D0081363

PROGRAMMABLE LOGIC CONTROLLERS

S. Brian Morriss
Conestoga College
Kitchener, Ontario

Prentice Hall
Upper Saddle River, New Jersey Columbus, Ohio

Library of Congress Cataloging-in-Publication Data

Morriss, S. Brian.
Programmable logic controllers / S. Brian Morriss.
p. cm.
ISBN 0-13-095565-5
1. Programmable controllers. I. Title.
TJ217.5.M67 2000
629.8'9—dc21 99-12358
CIP

Cover photo: The Stock Market
Publisher: Charles E. Stewart, Jr.
Production Editor: Alexandrina Benedicto Wolf
Production Supervision: WordCrafters
Cover Designer: Dan Eckel
Cover Design Coordinator: Karrie Converse-Jones
Production Manager: Deidra M. Schwartz
Marketing Manager: Ben Leonard

This book was set in Times Roman by Maryland Composition Company and was printed and bound by R. R. Donnelley & Sons Company. The cover was printed by Phoenix Color Corp.

Printed in the United States of America

10 9 8 7 6 5 4 3 2

ISBN: 0-13-095565-5

Prentice-Hall International (UK) Limited, *London*
Prentice-Hall of Australia Pty. Limited, *Sydney*
Prentice-Hall of Canada, Inc., *Toronto*
Prentice-Hall Hispanoamericana, S. A., *Mexico*
Prentice-Hall of India Private Limited, *New Delhi*
Prentice-Hall of Japan, Inc., *Tokyo*
Prentice-Hall (Singapore) Pte. Ltd., *Singapore*
Editora Prentice-Hall do Brasil, Ltda., *Rio de Janeiro*

PREFACE

This book begins by introducing the reader to programmable logic controllers (PLCs) and programming PLCs for simple Boolean operations. Unlike other texts, the remaining chapters describe the advanced features available in modern PLCs. The reader will learn how to program a PLC to manipulate whole data words and data files, how to program control of continuous analog processes, and how to configure a PLC so that it will interrupt its normal program scan cycle to control processes that require immediate attention. Communications, structured programming techniques, and the IEC 1131-3 standard for open structured programming are discussed. How to set up and troubleshoot a PLC are described.

Each chapter describes how a generic PLC performs particular operations and why a programmer might want to use those features. An understanding of the underlying computer operations is developed using language that is consistent with common programming languages and open interfacing of PLCs. Five PLC models from three major manufacturers are discussed in each chapter, using the proprietary terminology that each PLC manufacturer uses.

Rockwell Automation's Allen-Bradley PLCs are the industry leader in North America. As such, they provide a de facto standard that other PLC manufacturers must conform to in terms of features and performance. Allen-Bradley is active in organizations working to develop open standards.

1. **Allen-Bradley PLC-5**—A PLC-5 is a full-sized PLC. There is an enhanced series and a classic series. A classic PLC-5 will not be able to do everything described in this book.
2. **Allen-Bradley SLC 500**—An SLC 500 is a small PLC. SLC 500s have features similar to those of PLC-5s. The SLC 5/04 is the most recent version. Lower-level SLC 500s may not offer all the features described here, although the smallest SLC 500 (called the integrated model) has some features not offered by larger models. These features are also described. The programming language described here also is used with Allen-Bradley's MicroLogic line of PLCs.

Siemens AG's SIMATIC PLCs lead the European market and are making significant inroads into North America. They have been sold in North America under the Westinghouse and Texas Instruments brand names, but Siemens is now using its own name. Siemens PLCs typically offer the newest ideas in PLC programming first, so their PLCs are very powerful and sometimes a little intimidating to users. Siemens PLCs are consistently among the first to adopt open standards, and as a result Siemens has become a major force in the push toward open standards.

3. **Siemens AG's Simatic 5 (S5)**—The S5 line of PLCs is programmed in STEP 5. STEP 5 includes ladder logic that is as easy to use as other PLCs' ladder logic, but in this book we often use STEP 5's statement list (STL) programming language. Any reader who knows ladder logic will have no trouble learning STEP 5's ladder logic language, but some S5 features can only be programmed in statement list. Two versions of S5 PLCs are specifically discussed: the full-sized S5-115U type and the smaller S5-100U. They and the other members of the S5 family are programmed alike.
4. **Siemens AG's Simatic 7 (S7)**—Siemens S7 line of PLCs are programmed in STEP 7. STEP 7 is based on the IEC 1131-3 standard, a voluntary international standard describing a single set of common programming languages. Programs written in IEC 1131-3-compliant languages can be downloaded to any IEC 1131-3-compliant PLC. The S7 line of PLCs has just been released so they are the best examples of what future PLCs will probably be like. Two versions of S7 PLCs are specifically discussed: the S7-300 and the S7-400. They and the other members of the S7 family are programmed alike.

OMRON Electronics' CQM1—OMRON Electronic PLCs lead the market in Japan. OMRON is a force among sensor suppliers in North America, but they have only recently begun to actively push their PLC lines here. Largely because of their low cost, but also because their PLCs are good, OMRON has already attracted a large share of the PLC market.

5. **OMRON Electronics'** CQM1 is a small modular PLC. Recently introduced, the CQM1 offers many of the features of OMRON's full-sized C200H PLC, and its programming language is (almost) the same. Some CQM1 PLC models offer special-purpose features, which are also described here. OMRON's ladder logic language is very different from other ladder logic languages.

This single book does not contain all of the detail in the nearly 30 manuals it summarizes. Readers should have access to manual sets for the PLC(s) that they will be using for tabulated data, syntax information, and other specific information.

ORGANIZATION

Chapters 1, 2, 3, and 4 introduce PLC programming and describe the components and the unique operating system characteristics of a PLC. They cover the PLC's historical development, connecting of sensors and actuators, programming using Boolean logic, and use of timers and counters.

Chapters 5, 6, and 7 cover intermediate PLC programming topics such as memory structure, data types, addressing modes, and microprocessor basics. Instructions that manipulate data elements containing multiple bits (e.g., bytes, words, files, arrays, and structures) and math and logic operations are also covered.

Chapters 8 and 9 cover structured programming techniques from the early master control relay instructions, through jumps and subroutines, to the modern requirements for strict control of data access through the use of data declarations and parameter passing. The IEC 1131-3 standard's requirements for an open programming language are described.

Chapter 10 describes how to set up and configure a PLC.

Chapters 11, 12, 13, and 14 cover specialized advanced topics. The use of interrupts to customize the response characteristics of a modern PLC is discussed. Using a PLC for process control, and configuring and programming data communications between PLCs are covered. An example shows how a PLC and a robot can share control of a workcell.

Chapter 15 is dedicated to troubleshooting features.

Chapter 16 analyzes current trends to help readers understand the future of PLC technology.

ACKNOWLEDGMENTS

Thanks go to my students, who tested much of the material here. Thanks also go to the many helpful individuals at Rockwell Automation, Siemens AG, the Westinghouse division of CBS, and OMRON Electronics, without whom my college wouldn't even have a PLC lab. Finally, thanks to my wife, Linda, who spent most of the last summer without me as I tried to meet the completion deadline for this text.

I would also like to acknowledge the permission given by the following companies to reproduce their tables, figures, and graphs in this book: Allen-Bradley Company LLC, Siemens Corp., OMRON Electronics, Inc., and Westinghouse Electric Division of CBS.

I appreciate the valuable feedback from the reviewers of this edition: Tony Farrell, Michigan Technical University; Bill Maxwell, National State Technical Institute–Tennessee; and John Tielemans and Jeff Uniac, Conestoga College–Canada.

ABOUT THE AUTHOR

S. Brian Morriss has worked in automation for 19 years and was a design engineer at Automation Tooling Systems (ATS) Inc. of Cambridge, Ontario, Canada. ATS is one of the world's largest suppliers of customized automated manufacturing systems. Morriss currently teaches at Conestoga College in Kitchener, Ontario, Canada, and coordinates the content of a highly successful robotics and automation mechanical engineering technology program. He is in charge of the acquisition, installation, and configuration of several types of PLC training stations at Conestoga, and has been teaching PLC programming for over 10 years. Morriss is the author of *Automated Manufacturing Systems* (New York: Glencoe/McGraw-Hill, 1995), which was re-released in a Singapore International edition in 1996 and is being translated into Korean and Chinese.

Mr. Morriss is a Professional Engineer with a B.Sc. in systems design engineering (1977) from the University of Waterloo.

BRIEF CONTENTS

CONTENTS

1

WHAT IS A PLC?

OBJECTIVES

In this chapter you are introduced to:

- The hardware and operating characteristics that make a PLC significantly different from other computers.
- Ladder logic.
- Some options available in selecting a PLC.
- A brief history of the PLC's development.

A **programmable logic controller** (PLC) is a specialized computer, designed to be used for industrial control. PLCs are used extensively in industrial control because they are easy to set up and program, behave predictably, and are tough enough to keep working in even the dirtiest production environment.

Programmable logic controllers are sometimes called programmable controllers but are more commonly called PLCs. A PLC looks different from a standard personal computer. For a start, a PLC doesn't have a keyboard or a monitor. Figure 1.1 shows several PLCs (the PLCs we describe in more detail in this book).

PLC BASICS

Appearance is only a part of the difference between PLCs and other computers. PLCs differ from most computers in two major ways: (1) PLCs are constructed to make it easy for a user to put together a PLC-controlled system, and (2) PLCs come preprogrammed with an operating system and with application programs optimized for industrial control.

Construction of a PLC

Some PLCs are integrated into a single unit, whereas others are modular. **Integrated PLCs** are sometimes called **shoebox** or **brick PLCs** because of their small size. If an integrated PLC is

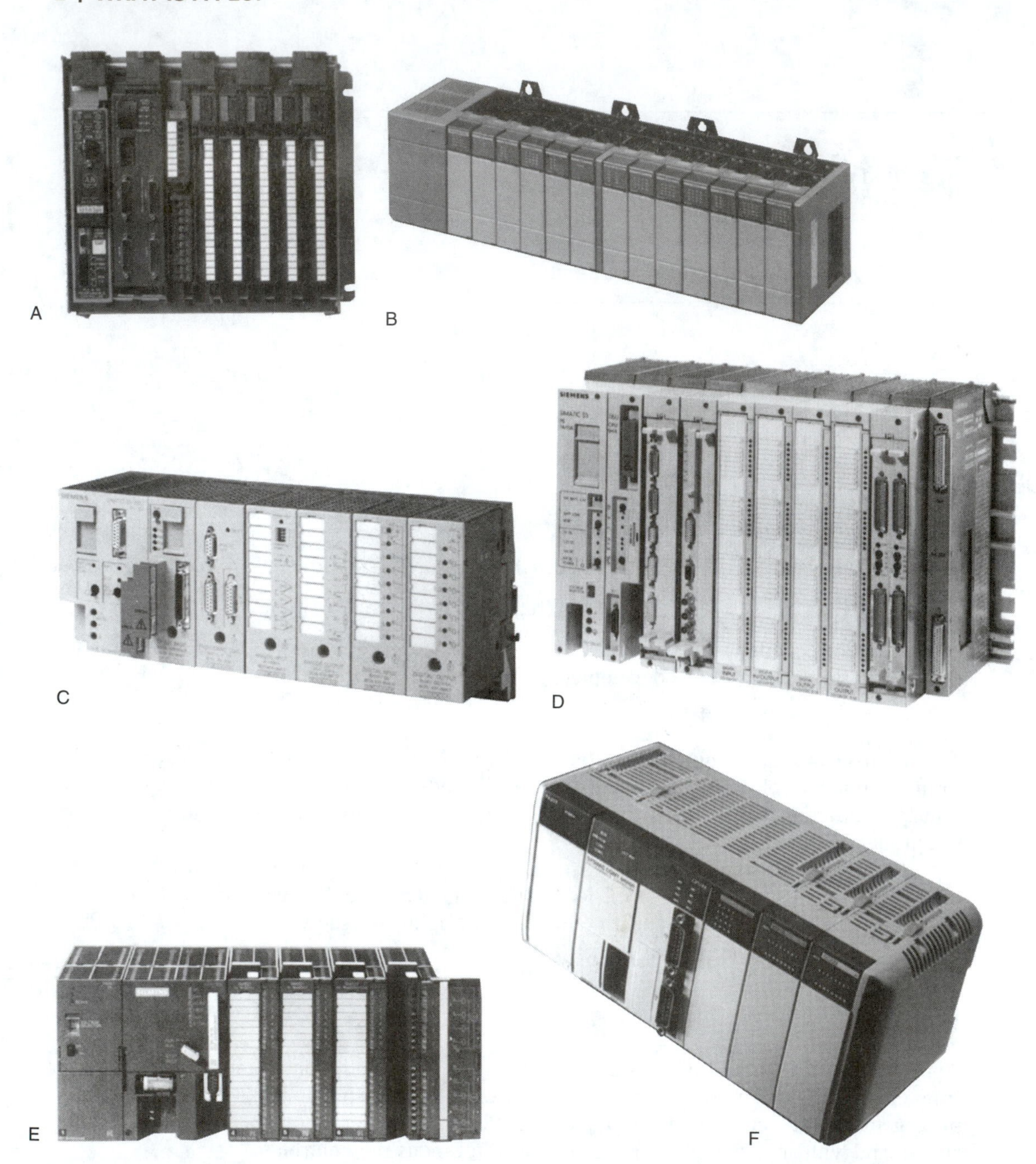

FIGURE 1.1
Programmable logic controllers: (a) PLC-5 and (b) SLC 500 by permission of Rockwell Automation; (c) S5-100U, (d) S5-115U, and (e) S7-300 by permission of Siemens AG; (f) CQM1 by permission of OMRON Electronics Inc.

available with the capabilities that a user needs, it is usually the most economical option. **Modular PLCs** consist of optional components required for a more complex control application, as selected and assembled by the user.

The PLC components that must be purchased separately when buying a modular PLC are covered in more detail in Chapter 2. They include the following components, as demonstrated in Figure 1.2:

- The **CPU** module, containing the central computer and its memory.
- **Input** and **output** modules (I/O modules), to allow the PLC to read sensors and control actuators. PLC manufacturers offer a wide variety of I/O module types.
- A **power supply** module, to provide power to the CPU and often to provide power to drive sensors and low-power actuators connected to I/O modules.
- A **rack** or **bus,** so that the CPU module can exchange data with I/O modules. In some PLCs, this component isn't required because each module plugs directly into its neighboring module.

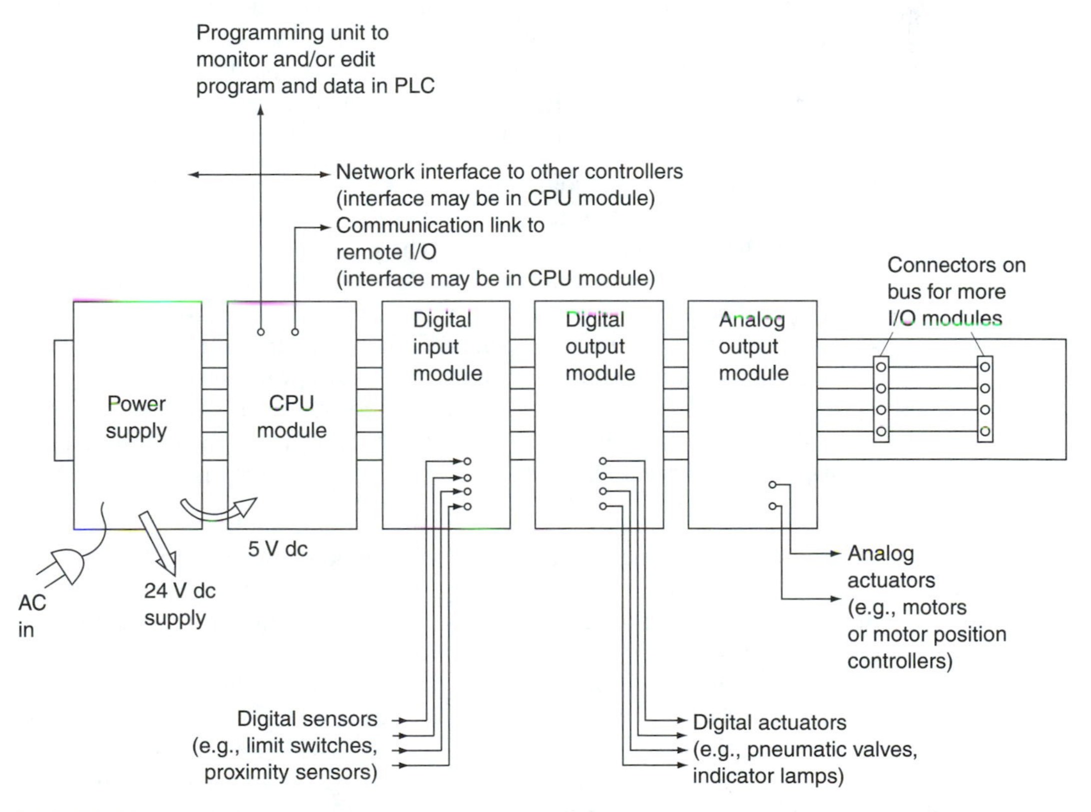

FIGURE 1.2
PLC in an automated system.

A PLC system with these components is all that is needed to control an automated system. Since a PLC must be programmed before it can be used, another component is required:

- A **programming unit** is necessary to create the user-program and send it to a PLC CPU module's memory.

Additional optional PLC components are often available, including:

- **Communication adapters for remote I/O,** so that a central controller can be connected to remote sensors and actuators.
- **Network interfaces** to allow interconnecting of PLCs and/or other controllers into distributed control systems.
- **Operator interface** devices to allow data entry and/or data monitoring by operators.

Operating System and Application Programs

The CPU module of a PLC comes with a very different operating system program than those used in most other computers, and comes complete with application programs preprogrammed into the CPU's memory. The operating system programs cause the PLC to start when power is turned on, to run the user-program when the PLC is switched into run mode, and to respond to user commands by running the appropriate application programs. The application programs allow the user to enter programs and data into the PLC's memory. Some parts of the user-accessible memory are retained even when the PLC's power is disconnected.

> Unlike a PLC, a standard personal computer comes with only a very small amount of **preprogrammed operating system** in its memory, and it has little or no memory for retaining data when power is lost. The personal computer's operating system performs a simple self-check when the computer is started, then must load **additional operating system programs** (such as DOS, Windows, or UNIX) from disk drives into memory. Even then, the personal computer can't do any useful work until it also loads an **application program** (e.g., a word processor, or CAD program, or a programming language such as Basic) from a disk drive into memory. The tiny bit of memory that retains user data when power is off is only adequate to remember configuration information (such as the type of hard drive installed).

A PLC retains its operating system, application programs, user-programs, and some data in retentive memory (sometimes called nonvolatile memory) while the PLC is turned off and even when disconnected from the power supply. A PLC can therefore resume running a user-program as soon as power is restored, although PLCs are often programmed to require some operator action before restarting (for safety reasons).

A PLC's operating system makes the PLC run user-programs very differently from the way other computers run user-programs. A PLC operating system executes an initialization step once each time it is put into run mode, then *repeatedly makes the PLC execute a scan cycle sequence as long as the PLC remains in run mode.* This basic scan cycle inherent in all PLCs is shown in Figure 1.3. Although there are some slight differences between PLCs, especially in what they do during the initialization step and in the terminology their manufacturers use to describe the scan cycle, *the three-step scan cycle is fundamental to how a PLC controls automated*

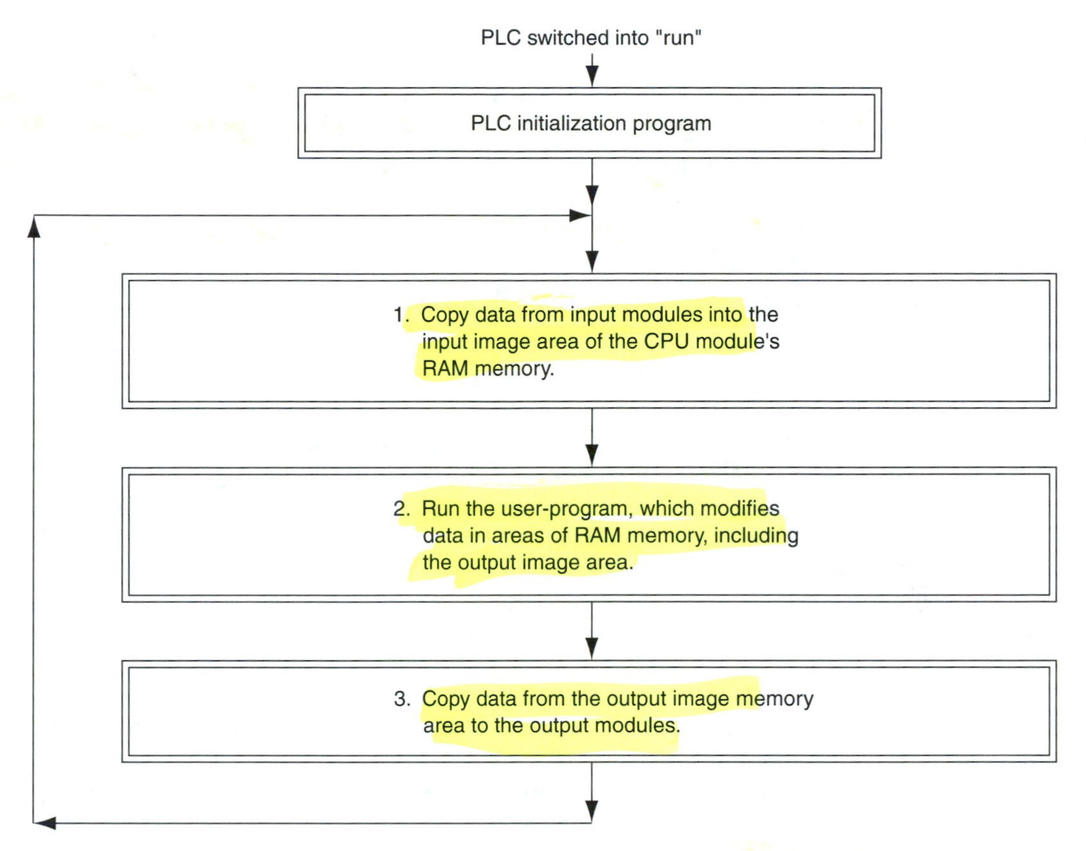

FIGURE 1.3
Standard PLC scan cycle.

systems. As Figure 1.3 shows, a PLC's operating system makes the PLC perform the following steps:

1. A preprogrammed **initialization step,** which is executed one time only, each time the PLC goes into run mode, before the three repeating scan cycle steps are executed for the first time.

2. The repeating three-step scan cycle, consisting of[1]:

(a) An **input scan.** The PLC reads data from all of its input modules (acquiring data from sensors attached to the input modules). This input data is placed into an area of the CPU module's memory reserved for images of input data.

(b) A **user-program scan.** The user-written control program is run *once,* from beginning to end. The program will contain instructions to examine input *image* data and

[1] More powerful PLCs may execute additional steps during this cycle. Some additional steps are preprogrammed as part of the operating system, often to control serial communications. Some PLCs can be programmed or configured by the user to perform additional operations.

to determine what values the PLC should output to the actuators. The PLC does *not* write the output data to the output modules yet, but saves it in an area of the CPU's RAM memory reserved for images of output data. The user-written program can examine and change all "addressable" areas of RAM memory. (This means that input image data can be changed by the user-program and output image data can be examined.) Some RAM memory isn't addressable, so it can't be changed by the user-program. The user-program, for example, is not in addressable memory.

(c) An **output scan.** During this step, the PLC copies all data from the CPU's output image area of RAM to the output modules.

Every time the PLC finishes one scan cycle and starts another, the operating system also restarts a **watchdog timer.** The watchdog timer runs while the scan cycle executes. If the watchdog timer reaches its preset value before being restarted (if a scan cycle takes unusually long to complete), the PLC will immediately **fault,** and stop running. After faulting, the PLC usually needs operator intervention before it can resume running. Operating system diagnostic features that help during the troubleshooting of faults are described in Chapter 15.

Other parts of a PLC's operating system program offer capabilities that a user takes for granted when buying any computer but which have a significant impact on the quality of the overall PLC control system. Always determine the quality of the operating system's application programs and driver routine programs before selecting a PLC.

All PLCs also come preprogrammed with **application programs** that run in response to commands the PLC receives from the programming unit, operator interface panels, or other computers connected to the PLC. Application programs allow users to do things such as writing and storing programs and data in the PLC's RAM memory, and allow the user to command that the PLC run programs and send status information to operator interface terminals, allowing monitoring of program execution and monitoring of data PLC's memory.

Driver routines are subroutines that other programs (the scan cycle program and sometimes the user-program) can call to operate the I/O control circuitry in the CPU module. Driver routines are preprogrammed in ROM memory by the PLC manufacturer. I/O functions that require driver routines include:

1. *Connecting to a programming unit.* All CPU modules include drivers and communication hardware to allow a programmer to monitor and change user-programs and working data via a programming unit. The programming unit is often connected to a PLC serial port dedicated for that purpose.

2. *Reading and writing of local I/O modules during the scan cycle.* All CPU modules for modular PLCs include driver software and hardware to exchange data with the local I/O modules, which are the I/O modules connected directly to the CPU module via parallel conductors in the rack or bus.

3. *Receiving and sending remote I/O data.* Remote I/O stations may be separate racks of I/O modules controlled by this PLC, may be independent I/O modules not connected into racks, or may even be independent sensors and actuators with built-in communication adapters. The PLC's CPU module may contain drivers and interface hardware to allow the exchange of data with remote I/O stations through serial communication links in the CPU module. If the CPU module does not have built-in remote I/O communications drivers and hardware, a separate communication module will be necessary (if available) to allow control of processes remote from the main PLC.

4. *Receiving and sending extended I/O data.* Extended I/O stations are separate racks of I/O modules controlled by this PLC but which are mounted close to the main rack so that faster data exchanges can take place using different driver programs and hardware than would be used for remote I/O communications.

5. *Receiving and sending data on a local area network (LAN).* If the PLC has the appropriate drivers and hardware, the CPU module can be connected directly to a system of other controllers (including other PLCs) via a shared serial communications link. The user-programs in the PLCs can then include programmed exchanges of data. Some PLCs come with a single communication port but with selectable drivers so that the user can connect the CPU module to the LAN of his or her choice. Programming units and operator interface panels are sometimes connected to a PLC via the local area network port and may be able to share the use of the LAN with the CPU. If the programming unit connects via the LAN port in this way, you may be able to use the programming unit to monitor and change the memory of all the PLC CPU modules in the LAN. Some CPU modules come with multiple communication ports. If a CPU does not have the required LAN driver or hardware, a separate communications I/O module may be available. Third-party vendors sell interface hardware and driver software to allow connecting to most of the popular LAN types so that you can (for example) connect a Siemens PLC to an Allen-Bradley LAN.

6. *Reading or writing slave devices* (such as a serial printer or a bar code reader). This is accomplished through a serial link using standard communications protocol such as RS-232.

PLC User-Programs

User-programs are not part of the preprogrammed set of programs purchased with the PLC. They must be entered into a PLC's RAM memory by a programmer using a programming unit, which can then be disconnected from the PLC. PLCs save user-programs in memory that is either unaffected by power loss or is maintained by a long-life battery. The user-program remains in the PLC's memory until a programming unit is used to change it.

PLC user-programs are usually written in **ladder logic.** Ladder logic programs are graphics based. They look a little like a drawing of a stepladder and a little like the relay logic circuit diagrams that industrial electricians use. Figure 1.4 shows a simple ladder logic program and the system that it controls. *Each rung in a ladder logic program consists of a logical statement that can be evaluated as being either true or false, and which controls whether the rung's output function is performed.* The first rung of the program in Figure 1.4, for example, controls a spray paint actuator. The spray paint is turned on whenever the system's control switch is on but only if a sensor detects a part in the paint booth at the same time. The next rung turns an alarm light on to call the operator if either the paint supply is empty or if the conveyor belt stops for any reason. If you can't quite see how the program works, don't worry. We examine how ladder logic programs work in Chapter 3.

A PLC repeatedly executes its scan cycle, which includes the user-program, at intervals measured in milliseconds. It should be obvious after examining the scan cycle diagram of Figure 1.3 that there will always be a short delay between when a sensor detects a change and when

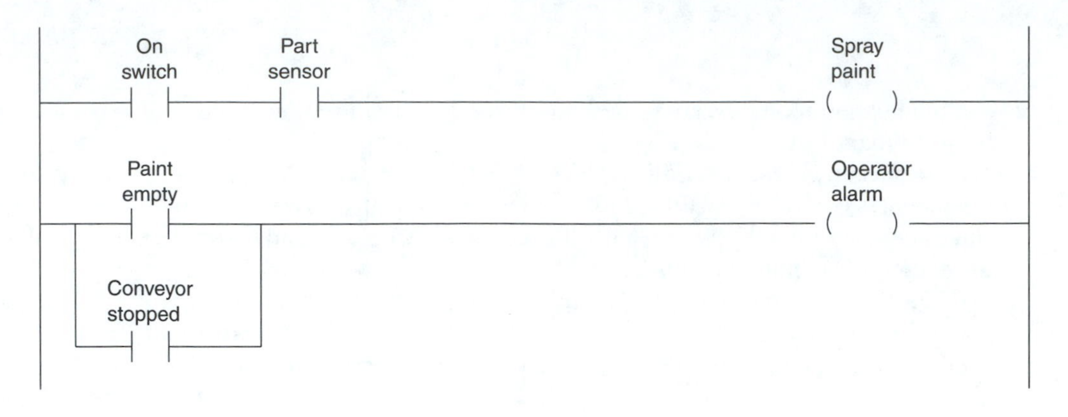

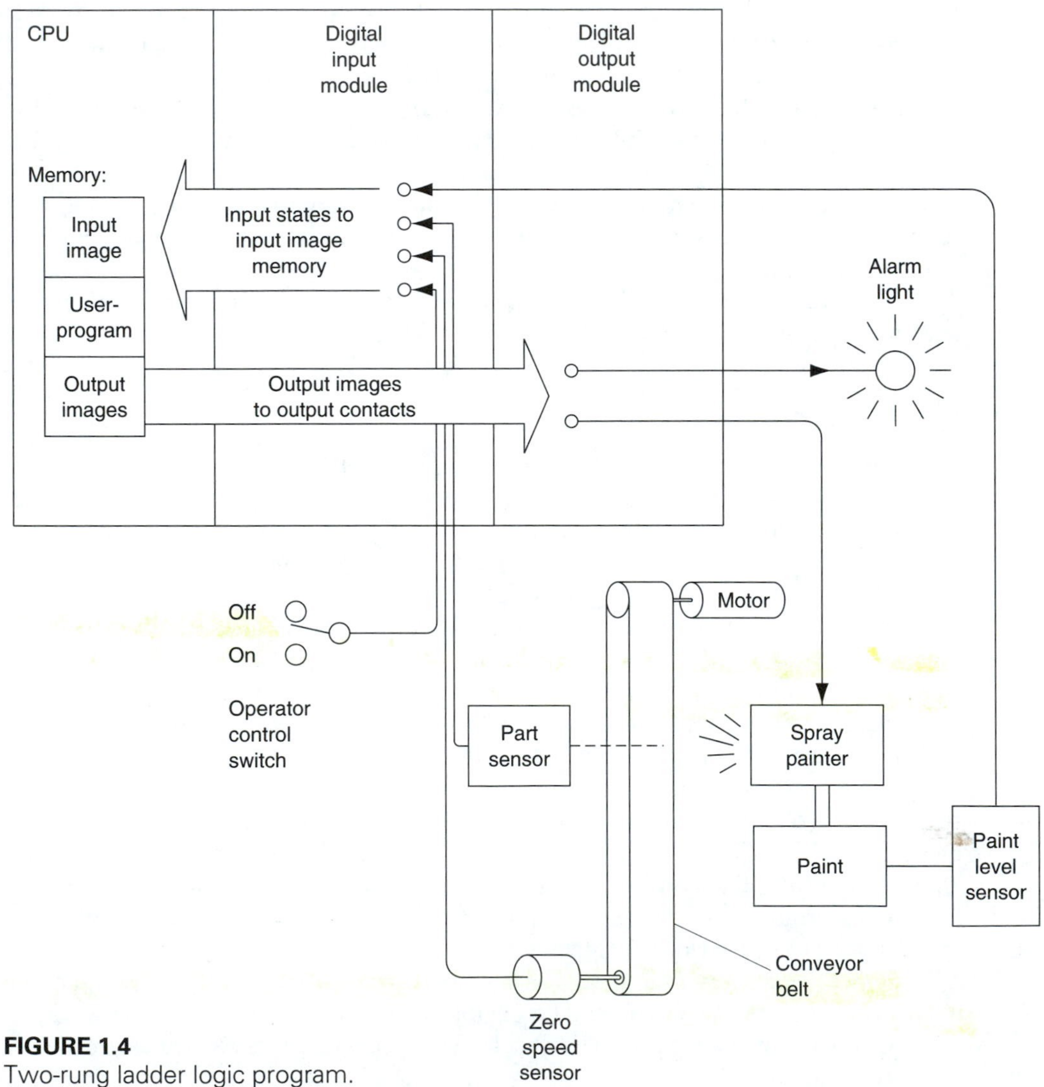

FIGURE 1.4
Two-rung ladder logic program.

an actuator starts to respond. In the worst case, if the part sensor in our example program detects a part immediately *after* its value has been read during a scan cycle, the spray paint actuator won't be turned on until the *next* scan cycle ends. Since PLCs execute their scan cycles at intervals as short as a few milliseconds, the delay usually isn't a problem.

The programmer should always remember that the conditions that a user-program is checking are only recent images of input conditions. Similarly, the user-program does not change the true outputs but only images that will eventually be written to the outputs. Remember, too, that earlier rungs in a user-program may already have changed the output or input image values that a rung is using. A common programming error that inexperienced programmers make is to program two separate rungs controlling the same output image. The second rung will always overwrite the result of the first instruction before the output image is actually copied to the output module.

Programmed correctly, the program of Figure 1.4 would ensure that every part that passed through a paint booth would be painted, with no *perceptible* delay between the part's arrival and the paint spray going on. No paint would be used when it wasn't needed. The program would control the paint booth for days or weeks or years, requiring operator attention only when the PLC detects one of the situations for which it has been programmed to monitor. This PLC-based automatic system would pay for itself faster than it would take to convince some accountants to buy it.

SELECTING THE RIGHT PLC

PLCs come in a variety of sizes and with a variety of capabilities. Choose the right one for the task. Most organizations start with an *integrated PLC* as the controller for their first simple automated system. Allen-Bradley's Micrologix 1000, Siemens' S7-100, or OMRON's CPM1 (shown in Figure 1.5) are typical small but reasonably powerful integrated PLCs. Most PLCs at this level even have a built-in 24-V dc power supply adequate to drive typical sensors. The built-in power supply isn't big enough to power any but the smallest actuator, so separate power supply(s) to drive the actuators controlled by the PLC are required. The user will also need a programming unit. PLC manufacturers sell programming software that can be loaded into a personal computer, to turn the personal computer into a programming unit, and a serial cable to connect the personal computer to the PLC. The total cost of the PLC, programming software, and connecting cable is usually much less than the cost of the personal computer. Allen-Bradley's Micrologix 1000 can be programmed using the same programming software as that used for the larger SLC 500 PLC; Siemens S7-200 can be programmed using the S7-300 programming software; and OMRON's CPM1 can be programmed with CQM1 programming software.

The simple PLCs described above are adequate for controlling a system that has, perhaps, 20 switched sensors and actuators. Some integrated PLCs can even provide analog output signals at one or two output contacts or can accept signals from one or two analog sensors or encoders, but they are intended primarily for use in systems where the sensor signals and actuators are digital (are either on or off). The programming languages used for these small PLCs are similar to the languages described in this book, although some of the instructions described here may not be available.

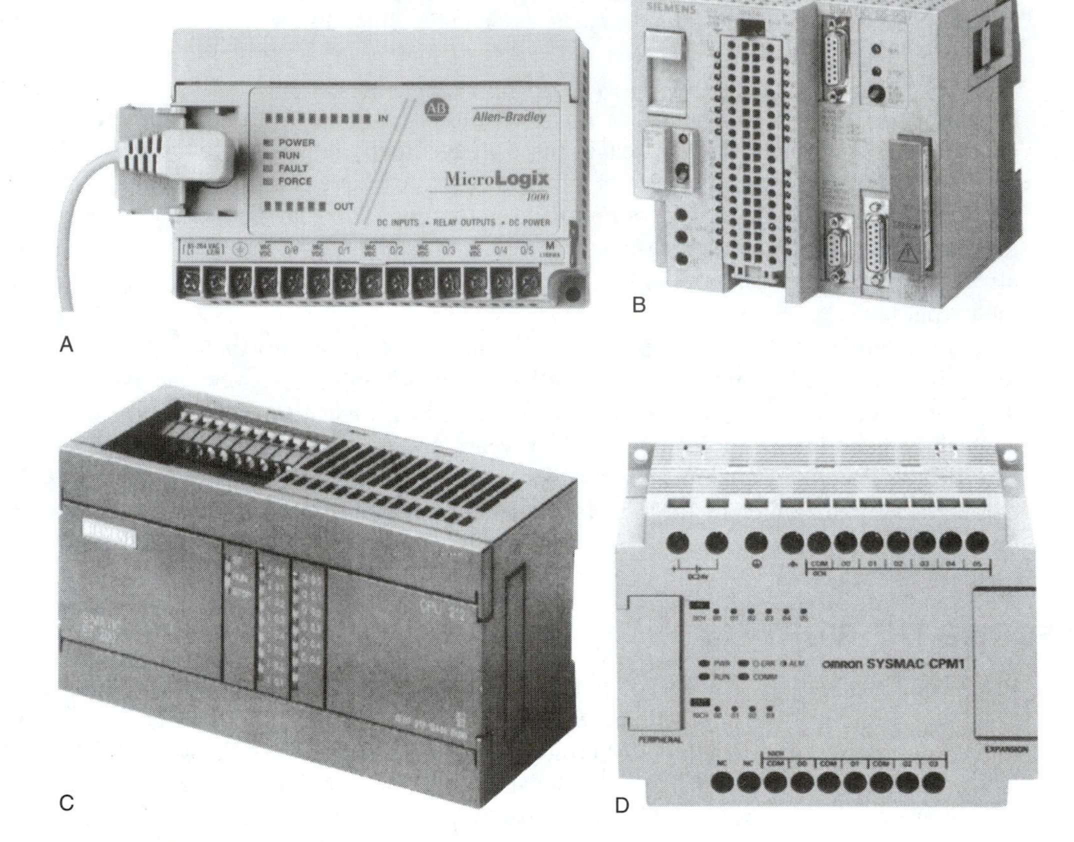

FIGURE 1.5
Small PLCs: (a) MicroLogix 1000 by permission of Rockwell Automation; (b) S5-95U and (c) S7-200 by permission of Siemens AG; (d) CPM1 by permission of OMRON Electronics Inc.

If the system being controlled needs a controller with only a *very* few inputs and outputs, all digital, it may make more sense to buy an even smaller *programmable logic module,* such as Siemens' LOGO logic module, shown in Figure 1.6. The user's purchase order can even include a program that Siemens will write to the logic module before they ship, so not only is the PLC controller cheap, the user doesn't necessarily even need to purchase a programming unit.

If one of the small PLCs described above isn't adequate, the user will want to purchase a *modular PLC* (described in detail in Chapter 2). In purchasing a modular PLC, the purchaser buys only the interface modules actually needed to connect the PLC to digital sensors and/or actuators, to nondigital sensors and/or actuators, or to other computerized controllers.

Some manufacturers have recently made software and interface cards available so that a personal computer can do the work of a PLC. These systems are often given names such as *soft*

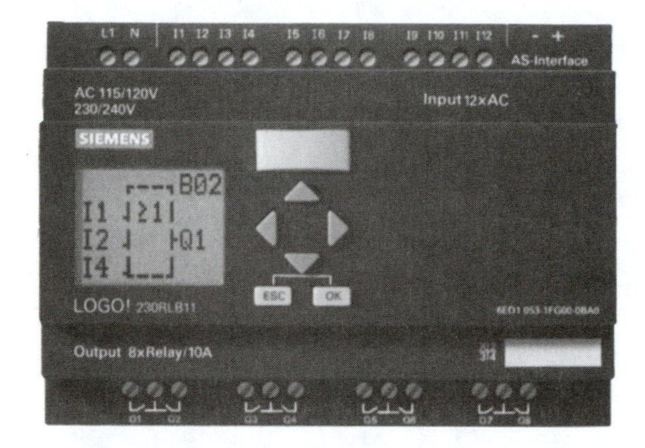

FIGURE 1.6
Siemens' LOGO programmable logic block. By permission of Siemens AG.

logic controllers. A soft logic controller must still be connected to sensors and actuators, so I/O modules that can communicate with the soft logic controllers are required. Soft logic controllers are discussed in Chapter 16.

EVOLUTION OF THE PLC

Before the PLC, automated manufacturing processes had to be controlled using hardware devices. A rotating camshaft, for example, would ensure that events occurred sequentially in a manufacturing cycle, just as camshafts are still used to ensure that automobile engines cycle correctly, but camshafts require mechanical linkages between the camshaft and the operations being controlled. Later control systems used electrical relays to control widely dispersed systems. Some circuits contained sensors that controlled relay switches. The relays controlled electric current in other circuits to control electrical actuator and/or other relays. With timers and counters in the system, systems of relays could control sequential manufacturing processes. The term **relay logic** was used to describe control systems based on interconnected relays. Programming a relay logic controller meant building it. To modify a relay logic program, the control system had to be rebuilt. Debugging a faulty relay logic control system required tedious electrical troubleshooting of the electrical components and contacts.

Computers were first used as industrial controllers in the form of **sequencers.** To program a sequencer, a programmer entered a series of binary patterns (data words) into the computer's memory. The sequencer controlled a manufacturing process by outputting the binary words, one at a time. Each binary bit of a data word controlled an actuator, so when one output data word was replaced by the next, some actuators would turn on and others would turn off. Two methods of controlling the changing of output words were used:

1. In a time-based sequencer, the programmer specified a time delay between steps.
2. In an event-based sequencer such as that shown in Figure 1.7, the programmer entered an *input* data word for each output data word. After outputting a data word, the sequencer compared its matching input data word against a pattern of on–off states of sensors. When the input

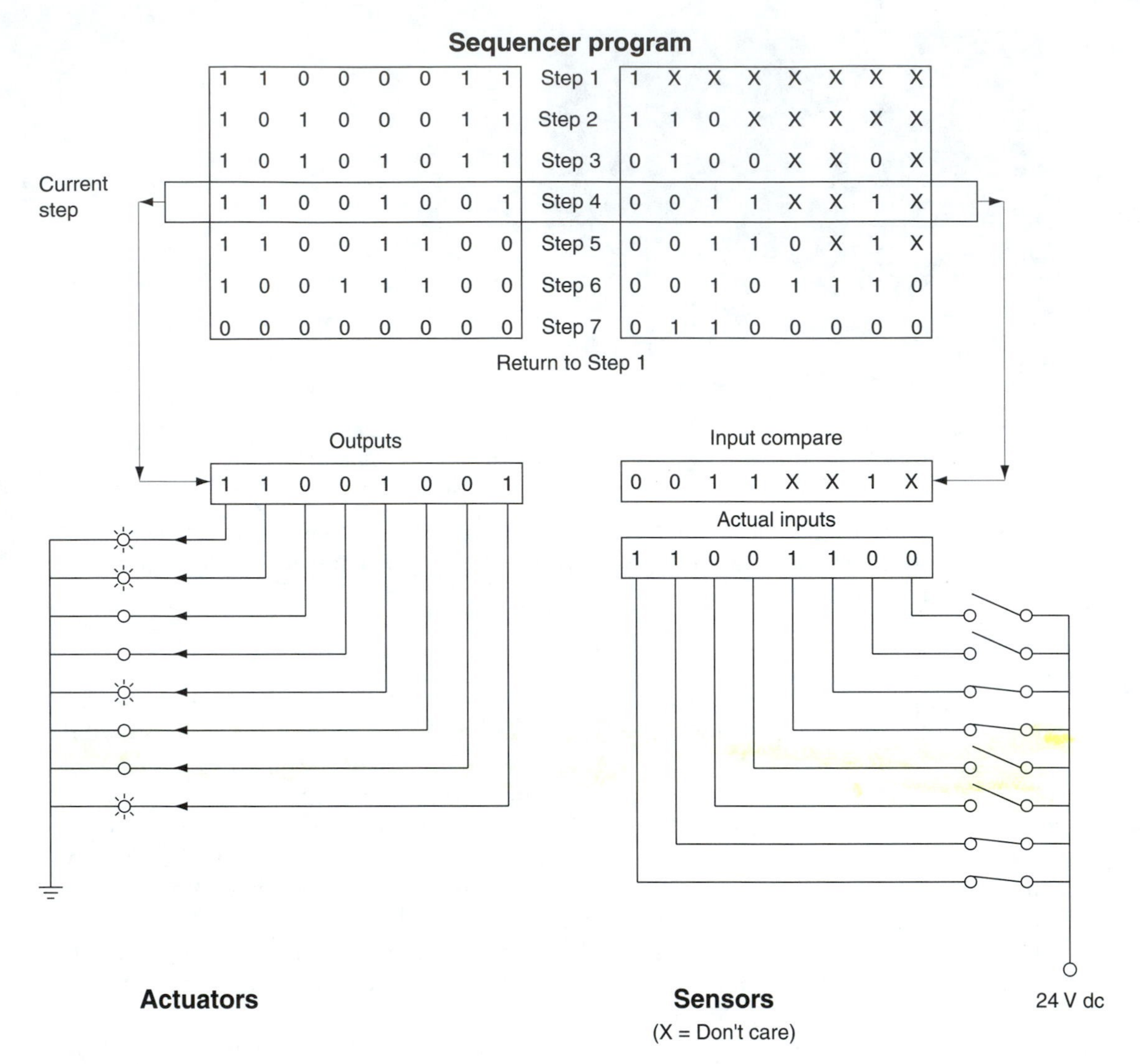

FIGURE 1.7
Event-based sequencer.

pattern matched the input data word, the next output data word was outputted. (A single bad input contact could cause an event-based sequencer to become stuck.)

Sequencers that combine event- and time-based sequencing did (and still do) exist. Some modern PLCs offer sequencer instructions, so that a PLC program can include sequential control. Sequencer instructions are covered in Chapter 7 when we discuss manipulation of data files.

Relay logic controllers were better than sequencers in several ways: With their multiple parallel circuits, each control circuit operated independent of other control circuits, so a single failed sensor couldn't bring the entire controlled system to a stop. Also, the multiple parallel circuits all performed their control functions simultaneously, so each could change an output immediately as soon as the change was required. Relay logic controllers were said to operate in **real time,** in that they provided control signals as soon as the control signals were required. Sequencers are not real-time controllers.

The first PLC, then called a **programmable controller,** was an early computer that was designed to operate in as close to a real-time mode as possible. Since a computer can actually do only one thing at a time, in the order that its programming dictates, the early programmable controller needed an operating system that made it look as if it was controlling several independent processes simultaneously. The early programmable controller operating system introduced the **three-step scan cycle.** First, the programmable controller read the state of all the input contacts into its input image area of memory. This effectively took a nearly instantaneous "snapshot" of the input conditions. Next, the programmable controller ran its user-program, which consisted of ladder logic rungs, each of which independently controlled the state of a single bit in the output image area of memory, in much the same way as each of the parallel relay logic circuits would control a single actuator. (Ladder logic programs were even designed to look like relay logic circuit diagrams so that industrial electricians could write programmable controller programs.) Finally, after the user-program finished executing, the programmable controller rapidly copied the contents of the output image area of memory to the output contacts, turning actuators on or off. The control wasn't actually "real time" because there was always a delay between the sensed need for a change and when the programmable controller made the output change, but the efficiencies in reading and writing large blocks of input and output information made the programmable controller operate quickly enough that the delay was short. The scan cycle also made the control **deterministic,** which means that the programmable controller's response was predictable and dependable.

The programmable controller was forced to undergo a name change in the 1980s. *Programmable controller* was usually shortened to *PC* by users. When IBM introduced its personal computer, it became known as a PC. Programmable controllers are now called **PLC**s, which stands for the awkward and rarely spoken *programmable logic controller.*

The PLC has undergone other evolutions as well. The earliest PLC could use only Boolean ladder logic elements to examine and control the on–off states of input and output image data. (Boolean logic is covered in Chapter 3.) It didn't take long before a new generation of PLCs offered **working data memory** in addition to input and output image memory, and offered **programmable timer and counter instructions,** which used whole data words in memory to store count values and time values. There are still many PLC applications that don't require the use of any advanced features beyond timers, counters, and working data bit storage. Chapter 4 covers timers and counters.

One improvement inevitably leads to others. Once PLCs could store working data, such as count values, users demanded **retentive memory** so that data wouldn't be lost when the PLC program was stopped. **Battery-backed memory** and **EEPROM storage** of data became available to retain the contents of selected memory locations (including the user-program) even if power is disconnected.

Timer and counter instructions are actually calls to function programs that are included in the PLC's operating system, and manufacturers invented **function block** graphic elements to represent these preprogrammed subroutine calls in ladder logic. Function block graphic elements had to make it clear what input data the function program needed and what output elements it controlled. Function blocks sometimes display a working data value (such as a counter's accumulated value) while the PLC's operation is being monitored using a programming unit. Chapters 4 and beyond include examples of the function block elements used in PLC programming languages.

Since PLCs now offered timer and counter instructions, which perform mathematical and comparison operations on **data words,** the next logical step was to offer instructions that would allow a programmer specifically to **store** data words in memory, to **retrieve** those data words, and to do other **mathematical, comparison,** and **logical operations** on those data words. I/O modules were developed to allow a PLC to read and write analog I/O values as digital words and to use serial communications to exchange data words with other computers. Chapters 6 and 7 cover the instructions that manipulate individual data elements and sets of data elements. In Chapter 5 we describe the data storage capabilities of Allen-Bradley, Siemens, and OMRON PLCs and describe how a user-written program can "address" data in those memory areas.

Most function block instructions and data manipulation instructions worked by calling preprogrammed functions or subroutines that were included as part of the PLC's operating system. Soon, PLCs began to offer programmers the power to write their own functions and subroutines and to program Boolean logic statement–controlled jumps to them. The use of instructions that conditionally included or excluded parts of the user-program meant that previously deterministic PLC control became more variable. If a function or subroutine was called, the scan cycle took longer to execute. Other PLC instructions were offered that further reduced the deterministic character of PLC control. **Master control relay** and **jump instructions** allowed programs to override (or completely skip past) parts of a user-program. **Immediate I/O instructions** were offered, which allowed the programmer to deviate further from the rigid restraints of the scan cycle by reading data from input modules and writing to ouput modules during execution of the user-program. Most modern PLCs even offer **interrupt** capabilities, so that a scan cycle can be interrupted at any point to execute important control programs in response to signals from I/O modules, or from timers, or to handle detected PLC fault conditions. After the interrupt, the PLC resumes the scan cycle as if it hasn't been interrupted. Chapter 8 covers structured programming techniques, and Chapter 11 covers interrupt use.

As PLCs and PLC programming languages became more powerful, the constraints of ladder logic, even with function block elements, became evident. PLC function blocks and methods of addressing memory were becoming less standardized as manufacturers disagreed on the best way to implement advanced features. Some PLCs even began to offer programming languages other than ladder logic. In an attempt to bring programming standards and memory-access standards to PLCs, the International Electro-technical Commission (IEC) developed a voluntary standard, called IEC 1131-3, which provides guidelines for (among other things) the allowable programming languages and their elements. IEC 1131-3 classifies PLC programming languages as *ladder logic, function block, instruction list, structured text,* and *sequential flowchart.* We examine these IEC-compliant languages in Chapter 9. (The IEC 1131 set of standards has recently been renumbered as the IEC **6-**1131 set of standards, but most suppliers still call them IEC 1131.)

Some PLCs now include more than one microprocessor, so that the PLC can perform several operations simultaneously. The additional microprocessors are sometimes found in intelligent I/O modules, complete with their own operating system programs and memory, or sometimes in the CPU module with the main microprocessor. The additional processors are *slave* processors, and they operate under the control of the single *master* processor. The master processor executes the scan cycle, so the user-program still controls what the PLC does. In at least one recent PLC, the I/O scan steps of the scan cycle are executed by one microprocessor while another microprocessor runs the user-program.

Communications capabilities of PLCs have also evolved. The earliest PLC could read input contacts and write to output contacts. All PLCs allow a **programming unit** to be connected, so the user can change the user-program and working data areas of memory and can monitor the program and memory as the PLC runs. Some PLCs can be connected to **remote racks** of I/O modules via serial communication handlers, which may be in separate communication modules or may be built right into the CPU module. Serial interfaces now allow PLCs to be connected to **sensor/actuator networks** so that the PLC can read and write groups of remote sensors and actuators that aren't attached to I/O modules of the PLC. PLC manufacturers offer **proprietary local area networks** through which PLCs and programming units can be interconnected. **Gateways** are sometimes available to connect PLCs and computers from other suppliers to a proprietary local area network (LAN). Some PLCs already implement standards for **fieldbus** LANs, which can be used to interconnect any PLC and any compatible sensor, actuator, and other control devices. Standard communication capabilities are discussed in Chapters 10 and 13, and communication advances are discussed in Chapter 16.

Finally (?), personal computers may be ready to replace PLCs in industrial control. (Just taking the name "PC" wasn't enough, I guess.) There are some very good reasons why this could happen and some very good reasons why it may not. In case you are now wondering if there is any point in reading any further in this book, I would like to point out that the advocates of personal computers have been predicting the death of the PLC for the last two decades, during which time PLC sales have continued to skyrocket. In Chapter 16 the cases for and against personal computers are presented and other trends that will affect PLC evolution are described in as objective a fashion as possible.

TROUBLESHOOTING

Difficulties in introducing PLCs to the plant floor are usually due to a lack of understanding of what a PLC can (and can't) do, or due to a resistance to change (which often stems, at least in part, from a lack of understanding). To ease the introduction of PLCs, it is a good idea to provide training to both operations *and* decision-making personnel so that they will understand what can (and can't) be done with a PLC. This training should include a hands-on component, where the trainee can experiment with a PLC.

There's nothing like learning on the job. The best way to overcome lack of experience is to pick a simple process that could be controlled by a PLC, then implement PLC control. This will force the implementation team to face their lack of knowledge and their resistance to change during all phases of the project: from planning, through hardware selection and assembly, programming and testing, to implementation on the plant floor and operator training. The process should be one that isn't too important, of course, because it may not work for awhile. It shouldn't be too unimportant, either, or there won't be any incentive to finish it if it proves more difficult than anticipated. Once a few simple projects have been completed, the implementation team will have learned enough that they will wonder why the first project seemed so difficult, and they will be anxious to implement their new skills.

PLC training should emphasize the scan sequence. Most users start by writing programs that contain conflicting logical statements, then wonder why one of the statements doesn't seem to do its job. The program in Figure 1.8a won't turn on the operator alarm if the paint runs out, for example; the alarm goes on only if the conveyor stops. Once the user grasps that the *entire* user-program is evaluated, from the first logical element on the first rung to the last logical element on the last rung, before any output actions are taken, the user will understand why perfectly logical statements controlling an output state won't have any effect if they are followed by another logical statement that controls the same output. The program in Figure 1.8a doesn't work because two rungs try to control the same output, and even if the first statement would normally turn on the operator alarm (it does try to), the second statement keeps the alarm off unless the conveyor is stopped. When the operator alarm is controlled by a single statement (it is shown correctly in Figure 1.8b), the alarm will operate when either alarm condition occurs.

Understanding the PLC's scan cycle will also help the programmer understand why the program in Figure 1.9 doesn't work. The PLC never pauses during execution of a user-program, even to wait for an actuator to complete its action, and never ignores any logical statements, even if their function is finished. In this program, when a part is detected the PLC will start the

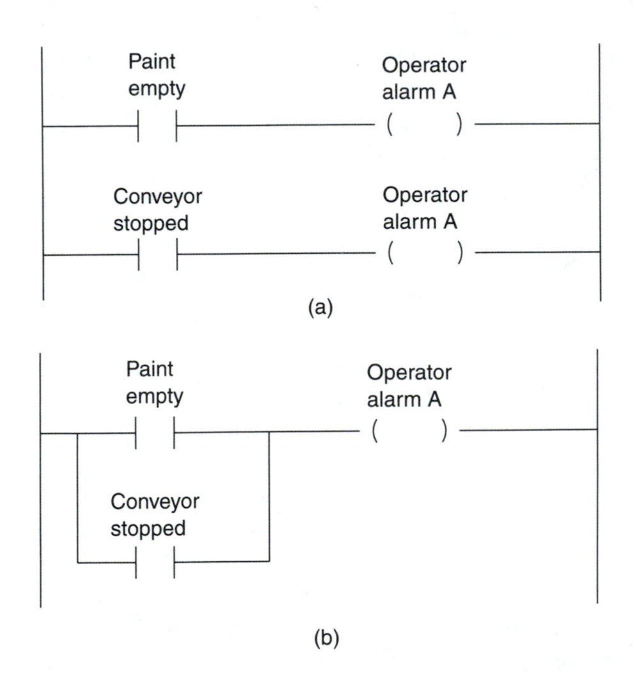

FIGURE 1.8
Conflicting logical statements: (a) incorrect version; (b) correct version.

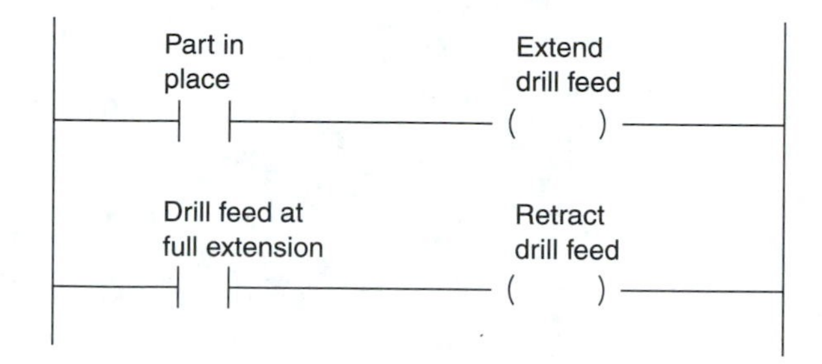

FIGURE 1.9
Another faulty program.

drill feed. For a while, the repeated program scans will verify that the part is in place (so the feed will stay on) and that the drill hasn't drilled far enough to reach its full extension. The second rung will have no effect. Eventually, the drill will reach its full extension and the second rung will initiate retraction. Since the retract output is in the second of the two feed-control rungs, it will override the first rung's extend output, so the drill will start to retract, *but* as soon as the drill has retracted slightly, the second conditional statement ceases to be true and the retract output goes off. The part is still in place, so the extend logical statement is still true. The drill feed will alternate between extend and retract as fast as the feed can actuate and deactuate the sensor at the end of its stroke.

In Chapter 3 we will see how to create sequencer programs that overcome the problem demonstrated in Figure 1.9, and in Chapter 8 we learn some structured programming instructions that can be used to cause the PLC to bypass selected rungs.

QUESTIONS

1. How is an integrated PLC different from a modular PLC?
2. What components (or types of components) are required as parts of a modular PLC? Page 3
3. What kind(s) of power does a power supply module provide? Page 3
4. (a) List and explain the steps in a PLC's three-step scan cycle. page 5 + 6
   (b) How long would a typical scan cycle take to execute?
   (c) What happens when a scan cycle ends?
5. If the sensors and actuators controlled by your PLC must be situated a long way from the PLC, you need to select a remote I/O option. What methods could you use?
6. What is an input image? page 5+6
7. Before the advent of PLCs and other computer-based controllers, what mechanical device was used to control automated processes? Page 11
8. What is the difference between an event-based sequencer and a time-based sequencer?
9. Why was ladder logic developed? (Why wasn't a more traditional computer language, such as Basic or Pascal or C, used?) Page 7
10. Identify the five programming languages allowed by the IEC 1131-3 standard.
11. Define *real-time control.* What is it about the PLC scan cycle that makes a PLC inherently good in real-time control applications? page 13

2

PLC COMPONENTS

OBJECTIVES

In this chapter you are introduced to:

- The operation of a CPU module and how it uses its memory.
- What a power supply module and a bus (or rack) are for.
- Digital, analog, and intelligent I/O modules:
  - How to wire sensors and actuators.
  - What they do.
  - How to address sensor and actuator data in a user-program.
- Programming units: what they do and options.

As we saw in Chapter 1, some PLCs are integrated into a single unit and some are modular. A modular PLC consists of several components that can be connected by being plugged into a common bus or rack. Every PLC needs:

- A CPU module
- A power supply module
- At least one I/O module

An **integrated PLC** contains all of those components in a single case, so the I/O capabilities of an integrated PLC are decided by the manufacturer, not by the user. Some integrated PLCs can be expanded by having additional I/O modules plugged into expansion sockets, making them somewhat modular.

Modular PLCs must contain a CPU module, a power supply, and I/O modules in components purchased separately. They may all plug together or they may all be plugged into the same rack. Some manufacturers offer CPU modules with a few built-in high-speed I/O capabilities as CPUs for modular systems, making the modular PLCs somewhat integrated. A typical modular PLC assembly is shown in Figure 2.1.

Modular PLCs come in a variety of sizes and are traditionally classed as *small, medium,* or *large*. The classification is based in part on the *power in the CPU* module and in part on the

FIGURE 2.1
Assembly of a modular PLC. Photo of CQM1 by permission of OMRON Electronics Inc.

amount of memory the CPU module offers for input and output image data. I/O image table size restricts the number of sensors and actuators that can be connected to a PLC. With increasingly more powerful processors, inexpensive memory, and high-speed data communication interfaces between PLCs, these classifications are becoming meaningless. Some of today's small PLCs make the large PLCs of only a few years ago look like toys. We distinguish only between integrated PLCs and modular PLCs in this book.

THE CPU MODULE

As demonstrated in Figure 2.2, the **CPU module** contains the **central computer** and its **memory.** The memory includes preprogrammed ROM memory containing the PLC's operating system, driver programs, and application programs, and the RAM memory, where user-written programs and working data are stored. PLC manufacturers offer various types of retentive memory to save user-programs and data while power is removed, so that the PLC can resume execution of the user-written control program as soon as power is restored. If the PLC has one of the following retentive memory options, it doesn't have to be reprogrammed each time it is turned on, so a keyboard and monitor don't need to be included as a part of every PLC.

1. In most PLCs, at least part of the RAM memory's contents are protected by a **long-life battery,** for years of use. Other PLCs have only **capacitor-based** power backup, so RAM memory is saved only for short periods of power outages (measured in hours). A PLC with only capacitor RAM backup must also offer at least one of the options noted below.

2. Many PLCs also offer **removable memory modules,** which are plugged into the CPU module. The user can make a copy of the user-program and data on **EEPROM** (electrically erasable programmable read-only memory) chips in the removable memory modules. The user may have to purchase an optional EEPROM writer, but some PLCs contain the special circuitry needed to write data to the memory modules at higher voltage levels than the CPU module would normally use. PLCs can be configured to copy the contents of a memory module into

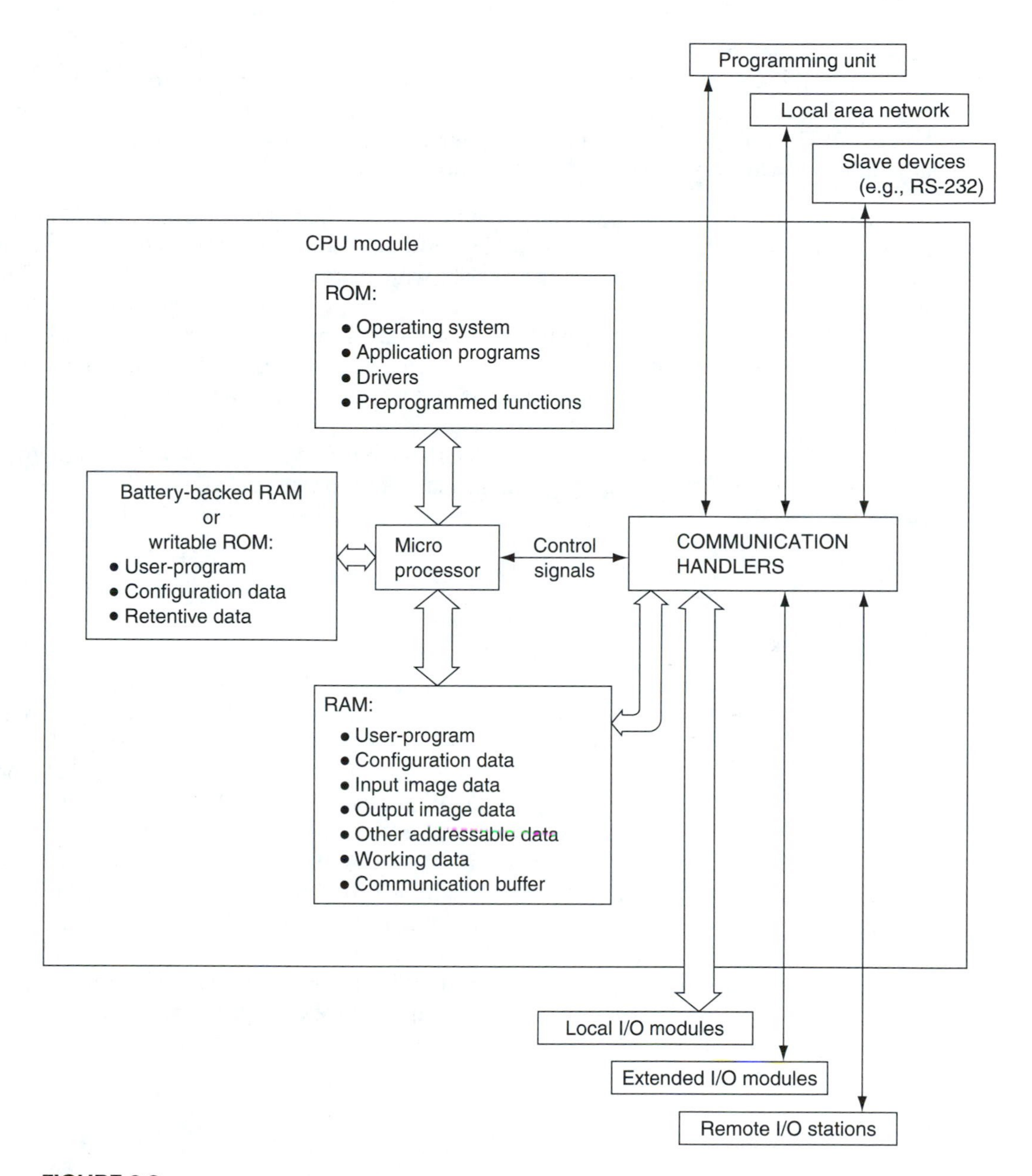

FIGURE 2.2
Memory and processors in a CPU module.

RAM memory whenever power is switched on. An EEPROM module can be plugged into any PLC of the same make, so they are also useful in copying programs and data from one PLC to the next.

3. Recently, PLCs have started to include **flash memory.** Flash memory is like EEPROM memory except that it can be written to faster and can be written to without needing special circuitry. Flash memory is used on some removable memory modules instead of using the older EEPROM memory chips, but flash memory is also sometimes built into the CPU module, where it automatically backs up parts of RAM memory even as the PLC runs. If power fails while a PLC with flash memory is running, the PLC will resume running without having lost any important working data after power is restored.

Modern CPU modules often contain more than one microprocessor. The main microprocessor chip's job is to execute the scan cycle while **slave microprocessors** handle the communications functions required to exchange data with increasingly powerful I/O modules, remotely located sensors and actuators, and with other controllers via local area networks. The slave microprocessors work in response to commands from the main microprocessor or in response to messages from microprocessors connected to this CPU module via serial communications links. Slave microprocessors may have direct access to data memory that they share with the main microprocessor (in which case, memory contents can be changed outside the main processor's control) and/or may have their own memory, which the main microprocessor reads and writes as part of its scan cycle or in response to interrupt signals that the slave processor generates. Use of slave microprocessors as communication handlers is discussed in more detail in Chapter 13 and in the communications interrupt section of Chapter 11.

THE RACK OR BUS

During every scan cycle, a CPU module reads and writes I/O modules that are part of the modular PLC. The CPU module is connected to each of those I/O modules via a set of parallel conductors called a **bus.** In some modular systems, the bus is in a backplane circuit card in a rack, and all PLC modules are plugged into slots in the rack. In other modular systems, I/O modules are plugged into the side of the CPU module or into the side of an I/O module that is already plugged into the CPU, so bus conductors are connected through the I/O modules.

Bus conductors are used for data that the CPU can send to or receive from the I/O modules, several bits at a time. The CPU must specify *which* of the I/O modules the CPU wants to read from or write to. I/O module addresses are assigned automatically according to how far the module is located away from the CPU module along the bus. Some bus conductors are used for miscellaneous control signals passed between the CPU module and I/O modules and to provide power to run the circuitry inside I/O modules. The bus does *not* provide power to operate the sensors or actuators attached to I/O modules.

THE POWER SUPPLY

As shown in Figure 2.3, a **power supply module** converts available power to dc power at the level(s) required by the CPU and I/O module internal circuitry. In North America the available power is typically 60 Hz 120 V ac or 220 V ac, although power supply modules are available for other input power characteristics. Output power must drive the computer circuitry at **5 V dc.** Power supply modules may be connected to the bus or may have to be wired to the CPU module in modular PLC systems.

Some—but not all—power supplies include power conversion circuitry that outputs **24 V**

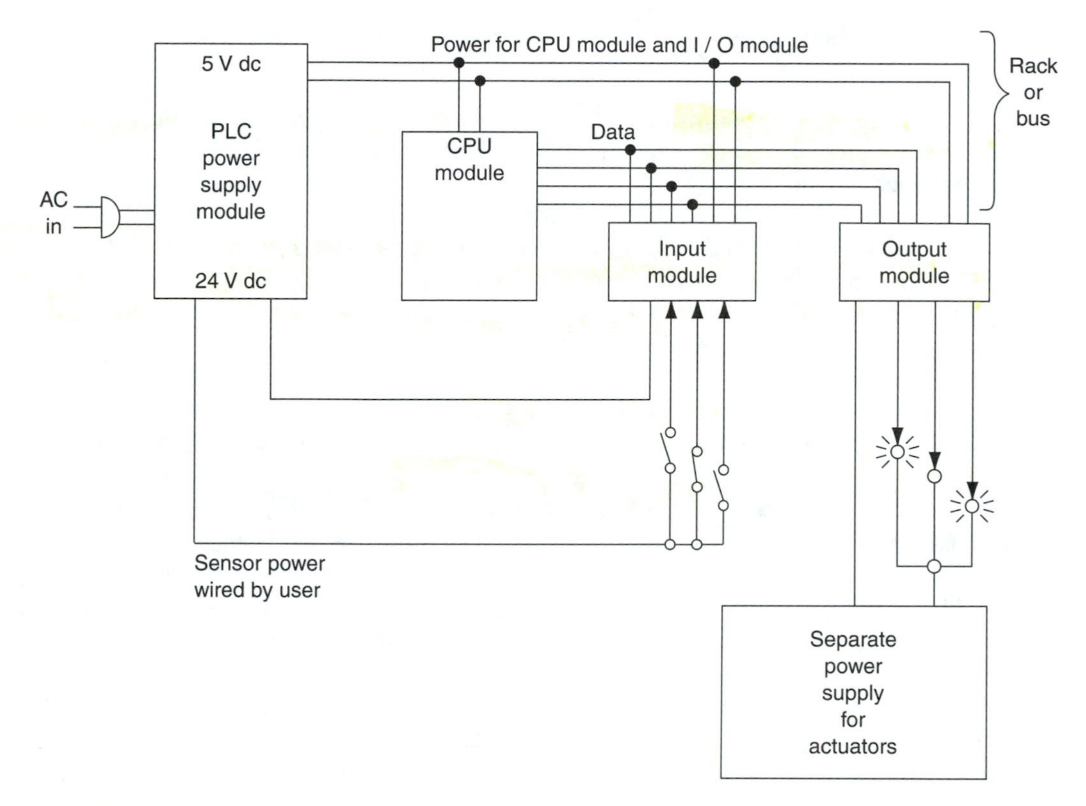

FIGURE 2.3
Power supplies in a PLC system.

dc via screw terminals on the power supply module. These PLCs provide only enough power to drive a few of the sensors and actuators that are connected to I/O modules.

If the PLC-based control system requires significant power to drive sensors and actuators, or needs dc levels other than 5 or 24 V dc, or needs other electrical signal characteristics, the user must provide additional power supplies and (for the high-power applications typical of some actuators) may need to supply relays, optical isolators, or other circuit-isolation devices.

I/O MODULES

Input and **output modules** (I/O modules) allow the PLC to be connected to sensors and actuators. The I/O modules isolate the low-voltage, low-current signals that the PLC uses internally from the higher-power electrical circuits required by most sensors and actuators. The user purchases the types of I/O modules that are needed for the sensors and actuators that need to be used, and the user can connect several different types (or several of the same type) of I/O modules to a PLC's bus. I/O modules offered by PLC manufacturers are designed to work with that manufacturer's CPU module, so the user can be confident that compatibility won't be a problem. Most manufacturers have a wide range of I/O modules that the user can select from, including:

1. **Digital I/O modules,** which are used to connect the PLC to sensors and actuators that can only switch on and off. Modules are available for a variety of dc and ac voltages and currents. Each module typically can be connected to several digital sensors and/or to several digital actuators of similar electrical characteristics.
2. **Analog I/O modules,** which are used to connect the PLC to sensors that can provide electrical signals that are proportional to a measured value or to actuators that vary their output proportionally with the electrical signals they receive from an output analog module. A single analog I/O module can typically only be connected to a few sensors or actuators of similar electrical characteristics.
3. Miscellaneous **intelligent I/O modules,** each with its own built-in microprocessor and memory. Intelligent I/O modules are designed for special purposes such as counting high-frequency signals or providing servo control of motors.
4. **Communication interface modules,** which are intelligent I/O modules that handle the exchange of data via a communication link. The user-program in the CPU writes data to the communication interface module, and the module ensures that it is placed on the communication network. Similarly, the communication interface module can accept data from other computers via the communications network and hold it until the CPU reads it from the module. Modern CPU modules can be connected directly to communication networks, so communication interface modules are needed only if communications requirements exceed the CPU's built-in capabilities. In this chapter we cover communications interface modules as intelligent I/O modules. Data exchanges via serial communications are discussed in Chapter 13.

When most PLCs power up, they perform a **self-check** that includes searching the bus to determine *how many* modules are present, in order to optimize the data exchange that will be performed each scan cycle. The PLC often exchanges different amounts of information with different types of I/O modules, so optimization of communication also requires the PLC to know what *type* of module is in each slot of the rack. In a truly optimized data exchange, a PLC will not waste time reading from an output module and will not try to write data to an input module. (The CPU module's memory will still contain an input image and an output image data word for each slot, but the data won't reflect sensor or actuator states.) Some I/O modules require different amounts of data to be exchanged, so the PLC must recognize these modules, too.

1. Some PLC manufacturers require the PLC user to set switches in the rack, depending on the type of module that will be plugged into that position in the rack. The PLC will read the rack switches to configure itself as the PLC is powered up.
2. Some PLC manufacturers build switches into I/O modules so that the I/O modules can be set up to operate in one of a few optional manners. When the PLC examines those I/O modules as it powers up, it will recognize the slightly different data exchange requirements.
3. Some PLC manufacturers prebuild identification features into I/O modules so that the PLC can examine each I/O module on the bus to configure itself as it is powered up.
4. Some PLC manufacturers require that the programmer enter configuration data into the PLC's memory before the PLC is put into run mode so that the PLC will perform the ap-

propriate data exchange functions. Programmer configuration is required only for the most sophisticated types of I/O modules.

Digital I/O Modules

Digital input modules allow a CPU module to read input image data words from the module. Each individual bit of a data word reflects the open or closed state of a single switch or switched sensor. Digital output modules accept output image data words from the CPU module. Each bit of the data word will turn a single actuator on or off.

Digital I/O modules primarily provide **electrical isolation** between the low-power internal circuits of the PLC and the (typically) higher-power circuits containing sensors or actuators. A digital output module also provides a **buffering** feature so that after the CPU writes a data word to the output module, the module will retain that data word (holding some actuators on and others off) until the next scan cycle, when a new output image data word will be written to the output module.

Digital Output Modules Some digital output modules contain **relays** as isolation devices (see Figure 2.4). Each bit of the data word sent to this output module will be either a 1 (5 V dc) or a 0 (0 V dc), and each bit's storage location is wired to control a relay coil. If the bit is a 1, the coil will actuate, closing a switch to allow current to flow in a circuit containing an actuator. The I/O module may have two external contacts per relay so that a separate actuator and power supply can be connected into each relay-controlled circuit. Each output circuit can then operate at a different power level (within the rated capacity for the relay) with different polarities, and some can even contain ac power supplies. More inexpensive relay output modules may have only one external contact per relay, with a common contact for a group of outputs (or for all the outputs), so that all circuits in that group must be wired to the same power supply. The common terminal of the output module must be connected to one terminal of the dc or ac power supply, and the other power supply contact is connected through individual actuators to the relay contacts.

Other output modules contain **transistors** that the PLC uses to switch external circuits on and off. (Additional components electrically isolate the internal circuits from the external circuits but are not shown in the following simplified diagrams.)

> Transistors can be used as electrically controlled switches. If a positive charge is applied to the base of an NPN transistor, the transistor switch will allow conventional current[1] to conduct into the collector contact and out through the emitter contact. A PNP transistor conducts conventional current from the emitter to the collector if the base charge is negative with respect to the emitter charge. Field-effect transistors (FETs) are also used to switch external circuits on and off by changing the charge at an FET's gate contact.

Transistor output modules can be either current sourcing or current sinking. A **current-sourcing** output module (see Figure 2.5) *provides* conventional dc current to actuators via the

---

[1] Conventional current direction is from positive to negative.

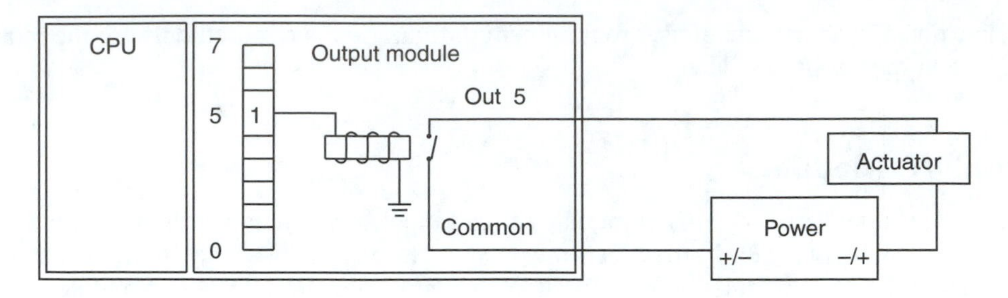

FIGURE 2.4
Relay output module (only one output circuit shown).

individual actuator output contacts. The current comes from the external power supply via a contact often labeled "common." The *PNP* transistors used in current-sourcing output modules are slightly more expensive and less abuse tolerant than NPN transistors. *NPN* transistors are used in the more common **current-sinking** transistor output modules, shown in Figure 2.6. In a current-sinking output module, the output module allows (conventional) current to flow *from the actuator into the individual output contact* and back to the power supply via the common contact. First-time PLC users often wire dc power supplies backward to current-sinking output modules.

As **triacs** drop in price, they are being used increasingly in place of transistors in output modules. As shown in Figure 2.7, a triac-controlled output circuit allows current to flow in either direction through the triac when the output module's circuitry applies a charge at the triac's gate contact. Power supply polarity is unimportant (ac power is even allowed), although the user must ensure a uniform polarity within each group of actuator circuits sharing a common contact at the output module.

Digital Input Modules Input modules usually contain **optoisolators** in each sensor-controlled circuit. When current from an external power supply flows through the optoisolator input contacts because an external sensor switch is closed, a light-emitting diode in the optoiso-

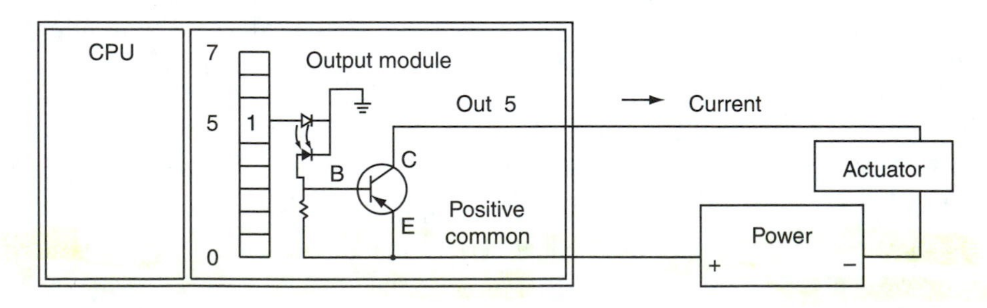

FIGURE 2.5
Sourcing transistor output module (only one output circuit shown). Sometimes called PNP output or open-emitter output.

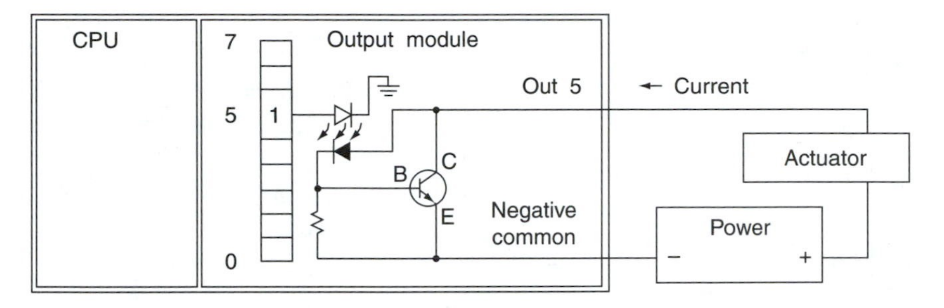

FIGURE 2.6
Sinking transistor output module (only one output circuit shown). Sometimes called NPN output or open-collector output.

lator generates light. A light-sensitive diode in the optoisolator allows current to conduct in a lower-power PLC internal circuit when it receives light. The optoisolator allows an external circuit to control a PLC's internal circuit without any electrical connections between the two. Because light-emitting diodes *are* diodes, the external circuit with the sensor must be connected to the digital input module with the correct polarity, so digital input modules are also classified as either current sourcing or current sinking, depending on whether they require (conventional) current to conduct out of or into the individual sensor contacts, as shown in Figures 2.8 and 2.9. Although it is reasonable to assume that an input module would receive current from a sensor, most input modules are current sourcing and must be connected to sensors so that closing the sensor switch will allow current to be conducted *from* the input module, through the sensor, to a power supply's negative contact. Current-*sourcing* input modules are more common only because *output* modules used in the PLC system are often current *sinking,* for reasons explained previously. Input and output modules are compatible with each other if a (current-sourcing) input module can receive signals from a (current-sinking) output module.

Optoisolators with two light-emitting diodes, as shown in Figure 2.10, are dropping in price, so more input modules are being built like this. Whichever direction current flows, one LED generates light, so power supply polarity is no longer important (except within groups of contacts sharing a common contact), and ac power supplies are also allowable.

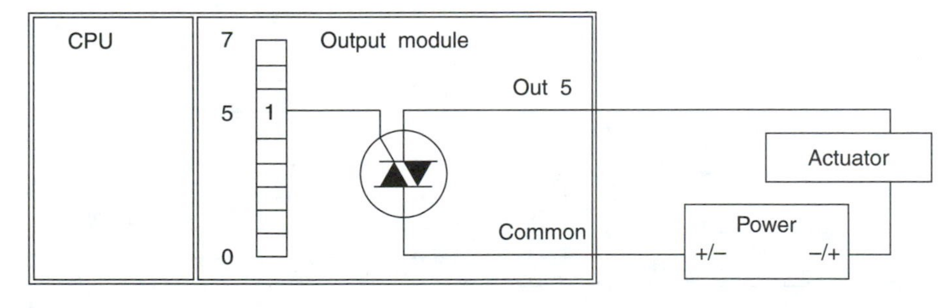

FIGURE 2.7
Triac output module (only one output circuit shown). Sourcing or sinking output.

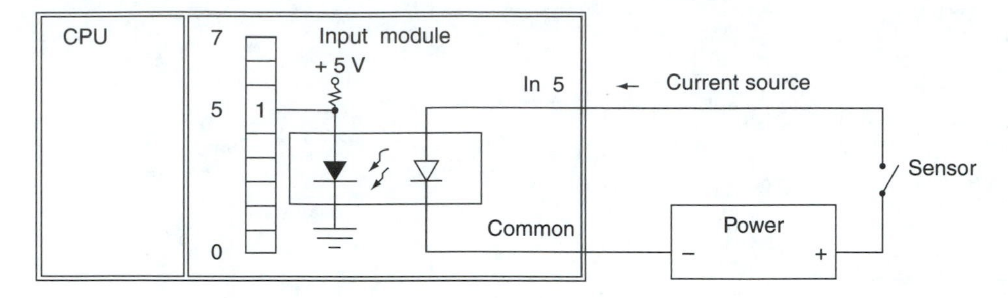

FIGURE 2.8
Sinking optoisolated input module (only one input circuit shown).

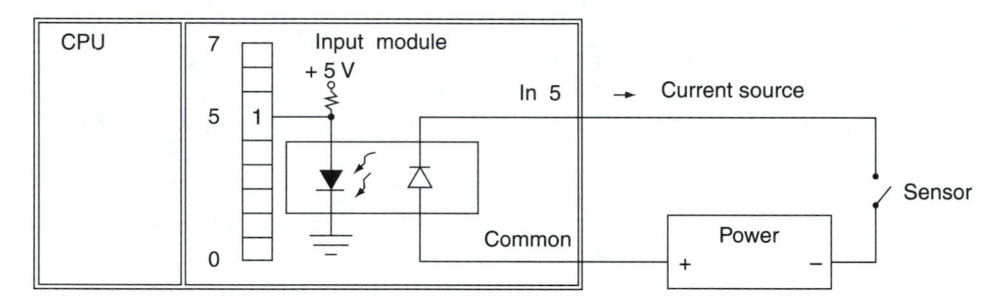

FIGURE 2.9
Sourcing optoisolated input module (only one input circuit shown).

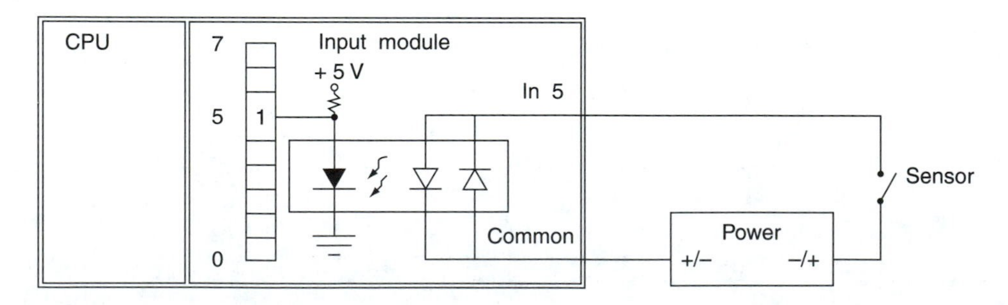

FIGURE 2.10
Sinking or sourcing optoisolated input module (only one input circuit shown).

Addressing of Digital I/O: Introduction to Addressing A PLC program instruction that examines a digital sensor or controls a digital actuator must include the address of the input image bit for the sensor or the output image bit for the actuator. The I/O image addresses reflect the actual locations of sensors and actuators. In this section we describe the relationship between a digital I/O address and the location of the I/O module and contact connected to the sensor or actuator. Addressing of I/O is discussed in more detail in Chapter 5.

ALLEN-BRADLEY PLC-5 DIGITAL I/O ADDRESSING

Digital I/O addressing for the Allen-Bradley PLC-5 contains four identifiers and two separating characters, as follows:

I:123/04

1. The leading letter is either the letter **I,** for input image bit, or **O,** for output image bit. All output image bits are saved in the *output image file,* sometimes called file 0 (zero). All input image bits are saved in the *input image file,* sometimes called file 1 (one). There are separate locations in memory for input and output data words. **I:**123/04 is a different address from **O:**123/04.

2. ":" must follow the leading letter.

3. The first two numerals after the ":" indicate which I/O rack contains the I/O module. A PLC-5 system may contain as many as 24 racks of I/O modules. The racks are numbered 00 to 07, 10 to 17, and 20 to 27. (Notice that these are octal numbers and do not contain the digits "8" or "9.") The CPU module must be located in either rack 00 or 01. An Allen-Bradley PLC-5 can be set up so that the *addressed* racks don't start and end at the ends of the metal rack structures. Because of this, Allen-Bradley calls a *physical* rack a **chassis** to distinguish it from the logical racks to which the addresses refer.

4. The third numeral indicates a **group** of I/O module(s). The leftmost group in a rack is group 0, and the rightmost group is group 7. (Again, no 8's or 9's are allowed.) A PLC-5 contains one 16-bit data word for the input images and one 16-bit data word for the output images from each group. A group therefore represents a maximum of 16 sensors and/or 16 actuators. Some of the bits in the data words may not be used. If, for example, only an *input* module is plugged into the rack position reserved for a group, none of the *output image bits* for that group are used.

The name *group* is used because early Allen-Bradley I/O modules had only eight input or eight output contacts, so *two* I/O modules in adjacent slots shared a single 16-bit input or output image word.

- Eight-bit I/O modules are still available, so a PLC-5 chassis *can* still be configured for *two-slot addressing* (see Chapter 10 for configuration). Any two digital I/O modules can be inserted into the adjacent slots as long as they do not contain more than 16 input contacts and 16 output contacts total.
- A modern I/O module often contains 16 sensor contacts and/or 16 actuator contacts. The PLC-5 system allows a chassis to be configured for *one-slot addressing,* in which every slot in the chassis is treated as a full group with a 16-bit input image word and a 16-bit output image word.

17	16	15	14	13	12	11	10	07	06	05	04	03	02	01	00

FIGURE 2.11
Allen-Bradley I/O image data bit addressing.

- Further miniaturization has made it possible to build 32 input contacts and/or 32 output contacts into a single I/O module, so Allen-Bradley also allows a chassis to be configured for *1/2-slot addressing*. Each physical slot in a chassis is treated as two groups. There are two input image data words and two output image data words representing the sensor and actuator states at the single I/O module.

5. A "/" character is typically used to separate the three-digit I/O image word address from the two-digit bit address. The programmer can substitute a "**.**" (period) for the "/" (so the example bit address can also be written as I:123.04). When entering a ladder logic program, neither the "/" nor the "**.**" will be displayed. Instead, the three-digit number will be displayed above the instruction and the two-digit number below, as in Figure 2.12. If the instruction uses a whole word, omit this "/" and the following bit identification number.

6. The final two numerals indicate which bit of the I/O image data word the instruction is using. As shown in Figure 2.11, the 16 bits are numbered from 00 to 07, and 10 to 17 (octal again).

Figure 2.12 shows a one-rung ladder logic program with PLC-5 digital addressing. It examines bit 6 in the input image data word representing the input module in rack 04, group 5. If that bit is on (it is), the program will turn bit 16 on in the output image data word representing the output module in rack 05, group 3. (None of the other I/O image bits are affected by this program, so their values aren't shown here.)

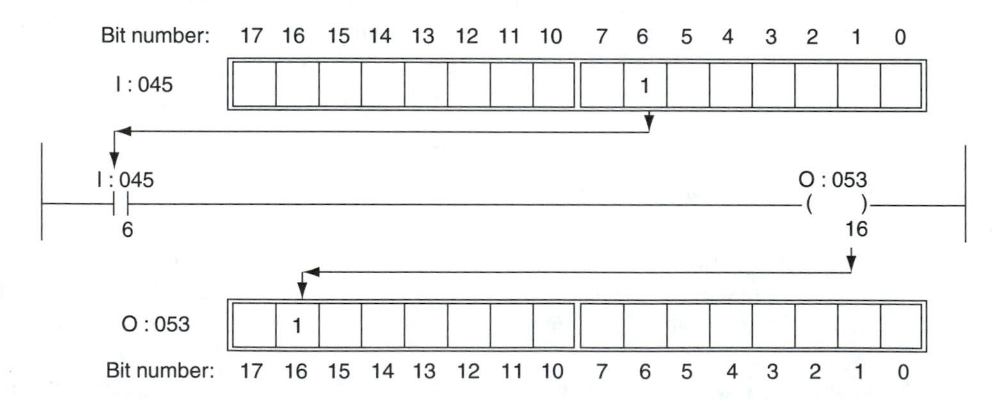

FIGURE 2.12
Ladder logic and I/O image file use in the PLC-5.

SLC 500

ALLEN-BRADLEY SLC 500 DIGITAL I/O ADDRESSING

Digital I/O addressing in **Allen-Bradley's SLC 500** line of PLCs is similar to, but simpler than, that used in the PLC-5 line. Words or bits can be addressed. I/O image bits are addressed in the form

I:12/04

1. The leading letter is either the letter **I,** for input image bit, or **O,** for output image bit. All output image bits are saved in the output image file, sometimes called file 0 (zero). All input image bits are saved in the input image file, sometimes called file 1 (one). The input image file is completely separate from the output image file, so I:12/04 is a different address from O:12/04.

2. ":" must follow the leading letter.

3. The two-digit number following the letter indicates the location that the I/O module occupies on the bus. I/O image data word number 0 is reserved for images of I/O capabilities built into the CPU module. I/O image words 1 to 30 are reserved for I/O modules beside the CPU (slot 1) to I/O modules 30 slots away.

4. A "/" character is typically used to separate the two-digit I/O image word address from the two-digit bit address. In some programming software, the programmer can substitute a "." (period) in place of the "/" (so the example bit address could also be written as I:12.04). If the instruction uses a whole word, omit this "/" and the following bit identification number.

5. The final two numerals indicate which bit of the I/O image data word the instruction is using. As shown in Figure 2.13, the 16 bits are numbered from 00 to 15.

Figure 2.13 shows a one-rung ladder logic program with SLC 500 digital addressing. It examines bit 6 in the input image data word representing the input module in slot 4. If that bit is on (it is), the program turns bit 14 on in the output image data word representing the output module in slot 5. (None of the other I/O image bits are affected by this program, so their values aren't shown here.)

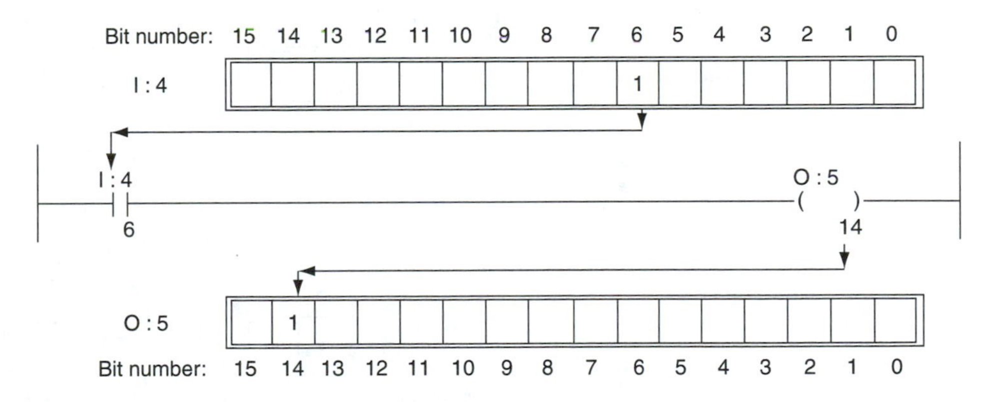

FIGURE 2.13
Ladder logic and I/O image file use in the SLC 500.

SIEMENS S5 DIGITAL I/O ADDRESSING

Digital I/O data can be addressed in STEP 5 programs for Siemens S5 PLCs as whole words, bytes, or as individual bits. For word or byte address formats, see the following section on addressing intelligent I/O. Bit addressing is usually used in Boolean logic, and would look like this:

I4.5

1. The first character is either an **I** for input image or a **Q** for output image (a Q is an O with a tail so that it doesn't look like a zero).

2. The number between the letter and the period indicates the location of the I/O module along the bus.

- In smaller S5 PLCs, such as the S5-100U, standard digital I/O modules cannot have more than eight input or output contacts, and each slot along the bus is represented by one process image input (PII) data byte and one process image output (PIQ) byte. I0 and Q0 are reserved for the I/O module in the slot closest to the CPU module, I1 and Q1 for the next module, and so on, up to I31 and Q31 for the thirty-second I/O module. The newer I/O modules, which have more than eight input or eight output contacts, are treated as analog I/O modules (see below).
- In larger S5 PLCs, such as the S5-115U, I/O modules can have as many as 32 input contacts and/or 32 output contacts, so *four* I/O image bytes are required to represent digital I/O values at a single I/O module slot. I0 to I3 and Q0 to Q3 are reserved for the slot closest to the CPU module. I4 to I7 and Q4 to Q7 are reserved for the next slot, and so on, up to I63 and Q63 for the sixteenth I/O module.

3. The "." (period) is necessary.

4. After the period, a number between 0 and 7 indicates which of the 8 bits is being addressed. (There are only 8 bits in one byte of an S5's memory.)

Figure 2.14 shows a one-rung ladder logic program with STEP 5 digital addressing. It ex-

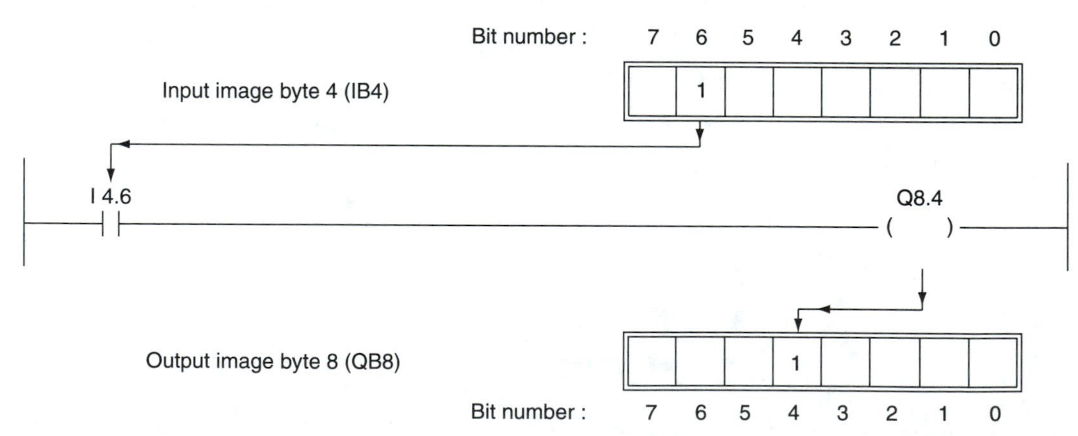

FIGURE 2.14
Ladder logic and I/O image use in the S5 PLC.

amines bit 6 in the input image data byte, which represents the input module in slot 4 (in a small S5 PLC). If that bit is on (it is), the program turns bit 4 on in the output image data byte representing the output module in slot 8. (None of the other I/O image bits are affected by this program, so their values aren't shown here.)

S7

SIEMENS S7 DIGITAL I/O ADDRESSING

Digital I/O addresses as used in STEP 7 (S7) programs for Siemens S7 PLCs are very similar to the system in STEP 5 programs. Words or bytes can be addressed, as described below in the section on intelligent I/O addressing, but a program would often examine or control a single bit. An address for an individual bit would look like this:

I4.5

1. The first character is either an **I** for input image or a **Q** for output image (a Q is an O with a tail so that it doesn't look like a zero).

2. The number between this letter and the period indicates the location of the I/O module along the bus. S7 I/O modules can have as many as 32 input contacts and/or 32 output contacts, so four I/O image bytes are required to represent digital I/O values at a single I/O module slot. I0 to I3 and Q0 to Q3 are reserved for the I/O slot closest to the CPU module, I4 to I7 and Q4 to Q7 are reserved for the next slot, and so on, up to I127 and Q127 for the thirty-second I/O module.

3. The "." (period) is necessary.

4. After the period, a number between 0 and 7 indicates which of the 8 bits of the byte of memory is being addressed.

Figure 2.15 shows a one-rung ladder logic program with STEP 7 digital addressing. It examines bit 6 in the input image data byte 4, which represents a sensor connected to the second I/O module from the CPU module. If that bit is on (it is), the program turns bit 4 on in the out-

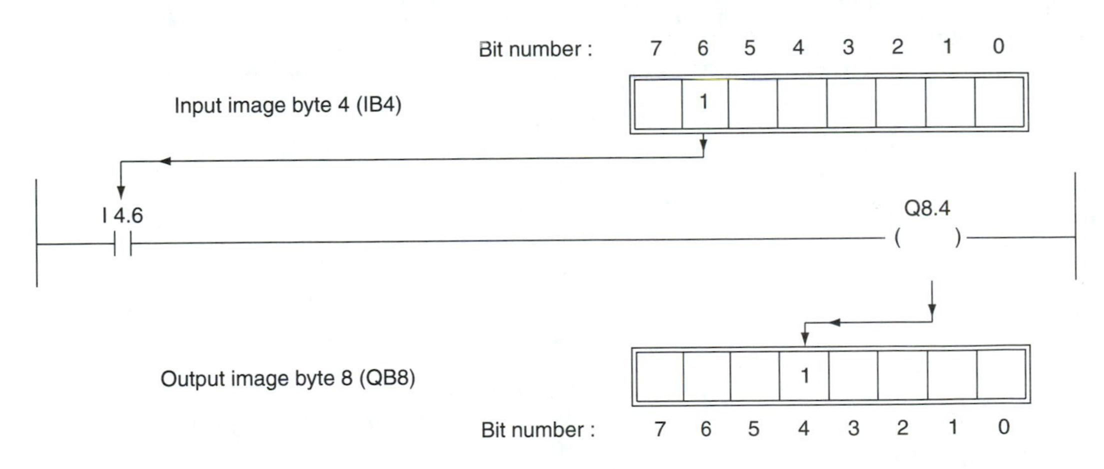

FIGURE 2.15
Ladder logic and I/O image use in the S7 PLC.

put image data byte 8 (to control a sensor attached at the third I/O module). (None of the other I/O image bits are affected by this program, so their values aren't shown here.)

OMRON CQM1 DIGITAL I/O ADDRESSING

Digital I/O addresses as used in OMRON's CQM1 PLC would look like this:

00405 which actually means **IR 00405**

1. The **IR** prefix indicates that the I/O register area of memory is being accessed. The programmer does not enter the IR prefix, because IR is usually assumed if no prefix is entered, so address 00405 is the same address as IR 00405.

2. The first numeral is a **0** (zero) to indicate an *input* image or a **1** (one) to indicate an *output* image. (Yes, you read that right.)

3. The second and third numerals indicate the location of the I/O module along the bus.

- x**00**xx indicates input image and output image data words reserved for I/O capabilities built into the CPU module. x**01**xx to x**11**xx indicate data words reserved for images for up to 11 input modules and 11 output modules.
- Input image address 001xx is reserved for the input module *closest* to the CPU, even if there are output modules closer. 002xx is for the second closest input module, and so on, to the eleventh input module.
- Output image addresses are assigned to output modules similarly. 101xx is for the *closest* output module, even if there are input modules closer, 102xx is the next closest, and so on, to the eleventh output module.

4. The final two numerals, numbers between 00 and 15, indicate which of the (maximum) 16 bits is being addressed. If an instruction is used that can manipulate whole 16-bit words, leave the bit identifier number out of the address (e.g., enter "004" only).

Figure 2.16 shows a one-rung ladder logic program with CQM1 digital addressing. It ex-

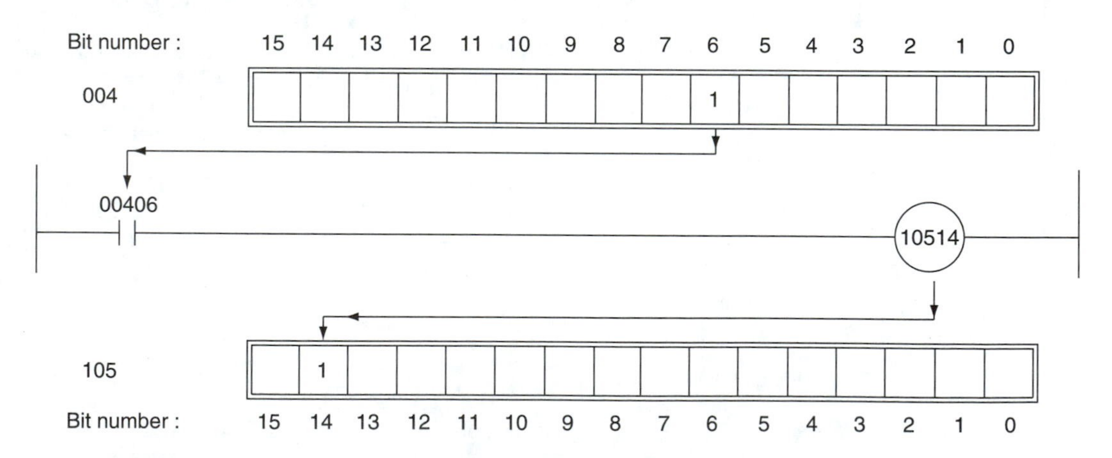

FIGURE 2.16
Ladder logic and I/O image file use in the CQM1.

amines bit 6 in the input image data byte representing the fourth input module from the CPU. If that bit is on (it is), the program turns bit 14 on in the output image data byte representing the fifth output module from the CPU. (None of the other I/O image bits are affected by this program, so their values aren't shown here.)

Analog I/O Modules

Analog output modules use *digital-to-analog (DAC)* chips to convert binary numbers to dc voltage or current signals with a magnitude proportional to the size of the digital value. **Analog input modules** contain *analog-to-digital (ADC)* chips, which convert analog dc signals to binary numbers. The Boolean ladder logic instructions that we have seen so far can be used to read and write binary numbers representing analog values one bit at a time, but in Chapter 6 we examine instructions that do the job more efficiently.

Analog I/O modules manipulate analog signals in several standard ranges, including (4 to 20) mA or (0 to 20) mA, (0 to 5) V or (0 to 10) V, and (−5 to +5 V) or (−10 to +10 V). Some analog I/O modules offer the user the ability to select which of those (or other) ranges to use by setting DIP switches on the analog I/O module, or sometimes through software configuration of the analog I/O module. (See Chapter 10 for I/O module configuration.)

The **resolution** of an analog I/O module is limited by the number of binary bits that its DAC or ADC chip uses. Each additional bit doubles the number of analog values the module can distinguish between, and therefore halves the difference between distinguishable values. Typical DACs and ADCs used in PLC systems use 12- or 13-bit binary values, so they have a resolution of 4096 or 8192 steps. If a 13-bit ADC is used, an input signal with a 20-V range (e.g., −10 to +10 V) can, in theory, be recognized as being within 8192 subranges of just under 0.0025 V each. In fact, most analog I/O modules reserve part of their range for over- and underrange values, so an input module for −10 to +10 V might actually accept and convert analog values of between −20 and +20 V, making each step of 13-bit resolution closer to 0.005 V in size. Since even this degree of resolution is better than a typical sensor's quality would justify, a 12-bit ADC is good enough for many applications. Analog output modules, similarly, rarely need DACs with more than 12 or 13 bits of resolution.

A PLC can only read or write binary numbers in standard sizes of 8, 16, 32, or 64 bits, so a 12- or 13-bit binary number representing an analog value must be manipulated as part of a 16-bit number. Some analog output modules ignore the low bits of the 16-bit values they receive (e.g., Siemens PLCs), and some analog output modules ignore the high bits of the 16-bit numbers they receive (e.g., Allen-Bradley's PLC-5). Analog input modules similarly put the relevant 12 or 13 bits into either the high or low bits of a 16-bit number, but the other bits are often used to report the status of the input module. Figure 2.17 shows the structure of a 16-bit binary value that a Siemens S5 analog input module might generate. The 13-bit binary value representing the analog input value is in the high bits (0000_0100_0000_0). The 13-bit number should not be used by the PLC program as a valid number, however, because one of the low two status bits indicates that the analog input module detects a problem in its analog input circuit.

The programmer should check the specifications for analog I/O modules being used to ensure that user-programs use binary values within the range that the I/O module is designed to handle, and to ensure that the analog I/O modules will interpret them in the way the program expects them to be interpreted:

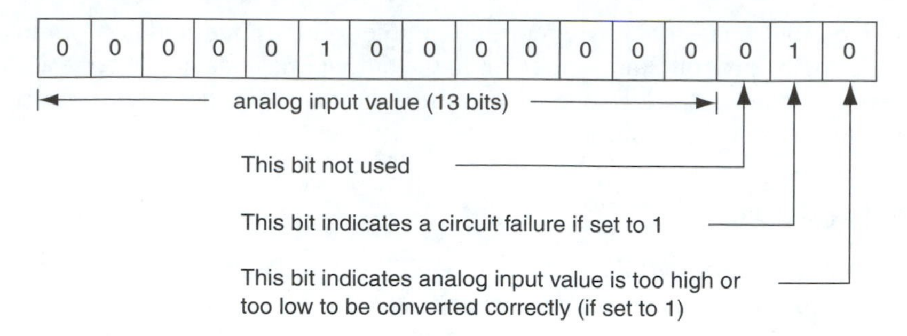

FIGURE 2.17
Sixteen-bit binary value representing an analog input value from a Siemens S5 analog input module.

1. Some analog I/O modules treat binary numbers as **unsigned binary** integer numbers. The 12-bit binary number range 0000_0000_0000 to 1111_1111_1111 would be interpreted as decimal numbers between 0 and 4095.

2. Other analog I/O modules treat binary numbers as **signed binary** integers. The 12-bit number range 1000_0000_0000 to 0111_1111_1111 would represent decimal values between −2048 and +2047.

3. Yet other analog I/O modules interpret binary numbers as **BCD code.** 16-bit binary BCD numbers 0000_0000_0000_0000 to 0100_0000_1001_0110 must be written to or read from the module to represent the decimals 0000 to 4096. The analog I/O module converts between the BCD format and the 12-bit natural binary format used by the DAC or the ADC.

Some manufacturers offer analog modules with input and output channels on the same module. Some analog I/O modules contain simple computers that can do more than just convert values between binary and analog. These analog I/O modules are discussed in the section on intelligent I/O modules.

Analog Output Modules An analog output module must store the binary value(s) that the CPU module writes to the module. Digital-to-analog (DAC) chips in the module continuously generate analog output signals proportional to the binary number's size, so as soon as the CPU module writes a changed number to the output module, the analog output value changes. Analog output modules often have between two and four output channels, which means that they can simultaneously convert that many digital numbers to analog signals and can provide each analog signal via a separate set of contacts. Each channel of an analog output module has three contacts. A power supply must be connected to provide power to the DAC, and the DAC allows a portion of the supplied voltage or current to be outputted via the third contact, as shown in Figure 2.18.

Analog Input Modules Analog input modules must convert a voltage or current signal to a digital number that the PLC's CPU module can read. Several types of ADC chips exist to do this. One type, the **flash** ADC, continuously generates binary values but is too expensive for use in most PLCs. The other types, of which the most common are the **successive approximation** and **dual ramp,** require time to perform analog-to-digital conversion, so they sample

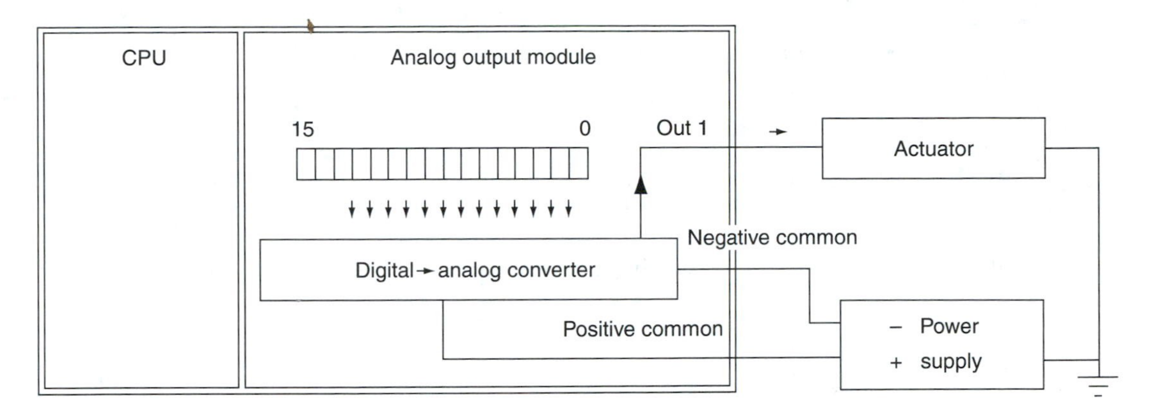

FIGURE 2.18
Analog output module (one output channel shown).

the analog signal repeatedly, then perform conversions on each sample. Changes in the analog input signal's value between samples are ignored. The result of the most recent conversion process is stored in the analog input module's memory for the CPU module to read. Binary values representing analog input signals are always slightly out of date, but if the user is careful in selecting an analog input module with an adequately high sampling frequency, the delay shouldn't be a problem in most applications.

Analog input modules often offer more channels than analog output modules, because a single ADC is often shared among several channels (making the time between updatings of an binary value that much longer). Figure 2.19 shows that each analog input channel requires at least three contacts: two for the power supply and one for the single-ended signal from the sensor, if the sensor is powered by the same power supply. Some analog input modules measure differential input signals, so each channel must have two sensor contacts across which there is a voltage potential or a continuous current loop.

Addressing of Analog Values All PLC languages now include instructions to manipulate whole data words at a time instead of manipulating them one bit at a time as Boolean ladder logic instructions do. These advanced instructions, which we examine in Chapter 6, must use addresses that represent whole data words. Analog I/O modules have their own memory and often have some built-in "intelligence." Addressing of analog I/O values will be discussed with addressing of other intelligent I/O modules, following the next section.

Intelligent I/O Modules

Some I/O modules contain complete computers. This book calls them all *intelligent* I/O modules, although each manufacturer has its own name for them.[2] Even some apparently simple

---

[2] Allen-Bradley's various **block transfer** modules for PLC-5s all contain at least rudimentary computers, as do the **specialty** modules they sell for SLC 500 PLCs. Siemens has **signal** modules, **CP** modules, and **SP** modules. OMRON offers **specialty** modules.

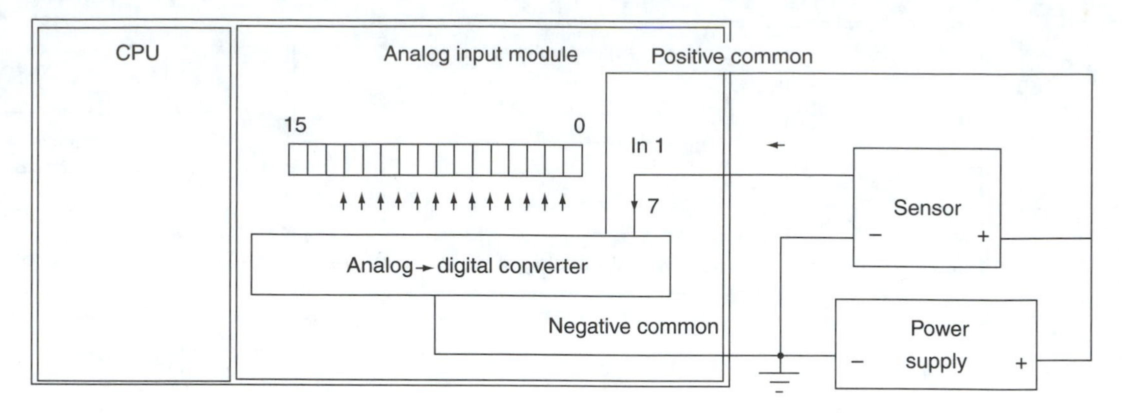

FIGURE 2.19
Single-ended analog input module (one input channel shown).

digital I/O modules now come with embedded processors, making them simple computers. It isn't easy to describe what an intelligent I/O module is because there are so many things that a computer can do, and because there are so many different sizes of computers.

In addition to just writing data to, or reading data from, an intelligent I/O module, a user-program must also include instructions to send **commands** and/or **configuration** data to the computers in intelligent I/O modules, and must contain instructions to read the results that these modules produce and/or **status** information describing the module's state.

Analog I/O modules are examples of simple intelligent I/O modules. Some analog I/O modules can be configured (by writing configuration data words to them) to scale analog values so that the user-program running in the CPU can be programmed to exchange engineering unit data values with the module. The user-program can, for example, send the number 25 to an analog output module to make a dc motor rotate at 25 rpm. Other intelligent I/O modules are available to perform functions as simple as counting rapidly changing signals from sensors such as optical encoders, or as complex as making a motorized conveyor execute a sequence of moves.

Addressing of Intelligent I/O Modules There is no universally accepted procedure for reading or writing intelligent I/O module data. Some PLCs automatically assign **several addresses** per intelligent I/O module. The user-program then reads and writes intelligent I/O module data using the same instructions used to read or write digital I/O modules. Other PLCs require the user-program to include instructions that transfer blocks of multiple data words between the I/O module and the CPU's data memory. In this second system, a user-program can manipulate the copy of the intelligent I/O module's data only while it is in the CPU's data memory. Some PLCs have systems that **combine** block transfers of data and automatic exchanges of data between the CPU and intelligent I/O modules. Analog I/O modules are often treated as intelligent I/O modules. Some digital I/O modules may be treated as intelligent I/O modules.

ALLEN-BRADLEY PLC-5 ANALOG I/O ADDRESSING

Allen-Bradley's PLC-5 offers a set of intelligent I/O modules that Allen-Bradley calls **block transfer (BT)** I/O modules. Analog I/O modules are BT modules. BT modules do not have individual addressable memory locations. Whole blocks of data words must be transferred to and/or from BT modules using block transfer instructions, which are described in Chapter 7. Data words and bits must be written into the CPU's working data memory, and then a block transfer instruction can be used to copy that block of data to a BT module. A block transfer instruction can be used to copy data from a BT module to any area of the CPU's working data memory, and then the user-program can read individual words or bits from the data memory. In Chapter 7 we describe how to address a PLC-5's data memory. The structure of the data block is important. Each BT module comes with a manual describing the structure of the data blocks it accepts from a CPU or writes to a CPU, and the programmer must understand the structure to know which of the words in the data block to operate on.

ALLEN-BRADLEY SLC 500 ANALOG I/O ADDRESSING

Allen-Bradley calls SLC 500 intelligent I/O modules **specialty I/O modules.** Analog I/O modules are included in the set of specialty I/O modules. Data must be read directly from a specialty I/O module or written directly to it. To do so, the module's memory must be addressed in this form:

M0:12.3/45

1. **M** indicates that the data exists in the memory of a specialty I/O module.

2. The digit after the M is a **0** (zero) or a **1** (one). 0 sometimes indicates an output value and 1 sometimes indicates an input value, but 0 and 1 may only indicate separate areas in a module's memory.

3. The "**:**" is required.

4. The number between the ":" and the "." indicates the location of the specialty I/O module on the bus. The location nearest the CPU is slot 1, and numbers up to 30 indicate the next 29 slot locations.

5. The "**.**" (period) is required.

6. The number following the period may be 0 to some higher number, and represents the address of the data word in the specialty I/O module's memory. Specialty I/O modules have differing amounts of memory.

7. The "/" is optional. Include it and the following number only if the program instruction is to operate on a single bit. Do not include a "/" if the instruction is to operate on a whole 16-bit data word.

8. The number following the "/" is a number from 0 to 15, indicating on which of the 16 bits this instruction is to operate.

SIEMENS S5 AND S7 ANALOG I/O ADDRESSING

In the S5 series of PLCs, there are two distinct classes of intelligent I/O modules. Some are treated as analog modules and are addressed in a form similar to how digital I/O data is addressed, except that data is usually read or written as whole data words in the following form:

PW 72 (or IW 72 or QW 72)

1. The first letter must be a **P** if it represents a memory location *in* the I/O module. In medium-sized S5 PLCs such as the S5-100U, analog I/O modules are *included in the scan cycle's reading and writing of I/O modules,* so the programmer *can* use an **I** or a **Q** to read or write the data values in the CPU's input or output image tables. In these PLCs, use of the P prefix is not allowed except in an interrupt service routine, as discussed in Chapter 11.

2. The next character must be a **W** to indicate a full 16-bit data word, or a **Y** to indicate an 8-bit byte. To address a single bit, omit this character and use the format described earlier for addressing digital I/O modules. *You cannot address individual bits in an analog I/O module if the P prefix is used.*

3. The following number indicates the location of the analog I/O module and the channel of that I/O module. Each channel represents storage for one word, consisting of two bytes of data. When an instruction uses a word of data, it operates on two bytes, as shown in Figure 2.20.

(a) Medium-sized S5 PLCs such as the S5-100U allocate four input image words and four output image words for each slot into which an analog I/O module can be plugged. Only the eight slots nearest the CPU module can have analog I/O modules plugged in. As shown in Figure 2.21, the closest slot includes analog input and output addresses that start at byte 64 and go to byte 71. The last byte in the last module's slot is byte 126.

(b) Large S5 PLCs such as the S5-115U have analog I/O modules that have 32 bytes of memory that the PLC program can read or write. Only the four slots nearest the CPU module can have analog I/O modules plugged in. As shown in Figure 2.21, the closest slot includes analog input and output addresses that start at byte 128 and go to byte 159. The last byte in the last module's slot is byte 255.

The STEP 5 programming language does not allow reading or writing of individual bits in an analog I/O module's memory. Some highly sophisticated intelligent I/O modules in the S5 system, especially the **communication processor (CP)** and the **interface module (IM)** types, have memory containing data that is *automatically* copied between the CPU module and the intelligent I/O module's memory. Two areas of the CPU's memory can be used for this purpose:

1. One area is called the **interface flag area.** When you configure an S5 PLC (see Chap-

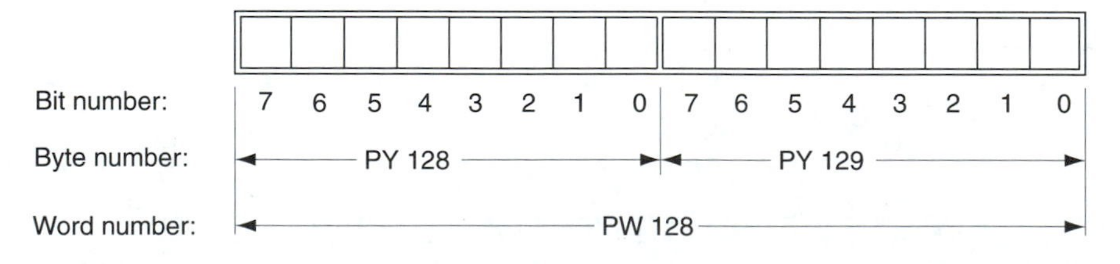

FIGURE 2.20
An S5 analog channel uses one word, which consists of two bytes.

Slot number	0	1	2	3	4	5	6	7
CPU		I/O modules		I/O modules			I/O modules	
Medium-size S5	IB 64 – 71	IB 72 – 79	IB 80 – 87	IB 88 – 95	IB 96 – 103	IB 104 – 111	IB 112 – 119	IB 120 – 127
	QB 64 – 71	QB 72 – 79	QB 80 – 87	QB 88 – 95	QB 96 – 103	QB 104 – 111	QB 112 – 119	QB 120 – 127
Large-size S5	PY 128 – 159	PY 160 – 191	PY 192 – 223	PY 224 – 255				

FIGURE 2.21
Analog address allocation in S5 PLCs.

ter 10), you specify which parts of the PLC's data memory will be used for this purpose. The user-program should be written to include Boolean instructions to monitor the interface flags containing status information from the I/O module, and to control other interface flags to control the I/O module. See Chapter 5 for addressing an S5's working memory.

2. The other area is the **interface area,** which is an area of the S5's memory reserved only for exchanging data with intelligent I/O modules. This area of memory does not contain memories that can be addressed in the conventional manner, but the S5 PLC includes **data handling function blocks,** prewritten subroutines that can be used to copy sets of data words into and out of interface memory. To control an intelligent I/O module, a program must write individual data words or bits to addressable working memory, then use a data handling function block to copy that area of working memory to the interface memory. Reading the I/O module's status requires using a data handling function block to copy data from the interface memory to working memory, then reading individual words and bits from the working memory. See Chapter 5 for addressing an S5's working memory. In Chapter 10 we describe how to configure an I/O module so that the CPU "knows" which section of interface memory it must use in copying data to and from the module. In Chapters 7 and 13 we describe how to use data handling function blocks. Each intelligent I/O module comes with a manual describing the structure of the interface data block that the module exchanges with the CPU. The programmer is responsible for using this information to be able to assemble data blocks for controlling the I/O module and for interpreting status data from the module.

OMRON CQM1 ANALOG I/O ADDRESSING

The OMRON CQM1's analog I/O modules and other intelligent I/O modules are addressed using the same methods as for addressing a digital I/O module, except that whole data words are usually manipulated as single units, using addressing in the form

004 which actually means **IR 004**

Notice that a word address is the same as a bit address except that it does not include the two digits that identify a bit.

Analog/intelligent I/O modules may use more than one input or output image word per module. Some CQM1 CPU modules have a pulse width modulation feature that can be used to output *analog* signals, but they require the use of a special set of instructions, which are described in Chapter 11.

PROGRAMMING UNITS

A PLC needs a user-program and may need initial data in working memory before it can be used to control an industrial process. The programmer therefore needs a way of entering the program and data into memory, despite the fact that the average PLC does not come with a keyboard or monitor.

PLC manufacturers sell various types of programming units for program and data entry. Programming units can also be used to monitor the execution of a program and to monitor the contents of working data memory as the PLC runs. The simplest type of programming unit is the small device often called a **pendant.** A pendant has a connecting cable so that it can be plugged into a PLC's programming port. Pendants usually have only a few special-purpose keys and a small display for a few lines of text or for one simple ladder logic rung. A pendant is easy to carry around, but programming and monitoring a sophisticated PLC is difficult with only a pendant.

At one time, most PLC manufacturers offered proprietary programming units, which looked like personal computers except that they could only be used to program and monitor PLCs. These types of programming units are rarely seen today. PLC manufacturers now offer sophisticated **programming software** that can be installed in a personal computer, and interface hardware so that the personal computer can be connected to the PLC's programming port. Sometimes the interface hardware consists of a serial cable, sometimes with an **RS-232C-to-current loop converter,** so that the user can use the personal computer's RS-232C serial port to exchange data with the PLC. For large amounts of data exchange at high rates, the RS-232C standard isn't adequate, so some manufacturers offer optional **high-speed communications interface cards** that can be installed in a personal computer and are connected to the PLC via a serial interface cable. To avoid the need to lug a programming unit of any sort to the work site, some programming software and interface hardware allow a personal computer type of programming unit to be connected to a PLC local area network so that the programming unit can be used to modify and/or monitor the contents of any PLC on the same LAN.

TROUBLESHOOTING

Selecting the right PLC modules and interconnecting them correctly is only part of building a PLC-controlled system. If the control system still doesn't seem to work, the following potential problem sources should be investigated:

1. Since the PLC I/O modules aren't usually connected to a power supply *through* the PLC, they must be connected to a power supply externally. Have you connected a power supply in series with the I/O module and the sensor or actuator? (Is it the correct power supply?) Are proximity switches connected so that the power supply provides power to the sensor and provides power to the PLC input circuit?

2. Is the dc power supply connected with the right polarity? What polarity (if any) does the I/O module require to be connected as common? (The author has seen current-sinking output modules discarded as being defective because they wouldn't provide current to actuators.)

3. Is the PLC program addressing correct? Especially with inexperienced programmers, it is easy to type an input address for an output instruction, and difficult to find this type of error. If the program seems to work (the program can be monitored as the PLC executes it) and the LEDs on the PLC's output modules don't go on when the program monitor says they are on, look carefully at the address to which the program is outputting.

4. Is the PLC program correct? If this rung looks like it should operate properly, is there another rung somewhere else in the program that conflicts with this one? Most PLC programming software packages offer a cross-reference feature that tells you every place where an address is referred to in a program. Examine each of those rungs to ensure that none of them is causing the problem.

QUESTIONS/PROBLEMS

1. What three types of components are required in a modular PLC?

2. Which component contains the PLC's main memory? What methods are used to prevent the contents of a PLC's user memory from being lost when a PLC's power is disconnected?

3. Is the scan cycle program in a PLC's RAM memory or ROM memory? Is the user program stored in a PLC's RAM memory or ROM memory?

4. What is a communication handler?

5. What type of output module would you use to:
- **(a)** Simply turn a dc motor on or off?
- **(b)** Control a dc motor's speed from the main program?
- **(c)** Allow the main program to initiate a move sequence without having to control the execution of the move sequence?

6. Sketch how you would connect a mechanical switch to a current-sourcing digital dc output module.

7. Sketch how you would connect an inductive proximity switch to a current-sourcing digital dc output module. Would the proximity switch have to be an NPN sensor or a PNP sensor?

8. If a digital sensor were attached to the third input contact of the second input module away from the CPU module (the fourth I/O module away from the CPU), what address would you use in a user-program to examine the input image for that sensor in each of the following?
- **(a)** A PLC-5 (single-slot addressing)
- **(b)** An SLC 500
- **(c)** An S5
- **(d)** An S7
- **(e)** A CQM1

9. What address would you use to address a 16-bit input image containing the states of *all* of the digital sensors attached at the module in Question 8?

10. If your PLC is using an intelligent I/O module to handle communicating messages between the PLC and other computers attached to a local area network (LAN), what different types of data must the CPU read or write from the module (in addition to the message data being sent or received)?

11. To read a PLC-5 analog input module, you need a special instruction, not just a special address. What instruction would you use?

12. You have just written a program that is supposed

to detect tiny variations in an analog input by examining the least significant bit of the input data word to detect small changes. The analog value is changing, but for some reason your program *never* detects the change. The low bit never seems to change from zero. *Why does this bit never change?* (Assume that your program is *not* the problem.)

13. What memory areas can be used to exchange data between a Siemens CPU module and a communications processor (CP) module?

14. What types of programming units are available?

3

PROGRAMMING IN BINARY LOGIC (BOOLEAN LOGIC)

OBJECTIVES

In this chapter you are introduced to:

- Boolean logic instruction elements and how to combine them into Boolean statements to control industrial processes.
- Ladder logic's graphical instructions and the instruction list (Siemens' STL language) equivalents.
- Data bits.
- Programming "tricks" such as one-shots, latching, and sequencer programs.

For many control applications, actuators need only to be turned on or off, depending on whether sensors are on or off. For example, a conveyor motor might be turned on while a box is detected on the conveyor belt but not if there is already a box waiting to be removed from the chute at the end of the conveyor, as demonstrated in Figure 3.1.

Boolean logic control programs examine and control on and off states. Each Boolean logic program can contain several conditional statements. An example of a Boolean statement might be:

```
If (a box is on the conveyor) AND (there is not a box in the chute) THEN
(turn the conveyor motor on).
```

which could be translated into language reflecting the existence of sensors and actuators as:

```
If (sensor_A is ON) AND (sensor_B is NOT ON) THEN (turn actuator_C on).
```

In Boolean logic, you think in terms of true and false. In the Boolean logic statement above, there are two conditional elements, each of which can be true or false. If sensor_A is on, the first element is true; otherwise, it would be false. If sensor_B is *off,* the second element is true; otherwise, it would be false. If an entire Boolean logical statement evaluates as true, the output element it controls is made true (the action it controls will be performed; e.g., actuator_C

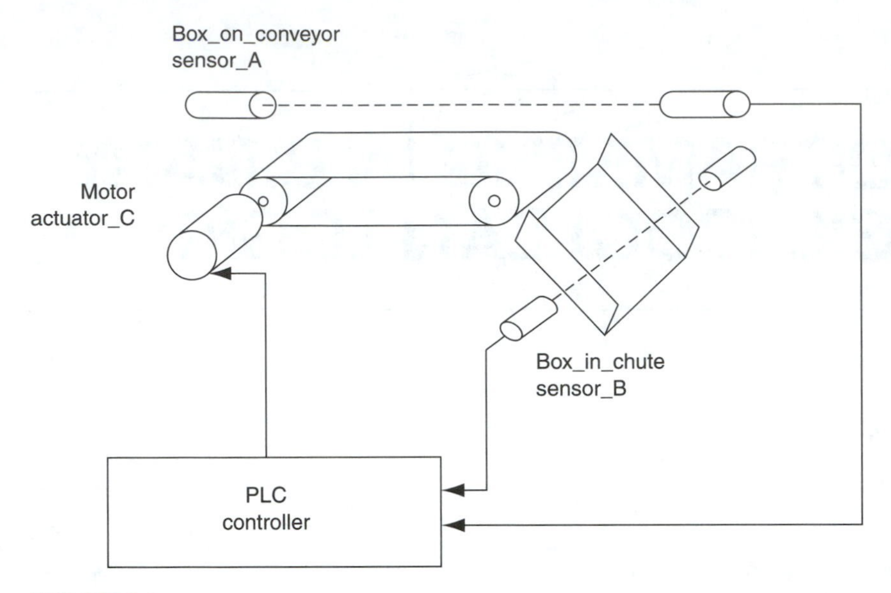

FIGURE 3.1
Conveyor motor control system.

will be turned on). Inside a computer, a binary **1,** which is sometimes referred to as **on,** is used to represent **true,** and **0 (off)** represents **false.**

In the Boolean logic statement we have been examining, there are four possible combinations of sensor on and off states, so there are four possible **results of the logical operations** (which Siemens calls the RLO). The Boolean AND operator will result in a final true only if both elements of the statement are true:

Sensor states		Boolean logic element states		
Sensor_A	Sensor_B	(Sensor_A ON?)	(Sensor_B NOT ON?)	(Actuator_C)
off	off	false	true	false (off)
off	on	false	false	false (off)
on	**off**	**true**	**true**	**true (on)**
on	on	true	false	false (off)

The on–off state of every sensor and actuator is represented by a binary digit in the PLC's input image or output image memory area of memory. In Chapter 1 we saw that PLCs execute scan cycles during which they read the states of sensors attached to input modules into input image memory and copy data from output image memory to the output modules attached to actuators. In Chapter 2 we examined the relationship between memory addresses and locations of I/O modules. Boolean logic programs can also read and write bits in other areas of a CPU module's memory.

The earliest PLCs could be programmed only in Boolean logic. Even today, Boolean logic remains the underlying control language in most PLC control programs. PLC programs are usually written in a graphics-based programming language called **ladder logic,** although other programming languages with Boolean logic instructions are sometimes used. Those other languages include assembler language–like **instruction list** programming languages, and higher-level **structured text** languages based on programming languages such as Basic, Pascal, and C. We examine the ladder logic and instruction list programming languages in this chapter.

Significant differences exist between PLC programming languages, even in ladder logic. The **IEC** (International Electrotechnical Commission), a sister body of the International Standards Organization (ISO),[1] has recently been making efforts to standardize the PLC programming languages. In fact, ladder logic (LD), instruction list (IL), and structured text (ST) are names assigned by the IEC to represent three of the five standardized languages. The **IEC 1131-3** standard (now renumbered as **IEC 6-1131-3**) is discussed in Chapter 9.

LADDER LOGIC FOR BIT MANIPULATION

Ladder logic programs and ladder logic elements are graphics-based. Ladder logic programs look similar to the relay logic circuit diagrams that industrial electricians use. A ladder logic program consists of horizontal **rungs** drawn between two vertical **rails,** so programs look a little like drawings of stepladders. In keeping with the electrical diagram theme, the left rail can be considered an electrical power rail and the right rail a common connection. Each rung contains instruction elements that examine memory bits and contains at least one output element that controls a memory bit. If the bit-examine elements on a simple rung are true, the result of the overall logic operation (RLO) is true, and the bit(s) controlled by that rung's output element(s) are turned on. The electrical analogy is that if a path exists for electrical current to flow from the supply rail through switches to the common rail, the current will turn actuators in the circuit on. Unlike an electric circuit, however, a ladder logic program's output elements can usually only be entered as the rightmost elements on each rung.

> As you read this section, please remember that a PLC *repeatedly* executes the ladder logic user-program, one rung at a time, from the first element at the top left to the last element at the bottom right. Remember, too, that execution of the user-program is the second of the three-step PLC scan cycle and uses only *images* of input and output states. Actual input conditions are read during the first scan cycle step, before the user-program starts executing, and actual output states are changed during the third scan cycle step, after the user-program ends.

Ladder Logic Elements

The basic set of ladder logic Boolean elements (instructions) consists of:

—| |— EXAMINE ON, sometimes called EXAMINE INPUT CLOSED, which is evaluated as true if the bit that is examined is a 1. (Other words used for 1 include *true, on,* and *set.*)

—| / |— EXAMINE OFF, sometimes called EXAMINE INPUT OPEN, which is evaluated as true if the bit that is examined is a 0. (Other words used for 0 include *false, off,* and *reset.*)

---

[1] ISO is not an acronym for International Standards Organization. *Iso* is a Greek word meaning "one" or "together."

—( )— ENERGIZE OUTPUT, sometimes called OUTPUT COIL, which controls a memory bit. The memory bit:
- Will be made 1 (or on) if the logical statement preceding this element is true, or
- Will be made 0 (or off) if the logical statement preceding this element is false.

Two additional variations of the ENERGIZE OUTPUT coil are in common use:

—(L)— OUTPUT LATCH, which controls a memory bit that will be made 1 (on) if the logical statement preceding this element is true, but will NOT be changed to 0 (off) if the logical statement becomes false.

—(U)— OUTPUT UNLATCH, which controls a memory bit that will be made 0 (off) if the logical statement preceding this element is true, but will NOT be changed to 1 (on) if the logical statement becomes false.

Each of these elements must include the address of the bit in the CPU's memory that is being examined or controlled:

1. EXAMINE ON and EXAMINE OFF elements can examine any data bit, including data bits in the input image data table, the output image data table, and bits in memory areas containing other working data. In some PLCs (e.g., Siemens S7), status bits in the CPU can be examined if the instruction contains a special address indicating that the instruction refers to the status bits (e.g., BR, OV, OS, >0, etc.; see Chapter 5).

2. The output elements (ENERGIZE OUTPUT, OUTPUT LATCH, and OUTPUT UNLATCH) can control data bits in the output image data table or in the working data areas of memory. Most PLCs also allow output instructions to change data bits in the input image data table, although the scan cycle will overwrite the input image data bits next time it reads new values from the input modules. Programs that change input image table data can have some interesting and chaotic results if later rungs examine the changed input images.

Every brand of PLC requires the programmer to use a specific **proprietary addressing convention.** The addressing of input and output image bits was covered in Chapter 2 and is reviewed in Chapter 4. In examining the proprietary addressing conventions, the reader may note a pattern: All the addressing systems contain three fields, as shown in Figure 3.2. The first field identifies the memory area, the second field identifies the I/O module, and the third field identifies a single sensor or actuator contact at that I/O module. The same form is used to identify other data bits in a PLC's memory.[2]

Boolean ladder logic is reasonably standard from one PLC to the next, so we can become familiar with ladder logic without relying on any one of the proprietary addressing systems. (In subsequent chapters we will be forced to cover the programming of each PLC separately.) Until it becomes necessary to use proprietary addressing, we will use the following simplified (imaginary) addressing system:

Identifiers:	**IN**	for input image data
	OUT	for output image data
	D	for working data memory

[2] In some PLCs, if an address is entered without a prefix, it is assumed that the input or output image area is being addressed.

Data type identifier:	Data word address	Data bit number
• Input • Output • Data	• Location of I/O module relative to CPU module • Unique memory word or byte address	• Location of sensor/actuator contact • Memory bit

FIGURE 3.2
Standard bit address format.

Data words: **0 to 7**
Data bits: **0 to 7**

For example:

`IN.3.6` means the input image data bit representing contact 6 of input module 3
`OUT.0.7` means the output image data bit representing contact 7 of output module 0
`D.2.4` means data bit 4 of data memory word 2 (which is not associated with any input or output data)

Constructing Ladder Logic Programs

Each ladder logic rung *must* have an output element as the rightmost element. EXAMINE instructions usually precede the output element, as shown in rung 1 of Figure 3.3. In this example rung, three conditions must be true simultaneously for the output to be turned on. This is known as a **Boolean AND statement** and is equivalent to a series electric circuit. (Notice that PLC-based Boolean programming is already more powerful than electrical circuits, because EXAMINE OFF instructions are available. Electrical switches can't conduct current when they are open!)

In some ladder logic languages, it is possible to program rungs without any EXAMINE elements, as shown on rung 2 of Figure 3.3. The bit controlled by the output element on such a rung will always be 1 (true). In other ladder logic languages, Boolean statements without con-

Rung 1: The following rung will only turn OUT.5.0 on while input image bit IN.1.0 is on, and output image bit OUT.6.0 is on, and data bit D.7.0 is off.

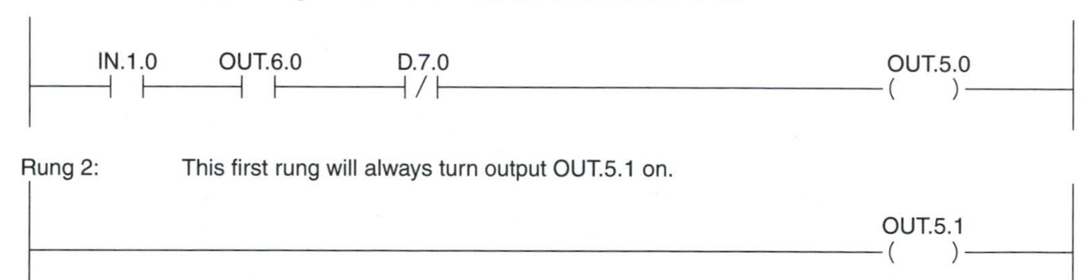

FIGURE 3.3
Simple ladder logic rungs.

ditions aren't allowed, so the manufacturers usually provide data bits that are always on, which can be examined to yield a statement that is always true. Yet other Boolean logic languages offer statements that can be used to set the Boolean "result of logic operation" to true.

A rung can contain branches from the main rung, similar to parallel electric circuits, so that there are alternative ways for an output bit to be turned on or so that more than one actuator can be controlled by the same rung. The first rung in Figure 3.4 shows how a ladder logic

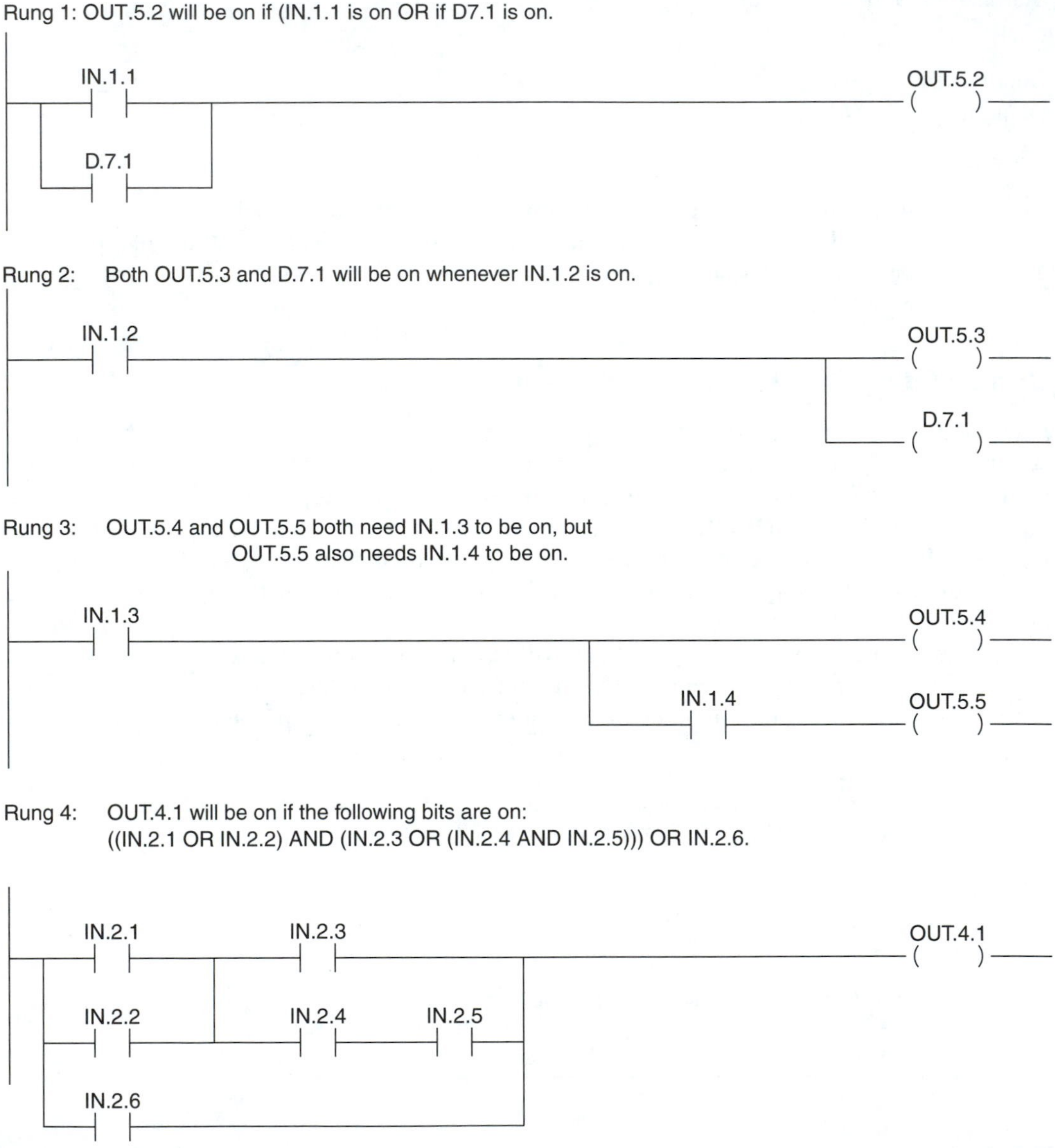

Rung 1: OUT.5.2 will be on if (IN.1.1 is on OR if D7.1 is on.

IN.1.1
D.7.1
OUT.5.2

Rung 2: Both OUT.5.3 and D.7.1 will be on whenever IN.1.2 is on.

IN.1.2
OUT.5.3
D.7.1

Rung 3: OUT.5.4 and OUT.5.5 both need IN.1.3 to be on, but OUT.5.5 also needs IN.1.4 to be on.

IN.1.3
OUT.5.4
IN.1.4
OUT.5.5

Rung 4: OUT.4.1 will be on if the following bits are on:
((IN.2.1 OR IN.2.2) AND (IN.2.3 OR (IN.2.4 AND IN.2.5))) OR IN.2.6.

IN.2.1
IN.2.3
OUT.4.1
IN.2.2
IN.2.4
IN.2.5
IN.2.6

FIGURE 3.4
More complex ladder logic statements

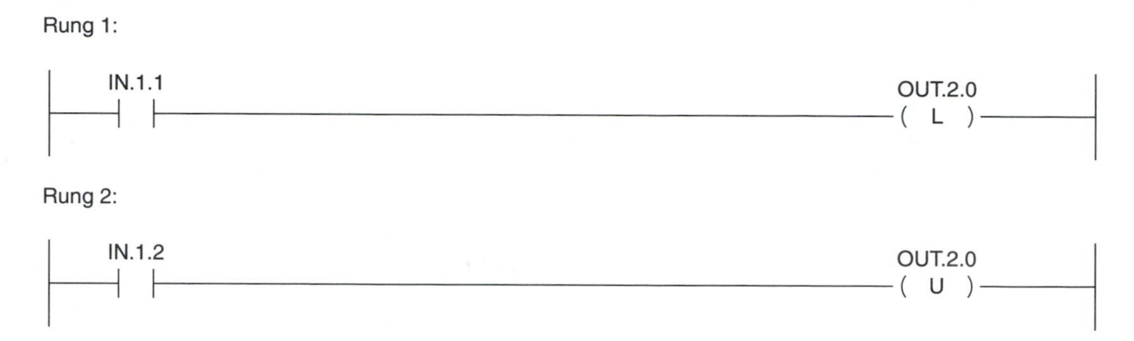

FIGURE 3.5
Output latch and output unlatch example.

Boolean statement can contain alternative conditions for turning an output bit on. This is known as a **boolean OR statement.**

Some ladder logic programming languages allow OUTPUT instructions to be on branches, but the output elements usually have to be programmed as the rightmost element on that rung. Rung 2 of Figure 3.4 shows how a single logical statement can control two output bits. Rung 3 shows how additional logic can be added to an output branch.[3]

Combined AND and OR logical statements can be constructed into complex **networked ladder logic statements,** but branches that don't end with output elements *must* recombine with the main branch to the left of the main branch's output element. Whenever a branch from the main rung recombines with the main rung (to the right of where it branched off), the PLC will evaluate the result of the OR logic *before* ANDing any additional logical statements farther to the right. Rung 4 of Figure 3.4 shows a network of AND and OR statements that are evaluated starting from the top left of the rung. For simplicity, the example's logic elements are all "examine on," and the input addresses were chosen to indicate the order in which the logic is to be evaluated.

The OUTPUT LATCH and OUTPUT UNLATCH instructions allow separate logical statements to control turning a binary bit on and turning it off. Figure 3.5 shows how a pair of rungs can be used to latch and unlatch a single output image bit. The two (or more) rungs that control latching and unlatching of this bit do not have to be programmed consecutively in a ladder logic program. If both logical statements are false during a scan cycle, the output bit is left in its previous state. If both logical statements are true during a single scan, the last-executed rung overrides the effect of the previous rung(s). In Figure 3.5, for example, if IN.1.1 and IN.1.2 are both on, this program will turn OUT.2.0 off, because the rung with the UNLATCH instruction is executed last.

A few PLCs allow output elements to be placed in locations other than at the farthest right-hand side of a rung.[4] The rung must still have a final output element at the rightmost side,

---

[3] Allen-Bradley and OMRON PLCs allow branches to contain outputs, but the Siemens ladder logic programming language does not.

[4] Siemens STEP 7 programming language uses a different symbol (#) for output elements that aren't rightmost on a rung.

OUT.5.5 will only be turned on if both IN.1.3 AND IN.1.4 are on.
OUT.5.4 will be turned on while IN.1.3 is on.

FIGURE 3.6
Output elements controlled by part of a rung's logic.

of course. Figure 3.6 shows a rung that controls OUT.5.5 dependent on the entire ladder logic rung, and controls OUT.5.4 dependent on only part of the rung. This ladder logic structure doesn't fit the electrical circuit analogy, because an actuator (OUT.5.4) can be turned on even if the circuit isn't complete (if IN.1.4 isn't on). Notice that this rung has the same effect as rung 3 of Figure 3.4.

INSTRUCTION LIST PROGRAMS FOR BIT MANIPULATION

Instruction list is the IEC 1131-3 name for programming languages that look a lot like assembler language programs. The instruction list language specified in the IEC 1131-3 standard is almost identical to **Siemens' statement list (STL)** programming language.

> True assembler language programs have separate instructions for each machine language instruction that a microprocessor chip can execute, and the instructions' names are chosen to sound (in English) like a description of what the microprocessor does in response to the machine language instruction. In true assembler, for example, the assembler instruction SUB means to subtract one number from another. The instruction would always include parameters telling the microprocessor what numbers to use.

Instruction list instructions aren't true assembler language instructions, but they do allow the programmer to specify what the PLC's microprocessor should do more exactly than if the PLC is programmed in ladder logic or one of the other PLC programming languages. Instruction list allows the programmer to write shorter (and faster) programs that don't need as much translation before the microprocessor can execute them. Siemens ladder logic programs are automatically translated into STL by a programming unit before they are sent to the Siemens PLC. The translated STL program may not be in the most speed-optimized form because the programmer has no control over the translation.

In all Boolean programs, the PLC must keep track of the logical result of each logical statement as the statement is evaluated from its first to its last logical element. In instruction list programming languages (Siemens' STL language, for example) the presence of a **result of logical operation (RLO)** status bit is acknowledged. The PLC sets the RLO bit to 1 (true) before beginning to examine an instruction list logical statement, and if the RLO bit is still 1 when the entire Boolean statement has been evaluated, the output statement executes. Boolean output instructions are said to be RLO-dependent, unlike some other instruction list data manipulation instructions, which we cover in later chapters. RLO-independent output instructions execute regardless of the state of the RLO and have no effect on the RLO from Boolean logic statements executed previously.

Instruction List Boolean Logic Elements (STL)

The Siemens STL Boolean logic instructions include:

A The **AND** instruction, which must be followed by the address of the bit that is being ANDed to the RLO. For example:

`A IN.4.0` is an STL language statement that will be evaluated as true if the bit in address IN.4.0 is a 1.

O The **OR** instruction, which also needs a bit address. This is the STL equivalent of a ladder logic branch recombining with the main rung.

AN **AND NOT,** which will be true if the bit in the specified address is a 0.

ON **OR NOT.**

STL Boolean output statements include:

= Causes the specified address bit to be turned on if the RLO is true, or turned off if the RLO is false. For example:

```
A IN.4.0
O IN.4.1
= OUT.5.0
```

will turn output image bit OUT.5.0 on while either input image bit IN.4.0 or IN.4.1 are on.

S The STL equivalent to a ladder logic OUTPUT LATCH instruction. It causes the specified address bit to be turned on (set) if the RLO is true, but won't turn the bit off if the RLO is false.

R Causes the specified address bit to be turned off (reset) if the RLO is true, but won't turn the bit on if the RLO is false.

For complex logical statements, equivalent to complex branched rungs in ladder logic, brackets are necessary to tell the PLC to save the RLO from one part of the statement so that it can be combined with the RLO from another part. The following **bracket instructions** are available and are entered without bit addresses:

A(Causes the PLC to evaluate the RLO of the logical statements starting from the next program line up to the closing bracket instruction, then to AND that RLO with the RLO from logic preceding this statement.

O(Works out the RLO of the following statements, then ORs that RLO with the RLO developed previously.

) Closes a bracket opened with an A(or an O(instruction.

The end of an STL program must be marked with a

BE BLOCK END instruction. (BE is added automatically by some program editors.)

Instruction list languages such as STL usually include instructions that perform operations not available in ladder logic. STL, for example, offers TEST BIT (TB and TBN) instructions for examining memory bits that cannot be examined using ladder logic or even using STL's A and O instructions. STL also contains output instructions that can set or reset bits unconditionally (SU and RU), so that a Boolean logic statement isn't even necessary.

Constructing Instruction List Programs (STL)

In ladder logic each logical statement must be a separate rung, which Siemens calls *segments.* In STL it is possible, *but not necessary,* to break Boolean logic programs into segments. Whenever the PLC encounters an instruction that can affect the RLO (the A, O, AN, ON, A(, or O(Boolean logic instructions) *after* an instruction that outputs the RLO to a bit (the =, S, R Boolean logic instructions), the PLC will automatically reset the RLO to 1 so that the following logical statement is evaluated independent of the result of the preceding statement. Since the RLO is not reloaded after an output instruction UNTIL a new input instruction is encountered, it is possible to use a single Boolean statement to control multiple output bits. Note that you can't write a Siemens ladder logic segment to control multiple outputs, but you can write an STL program to do so.

The following example STL program executes the same Boolean logic statements that we have seen in the ladder logic examples earlier in this chapter (and still uses the same imaginary addressing system). Blank lines have been added for readability, but are not required. The comments following the semicolons (;) are shown exactly as a programmer would enter comments into a STEP 5 STL program.

```
A       IN.1.0      ; Equivalent to Figure 3.3, rung 1:
A       OUT.6.0     ; A simple AND statement
AN      D.7.0
=       OUT.5.0

A       D.7.0       ; Equivalent to Figure 3.3, rung 2
ON      D.7.0       ; A logical statement that can
=       OUT.5.1     ;   only result in a true RLO is needed
                    ;   following any other logical statement

A       IN.1.1      ; Equivalent to Figure 3.4, rung 1
O       D7.1
=       OUT.5.2

A       IN.1.2      ; Equivalent to Figure 3.4, rung 2
5       OUT.5.3     ; One logical statement controls two
=       D.7.1       ;   outputs

A(                  ; Equivalent to Figure 3.4, rung 3, or to
A       IN.1.3      ;   the single rung of Figure 3.6
=       OUT.5.4     ; Note that brackets can be used to save
)                   ;   an RLO for use even after that RLO is
A       IN.1.4      ;   used to control and output
=       OUT.5.5

A       IN.2.1      ; Equivalent to Figure 3.4, rung 4
```

```
O       IN.2.2      ; Note that the development of a final RLO
A(                  ;   doesn't necessarily require as many
A       IN.2.3      ;   brackets as one might expect
O(
A       IN.2.4
A       IN.2.5
)
)
O       IN.2.6
=       OUT.4.1

A       IN.1.1      ; Equivalent to the two rungs in Figure 3.5
S       OUT.2.0
A       IN.1.2
R       OUT.2.0

BE                  ; Required at the end of an STL program
```

SOME COMMON BINARY LOGIC PROGRAMMING TRICKS

Boolean logic PLC programmers have developed a few tricks that take advantage of the three-step scan cycle's effects. Three of those tricks are described next.

One-Shots

The two-rung program in Figure 3.7 will cause the bit that is controlled by rung 1 to go on for one scan cycle's duration each time the input bit (IN.1.0) changes from off to on. The other rung controls a data bit and *must* follow the first rung (not necessarily immediately after the first rung). The second rung is needed to allow the one-shot to execute. Here is how it works:

1. Assume that the input is off (i.e., IN.1.0 is 0). As a result, both OUT.3.0 and D.2.0 will be off (0).

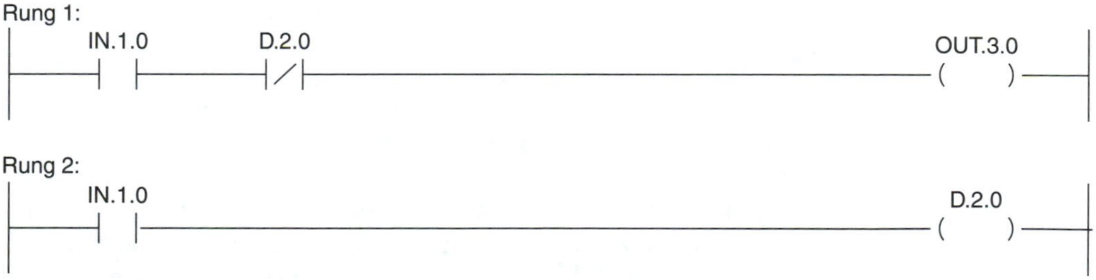

Rung 1:
IN.1.0 D.2.0 OUT.3.0

Rung 2:
IN.1.0 D.2.0

FIGURE 3.7
One-shot Boolean logic construct.

2. During the first scan cycle after IN.1.0 becomes a 1, the logical statement in rung 1 will be completely true, so the output image bit (OUT.3.0) will become a 1. After finishing rung 1, the PLC will evaluate rung 2, making the data bit (D.2.0) a 1. During the third step of the three-step scan cycle, the PLC will turn the actuator associated with output image bit OUT.3.0 on.

3. During the second scan after IN.1.0 becomes a 1, the PLC will reevaluate rung 1, and since D.2.0 is no longer a 0, the logical statement will be false, so the output image bit (OUT.3.0) will be turned off. At the end of this scan cycle, the actuator associated with OUT.3.0 will be turned off, after having been on for only one PLC scan cycle.

4. The one-shot will work again only after the input bit has gone off (turning D.2.0 off), then back on again.

Most PLC manufacturers have acknowledged that some programmers have difficulty learning the concept of the one-shot as described above and have added Boolean **one-shot instructions** to their ladder logic instruction sets. Figure 3.8 shows the very different graphical depictions of one-shots as offered by the three PLC manufacturers covered in this book. (All examples use the bit address format invented for this chapter.) All one-shot instructions control a user-specified bit (which doesn't have to be an output image bit) and all require a data bit to remember the state of the controlling logical statement from the previous scan cycle. Allen-Bradley and Siemens' PLCs require that the programmer specify a data bit, but the OMRON DIFU (13) one-shot instruction makes use of an internal working bit so that the programmer doesn't have to specify a data bit address. Most PLCs (including those covered in this book) offer additional one-shot instructions that provide a one-shot when a Boolean logic logical statement goes from true to false.

Latching and Sealing

The use of Boolean logic programming to *seal* an output predates the OUTPUT LATCH and OUTPUT UNLATCH instructions discussed earlier in the chapter. The Boolean logic seal construct is shown in Figure 3.9. It consists of:

1. A two-part ANDed logical statement that can turn an output on if the statement is true.
- **(a)** The first part of the logical statement is normally false but if IN.1.0 goes on, then the output goes on.
- **(b)** The second part of the ANDed statement is normally true but if IN.1.1 goes off then the output will go off.

2. A branch that ORs the state of the output bit with the state of the (normally false) first part of the ANDed statement.

Once the output has been turned on, the top branch of the ORed statement can return to its normally false state, and the output that is examined on the lower branch will hold (or seal) itself on for as long as the normally true ANDed element of the main branch's logic remains true. The seal logic is similar to the standard wiring of an on–off switch circuit, in which a normally open ON switch can be closed to actuate a relay turning an actuator on, and the relay will hold itself actuated until a normally closed OFF switch is opened to interrupt power to the relay.

We have already seen the OUTPUT LATCH and OUTPUT UNLATCH instructions that perform the same function as the seal construct. Some programmers prefer to use the seal construct since

Allen-Bradley:

IN.1.0 D.2.0 OUT.3.0

[ONS] ( )

Siemens S5:

IN.1.0 OUT.3.0

One-shot F-T

PS : D.2.0

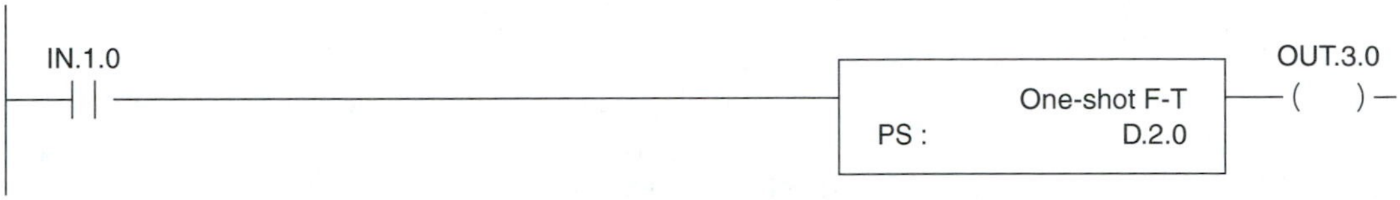

Siemens S7:

In ladder logic:

IN.1.0 D.2.0 OUT.3.0

(P) ( )

In STL:

```
A    IN.1.0
FP   D.2.0
=    OUT.3.0
```

OMRON:

IN.1.1

DIFU(13) OUT.3.0

FIGURE 3.8
One-shot instructions.

it brings all the logic that controls an output bit together onto a single rung, making troubleshooting easier. Other programmers prefer to use the OUTPUT LATCH and OUTPUT UNLATCH instructions, because they allow the programmer to enter the rungs that latch the output with other logic that turns other actuators on, and to enter rungs that unlatch the output with rungs that turn other outputs off. With OUTPUT LATCH and OUTPUT UNLATCH, the programmer can also enter several independent rungs to turn an actuator on or off under completely different conditions.

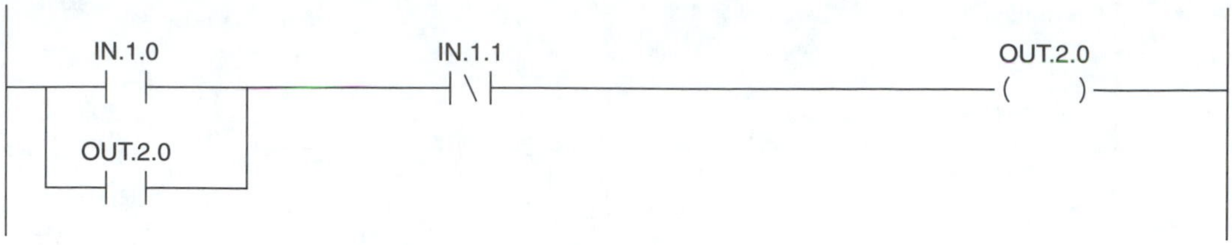

FIGURE 3.9
Sealing ladder logic construct.

Some PLCs offer a single ladder logic element that combines the functions of the OUTPUT LATCH and OUTPUT UNLATCH instructions, performing the same function as the seal construct. OMRON, for example, offers a KEEP instruction, as shown in Figure 3.10. When the logic on the input rung labeled with an S (set) goes true, the bit addressed in the KEEP instruction is turned on and will stay on until the logic on the input rung labeled with an R (reset) goes true. (OMRON also offers SET and RESET instructions that perform OUTPUT LATCH and OUTPUT UNLATCH functions.)

Sequencers

Latching (or sealing) can be used in a PLC program to control a process that must be performed one step at a time in a predefined order. Data bits are often used to keep track of which step(s) are active at the current time. Figure 3.11 shows a three-step sequencer program. Remember that a PLC program executes many times per second, but each step in a manufacturing sequence controlled by a PLC may take minutes or hours.

> The first rung can execute its output statements only if none of the three data bits are on, which will only be true if the sequenced process is not already active. When the other condition is also right to start the sequenced process (IN.1.1 is on), this rung will start an action (i.e., latch OUT.2.1) and will latch a data bit (D.3.1) to indicate that the sequence's first step is active.
>
> The next rung can execute its output statements only if the first step of the sequence has begun. When the conditions are right to end the first step and begin the

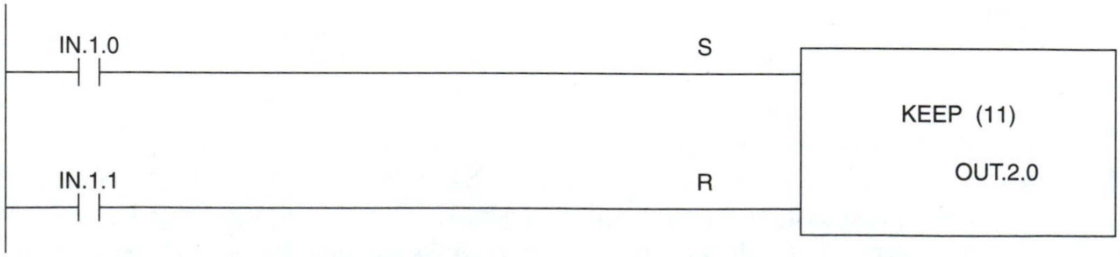

FIGURE 3.10
OMRON KEEP instruction for sealing an output bit.

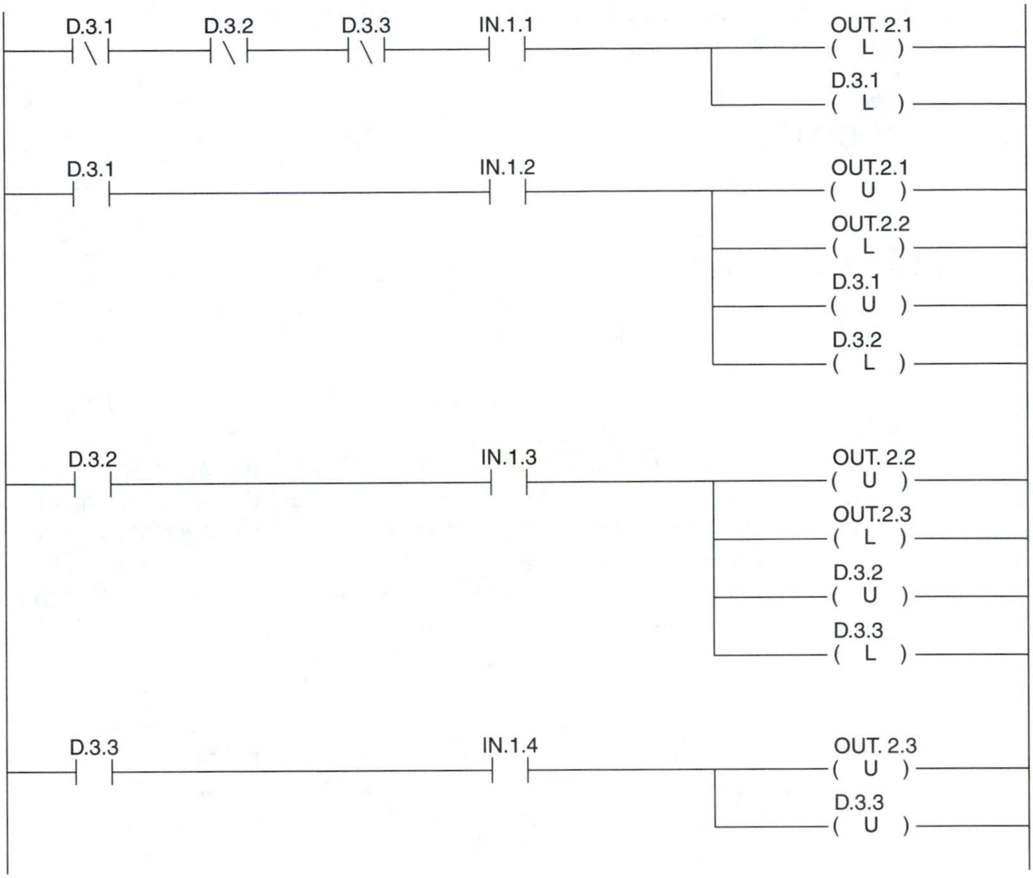

D.3.1 D.3.2 D.3.3 IN.1.1 OUT. 2.1 (L) D.3.1 (L)

D.3.1 IN.1.2 OUT.2.1 (U) OUT.2.2 (L) D.3.1 (U) D.3.2 (L)

D.3.2 IN.1.3 OUT. 2.2 (U) OUT.2.3 (L) D.3.2 (U) D.3.3 (L)

D.3.3 IN.1.4 OUT. 2.3 (U) D.3.3 (U)

FIGURE 3.11
Three-step sequencer program in ladder logic.

next (when IN.1.2 goes on), the second rung terminates the action begun by rung 1 and initiates a new action (OUT.2.1 unlatches, and OUT 2.2 latches), and the states of data bits are changed to reflect that the sequence is now at step 2.

The third rung is similar to the second. It controls the transition from step 2 to step 3.

The final rung controls the termination of step 3 and unlatches the last data bit so that the first step can be executed again when the conditional statement on rung 1 becomes true again.

When the sequence order must be variable, or if the conditions for beginning or ending steps become complex, the ladder logic program can become quite cumbersome. Some PLCs have advanced programming capabilities that can be used to make programming of a complex sequencer simpler. We discuss two of those methods later in this book: The use of special sequencer instructions is discussed in Chapter 7 and programming in a sequential flowchart (SFC) language is discussed in Chapter 9.

TROUBLESHOOTING

There are some binary programming errors that are often made by new PLC programmers:

1. The programmer enters the wrong address prefix. Instead of entering an EXAMINE ON instruction to examine IN.2.1, for example, the programmer has entered the address OUT.2.1, which is a valid address but isn't in the part of memory the programmer intended. Similarly, an output instruction might accidentally control IN.2.3 instead of OUT.2.3. In the second example, not only does the actuator associated with OUT.2.3 fail to respond, but if a later rung in the program tries to examine the state of a sensor associated with bit IN.2.3, that input image bit's state will be wrong because it has been overwritten.

2. Two conditional statements have been entered to control the same bit:

(a) If one statement controls an OUTPUT LATCH and the other statement controls an OUTPUT UNLATCH for the same output image bit, the associated actuator will always stay off while both conditional statements are true. Here the order of the statements in the program is important because the last instruction executed will control the actuator's state. Write these logical statements in the order that gives you the default state you want, or include logic in the statements to disable one statement if the other statement is true.

(b) If both statements control the same bit through OUTPUT ENERGIZE instructions, and if the first rung contains true logic and the other rung contains false logic, the memory bit will be turned on, then back off, during execution of the user program. If the bit is an output image bit, the PLC will turn the associated actuator *off* during the third step of the PLC scan cycle even though one logical statement indicates that the actuator should be on. All logic to control a single actuator should be combined into a single statement. If there are two different conditions under which the actuator should be on, OR the conditional statements on the same rung.

(c) If one rung contains an OUTPUT ENERGIZE instruction and another contains an OUTPUT LATCH or OUTPUT UNLATCH controlling the same bit. Again, the last active rung that the PLC executes will control the actual state of the output. If the conditions controlling the LATCH or UNLATCH instructions are *false,* those rungs won't be active, but the rung with the OUTPUT ENERGIZE instruction will be active every program scan. This construction could lead to intermittent failure of the control system.

3. After switching a PLC programmed with a sequencer program out of run mode to repair a mechanical malfunction, then restarting the PLC, the sequence doesn't resume executing properly. The PLC will have set some data bits and will have latched some output image bits prior to being switched out of run mode, depending on the step it was executing. The data bits may still be on, and in some PLCs even the output image bits will still be on. The PLC will try to resume the sequence at the step it was performing when it last ran. If the operator has cleared workpieces from the mechanical system, resumption of the sequence may not be possible. The operator, or the program, should ensure that the mechanical system and/or the contents of data memory are in the appropriate state prior to switching the PLC back into run mode. See Chapter 4 to learn more about data bits. See Chapters 10 and 11 (specifically the initialization interrupts and fault interrupts sections) to see how a PLC can be programmed or configured for proper automatic restart.

4. Upon restarting a PLC, the one-shots execute even though the conditions that would normally make them execute haven't changed. Most (but not all) PLCs clear their input image and output image memory automatically upon being put into run mode. A PLC, therefore, may "see" conditional statement logic changing from false to true during the first or second program scan even though the actual conditions in the workcell haven't changed. The program may have to contain logic to disable one-shot logic from executing for the first, and possibly for the second, scan cycle. See the initialization interrupts section in Chapter 11.

A few PLCs allow output elements to be placed in locations other than at the farthest right-hand side of a rung.

QUESTIONS/PROBLEMS

1. What value represents the "true" state in a PLC? Which represents the "false" state?

2. What are the three graphical elements that operate on bit values in ladder logic?

3. Which of the following memory areas of a CPU module's memory can be changed by a user-program?
- Operating system program area
- User-program area
- Input image area
- Output image area
- Data area

4. While a PLC is in run mode, it executes a scan cycle. Describe, *in order, exactly* what the PLC does during each scan cycle step and what it does when each step is completed. (Use a diagram if it helps.)

5. If the following program is running, what will the state of the output be after the program executes to completion if the input and data bits change as described in the table?

IN 1 — OUT 3
DATA 2 — OUT 3

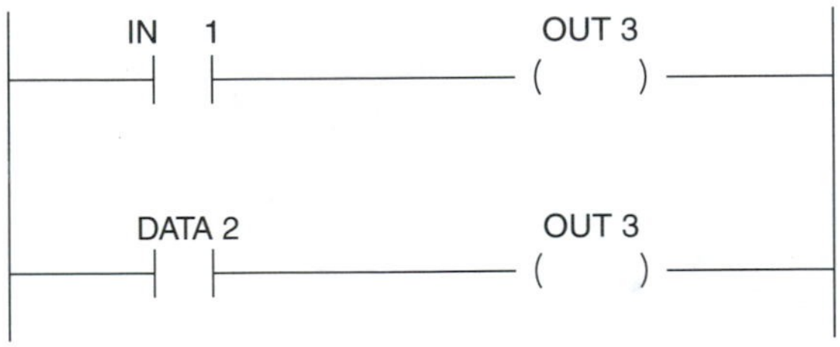

IN 1	DATA 2	OUT 3
OFF	OFF	
OFF	ON	
ON	OFF	
ON	ON	

6. Why should you avoid using ENERGIZE OUTPUT instructions in more than one rung to control the same output image bit?

7. If the following program is running, what will the state of the output be after the program executes to completion as the sensors that provide the input image states change as described in the table on p. 62?

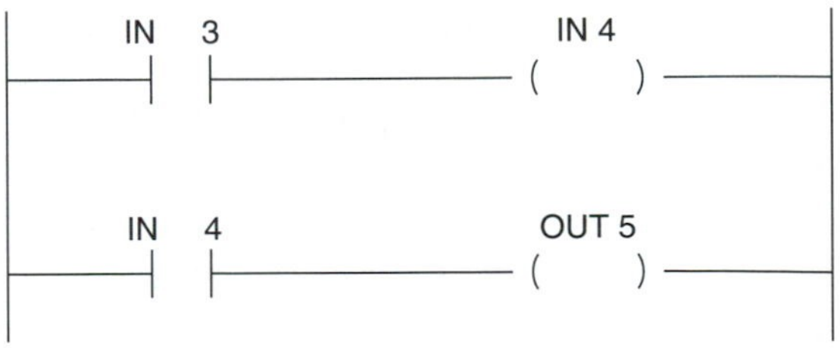

IN 3 — IN 4
IN 4 — OUT 5

SENSOR FOR:		OUT 5
IN 3	IN 4	
OFF	OFF	
OFF	ON	
ON	OFF	
ON	ON	

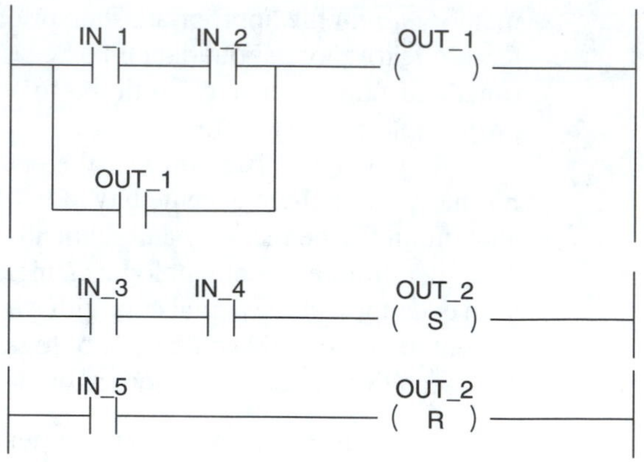

8. Examine the following ladder logic rungs. Describe what they do.

IN_1 IN_2 OUT_1 ( )
OUT_1
IN_3 IN_4 OUT_2 (S)
IN_5 OUT_2 (R)

PROGRAMMING EXERCISES (NO PLC REQUIRED)

1. Write the two-rung program for a one-shot, using only the simple EXAMINE ON, EXAMINE OFF, and OUTPUT instructions. (Write in ladder logic and in STL if programming a Siemens PLC.)
2. Write a ladder logic program that will:
   - **(a)** Turn OUT.1.0 on while IN.2.0 is on, but only if IN.2.1 is off at the same time.
   - **(b)** Turn OUT.1.1 on each time the conditions for turning OUT.1.0 on become true, but OUT.1.1 must remain on even if those conditions cease to be true.
   - **(c)** Turn OUT.1.1 back off again each time IN.2.3 goes on. The conditions for turning OUT.1.1 on must override the conditions for turning OUT.1.1 off if both conditions are true at the same time.
3. Rewrite the program for programming exercise 2 so that OUT.1.0 also goes on when IN.2.2 is on but the rest of the program is unaffected.
4. Write an instruction list (Siemens STL) program that performs the task described in programming exercise 2.
5. Rewrite the instruction list program for programming exercise 4 so that OUT.1.0 also goes on when IN.2.2 is on, but the rest of the program is unaffected.
6. Using only the standard Boolean logic statements, write a ladder logic program that turns OUT.3.0 on for one cycle whenever IN.4.0 goes from false to true (goes on).
7. Modify your program from programming exercise 6 so that OUT.3.0 goes on for one scan cycle each time IN.4.0 goes on or goes off.
8. Write a ladder logic sequencer program, using the seal structure, so that a momentary contact start button causes a conveyor to start, taking a car into a paint booth, where a stationary paint gun will start when the car is detected, but the car will continue to move past the gun. The paint will shut off after the car has passed, and the conveyor will retract fully. (The painted car is automatically replaced by an unpainted car when the conveyor has retracted. You do not have to program this part.) Repeat to paint the next car(s). When a momentary-contact stop switch is pressed, the current cycle should finish, but a new cycle won't start.

SUGGESTED PLC LABORATORY EXERCISES

For a system with:

- Four control panel switches: inputs 0 to 3
- Four indicator lamps or visible output module LEDS: outputs A to D
- Two spring-return-valve-controlled cylinders: outputs E and F
- One detent-valve-controlled cylinder: outputs G and H
- Three sensors to detect extension of each of the cylinders

1. Write a ladder logic program that will:

(a) Extend cylinder E if switch 0 is on *and* switch 1 is off, or if switch 2 is on regardless of the condition of switches 0 and 1.

(b) Extend cylinder F while cylinder E is extended or extending. This rung should examine the *output* for cylinder E to determine whether it is supposed to be extended, and the sensor at cylinder E to see if it has extended. (Adjust the airflow to the cylinder so that extension is delayed.)

(c) Extend cylinder G–H *only* during the time when both cylinders E and F are extended. Examine the sensors to determine if the cylinders are extended. (Getting the cylinder to extend is only half the job. You will need another rung to retract it when it isn't supposed to be extended.)

2. Write a (Siemens) STL language program, in one network, to do the same as the ladder logic program above.

3. (a) Before writing a program, actuate the switches and sensors while watching the input module LEDs. Note which switch corresponds to which contact at the input module.

(b) Program the PLC to:

(1) Turn indicator light A on when control switch 0 is on.

(2) Turn indicator light B on if switch 1 is on, but only if switch 2 is *not* on at the same time.

(3) Turn indicator light C *off* if either indicator A or B is on. *Examine the outputs.*

(4) Turn the first available data bit in memory on if switch 1 is on.

(5) Turn both indicator lights C and D on if the data bit is on while either, but not both, of switches 2 and 3 is on.

4. Write a program, using latching (or sealing), to execute the following sequence:

- Extend cylinder E when switch 0 goes on.
- After cylinder E extends (use the sensor) and switch 1 is turned on, extend cylinder F.
- After cylinder F extends and switch 2 is turned on, retract cylinders E and F.
- When cylinders E and F are retracted, extend cylinder G–H but only if indicator lamps A and B are on.
- Turn indicator lamps A and B on at any time when switch 2 is on (regardless of which sequenced step the rest of the program is executing).
- Retract cylinder G–H when lamps A and B go off, but only if it has been extended as part of the normal sequence.

5. Modify the program from laboratory exercise 4 so that cylinder E can extend again before the entire sequence ends, provided that it has retracted fully after its previous extension, but cylinder F cannot extend again until cylinder G–H has retracted.

4

COUNTERS AND TIMERS

OBJECTIVES

In this chapter you are introduced to:

- Counting events and to timing by counting time units.
- Counting directions and preset values.
- Accumulated values, status bits, memory requirements, and some data manipulation possibilities.
- Up and down counting.
- On-delay, off-delay, and retentive and pulse timers.
- Detailed descriptions of Allen-Bradley, Siemens, and OMRON counters and timers.

Controlling a process often requires a PLC that can count events (such as the number of bottles that have been put into a beer case) or a PLC that can time events (such as how long to delay before feeding another bottle, while the full beer case is being replaced by an empty one). Figure 4.1 shows a (simplified) program containing a counter and a timer to control the beer case filling system.

All PLC programming languages include counter and timer instructions. **Counters** count the number of times logical statements go from false to true. A logical statement might, for example, examine the sensor that detects beer bottles passing on a conveyor, so that bottles can be counted. **Timers** count time units. A timer might be started whenever a full beer case is moved away from a sensor. Counters and timers can be used to control Boolean operations, such as to start conveying a full beer case away or to stop the movement of beer bottles toward the boxing station. Counter and timer instructions affect data words in memory locations where time values or count values are kept, so although timers and counters are used in Boolean programs, counters and timers are not simple Boolean programming elements.

Counter and timer instructions are entered as output elements that are controlled by Boolean logic statements. Each counter or timer manipulates status bits that can be examined in other Boolean logic statements. One status bit indicates whether the counting or timing has

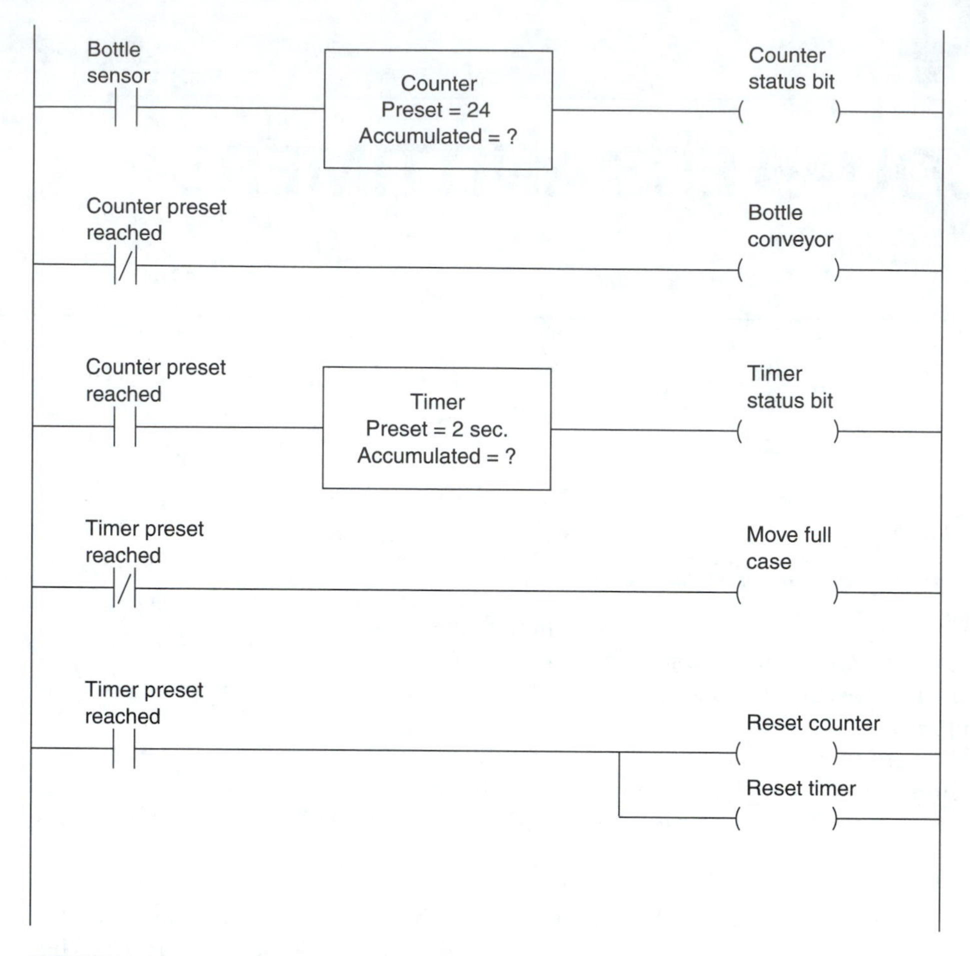

FIGURE 4.1
Simplified control program showing a counter and a timer.

reached the end. The PLC must store counter and timer *accumulated* values as separate data words and must change the accumulated values as counting or timing instructions are executed. For advanced programming operations, Boolean logic statements can be used directly to control counter and timer status bits, and instructions that manipulate data words can be used to examine or change accumulated values.

COUNTER INSTRUCTIONS

Some PLCs have counters that are used to *count down toward zero.* These counters have status bits that indicate when the accumulated value is equal to zero. The PLC program must include an instruction to initialize the accumulated value to a programmer-selected preset value and instructions that then decrement the accumulated value down toward zero. Instructions are also available to count up and to reset the accumulated value to zero, although they are not always used.

Other PLC counters are used to *count up to a preset value.* Each time the count changes, the counter instruction compares the accumulated value with the programmer-selected preset value, changing a status bit when the accumulated value reaches or exceeds the preset value. These PLCs need to store the preset value as well as store the accumulated value. The PLC program needs an instruction to reset the accumulated value to zero and instructions that count up toward the preset. Instructions are available to count down and to set the accumulated value to a nonzero value, although they are not always used.

TIMER INSTRUCTIONS

Timer instructions are actually counters that **count time units** while their input logic statement remains true (or in some cases, while the statement remains false). Like counters, each timer requires (at least) one data word of memory to hold the timer's accumulated value and (at least) one status bit in memory for the timer's "done timing" bit.

Unlike counters, most types of timers *restart automatically whenever their control logic disables then reenables the timing,* so timer reset instructions aren't always necessary. Other timers are *retentive,* which means that they stop timing when disabled by their control logic, and resume timing when reenabled. Retentive timers do need a timer reset instruction. Some PLCs offer timers that run to completion once started, even if their control logic changes state after it starts the timing.

As with counters, some PLC timer instructions decrement an accumulated-time value *toward zero,* and other PLC programming languages only offer timers that increment an accumulated-time value *toward a preset value.* All PLC timers change a status bit when the accumulated value reaches zero or the preset value. Some PLCs offer additional status bits and/or additional timer types that use their status bits to indicate (for example) when the *timer is running.*

Most PLC timer instructions allow the programmer to select the *size of the time unit,* in addition to specifying the *number of time units* to count. A timer will change its accumulated value only after one full unit of time has expired, so selecting a large time unit reduces the timer's accuracy. Timer instructions can change their timer status bit(s) only when the instruction executes, so long programs can also lead to low timer precision.[1] If, for example, a program takes 50 milliseconds (ms) to execute one scan cycle, the timer can be inaccurate by ± 50 ms even if the timer uses time units smaller than 50 ms!

Many PLCs offer **real-time clock** features, where the PLC keeps track of time without requiring instructions from the user-program. The user-program can include instructions to examine and even change accumulated real-time clock values.

---

[1] Timers and counters that count more quickly, independent of the scan cycle time, are offered by most PLC manufacturers. High-speed counters and timers are discussed in Chapter 11.

PLC-5

SLC 500

ALLEN-BRADLEY COUNTERS AND TIMERS

In the PLC-5 and in the SLC 500 series of Allen-Bradley PLCs, a *three-word data structure* is set aside for each counter or timer. The structure, shown in Figure 4.2, contains two 16-bit signed-binary numbers representing the accumulated value and the preset value, and also contains status bits. Each status bit or data word can be addressed separately in a user program.

There are three **counter** instructions available in the PLC-5 and SLC 500[2] line of PLCs. A **COUNT UP** instruction, shown in the program in Figure 4.2, causes the accumulated value to *increment* once each time the statement that controls the COUNT UP instruction *goes from false to true.* A **COUNT DOWN** instruction (not shown) *decrements* the accumulated value every time its control logic goes from false to true. The **RESET** instruction returns a counter's accumulated value to zero and resets all its status bits. Multiple COUNT UP and COUNT DOWN instructions *can share the same counter data structure,* so that they manipulate the same accumulated value and status bits.

The programmer must enter a counter preset value as a constant while entering a COUNT UP or COUNT DOWN instruction into a program. The preset value can be any number in the range −32,768 to +32,767. The accumulated value can be counted up or down within the same range. If a counter counts up past +32,767, it will *overflow* to −32,768. If it counts down past −32,768, it will *underflow* to +32,767. COUNT UP and/or COUNT DOWN instructions that manipulate the same counter data element should all be entered with the same preset value because the preset value in the counter structure is changed each time the programmer enters a counter instruction.

There are five **status bits**[3] in a counter data structure. Status bits can be examined by Boolean logic program instructions, as shown in Figure 4.2. The *DN (done)* bit is on whenever the accumulated value is equal to or greater than the preset value. The *CU (count up)* and *CD (count down)* bits reflect the state of control logic on the most recently executed COUNT UP and/or COUNT DOWN instructions that use this data structure. The *OV (overflow)* and *UN (underflow)* bits are latched on whenever the accumulated value overflows or underflows. If the OV or UN bits are on, the DN bit's state may be wrong because the accumulated value is invalid. Boolean logic output instructions also can control these status bits, but the programmer should avoid using this capability because doing so may make the counter ineffective.

> *Be careful:* Allen-Bradley counter instructions also increment (or decrement) their accumulated values when the PLC is switched into run mode while the counter instruction's control logic is true, after resetting the counter while the control logic is true, and when execution of a structured program makes it look as if the control logic has gone from false to true (as we will see in Chapter 8).

Figure 4.2 demonstrates a COUNT UP instruction that is controlled by a very simple logic statement, which examines a single input image bit (the SLC 500 address I:4.3). Another rung turns a bit file bit (B3/2) on if the counter's DN bit is on (DN will be on if the accumulated value is equal to or greater than the preset value), but only if the accumulated value hasn't overflowed or underflowed since the last time the counter was reset. A RESET instruction will set the accumulated value back to zero and will reset the counter's status bits if another input image bit

---

[2] The fixed SLC 500 and the SLC 5/01 also offer a high-speed counter instruction. The high-speed counter's operation is described in the timed interrupt section of Chapter 11.

[3] There is also a sixth status bit, the UA bit, for high-speed counters only. See Chapter 11.

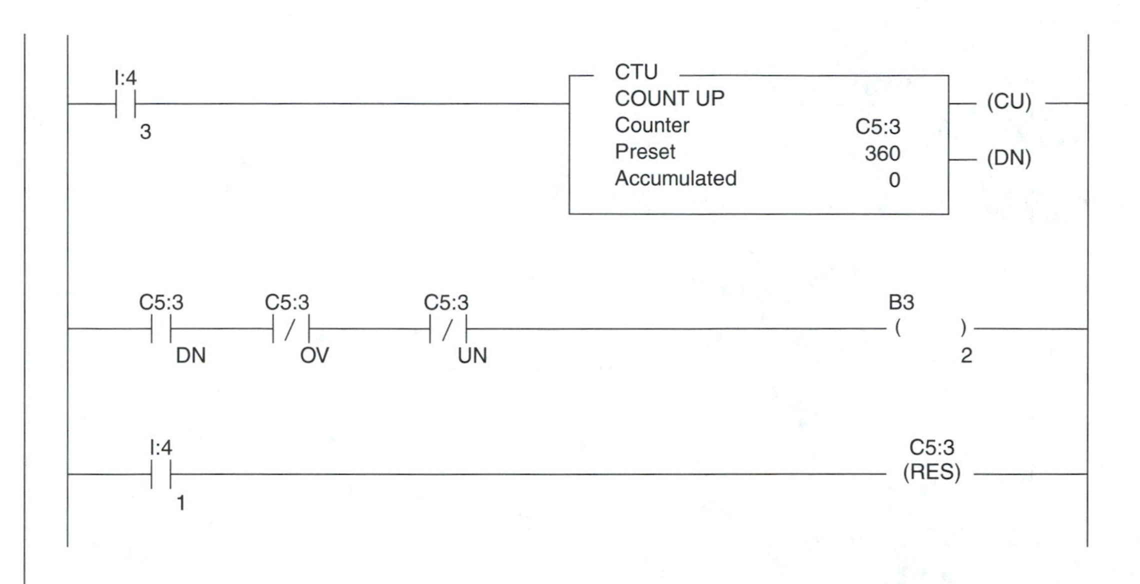

Structure C5:3

	CU	CD	DN	OV	UN	*	
C5:3.0 (status)	1	0	0	0	0	0	
C5:3.1 (.PRE)	+360						
C5:3.2 (.ACC)	17						

* SLC 500 high-speed counters' counter elements have an additional status bit (UA) in bit 10 of word 0.

FIGURE 4.2
Allen-Bradley counter instructions and the counter element data structures they manipulate.

(I:4/1) is on. Other data manipulation instructions can operate on counter preset and accumulated values. The MOVE instruction, math operations, and the COMPARE instruction, which are examined in later chapters, are important examples. Figure 4.2 also shows the format in which the timer's three-word data structure stores the status bits, preset value, and accumulated value.

Allen-Bradley **timers** are actually counters that count time units. The preset value, entered with the timer instruction, must be a positive value between 0 and +32,767. This value represents a number of time units. In lower-level SLC 500s, the time base unit is always 0.01 s, but in better SLC 500 models and in the PLC-5, the programmer can select 0.01- or 1.0-s time base units. The timer can increment only the accumulated value and stops incrementing when it reaches the preset value. Only three status bits exist in the timer's data structure: the *DN (done)* bit, the *TT (timer timing)* bit, and an *EN (enabled)* bit. The EN bit reflects the state of the timer's control logic, and Figure 4.3 shows how the other status bits act.

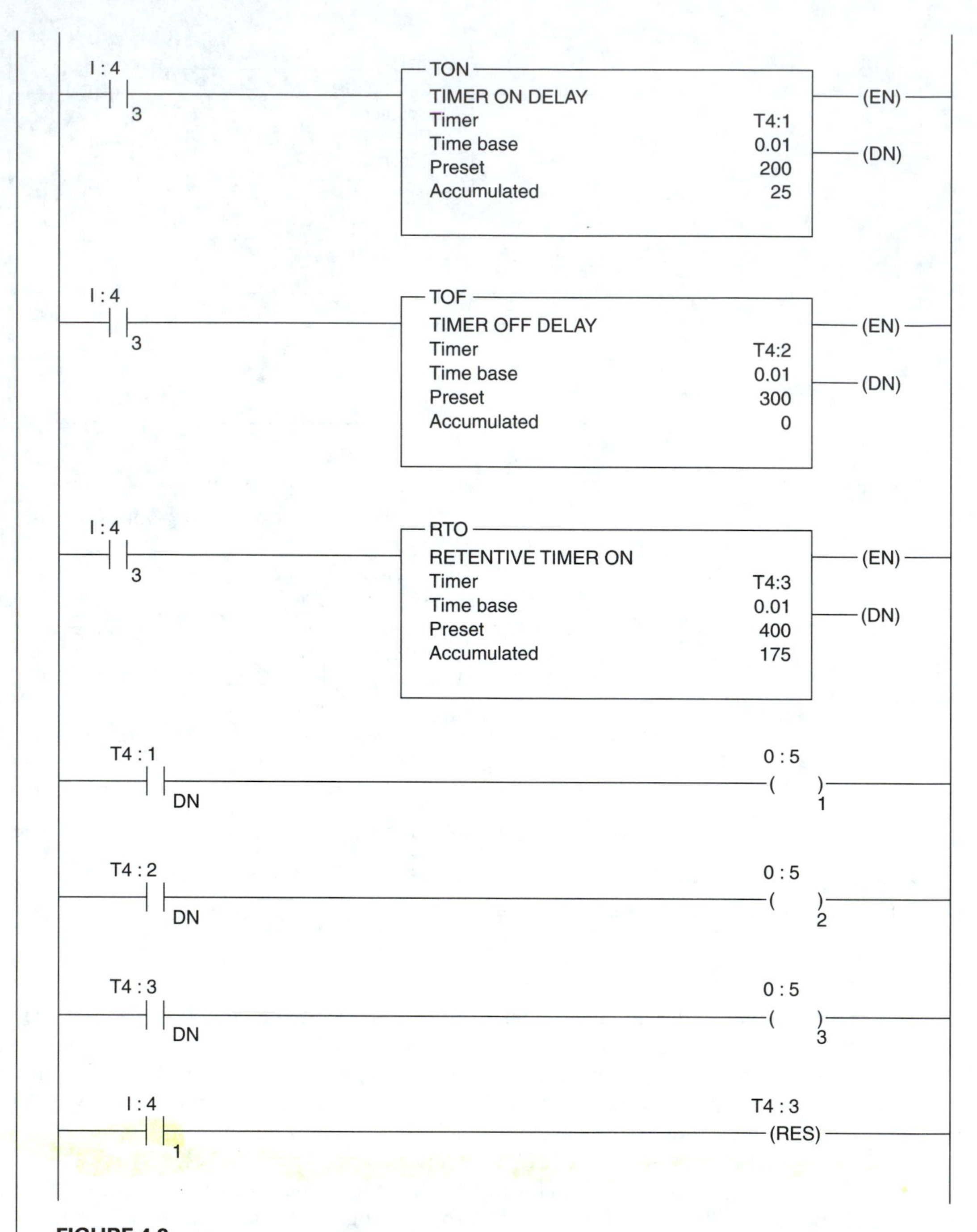

FIGURE 4.3
Allen-Bradley timer instructions, timer element structure, and timer operation table.

The T4 timer element data file for this program contains three data words for each of the three timers:

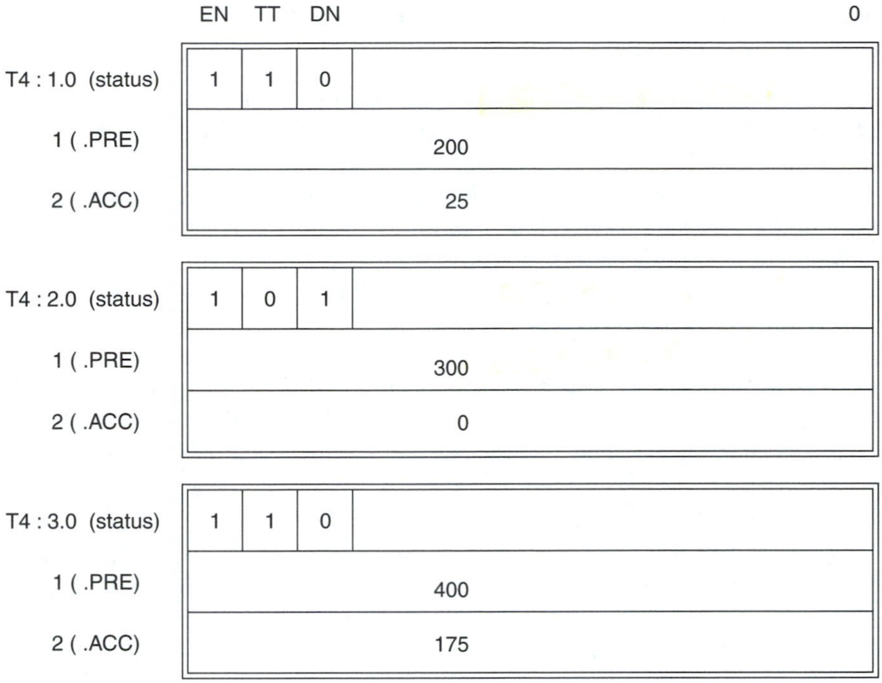

	EN	TT	DN	0
T4 : 1.0 (status)	1	1	0	
1 (.PRE)	200			
2 (.ACC)	25			
T4 : 2.0 (status)	1	0	1	
1 (.PRE)	300			
2 (.ACC)	0			
T4 : 3.0 (status)	1	1	0	
1 (.PRE)	400			
2 (.ACC)	175			

Timer type	Resets accumulated value and status bits when:	DN bit while reset	Times while control logic is:	DN bit while timing toward preset	TT bit while timing toward preset	DN bit when accumulated equals preset
TON	Control logic goes false	0	True	0	1	1
TOF	Control logic goes true	1	False	1	1	0
RTO	RES instruction executes	0	True	0	1	1

There are three Allen-Bradley timer instructions and a timer reset instruction, which are used as shown in Figure 4.3. The three timer instructions are affected by their control logic as shown in the table in Figure 4.3.

1. The **timer on delay (TON)** automatically resets its accumulated value and all status bits whenever its control logic goes false, and restarts timing when the input logic goes true. The accumulated value is updated only when the timer instruction executes, but the same instruction can be programmed more than once in a user-program to cause more frequent updating. The TT status bit is on as the timer times toward its preset, then goes off. The DN bit goes on when the accumulated time reaches the preset value.

2. The **timer off delay (TOF)** is almost the exact inversion of the TON timer except that like the TON timer, it turns its TT bit on while it times.

3. The **retentive timer on (RTO)** timer is identical to the TON timer except that it does not reset its accumulated value when its control logic goes false. The TT bit goes off, while the RTO timer "pauses," when its control logic is false. Since the RTO doesn't reset automatically, the RTO timer's accumulated value must be reset by executing a RESET (RES) instruction.

> Watch out! Timers act as if their control logic has gone from false to true if the PLC is switched into run mode when the timer instruction's control logic is true. Timers affected by structured programming can act as if their control logic has gone false or has gone true even if the logical statement hasn't actually changed state. See Chapter 8 for details.

Figure 4.3 demonstrates some differences among the three timer types. All timers in this example are controlled by the same simple control logic, which is now true (as the EN bit shows) and has been true for 0.25 s (as the accumulated value in T4:1.ACC shows). The TON is running because its control logic is true but its DN bit hasn't gone on yet. The TOF is reset because its control logic is true and its DN bit is on (it will start to time when its control logic goes false and will turn its DN bit off when the time has expired). The RTO didn't reset its accumulated value the last time its control logic went false, so it has timed an additional 0.25 s over its previously accumulated 1.50 s. Its DN bit isn't on yet. When I:4/1 goes on, the RTO instruction's accumulated time at T4:3.ACC will reset (become 0) even if the RTO is still timing.

S5 SIEMENS S5 COUNTERS AND TIMERS

When a Siemens counter or timer is started, an initial preset value is stored as the counter or timer's accumulated value. The accumulated value must then be *decremented toward zero,* because Siemens counter and timer status bits change only after their accumulated value has been decremented to zero.

> To be able to understand the STL language programs that store initial values into timer or counter memory locations, it is important to understand the LOAD (L) instruction, although LOAD won't be covered formally until a later chapter. The LOAD instruction supplies a value that can then be copied into a counter or timer's accumulated value. For example:

```
L FW004
SP T000
```

gets a value from memory address FW004 and copies it into the memory address reserved for timer 0 (T000). Another example:

```
L KC+15
S C001
```

places the counter-format constant (KC) 15 into the memory reserved for counter 1 (C001).

To program a STEP 5 **COUNTER** instruction, at least two Boolean statements are usually used. As shown in ladder logic and in STL in Figure 4.4, one Boolean statement controls a **COUNTER SET (S)** input to place a preset value into the counter's accumulated value storage address. Another statement controls a **COUNT DOWN (CD)** input to decrement the accumulated value. Boolean statements can also be entered to control the ladder logic counter element's **COUNT UP (CU)** and **COUNTER CLEAR (R)** inputs, but aren't necessary. Each of these four counter controls can be programmed separately in STL, and the STL part of the example in Figure 4.4 does not include statements to control the inputs that aren't used in the ladder logic example.

In at least one other programming software package for the S5, the COUNT UP, COUNT DOWN, COUNTER SET, and COUNTER CLEAR functions are implemented in separate lad-

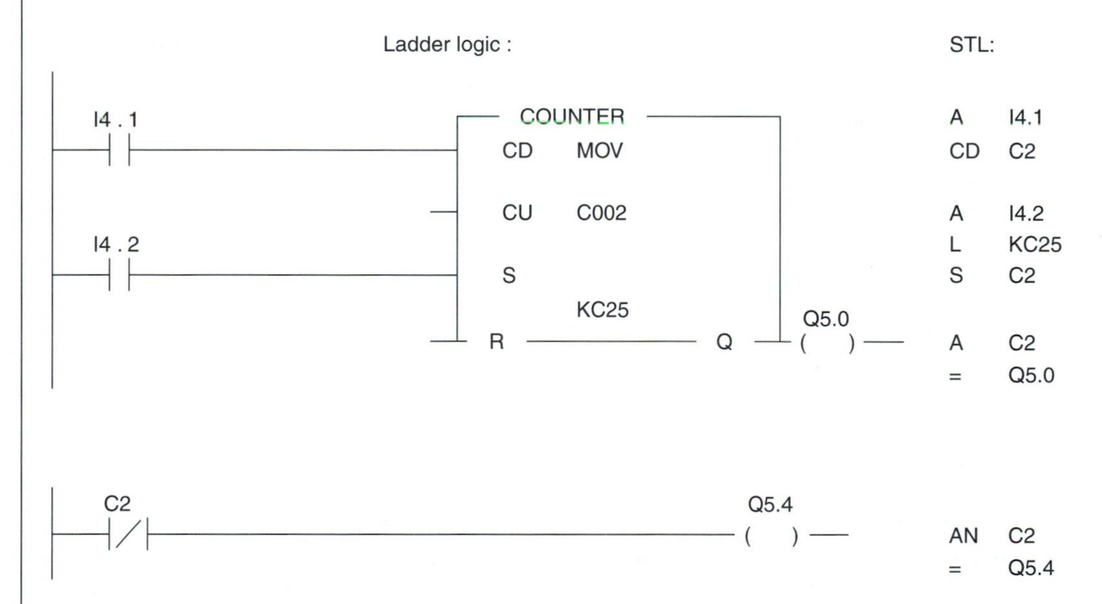

FIGURE 4.4
STEP 5 COUNTER SET and COUNT UP instructions.

der logic blocks, to reflect that they aren't always all required, and to reflect that they don't need to be programmed into sequential program memory locations.

Figure 4.4 shows two methods of programming logical statements using the counter's status bit called a **counter coil (Cx).** The counter coil is on when the accumulated value is greater than zero, and goes off when the accumulated value reaches zero. In ladder logic, an address (e.g., Q5.0) can be entered at the output contact (Q) of the counter element, or a separate rung can examine the same counter status bit as part of a Boolean logic statement. In STL, the two methods are programmed the same way.

Additional counter control instructions can *only* be programmed in STL, and will be described later. They include the **ENABLE TIMER/COUNTER (FR), SET BIT UNCONDITIONALLY (SU),** and **RESET BIT UNCONDITIONALLY (RU)** instructions.

Some counter control instructions execute like one-shots. They execute once each time the RLO (result of logic operation) of their Boolean control logic statement *goes from false to true.* The **COUNTER SET (S), COUNT DOWN (CD), COUNT UP (CU),** and **ENABLE TIMER/COUNTER (FR)** instructions work this way. The COUNTER CLEAR (R) instruction, on the other hand, will execute as long as every scan cycle while its *RLO remains true,* and the SET and RESET BIT UNCONDITIONALLY (SU and RU) instructions *execute despite the RLO.*

The initial preset value that the COUNTER SET instruction writes to the counter's memory address must be between 0 and 999. In ladder logic, the programmer enters a counter constant value into the ladder logic box or enters the address of a memory location that contains a preset value. In STL, a LOAD instruction (L) must precede the COUNTER SET instruction (S), and the LOAD instruction must include the counter constant or the memory address. Counter constants are entered as **KCx,** where "x" is a number between 0 and 999. If an address is entered, that address must contain a number in BCD format. (In BCD format each of the low three binary nibbles represents one decimal digit between 0 and 9, and the high nibble is all zeros.) The PLC translates preset numbers into binary and stores them in the counter's memory location (Cx, where "x" is the counter's number). COUNT UP (CU) or COUNT DOWN (CD) instructions increment and decrement the memory location containing this value. A LOAD (L) instruction can read the current accumulated value from the counter memory. If the counter's accumulated value is read using the **LOAD COUNTER/TIMER (LD)** instruction, the PLC converts the accumulated value back into BCD as it is read.

The ENABLE TIMER/COUNTER (FR) instruction (only programmable in STL, and only in a function block) executes once each time its RLO goes true, *causing a counter to reset other status bits that contain the most recent state of the COUNT UP, COUNT DOWN, and COUNTER SET control logic, so that those instructions will execute again next time they are run with true RLO,* even if their control logic statements haven't actually gone false first.

A Siemens **TIMER** is started by its Boolean start-control logic. The timer in Figure 4.5 reloads its preset value and starts timing down each time its *start* control RLO goes from false to true. After a timer is started, the PLC must repeatedly execute the timer instruction to decrement the accumulated value and to change the timer's **timer coil (Tx)** when the timer has run down to zero. An address for the timer coil (Tx) to be copied to can be entered at the ladder logic timer element's output (Q), or the timer coil (Tx) can be examined as part of another Boolean logic statement, as demonstrated in the second network of the program in Figure 4.5.

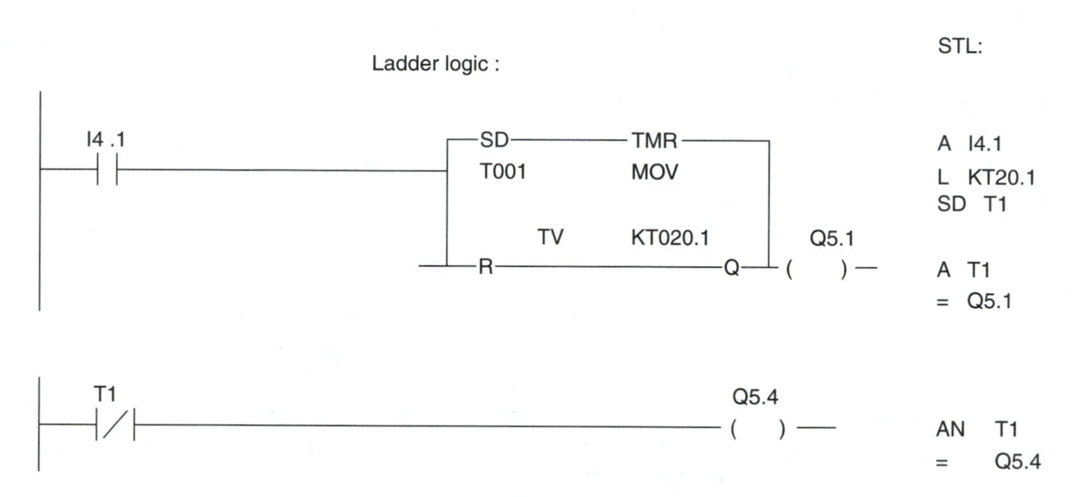

FIGURE 4.5
Program containing a STEP 5 timer, shown in ladder logic language and in STL language.

The initial preset time value included with a timer start instruction must be entered as a **timer constant** value or as the address of a memory location that contains a 16-bit number in **timer format.** Timer constant values are shown in Figure 4.6. A timer constant number must be entered with the prefix KT, followed by a three-decimal-digit number indicating the number of time units, then a period followed by a decimal-digit code for the size of the time unit. If the program contains an address for the PLC to get the value from, the 16-bit binary word in that address must be in timer format. As demonstrated in Figure 4.6, in timer format the low 3 bytes of the 16-bit value must each be a BCD code for decimal digits 0 to 9 (maximum time setting is 999 time units), the highest 2 bits must be zeros, and the two remaining binary digits must contain one of the four codes for time unit sizes. The PLC translates timer constant and timer format numbers into natural binary and stores it into the timer's memory location (addressed as Tx, where "x" is the timer's number). Siemens does not make it clear where the time unit size code is stored. If your program includes a LOAD instruction to read the contents of a timer accumulated value, the PLC reads only the number of time units; it does not reconvert the number back into timer constant or timer format form. If the timer value is read using the **LOAD TIMER COUNTER (LD)** instruction, the PLC will convert the binary number back into BCD code for a three-decimal-digit number, but the code that identifies the time unit size is lost.

Timers are accurate only to within ± 1 time unit. Additional inaccuracies can be caused by delays between when a timer runs out and when the timer instruction is executed to affect the timer's status bit, delays between when the status bit changes and when the user-program examines the bit to control an output, and delays inherent in the scan cycle's reading of inputs and writing of outputs.

Time Unit		
Code		Size (s)
Decimal	Binary	
0	00	0.01
1	01	0.1
2	10	1
3	11	10

Examples:

Timer constant	Timer format	Time setting (s)
KT205.0	0000_0010_0000_0101	2.05
KT555.1	0001_0101_0101_0101	55.5
KT125.2	0010_0001_0010_0101	125
KT999.3	0011_1001_1001_1001	9990

FIGURE 4.6
Timer codes, timer constant, and timer format numbers, and the times they represent.

The ladder logic **on-delay (SD)** timer instruction for an S5 PLC is shown in the example program in Figure 4.5. (The STL program beside the ladder logic program is *not* a direct translation.) In the example, I4.1 is used to control the start of the timer. When I4.1 goes from off to on, the timer instruction places an initial time value into the memory location used by timer 1 (T001), then starts decrementing that number. The timer constant value KT20.1 in Figure 4.5 sets the timer to run for 20 time units of 1/10 s each. The ladder logic timer instruction offers a **TIMER RESET (R)** input, which can be used to clear the timer's memory location while the (optional) control logical statement is true (reset control logic is not used in this example). An on-delay timer turns its timer coil status bit on after the timer has timed down to zero. The timer coil status bit will be copied to another address by the ladder logic instruction if the programmer enters the other bit address at the Q contact. The timer coil bit for timer 1 (T001) can be examined in Boolean statements elsewhere in the program as bit address T001.

The **RESET (R)** instruction, as noted above, can be used to clear a timer's memory location and status bit and to keep them at zero while the reset control logic stays true. The **ENABLE TIMER/COUNTER (FR)** instruction causes a timer to restart its timing from the initial value *if the timer hasn't yet run out.* The FR instruction can only be programmed in STL, only in a function block, and executes only when the RLO of the Boolean logic controlling it goes from false to true. The R and FR instructions have different effects on how different timers work, as described below.

Siemens offers five different types of timers:

1. The **on-delay (SD)** timer. Turns its timer coil bit off when its SET (S) RLO is false. Sets its time value and starts timing (down) when the S logic goes true. Turns its timer coil bit on after timing finishes. While the on-delay timer's start logic remains true but the timer has finished running, an R pulse followed by an FR execution will restart the timer.

2. The **retentive on-delay (SS)** timer. Sets its time value and starts timing when its SET (S) logic goes true. Continues to run, even if the S logic goes false, until it finishes timing, when it turns its timer coil bit on. If allowed to time to completion, S control logic can't restart the timer until after the timer has been reset. While the retentive on-delay timer's S logic remains true but the timer has finished running, an R pulse followed by an FR execution will restart the timer.

3. The **pulse (SP)** timer. Turns its timer coil bit off when its SET (S) RLO is false. Sets its time value, turns its timer coil bit on, and starts timing when its S logic goes true. Turns its timer coil bit off when it finishes timing. The FR instruction will restart a pulse timer while the timer's S logic remains true, even if the timer has finished timing.

4. The **extended pulse (SE)** timer. Resets, turns its timer coil bit on, and starts timing when its SET (S) RLO goes true. Continues to run, even if its S logic goes false, until it finishes timing, when it turns its timer coil bit back off. The FR instruction will restart an extended pulse timer while the timer's S logic remains true, even if the timer has finished timing.

5. The **off-delay (SF)** timer. Turns its timer coil bit on when its SET (S) RLO goes true. Sets its time value and starts timing when its S logic goes false. Turns its timer coil bit off when it finishes timing. The off-delay timer cannot be restarted by the FR instruction after it has finished timing while its S logic remains false.

The Boolean logic instructions: **SET BIT UNCONDITIONALLY (SU)** and **RESET BIT UNCONDITIONALLY (RU)** can be used to turn certain selected memory bits, including counter and timer status bits, on or off. These instructions execute despite the RLO, but they can only be entered in STL and only in a function block.

S7

SIEMENS STEP 7 COUNTERS AND TIMERS

Counters and timers in STEP 7, for the S7 line of PLCs, are the same as STEP 5 timers, except in two ways:

1. The ladder logic elements appearances have been changed to conform to the IEC 1131-3 standard, as shown in Figure 4.7. There are now three separate ladder logic COUNTER instructions instead of one: the **up-down counter** plus a new **up counter** and a **down counter.** Each instruction has a SET (S) and a RESET (R) input as well as the count-control inputs (CU and/or CD). The programmer can program control statements at any of the input(s). The counter's PRESET VALUE (PV) is programmed by entering an address or a constant at that input. (A BCD number, indicated by STEP 7's BCD prefix, has been entered for the timer in Figure 4.7.) There are additional outputs for the CURRENT VALUE (CV) and for the CURRENT VALUE in BCD (CU_BCD). If the programmer enters addresses at these outputs, the instruction will copy the counter's current value to those addresses.

Ladder logic *timer instructions* have also been changed to allow the timer to output its time-remaining value to specified addresses in natural binary (BI) or in BCD format. The timer

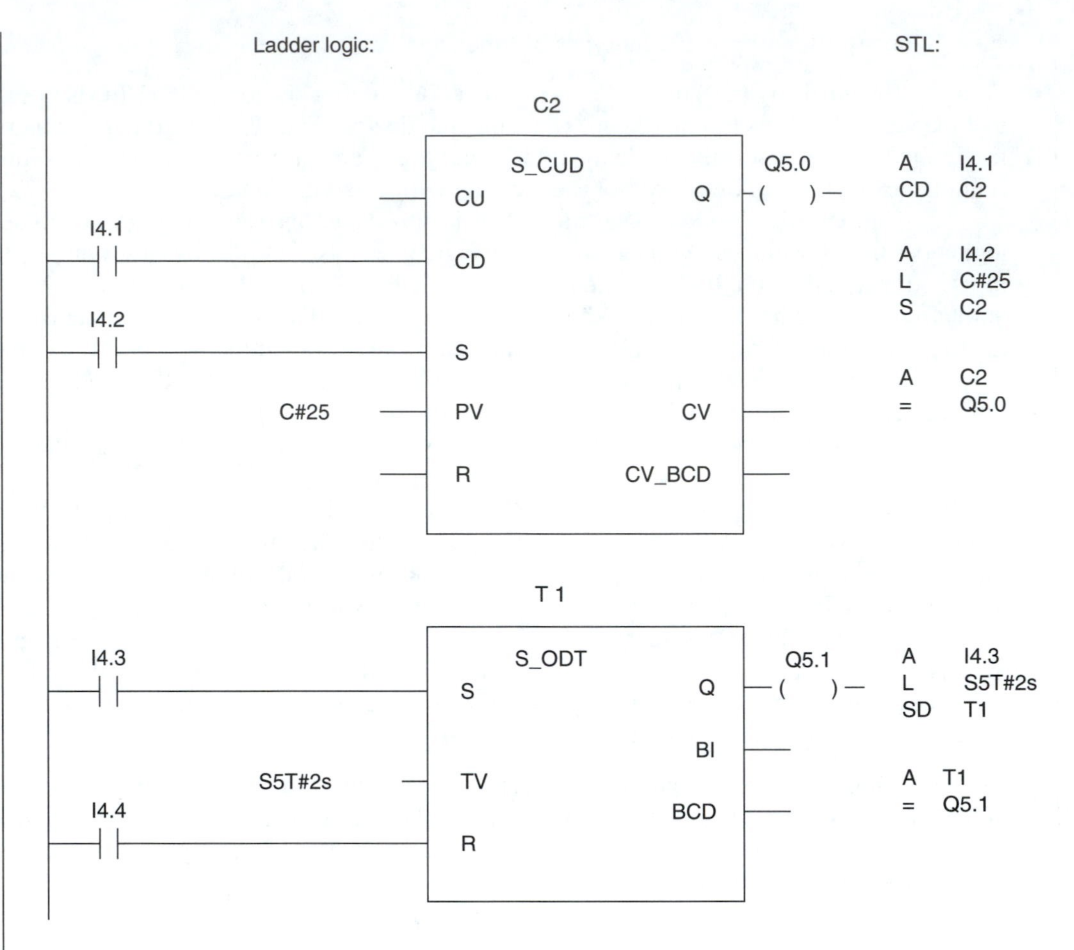

FIGURE 4.7
S7/IEC 1131-3 ladder logic counter and timer instructions.

element still allows programming of logical statements to control the START (S) input and the RESET (R) input, and for specification of an OUTPUT (Q) bit. A constant or an address is entered at the TIME VALUE (TV) input. In Figure 4.7, 2 s is entered in STEP 7's constant format for S5 timers.

2. The FR instruction can be programmed in an organization block or in a function as well as in a function block, although it is still available only in the STL language.

CQMI OMRON CQM1 COUNTERS AND TIMERS

There is a single set of memory addresses (TC 0 to TC 511) set aside for both CQM1 timers and counters so that a timer and a counter can't have the same number (e.g., if there is a timer named

TIM 6, there can't be a counter named CNT 6). The CQM1 offers regular timers and counters and high-speed timers and counters. The high-speed timers and counters are discussed in Chapter 11, with related topics such as the comparison table instructions and pulse output instructions. In this chapter we discuss only the two types of standard counters and the single standard timer.

One type of CQM1 counter is the **reversible counter CNTR (12),** shown in Figure 4.8. When the control logic on its **RESET (R)** input goes true, the value **0** is copied into the TC area memory location reserved for this counter's **PRESENT VALUE (PV).** The PV is held at zero until the R control logic goes false. If the RESET (R) logic is off, the counter's PV will be incremented each time the control logic on the **II** input goes true, and decremented each time the control logic on the counter's **DI** input goes true. (If the II and DI input logic statements both go true together, the PV does not change.) The CNTR (12) instruction's PV is said to be *circular:* It can be decremented below 0, in which case it reloads the **SETPOINT VALUE (SV)** automatically and counts down from there, or it can be incremented above the SV value, in which case it jumps to 0 and counts up from there. The CNTR (12) instruction's **completion flag** (CNT 6) goes on when the count has been decremented past 0 to the SV, or has been incremented past SV to 0, but stays on only until the next time the count changes. The counter's SV value is entered as a parameter of the CNTR (12) instruction. It can be entered as a *four*-digit-decimal constant (e.g., #0025, as in Figure 4.8) or as the address where a 16-bit BCD value is stored.

Other program instructions can read or write to the counter's PV by addressing the counter's memory location as **CNT x,** where "x" is the unique number for this counter. (PV values are in BCD format.) The counter's completion flag is also addressed as **CNT x.** The CQM1 recognizes whether CNTx means the PV or the completion flag by the type of instruction.

The much simpler **COUNTER (CNT)** instruction is a down counter. This counter will only count down, but won't count down past 0. When the control statement on the **RESET (R)** input goes true, the **SETPOINT VALUE (SV)** is copied into the counter's PV memory location (and won't change while R is true). The CNT instruction's **completion flag** goes on when the PV is 0.

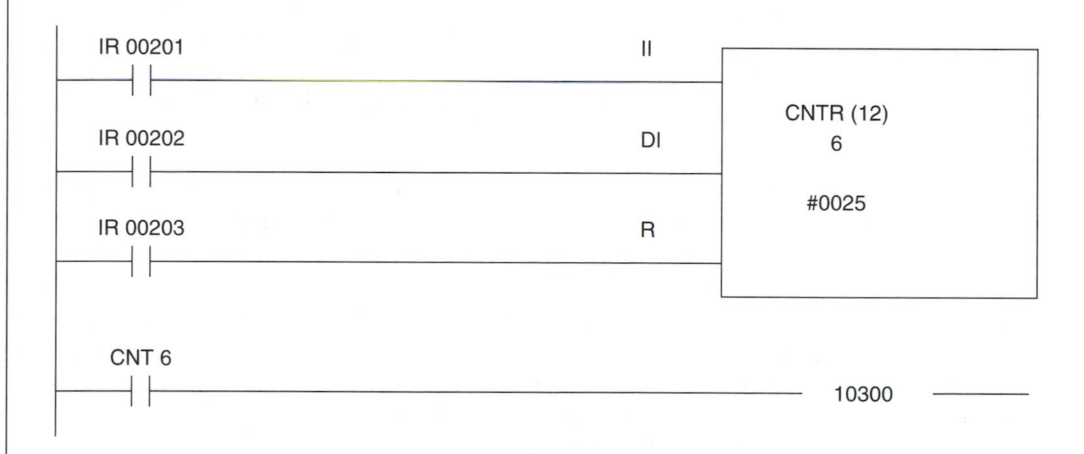

FIGURE 4.8
CQM1 reversible counter and an instruction using its completion flag.

For a timer, the CQM1 offers only the on-delay type. The **timer (TIM)** instruction is entered with a unique timer identifying number, and a **SETPOINT VALUE (SV)** is entered as a parameter with the instruction. The SV can be entered as a four-digit-decimal constant or as the address where a 16-bit BCD value is stored. The value represents the timer's running time in tenths of a second (e.g., #0125 is a constant meaning 12.5 s). Each time the control logic at the timer's single input goes false, the SV value is copied to the memory location reserved for the timer's **PRESENT VALUE (PV),** and the timer's completion flag is turned off. When the control logic goes true, the timer's PV begins decrementing. When the PV reaches zero, the timer's **completion bit** is turned on (even if the timer instruction hasn't been executed since its PV ran down to zero). The time remaining in the timer's PV can be addressed as **TIM x** in other CQM1 instructions, and the timer's completion bit can also be addressed as TIM x, where "x" is the timer number assigned when the timer instruction was entered.

TROUBLESHOOTING

One of the most common reasons that timers or counters fail to work is because they have been programmed so that the *timer/counter instruction is never executed with its control logic false,* so that it never sees that logic go true (or vice versa), and so that the timer never starts or the counter never counts. If you observe that the timer/counter instruction does execute once after each time the PLC is put into run mode, but refuses to work again until the PLC is switched out of run mode, this is very likely the programming error. There are several possible reasons why a timer/counter instruction might only ever "see" its control logic as true:

1. The programmer may not have entered any control logic, so there is no way the control logic can be false.

2. The control logic is always true, perhaps because of a programming error somewhere else. Monitor the control statement while the PLC is executing. Consider using the data monitor screen instead of just watching the program monitor.

3. There are structured programming techniques that can be used to cause the PLC to skip past parts of a program under certain conditions. Structured programming techniques are covered in Chapter 8. Perhaps the timer/counter's control logic does change state, but your structured program executes the timer/counter instruction only while that control logic is true (or only while false).

If you still believe that the control logic is changing state as it should but suspect that it is changing too fast to see on your slow programming unit display, temporarily add a unique new counter instruction that uses the same logic. Add this element immediately before or after the timer/counter that isn't working! If the new counter's accumulated value doesn't change, the control logic isn't changing at this point in your program. Try adding the same logic to control *two* new counters on a temporarily added first rung and last rung in your program. If neither counter changes, your program is looking for a change of state that isn't happening. (Could there be a faulty sensor or a broken circuit outside the PLC?)

If one or both temporarily added counters do change their accumulated values, you are looking for another type of error:

1. Did you use the same address for what should be two different timers or two different counters? Your programming software may have a cross-referencing feature that will tell you where else you have used this timer address or counter address. Remember that some PLCs (e.g., the OMRON CQM1) use the same set of addresses for timers and counters, so you can't use the same address number for a timer that you have already used for a counter!

2. Look for Boolean logic statements that affect the timer/counter status bits. (They may be overwriting your timer/counter's output bits.)

3. Look for statements that reset the timer. Are they preventing the timer from running? Try adding some logic temporarily so that you can make the reset logic false to see if that allows the timer/counter to work. Remember, resetting a timer/counter's status bit and accumulated value does *not* enable the timer/counter to time/count again; the counting logic or timer start logic must change state (or the Siemens FR STL instruction must be used to make the timer/counter think its control logic has changed state).

4. Look for instructions that may be affecting the timer/counter's accumulated value or its preset value (if a preset value can be changed). There are several instructions that can change these values, including MOVE instructions.

5. If your PLC is designed to decrement timer/counter values toward zero, are these values being properly set to nonzero values to allow timing/counting to execute? Are the values being set repeatedly to their starting value while the timer/counter is trying to decrement them? Monitor the accumulated values as your PLC runs.

6. If your PLC is designed to increment count/time value toward a preset value, are the counter/timer accumulated values being reset to zero to allow the counter/timer to work? Are they being reset repeatedly as your counter/timer is trying to increment them?

If it is a timer that you are having trouble with, perhaps you are just using the wrong type of timer or the wrong timer status bits. Remember, most timers reset themselves automatically when their control logic goes false. If the control logic isn't allowing the timer to run long enough, add a latch to hold the logic for an Allen-Bradley timer or an OMRON timer true, or use a Siemens timer that doesn't require its control to stay true. Unlatch the logic when the timer's status bit indicates that the timer has finished running. (Some of my students try programming off-delay timers when they first have trouble with on-delay timers, but this doesn't help at all, because they end up with the same problems and more.)

Another common problem with timers and counters is that their accumulated values change when they shouldn't be changing. If you remember that each counter/timer has a single memory location where it keeps its accumulated value and has a few status bits at reserved memory locations, you will probably not make these mistakes:

1. Each timer or counter must have its own unique identifying number, which is used by the PLC to identify the address(es) where its accumulated value and status bits are kept. Do not use that timer/counter address for another timer/counter, or both instructions will try to change the contents of the same accumulated value and will use the same status bits! Most PLCs maintain separate memory areas for timers and for counters, so you can have a timer 12 and a counter 12, but the OMRON PLC has a single area of memory for timers and counters. If an OMRON program has a timer 12, there shouldn't be a counter 12.

2. Do not enter more than one instruction to count up a single counter (or more than one count-down instruction for a counter), because only one status bit is used to recall the most recent count-up status (and one for count-down status). If you have entered two logical statements to count up, and one logical statement is true while the other is false, the PLC will see the count-up status bit changing from false to true *every* program scan and will increment the accumulated value for that counter every program scan.

QUESTIONS/PROBLEMS

1. What does a timer count?

2. Depending on the type of PLC, a preset value may or may not have to be stored, but *all* PLCs need to store a value for a timer or counter's __________ value.

3. Under what Boolean conditions does:
- **(a)** An up counter count up once?
- **(b)** An on-delay timer start timing and continue to time?
- **(c)** An off-delay timer start timing and continue to time?

4. How is a retentive on-delay timer different from a standard on-delay timer? (Provide at least two differences.)

5. If a (retentive) timer has been programmed to count 1-s time units, and the timer's logic has been true for one interval of 3.5 s and 10 intervals of 1/2 s, how much time is in the timer's accumulated value register?

6. If a counter stores a 16-bit signed value as its accumulated value, what is its counting range?

7. Complete the following table for nonretentive on-delay timers and off-delay timer outputs (on or off?). All timers are set to run for 5 s.

	Timer outputs	
Input conditions	On delay	Off delay
Went true 10 s ago		
Went false 1 s ago		
Went false 6 s ago		
Went true 1 s ago		
Went true 6 s ago		

PROGRAMMING EXERCISES (NO PLC REQUIRED)

1. Draw a STEP 7 ladder logic network containing a *complete* up-down Counter that could be used to turn an output (Q5.0) off after an input (I4.0) goes on four times. Add any other components to the network that are necessary, but *only if they are absolutely necessary.* Describe what each of the *unused* contacts in the counter in your network could be used for.

2. Write a program for the PLC you are using, with appropriate addressing. The program should:
- **(a)** Count the number of times that objects are detected passing a sensor on a conveyor.
- **(b)** Time while the sensor is blocked each time, and turn an output on whenever the sensor is blocked for longer than 2 s.
- **(c)** Count the number of times the timer reaches 2 s.
- **(d)** Reset the timer to repeat timing after each time the timer exceeds 2 s [so the counter in part (c) will count every 2-s interval].
- **(e)** Turn on an alarm when the counter in part (c) reaches 100.

SUGGESTED PLC LABORATORY EXERCISE

For a system with:

- Four control panel switches: inputs 0 to 3
- Four indicator lamps or visible output module LEDS: outputs A to D
- Two spring-return-valve-controlled cylinders: outputs E and F
- One detent-valve-controlled cylinder: outputs G and H
- Three sensors to detect extension of each of the cylinders

1. Write a PLC program to:

(a) Extend cylinder E for exactly 4 s every time that switch 0 goes on. The cylinder should extend for 4 s, regardless of how long the switch stays on.

(b) Extend cylinder F exactly 2 s after cylinder E extends, and retract when cylinder E retracts. Use the sensor at cylinder E to control this timer.

(c) Use a retentive timer to keep track of how much time cylinder F is on while switch 0 is off. Turn indicator light A on when the time exceeds 10 s. Reset the time when switch 1 goes on.

(d) Program a counter that will count actuations of switch 0. Switch 1 must cause the counter to count in the opposite direction. Switch 2 must reset the count to the beginning.

(e) After switch 0 has actuated three more times than switch 1, cause cylinder G–H to extend for 3 s, then retract. (Cylinder G–H requires actuation of separate solenoids to extend and retract. You shouldn't output to both solenoids at the same time. Actuate each solenoid for only 1 s.)

5

MEMORY ORGANIZATION AND DATA MANIPULATION

OBJECTIVES

In this chapter you are introduced to:

- Types and sizes of data units that a PLC must store, from bits to structures and files.
- Constant, absolute, symbolic, indexed, and indirect addressing modes possible in user-programs.
- Pointers and parameter passing.
- Addressing the memory of Allen-Bradley, Siemens, and OMRON PLCs, with examples.
- IEC 1131-3 data types, variable declarations, and "instances" of data for function blocks (as implemented in Siemens STEP 7).

The earliest PLCs could only manipulate on–off states of input and output image data bits. Modern PLCs can manipulate whole data words and sets of data words. In this chapter we discuss how data is stored and how it can be addressed in a modern PLC. Instructions to manipulate that data are examined in the following chapters.

MEMORY OVERVIEW

The memory in a PLC's CPU module must contain the programs and data that the PLC needs to operate. Memory is divided into three types:

1. One part of memory is not accessible to the user. This area of memory contains the PLC's proprietary operating system programs and data, as provided with the PLC.

2. A second area of memory, used to store the user-program and specific types of configuration data, can only be changed using a programming unit and usually can't be changed while the PLC is in run mode. In a typical PLC, little of this area of memory is addressable from within a user-program. Areas that are addressable, if any, can only be read by a user-program but cannot be changed.

3. A third section of memory is *fully addressable,* which means that instructions in the user-program can read and write to it. Output image memory must be addressable, of course, and some PLCs allow user-programs to change the contents of the input image area of memory (until the scan cycle overwrites user-program changes). Additional addressable memory areas are available for the storage of working data, including counter and timer values and other data values. Some parts of the PLC configuration memory are addressable, so that a PLC can be programmed to change its working characteristics dynamically (as it runs).

Besides memory in a CPU module, some more recent PLCs have I/O modules containing data memory that is directly addressable from a user-program. Data memory chips in the CPU module may be standard dynamic RAM type, which lose their data when power is lost. Modern PLCs often have batteries that provide power to maintain the data in dynamic RAM when power is interrupted, but there may be parts of memory that are not protected by battery power. Some PLCs have capacitors that can hold enough charge to prevent data loss during short power failures, so that no data is lost. The most recent PLCs contain flash memory chips, which can retain data without requiring capacitors or batteries. Memory that does not lose its data while a PLC's power is shut off is called *nonvolatile* or *retentive memory.* Although it is now economically feasible to make all of a PLC's memory retentive, some PLC manufacturers still designate some memory areas or parts of memory areas as *nonretentive.* Whenever the PLC is switched on, or sometimes whenever the PLC is put into run mode, the PLC clears those memory areas. We discuss this clearing of memory in the initialization interrupts section of Chapter 11. We discuss memory-use options in Chapter 10.

Most PLC programming units include a **data monitor screen** that the programmer or PLC operator can use to monitor the contents of data memory, even as the PLC program runs. In many PLCs, simply typing [ALT] [D] calls the data monitor screen. While the scan cycle controls some data (e.g., input image data and memory locations written to by the user-program), other data used by the program can be changed through the data monitor screen as the PLC runs! The user-program's behavior will be dynamically adjusted as the operator modifies the working data.

DATA TYPES

Each **addressable memory location** in a PLC is addressed using notation that identifies the type of data for which that area of memory is reserved. The address notation also indicates how many data bits data items of that data type require.

Data that is entered as a constant in a program usually includes a prefix identifying the format in which it is entered so that the PLC will know how to translate it into binary.

1. A **bit** consists of one **b**inary dig**it.** A bit is often described as a **Boolean** type of data element. Modern computers don't read and write individual bits of memory. Bits are stored in sets (*bytes* or *words*) in memory, so the address of each bit must indicate the byte or the word address to be read or written to, and must indicate which bit is to be used.

2. A **byte** consists of 8 bits and could be used to store a set of 8 independent bits but is often used to store an **unsigned integer** value of between 0 and 255. A ***word*** of data usually means 16 bits. A 16-bit memory location can be used to store 16 independent bits or an unsigned binary number representing an unsigned integer value of between 0 and 65,535, or to store a signed binary number representing a **signed integer** value between −32,768 and +32,767.

Newer PLCs offer **double-word** data elements of 32 bits, which can be used to store signed integer values of between (approximately) −2 billion to +2 billion, or unsigned integer values of between 0 and (approximately) 4 billion.

3. Several PLCs reserve 16-bit words in memory for storage of data in timer or counter format, as used by that PLC's timer and counter instructions. PLC manufacturers have each developed their own systems, but a typical timer word would be used to store an accumulated value in part of the 16-bit word, with the other bits used for status indicators, such as whether the timer/counter has reached its preset value. There is a movement toward standardization of counter and timer value formats.

4. Some PLCs reserve 32 bit-memory locations for the storage of **floating-point numbers.** The IEEE 754 format is commonly used to allow 32-bit binary numbers to represent real numbers in the range $\pm 1.1754944 \times 10^{-38}$ to $\pm 3.4028238 \times 10^{+38}$. Still other PLCs reserve 64-bit memory locations for even larger scientific notation numbers.

5. An **ASCII code** represents a single keyboard character and is often called a data unit of type **CHAR**. One ASCII code requires 8 bits of memory (1 byte).

6. Several recent PLCs now allow the programmer to store a set of similar data elements (e.g., integers or floating-point numbers) in a data element known as an **ARRAY**, then manipulate the array as a single data element. An array of ASCII codes is usually called a **STRING**.

7. A set of data elements that do not have to be of the same type, called a **STRUCTURE**, can be manipulated as a single data element in a few modern PLCs. Structures may contain arrays or even other structures. In some PLCs, timer and counter data is stored in structures.

8. A **pointer** is a memory *address* that is stored in data memory in a format that the PLC can interpret. Pointers are used in indirect addressing, described in the next section.

Just because a PLC manufacturer has provided memory areas intended for storage of specific types of data doesn't mean that a programmer always has to use that memory for that purpose. There is nothing to prevent a user-program from copying a full word of input image data to an area of memory intended for signed integer data, nor is there anything to prevent a user from programming Boolean instructions to examine or change individual bits of a memory location containing an integer number or an ASCII code. (PLC manufacturers have often included protection features in their programming software to, for example, prevent programs from changing input image bits or parts of ASCII codes. Users have generally demanded that these safeguards be removed.)

Some *instructions* are only capable of working on specific types of data, and won't work if the wrong address notation is entered (in fact, most programming software won't even allow entry of the wrong type of data). For example, the Boolean instruction EXAMINE ON requires the address of a single bit, whereas a MOVE instruction (available in most PLC programming languages) requires the destination memory address for a whole data byte, word, double word, or floating-point number. Neither Boolean instructions nor MOVE instructions work on arrays or structures.

ADDRESSING MODES

PLCs now allow the programmer to select from several addressing methods in entering data memory addresses in programs.

1. A **constant** is a number included with the program, and the programmer doesn't need to know what address it occupies in the program memory. Nonetheless, since it does occupy a memory location in the computer, a constant is said to have an *addressing mode,* and that mode is constant.

2. The most common addressing mode used in a program to indicate a data memory location is the **absolute address** mode. An absolute address is the actual address where the data is stored, usually entered as an alphanumeric prefix identifying the type of data stored in this area of memory, followed by numbers identifying exactly which memory location is being addressed in that area. All PLCs allow absolute addressing of data.

3. **Symbolic addressing** is a variation of absolute addressing. The programmer uses the programming unit to enter a table that identifies alphanumeric names for individual absolute addresses. After entering the table, the programmer can then enter PLC programs referring to memory addresses by their alphanumeric names. The programming unit will substitute the actual absolute addresses before sending the program to the PLC.

4. Several PLCs now allow **indexed addressing,** or allow the programmer to use interchangeable **data blocks.** These two addressing techniques are very similar when examined closely. To use indexed addressing, the user-program must first have placed an offset value into a memory location reserved for offset values. (Some instructions do this automatically.) Whenever the PLC encounters an instruction containing a *base address* in indexed addressing notation, the PLC will *add* the offset value to the base address, to calculate the *absolute address* of the data that is to be manipulated. (In an Allen-Bradley PLC, for example, any address preceded by a "#" is treated as an indexed address.) To use data blocks, the user-program must first "call" one of the data blocks (thereby establishing an offset value). Addresses of data words in the program are then assumed to be numbered from the start of that offset area in memory.

5. Some PLCs now offer **indirect addressing.** Whenever the PLC encounters an address entered in indirect addressing format, the PLC reads the contents of the specified address and treats that value as the absolute address of the data item to be manipulated. The first memory location is said to contain a **pointer** to the second memory location. (In a Siemens S7 PLC, preceding an address with "#P" is one way to make the PLC recognize the memory's contents as an indirect address.)

6. Some PLC programming software packages allow **formal operand** names to be defined to represent values that are supplied by one program to another program (to a "subroutine" or "function," etc.) that it calls. The alphanumeric formal operand names can then be used instead of actual addresses in the program that is called. The IEC 1131-3 standard includes the requirement that each program or function should have an area of memory that is available only to that program or function, not to any other program or function. As this IEC 1131-3 requirement is implemented, more PLC programming languages will allow the use of formal operands.

ADDRESSABLE DATA STORAGE IN PLCS

In Chapter 2 we discussed input and output images addressing, and in Chapter 4 we discussed addressing of counters and timers, so we have already covered some addressing conventions. In this section we review those addressing formats and cover the other types of addressable memory and the addressing modes available in selected popular PLCs.

The MOVE instruction is used in this section to demonstrate the capabilities of selected PLCs to store data, although the MOVE instruction will not be covered formally until Chapter 6. A MOVE instruction copies a data value from one memory location into another, without changing the contents of the source memory location. (A MOVE is a really a copy instruction.)

PLC-5 SLC 500

DATA FILES AND ADDRESSABLE DATA IN THE ALLEN-BRADLEY PLCS

Allen-Bradley's PLC-5 and SLC 500 PLCs have very similar memory structures. In this section we cover the similar data addressing characteristics. Where there are significant differences, they are covered in the following sections, which deal specifically with the PLC-5 or the SLC 500.

In both Allen-Bradley PLCs, the CPU memory contains a **processor file** that consists of the **program files** and the **data files** that the PLC needs when it is running. Program files contain the user-program(s) and some operating system configuration information that the PLC needs to be able to run. (Program files are not discussed in this chapter.) The data files contain user data and system configuration data that the program files require. Data stored in data files is retained by an Allen-Bradley PLC until specifically changed or cleared, even if the PLC is switched out of run mode and even if it is disconnected from its power source, as long as the memory-backup battery works. Allen-Bradley PLCs differ from the other PLCs covered in this textbook in that those other PLCs have data memory designated as *nonretentive.* Nonretentive memory is cleared automatically by those other PLCs when they are switched out of run mode or when they are disconnected from power.

Allen-Bradley Constant Data Values

Constants can be entered directly into some user-program instructions instead of entering addresses for data values. The following values can be entered. In some cases, an identifying prefix must be used:

1. Signed decimal integer values are accepted in the range -32768 to $+32767$.

2. Larger decimal numbers can be entered, with or without decimal points, if the PLC (and the instruction) is capable of using floating-point numbers (not all SLC 500s can). For example, 9876543210 is acceptable, as is -10.5. The size of the acceptable number is subject to the same limit, as are scientific notation numbers (see below).

3. Very large or very small decimal numbers can be entered in scientific notation format (e.g., 9.87E25, which means 9.87×10^{25}, or −9.8E-25, which means -9.8×10^{-25}) but only within the range that a 32-bit IEEE 754 format floating-point number can represent ($\pm 1.1754944 \times 10^{-38}$ to $\pm 3.4028238 \times 10^{+38}$).

4. Hexadecimal, octal, binary, and ASCII constant values are accepted in some instructions if prefixed correctly. The prefixes are:

&H for **hexadecimal** values from 0 to FFFF (e.g., &HFF12)
&O for **octal** values from 0 to 177777 (e.g., &0177000)
&B for **binary** (e.g., &B1111000011110000)
&A for **ASCII** (e.g., &Ahi)

Allen-Bradley Data Files

Data that is not entered into the user-program as constants must be in addressable memory in the data files. Each data file is used to store a single type of data. Three of the data files are essential to the PLC:

- **O** **output image data words;** covered in the sections dedicated to the PLC-5 and SLC 500 processors
- **I** **input image data words;** covered in the sections dedicated to the PLC-5 and SLC 500 processors
- **S** **status words;** contains configuration data for the PLC and status data describing the conditions of the PLC as it operates

Additional data files are available for storage of specific types of user data. Some of these additional data files are intended for specific types of data by default. The default data files include:

- **B3** **bit storage** in 16-bit data words
- **T4** **timer** data structures of three data words each
- **C5** **counter** data structures of three data words each
- **R6** **control** data structures of three data words each
- **N7** **integer** data word storage
- **F8** **floating-point** number storage in the PLC-5 and better versions of the SLC 500

Data files **10 to 255** in the SLC 500, or files **9 to 999** in the PLC-5, are also available, but these data files do not have default data types. They can be used for additional storage for bits, timers, counters, control elements, integers, floating-point numbers, or for any of the other data types described below or in the sections devoted to the SLC 500 and PLC-5. Entering the file number of an available data file during programming automatically reserves that data file for storage of the data type specified with the file number. For example, to create data file 12 as an additional data file for bit storage, write a program that uses the following address:

`B12/x` or `B12:x` or `B12:x/x`

This will reserve file 12 for bit storage use. ("x" must be a valid bit or word address. The bit file address structure is described below.) Data files can also be reserved for specific data types by using the file-creation features in the programming unit.

Allen-Bradley Status Files

The default status file is the only status file that can exist. It contains Allen-Bradley–defined status words. Each 16-bit status word contains either configuration data for the PLC or status data describing the conditions of the PLC as it operates. Any of the status words can be examined by a user-program. Most status words are *dynamic,* which means that a user-program can write to them, perhaps changing the PLC's configuration, as the user-program runs. Other status words are *static,* which means that they can only be changed while the PLC is in program mode or, occasionally, that a user-program can change them but the PLC will only recognize that they have been changed the next time the PLC is switched into run mode. Status *words* are addressed in the following form:

S:e where "e" is the number of a 16-bit status word *element* (e.g., S:1 is a data word that contains 16 processor status bits)

The first status word is numbered 0, and the last depends on the model of SLC 500 or PLC-5. The status words for the SLC 500 and the PLC-5 are listed in Appendices A and B. Notice that the PLC-5 and the SLC 500 use the same status word address for the same function where possible, but the status files are not identical.

Individual status bits can be addressed in Boolean logic statements using the following form:

S:e/b[1] where "b" is a number between 0 and 15 indicating which bit to operate on (e.g., S:1/15 is the highest bit of word 1; this important status bit goes on for one scan cycle each time the PLC is put into run mode)

Allen-Bradley Bit Files

A bit file is intended for **bit storage.** File 3 is, by default, reserved as a bit file and is addressed with the prefix "B3." The prefix "B" can be used with any unused file number during programming to reserve that data file for additional bit storage. When used for bit storage, individual bits in bit file 3 are usually addressed in this form:

B3/b where "b" is a number between 0 and 4096 in the SLC 500, or between 0 and 15999 in the PLC-5

Allen-Bradley allows an alternative way of addressing bits in the bit storage file. Recognizing that all bits are actually parts of 16-bit data words, they can be addressed in this form:

B3:e/b where "e" is the word element, numbered from 0 to 255 in an SLC 500, or 0 to 999 in a PLC-5; "b" is a number between 0 and 15, indicating one of the 16 bits of that word

Figure 5.1 shows memory locations in the bit file memory, and demonstrates the two ways of addressing the *same* data bit. In the example, "B3/20" means the twenty-first bit from the start of file B3. The same bit can also be described as "B3:1/4," which means the fifth bit of the second data word.

Bit files consist of 16-bit data words, which can be manipulated as whole words. *Whole "bit" data words* can be manipulated if addressed in this form:

B3:e where "e" is a word element number between 0 and 255 for the SLC 500, or between 0 and 999 in the PLC-5

---

[1]In the PLC-5 only, the programmer can enter a period (.) in place of a slash.

	15															0
B3 : 0	15	14	13	12	11	10	9	8	7	6	5	4	3	2	1	0
B3 : 1												20	19	18	17	16
												4	3	2	1	0

FIGURE 5.1
Two ways of addressing the same Allen-Bradley bit file data bit.

Allen-Bradley Timer Structure Files

File 4 is, by default, reserved for timer data *structures,* and all data in this file must be addressed with the prefix "T4." Each timer element structure is made up of three 16-bit data words, arranged as shown in Figure 5.2.

When programming a timer instruction, the programmer enters the address of a *whole timer structure* in the form

T4:e where "e" is the number of the three-word structure element (0 to 255 in an SLC 500, 0 to 999 in a PLC-5)

Each *data word* of a timer data structure can be addressed as a word, in the following form:

T4:e.m where "e" is the number of a three-word structure element number (0 to 255 in the SLC 500, 0 to 999 in a PLC-5); "m" is the Allen-Bradley–defined mnemonic for one of the data words in the timer structure (e.g., T4:3.ACC is a way of addressing the accumulated value for timer 3)

In earlier Allen-Bradley PLCs, the programmer couldn't use mnemonics, but had to address data words numerically, as follows:

T4:e.s where "s" is the number of one of the three-word subelements (0 to 2) (e.g., T4:3.2 is the address for the accumulated value for timer 3)

Some programming software allows the programmer to choose either numerical or

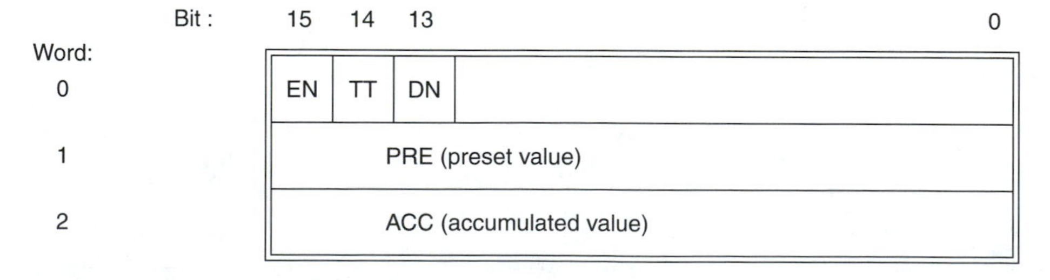

FIGURE 5.2
Timer data structure in Allen-Bradley PLCs.

mnemonic addressing. The programmer should learn the mnemonics method because some programming software requires the use of mnemonics in timer structure addressing.

Individual bits of the three-word structure can be addressed in the following formats:

T4:e/m where "m" is the Allen-Bradley–defined mnemonic for one of the addressable status bits in word 0 (e.g., T4:3/TT is another way of addressing the TT bit of timer 3)

T4:e.m In the PLC-5, use of a period (.) instead of a slash (/) is allowed and even recommended by Allen-Bradley (the period can't be used in programming an SLC 500)

Allen-Bradley Counter Structure Files

File 5, by default, is reserved for counter data structures. When addressing data in this file, the prefix "C5" must be used. Each three-word structure is saved in the format shown in Figure 5.3.

Addressing of counter data is similar to addressing of timer data, except that "C5" replaces "T4" as follows:

C5:e for a three-word counter element
C5:e.m for one of the data word subelements of a structure
C5:e/m for a single status bit of word subelement 0 (in an SLC 500)
C5:e.m for a single status bit of word subelement 0 (in a PLC-5)

Numerical values instead of mnemonics can be used in an SLC 500 program, but the programmer should learn the mnemonics method.

Allen-Bradley Control Structure Files

File 6, by default, is reserved for control data structures. Data in the control structure file must by addressed with the prefix "R6." (The "R" prefix is used because the first four letters in "control" were already used for **c**ounters, **o**utputs, **in**tegers, and **t**imers before Allen-Bradley introduced cont**r**ol elements.) The three-word structure format is shown in Figure 5.4.

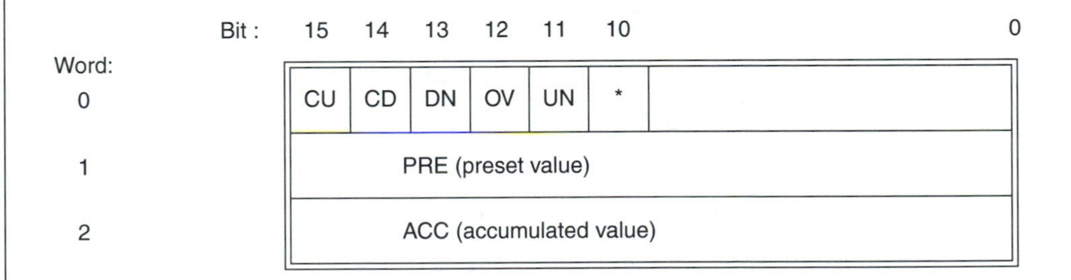

FIGURE 5.3
Counter data structure in Allen-Bradley PLCs.

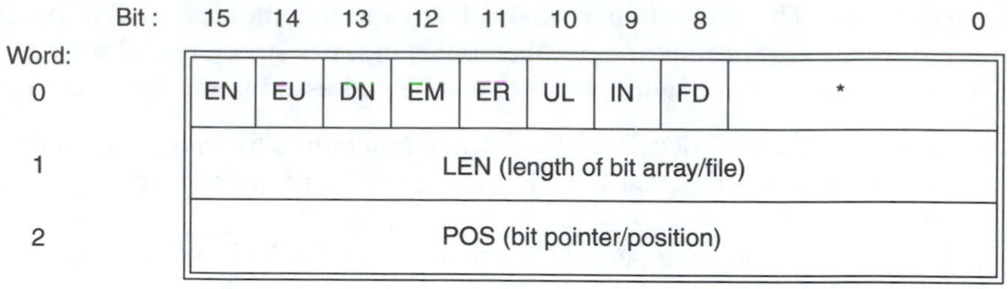

Word	Bit: 15	14	13	12	11	10	9	8	7–0
0	EN	EU	DN	EM	ER	UL	IN	FD	*
1	LEN (length of bit array/file)								
2	POS (bit pointer/position)								

* Bits 0 to 7 used in the SLC 500 to store an error code.

FIGURE 5.4
Control data structure in Allen-Bradley PLCs.

Control data can be addressed as follows:

R6:e for a three-word counter element
R6:e.m for one of the data word subelements of a structure
R6:e/m for a single status bit of a word subelement 0 (in an SLC 500)
R6:e.m for a single status bit of a word subelement 0 (in a PLC-5)

Numerical values instead of mnemonics can be used in an SLC 500 program, but the programmer should learn the mnemonics method.

Allen-Bradley Integer Files

File 7, by default, is reserved for storage of integer data values between −32,768 and +32,767. Each integer value is stored as a 16-bit signed binary number. Individual data words in this file can be addressed as follows:

N7:e where "e" indicates one of the data words in file 7 (0 to 255 in an SLC 500, 0 to 999 in a PLC-5)

Individual bits of data words can be addressed as follows:

N7:e/b where "b" is a bit number between 0 and 15

Allen-Bradley Floating-Point Structure Files

File 8, by default, is reserved for floating-point number storage in the PLC-5 and in better versions of the SLC 500. Floating-point numbers in the range $\pm 1.1754944 \times 10^{-38}$ to $\pm 3.4028238 \times 10^{+38}$ are coded into binary using the IEEE 754 floating-point-number format, which requires 32 bits, so each floating-point-number element is a structure of two 16-bit data words. Individual bits and 16-bit data words cannot be addressed, so floating-point-number storage elements can only be addressed as follows:

F8:e where "e" is an element number (0 to 255 in the SLC 500, 0 to 999 in the PLC-5)

Allen-Bradley ASCII Character Files

There is no default file for ASCII characters, but any unused file can be reserved for ASCII characters. ASCII characters include the original standard teletype keyboard character set. For each character, a binary *ASCII code of 7 bits* exists. (ASCII code remains the most commonly used format for binarizing keyboard characters.) In an Allen-Bradley ASCII character file, each code is stored in half of a 16-bit data word, so one word can store two characters. Since Allen-Bradley doesn't allow the addressing of bytes (8-bit elements), each addressable ASCII file element contains a pair of ASCII-coded characters. ASCII file data is addressed as follows:

Af:e where "f" is the number of the file and "e" is the number of a word in that file (each word is a 16-bit number that contains two ASCII codes representing two ASCII characters)

Individual bits of ASCII data file elements can be addressed in the following form:

Af:e/b where "b" is a bit number (0 to 15) in the two-ASCII-code (0 to 255 in an SLC 500, 0 to 999 in a PLC-5) of data file "f" (one of the extra data files)

Af:e.b (the period being recommended by Allen-Bradley in programming the PLC-5)

Allen-Bradley String Array Files

No default files are reserved for string data storage, but any unused data file can be reserved. Each string array consists of forty-two 16-bit data words, of which the first data word contains a number describing the actual number of ASCII characters currently stored in the string. The following 16-bit data words each contain two 8-bit ASCII codes representing ASCII characters, as shown in Figure 5.5. (ASCII characters are shown in Figure 5.5, although ASCII codes would actually be stored.)

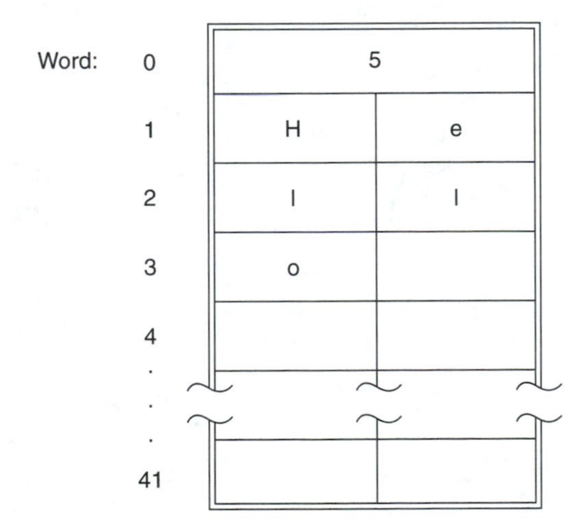

FIGURE 5.5
ASCII string storage in an Allen-Bradley string data file.

Allen-Bradley PLCs offer instructions for manipulating strings. String element words and bits can only be manipulated using those instructions, because Allen-Bradley PLCs do not allow addressing of individual string data words or bits, except for the first word, which contains the length of the string. This first word can be accessed using the address

STf:e.LEN where "f" is the file number and "e" is the string element number

Figure 5.6 demonstrates the use of some of the Allen-Bradley addressing conventions covered so far. It shows an Allen-Bradley ladder logic rung that examines the seventh bit from a bit file (the first bit is bit 0). If the bit is on, the rung:

1. Moves a constant decimal value (0) to a 16-bit bit file word that contains B3/16 to B3/31, turning them all off.

2. Copies the third data word (the accumulated value) of timer structure 8 of file 4 to an integer data file (N7:12).

Allen-Bradley Indexed Addressing

Allen-Bradley PLCs can also be programmed using indexed addressing for data memory. If an address is entered with a "#" preceding it, the PLC will perform a calculation to generate an element number before operating on the address. The calculation consists of adding the number from status file word S:24 (called the *offset number*) to the element number entered in indexed addressing format (the "#" and a *base number*). This addition generates the absolute address of the data element to be operated on. The sample program in Figure 5.7 includes a MOVE instruction that puts an offset value into S:24, and a second MOVE instruction with a base number in indexed addressing format. The table in Figure 5.7 shows how the absolute address of the data to be moved to B3:1 is calculated, and shows other indexed addressing examples.

Since the offset number in status file word 24 (S:24) affects the actual address to which an indexed address refers, the programmer *needs to ensure that S:24 contains the appropriate value,* usually by using a MOVE instruction in the user-program to put a value into S:24 (as in

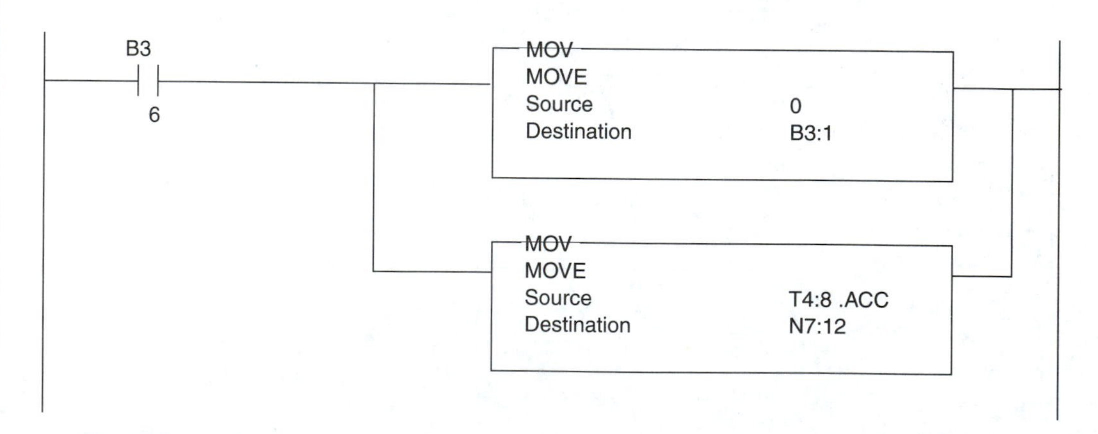

FIGURE 5.6
PLC-5 or SLC 500 program rung that conditionally performs two moves.

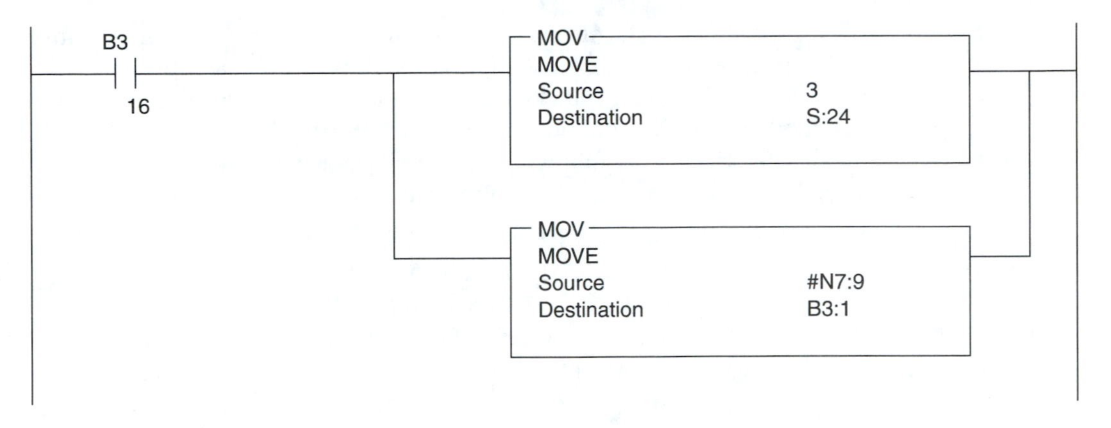

Indexed address	S : 24 contents	Actual address
#N7 : 9	3	N7 : (3+9) = N7 : 12
#N7 : 9	–3	N7 : (–3+9) = N7 : 6
#N7 : 9	–15	See text
# C5 : 2 . ACC	5	C5 : (5+2) . ACC = C5 : 7 . ACC

FIGURE 5.7
Allen-Bradley indexed addressing examples.

Figure 5.7). Positive or negative values can be entered (within the limits dictated by 16-bit signed binary numbers) and S:24 can be changed before repeating an instruction with an indexed address. Care should be taken that elements which are addressed do, in fact, exist. There is no address N7:**−6,** so adding the S:24 offset value **−15** to a base address #N7:**9** will generate an invalid address. N7:12 could also be an invalid address if N7:10 was the last data word in N7. Allen-Bradley calls generation of nonexistent addresses **crossing data file boundaries** because N7:−6 *does*, in fact, exist. N7:−6 is a data word in the previous data file: R6. SLC 500 PLCs can be configured to allow crossing of data file boundaries or to fault when a program tries to do so. PLC-5 PLCs will fault if a data file boundary is crossed.

Some instructions automatically use indexed addressing and modify the contents of S:24 as they execute. These instructions require the programmer to enter addresses preceded by the "#" symbol but don't require the user to program instructions to write values to S:24. When writing a program that uses indexed addresses in instructions that do *not* automatically control the contents of S:24, it is good programming practice always to move desired offset values into S:24 immediately preceding those instructions, especially in programs that contain instructions which automatically change the contents of S:24.

Allen-Bradley Indirect Addressing

The PLC-5 and better SLC 500s offer indirect addressing, which is what Allen-Bradley calls the use of **pointers.** In indirect addressing, the numerical part of an address specification can be entered as the address of a data memory location that contains the absolute number. (Confused?) For example, the indirect-format address "N**[B3:4]**:1" actually means "N**7**:1" if **B3:4** contains the number **7.** The square brackets are essential so that the PLC-5 will recognize indirect addressing. File numbers, word numbers, and binary file bit numbers can all be indirectly referenced in an address, as in the following examples. Data words in structures can only be referenced indirectly in timer, counter, and control structures.

`N[N7:1]:5`	means N**16**:5 if N7:1 contains "16."
`B3:[N7:1]`	means B3:**16** if N7:1 contains "16."
`B3/[N7:1]`	means bit B3/**16** if N7:1 contains "16."
`N9:5/[N7:1]`	*won't work.* Indirect addressing of bits is possible only for bits in bit files.
`O:[N7:1]`	means O:**16** in an SLC 500 if N7:1 contains "16."
	means O:**020** in a PLC-5 if N7:1 contains "16." Indirect address references for PLC-5 I/O image addresses are converted to octal. (Decimal "16" = octal "20.")
`C5:[B3:1].ACC`	means C5:**2**.ACC if B3:1 contains "2."
`C5:2.[B3:1]`	means C5:1.**2** if B3:1 contains "2"; *won't work in a PLC-5,* because mnemonic addressing must be used to address words.
`ST9:[B3:1].5`	*won't work.* Indirect references to a string element aren't allowed.
`ST9:1.[B3:1]`	*won't work.* Indirect references to a word of a structure other than a timer, counter, or control element structure aren't allowed.
`SC10:[B3:1].TIM` (PLC-5 only)	means SC10:**2**.TIM if B3:1 contains "2."

Allen-Bradley Symbolic Addressing

Allen-Bradley PLCs can also be programmed using symbolic addressing. The programming software allows the programmer to define alphanumeric symbolic names to represent absolute addresses. After defining symbolic names, the symbolic names can be entered in the user-program instead of entering the absolute addresses.

Allen-Bradley Formal Parameters

Alphanumeric formal parameters cannot be entered in Allen-Bradley programs (at least not when using the PLC-5's 6200 programming software or the SLC 500's APS programming software). By the time you read this, formal parameters probably will be available.

When an Allen-Bradley PLC program contains a JUMP TO SUBROUTINE (JSR) instruction, the JSR instruction allows the programmer to specify *input parameter* addresses. The matching SUBROUTINE (SBR) instruction also allows addresses to be entered as input parameters. Whenever the subroutine is called, the PLC copies data from the input parameter addresses listed in the SBR instruction to the input parameter addresses listed in the SBR label.

Similarly, output parameter addresses can be listed in a JSR instruction and in the RETURN (RET) instruction at the end of a subroutine. Whenever a subroutine ends, the PLC copies data from the output parameter addresses listed in the RET instruction to the output parameter addresses listed in the JSR instruction.

As the subroutine executes, it can manipulate data that was copied to its input parameter addresses without affecting the original values, and can write results to its output parameter addresses without prematurely changing the eventual destination addresses for that data.

ADDRESSABLE DATA IN THE ALLEN BRADLEY PLC-5

In the preceding section we discussed data storage and addressing features common to the PLC-5 and the SLC 500. In this section we describe data memory characteristics unique to the PLC-5.

Allen-Bradley PLC-5 I/O Image Data Files

Data file 0 (zero) is reserved for output image data, and addresses must have the prefix "O:" (oh). Data file 1 (one) is reserved for input image data, which must be addressed with the prefix "I:" (eye). The output image and input image data files are arranged as shown in Figure 5.8. Notice that the octal numbering system is used, to be consistent with earlier Allen-Bradley PLCs.

Addressing for 16-bit output image and input image file *data words* must be in the following format:

p:e where the prefix "p" is the letter O (for *output*) or I (for *input*); "e" is a three-digit octal[2] number of which:

- The first two digits indicate the logical rack.
- The last digit indicates the I/O group where the I/O module that corresponds to the image is actually located (PLC-5 I/O groups were described in Chapter 2).

[e.g., I:123 is the address of the 16-bit input image word corresponding to the digital input module(s) in I/O group 3 of rack 12.
e.g., O:027 is the address of the 16-bit data word corresponding to the digital output module(s) in the last I/O group (group 7) of rack 02]

Individual bits of I/O image data words are addressed in the form

p:e/d where "d" indicates (in octal) which of the 16 bits is being operated on [the 16 bits of an I/O image data word are numbered 00 to 07 and 10 to 17;
e.g., I:123/04 is the address of bit 4 of input image data word 123; it represents the state of a sensor attached at the fifth screw terminal of an input module in I/O group 3 of logical rack 12)

Allen-Bradley PLC-5 BCD/Hexadecimal Data Files

Unused data files can be assigned for storing 16-bit numbers representing BCD/hexadecimal values. Four-digit BCD/hex data values [between decimal 0000 and 9999 for *binary-coded decimal* (BCD), or hexadecimal 0000 to FFFF] can be stored. Each of the four digits is stored as a separate 4-bit code in the 16-bit number (4 bits = 1 nibble). When using a programming unit to

---

[2]Octal numbers use the digits 0 through 7 only. Racks are therefore numbered 00 through 07, 10 through 17, and so on. The largest current PLC-5 system can have racks numbered to 27. I/O groups are numbered 0 through 7.

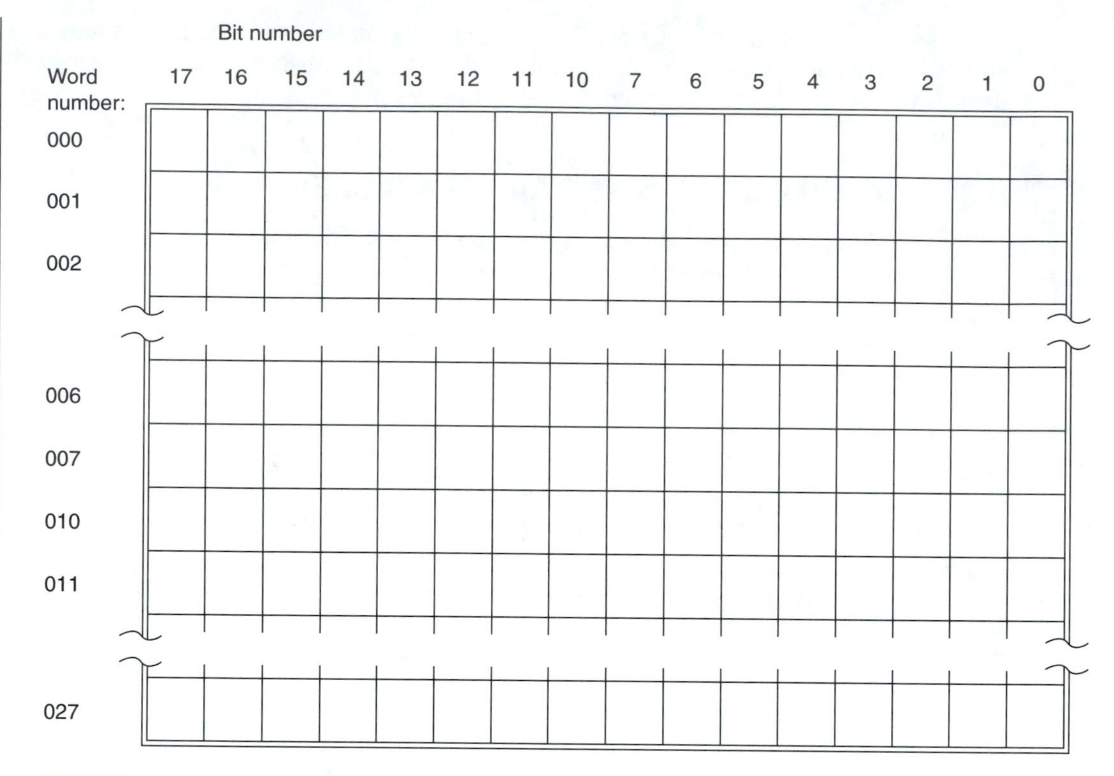

FIGURE 5.8
Allen-Bradley PLC-5 I/O image file format.

monitor programs containing addresses of BCD/hex data words, the programming unit displays a four-character decimal/hex number to show the address's contents. The 16-bit binary numbers are not treated as signed binary numbers when being converted for display purposes. Other than how numbers are displayed on programming units, the PLC-5 does not treat BCD/hex numbers differently from integer numbers. PLC-5 math operations treat BCD/hex numbers as if they were signed binary numbers representing integers.

When some input and output interface devices accept or generate BCD values, they use a special convention to enable representation of signed binary numbers in BCD. Each of the low 3 nibbles of the 4-nibble binary number will contain a binary representation of the decimal characters 0 through 9. The high nibble uses binary 0000 to represent a positive sign and a binary 1111 for a negative sign. Decimal values between −999 and +999 can thus be represented. On the PLC-5 programming unit, these numbers will appear as 0999 and F999.

Hexadecimal values are often used as a compact way to represent binary values, so applications using 16-bit binary numbers are sometimes easier to understand if those numbers are displayed in hexadecimal rather than in binary or in decimal forms.

BCD/hex *data words* are addressed as follows:

Df:e where "f" is any available data file (3 to 999) and "e" indicates a 16-bit word element (0 to 999)
(e.g., D9:4 is the address for a decimal number stored in BCD format in data word 4 of data file 9)

BCD/hex *data bits* can be operated on if addressed in the form

Df:e/d where "d" indicates bits numbered 0 through 15
(e.g., D9:4/15 is the most significant bit of the 16-bit value stored as word 4 in file 9; if it is a 1, the decimal number is negative)

Allen-Bradley PLC-5 Block Transfer Structure Files

In an *enhanced* PLC-5, unused data files can be reserved to store block transfer data elements consisting of six-word structures. A block transfer element is used for control purposes while the CPU transmits data to/from intelligent I/O modules, which Allen-Bradley sometimes calls block transfer modules. In the *classic* line of PLC-5s that predated enhanced PLC-5s, the programmer had to use a set of five data words in an integer file for control of each data exchange with an intelligent I/O module.

There is no diagram of the format of a block transfer structure in this book because the programmer does not need to know its format. Individual data bits and words of a block transfer data element cannot be addressed numerically. They can *only* be addressed by the Allen-Bradley mnemonics. Block transfer elements are addressed as follows:

BTf:3 for the whole six-word structure, where "f" is one of the available data files (3 to 999) and "e" is an element in that file (0 to 999)

BTf:e.m for a 16-bit data word or a single status bit from the element ("m" mnemonics for words include RLEN, DLEN, FILE, and WORD; "m" mnemonics for individual status bits include EN, ST, DN, ER, CO, EW, NR, TO, and RW)

Allen-Bradley PLC-5 Message Control Structure Files

In an enhanced PLC-5, unused data files can be assigned to store 56-word structures used for message control data. When a user-program includes a MSG (message) instruction to send or receive data via one of Allen-Bradley's *data highway* local area networks, the MSG command must specify a message control element in the following form so that the PLC can use the message control structure in handling the data exchange:

MGf:e where "f" is an available data file (3 to 999) and "e" is the element number (0 to 999) of a 56-word message control element in that file

The 56-word structure includes status bits and words that can be addressed in other instructions in the user-program in this form:

MGf:e.m where "m" is a mnemonic for either a status bit or a 16-bit data word

Allen-Bradley PLC-5 PID Structure Files

In an enhanced PLC-5, unused data files can be assigned for the storage of 82-word structures of PID data used by the PID instruction. The PID instruction and the PID data structure element are described in detail in Chapter 12. Before development of the enhanced PLC-5, programmers had to set aside a block of 23 data words in an integer file for storage of data important to the PID instruction, but now a PID instruction can reserve and use the improved 82-word PID element by addressing it in the following ways:

PDf:e where "f" is an available data file (3 to 999) and "e" is the element number (0 to 999) of an 82-word PID data element in that file

Allen-Bradley's new PID data element's structure includes data words and status bits, and also floating-point values. Each value can be addressed using mnemonics in this form:

PDf:e.m where "m" is a mnemonic for either a status bit, a 16-bit data word, or a 32-bit floating-point number

Allen-Bradley PLC-5 SFC Step Timer Structure Files

In an enhanced PLC-5, unused data files can be assigned to store SFC step timer data elements, consisting of three data words each. SFC step timers can be used so that a programming unit can monitor a program written in structural flowchart (SFC), displaying the timing status of each SFC step as it is being monitored. The user-program can also address individual words and status bits of the SFC timer elements set up for each SFC step, in the following forms:

SCf:e for a three-word counter element

SC5:e.m for one of the data word subelements or status bits of a structure (mnemonics for words include PRE and TIM; mnemonics for status bits include SA, FS, LS, OV, ER, and DN)

SLC 500

ADDRESSABLE DATA IN THE ALLEN-BRADLEY SLC 500

We have already discussed the data storage and addressing features that are common to both the PLC-5 and the SLC 500. In this section we describe data memory characteristics unique to the SLC 500.

Allen-Bradley SLC 500 I/O Image File

Output image data words are maintained in data file 0. Input image data words are maintained in file 1. Whole 16-bit words in these memory areas are addressed in the following format:

p:e or p:e.s

where:

- The prefix "p" is the letter O (oh) for an output image data word, or I (eye) for an input image data word.

- The element number "e" indicates the slot that contains the input or output module that the image word represents. Number 0 means the CPU module, which contains I/O contacts only if the SLC 500 is a fixed controller. Numbers 1 through 30 represent slots numbered from the CPU module.
- The subelement "s" is required only if the I/O module needs more than a single 16-bit input image or output image. 0, 1, 2, and so on, would indicate the first, second, third, and so on, 16-bit data word of image data. The first data word (word 0) is *assumed* if no "s" value is specified. The maximum value for "s" is 255, which makes very powerful *specialty* I/O modules possible.

Individual I/O image *data bits* can be addressed in the following format:

p:e/d or **p:e.s/d** where "d" is the address of the binary bit to be operated on (0 to 15)

Allen-Bradley SLC 500 Module Data Files (M0/M1)

The SLC 500 allows the addressing of data files that are actually contained in I/O modules. This type of data file addressing is only possible, of course, if a specialty I/O module with its own memory has been installed. If so, the module data files (M0 and/or M1) can be addressed as follows:

Mf:e.s where "f" is the number 1 or 0, "e" is the slot number where the module is located (1 to 30), and the subelement number "s" indicates which word of the module's memory is being operated on (0 to the module's maximum)

M0 or M1 data files can also be addressed at the bit level in the following form:

Mf:e.s/b where "b" is the bit number (0 to 15)

Configuration data can also be written to specialty I/O modules but only while the PLC is not running. Since this type of I/O module data memory (called the *G file*) is not addressable from a user-program, we do not discuss it here.

Allen-Bradley SLC 500 Data File 9

The available data files, 10 to 255, can be used to store any type of data (except I/O images, status, or M0/M1 data). The reader may ask why data file 9 isn't available. In Chapter 13 we will see that file 9 is sometimes needed for certain types of communication data, so it should be reserved for that purpose.

In Figure 5.9, an SLC 500 ladder logic rung examines a bit in a binary file. If the bit is on, the rung:

1. Moves a 16-bit value from the input module in slot 8 to an integer file data word (N7:2). The input module doesn't use more than 16 input image bits, so no "s" parameter is required.

2. Moves a two-ASCII code data word (a 16-bit value representing the ASCII characters "H" and "i") to the fourth data word (word 3) of file 0 in a specialty I/O module's memory. The I/O module is in slot 12.

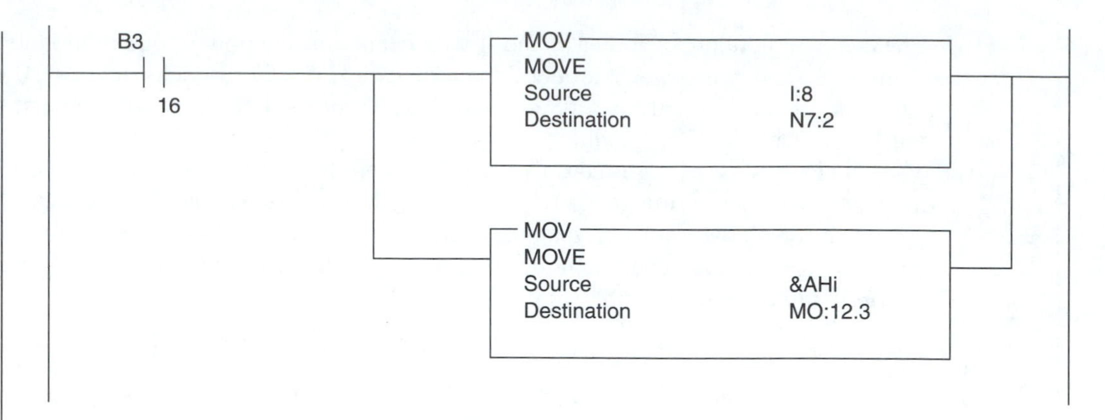

FIGURE 5.9
SLC 500 program rung that conditionally performs two moves.

S5

ADDRESSABLE DATA IN SIEMENS STEP 5

The STEP 5 operating system allows the programmer to enter constants or addresses for data in I/O image, timer, counter, flag, data blocks, and reserved system memory. All data can be entered and addressed as bytes (8 bits) or words (16 bits).

Siemens STEP 5 Constant Data Values

Constant values can be entered in user-programs. Prefixes with each data entry identify the type of data entered, so that the PLC can convert that data to binary correctly. Data types are identified using the following prefixes (underscore characters have been used in the binary numbers for readability—don't use them in programs):

KM for *binary* values from 0 to 1111_1111_1111_1111 (e.g., KM 1001_1111_0011_1110 will be stored as 1001_1111_0011_1110)

KH for *hexadecimal* values from 0 to FFFF (e.g., KH 9F3E will also be stored as 1001_1111_0011_1110)

KF for *signed decimal* values from −32768 to +32767 (e.g., KF −24771 will also be stored as 1001_1111_0011_1110)

KY for *pairs of unsigned decimal* values from 0 to 255 each (e.g., KY 12,255 will be stored as 0000_1100__1111_1111)

KS for *pairs of ASCII characters* (e.g., KS 's', 'F' will be stored as 0111_0011__0100_0110)

KC for the special coding form for *counter-accumulated values* [e.g., KC 50 will be stored in counter format (BCD) in the program (i.e., 0000_0000_0101_0000); and in natural binary in the counter's memory (i.e., 0000_0000_0011_0010[3])]

KT for the special coding form for *timer time-remaining values* [e.g., KT 50.2 will be stored in timer format (modified BCD) in the program (i.e., 00_ 10_

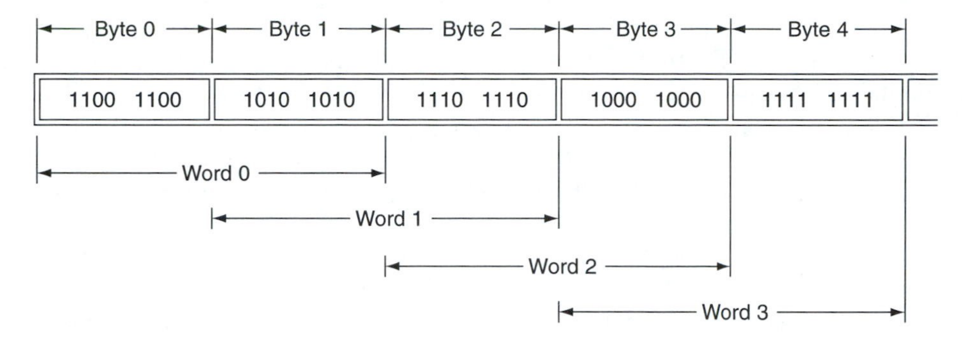

FIGURE 5.10
Relationship between bytes and words in an S5 PLC.

0000_0101_0000); and in natural binary in the timer's memory (i.e., 0000_0000_0011_0010[3])]

STEP 5 does *not* offer floating-point-number capabilities.

Siemens STEP 5 Addressable Data Memory

Data memory of the S5 series of PLCs is arranged in separately addressable bytes. The STL language LOAD (L) instruction copies data from memory into a 16-bit accumulator, and the TRANSFER (T) instruction copies data from the 16-bit accumulator into memory. (The ladder logic MOVE instruction consists of an STL-language LOAD followed by an STL TRANSFER instruction.)

Data *words* can be addressed in LOAD or TRANSFER instructions, but each word consists of data from two separate bytes, as demonstrated in Figure 5.10. A LOAD instruction that gets data from word 2, for example, will read an 8-bit value from address 2 and another 8-bit value from address 3. Word 3 consists of byte 3 and byte 4. Usually, a programmer will use only even numbers in addressing words, so that each word is a unique set of 16 bits. If a LOAD instruction includes the address for a *byte*, the S5 PLC copies a single byte of data from a single address to the *low* byte of the accumulator *and clears the high byte of the accumulator.* When a TRANSFER instruction includes a byte address, it copies the contents of the *low* byte of the accumulator to the specified address, but doesn't copy the accumulator's high byte to memory.

Siemens STEP 5 I/O Data

Input and output images are stored in memory areas that Siemens calls **process image input (PII)** and **process image output (PIQ)** tables. Each slot where a digital I/O module can be plugged in has both PII memory and PIQ memory assigned to represent possible I/O data. S5

---

[3]Siemens says that the low 10 bits of a timer/counter memory address are used to store the counter/timer accumulated value. The high bits are probably used for status bits, but when a user-program reads those addresses, the S5 makes the high bits zeros.

PLCs have 128 bytes of addressable PII and another 128 bytes for PIQ memory. A medium-sized S5 PLC (e.g., an S5-100U) reserves 1 byte of PII and 1 byte of PIQ memory space for each digital I/O module, and 8 bytes of PII and PIQ memory for each analog I/O module. A large S5 PLC (e.g., an S5-115U) reserves 4 bytes of PII and PIQ memory space for each digital I/O module, but no PII or PIQ space for analog modules. These large PLCs reserve 32 *direct access addresses* for reading bytes of input data from analog I/O modules or for writing bytes of data to analog I/O modules. When the user-program reads from or writes to one of these addresses, the PLC actually reads or writes memory in the I/O module. (Analog I/O modules don't always contain the full 32 bytes of input or output memory.) Figure 5.11 shows the byte addresses that correspond to I/O module locations for medium-sized (S5-100U) and large (S5-115U) PLCs. Notice that when an analog I/O module is plugged into a slot, it has different byte addresses than a digital I/O module plugged into the same slot would have.

In STEP 5, I/O image *data bytes* are addressed in the following absolute addressing formats:

IB x for a process image input (PII) byte, or
QB x for a process image output (PIQ) byte, where "x" is a number between 0 and 127, corresponding to the I/O module location, or

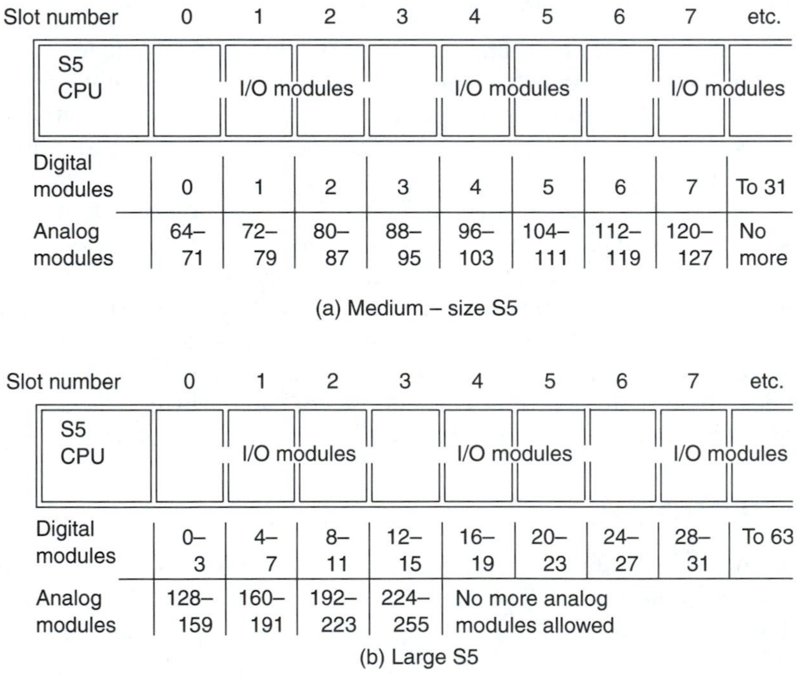

S5 CPU — I/O modules — I/O modules — I/O modules

Slot number	0	1	2	3	4	5	6	7	etc.
Digital modules	0	1	2	3	4	5	6	7	To 31
Analog modules	64–71	72–79	80–87	88–95	96–103	104–111	112–119	120–127	No more

(a) Medium – size S5

S5 CPU — I/O modules — I/O modules — I/O modules

Slot number	0	1	2	3	4	5	6	7	etc.
Digital modules	0–3	4–7	8–11	12–15	16–19	20–23	24–27	28–31	To 63
Analog modules	128–159	160–191	192–223	224–255	No more analog modules allowed				

(b) Large S5

FIGURE 5.11
I/O address allocation in S5 PLCs: (a) addressing in a medium-sized S5 PLC system; (b) addressing in a large S5 PLC system.

PY x for direct access to an I/O module's memory in a large S5 PLC, where "x" is a number between 0 and 255 (the S5 does not maintain PII or PIQ memory in the CPU module for byte addresses over 127, so direct access *must* be used in addressing I/O modules at these addresses)

Nearly direct access to I/O modules of medium-sized S5 PLCs is also possible using the "PY" prefix under certain conditions (see the section on immediate addressing in STEP 5 in Chapter 11).

STEP 5 also allows the same I/O image data memory to be addressed as 16-bit words in the following formats:

IW x for a process image input word, or
QW x for a process image output word, where "x" is a number between 0 and *126*, or
PW x for direct access to an I/O module's memory in a large S5 PLC, or for nearly direct access to I/O modules in a medium-sized S5 PLC ("x" is an even number between 0 and *254*)

Usually, a LOAD (L) instruction addressing memory *words* gets the accumulator's high byte of data from the specified address and the accumulator's low byte from the next address. L IW072, for example, gets data from IB072 and from IB073, as shown in Figure 5.12a.

In *small S5 PLCs* where only one input image (PII) byte of memory is reserved for each digital input module, the PLC prevents the LOAD WORD instruction from reading data from more than one input module and prevents the TRANSFER WORD instruction from writing data to more than one output module. As shown in Figure 5.12b, LOAD and TRANSFER instructions addressing data as *bytes* cause the S5 PLC to use only the low byte of the accumulator (clearing the high byte during a load). Figure 5.12c shows that a small S5 PLC's LOAD instruction with a word address reads a byte of data from the specified input image address into the high byte of the accumulator and puts zeros into the low byte. Similarly, a TRANSFER WORD instruction will copy only the high byte from the accumulator to the specified output image byte, ignoring any data that might be in the accumulator's low byte. Figure 5.12d shows what might happen if the programmer were to accidentally mix word and byte addressing.

Figure 5.12d and e show that since larger S5 PLCs have digital I/O modules that aren't limited to 8 bits of I/O capabilities, whole words can be read from and written to those modules' PII/PIQ table addresses. Figure 5.13f is to remind the reader that there is no PII/PIQ memory for analog modules in a large S5 PLC.

Individual bits of I/O data can be addressed in the following format:

Ix.b or Qx.b where "x" is the byte number and "b" is the bit number from 0 to 7

Direct addressing cannot be used to read or write individual bits, so entering the address "Px.b" will result in an error.

Siemens STEP 5 System Data

PLC status and configuration data is stored in the PLC's reserved system data area of memory, some of which is addressable in a user-program, but most of which is not. By changing the con-

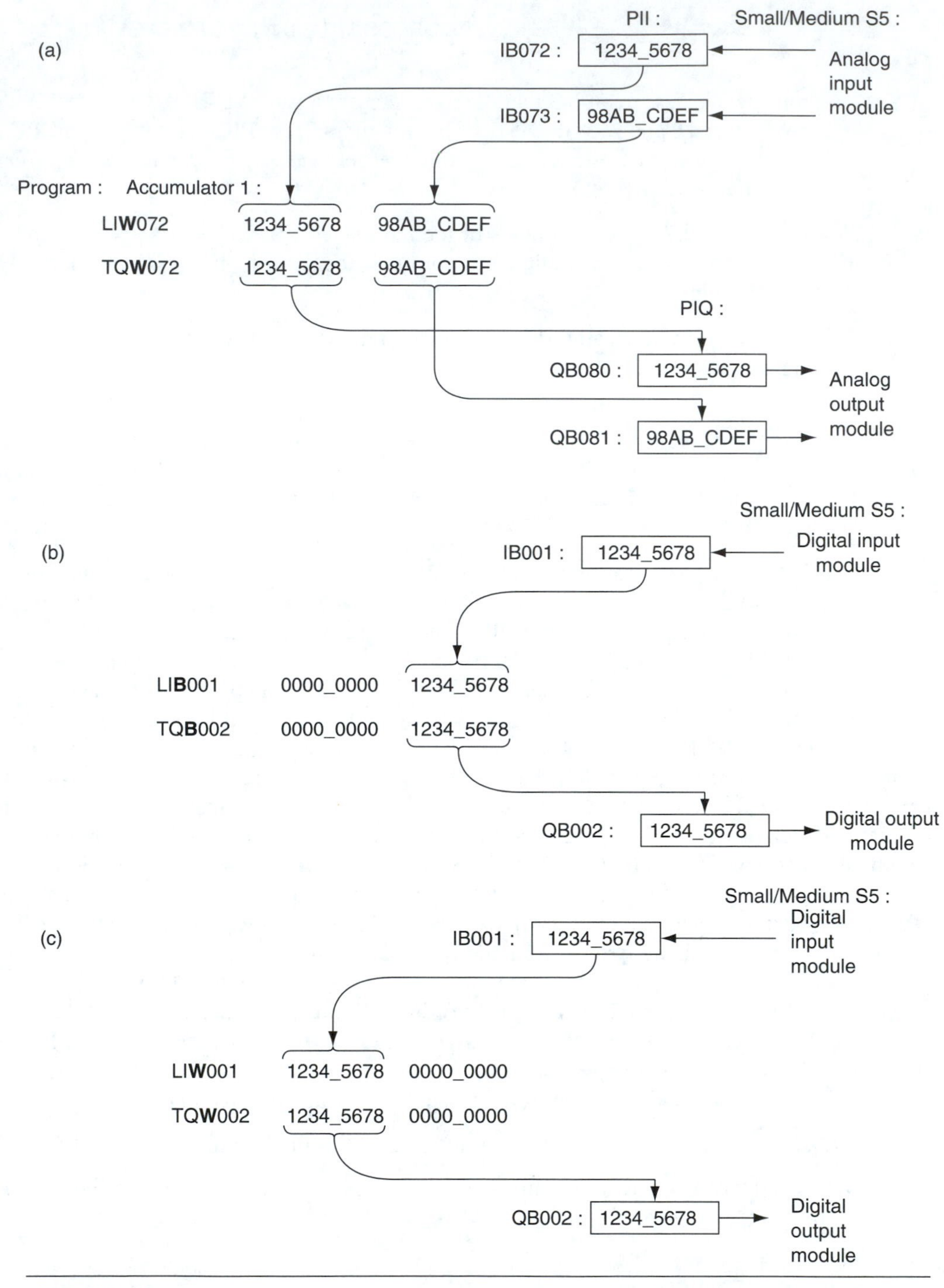

FIGURE 5.12
Loading input data and writing output data as bytes and words in Siemens S5 PLCs. Small and medium-sized S5 PLCs: (a) addressing analog module PII/PIQ data as words; (b) addressing digital module PII/PIQ data as bytes; (c) addressing digital module PII/PIQ data as words; *(continued)*

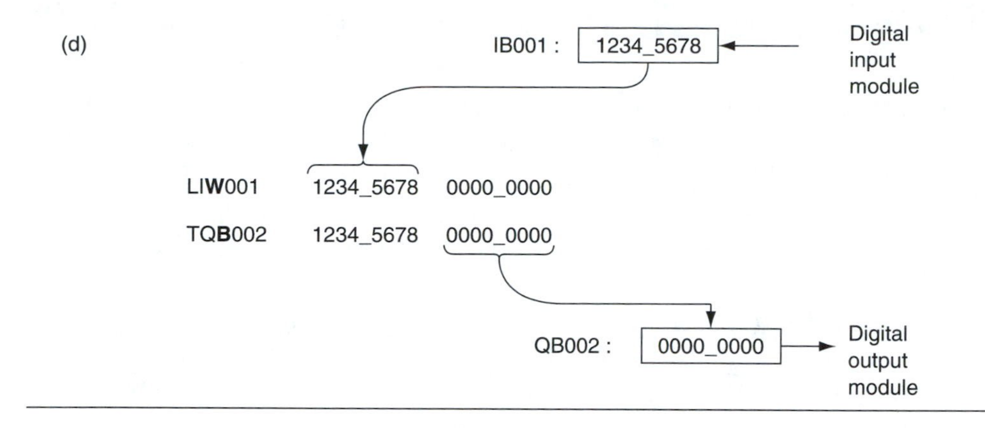

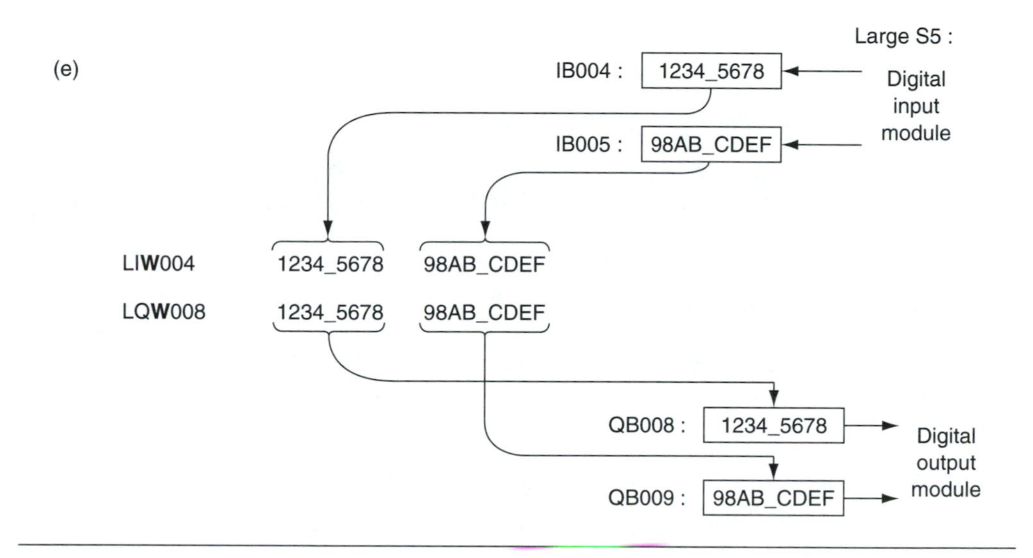

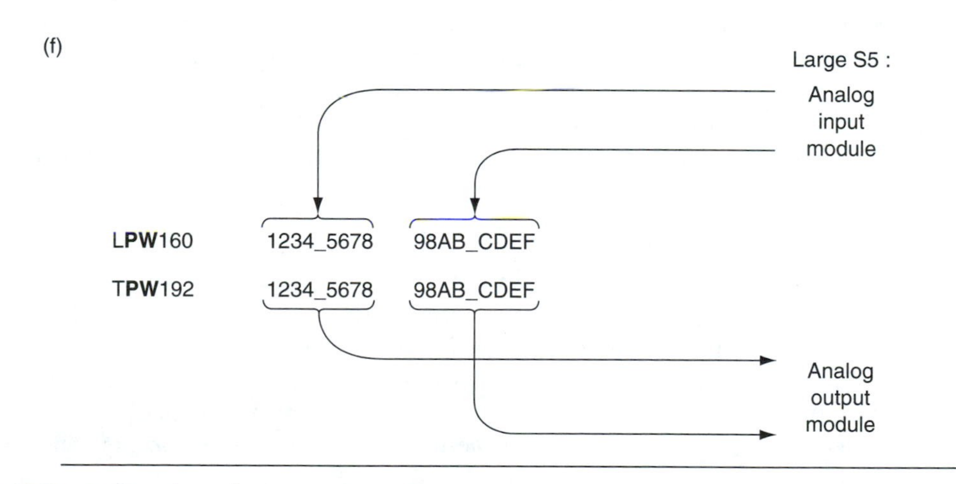

FIGURE 5.12 *(continued)*
(d) mixing word and byte addressing of digital module PII/PIQ data. Large S5 PLCs: (e) addressing digital module PII/PIQ data as words; (f) addressing analog module data as words.

tents of the appropriate system data words, the user-program can change the PLC's configuration. In Chapter 8 we will see that system data words are not as freely addressable as other data memories.[4] When available, they are addressed in the following form:

RSx where "x" is one of the addressable memory addresses in the range 0 to 255 (see your programming manual for addressable reserved system memory addresses and their uses)

Individual bits of system data words can be addressed using the TEST BIT (TB) or TEST BIT NOT (TBN) instructions and can be changed using the SET UNCONDITIONAL (SU) and the RESET UNCONDITIONAL (RU) instructions, subject to conditions similar to addressing system data words, in the following form:

RSx.d where "d" is the bit number (0 to 15)

Siemens STEP 5 Counter Data

Each counter uses the low 10 bits of a single data word of memory to store its accumulated count value in natural binary format. Additional memory bits are used to store status bits for each counter.

When setting a count value, the program must supply a counter-constant (KC) number or must supply the address of a memory location that contains a number in counter format. (See Chapter 3 for a description of counter constants and counter format.) Counter constant and counter format numbers are automatically converted to 10-bit natural numbers as the PLC transfers them to a counter memory location. Counter accumulated values can be addressed as follows:

Cx where "x" is a number between 0 and 127 (e.g., C25)

If the accumulated value is read from a counter address using the standard LOAD (L) instruction, the PLC will provide the value as an unsigned 10-bit natural binary number (the high 6 bits will be zeros). If the LOAD BCD (LC) instruction is used to read the counter accumulated value, the PLC will convert the natural binary number into a 12-bit BCD number.

The *bits* of the counter's accumulated value can be addressed individually using the TEST BIT instructions (TB and TBN) and the SET/RESET UNCONDITIONAL instructions (SU and RU). These instructions can only be programmed in STL and only in a function block. The addressable bits are addressed as follows:

Cx.b where "x" is the timer number and "b" is one of the bits 0 to 15

Low-numbered counter accumulated values are retentive, while other accumulated values aren't (see your manual for specific addresses). Retentive counters' accumulated values are retained even if the PLC's power is disconnected.

Counter status bits are also maintained in the S5's memory, to keep track of things such as when the count has reached zero and whether the counter's count-up and count-down logical statements were true last time they were evaluated, so that the counter can detect when the

---

[4]RS data addresses can only be entered in function block programs.

logic goes from false to true. Of these bits, only the counter coil status bit is addressable. This bit goes off when the counter's accumulated value is zero. This bit is addressed as follows:

Cx where "x" is a number between 0 and 127 (e.g., C25)

Yes, the same address is used for the accumulated value and for the counter coil. The S5 PLC knows which of those uses the programmer means by the type of instruction the address is used with. If "C005" is used in a Boolean EXAMINE ON instruction, the PLC will supply the Boolean state of the counter coil for counter 5, whereas if "C005" is used in a LOAD instruction, the PLC will supply the 10-bit natural binary accumulated value number.

Siemens STEP 5 Timer Data

Timers each use one data word of memory to store the amount of time that they have left to run, once started. Memory use is similar to that of counters: A time-remaining value is stored in the low 10 bits of each timer's data word, and timer status bits are maintained in the PLC's memory. When a timer is programmed, the program must either include a preset time as a constant (in timer constant format) or specify which address to read to get the preset value. If the preset value is to come from a memory location, that value must be in timer format in memory. In Chapter 3 we describe timer constant and timer format numbers.

The S5 PLC converts timer constant and timer format numbers to 10-bit natural binary numbers for storage in the low 10 bits of the memory address reserved for the timer's time-remaining value. The code that indicates the size of time units that the timer uses is stored in bits 12 and 13 of the same data word. The time-remaining value can be read using the LOAD (L) or LOAD BCD instruction, which causes the PLC to convert the 10-bit natural binary number to a 12-bit BCD number as it is being read. Both instructions strip the time-unit code from the number as it is being read. The timer's *time-remaining value* can be addressed in the form

Tx where "x" is a number between 0 and 127

The *individual bits* of the timer's time-remaining value and time-unit code can be examined using the TEST BIT instructions (TB and TBN), and can be changed using the SET/RESET UNCONDITIONAL instructions (SU and RU) in a function block. The bits would be addressed in this format:

Tx.b where "x" is the timer number and "b" is one of the bits 0 to 15

Timer status bits are also maintained in the S5's memory so that the timer can be started and run to completion, and to indicate when timing is over. Of these bits, only the timer coil status bit is addressable. This bit may go on or off when the timer is started and will change state when the timer stops running or when it's time-remaining value reaches zero, depending on the type of timer. This bit is addressed as follows:

Tx where "x" is a number between 0 and 127 (e.g., C25)

The same address is used for the timer's time-remaining value and for the timer coil. The S5 PLC knows which of those uses the programmer means by the type of instruction the address is used with. If "T005" is used in a Boolean EXAMINE ON instruction, the PLC will supply the Boolean state of the timer coil for timer number 5, whereas if "T005" is used in a LOAD instruction, the PLC will supply the 10-bit natural binary time-remaining value.

Siemens STEP 5 Flag Data

Flag data bytes are available in an S5 PLC for the storage of working data. Flag bytes with low addresses are retentive, whereas others are cleared each time the PLC is put into run mode. (See your manual to determine how much flag memory your S5 PLC has and how much is retentive.) Flag bytes can be addressed in this format:

FY x where "x" is a number between 0 and (up to) 255

Flag memory can also be addressed as words in this format:

FW x

Like I/O memory, a flag word consists of the flag byte of the same number and the next flag byte, as shown in Figure 5.13.

Individual bits of flag bytes can be addressed in the following format:

F x.b where "x" is the byte number and "b" is the bit number from 0 to 7

Siemens STEP 5 Data Blocks

S5 PLCs can store user data in data blocks. Data block addresses can't be used in a program until a data block containing data words has been created. Data blocks can be created using a programming unit, or a user-program can contain instructions to create a data block. Each data word in the data block must also be created during creation of the data block. A data block can contain from 0 to 255 data words.

In a ladder logic program, data words in data blocks can be addressed as follows:

DBx:DWn where "x" is the number of an existing data block. Numbers 2 to 255 are available (the STEP 5 operating system uses data blocks 0 and 1). "n" is the number of an existing data word in that data block. Numbers 0 to the actual last word that was created (maximum number 255) are allowed.

The high and low bytes of each data word can be addressed separately in the following forms:

DBx:DRn to address the low (right) byte
DBx:DLn to address the high (left) byte

In programs written in the STL programming language, data word and data byte addresses do not include the "DBx:" but the program must include an instruction to "call" a data block be-

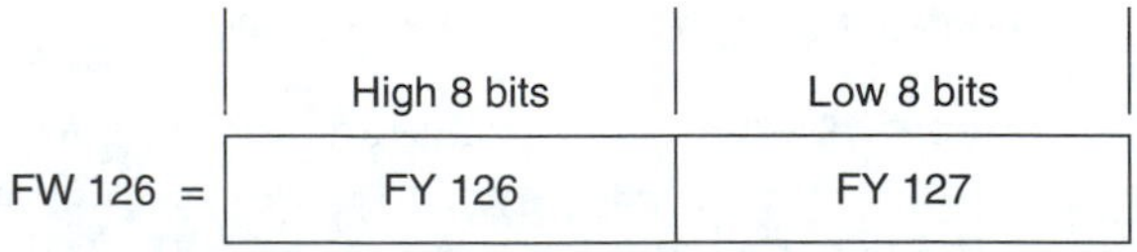

	High 8 bits	Low 8 bits
FW 126 =	FY 126	FY 127

FIGURE 5.13
An S5 flag word consists of two flag bytes.

fore the program tries to use a data word or data byte. The following STL instruction calls a data block:

C DBx

After calling a data block, an STL program can address data words or bytes in that data block in the following forms:

DWn `or DRn or DLn`

Individual data word *bits* can only be examined using the TEST BIT instructions (TB and TBN), and can only be changed using the SET/RESET UNCONDITIONAL instructions (SU and RU) in a function block. The bits of a data word would be addressed in this format:

Dx.b where "x" is the data word number (in the currently open data block) and "b" is one of the bits 0 to 15

Only one data block can be opened (using the CALL instruction) at a time. CALLING a data block automatically closes a previously called data block. Structured programs may include jumps to other programs which may include CALL instructions for their own data blocks, but when that other program ends, the S5 PLC automatically re-CALLS the data block that was open before the other program was jumped to.

While creating data blocks, the programmer must provide an initial data value and must indicate the default data form for each data word in that data block. The default data form can be any of the allowable formats for constant data, as described earlier in this section. Programming units will display each word of a data block in its default data format, although a user-program can manipulate data words, bytes, or bits in any format.

Use an "=" sign between type identifier and initial data value while creating data block data words (e.g., KH = 1CF3).

Figure 5.14 shows a STEP 5 ladder logic program, and the equivalent statement list (STL) program, which unconditionally:

1. Moves a byte of digital input image data (IB3) from the process image input table to the first flag data byte (FY0).

A ladder logic MOVE instruction does the same as an STL LOAD (L) and TRANSFER (T) instruction pair.

2. Moves the second data word (DW1) from data block 2 to the accumulated value memory location for counter 18 (C18).

3. Moves a constant value, entered in signed binary (KF 25), into one of the system data words (RS 97). System data words that are addressable from a user-program can only be addressed in one type of user-program: a function block.

Siemens STEP 5 Symbolic Addressing

STEP 5 allows the programmer to make entries into a symbol library. The symbolic addresses cannot be used in entering addresses during programming. They can only be caused to appear

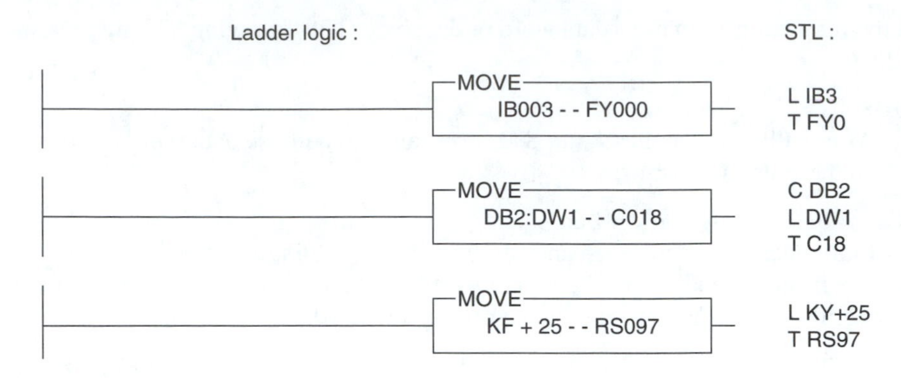

FIGURE 5.14
S5 PLC program demonstrating use of data words. Shown in ladder logic and in STL.

on the programming unit screen alongside the absolute addresses and block identifiers that the programmer must still enter.

Symbolic names are *not* part of the program in an S5 PLC. They are only maintained in the STEP 5 programming unit and in STEP 5 programs that are saved to disk files using that programming unit. If another programming unit is used to read a program from a PLC, the alphanumeric names defined using the original programming unit will not be present.

Siemens STEP 5 Indexed Addressing

STEP 5 allows the programmer to use indexed addressing in two ways:

1. Data block addressing, described above, is actually a form of indexed addressing. When a data block is called, the starting address of the memory area reserved for the data block becomes a base address and data word addresses are effectively offsets from the start of the data block. If a program is written to call one data block under a specific set of conditions or a different data block under a different set of conditions, the data word addresses (offsets) in the program instructions can refer to data in either data block (base).

2. There is a PROCESSING OPERATION (DO) instruction (only programmable in STL and only in a function block) that is used in indexed addressing. As shown in the following STL language example, the DO instruction must include a flag word (FW) or a data word (DW) address, which must contain the offset part of an address. The DO instruction must be followed by an instruction indicating what operation to perform using the indexed address and an address identifier indicating the base part of an address. The base address identifier must indicate one of the Siemens S5 memory areas and must indicate the size of the data unit required (e.g., I indicates an input image *bit*, IW indicates an input image *word*, and C indicates either a counter word or a counter status bit, depending on the instruction that uses it).

STEP 5 requires that a numerical address be entered with each address identifier, so the address 0 or 0.0 must be entered with the address identifier in the instruction following the DO. The 16-bit offset number stored in a flag or data word identified in the DO instruction must consist of a number in the low byte indicating a byte or word address number and (if required) a bit number in the high byte. If a bit address isn't required, the high byte should be zero. For example:

```
L KH0203    ;  Put byte address "03" into the low byte and
T FW4       ;     bit address "02" into the high byte of FW4
DO FW4      ;  Use the number in FW4 as offset
S Q0.0      ;  Set the bit in output image memory that is
            ;     offset from its start by 3 bytes and 2 bits
            ;     (= Q3.2)
```

and

```
L +21     ;  Put the number "21" into DB5:DW6 (a
C DB5     ;     number below 256 uses only the low byte)
T DW6
DO DW6    ;  Use the number in DB5:DW6 as an offset
L T0      ;  Load a number from the timer memory area,
          ;     offset from the start of the timer area by
          ;     21 (= T21)
```

Siemens STEP 5 Indirect Addressing

Two forms of indirect addressing are available in S5 PLCs. One method uses a LOAD REGISTER INDIRECT (LIR) or TRANSFER REGISTER INDIRECT (TIR) instruction and an accumulator to indicate (point to) the address to be used. The other method uses the PROCESS INDIRECTLY (DI) instruction.

To use the REGISTER INDIRECT instructions, a 16-bit number representing an *address* in the PLC's memory must first be loaded into accumulator 1. The LOAD REGISTER INDIRECT (LIR) instruction can then be used to load a number from the indicated address into an accumulator, or the TRANSFER REGISTER INDIRECT (TIR) instruction can be used to transfer a number to the indicated address from an accumulator. These instructions can only be programmed in STL, and only in function blocks. The instruction also must indicate which accumulator to load to or to transfer from, in the following form:

LIR 0 loads accumulator 1 with a value from memory (overwriting the indirect address number that was in accumulator 1 before LIR is executed)

LIR 2 loads accumulator 2 with a value from memory (the indirect address number in accumulator 1 isn't overwritten)

TIR 2 transfers the contents of accumulator 2 to the address indicated by the address in accumulator 1

To use these instructions, the programmer must know the actual 16-bit memory address that is to be read from or written to, not the address in its usual prefix-number form. The S5 PLC manuals identify the actual address ranges that are used for I/O images, flags, counters, system data, and so on.

The **PROCESS INDIRECTLY** (DI) instruction is also a form of indirect addressing and is only programmable in STL and only in a function block. (The reader may wish to skip this section until after reading Chapter 8.) The PROCESS INDIRECT (DI) instruction is similar to the PROCESSING OPERATION (DO) instruction, in that it affects which address is used by the instruction that follows the DI instruction. The DI instruction differs from the DO instruction in that when the DI instruction executes, it uses a pointer from accumulator 1 to select an address from the list of parameters that were passed to this function block when the block was called.

The following example illustrates the process. In the example, a calling program passes the addresses of three words (IW4, FW6, and FW40) to FB4, as it is called. FB4 includes a declaration section that identifies those three addresses as input parameters (IWs) for the function block. In the simplified example program, FB4 puts the number "3" into accumulator 1 before the DI instruction executes. The TRANSFER instruction that follows the DI instruction in this example would normally copy the contents from accumulator 1 (the number 3) to QW6 if no DI instruction were preceding it. But since the DI instruction does precede the TRANSFER instruction, the PLC knows to use the number in accumulator 1 (3) to indicate which of the input parameters to use as the source of the data to be transferred to QW6. Since the third input parameter is FW40, the contents of FW40 are loaded, then transferred to QW6. In our example, the calling program is

```
JU FB3
p1:IW4
p2:FW6
p3:FW40
```

and the function block (FB4) is

```
DECL: p1 IW
DECL: p2 IW
DECL: p3 IW

L KF +3
DI
T QW6
BE
```

Siemens STEP 5 Formal Operands

User-defined names for absolute memories or blocks can be defined as **formal operands,** but only in function blocks. Once defined, they can be entered in place of absolute addresses while entering a program into a function block. In Chapter 8 we show how to declare formal operand names while programming function blocks, and how to use formal operand names in the function block.

Formal operands are already defined in the preprogrammed function blocks that come in an S5 PLC. Formal operand names are *not* part of the program in an S5 PLC. They are maintained only in the STEP 5 programming unit and in STEP 5 programs that are saved to disk files using that programming unit. If another programming unit is used to read a program from a PLC, the alphanumeric names defined using the original programming unit will not be present (unless they are also defined in the other programming unit).

S7

ADDRESSABLE DATA IN SIEMENS STEP 7

The S7's memory is organized to allow highly structured programming, where a control program will actually consist of several user-programs. Some of the memory is shared by all user-programs, but other areas are reserved for individual programs. STEP 7 offers a wide range of data types and encourages programming using symbolic and variable names.

Siemens STEP 7 Constant Data Values

Data values can be entered as STEP 7 program constants using the data types and constant formats that are specified in IEC 1131-3 and some formats that are not included in the IEC 1131-3 standard. *The IEC 1131-3 defined data element types include:*

- **BOOL:** a 1-bit Boolean value, which is entered either:

```
In binary:  0 or 1, or
In text:  true or false
```

- **BYTE:** an 8-bit binary number representing values from 0 to 255, entered as a constant in hex (base 16) as follows:

```
B#16#0 to B#16#FF
```

- **WORD:** a 16-bit unsigned binary number representing values 0 to 65,535, or a 16-bit BCD data representing values from −999 to +999.[5] Constants can be entered in:

```
Binary:  2#0 to 2#1111_1111_1111_1111
Hex:  W#16#0 to W#16#FFFF
BCD:  C#−999 to C#999
Unsigned decimal:  B#(0,0) to B#(255,255)
```

- **DWORD:** a 32-bit unsigned binary number representing values 0 to (approximately) 4 billion, entered in:

```
Binary:          2#0          to
       2#1111_1111_1111_1111_1111_1111_1111_1111
Hex:  D#16#0 to D#16#FFFF_FFFF
Unsigned decimal:  B#(0,0,0,0) to B#(255,255,255,255)
```

- **INT:** a 16-bit signed binary number, representing integer values −32,768 to +32,767. Constants can be entered as:

```
Integer values:  −32768 to 32767
```

- **DINT:** a 32-bit signed binary number, representing integer values from −2,147,483,648 to +2,147,483,647, entered as:

```
"Long" integers:  L#-2147483648 to L#2147483647
```

- **REAL:** a 32-bit ANSI/IEEE format binary number that can represent floating-point numbers from $\pm 3.402823 \times 10^{38}$ to $\pm 1.175494 \times 10^{-38}$. Constants are entered with decimal points or in scientific notation (e.g., −1.553 or 4.7e-25)

---

[5]If the high nibble of a BCD number is all 1's, the number is negative.

- **CHAR:** an 8-bit ASCII code for an ASCII character. ASCII characters must be entered in single quotes (e.g., 'R')
- **TIME:** a 32-bit signed integer value representing the number of milliseconds after midnight, entered in day/hour/minute/second/millisecond code form:

  `T#-24D_20H_31M_23S_648MS to T#24D_20H_31M_23S_648MS`

- **DATE:** a 16-bit binary number representing the number of days since January 1 1990. Constants must be positive values and can be entered in year/month/day form:

  `D#1990-1-1 to D#2168-12-31`

The non-IEC 1131-3 single-element data types allowable in STEP 7 include:

- **TIME_OF_DAY:** a 32-bit hour/minute/second/millisecond time system that is stored in the same binary format as an IEC 1131-3 TIME value, but constants can be entered more simply, in this form:

  `TOD#0:0:0.0 to TOD#24:59:59.999`

- **Count:** not a declarable data type; however, there is a counter type constant that consists of a 16-bit number stored as three BCD digits in the low 12 bits, with the high 4 bits unused. A count value would typically be loaded into the low word of the S7 PLC's 32-bit accumulator during program execution, then used to set a counter. Although the S7 PLC always converts the 12-bit BCD number to a 10-bit natural binary number as it is written into the counter area of memory, it must be loaded into the accumulator in BCD. It is entered as a constant in this form:

  `C#000 to C#999`

- **S5TIME:** a 16-bit number stored as three BCD digits in the low 12 bits, a code identifying the time unit size in bits 12 and 13, and unused bits in the 2 high bits. The value can be entered in hour/minute/second/millisecond time form:

  `S5T#10ms to S5T#2h_46m_30s (e.g., S5T#2m_10s)`

  or can be entered as a three-digit hex value following a time-unit code value. Allowable time-unit codes are 0, 1, 2, 3, meaning 0.01-, 0.1-, 1-, or 10-s time units. An S5TIME value in hex would appear as follows:

  `w#16#2130 (meaning 130 seconds, or 2 minutes + 10 seconds)`

- **BLOCK, COUNTER,** and **TIMER:** block, counter, or timer numbers are stored as 16-bit binary numbers as the programmer creates those elements. Blocks, counters, and timers are entered by name and number when required in a program (for passing counter data to another program, for example).

STEP 7 also allows additional *non-IEC 1131-3 complex data* constants. A complex data constant consists of an array or structure of other data elements. S7 complex data types include:

- **DATE_AND_TIME:** an 8-byte set of BCD data values representing a year/month/day/hour/minute/second/milliseconds/weekday specification, stored in memory as shown in Figure 5.15. A DATE_AND_TIME constant is entered in the following form:

```
DT#1999-12-31-24-59-59-999-7
```

- **POINTER:** a data element that identifies the location in memory of other data elements. STEP 7 offers two very different ways of using pointers. Both ways allow some optional pointer forms:

(a) One type of pointer data is used in the following forms of indirect addressing. (STEP 7 indirect addressing is described elsewhere in the chapter.)

(1) The **memory indirect** type of indirect addressing requires the program to place a pointer into a memory location before indirect addressing can be used. The memory indirect pointer can be:

- A 16-bit unsigned binary value (representing a decimal value between 0 and 65,535) to point to a timer, counter, data block, function, or function block.
- A 32-bit binary double word, of which the low 3 bits indicate a bit number (0 to 7), the next-lowest 16 bits represent a memory byte address (0 to 65,535), and the high 13 bits are zeros. This type of pointer is used to point to memory addresses other than timers or counters.

(2) The **register indirect** type of indirect addressing uses a 32-bit pointer, which the program must load into an address register before indirect addressing can be used. When entered as a constant, the pointer must be in one of these forms:

p#x.d for area-internal pointers; "x" is a number from 0 to 65,535 and "d" must be a number from 0 to 7

p#ax.d for area-crossing pointers; "a" is a prefix (P, I, Q, M, DBX, or DIX) indicating which memory area is being pointed to, "x" is a number from 0 to 65,535, and "d" must be a number from 0 to 7

When stored in memory, an area-internal pointer is a 32-bit number in the same form as the 32-bit memory indirect pointer described above. The area-crossing pointer data value is stored in memory in the same 32-bit data format, except that the most significant bit is a 1, and bits 24, 25, and 26 contain a 3-bit code identifying the memory area being pointed to (see below).

Byte n	Year (00–99)	Month (00–12)	Byte n + 1
Byte n + 2	Day (00–31)	Hour (00–60)	Byte n + 3
Byte n + 4	Minute (00–60)	Second (00–60)	Byte n + 5
Byte n + 6	MS (3 digits : 000–999)	Day (0–7)	Byte n + 7

FIGURE 5.15
S7 DATE_AND_TIME element structure.

(b) The second type of pointer is used to identify the start of a series of memory locations containing a complex data element such as a DATE_AND_TIME element, or an ARRAY, or a STRUCT. These types of pointers are used in passing complex data elements between programs. Instead of copying the entire complex data element from one program's memory space to the other program's space, only the pointer is passed. The two programs then use the same complex data element. A pointer constant for this purpose is entered in this format:

```
p#area_identifier byte_number ? bit_number
```

For example, for bit arrays:

```
p#M25.4
p#DB30.DBX25.4
```

and for bytes/words/double words:

```
p#MB25.0
p#DB30.DB25.0
```

A bit number *must* be entered even if the pointer points to a complex data item consisting of bytes, words, or double word. The bit number should be zero unless the pointer is used to point to a bit. The S7 PLC stores pointers in the 6-byte format shown in Figure 5.16, where:

"O" means bit not used
"b" means a byte address
"x" means a bit address

and memory area codes (shown in hex) include:

81 for I (input image memory)
82 for Q (output image memory)
83 for M (shared data memory)
84 for DB (shared data block memory)
85 for DI (instance data block memory)
85 for L (local data memory)
86 for the local data memory previously in use

<table>
<tr><th>Byte :</th><th colspan="16"></th><th>Byte :</th></tr>
<tr><td>0</td><td colspan="16">DB number (0 if not a DB)</td><td>1</td></tr>
<tr><td>2</td><td colspan="8">Code for memory area</td><td>0</td><td>0</td><td>0</td><td>0</td><td>0</td><td>b</td><td>b</td><td>b</td><td>3</td></tr>
<tr><td>4</td><td>b</td><td>b</td><td>b</td><td>b</td><td>b</td><td>b</td><td>b</td><td>b</td><td>b</td><td>b</td><td>b</td><td>b</td><td>b</td><td>x</td><td>x</td><td>x</td><td>5</td></tr>
</table>

FIGURE 5.16
S7 pointers to complex data elements.

- **STRING:** a series of CHAR elements stored in sequential memory bytes following a 2-byte header. The first byte in the header is a number indicating the number of memory bytes reserved for the string (including the header), and the second byte indicates the number of memory bytes currently containing valid data. Following the header, each memory byte contains one ASCII code. A STRING's default maximum size is 256 bytes, which includes the header, but shorter strings can be created, as we will see. A *string constant* can be included in a program inside single quotation marks, as follows:

'This 50-character string will need 52 memory bytes'

- **ARRAY:** a one- to six-dimensional set of similar data elements. Data elements are stored in sequential memory locations. Data for an array cannot be entered as constant data value(s) in a program. A program may contain a pointer constant to identify the start of the set of memory locations reserved for the array, and an array with initial values can be declared in variable declaration tables and/or in data declaration tables, which are described later in the chapter. When an ARRAY is created as a named variable, the S7 PLC saves a 6-byte pointer in memory to identify the start of the memory reserved for the named ARRAY.
- **STRUCT:** a structured set of any number of data elements of *any type* (even other STRUCT elements). As with ARRAYS, data for structures cannot be entered in programs as constant data value(s), but a program may contain a *pointer* constant that identifies the start of the memory reserved for the structure. Structures can be created and assigned initial values in variable declaration tables and/or in data declaration tables, which are described later in the chapter. Like ARRAY data, STRUCT data is stored in sequential memory locations, except that if a structure contains other complex data elements (e.g., arrays or other structures), the structure will actually contain only a pointer to each of those other data elements. Structures can contain other structures to a nesting depth that cannot exceed eight levels. When a structure is created as a named variable, the S7 PLC saves a 6-byte pointer in memory to identify the start of the series of memory locations in the STRUCT.
- **ANY:** a complex data element that is like an ARRAY in that it can contain multiple data elements of the same type, but the elements can be of any type, even STRUCT elements or other ANY elements. The ANY data element will be stored in a sequential series of memory locations containing basic data elements or pointers to complex data elements. When an ANY data element is created as a named variable, a 10-byte data set, similar to a pointer, is saved in memory. This data set identifies the starting address of the data in the ANY data element and includes a description of the type and number of elements it contains.

  A program can contain a pointer constant to identify an ANY data element. Slightly different from the pointer constant described previously, the ANY-pointer format is as follows:

```
p#area_identifier_and_byte.bit data_type how_many
```

For example:

`p#M25.0 byte 5`	for an array of five bytes starting at M25 (elementary data types must be entered in absolute address notation)

p#M30.0 struct 3 for an array of three structure data elements starting at M30 (complex data starting addresses can be entered using their symbolic names if the programmer prefers)

(Bit is always 0 unless data_type is BOOL.)

The 10-byte ANY pointer is stored in a format similar to that of the 6-byte standard pointer. The actual format depends on what types of elements are in the ANY array (see your manual set). When the function program uses an ANY parameter, the PLC refers to the 10-byte ANY pointer to determine which memory location(s) should actually be used.

Siemens STEP 7 Addressable Data Memory

The addressable memory of an S7 PLC contains:

- I/O image data
- Bit memory
- Timer data
- Counter data
- System data
- Data blocks
- Peripheral I/O memory

Peripheral I/O memory is actually located in I/O modules, but can be used in programs if it was in the CPU module.

- Local data stack (L stack) memory

Every program in an S7 PLC can have its own local data memory, and each program's local data memory is protected from being changed by other programs.

All memory areas can be addressed using **absolute addressing.** An absolute address consists of a prefix to identify the memory area and the data unit size, then a unique number identifying the specific memory. STEP 7 also allows indirect addressing and indexed addressing as well as the use of pointers to identify memory.

In compliance with the IEC 1131-3 requirement that every PLC program should have an area of memory that no other program can access, STEP 7 offers a new **instance data block,** which was not available in STEP 5, and the **local data stack** memory mentioned above. To give a programmer slightly more power than the writers of the IEC 1131-3 standard probably would like, pointer and/or ANY data types can be used to allow programs to access local stack memory that actually "belongs" to other programs.

STEP 7 encourages the use of **symbolic addressing.** Symbol tables and data declaration tables can be used to define symbolic names to represent absolute addresses. Individual STEP 7 programs include variable declaration tables in their headers, which can be used to declare (local) symbolic names that can be used in that program instead of absolute addresses. Use of symbolic names allows programs to be copied more easily to other PLCs, even if the other PLCs have different memory arrangements, because the programmer can easily modify the symbol table, data declaration tables, and variable declaration tables instead of having to rewrite the entire program with different addresses. The easy transportability of a program from one PLC to another is a requirement in the IEC 1131-3 standard.

Siemens STEP 7 Input Image Data

Input image data is stored in the process-image input table (PII), which contains 128 bytes of input data from digital input modules. PII data can be addressed in the following absolute address forms:

IB x for addressing 8-bit *bytes,* where "x" is a number from 0 to 127, reflecting the location of a digital input module. Addresses are automatically assigned to I/O modules as the PLC is configured to recognize I/O modules. Lower-numbered bytes represent modules closer to the CPU module. Siemens reserves additional IB addresses up to 65535, for future expansion.

I x.d where "d" is the number of *one bit* in byte "x," numbered 0 to 7.

IW x for addressing 16-bit data *words,* where "x" is a number from 0 to **126,** indicating the address of the first of two consecutive bytes making up the 16-bit word. The two bytes are combined into a word as shown in Figure 5.17a.

ID x for 32-bit *double words,* where "x" is a number from 0 to **124,** indicating the first of 4 consecutive bytes making up the 32-bit double word. The 4 bytes are combined into a double word as shown in Figure 5.17b.

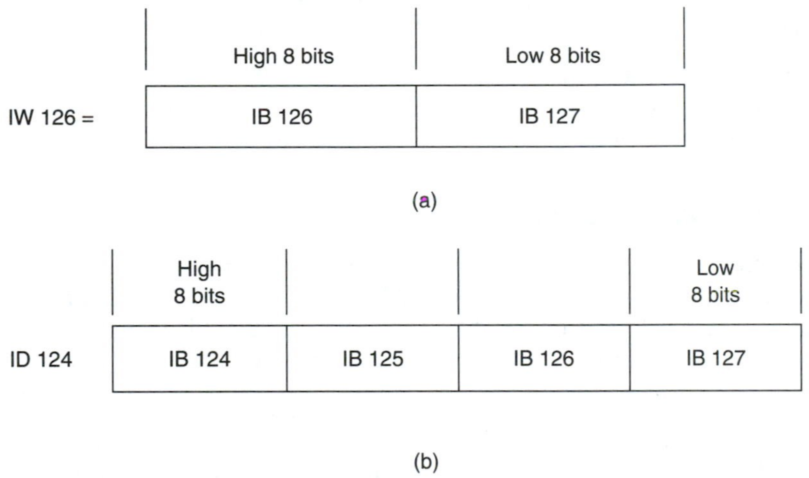

	High 8 bits	Low 8 bits
IW 126 =	IB 126	IB 127

(a)

	High 8 bits			Low 8 bits
ID 124	IB 124	IB 125	IB 126	IB 127

(b)

FIGURE 5.17
S7 word and double-word structure: (a) one word = two bytes; (b) one double word = four bytes.

Siemens STEP 7 Output Image Data

Output image data is stored in the process-image output table (PIQ), which contains 128 bytes of output data for digital output modules. PIQ data can be addressed like PII data, in the following absolute address forms:

QB x for 8-bit *bytes*, where “x” is a number from 0 to 127 (Siemens reserves additional QB addresses up to 65,535, for future expansion)
Q x.d where “d” is the number of one *bit* in the byte, 0 to 7
QW x for addressing 16-bit data *words*, where “x” is a number from 0 to **126**
QD x for 32-bit *double words*, where “x” is a number from 0 to **124**

Siemens STEP 7 Peripheral Input Data

Peripheral input data is used like PII data but is actually in the memory of an input module. Peripheral input data reflects the current input status, not the input status as it was when the scan cycle read the input module data into input image memory. Individual peripheral input data bits cannot be addressed, but addresses in the following absolute address forms are allowed:

PIB x for 8-bit *bytes*, where “x” is a number from 0 to 65535. Note that input modules with byte addresses 0 to 127 can be read directly, even though they have input image data memory that is updated every scan cycle. Instructions with peripheral I/O addresses take longer to execute than those with PII or PIQ addresses.
PIW x for addressing 16-bit data *words*, where “x” is a number from 0 to **65,534.**
PID x for 32-bit *double words*, where “x” is a number from 0 to **65,532.**

Siemens STEP 7 Peripheral Output Data

Peripheral output data is used like peripheral input data. Individual peripheral output data bits cannot be addressed, but addresses in the following absolute address forms are allowed:

PQB x for 8-bit *bytes*, where “x” is a number from 0 to 65535
PQW x for addressing 16-bit data *words*, where “x” is a number from 0 to **65,534**
PQD x for 32-bit double words, where “x” is a number from 0 to **65,532**

Siemens STEP 7 Bit Memory

Bit memory (sometimes called *marker memory*) is used to store bits, bytes, words, or double words of user data. Sixteen bytes of bit memory (MB0 to MB15) are retentive by default, but the programmer can increase or decrease the number of retentive bit memory addresses. Bit memory can be addressed in absolute addressing form as follows:

MB x for 8-bit *bytes*, where “x” is a number from 0 to 255
M x.d where “d” is the number of one *bit* in the byte, 0 to 7
MW x for addressing 16-bit data *words* consisting of two consecutive bytes, where “x” is a number from 0 to **254**
MD x for 32-bit *double words* consisting of four consecutive bytes, where “x” is a number from 0 to **252**

Siemens STEP 7 Timer Data

Like S5 timer data words, timer data words contain 10-bit natural binary numbers in the low 10 bits, representing the timer’s time remaining. Siemens doesn’t tell us what the high bits are used for, but they are probably used for timer status bits. Unlike in S5 PLCs, none of the timer data values are retentive unless the programmer specifically configures some timer words as retentive. The binary time-remaining value can be read using a ladder logic instruction or an STL LOAD (L). The STL LOAD BCD (LC) instruction can also read a timer's time-remaining value and convert it back into BCD. Whenever a time-remaining value is read, the PLC will only read the value from bits number 0 to 9 into the accumulators. The *time-remaining value* is addressed in absolute address form as follows:

T x where "x" is a number from 0 to 255

The timer's *status bit* can also be addressed in absolute addressing mode as follows:

T x where "x" is a number from 0 to 255

The absolute address of the status bit is the same as the address for the time-remaining value, but the PLC can determine which value to use by the type of instruction that contains the address.

The S7 line of PLCs also offers a real-time timer, which keeps track of the date and time, as users expect of modern computers. During configuration (see Chapter 10), the programmer specifies an address to contain real-time data. The real time can then be examined in programs to cause functions to be executed at program-specified times or intervals.

Siemens STEP 7 Counter Data

As in S5 PLCs, counter data words in the S7 line contain 10-bit natural binary numbers in their lowest bits. This number represents the counter's accumulated value, which can be read by ladder logic instruction or in STL using the LOAD (L) instruction. The STL LOAD BCD (LC) instruction converts the accumulated value back into BCD as it reads it into the accumulator. By default, only eight counter data words (C0 to C8) are retentive, but more retentive timers can be configured. Counter accumulated value data words can be addressed in absolute address form as follows:

C x where "x" is a counter number from 0 to 255

The counter's *status bit* can also be addressed in absolute addressing mode as follows:

C x where "x" is a number from 0 to 255

The absolute address of the status bit is the same as the address for the accumulated value, but the PLC can determine which value to use by the type of instruction that contains the address.

Siemens STEP 7 System Data

The S7 line of PLCs makes system data available to the programmer in several ways:

1. By automatically passing system data values as parameters to user-written organization block programs every time the S7 operating system calls one of those programs. The programmer must know what system values the S7 PLC passes, and the absolute local stack addresses or symbolic names for the parameters passed. For example, a value with the symbolic name **Event_ID** is passed to every organization block the S7 operating system calls (organization blocks cannot be called by user-programs, only by the operating system). The Event_ID number identifies the reason the organization block was called, and the program in the organization block can check the value in Event_ID before responding. Your manual set contains more information on Event_ID numbers.

2. By allowing user-written programs to include a call of a built-in system function program, SFC 51, which also goes by the name RDSYSST. RDSYSST can be used to copy any one of a variety of **system status lists** from the system data area of memory to memory locations specified by the programmer. A table showing the list of the "partial" system status lists that can be read, and the **system status list number (SZL-ID)** that must be included with the call to RDSYSST to read that data, is included in Figure 15.6. More complete information is contained in the STEP 7 manual set.

3. By absolute addressing of a few of the **MPU registers.** The addressable MPU registers include:

(a) The 16-bit **status word,** shown in Figure 5.18, is addressable as

STW `(e.g., L STW reads the current status word)`

Some individual status bits of the status word can be addressed independently, so they can be examined in Boolean logic statements. Other status word bits are not separately addressable, but are used automatically by some instructions.

- **CC0, CC1, BR, OV,** and **OS** are condition code bits that are affected by data values that the MPU most recently manipulated or generated. (See your manual set for how various instruction types affect these bits.)
- **FC, RLO, STA, OR,** and **BR** (BR is sometimes addressed as **ENO**) are bits that are affected by the interim or final result of logical operations performed by the PLC as it executes instructions. If the **RLO** bit is 1, conditional output statements will execute. The **BR/ENO** bit is made 1 if a built-in word manipulation instruction, built-in function, or function block program executes correctly. The BR/ENO bits can also be written to by user-written functions or function blocks. The other bits are used by the S7 PLC during logical statement execution, in developing the final RLO result. (See your manual set.)

(b) The MPU's **address registers** contain data that can be used in register indirect addressing (described later in this section) if the instruction addresses them as **AR1** or **AR2.**

Bit number :	15 ... 9	8	7	6	5	4	3	2	1	0
		BR	CC1	CC0	OV	OS	OR	STA	RLO	$\overline{FC}$

FIGURE 5.18
S7 status word.

Siemens STEP 7 Shared Data Blocks

S7 PLCs offer two types of data blocks: shared data blocks, addressable from any S7 PLC program, and instance data blocks, intended to be reserved for the use of the function block program that the instance data block references. In this section we discuss mainly **shared data blocks,** leaving instance data blocks until later.

During creation of data blocks, the programmer can create a **user data type (UDT)** element, which looks like a data block and is created in the same way as a data block is created but is actually just a template for a complex data element. UDTs are *not* data elements and cannot be used as data elements. UDTs exist only in a programming unit, where they are used during the creation of shared or instance data blocks, or during the creation of STRUCT data elements.

Data blocks are sets of data elements. Each data block must be created by the programmer or by instructions in a user-program before the data block can be used in a program. Once created, the user-program must *open* that data block before its contents can be read or written to. A ladder logic instruction to open an existing data block would look like Figure 5.19. Figure 5.19 also includes the STL language instruction for opening the data block. Only one shared data block can be open at a time, so opening a shared data block automatically closes the shared data block opened previously. By default, only one data block (DB1) is retentive, but the programmer can increase or decrease this number within practical limits.

Data Declaration and Data Types for Shared Data Blocks

Shared data blocks can be created using the STEP 7 programming software to *declare* a data block as a structure of data elements as shown in Figure 5.20. The header in Figure 5.20 indicates that the programming software is creating this data block for a PLC identified in the programming software as "Stn_3", which is in a system of networked PLCs that the programming software knows as "Plant_2", for which this programming unit saves data on disk C:\. The data block being created is DB4. DB4 can be assigned a symbolic name elsewhere in the programming software, but the symbolic name will not appear on this programming screen.

The data declaration table includes the key words STRUCT and END_STRUCT to define the beginning and end of the set of data elements making up the data block. The programming software inserts these key words. Between these key words, each record (row) in the data declaration table defines one data element, indicating:

1. The absolute **address** of each data element created in the data block (the Addr. column in Figure 5.20). The programmer doesn't make entries in this column; the STEP 7 programming software assigns addresses as the programmer finishes each table row. Each absolute address identifies where that data element is stored, in addresses measured from wherever the data block

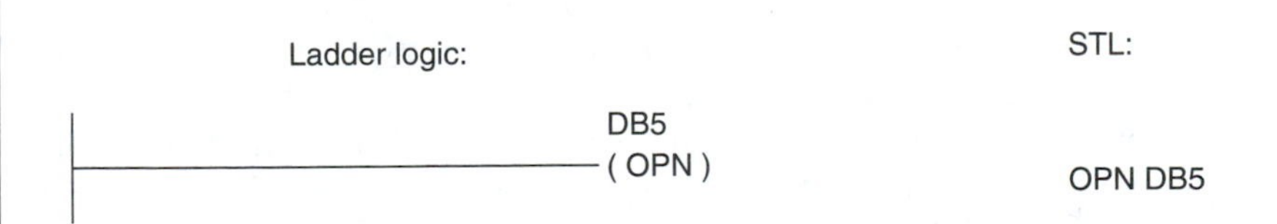

FIGURE 5.19
STEP 7 instructions for opening a data block.

Data block c:\Plant_2\Stn_3\DB4

Address	Name	Type	Initial value	Comment
0.0		STRUCT		
+0.0	latch_1	BOOL	0	
+0.1	latch_2	BOOL	0	
+2.0	count-up	WORD	20	
+4.0	big_num	DINT	0	
+8.0	ratio	REAL	9.5	
+12.0	message	STRING [20]		
=32.0		END_STRUCT		

FIGURE 5.20
Step 7 data declaration table, as seen when using a programming unit to create a data block.

starts in memory. In the example, two BOOLean bits are reserved for the first two elements, and STEP 7 assigned the lowest 2 bits of the first available data byte. The next data element, a WORD, is assigned two bytes (starting at byte 2). Additional space is reserved for a DINT number, a REAL number, and a STRING. Notice that the string declaration includes a size declaration so that it will only contain space for 20 ASCII codes (plus a 2-byte header) instead of the default 254 ASCII codes. The total data block uses 32 bytes. The program can operate on bits, bytes, words, and double words in a data block using the absolute addresses. Once the data block has been created and opened, data in the shared data block can be addressed in the following absolute addressing formats:

DBX x.b for an individual *bit*, where "x" is the byte number (0 to up to 65,535, less for current S7 PLCs) and "b" is the bit number (0 to 7)
DBB x for *byte* number "x"
DBW x for *word* number "x"; includes two consecutive bytes
DBD x for 32-bit *double words* consisting of 4 bytes

2. A **symbolic name** for each element, which is optional but advised. If a symbolic name is entered, the data item can be addressed in the user-program by using that name. Symbolic names can be used to address data elements larger than bits, bytes, words, or double words.

3. The **data type** column defines how much memory must be reserved for a data item, and the next column allows the programmer to (optionally) assign initial values (in constant form) for each data item. Almost any type of data can be declared as a data element in a shared data block, including timer constants, but not including timer, counter, or block identifying numbers; or pointer data types; or ANY data types.

Complex data types, such as ARRAYs and STRUCTures, can also be declared as elements in data blocks and can even be defined in the data block if they are not already defined as a UDT. The complex data type element in a data block can then be manipulated as a single data item in a user-program, or the simple data elements that make up the complex data element can be operated on separately. Figure 5.21 shows a data declaration table that reserves memory in a data block for four elements of complex data types: a DATE_AND_TIME element, a STRING element, an ARRAY, and a STRUCT element. Notice that variable names can be declared for a complex data element and sometimes also for the data elements in the complex data item. Use of symbolic names is often the only practical way of addressing complex data items and their components.

In the data declaration table of Figure 5.21, symbolic names have been assigned for a DATE_AND_TIME element, a STRING element, an ARRAY element, a STRUCT element, and for the simple elements in the STRUCT, but not for the simple elements in the ARRAY. Once the data block that this data declaration defines is opened by a program, the declared symbolic names can be used in place of the data elements' absolute addresses.

Data Block c:\Plant_2\Stn_3\DB5

Address	Name	Type	Initial value	Comment
0.0		STRUCT		
+0.0	today	DATE_AND_TIME		
+8.0	message	STRING [20]		
+28.0		STRUCT		
+0.0	matrix	ARRAY [0 . . 4 , 0 . . 9]		
*2.0		WORD		
=100.0		END_STRUCT		
+128.0	misc_data	STRUCT		
+0.0	light	BOOL		
+2.0	sum	WORD		
=4.0		END_STRUCT		
		END_STRUCT		

FIGURE 5.21
Step 7 data declaration table declaring complex data elements.

A data declaration table also defines the amount of memory to be reserved for the declared elements. The STRING element is declared to be large enough to contain only 20 ASCII CHAR elements. The ARRAY is declared as a two-dimensional, five-row (0 to 4) by 10-column (0 to 9) set of similar data items (WORDS). The key words STRUCT and END_STRUCT must precede and end the ARRAY declaration. The size of the array is declared after the STRUCT key word, followed by the type of data elements in the array. The data block declaration table also defines a STRUCT (named misc_dat) containing two simple elements (a BOOL element named *light* and a WORD named *sum*). The keyword END_STRUCT indicates that the STRUCT's definition is complete. (And the next END_STRUCT completes the data block declaration.)

Once a data block is opened in a user-program, the elements of the ARRAY in the data block can be addressed using the array's symbolic variable name, followed by the array indices' numbers for a single element. For example:

`matrix[0, 9]` is the address of the WORD that occupies the first row, last column, of the array declared in Figure 5.21

To address individual subelements of a STRUCT data element, use the variable names separated by periods, starting with the highest-level data element and working down to the lowest. For example,

`misc_data.sum` is the address of the WORD subelement of the STRUCT

If a STRUCT (e.g., more_data) was defined as including another STRUCT (e.g., sub_data), which in turn contained a three-dimensional ARRAY (e.g., set) of BOOL elements, one bit of the array could be addressed as follows:

```
more_data.sub_data.set [2,5,3]
```

If STRUCT and ARRAY data types don't give the user as much flexibility as required (demanding user!), a programmer can use the STEP 7 programming software's data type editor to define additional user data types (called **UDT** data types) in exactly the same way that the programmer creates a shared data block. The creation process is identical except that the STEP 7 programming software option, **Data Block Referencing a User Data Type,** must be selected instead of the **Data Block** option in the dialog box before actually creating the data block. If a UDT has been defined before creation of a shared data block, the UDT can be declared as an element in the data block, thereby reserving memory for all the elements of a UDT data element in the data block. Symbolic names, data types, and initial values for the components of the UDT-type element are declared during creation of a UDT and cannot be changed during creation of the shared data block that uses the UDT.

Siemens STEP 7 Instance Data Blocks

Instance data blocks are used for storage of data that is available only to a specific function block program.

Function blocks are subprograms that can be called by other programs. When a function block is called, the call must include the number of an instance data block that is to be opened, as in Figure 5.22. (Note that STEP 7 uses the notation "DB" for both instance data blocks and shared data blocks.)

Instance data blocks are *created* very differently from how shared data blocks are created. When the programmer uses the STEP 7 programming software to create an instance data block, the menu option that reads "**data block referencing a function block**" must be selected from the dialog box used to create data blocks. The programmer then enters the number of the function block that will use this instance data block. That's it! STEP 7 automatically creates an instance data block containing the data elements and the initial values that are declared in the **function block's variable declaration table** (except for data elements declared as **temp**). The function block *must* be programmed, with a complete variable declaration table, before the programmer creates the instance data block. Several instance data blocks can be created for each function block. Figure 5.23 shows an example of a variable declaration table in a function block. Instance data blocks that reference this function block will contain data elements with the symbolic names, data types, and initial values defined in this variable declaration table, except that the variables declared as being of type "temp" would *not* be included in the creation of the instance data block. When viewing the contents of an instance data block, the programming unit shows the declaration type column.

The variable declaration table in a function block indicates:

1. The absolute address (the Addr. column in Figure 5.23) for each data element in the instance data block or in the local stack memory (local stack memory is covered in the next section of this chapter). The programmer doesn't make entries in this column; the STEP 7 programming software assigns these "absolute" addresses after the programmer finishes entering each table row. In the example, 2 bits are reserved in the instance data block for two BOOL data units to represent the states of a switch and an actuator. Two more bytes (starting at bit 2.0) are reserved for an INT (integer value) and another 2 bytes starting at bit 4.0 are reserved for another INT. The following absolute addressing forms can be used for addressing instance data block contents:

DIX x.b for an individual *bit,* where "x" is the byte number (0 to up to 65,535, less for current S7 PLCs) and "b" is the bit number (0 to 7)

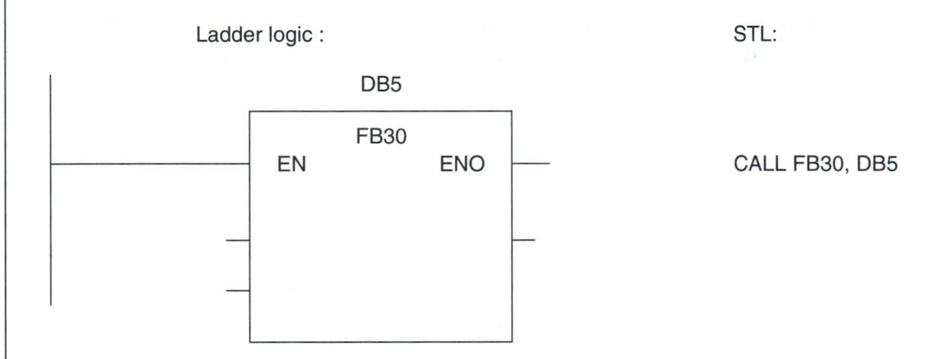

FIGURE 5.22
STEP 7 instruction (in ladder logic and in STL) to call a function block program. It includes specification of an instance data block.

Address	Declaration type	Name	Data type	Initial value	Comment
0.0	in	switch_1	BOOL		
0.1	out	cylinder	BOOL	False	
2.0	in_out	loop_cnt	INT	20	
4.0	stat	work_dat	INT	0	
6.0	temp	push_cnt	INT	0	

FIGURE 5.23
Form and content of a simple variable declaration table, part of a STEP 7 function block program.

DIB x for *byte* number "x"
DIW x for *word* number "x"; includes two consecutive bytes
DID x for 32-bit *double words* consisting of 4 bytes

Notice that the absolute addresses restart at 0.0 for the element with declaration type "temp." This reflects the fact that this variable will be stored in an area of memory other than in the instance data block. Variables declared as "temp" are only stored in the local stack area of memory, which we discuss in the next section.

2. The **declaration type** column identifies the source, destination, and lifetime of each data value. Any number of data elements can be declared for each declaration type, but they must be declared in the following order in the variable declaration table:

(a) "**in**" parameters followed by "**out**" parameters followed by "**in-out**" parameters. These parameter declaration types define whether a data element's value is to be provided to the function block ("in"), written to by the function block ("out"), or both ("in-out"). We discuss parameter passing in Chapter 8.

(b) "**stat**" parameters are next. These data elements are for working values for use by the function block. "Stat" parameter values are retained in the instance data block until the next time the function block is called with the same instance data block.

(c) "**temp**" parameters are listed last. Their value is lost when this function block ends, although the function block's variable declaration table can include initial values that the "temp" variables are to be set to next time the function block is executed with any instance data block.

3. Block-local symbolic names can be, but do not have to be, entered. If entered, this function block can address data items using a "#" followed by a block-local symbolic name, instead of having to address the data item using absolute addressing. Using symbolic names allows the programmer to address some data items more easily. The following examples show how data from the variable declaration table in Figure 5.23 could be addressed using their block-local symbolic names:

```
#cylinder     means the same as DIX 0.1
#work_dat     means the same as DIW 4
```

STEP 5 allows the programmer to use a symbolic table to define *global* symbolic names for shared memory addresses. These names can also be used in function blocks. Be sure to include the "#" when addressing *block-local* symbolic named data, or your program may accidentally address a global variable with the same name.

4. The **data type** column defines how much local stack memory must be reserved for a data item, and the **initial value** column allows the programmer to assign initial values in constant form for each data item. Allowable data types and constant formats include all the data types described so far, plus some types described as "parameter" types by Siemens, including:

(a) **TIMER:** causes the function program to reserve two bytes of memory and to accept parameter values consisting of the letter "T" and a number between 0 and 255, identifying a timer, from the calling program.

(b) **COUNTER:** like the timer parameter, except that the calling program must pass the letter "C" and a number between 0 and 255, identifying a counter.

(c) **BLOCK_FB, BLOCK_FC, BLOCK_DB, BLOCK_SDB:** each causes the function to reserve 2 bytes of memory and accept the address for a logic block (FB or FC) or data block (DB or SDB) from the calling program. The calling program can either pass the parameter by its full name—the block type identifier (FB, FC, DB, or SDB) followed by a number from 0 to 255—or can specify a block by its symbolic name.

(d) **POINTER:** used to pass an address specification to a function or function block program so that the function/function block program can read or write to that address. Pointers can always be passed to functions and function blocks but cannot necessarily be declared as an "out" data declaration type. (See your manual set.) Remember that the pointer-type parameter is stored in 6 bytes of memory, as was shown in Figure 5.16.

The allowable codes for memory areas include a code for the previous local stack memory. This means that a program can call a function (or function block) and can pass a pointer parameter that will allow the function (or function block) to access the local stack memory that belongs to the calling program.

(e) **ANY:** in which case the instance data block will set aside 10 bytes of space to contain an ANY pointer to an ANY data set. As with a standard pointer parameter, an ANY pointer parameter can give a function block access to memory areas containing one or more data elements of simple or complex data types, including memory areas in the calling program's local stack memory.

Since more than one instance data block can be created with reference to the same function block, program calls to a function block must include the address of the instance data block that is to be opened with the function block. A function block can close that instance data block and open a different instance data block. Figure 5.24 shows how to program the opening of a new instance data block in ladder logic and in STL. Notice that the open instruction *does* use addressing differentiating a shared data block (DBx) from an instance data block (DIx).

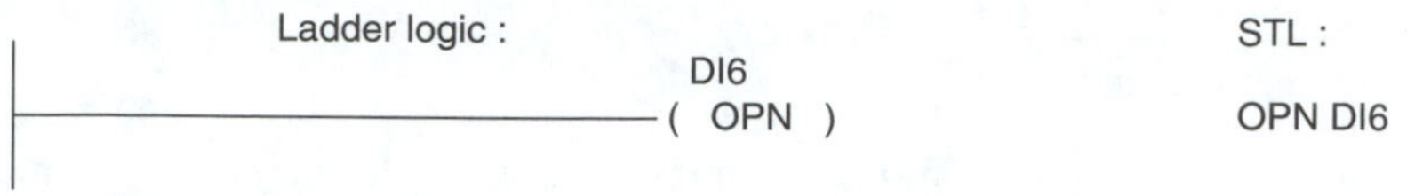

Ladder logic :

DI6
(OPN)

STL :

OPN DI6

FIGURE 5.24
STEP 7 instructions for opening an instance data block.

Siemens STEP 7 Local Stack Data

Local stack memory is used to store values that a STEP 7 program is using, but won't be needed when that STEP 7 program ends. To understand, we need an introduction to structured programming. Structured programming is covered in more detail in Chapter 8.

> The STEP 7 operating system causes **organization block** programs, written by the user, to be executed during the scan cycle. One of those organization blocks (OB1) contains the main user-program that is executed every scan cycle. The other organization blocks can contain user-programs that are run only under specific circumstances. OB1 has a lower priority than the other organization blocks, so the operating system interrupts OB1 whenever conditions require another organization block to run and resumes executing OB1 when the other organization block finishes. STEP 7 contains provisions for as many as 80 different **priority classes** of organization block programs. Each priority class of program has a fixed amount of *local stack* memory available for its use, which is separate from the local stack memory that other priority classes of programs can use.[6]
>
> The program in any organization block can call subprograms called *functions* and *function blocks.* Every organization block, function, and function block program is assigned some local stack memory, as it is called, and releases that local stack memory as it finishes executing. Some local stack memory is used automatically, for purposes defined by the STEP 7 operating system.

Every organization block, function, and function block addresses its local stack area starting from address L 0.0, although several programs may be assigned memory from the local stack at the same time. Every user-written program must include a variable declaration table that indicates how much local stack memory the program needs. Figure 5.23 demonstrated the form and content of a simple variable declaration table that might be programmed for a function block. Organization blocks and functions also include variable declaration tables.

The variable declaration table in a *function block* program is used in part as a template for its instance data block(s) *and in part to declare the function block's required "temp" variables.* Whenever a function block is called, STEP 7 reserves local stack memory for the "temp" variables. "Temp" variables (such as the INT variable defined in Figure 5.23) can be addressed using an absolute address (e.g., `LW 0`) or its symbolic address (e.g., `#push_cnt`).

[6]The amount of available memory varies between S7 PLCs. Refer to your manual set. The PLC will *fault* (stop running) if a program tries to use too much local stack memory in any one priority class or if the total available local stack memory is exceeded by all programs at all priority classes combined.

Address	Declaration Type	Name	Data type	Initial value	Comment
0.0	in	switch_1	BOOL		
0.1	out	cylinder	BOOL	0	
2.0	in_out	loop_cnt	INT	20	
4.0	temp	push_cnt	INT	0	

FIGURE 5.25
Form and content of a STEP 7 function's variable declaration table.

A *function*'s variable declaration table looks similar to a function block's variable declaration table, as shown in Figure 5.25. Notice, however, that the function's address column starts at 0.0 and does *not* start numbering again when it encounters the first "temp" variable. *All* the variables for a function are stored in local stack memory. Since a function can't have an instance data block, its variable declaration table cannot include variables with the declaration type "stat".[7] There are a few differences between the data types of variables that can be declared in function blocks versus in functions, but the rules for declaring variables and for addressing variables are generally the same as in a function block.

An organization block's variable declaration table can only contain variables of declaration type "temp", and all its working data must be stored in local stack memory (unless the organization block opens a shared data block). Whenever the STEP 7 operating system calls an organization block, it automatically opens 20 bytes of local stack memory and places data into that memory for the program in the organization block to use. Every STEP 7 organization block comes preprogrammed with a variable declaration table defining the 20 bytes of local stack memory it uses, but the programmer can add to the variable declaration table to declare more "temp" variables.

Siemens STEP 7 Memory Indirect Addressing

Memory locations can be addressed using memory indirect addressing format, in which pointer data in memory consists of an address indicating where data is to be read from, or where to put data.[8] The pointer used in memory indirect addressing is different from the 6-byte pointer data type described in the data constant section and in the instance data block sections in this chapter. The pointer that is used in memory indirect addressing is a number that must be stored in the 16- or 32-bit formats shown in Figure 5.26, where:

---

[7]Any program (function, organization block, or function block) can open a shared data block and can read or write data in that data block, so shared data blocks can be used to retain working data even after a program has ended.
[8]Because the local stack memory is sometimes used automatically by the STEP 7 operating system, Siemens advises against using indirect addressing to areas in the local stack memory.

Byte :			1	0
Bit :			15	0
			bbbb bbbb	bbbb bbbb

Byte :	3	2	1	0
Bit :	31			0
	0000 0000	0000 0bbb	bbbb bbbb	bbbb bxxx

FIGURE 5.26
S7 pointers as used in memory indirect addressing.

- **bbb bbbb bbbb bbbb b** is an unsigned 16-bit number indicating a byte address (0 to 65,535).
- **xxx** is a 3-bit value indicating a bit number (0 to 7); must be 0 if the pointer "points" to a byte, word, or double word.

The 16-bit pointer format is used to point to a counter, timer, data block, function, or function block. It consists of an unsigned binary number representing a number between 0 and 65,535. The 32-bit pointer format is used to point to a byte in memory or may point to 1 bit of that byte.

The memory indirect addressing format is used in instructions in the following format:

instruction $p_1s_1[p_2s_2x]$

where:

- "instruction" is any instruction that can use memory indirect addresses.
- "p_1" is the prefix that identifies the area of memory that contains the data value to which the pointer points.
- "s_1" indicates the size of the data value (if it is necessary to specify a size).
- The square brackets tell the PLC that this is memory indirect addressing, and the address inside the brackets is the address where the pointer is stored.
- "p_2" is a prefix indicating which part of memory the pointer is stored in (M, L, DB, or DI).
- "s_2" indicates the size of the pointer data (W or D).
- "x" is the address number of the pointer.

For example:

```
L +25
T MW4        //   Places "25" into MW4 as a pointer.
L C [MW4]    //   Causes the PLC to read the pointer (+25) from MW4, then load the ac-
                      cumulated value from the counter "pointed" to (C25). Since a
```

counter's accumulated value is always 16 bits, it isn't necessary to indicate its size.

and

```
L PW [MD8]     //
```
Causes the PLC to read a 32-bit pointer double word from MD8, then load a 16-bit value from the peripheral I/O word that the pointer "points" to.

The 32-bit pointers for memory indirect addressing must be created using a pointer constant in the format described in the data constant section of this chapter. The pointer constant is stored as a 32-bit number. The process of creating them using a 32-bit pointer is as follows:

```
L p#24.6
T MD8          //   Places the 32-bit pointer into MD8
A I [MD8]      //   Examines bit 6 of byte 24 of input image memory
```

Symbolic addresses can be used in place of the absolute addresses used in the examples in this section, of course. If the symbolic name My_Pointer has been declared to represent the address MD8, the program shown above could be written as

```
L p#24.6
T "My-Pointer"
A I ["My-Pointer"]
```

Siemens STEP 7 Register Indirect Addressing (Indexed Addressing)

Register indirect addressing is an extremely powerful form of indexed addressing. It uses pointers for both the base and the offset of the address. The S7's CPU contains two 32-bit address registers that can be used to store the 32-bit pointers used as the base value for register indirect addressing. Pointer constant values are stored into address registers using the instructions

LAR1 or **LAR2** to load a constant or the contents of a memory location into address register 1 or 2

The format of the 32-bit number used in register indirect addressing is slightly different from the format of the 32-bit number that is used in memory indirect addressing. It must be in the form shown in Figure 5.27, where:

- **bbb bbbb bbbb bbbb b** and **xxx** are byte and bit numbers, as in memory indirect addressing.
- **c** is:
  - 0 if the pointer is meant to be **area-internal,** which means that the instruction that uses this pointer also must include a specification of the area of memory to be accessed.
  - 1 if this pointer is meant to be **area-crossing,** which means that this *pointer* indicates the area of memory to be accessed.
- **rrr** is a 3-bit code indicating which area of memory to use if the pointer is an area-crossing type, if c = 1. The codes include:

Byte :	3	2	1	0
Bit :	31			0
	c000 0rrr	0000 0bbb	bbbb bbbb	bbbb bxxx

FIGURE 5.27
S7 register indirect pointer format.

```
000 for P    010 for Q    100 for DBX
001 for I    011 for M    101 for DIX
111 for the local memory reserved for the program that called this
program
```

If the pointer's most significant bit is a "0," indicating an area-internal pointer, the instruction that uses register indirect addressing must include the address in the following form:

`instruction` **ps[ARn,o]**

where:

- "p" is the area identifier (I, Q, P, M, L, DB, or DI) for the data value on which this instruction will operate.
- "s" is the size of the data value on which this instruction will operate (B, W, or D). A size value isn't needed if the instruction uses a bit value.
- "n" is 1 for address register 1 (AR1), or 2 for address register 2 (AR2). AR1 or AR2 must contain a 32-bit register indirect pointer to be used as the base port of the indexed address.
- "o" is the indexed address offset value. This can be entered in 32-bit memory indirect pointer constant format, or as an address where a 32-bit memory indirect address pointer is stored. If entered as a pointer constant, it would be in this form:

  P#x.n

  where "x" is the byte number of the base address and "b" is the bit number of the base address. "b" must be 0 if the pointer points to a byte, word, or double word.

If the pointer's most significant bit is a "1," indicating an area-external pointer, the instruction that uses register indirect addressing must include the address in the following form:

`instruction` **s[ARn,b]**

which is the same as in area-internal addressing, except that the memory area specifier isn't used.

The following examples demonstrate the use of register indirect addressing. (The instructions that load pointer constants and transfer them to the address registers do not have to immediately precede the instructions using register indirect addressing, of course):

Example 1: Using two pointer constants:

```
LAR1 p#24.6          //   Loads a constant for an area-internal pointer to byte 34, bit 6, into
                            address register 1
A I [AR1,p#2.1]      //   Examines input image byte 26 (24 + 2 = 26), bit 7 (6 + 1 = 7)
```

Example 2: Using an offset pointer from memory and a pointer constant for a base address

```
L p#Q5.2             //   Loads a constant for an area-crossing pointer to output image bit
                            Q5.2 into accumulator 1
T MD4                //   Transfers it to data memory address MD4
LAR2 MD4             //   Copies the pointer from MD4 to address register 2
S [AR2,p#1.2]        //   Sets output image bit Q6.4
```

Example 3: Using two pointers from memory

```
L p#M6.0             //   Loads a constant for an area-crossing pointer to data memory byte 6
T LD8                //   Transfers it to local stack memory LD8
L p#1.0              //   Loads a 32-bit memory indirect type pointer constant for a pointer
T LD12               //   Transfers it to local stack memory LD12
LAR1 LD8             //   Loads address register 1 with the pointer from LD8
L D [AR1,LD12]       //   Loads a double word from data memory byte addresses MD7.
```

Two instructions, **+AR1** and **+AR2,** can be used to increase or decrease the pointer values in the address registers. +AR1 and +AR2 are covered in the section on the Siemens STEP 7 math/logic instructions in Chapter 6.

CQMI

DATA REGISTERS AND ADDRESSABLE DATA IN THE OMRON CQM1

The CQM1 has the least complex working data memory structure of the PLCs discussed in this book and the least number of ways of entering constants.

OMRON CQM1 DATA CONSTANT VALUES

Data constants can only be entered in a CQM1's program as decimal or hexadecimal values, and must be prefixed with a "#". Some CQM1 instructions that operate on data in memory assume that the data is in BCD form before using it, while others assume unsigned binary, and others assume signed binary. Some instructions automatically use 16-bit data words, and others automatically use 32-bit double words. The data constants that are entered must be in the decimal or hexadecimal form that the program requires and must contain an appropriate number of decimal or hexadecimal digits.

OMRON CQM1 Memory Register Areas

There are eight separately addressable memory areas.

OMRON CQM1 I/O Image Data An **internal register (IR)** area of memory is reserved for 16-bit data words containing image and output image data, but includes more memory than is required for input and output images. The rest of this memory area is available for the storage of other working data. The MACRO instruction automatically uses part of this memory area. In some models of the CQM1, parts of the additional memory are reserved for special features of that PLC model. OMRON's literature indicates that OMRON intends to use part of the extra internal register memory for other special features in future CQM1 models, so the programmer should avoid those addresses. Data words in the internal register (IR) area of memory can be addressed in the following form:

000 to **243**

(The "IR" prefix is assumed if no prefix is entered.)

- Addresses 000 to 011 are needed for input images.
- Addresses 100 to 111 are needed for output images.
- Other addresses are available for special functions or for working data.

Individual data bits can be addressed in the following form:

xxx00 to **xxx15** where "xxx" is 000 to 243

OMRON CQM1 Timer/Counter Registers The **timer/counter register (TC)** area of memory is used to store 16-bit BCD numbers representing the present value (PV) of either timers or counters (i.e., there cannot be both a counter 5 and a timer 5). Address a present value word using one of the following addressing formats, depending on whether an address contains a timer (TIM) PV or a counter (CNT) PV:

```
TIM 000 to TIM 511   or   CNT 000 to CNT 511
```

To examine the single-bit completion flag for a timer or counter, use the same address. The instruction implies that the present value or the completion flag should be used, and the PLC will access the correct data entry. For example:

`TIM 025` can be used as the address for timer 25's present value *and* as the address for timer 25's completion flag

OMRON CQM1 Data Memory The **data memory (DM)** area contains 16-bit data words, some of which are automatically used by the CQM1 for configuration and status, and the rest of which are available for user data storage. Data memory words can be addressed as:

```
DM 0000 to DM 6655
```

of which DM 6144 to DM 6655 are used for configuration and status information and cannot be changed by a user-program. Some configuration words can be read and changed using a programming unit while the PLC is not in run mode.

The CQM1 does not allow the addressing of individual bits in the DM area of memory.

OMRON CQM1 Holding Registers The **holding register (HR)** area of memory is EEPROM memory, so the 16-bit user-data words they can contain will not be affected by power loss. Words are addressable in the following range:

HR 00 to **HR 99**

and individual bits can be addressed as

HR xx00 to **HR xx15** where "xx" is 00 to 99

OMRON CQM1 Special Registers The **special register (SR)** area of memory contains 8-bit bytes, of which individual bits are used by the CQM1 for status and control purposes. Bytes are addressed as follows:

244 to **255**

The numbers start where the numbers of the IR memory area end. The "SR" prefix is assumed and does not have to be entered.

The programmer will often address individual bits of SR bytes in the following form:

xxx00 to **xxx07** where "xxx" is 244 to 255

Appendix C contains a description of the SR bits.

OMRON CQM1 Auxiliary Data Memory Registers The **auxiliary data memory register (AR)** area contains additional 16-bit data words containing **control and status** bits, which can be addressed as words:

AR 00 to **AR 27**

or as bits:

AR xx00 to **AR xx15** where "xx" is 00 to 27

Appendix C contains a description of the AR bits.

OMRON CQM1 Temporary Register The **temporary register (TR)** contains only 1 data byte, which can *only* be addressed as bits in this form:

TR 0 to **TR 7**

These bits are used by the CQM1's operating system to contain temporarily saved result-of-logic-operation states during execution of complex Boolean ladder logic statements. All PLCs must store temporary result-of-logic states while they evaluate branched rungs, but OMRON allows these stored states to be addressed so that they can be copied to other memory bits for use later.

OMRON CQM1 Link Registers The **link register (LR)** area is used during data communications via the RS-232 port, as we will see in Chapter 13, but this memory area is available for other data storage if it isn't required for data communications. The 16-bit words can be addressed as

LR 00 to **LR 63**

and individual bits are addressed as

LR xx00 to **LR xx15** where "xx" is 00 to 63

Figure 5.28 shows a rung that examines bit 15 of special register word 253 (this bit goes on for one cycle each time the PLC starts running). If the bit is on, the rung performs two move operations:

1. It copies the 16-bit number from holding register 10 (containing a value saved from the last time the PLC ran) to the output image data word for output module 102 (so 16 digital output actuators start in a predefined state).
2. It copies the constant value 2000 to data memory location 0011.

OMRON CQM1 Indirect Addressing

Two data memory (DM) memory locations are required for indirect addressing. One DM memory location must contain a *BCD number* indicating the address of the other DM location. The "other" DM memory contains the data value on which the instruction will operate. For example, if a MOV (21) instruction intends to write a value to DM 2000 using indirect addressing, the destination address number (2000) must exist in another DM memory location (e.g., in DM 0010). The MOV (21) instruction will contain the destination address specification in the form ***DM 0010,** where the "*" tells the CQM1 that this is indirect addressing. The CQM1 will use the BCD "2000" value from DM 0010 as the actual DM destination address (DM 2000), to

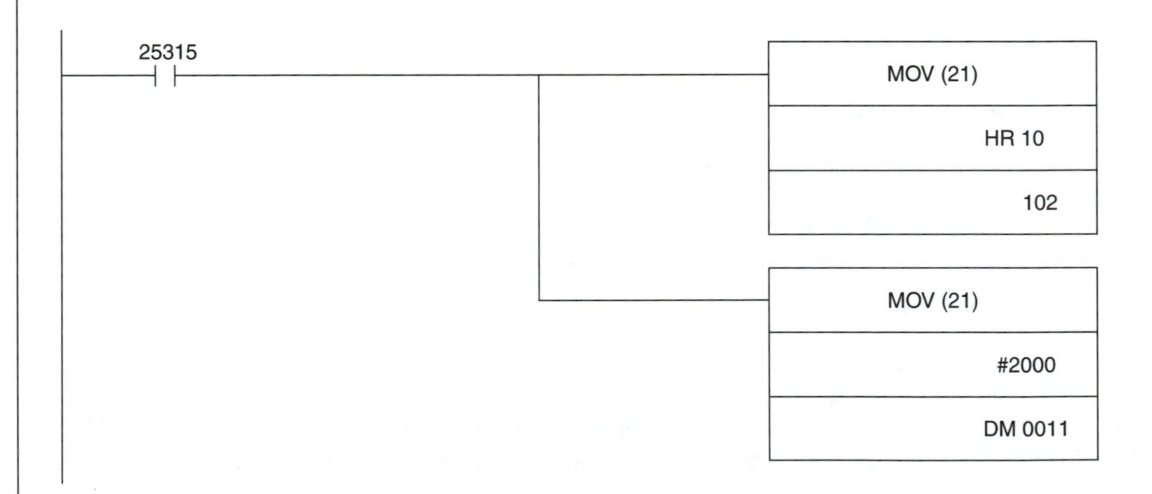

FIGURE 5.28
CQM1 program that moves two data words.

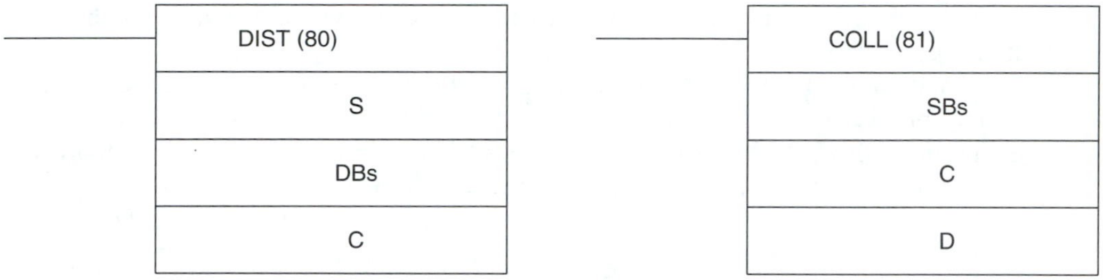

DIST (80)
S
DBs
C

COLL (81)
SBs
C
D

FIGURE 5.29
Instructions allowing CQM1 indexed addressing in moving data words.

which it will write the value being moved. Indirect addressing can only be used to access a memory location in the DM area of memory.

OMRON CQM1 Indexed Addressing

Only two of the CQM1 instructions allow indexed addressing. The SINGLE WORD DISTRIBUTE instruction, **DIST(80),** copies a 16-bit data word from a source address and copies it to a destination address. The destination address can be specified using a base address and an offset. Similarly, the DATA COLLECT instruction, **COLL(81),** copies a 16-bit data word from a source address to a destination address, but this time the source can be specified with a base address and an offset. The two instructions are shown in Figure 5.29. Note that both require a control (C) parameter. If the control parameter value is under 9000 in the DIST (80) instruction, or under 8000 in the COL(81) instruction, it is used as an offset value.[9] The control parameter value can be entered with the instruction in constant form (e.g., #0045), or the instruction can include the address where the actual offset is stored (in BCD form), so that other instructions in the program can manipulate the offset. The base address [the DBs parameter in DIST(80), or the SBs parameter in COLL (81)] can only be entered *as* an address, so the BCD value at that address can also be manipulated by other program instructions.

TROUBLESHOOTING

How can you have trouble using memory? Easy! And sometimes the causes of those troubles are hard to see. A valuable troubleshooting tool is the **online data monitor feature** (which Siemens calls a *variable access table*). All good programming software offers data monitoring. The programmer can use this screen to monitor (and change) the contents of memory as the program runs. Sometimes the programming software requires the programmer to use other screens (e.g., system status screens or system configuration screens) to access memory that the data screen can't access. Sometimes a program can be written to copy data from one part of mem-

[9]If the control word value exceeds these limits, the DIST(80) or COL(81) instructions do something slightly different than simply moving a data word (see Chapter 7).

ory to another to make troubleshooting easier (i.e., to allow seeing data from different memory areas on the same data screen).

A **histogram feature,** offered in some PLC programming packages, can be helpful in finding problems related to data. A programmer can use the histogram feature to have a programming unit track the value in an address (or sometimes in multiple addresses simultaneously) while the PLC runs, and then display a value-time graph showing how the value changed during the period when it was being monitored. If multiple histograms can be generated for the same time period, the programmer can examine how related data values changed during the time period. Histograms are described in Chapter 15.

A common programming error is to accidentally enter the wrong address into your program. If you write data to an incorrectly addressed memory location, you may not be able to find that data later, but the problem is more complex if you accidentally write that data into a memory that contains data you didn't want to lose!

Remember that timers and counters manipulate memory locations. If you accidentally use the same timer address for two timers, you have two instructions in your program that will both try to manipulate the same memory, one of which might be trying to keep track of time while the other is trying to reset the time to zero. If you use the same address for two counters, you might find that the counter increments every scan cycle, because one counter instruction has false control logic while the other has true control logic, so the counter thinks its logic is going true every scan cycle.

Memory that is addressed in a program must exist, of course, so it is important to check your manual set to determine which memories are available and which can be changed. In some PLCs (e.g., Allen-Bradley), data memory must be "created" before it can be used. The memory isn't actually created in this process, of course, it is just reserved for a specific type of data. Allen-Bradley programming software usually reserves memory as the programmer writes address references in a program, but some rules must still be followed (e.g., if a data file is used in one part of the program to store one type of data, the same program file can't be used elsewhere in the program to store a different type of data). Some memory areas, often those areas containing system configuration and status, cannot be written to as a program runs, or can only be written to using advanced program structure or instructions. The programming software may not warn the programmer when he or she enters a reserved memory address, so the programmer may have to monitor a program as it executes to discover that the program isn't changing that memory's contents.

Some memory locations, in some PLCs, are retentive, which means that their contents are retained while the PLC is out of run mode or even when power is removed. When the PLC resumes running its program, the values in those memories are the same as they were the last time the PLC was running. In an Allen-Bradley PLC, all memory (even output image) is retentive. Some Allen-Bradley programming elements (e.g., counters, on-delay timers, etc.) may change the memories that they control as soon as the PLC goes into run mode, because they may think their control logic has gone true. Whenever an Allen-Bradley program is uploaded from the PLC to a host computer, it is uploaded with the data files containing the contents of all data memory and all of the data is reloaded to the PLC when the program is later downloaded to the PLC. Other PLCs have memories areas that are nonretentive, which means that the PLC will clear them every time it is switched on or switched into run mode, or may have areas of memory that are partly retentive and partly nonretentive. The Siemens PLCs have counter, flag/bit, and data block memory, of which part is retentive and part isn't. It is the programmer's responsibility to read the manuals to determine which memories will be cleared automatically when the PLC is switched on, and which ones won't be cleared.

The advanced addressing modes, indexed addressing and indirect addressing, can lead to addressing problems because the program can change the address references as the program executes (by changing address pointers or offset values). *The program may generate an address that doesn't exist or which shouldn't be written to.* Some PLCs will fault automatically if a program tries to access a memory that doesn't exist. Other PLCs will allow the program to access data "across memory boundaries." If, for example, a program tries to reset the preset value of a counter address higher than the highest-allowable counter address, a memory address outside the counter accumulated value storage location might be cleared. If the PLC can access data across memory boundaries this way, it will usually (at least) set a fault bit that the program should check before the scan cycle ends. (You can see why most PLC manufacturers would rather have their PLCs fault.)

QUESTIONS/PROBLEMS

1. What is an output image table? During which parts of the standard PLC cycle does the PLC use this data?

2. If your PLC is in run mode and you use the data screen to change the value at an input image address, will the PLC change the memory as commanded? Explain your answer,

3. How many bits are there in a:
   (a) Word?
   (b) Byte?
   (c) Floating-point number (most common size)?
   (d) Bit?
   (e) ASCII code?

4. Some PLCs allow the use of floating-point numbers. What is a floating-point number?

5. What is a pointer? Is it used in indexed or indirect addressing?

6. Briefly describe the advantages of using symbolic addressing.

7. If your PLC has an analog input module and a full range of signed 16-bit digital values is used to represent a +10 to −10 V dc range:
   (a) What value would represent +10 V?
   (b) What value would represent −10 V?

8. Is all RAM in a PLC addressable? (Yes or no; do not explain.)

For the Type of PLC You Are Learning to Use

9. Data values included in programs are often entered in hexadecimal format. What other formats can we use to represent data when entering constants? (Answer for the PLC that you are learning to use, and give one example of each format.)

10. What is the first available address in the memory area that contains:
   (a) Output image data?
   (b) Timer accumulated values?
   (c) Timer done bits?
   (d) Signed integer words?

11. List the other memory areas, with an example address for each, and indicate what each is intended to store. If the address does *not* represent memory in the CPU module, say so.

12. Fill in the table that follows.
   (a) If the table entry is an address that indicates a digital input or digital output contact, use the right column to describe where in the PLC system you would find the screw contact connected to the sensor or actuator:
   (b) If the table entry is a constant, use the right column to describe what type of data it represents.

(c) If the table entry is any other type of address, indicate what its area of memory is intended to contain.

PLC address/ constant	Data size	Type of memory area/ I/O location/constant type
PLC-5		
I:223/14		
O:020		
B3/20		
N7:20.2		
T4:0		
T4:0.ACC		
T4:0.TT		
F8:6		
5.3		
&H1234		
SLC 500		
I:12/14		
O:10		
B3/20		
N7:20.2		
T4:0		
T4:0.ACC		
T4:0.TT		
M0:2.6		
5.3		
&H1234		
Siemens S5		
F0.1		
FY2		
FW2		
IB4		
Q2.14		
PY4		
12		
C2		
KH12		
KT 12.2		
Siemens S7		
M10.1		
MW10		
IB4		
Q2.14		
PIB4		
>0		
B#16#12		
T#2H5M		
P#M10.0		
OMRON CQM1		
10304		
003		
HR 04		
DM 1304		

If You Are Learning to Use an Allen-Bradley PLC

13. In an Allen-Bradley PLC: B3/4 is the same bit as B3: ___, and B3/___ is the same bit as B3:1.2.

14. There are ___ data words set aside in memory for each counter in an Allen-Bradley PLC. What do the data words contain?

15. There are ___ data words set aside in an Allen-Bradley PLC's memory for each timer. What do the data words contain?

If You Are Learning to Use a Siemens S5 PLC

16. If analog addresses for modules in a Siemens S5 PLC start at 64, what address would you use to write (transfer) a 16-bit value to the first channel of an intelligent I/O module in the second slot? If you wanted to check the status of that I/O module by examining the lowest bit of the first word of data from that module, what bit address would you use?

17. What two data items are represented by the Siemens S5/Westinghouse PC503 address T001?

If You Are Learning to Use a Siemens S7 PLC

18. How many S7 data blocks can be open at one time? What types of data blocks?

19. Where would you declare symbolic names that would only be valid within a single (local) program?

20. What are the complex data types? Use a single sentence to describe what type of data each contains.

21. Where do you declare the structure of an instance data block?

22. What is an "in-out" parameter type as it might be declared in the data declaration section of a function block?

If You Are Learning to Use an OMRON CQM1 PLC

23. Identify three data memory areas in an OMRON CQM1 PLC, indicating the memory area prefix and the type of data that memory area is supposed to hold.

SUGGESTED PLC LABORATORY EXERCISES

For a system with:

- Four control panel switches: inputs 0 to 3
- Four indicator lamps or visible output module LEDS: outputs A to D
- Two spring-return-valve-controlled cylinders: outputs E and F
- One detent-valve-controlled cylinder: outputs G and H
- Three sensors to detect extension of each of the cylinders

1. Clear the program memory of your PLC, then put the PLC in run mode and call up the data screen.
   - **(a)** Use the data monitor screen to display the contents of the input images reflecting the states of all input modules (if using a PLC-5, analog inputs cannot be monitored until after they are configured). Change some switch and sensor positions while watching the data screen. Change the display's data formats.
   - **(b)** While the PLC remains in run mode, try to change the input image bits using the programming unit's data write capabilities. Why can't you change them?
   - **(c)** Display the contents of the output images for the output modules (PLC-5 analog output modules must be configured first). Use the data write feature to change the outputs at the digital output modules while observing the actuators. Change the output values for the analog output module, and use a voltmeter to measure the module's output (for current output, measure the voltage drop across a resistor).
2. Write a program that includes a timer, a counter, and a rung that monitors one data bit to control the output bit controlling cylinder E.
   - **(a)** Use the data monitor to monitor the data values in timer, counter, and data memory while your program is running. Determine which values can be changed from the programming unit.
   - **(b)** Display and change the contents of each of the other types of data memory that your PLC offers. You may have to create data elements (in Allen-Bradley PLCs) or data blocks (in Siemens PLCs) before you can monitor their contents.
   - **(c)** Write a program to move the data word or byte from the control switches to the output lights AND to a data word, but only if cylinder E is extended. Change the switch positions, and use data write to change the state of the data bit that controls cylinder E.
3. Write a program with a timer.
   - **(a)** Include a rung that copies the timer's accumulated value to the third data word in a set of data values. Use indexed addressing with an offset of three for the destination address.
   - **(b)** Include a rung that copies a timer's accumulated value to the fifth memory available for storage of data words. Use indirect addressing for the destination address.

6

MANIPULATING SIMPLE DATA ELEMENTS

OBJECTIVES

In this chapter you are introduced to:

- The use of a microprocessor's accumulators, status word, and other registers in data manipulation operations.
- Allen-Bradley, Siemens, and OMRON instructions (or function blocks) for:
  - Moving single data units (bytes, words, double words, or part words).
  - Comparing single data units in Boolean logic statements.
  - Performing mathematical operations on single data units.
  - Performing logical operations on single data units.
  - Converting data types.
- Indirect and indexed addressing of data in data manipulation instructions.

If you are using a Siemens PLC, you should read Chapter 8 before reading Chapters 6 and 7.

Controlling industrial processes often requires manipulation of data. Operations as simple as adding the accumulated values from two counters may be required to sum the number of good parts made on two assembly lines. Math operations may need to be more complex if, for example, you want to perform statistical process control operations in the PLC in order to calculate whether a manufacturing process is still safely within specifications. A modern PLC can perform the same math operations that any other computer can perform, but the programmer must be careful to ensure that the amount of data manipulation being performed doesn't take too much time, or it will interfere with the PLC's prime job: fast sampling and program execution to control a manufacturing process.

The key to understanding how a PLC manipulates data is to remember that the PLC is only a special-purpose computer. It uses a microprocessor chip like any other computer, but its ROM memory contains programs that allow it to act differently from other types of computers.

MICROPROCESSOR BASICS

A microprocessor is sometimes called a **microprocessor unit (MPU).** It is the central component in a PLC's **central processing unit (CPU) module.** PLCs generally use the same MPU chips that are used in other small computers (i.e., they are often made by Intel, as are the MPUs in IBM-compatible computers; or by Motorola, as are the MPUs in Apple computers and in the PowerPC). PLC manufacturers don't usually identify the make of the MPU they use.

Some PLCs (e.g., Allen-Bradley) can easily be programmed by users who know little about how an MPU works. Other PLC manufacturers (e.g., Siemens and OMRON) believe that a PLC can only be used to its full potential if the user understands how the MPU operates. Even if you intend to use Allen-Bradley PLCs, it is still a good idea to learn a little about how a microprocessor operates, because Allen-Bradley PLCs are increasingly making it possible for the user with MPU familiarity to write better control programs.

A microprocessor contains several types of registers in which it stores information that it uses as it executes programs one step at a time. These registers include:

1. Accumulators store the data words that the MPU is manipulating. PLC programming languages typically allow the direct use of only one or two accumulators. The number of bits that an accumulator can hold dictates the maximum-size data element that the PLC can manipulate at one time. The Allen-Bradley PLC-5 and SLC 500 PLCs have 16-bit accumulators, as do the Siemens S5 PLC and the OMRON CQM1. The Siemens S7 line of PLCs have 32-bit accumulators.

2. Status registers contain status bits which are affected by data that the microprocessor manipulates. Some PLC data manipulation instructions automatically copy status register bits into addressable memory so that the user-program can examine them. Status bits typically available for the PLC programmer's use include:

(a) The **zero** status bit, which is set if the most recently manipulated number was (or became) a zero.

(b) The **sign** bit, which is set if the most recently manipulated number had a "1" in its most significant bit (MSB) position. In signed binary notation, a "1" in the MSB indicates a negative number.

(c) The **carry** bit, which is set if the most recent math operation generated a binary number with more bits than the microprocessor could hold (so the result is an incorrect answer).

(d) The **overflow** bit, which will be set whenever an attempt to increase the value of a number resulted in its most significant bit changing to a "1" or if decreasing a value caused its most significant bit to change to a "0." This would indicate an unintentional sign change, such as when 1 is added to the 16-bit signed binary number 32,767, which would result in the 16-bit signed binary number representing −32,768.

(e) The **result of logic operation (RLO)** bit, which a PLC uses as it works its way through complex Boolean logic instructions (true is 1 and false is 0). In the S7 PLC, the RLO status bit is accessible to the program. The S7 also allows the programmer to address a **binary result (BR)** status bit, also called the **ENO** bit, which indicates the success (1) or failure (0) of a subprogram to execute. The OMRON CQM1 stores RLO bits in its *truth register.*

3. Address registers (sometimes called **offset registers**) are used by the MPU to remember the starting addresses of the various sections of memory it is currently using, including the user-program memory area and the various addressable data memory areas. Increasingly powerful and complex memory-access techniques require additional address registers.

DATA MANIPULATION INSTRUCTIONS

Although most memory locations can be manipulated one bit at a time using Boolean instructions, modern PLCs now offer a far more powerful set of instructions. These more powerful instructions allow a program to manipulate whole multiple-bit data elements, as single-element data types (i.e., as bytes, characters, 16-bit data words, floating-point numbers, time values, etc.). In this chapter we describe what you can program a PLC to do, but provide detailed explanations for a only few examples of typical data manipulation instructions. Refer to your programming manuals for more detailed descriptions of your PLC's complete instruction set.

As we examine data manipulation instructions, remember that a PLC operates by repeatedly executing a three-step scan cycle. Remembering this characteristic makes it easier to understand the effect that data manipulation instructions have on memory and on processes that the PLC is controlling. The three-step scan cycle again:

1. Read input values from input modules into input image memory.
2. Execute the user-program one instruction at a time from start to end.
3. Copy data from the output image area of memory to output modules.

SIMPLE DATA ELEMENTS

Bytes, words, and **double words** are individual simple data elements.[1] Binary values stored as "simple" data elements may represent signed or unsigned integer values, real numbers, timer/counter accumulated values, or may be specially coded data units for use as characters, as timer constants, as pointers for indirect addressing, or for some other purpose. A multibit data element may, alternatively, only represent the on–off states of a set of digital sensors or digital actuators. Despite what a binary data element may represent, to the PLC it is just a binary number, and it can be manipulated by the instructions described below.

Moving Simple Data Elements

We have already seen some examples of PLC instructions for moving individual data values; they were used in Chapter 5 to show the addressing of data elements. A MOVE instruction might be used, for example, to copy a 16-bit value from working data memory to an output image data word, thereby controlling the state of 16 digital actuators with a single instruction.

Some MOVE instructions are *conditional,* which means that they execute only when the Boolean logic preceding the MOVE instruction is true. In other PLCs, MOVE instructions are executed *unconditionally,* which means that the PLC will always execute the move, even if the MOVE instruction is programmed with control logic that is false.

---

[1] A bit is also a simple data element, but bit manipulation was covered in Chapter 3.

In general, a MOVE instruction *copies* data from a memory location to another memory location of the same size. Being a computer, a PLC *must actually copy a data value twice* to move it, once from its source into the microprocessor's accumulator register, then from the accumulator to the destination address. Ladder logic instructions make this two-step operation look like a single step, but some PLC programming languages (such as Siemens' STL language and the instruction list (IL) language described in the IEC 1131-3 standard) require the programmer to program the two copy operations separately. These languages also allow the programmer to perform operations on the data value after it is copied into the microprocessor's accumulator, before copying it to its destination address. Operations that can be performed to change a data value are covered in the following sections.

Several PLCs offer ladder logic MOVE instructions that move only parts of binary data values, leaving part of the destination address unaffected by the move. Some PLCs also offer MOVE instructions that can convert data values from one data type to another during the move.

In Chapter 7 we discuss more powerful instructions for manipulating whole sets of simple data elements,

PLC-5

SLC 500

ALLEN-BRADLEY MOVE Instruction

In the Allen-Bradley PLC-5 and the SLC 500, the **MOVE** (MOV) instruction (shown in Figure 6.1) can be used to copy a source value (a constant, or the contents of a memory containing a 16-bit data word, or a 32-bit floating-point element) to a destination address. When moving a 16-bit number to/from a floating-point-number element, the MOV instruction converts the number between signed 16-bit binary and 32-bit floating-point number format. Decimal parts of floating-point numbers are rounded to integer values. Figure 6.1 shows the MOV instruction programmed to execute unconditionally, but Boolean logic can be programmed to the left of the MOV instruction, making the MOV conditional.

Both Allen-Bradley PLCs offer a **MASKED MOVE** (MVM) instruction that can be used to copy *part of a 16-bit number* to a destination address without changing the rest of the destination address's contents. A *mask* must be specified as a hexadecimal constant (as in Figure 6.2) or as an address from which the 16-bit mask value is to be retrieved. The 16-bit mask must contain a 1 in each bit position in which a bit from the source will overwrite a bit in the destination address. In the example, the constant hexadecimal mask 00FF has 1's in the low 8 bits only, so only the contents of the low 8 bits of the destination address are overwritten with the bit values from the low 8 bits of the source address.

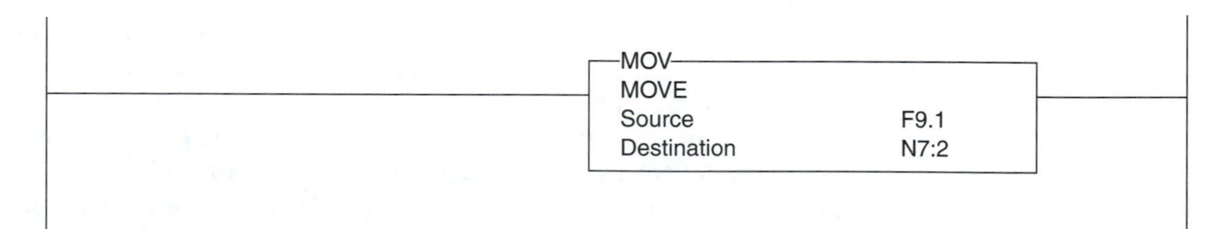

FIGURE 6.1
Allen-Bradley MOVE instruction.

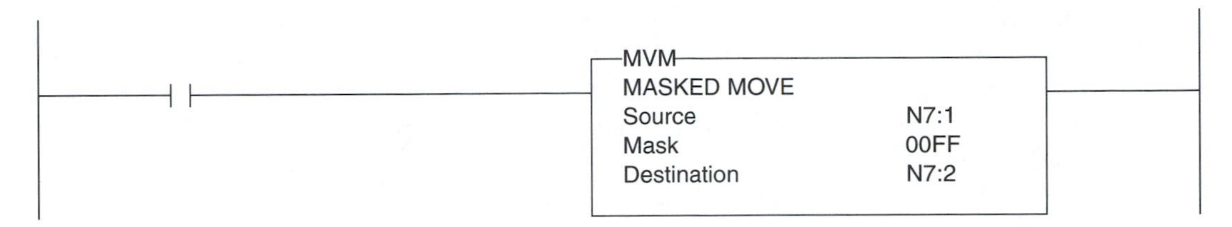

FIGURE 6.2
Allen-Bradley MASKED MOVE.

The PLC-5 also has a **BIT FIELD DISTRIBUTE** (BTD) instruction that allows the programmer to copy a series of consecutive bits from one 16-bit data source address to anywhere in a 16-bit data destination address. The programmer specifies how many consecutive bits to move, the starting bit number in the source address, and the starting bit number in the destination address.

The PLC-5 and SLC 500 PLCs have status bits[2] that are affected by the resultant number in the destination address after a MOV, MVM, or BTD.

S5

SIEMENS STEP 5 MOVE Instructions (with LOAD and TRANSFER)

Before learning about the data manipulation capabilities of the Siemens S5, you should know how to call STEP 5 programs from other programs, and you should know about function block (FB) programs. Some instructions can only be programmed in function blocks, and some addresses can only be accessed by programs in function blocks. (You should have read Chapter 8 before you started this chapter.)

STEP 5 user-programs can be written in ladder logic or in statement list (STL) language. STL is closer to the machine-level language that the PLC's microprocessor actually uses. All ladder logic instructions have STL equivalents, often consisting of several STL statements per ladder logic instruction. In fact, ladder logic instructions are always translated into their STL equivalents by the programming unit before they are sent to a PLC's memory. Some STL instructions perform operations that cannot be programmed in ladder logic.

The STL **LOAD** (L) and **TRANSFER** (T) instructions and their variations described in this section are not conditional! They are executed whenever the PLC encounters them in a program, regardless of the state of the result of logic operation (RLO). You must use structured programming techniques if you need to make LOAD and TRANSFER instructions execute conditionally. You can program a conditional ladder logic MOVE instruction in a function block, but the programming unit automatically translates it into a structured STL program containing unconditional LOAD and TRANSFER instructions.

Except in a function block, the STEP 5 ladder logic **MOVE** instruction is a direct equivalent of an STL **LOAD** (L) instruction followed by an STL **TRANSFER** (T) instruction, as shown in

---

[2] S:0/2 (the zero bit) or S:0/3 (the sign bit) could be set depending on the value in the destination address after the move. S:0/1 (the overflow bit) will be set if the floating-point number is too big to convert to an integer. S:0/0 (the carry bit) is cleared by a move instruction.

the program in Figure 6.3. The number being loaded can come from an address (e.g., from IW2 in Figure 6.3) or may be a constant value. The STL example also shows the data values that are being manipulated, as they would appear (in hex) on a programming unit's monitor as the program was being monitored. The LOAD instruction copies a number ("B500") into accumulator 1 of the MPU, pushing the old contents of accumulator 1 (shown as "xxxx") into accumulator 2. The old content of accumulator 2 (shown as "YYYY") is lost. The TRANSFER instruction then copies data from accumulator 1 to the memory address specified with the instruction (FW4) without affecting the contents of accumulator 1 or 2. (This example contains a potential flaw, which we discuss next.)

The LOAD and TRANSFER instructions (or the MOVE instruction) must include the address of the memory location being read from or written to, or the constant value being loaded. Remember that most addressable memory locations in an S7 PLC contain only *one byte* (8 bits). If a LOAD or TRANSFER instruction uses an address specification for a byte, the value is copied between that address and the *low* byte of accumulator 1. If, on the other hand, a LOAD or TRANSFER instruction uses an address specification for a word (16 bits), then 8 bits will be copied between the *high* byte of the 16-bit accumulator and the address whose number is specified in the instruction, and an additional 8 bits will be copied between the low byte of the accumulator and the next memory location (if any).

> The program in Figure 6.3 is running in a medium-sized (e.g., S5-100U) PLC. It uses word addressing, so the LOAD instruction gets the 8-bit value (hex: B5) from IB2 for the high byte of the accumulator, and the accumulator's low bits are made zeros because a Siemens S5-100U digital input module can provide only 8 bits of data. The TRANSFER instruction then copies the accumulator's high byte to the 8-bit memory location FY4 and copies the zeros from the low 8 bits of the accumulator to FY5, overwriting any useful data that was there. This is obviously a potentially faulty program.
>
> A better program would have addressed the 8-bit memories as bytes, as in Figure 6.4. This program LOADS the 8-bit input image value (B5) from IB2 into the accumulator's *low* byte and clears the high byte (the high byte is always cleared when

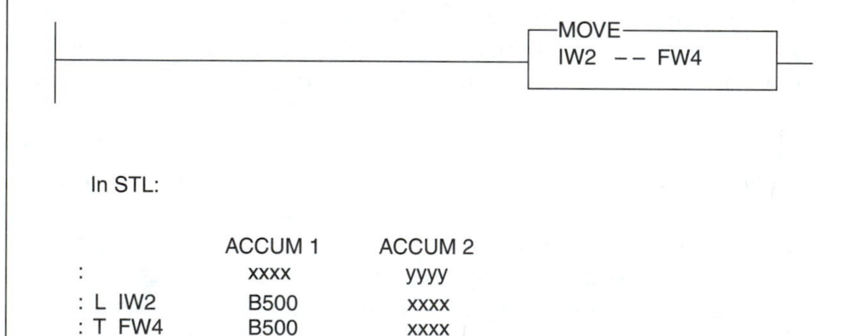

FIGURE 6.3
STEP 5 ladder logic MOVE instruction, with STL equivalent showing accumulator contents as the program executes.

```
                         ACCUM 1      ACCUM 2
:                        xxxx         yyyy
: L   IB2                00B5         xxxx
: T   FY4                00B5         xxxx
```

Memory	FY4	FY5
Before program execution	aa	bb
After program execution	B5	bb

FIGURE 6.4
Improved STL program for manipulating bytes.

a byte is loaded), then TRANSFERS the *low* byte from accumulator 1 to FY4, leaving FY5 unaffected.

The faulty program could have been worse. Bytes and words could have been mixed. In Figure 6.5a, the input image data is LOADed as a word, so it goes into the accumulator's high byte. The TRANSFER instruction uses byte addressing, so the zeroed low bits of the accumulator are written to FY4 and the valid data isn't copied anywhere. In Figure 6.5b, the input image byte is LOADed as a byte, so it goes into the low 8 bits of the accumulator, then it is TRANSFERred as a word. The cleared 8 high bits are transferred to FY4 and the valid data is sent to the wrong address, FY5.

If a LOAD or TRANSFER instruction addresses data in a data block, the STL program must include a third type of STL instruction: A **CALL DATA BLOCK** (C) instruction must precede the LOAD and/or the TRANSFER instruction, as shown in Figure 6.6. In ladder logic, the data block call is included in the ladder logic instruction, so the address must specify the data block and the data word. A ladder logic MOVE instruction with the source address DB4:DW5 is equivalent to an STL program that calls data block 4, then loads data word 5.

The **LOAD IN BCD** (LD) instruction can be used in place of the LOAD (L) instruction in STL, but only if the instruction is loading a value from a timer or counter memory. This instruction causes the PLC to translate the natural binary value in the low 10 bits of the timer or counter's address into BCD in the low three nibbles (and clear the top nibble) of the accumulator. A MOVE BCD ladder logic instruction consists of a LOAD IN BCD, then a TRANSFER instruction.

The **SWAP ACCUMULATOR CONTENTS** (TAK) instruction exchanges the data values in accumulators 1 and 2. It can only be programmed in STL, and only in a function block.

LOAD and TRANSFER instructions (and MOVE instructions) can read/write the **system area of memory (RSn)** only if they are programmed in function blocks. In smaller and medium-sized S5 PLCs, LOAD and TRANSFER instructions using direct addressing of **peripheral memory (PWn, PYn)** are allowed only in an interrupt program (OB2 or OB13; see Chapter 11).

```
                  ACCUM 1    ACCUM 2
:                 xxxx       yyyy
:  L   IW2        B500       xxxx
:  T   FY4        B500       xxxx
```

Memory	FY4	FY5
Before program execution	aa	bb
After program execution	00	bb

(a)

```
                  ACCUM 1    ACCUM 2
:                 xxxx       yyyy
:  L   IB2        00B5       xxxx
:  T   FW4        00B5       xxxx
```

Memory	FY4	FY5
Before program execution	aa	bb
After program execution	00	B5

(b)

FIGURE 6.5
Siemens load and transfer instructions mixing word addressing and byte addressing.

Moving Data Using Indexed and Indirect Addressing in STEP 5

The addressing techniques for the following instructions have been described in detail in Chapter 5 in the section on indexed addressing in STEP 5.

Indexed addressing can be used with the standard STL-language LOAD and TRANSFER instructions if the LOAD or TRANSFER instruction follows a **PROCESSING OPERATION** (DO) instruction that specifies an offset value and if the LOAD or the TRANSFER instruction specifies the memory area that the data is to be LOADed from or TRANSFERred to.

Special LOAD and TRANSFER instructions can use indirect addressing in the S5 PLCs. A 16-bit pointer value must be loaded into accumulator 1; then a **LOAD REGISTER INDIRECT** (LIR) instruction can be used to load a number from the address being "pointed" to into accumulator 1 or 2, or the **TRANSFER REGISTER INDIRECT** (TIR) instruction can be used to transfer a number from an accumulator to the address being pointed to. These instructions can only be programmed in STL, and only in function blocks.

FIGURE 6.6
Moving data between two Siemens data blocks.

```
C  DB4
L  DW5
C  DB6
T  DW7
```

The standard STL-language LOAD and TRANSFER instructions can also be used with indirect addressing of data if they follow a **PROCESS INDIRECTLY** (DI) instruction, but only in function block programs, and only if a set of possible addresses has been passed to the function block as parameters.

S7

SIEMENS STEP 7 MOVE Instructions (with LOAD and TRANSFER)

A basic understanding of STEP 7 structured programming is necessary to understand STEP 7 MOVE instructions. A STEP 7 structured user-program is made up of one or more main programs, called **organization blocks (OBs),** which can include instructions to call **function block (FB) programs** or the slightly simpler **function (FC) programs.** (You should have read Chapter 8 before reading this chapter.)

The *ladder logic* **MOVE** instruction consists of the equivalent STL language **LOAD** instruction (which copies a number into accumulator 1) followed by a **TRANSFER** instruction (which copies the number from accumulator 1 to a memory location). The LOAD and TRANSFER instructions execute unconditionally unless a structured program causes the PLC to bypass them.

Siemens has not made any significant changes to the MOVE instruction, or to its STL-equivalent LOAD and TRANSFER instructions for the STEP 7 programming language. The previous discussion describing STEP 5 for S5 PLCs also applies to STEP 7 for S7 PLCs, except as follows:

1. S7 PLCs have 32-bit accumulators, which otherwise act like the S5's 16-bit accumulators. Figure 6.7 shows LOAD instructions with *three* different source address size specifications:

- **(a)** Addressing memory as *bytes* causes the PLC to copy 8-bit data values between that address and the lowest 8 bits of accumulator 1. Loading a byte value causes the upper 24 bits of accumulator 1 to be cleared. Transferring a byte writes accumulator 1's lowest byte to the specified byte of memory.
- **(b)** Addressing memory as *words* in a LOAD or TRANSFER instruction causes the PLC to copy two bytes of data between the specified address and the next address, and the lowest 16 bits of the accumulator 1. Loading a word value causes the upper 16 bits of accumulator 1 to be cleared.
- **(c)** Addressing memory as *double words* causes the PLC to copy a 32-bit data value between the specified address and the next three addresses, and the 32-bit accumulator.

2. In addition to STEP 5's **SWAP ACCUMULATOR CONTENTS** (TAK) instruction (called **TOGGLE ACCUMULATOR** in STEP 7), there are four additional instructions that move data between accumulators only. **Push (PUSH)** copies the contents of accumulator 1 to accumulator 2, without changing accumulator 1. **Pop (POP)** copies accumulator 2 into accumulator 1 without

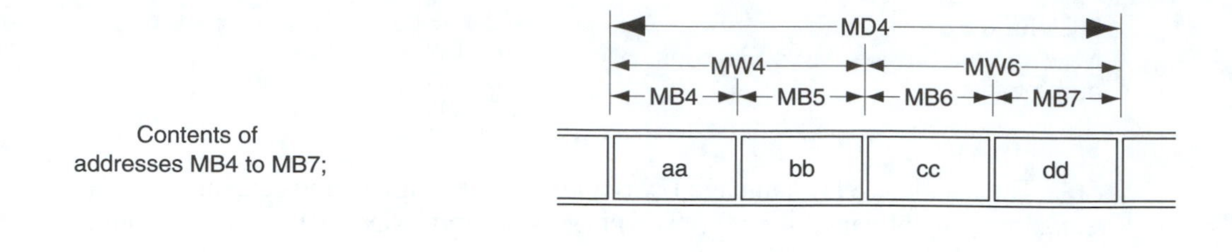

This instruction	Puts this value into accumulator 1: 31–24	23–16	15–8	7–0 (bit numbers)
L MB4	00	00	00	aa
L MW4	00	00	aa	bb
L MD4	aa	bb	cc	dd

FIGURE 6.7
Effects of addressing S7 memory as bytes, words, or double words.

changing accumulator 2. There are two CHANGE BYTE SEQUENCE instructions: **CAW** switches the two bytes of the low word in accumulator 1, and **CAD** reverses the order of all four bytes of the double word in accumulator 1. (After CAD executes, the lowest byte is switched with the highest, and the second lowest byte is switched with the second highest.)

3. The STL instruction LOAD IN BCD, which loads a timer or counter value and translates it into BCD in the process, is now **LC,** not LD.

4. Memory can be addressed using **symbolic names,** and in additional indirect addressing methods now called **memory indirect addressing** and **register indirect addressing** (explained in Chapter 5) as well as in direct (absolute) addressing format. STEP 5's rudimentary indirect addressing using the LIR and TIR instructions is no longer available.

5. Additional STL instructions are available for loading 32-bit pointer values into the MPU's address registers, or for transferring pointer values from the address registers. **LAR1** or **LAR2** are used to load a pointer into address register 1 (AR1) or into address register 2 (AR2). The pointer can be entered as a constant with the LAR instruction, or the LAR instruction may include an address specification if the 32-bit pointer is in memory, or the parameter "AR2" may be used with the LAR1 instruction if the pointer from AR2 is to be loaded into address register

1. Unless the instruction specifies another source, LAR1 and LAR2 use accumulator 1 as the source of the pointer.

The **TAR1** or **TAR2** instructions copy a pointer from AR1 or AR2 into a specified memory location, or into accumulator 1 if an address isn't included with the TAR1 or TAR2 instruction. The **CAR** instruction exchanges the pointers in AR1 and AR2.

In the section on Siemens STEP 7 math/logic instructions later in this chapter, we examine the **+AR1** and **+AR2** instructions, which can be used to increase or decrease pointer values in AR1 and AR2.

6. The number and size of the currently active data blocks can be read and loaded into accumulator 1.

`L DBNO`	loads the shared data block number
`L DBLG`	loads the shared data block size (in bytes)
`L DINO`	loads the instance data block number
`L DILG`	loads the instance data block size (in bytes)

CQMI

OMRON CQM1 MOVE Instructions

Several CQM1 programming instructions can be used to copy data from one memory location (or from a constant) to another memory location, but only three of them are intended to copy single data values without changing them. The instructions for changing data as it is being moved are discussed elsewhere in this chapter.

The **MOVE** instruction, **MOV (21),** shown in Figure 6.8, copies a source data word to a destination address whenever its control logic is true. The source is entered as the first parameter in the MOVE instruction, and can be any 16-bit constant or any address specification indicating where the number is to be copied from (in Figure 6.8, the constant #1234 is used.) The second parameter indicates the destination, and can be any address except in the TC area of memory (in Figure 6.8, an output image word address is specified).

Indirect addressing can be used with the MOV (21) instruction, as described in Chapter 5 in the section on OMRON CQM1 indirect addressing.

The other two **MOVE** instructions can be used to do the same work as the MOV (21) instruction, but they allow the use of indexed addressing to specify the source or the destination address in moving a single 16-bit data word. The SINGLE WORD DISTRIBUTE instruction, **DIST (80),** can copy a single 16-bit word to an indexed destination address, and the DATA COLLECT in-

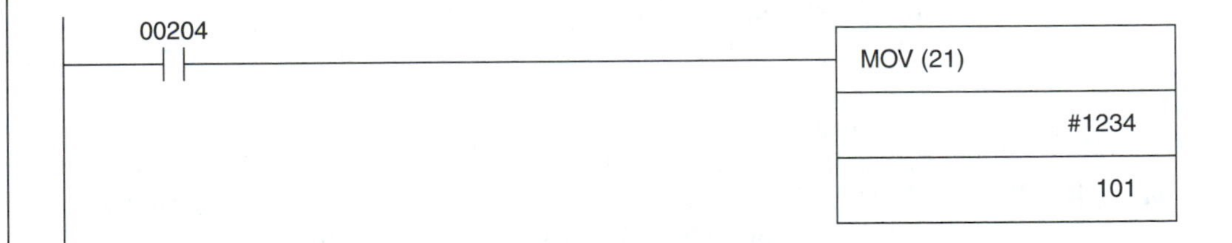

FIGURE 6.8
CQM1 MOVE instruction.

struction, **COLL (81),** can copy a data word from an indexed source address. In the section on OMRON CQM1 indexed addressing in Chapter 5 we describe how these instructions use indexed addressing. How these two instructions can be used for more than just single-word moves is described in Chapter 7.

Comparing Simple Data Elements

Most PLCs offer instructions that can compare two simple data values or compare a data value to a constant. The two data values are typically compared to see if they are of equal size or to see if one is greater than the other. The result of the COMPARE instruction is usually true or false and can be used in a Boolean logic statement that controls another PLC instruction. For example, two counter accumulated values, reflecting the number of items on separate conveyors, can be compared and the result can be used to control an output image bit so that an actuator pushes parts onto the conveyor with the fewest items.

PLC-5

SLC 500

ALLEN-BRADLEY COMPARE Instructions

Allen-Bradley's COMPARE instructions yield Boolean results. They can be included with other Boolean logic elements on ladder logic rungs that control a rung's output instruction. A rung may contain several COMPARE instructions, allowing the construction of Boolean logic statements with complex comparison requirements.

Allen-Bradley COMPARE instructions allow 16-bit *signed* binary numbers and/or floating-point numbers from memory to be compared to each other or to be compared against constant values. Constants can be entered in decimal, or in hexadecimal, binary, or octal with a prefix identifying its form (&H, &B, or &O, respectively).

When using floating-point numbers in CMP instructions, remember that rounding errors often occur during computerized mathematical operations with floating-point numbers, so that an expression as simple as (F9:1 = 0) might not actually be true when it should be.

PLC-5 COMPARE Instruction

The PLC-5 allows programmers to write their own logical statements. The expression is written in a **COMPARE** (CMP) instruction. The more recent enhanced line of PLC-5s even allows a logical expression to include brackets and mathematical expressions for manipulation of word values *before* the comparison operation. (Appendix D includes a complete list of operators that can be used. Note that the "|" character is used for dividing.) Brackets in the COMPARE expression cannot be used to build complex comparison statements. Only one comparison can be performed per CMP instruction.

Figure 6.9 demonstrates how a typical CMP instruction might be used. Note that the brackets force the mathematical operations (division of N7:1 by 4) and (summing of the 16-bit words in N7:1 and N7:3) to be performed before the logical comparison of the two results. If the 16-bit signed integer result of the division is greater than the 16-bit signed integer value result of the summing operation, the CMP instruction's result is a Boolean TRUE. The CMP instruction in this example is part of a complex Boolean logic statement (its result is Boolean

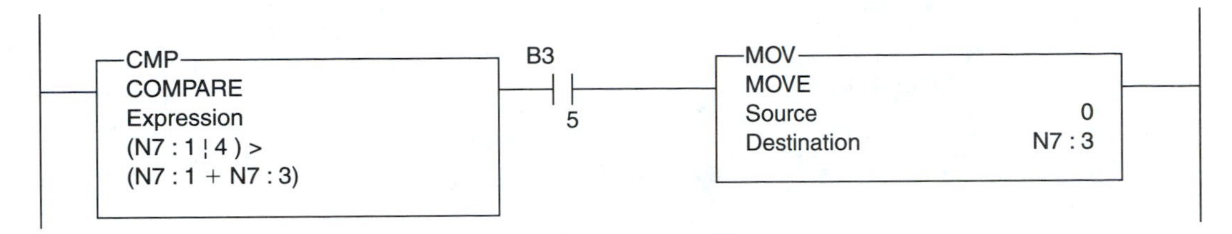

FIGURE 6.9
A PLC-5's COMPARE instruction as part of a statement controlling a move.

ANDed, with the result of examining bit B3/5), and the final Boolean logic result controls an output element (the MOVE instruction).

COMPARE expressions can include Boolean logical statements that perform the same logical operations than can be written in Ladder Logic. Operators such as:

> for "greater than"
= for "equal to"

can be used.

The boolean logical operators cannot be combined with operators which perform word logic. Operators such as

#not# to perform NOT operation on all bits of a data word
#AND# to perform an AND operation on all bits in two data words

cannot be used in the COMPARE (CMP) instruction expression.

The COMPARE (CMP) instruction takes significantly longer to execute than the simpler compare instructions explained below.

Other Allen-Bradley Comparison Instructions

SLC 500 PLCs do not yet offer the COMPARE (CMP) instruction. They require the programmer to use the older, simpler comparison instructions, which can only compare two source input values. Math and word-logic operations cannot be done in these compare instructions. The PLC-5 also allows the use of the simpler compare instructions.

The simple comparison instructions, used as shown in Figure 6.10, include six simple comparisons:

- **EQUAL** (EQU): true if the two source values are equal
- **NOT EQUAL** (NEQ): true if the two source values are not equal
- **LESS THAN** (LES): true if source A is less than source B
- **LESS THAN OR EQUAL** (LEQ): true if source A is less than or equal to source B
- **GREATER THAN** (GRT): true if source A is greater than source B
- **GREATER THAN OR EQUAL** (GEQ): true if source A is greater than or equal to source B

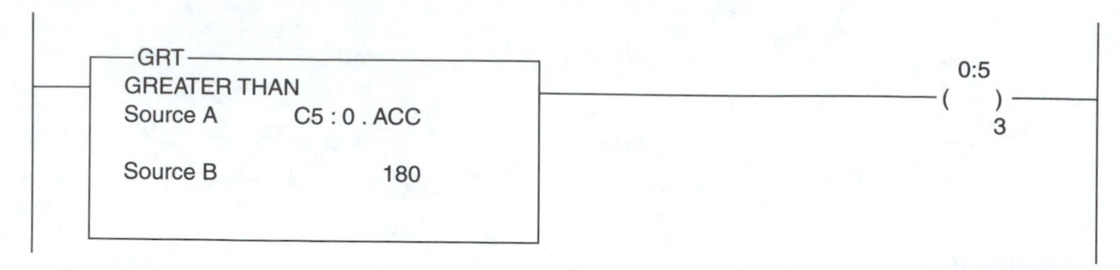

FIGURE 6.10
Allen-Bradley comparison instruction, shown controlling an output energize instruction.

Two slightly more powerful instructions are also available:

- **LIMIT TEST** (LIM): true if the test value is within the range entered as a "low limit" and a "high limit."
- **MASKED COMPARISON OF EQUAL** (MEQ): true if selected bits of the source 16-bit number are all equal to the same bits of the compare 16-bit number. A 16-bit mask indicates which bits to compare. Bits containing a 1 in the mask, entered as a hexadecimal constant or as a memory address reference, are compared.

S5

SIEMENS STEP 5 COMPARE Instructions

Siemens' compare instructions compare the value in accumulator 2 against the value in accumulator 1, treating both values as 16-bit *signed* binary numbers. A LOAD instruction puts a value into accumulator 1 and shifts the value previously in accumulator 1 into accumulator 2, so a pair of LOAD instructions will fill both accumulators.

The result of a STEP 5 comparison instruction is either true or false, and compare instruction can be used anywhere that other Boolean logic input instruction can be used in conditional statements to affect the RLO (to affect whether a conditional output instruction will execute). Compare instructions also affect two of the condition code bits (CC0 and CC1) in the MPU's status register. For more powerful programming, CONDITIONAL JUMP instructions can be used to check the CC0 and CC1 bits following a COMPARISON instruction. (Conditional jumps are discussed in Chapter 8.)

Figure 6.11 shows a conditional statement (in ladder logic and in STL) which loads two values, compares to see if the first (from IW64) is greater than the second (from data word 3 of data block 4), and turns an output on if it is.

Values can be compared in ladder logic or in STL to see if the first-loaded value is:

!=F equal to the second-loaded value
<>F not equal to the second-loaded value
>F greater than the second-loaded value
>=F greater or equal to the second-loaded value
<F less than the second-loaded value
<=F less or equal to the second-loaded value

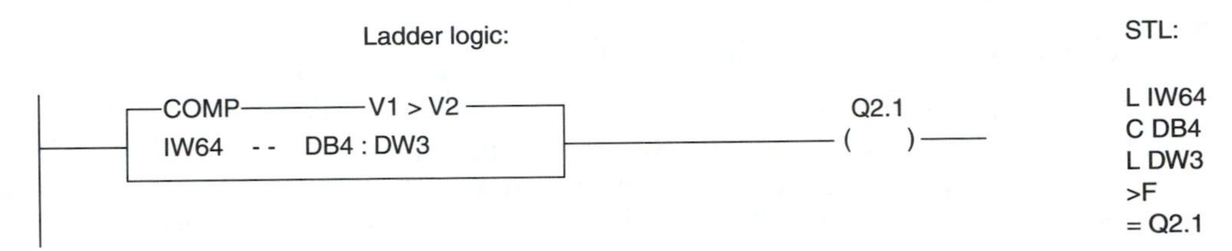

FIGURE 6.11
STEP 5 comparison instruction in a conditional statement.

The "F" in the STL compare instructions listed above is a reminder to the programmer that the PLC will treat the 16-bit numbers in the accumulators as signed binary notation representing (signed) ***F***ixed-point integer values.

S7 SIEMENS STEP 7 COMPARE Instructions

STEP 7 offers the same set of six comparison operators that are offered in STEP 5. STEP 7 changes include:

1. The "equals" operator has been changed from "!=" to **"=="**, and the "not equals" has been changed to "<".

2. The symbol "**I**" is used instead of "F" in the comparison instructions, to compare 16-bit signed fixed-point *I*ntegers (e.g., ==I is the instruction to compare the two 16-bit signed binary values in accumulators 1 and 2 to see if they are equal).

3. Double words and floating-point numbers can be compared. A "**D**" replaces the "I" (e.g., ==D) if double words (signed 32-bit binary numbers representing signed integers) are to be compared. An "**R**" replaces the "I" (e.g., >R) if floating-point numbers (representing real numbers) are to be compared. *Beware:* Quirks in the floating-point-number system sometimes make floating-point numbers that should be equal to each other unequal! It is always better to compare to see if a floating-point number is significantly larger or smaller than another floating-point number (or a constant) rather than comparing to see if they are equal.

4. The ladder logic instruction appearance has been changed slightly, as shown in Figure 6.12, to conform to IEC 1131-3 (which we examine in Chapter 9). The CMP instruction can be connected into Boolean logic rungs via the top connecting lines at the block's left and right. The addresses or constant values that will be compared are entered outside the block to the left of IN1 and IN2. IN1 represents the value that will be placed into accumulator 2, and IN1's value will go into accumulator 1 before the comparison is executed. The result of the statement in Figure 6.12 will be true if IN1 is greater than IN2.

FIGURE 6.12
STEP 7 ladder logic COMPARE instruction.

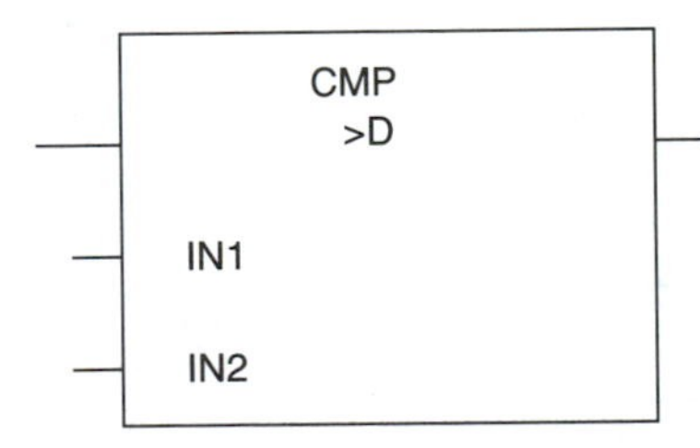

CQMI

OMRON CQM1 COMPARE Instructions

Unlike COMPARE instructions in the other PLC programming languages covered in this book, the CQM1's **COMPARE** instruction, **CMP (20),** is considered to be an *output* instruction. It must be the rightmost instruction on a rung, and is executed only if the Boolean control logic statement to its left is true. If executed, it affects bits in the SR area of memory. It does not generate a true or false result that can be used directly in a Boolean logic statement; the SR memory bits that it affects must be examined after the comparison.

As shown in Figure 6.13, a CMP (20) instruction compares a *first compare value* (Cp1, the first parameter) against a *second compare value* (Cp2, the second parameter) each time the CMP (20) instruction executes. The two compare values specified in entering the instruction can be addresses of memory locations containing values or can be constants entered as part of the compare instruction. Values are always treated as unsigned binary unless one value is a timer or counter present value, in which case both values are assumed to be in BCD format. When a CMP (20) instruction executes, it turns one of the following three status flags on (and turns the others off). The three status flags in the status register (SR) area of memory include:

1. ***25505*** *(the GR flag):* on if first compare value > second compare value
2. ***25506*** *(the EQ flag):* on if first compare value = second compare value
3. ***25507*** *(the LE flag):* on if first compare value < second compare value

Subsequent instructions can examine the status bits, as shown in Figure 6.13, where an output image bit (10300) is turned on if counter 8's present value (CNT 8) is over 5. Note in this example that the output (10300) can go on only if the logic that causes the CMP (20) instruction to execute is true. This construction prevents other instructions, which might affect SR 22505, from accidentally causing output 10300 to actuate.

The CMP1 offers some additional instructions for comparing other types of binary numbers. Like the CMP (20) instruction, they are all considered output instructions that execute when their Boolean logic control statements are true, and they all affect the GR, EQ, or LE flag bits.

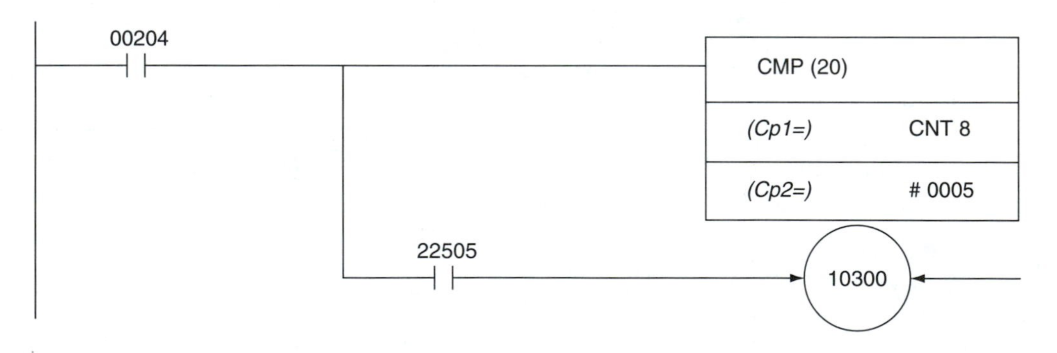

FIGURE 6.13
CQM1's COMPARE instruction, shown conditionally executed and with a branch that uses the result.

1. The **DOUBLE COMPARE** instruction, **CMPL (60),** compares two *32-bit unsigned* binary values (or BCD values). The Cp1 and Cp2 parameters must be addresses, and each indicates where the low 16 bits of the 32-bit numbers are. The high 16 bits come from the addresses above the addresses entered as Cp1 and Cp2.

2. The **SIGNED BINARY COMPARE** instruction, **CPS(−),** and the **DOUBLE SIGNED BINARY COMPARE** instruction, **CPSL (−),** are similar to the CMP (20) and CMPL (60) instructions except that the 16- or 32-bit numbers are interpreted as *signed* binary numbers. OMRON requires the programmer to enter a third parameter, which must be 000. The 32-bit version does not allow the parameters to be entered as constants.

3. An **AREA RANGE COMPARE** instruction, **ZCP (−),** is used to set the GR, EQ, or LE status bits depending on whether the compare data value (CD, first parameter) is over/in/under the range specified using a lower limit value (LL, second parameter) and an upper limit value (UL, third parameter). A **DOUBLE AREA RANGE COMPARE** instruction, **ZCPL (−),** works the same way but compares 32-bit numbers, which can be entered only as address specifications.

Math, Logic, and Conversion Operations on Simple Data Elements

PLCs are offering increasingly powerful instructions for performing mathematical operations and logical operations to change whole data words (unlike Boolean instructions, which affect only individual data bits). Early PLCs could add two data values, but instructions for subtraction, multiplication, and division are more recent, and trigonometric functions, logarithms, and so on are now available. Instructions for performing logical AND and OR operations on whole data words were added, followed by other word-logic instructions. PLC manufacturers soon began to wish they had simply allowed programmers to write their own mathematical/logical formulas instead of adding separate instructions for each operation, so some PLCs now offer a "compute"-type instruction that does allow the programmer to enter formulas, making better use of the PLC, which is, after all, a computer.

For those readers who are not familiar with Boolean logic at the word level, the NOT, AND, OR, and Exclusive OR word-logic operations are discussed here.

1. NOT inverts the state of each bit in a binary data element. A NOT instruction could, for example, get a data word from the output image table, perform a NOT operation on it, then put the result back into the same address. The result would be to turn all actuators off that were on, and to turn all actuators on that were off:

```
NOT  1111_0000_0000_1010
=    0000_1111_1111_0101
```

2. AND operations combine two binary data elements to generate a third binary element. Each bit of one source value is ANDed with the corresponding bit of the other source value, with the result placed into the corresponding bit of the result. A result bit is 1 only if *both* corresponding source bits are 1's (i.e., if source A is true AND source B is true). ANDing is often used to mask part of a data word off! One source value might, for example, come from an output image data word, the other source might be a constant (called the mask) supplied with the instruction, and the result might go back to the original output image data address. The programmer can use the mask constant to force some bits of the output image off (0's in the mask) without affecting the other bits (1's in the mask).

```
     1111_0000_1010_0101   (source A)
AND  0011_0011_0011_0011   (source B)
=    0011_0000_0010_0001   (result)
```

3. OR operations combine two binary data elements. The result data value will have a 1 in each bit position where *either* source value A **OR** source value B had a 1. (If two 1's are ORed, the result is still 1). OR operations can be used in masking operations to force some result bits on (where the mask constant has a 1) while leaving the other bits of the source alone (where the mask has 0's).

```
   1111_0000_1010_0101   (source A)
OR 0011_0011_0011_0011   (source B)
=  1111_0011_1011_0111   (result)
```

4. Exclusive OR operations are similar to OR operations except that a result bit will only be 1 if either, *but not both,* of the source values had 1's in that position. Exclusive OR operations are not used in control applications as frequently as are AND or OR.

```
             1111_0000_1010_0101    (source A)
Exclusive OR 0011_0011_0011_0011    (source B)
=            1100_0011_1001_0110    (result)
```

ALLEN-BRADLEY MATH, LOGIC, AND CONVERSION INSTRUCTIONS

Allen-Bradley COMPUTE Instruction

The PLC-5 and better SLC 500s offer a COMPUTE (CPT) instruction. In a CPT instruction, the programmer can enter a *mathematical/logical expression* containing memory references and/or constants (with the appropriate prefixes). The programmer must also enter an address destination for the floating-point or integer result. The CPT expression can perform four-function math and trigonometric and logarithmic operations and can perform logical operations on whole data words. The CPT instruction can also convert numbers between 16-bit signed integer and floating point, between binary and BCD numbers, and between radian and degree values. As shown in Figure 6.14, the CPT instruction is considered an output instruction in that it executes if its (optional) control logic statement is true. Figure 6.14 shows a math/logic expression that ANDs a constant with the value from address N7:1, then multiplies the result by the value in address F9:1 after increasing the value from F9.1 by 3. The final result is stored in an integer file element (N7:100). Appendix E includes a full list of math/logic operators and precedence for the CPT instruction.

Execution of the CPT instruction affects the math flags in status word 0 (S:0).[3] Three of the status bits (carry, zero, sign) will only reflect the size of the final result generated after the CPT instruction finishes, but one status bit (overflow) will remain set if any operation in the expression caused it to be set. The example in Figure 6.14 includes a rung that latches an SLC 500 output on if the calculation generated a wrong answer due to an overflow. The CPT instruction

[3] S:0.0 = carry, S:0.1 = overflow, S:0.2 = zero, and S:0.3 = sign.

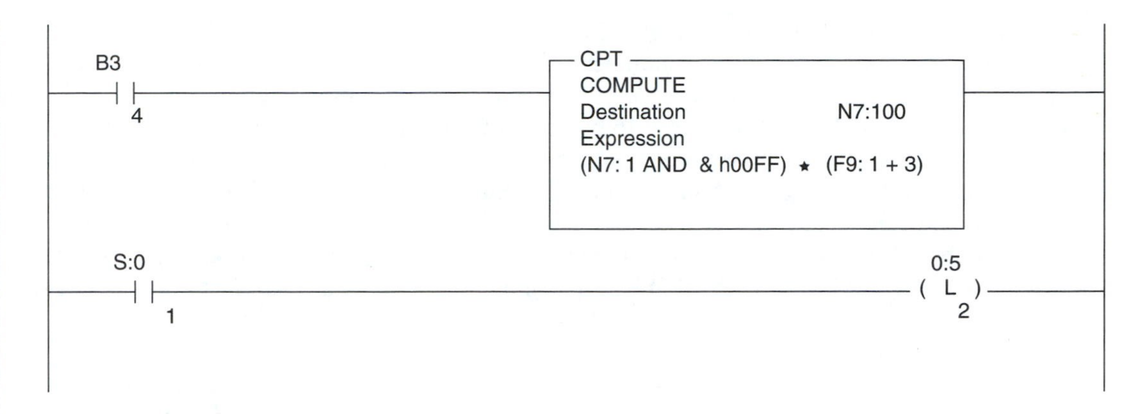

FIGURE 6.14
Allen-Bradley program with a COMPUTE instruction.

takes significantly longer to execute than the simpler single-operation math operations described in the following section.

Single-Operation Allen-Bradley Math and Logic Instructions

Single-operation math and logic operation instructions can be programmed in any PLC-5 or SLC 500 program. They include the basic **ADD** (ADD), SUBTRACT (SUB), MULTIPLY (MUL), and **DIVIDE** (DIV) instructions. Each uses two source operands (use address references or constants) to generate a result that is stored in memory at a destination address, as shown in Figure 6.15, where a multiplication is performed unconditionally (without the optional conditional control statement).

Both the PLC-5 and the SLC 500 offer additional math instructions to **CLEAR** (CLR) or to **NEGATE** (NEG) a number and to perform logarithmic (**SQR, LOG, LN, XCP**) and trigonometric functions (**SIN, COS, TAN, ASN, ACD, ATN**) on a number. Both PLCs have **NOT** (NOT), AND (AND), OR (OR), and **EXCLUSIVE OR** (XOR) instructions for performing logical operations on whole data words.

Some SLC 500s don't offer floating-point math capabilities, so SLC 500s have a feature that makes 32-bit integer addition and subtraction possible. The SLC 500 can only manipulate 16-bit integers, but won't store the actual 16-bit result in the destination address if an addition

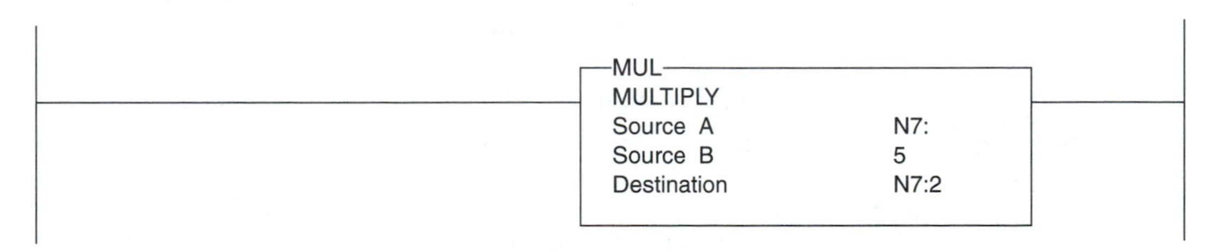

FIGURE 6.15
Allen-Bradley unconditional multiplication instruction.

or subtraction causes an overflow. Instead, the SLC 500 usually stores the largest or smallest possible signed 16-bit binary number (32, 767 or −32, 768). If, however, the **math overflow selection bit** (S:2/14) is set before the scan cycle with the math operation, the SLC 500 will store the actual result as the low 16 bits of the 32-bit math operation in the destination address. Regardless of the overflow selection bit status, whenever a math overflow occurs the SLC 500 latches the overflow bit (S:0/1) and the **overflow trap bit** (S:5/0). For 32-bit addition (or subtraction), the program should include a second math instruction to add the overflow trap bit to another 16-bit value representing the *high* 16 bits of the 32-bit math operation result (or to subtract it from the high 16-bit number, depending on the type of math instruction that caused the overflow). The overflow trap bit and overflow bit should then be unlatched after every 32-bit math operation.

The SLC 500 offers additional instructions for double-precision division (**DDV**), and instructions to find the absolute value of a number (**ABS**) or to scale a number. Instructions for scaling a number require the programmer to enter the coordinate values of two points along a scaling trend line (for the **SCP** instruction), or enter an offset and rate multiplier (for the **SCL** instruction).

The PLC-5 offers additional instructions to sort a series of numbers (**SRT**) and to find the average (**AVG**) and standard deviation (**STD**) of a series of numbers. These instructions operate on data in multiple memory locations, so they are covered in Chapter 7.

Math and logic operations affect the carry, overflow, zero, and sign bits in S:0.

S5

SIEMENS STEP 5 MATH, LOGIC, AND CONVERSION INSTRUCTIONS

Siemens' S5 PLCs only offer two types of STL-language *math* instructions: addition and subtraction. Instead of offering more sophisticated math and logic instructions, the S5 PLC comes with preprogrammed function blocks to perform operations such as multiply, divide, scale numbers, and convert binary number formats. Most STEP 5 math and logic instructions operate on the values in the two accumulators, leaving a result in accumulator 1. The ladder logic versions of the math instructions include the required LOAD and STORE instructions so that ladder logic programmers don't need to be aware of the accumulator use. The numerical results of math and logic instructions affect the status bits, as described in Appendix F.

Math instructions can only be programmed in function blocks. They execute unconditionally, which means that they change the contents of accumulator 1 and sometimes the CC0 and CC1 status bits every time they are encountered in a PLC program! Structured programming techniques can be used to make the PLC bypass these instructions. If programmed in ladder logic, then rungs with conditional logic controlling math instructions are automatically translated into structured STL equivalents. Math instructions usually treat 16-bit values as signed binary numbers, which Siemens calls *fixed-point integers,* as the "F" in the instructions implies.

Math or logic instructions in ladder logic are equivalent to one or two STL LOAD instructions, followed by the appropriate STL math/logic instruction, then an STL TRANSFER instruction to store the result. Figure 6.16 shows a **SUBTRACT** (−F) instruction, which subtracts the second-loaded value (in accumulator 1) from the first-loaded value (in accumulator 2), putting the

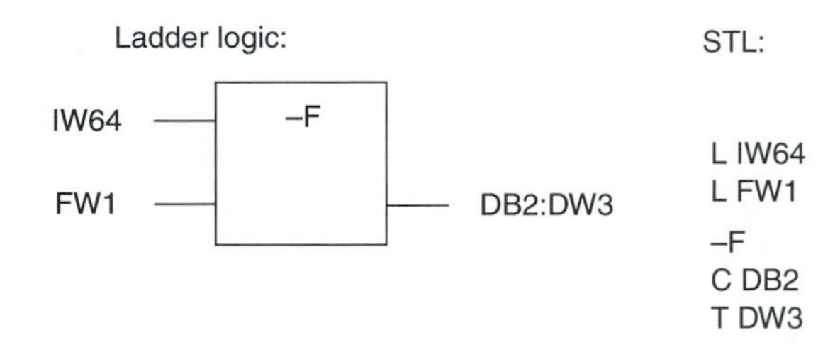

FIGURE 6.16
STEP 5 SUBTRACT instruction.

result into accumulator 1. The **ADD** (+F) instruction adds the numbers from the two accumulators, leaving the result in accumulator 2.

STEP 5 also offers **INCREMENT** (I) and **DECREMENT** (D) instructions, used to increase or decrease the value in the low byte of accumulator 1. The byte is treated as an 8-bit unsigned number. The high byte of the accumulator is not affected. The instruction must include a positive value to be added to or subtracted from accumulator 1. For example, the STL instruction "D 13" would subtract 13 from the low byte of accumulator 1 but would not affect the high byte.

An **ADD** (ADD) instruction is also available to add a signed constant value to the low byte (BF) or to the whole 16-bit number (KF) in accumulator 1. The ADD instruction treats the value in the accumulator as a signed number. For example, the STL instruction "ADD BF −13" would add −13 to the low byte of accumulator 1.

A **TWO'S COMPLEMENT** (CSW) instruction changes the sign of the signed 16-bit value in accumulator 1. INCREMENT (I), DECREMENT (D), ADD (ADD), and TWO'S COMPLEMENT (CSW) instructions can only be programmed in STL and only in function blocks.

STEP 5 offers four instructions to perform *logical operations*. Three of these instructions logically combine the 16-bit values from accumulators 1 and 2, leaving the result in accumulator 1. The instructions are **AND WORD** (AW), OR WORD (OW), and **EXCLUSIVE OR WORD** (XOR). A fourth, **ONE'S COMPLEMENT** (CFW), is equivalent to performing a NOT operation on the bits of the word in accumulator 1. These instructions all execute unconditionally. They can only be programmed in STL and only in function blocks.

S5 PLCs have a few preprogrammed function blocks[4] (and one organization block) that can be called by a user program. Some are called automatically by ladder logic instructions. The preprogrammed blocks include:

1. FB 240 to convert a number from *BCD to binary* and **FB 241** to convert a number from *binary to BCD.* Parameters include address or constant values for:

BINARY	the 16-bit binary number
BCD	the 16-bit BCD number for FB240
BCD1	the low 16-bit BCD output from FB241
BCD2	the high 8-bit BCD output from FB241
SBCD	sign bit for the BCD number (1 means negative)

[4] In Chapter 8 we describe how to use and how to write function blocks.

2. FB 242, which performs *multiplication,* accepting two signed 16-bit binary numbers as inputs and generating a 32-bit signed binary result (in two 16-bit words). A bit output also indicates whether the multiplication result was zero. Parameters include:

`Z1 and Z2`	the two 16-bit numbers to be multiplied
`Z31`	the low 16 bits of the result
`Z32`	the high 16 bits of the result
`Z3 = 0`	a bit indicating a result of zero (1 if not)

3. FB 243, which performs *division,* using signed 16-bit input values to generate a signed 16-bit integer quotient output and a 16-bit remainder output. Four output status bits are generated. Parameters include:

`Z1`	the number to be divided (dividend)
`Z2`	the number it is to be divided by (divisor)
`Z3`	the 16-bit integer result (truncated, not rounded)
`Z4`	the remainder (16 bits)
`Z3 = 0`	a bit indicating a result of zero (1 if not)
`Z4 = 0`	a bit indicating a remainder of zero (1 if not)
`FEH`	a bit indicating division by zero (1 means yes)
`OV`	a bit indicating overflow during division (1 means yes)

4. FB 250, which *scales an analog input value* according to scaling parameter values supplied by the calling program, and **FB 251,** which *scales an output analog value.* Parameters include:

`BG`	the 16-bit numerical part of the analog input or output module's address
`KNKT`	a constant consisting of two numbers, separated by a comma, indicating the analog channel number, and then a code identifying the type of analog I/O module (see your manual)
`OGR and UGR`	constants (in signed 16-bit range) indicating the upper limit (OGR) and lower limit (UGR) of the scaled value range
`XE`	16-bit value for FB251 to scale for output
`XA`	16-bit scaled result from FB250
`EINZ`	a bit that causes FB250 to execute once when it becomes 1
`TBIT`	a bit that FB250 holds on while it is performing a conversion ordered by EINZ
`BU`	a bit indicating that the scaled value is outside the range defined by OGR and UGR (1 = out)
`FEH`	a bit indicating that an error in KNKT, OGR, or UGR parameters was detected by FB251 (1 = error)
`FB`	a bit indicating that FB250 has detected a hardware error or an illegal address (1 = error)

5. OB 251, which calculates the result of a *servo-control* **(PID)** equation. OB 251 is covered in detail in Chapter 12.

S7

SIEMENS STEP 7 MATH, LOGIC, AND CONVERSION INSTRUCTIONS

All STEP 7 STL-language math and logical instructions operate unconditionally, using the contents of the accumulators as operands, leaving the result in accumulator 1 and affecting status register bits (the CC 0, CC 1, OV, and OS status bits, as detailed in Appendix F).

Ladder logic math/logic instructions include STL LOAD and TRANSFER instructions in addition to the STL math/logic instructions. Ladder logic math instructions (but not logic instructions) also include STL instructions to check the status register's OV bit after the math operation, to determine if the result is valid and to turn on an output (ENO) bit if it is. Figure 6.17 shows a ladder logic instruction that adds two 16-bit integer values, turns an output bit on if the result is valid (if the status register's OV bit is not set), and stores the result.

In STL programs, the programmer must use structured programming if he or she wants to make math/logic instruction conditional, and must program the checks of the status bits that are deemed to be required. The equivalent STL-language program beside the unconditional ladder logic ADD_I instruction in Figure 6.17 does *not* include any STL instructions that would be needed to make the instruction conditional, but does check the OV status bit.

In the remainder of this chapter we do not identify Siemens ladder logic instructions by name. There are ladder logic equivalents for each STL instruction that is discussed, unless otherwise noted. Since STEP 7 ladder logic instructions all incorporate STL instructions, the following descriptions also describe ladder logic capabilities.

STEP 7 allows *math operations* to be performed using signed 16-bit numbers representing integers **(I),** signed 32-bit double word numbers representing integers **(D),** or 32-bit floating-point numbers representing real numbers **(R).** New unconditional instructions have been added to enable converting the binary number format of the number in accumulator 1:

1. To convert a signed 16-bit integer to a signed 32-bit double-word integer **(ITD).**

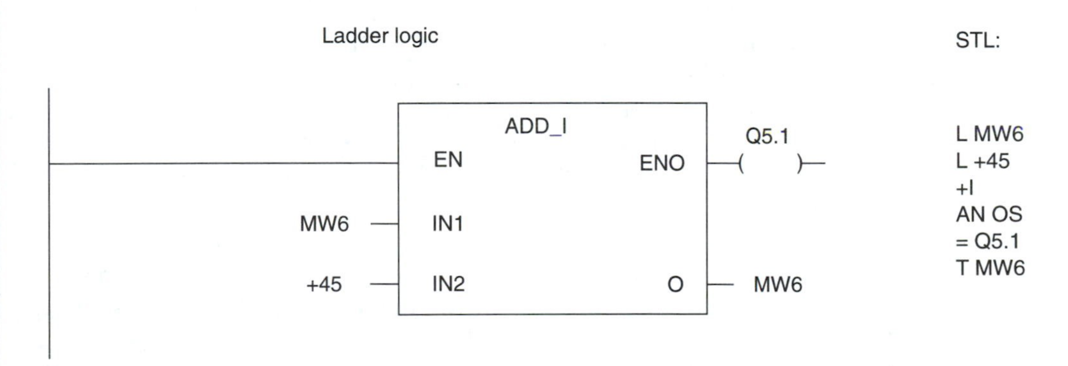

FIGURE 6.17
STEP 7 integer add instruction.

2. To convert BCD to 16-bit signed binary integer **(BTI)** or the reverse **(ITB),** or to convert BCD to a 32-bit signed binary double word **(BTD)** or the reverse **(DTB).** Remember that the LOAD BCD **(LD)** instruction can also be used to convert a number from timer or counter format to BCD as the number is loaded into accumulator 1.

3. To convert a signed 32-bit binary double word to a floating-point real number **(DTR),** or to convert from floating point to a 32-bit signed binary number with normal rounding (**RND**), with rounding up or down forced (**RND**+ and **RND**−) or with the decimal portion truncated **(TRUNC).**

After numbers are loaded into accumulators 1 and 2, they can be operated on by the *four basic math operations:*

- **Add** (e.g., +I)
- **Subtract** (e.g., −I)
- **Multiply** (e.g., *I)
- **Divide** (e.g., /I)

The second number that was loaded (in accumulator 1) is added to, subtracted from, multiplies, or divides the previously loaded number (in accumulator 2). The result is put into accumulator 1.

1. Basic math operations using 32-bit signed numbers (e.g., +D and +R) result in 32-bit signed numbers, of course. If the result is wrong due to needing a thirty-third bit for the true result or due to an illegal sign change, the appropriate status bits in the status word will be set (see Appendix F). Programs should include statements that monitor status bits after math operations.

2. Addition and subtraction of 16-bit signed numbers (e.g., +I) use the low 16-bit numbers from accumulators 1 and 2. The resulting signed 16-bit number does not affect the high 16 bits of accumulator 1 even if the operation should result in a 17-bit number, so the program should include statements to monitor the status bits.

3. Multiplication of 16-bit signed numbers (e.g., *I) causes a 32-bit signed binary result to be placed in accumulator 1, so the TRANSFER instruction storing the result should transfer the result as a double word. If the result fits into a 16-bit signed value, the low 16-bits of accumulator 1 will be correct and can be transferred as a word. The S7 PLC sets the OV and OS status bits if the result is too big for the low 16 bits of accumulator 1.

4. Division of 16-bit numbers (e.g., /I) results in a 16-bit number in the low 16 bits of accumulator 1, representing the truncated integer result, but the remainder of the division is placed into the high 16 bits of accumulator 1. If the result of the division is transferred to memory as a double word, other instructions in the program should read the result and the remainder as single words.

5. Division of 32-bit double words (e.g., /D) results in a truncated 32-bit result in accumulator 1. The remainder is lost. STEP 7 offers a fifth math instruction, **DIVIDE AND STORE A REMAINDER** (MOD), which performs the same 32-bit division, but MOD stores the remainder in accumulator 1 and discards the result. /D instructions can be followed by MOD instructions (after reloading the operands) so that 32-bit results and 32-bit remainders can be generated and stored separately, as follows:

```
L MD4
L MD8
/D        //    Generates integer result of MD4 / MD8
T MD12

L MD4
L MD8
MOD       //    Generates the remainder of MD4 / MD8
T MD16
```

Some instructions manipulate only *integer values:*

1. **INCREMENT** (INC) and **DECREMENT (DEC)** instructions only operate on the lowest byte of accumulator 1, adding or decrementing that byte by the unsigned value included with the instruction, and treating the byte as an unsigned integer, as in the following example:

```
DEC 25    subtracts 25 from the low byte of accumulator 1
```

2. The **ADD AN INTEGER** (+) instruction can be used to add a *signed* constant to the low, signed 16-bit word of accumulator 1 if the constant is entered in decimal, or to the signed 32-bit double word of accumulator 1 if the constant is entered in long signed integer constant format. For example:

```
+ -45     adds -45 to the low word in accumulator 1
+ -L#45   adds -45 to the double word in accumulator 1
```

3. **TWO'S COMPLEMENT INTEGER** (INVI) changes the sign on the low 16-bit word in accumulator 1, while **TWO'S COMPLEMENT DOUBLE WORD** (INVD) changes the sign of the whole 32-bit double word.

Floating-point numbers (R) in accumulator 1 can be operated on (leaving the result in accumulator 1) by several new unconditional instructions, including:

1. Instructions to change the number's sign: **NEGATE REAL NUMBER** (NEGR) and **ABSOLUTE VALUE** (ABS). The ladder logic versions of these instructions have an ENO output that is turned off if the overflow status bit is on, even though these instructions don't affect the OV bit. Since previously executed instructions can set the OV bit, the ladder logic programmer should be careful about using the ENO output to control subsequent operations. The following example shows how to use the NEGR instruction:

```
L MD4     //    Changes the sign of the 34-bit floating-point
NEGR            number stored in MD4
T MD4
```

2. Logarithmic and trigonometric math instructions, including **SQR, SQRT, LN, EXP, SIN, COS, TAN, ASIN, ACOS,** and **ATAN.**

STEP 7 offers the four basic *logic operators*, all of which execute unconditionally, and all of which affect the status register. The **AND WORD** (AW), OR WORD (OW), EXCLUSIVE OR WORD (XOW), and **ONE'S COMPLEMENT INTEGER** (INVI) instructions operate on 16-bit words, leaving the 16-bit result in the *low* word of accumulator 1. (The ONE'S COMPLEMENT instruction performs a NOT operation.) The 32-bit versions of the same instructions include **AND DOUBLE**

WORD (AD), OR DOUBLE WORD (OD), EXCLUSIVE OR DOUBLE WORD (XOD), and **ONE'S COMPLEMENT DOUBLE INTEGER** (INVD). The AND, OR, or EXCLUSIVE OR instructions can include an *address* for the 16- or 32-bit value to be combined with accumulator 1, or can include a 16- or 32-bit constant[5] to be combined with accumulator 1 (e.g., AW #16#12FF). If no address or constant is included with the instruction, the value in accumulator 2 (the low 16 bits or the whole 32 bits, depending on the instruction) is combined with accumulator 1.

In the ladder logic versions of logic instructions, the ENO output does *not* reflect the status of any status register bits. It simply outputs the RLO that existed before the logic instruction executed. The programmer must program any status bit checks that may be required.

Addition Instructions for STEP 7 Register Indirect Addressing

STEP 7 offers two instructions that can be used for advanced data addressing purposes. The **ADD A CONSTANT TO REGISTER** 1 (+AR1) and **ADD A CONSTANT TO REGISTER** 2 (+AR2) instructions can be used to change the pointers in address registers AR1 and AR2. Modifying a pointer makes the PLC operate on a value from a different address the next time an instruction that uses register indirect addressing[6] executes. (Register indirect addressing was described in Chapter 5.) If the **+AR1** and **+AR2** instructions are executed without any additional parameters, the 16-bit value from the low byte of accumulator 1 is added to the 16-bit pointer in the address register. If the **+AR1** or **+AR2** instruction includes a 32-bit pointer constant in the form **P#byte.bit,** that constant value will be added to the 32-bit pointer[7] in the address register.

OMRON CQM1 MATH, LOGIC, AND CONVERSION INSTRUCTIONS

The CQM1 has several sets of the basic four-function math operation instructions (add, subtract, multiply, and divide), one set for each way that the CQM1 can code numbers into binary. As shown in Figure 6.18, there are sets of instructions for 16- and for 32-bit BCD numbers, sets for 16- and 32-bit unsigned binary numbers, and sets for 16- and 32-bit signed binary numbers. The CQM1, of course, offers instructions for converting between these binary coding systems. A few specialized math instructions are also available for such operations as finding roots, averaging, trigonometry, and scaling numbers.

OMRON's add and subtract instructions all affect the **carry (CY)** bit in the status register area of memory (bit 25504), and they all include the previous CY bit status in calculating the result, in this way:

```
Addition:    R = Au + Ad + CY
```

After the addition operation, CY is set if the result is too big for storage in R.

```
Subtraction:    R = Mi − Su − CY
```

After subtraction, CY is set if the result is negative (i.e., if Su − CY was greater than Mi).

---

[5] Binary constant formats aren't allowed. Use hexadecimal constant format.

[6] AR2 is also used automatically by the STEP 7 operating system as functions and function block programs execute (probably as an offset for local addressing), so the programmer should be cautious in how AR2 is changed.

[7] Notice that the pointer constant does not include a memory area identifier. You cannot use +AR1 or +AR2 to change the area-crossing memory area identifier of the pointer in AR1 or AR2.

Math I Instructions and Parameter Requirements

Instruction :	Add	Subtract	Multiply	Divide
16-bit BCD	ADD (30)	SUB (31)	MUL (32)	DIV (33)
32-bit BCD	ADDL (34)	SUBL (35)	MULL (36)	DIVL (37)
16-bit unsigned binary	ADB (50)	SBB (51)	MLB (52)	DVB (53)
32-bit unsigned binary	ADBL (-)	SBBL (-)	–	–
16-bit signed binary	ADB (50)*	SBB (51)*	MBS (-)	DBS (-)
32-bit signed binary	ADBL (-)*	SBBL (-)*	MBSL (-)	DBSL (-)

* Can be used to operate on unsigned or on signed binary numbers. When used with signed values, the overflow and underflow flag bits must be used, as discussed later when flag bits are described.

Parameters used in math instructions	Add	Subtract	Multiply	Divide
First	Au : Augmend	Mi : Minuend	Md : Multiplicand	Dd : Dividend
Second	Ad : Addend	Su : Subtrahend	Mr : Multiplier	Dr : Divisor
Third	R : Result	R : Result	R : Result	R : Result

FIGURE 6.18
Math I instructions and parameter requirements.

Automatic inclusion of the CY bit in addition and subtraction allows the programmer to program sequential addition and subtraction so that binary numbers longer than 32 bits can be operated on! There are instructions to specifically set the carry bit: **STC (40),** or to reset the carry bit: **CLC (41).**

The parameters of a math instruction, such as the ADD (30) instruction shown in Figure 6.19, can be entered as constants or as addresses.

Addition and subtraction instructions generate result numbers with the same number of bits as their operands. Multiply instructions generate result numbers that are *twice* the size of the operands. Only one address is entered for each parameter, even if the instruction uses data words greater than 16 bits. If numbers larger than 16 bits are required as operands, or if results greater than 16 bits are generated, the math instructions automatically use the next sequential memory location(s) in addition to the memory locations specified. The address specified as parameters in the instruction will hold the *lowest* 16 bits of the larger number, and each higher address contains the next higher 16 bits. Figure 6.20 shows a MULL (56) instruction. This instruction multiplies two 32-bit BCD numbers (eight decimal digits) and generates a 64-bit BCD number (16 decimal digits). One eight-decimal-digit number comes from DM2100 (the low four digits) and from DM2101 (the high four decimal digits), and the other from DM2102 and DM2103. The 16-decimal-digit result will be stored in four sequential memories: DM2104 for the lowest four decimal digits to DM2107 for the highest four decimal digits.

Division instructions generate results (quotients) that are the same size as the operands, but they also generate a remainder. The remainder is the same size as the operands, and is automatically stored in the memories just above the result. For example, a DOUBLE SIGNED BINARY DIVIDE instruction that includes the address DM1000 for the result would put the low 16 bits of the 32-bit result in DM1000, the high bits of the result in DM1001, and would also store the remainder of the division in DM1002 (low 16 bits) and in DM1003 (high 16 bits). Be sure to leave enough memory space for multiplication and division results.

Some additional CQM1 math instructions are also available:

1. **INC(38)** and **DEC(39)** can be used to increment or decrement the BCD numbers in the address specified with the instruction.
2. **NEG(−)** and **NEGL(−),** available in the CQM1-CPU4x only, will find the two's complement of a binary 16- or 32-bit number, storing the result at another memory location.
3. **ROOT(72)** will find the root of a 32-bit BCD number, generating a 16-bit BCD result stored in another memory location.

FIGURE 6.19
CQM1 math instruction that adds the BCD number 100 to the BCD number representing a timer or counter present value, and stores the result in data memory.

FIGURE 6.20
A CQM1 math instruction that uses numbers larger than 14 bits.

4. MAX(−) and **MIN(−)** can be used to find the largest or smallest number in a sequence of data words in memory, storing that value in a destination address. Three parameters must be entered with these instructions. The first, a control parameter (C), indicates whether data values are to be treated as signed or unsigned binary, and indicates the number of data words to include in the search and whether to output the address at which the maximum/minimum value was found. Next, a range (R1) parameter is the address where the list of data entries to be searched through starts. Finally, a destination (D) parameter indicates where to write the maximum or minimum value, followed by the address where that data value was found, if that feature was selected.

5. SUM(−) can be used to add a series of words or bytes from a series of memory locations. A control (C) parameter indicates whether the data are BCD, unsigned, or signed binary numbers, indicates whether to sum the whole 16-bit data words or just the right bytes or just the left bytes, and indicates how many words to include in the sum. Next, a range (R1) parameter indicates the starting address of the values to be summed. Finally, a destination (D) parameter indicates the first of the two addresses where the 32-bit result will be stored.

6. AVG(−) calculates the average value of a single source word over as many as 64 scan cycles while this instruction's control logic is true. Each time the control logic goes from false to true, a new averaging sequence is begun. Three parameters are required. First, a source (S) parameter indicates the single address from which a value is read each scan cycle for inclusion in the average. The values are treated as natural binary numbers in calculating the average. Next, a number (N) parameter (address or constant) indicates (in BCD) the maximum number of scan cycles to include in the averaging (maximum = 64, even if a larger number is entered). Finally, a destination (D) parameter indicates the address where the working average is kept. Up to 65 addresses following the average result are also used. The address immediately following the destination address is used by the CQM1 for working data, and the following (up to 64) addresses are used to store the individual data values as they are read from the source address.

7. APR(−) is called the *arithmetic process* instruction because it can be used for two different mathematical purposes. It requires a control (C) parameter, a source (S) address parameter, and a destination (D) address parameter (in that order).

APR(−) can be used to find the **sine or cosine** of a number. If the control parameter is the constant #0000, the value in the source address will be treated as an angle (in BCD) in tenths of degrees, and the sine of that angle will be calculated, rounded, and stored (in BCD) in the destination address. Control constant #0001 will cause the APR(−) instruction to calculate the cosine.

APR(−) can be used to perform a **linear approximation** of a series of X–Y coordinates. If the control (C) parameter is an address, APR(−) will calculate *two* points along a linear approximated line. The value *in* the address specified by the control parameter is treated as a control word that indicates whether input and/or output numbers will be in BCD or binary, and how many X–Y coordinates are to be included in the calculation. In the address immediately following the control word, the programmer must enter an X_M value for which the APR(−) instruction is supposed to calculate the Y_M value as one of the two points on the linear approximation line that it will calculate. The next addresses must contain the X and Y coordinate data to be used in calculating the approximated line: X_0 is assumed to be 0, so it doesn't need to be entered and the first value in the list is the value of Y_0, followed by X_1, then Y_1; X_2, then Y_2; and so on, to the end of the list. (X values must be entered from lowest to highest and cannot be greater than X_M.) The source (S) parameter indicates the address where the X value for the other of the two calculated points has been entered, and the destination (D) parameter indicates the address where the APR(−) instruction is to store the calculated Y value for that X value.

8. The scaling instruction, **SCL(66),** uses two X–Y points on a straight line to calculate a third Y value for a third X value along the same line. For reasons known only to OMRON, X values are in unsigned binary while the Y values are in BCD. The first parameter, source (S), indicates the third (binary) X value. The second parameter, parameter (P1), is the address of the first of the four known point values (Y_0, X_0, Y_1, then X_1). Finally, a result (R) parameter indicates where the CQM1 is to write the third (BCD) Y value.

9. In the CQM1-CPU4x and higher, there is also a SIGNED HEX TO BCD SCALING instruction, **SCL2(−),** and a BCD TO SIGNED HEX SCALING instruction, **SCL3(−).**

Math instructions can affect flag bits in SR 254 and SR 255:

1. The **carry flag bit (SR 25504)** is affected by addition and subtraction, and can be used in addition and subtraction of numbers larger than the standard 16- and 32-bit numbers! If two 64-bit numbers, for example, are stored in addresses DM 2000 (to 2003) and DM 2004 (to 2007), a 32-bit addition of the numbers in DM 2000 (to 2001) and DM 2004 (to 2005) could be followed immediately by another 32-bit addition of DM 2002 (to 2003) with DM 2007 (to 2007). The second addition would automatically include the carry (if any) from the first addition. *Be sure to clear the carry flag before any addition and subtraction in which you don't want it to affect the result.*

The carry flag is also set if a large unsigned number (BCD or binary) is subtracted from a smaller unsigned number, so it indicates that the result is supposed to be negative, even though BCD math and unsigned math do not correctly generate negative numbers. The incorrect "negative" result can be converted to a valid BCD or unsigned binary number (its absolute value) if it is subtracted from zero[8] (after clearing the carry flag so that the carry flag doesn't affect this subtrac-

[8] In a CQM1-CPU4x or higher, a negative unsigned binary number can also be converted to a valid unsigned number (absolute value) by one of the two's complement instructions: NEG(−) or NEGL(−).

tion). In either case, the program must include some way to remember that the BCD or unsigned binary number actually represents a negative value (e.g., by setting a user-selected status bit).

The carry flag can also be set by multiplication. There is no easy way to recover from multiplication that generates a result too big for the memory reserved for it, but at least the program can recognize that the result is incorrect.

2. If unsigned addition or subtraction generates a result that is over or under the allowable range for a signed binary number of the same size, an **overflow bit (SR 25405)** or an **underflow bit (SR 25404)** is set. CQM1s below the CQM1-CPU4x do not offer these status bits, so sign overflows are trapped in these PLCs only if they cause carry errors. If the programmer of one of these PLCs wants to use an instruction marked with a * in Figure 6.18 to perform addition or subtraction using signed binary numbers, the program should include instructions watching for changes to the most significant bit of the result number.

3. The **error bit (SR 25503)** is set if the PLC can't read or write an operand address (perhaps because the address doesn't exist), if an indirect addressing offset includes bit patterns not recognizable as BCD, or if a division instruction divides by zero.

4. The **equals zero bit (SR 25506)** is set if the result is equal to zero.

Data conversion instructions include:

1. BIN(23) and **BCD(24).** BIN(23) converts a 16-bit BCD value, indicated by the source (S) parameter (a constant or an address), to a 16-bit unsigned binary value and places the result in a destination (D) address. BCD(24) converts a source 16-bit unsigned binary number to a 16-bit BCD destination number *unless* the BCD value would exceed 9999, in which case the conversion is not performed, but the CY status bit (SR 25504) is set.

2. The CQM1-CPU4x and higher also offer **BINL(−)** and **BCDL(−),** which will convert between 32-bit BCD and 32-bit unsigned binary.

3. 32-bit unsigned binary values representing a number of seconds (0 to 35,999,999) can be converted to hours/minutes/seconds format using **HMS(−).** A source (S) parameter indicates where the 32-bit number is stored, and a destination (D) parameter tells the CQM1 where to place the BCD hr/min/sec result. Hours (0 to 9999) are stored in the address above D, minutes (0 to 59) in the high byte of D, and seconds (0 to 59) in the low byte of D. **SEC(−)** can be used to convert from hr/min/sec format at the source (S) address to a 32-bit seconds-only number at the destination (D) address.

4. MLPX(76) reads the value (0 to 15) of an individual 4-bit "character" (nibble) in a 16-bit source (S) word and sets the bit of the same number (0 to 15) in a 16-bit result (R) word. **DMPX(77)** does the opposite, converting up to four 16-bit source (S) values, each with at least one bit set, to 4-bit nibble values in a single destination (D) word. (See your manual.)

5. The ASCII CONVERT instruction, **ASC(86),** and the ASCII-TO-HEX CONVERT instruction, **HEX(−),** are similar to MLPX(76) and DMPX(77), except that they convert between nibble values (hexadecimal 0 to F) and the 8-bit ASCII codes for each of those hexadecimal values. (See your manual.)

6. A SEVEN-SEGMENT DECODER instruction, **SDEC(78),** can be used to convert the nibble values of a 16-bit source word to 8-bit values suitable for driving a standard seven-segment display.

The CQM1 offers the standard set of logic instructions to operate on 16-bit data words. There is a COMPLEMENT instruction, **COM(29),** which inverts all 16 bits (performs a NOT operation) of a word at an address specified as the word (Wd) parameter (changing the value in that address). The LOGICAL AND, **ANDW(34)**; LOGICAL OR, **ORW(35)**; and EXCLUSIVE OR, **XORW(36)** instructions use two input words (I1 and I2) in generating a logical result word (R). EXCLUSIVE NOR, **XNRW(37),** is the equivalent of an XORW(36) followed by a COM(29). Input parameters (I1 and I2) can be entered as addresses or as hexadecimal constants. The result parameter (R) must be an address.

TROUBLESHOOTING

A PLC is a computer. It is designed to do math and logic operations quickly and accurately. So why is your program getting the wrong answers? Unfortunately, computers will get the wrong answer if they are programmed to, and they will happily attempt to perform operations they can't handle if they are programmed to.

A relatively easy-to-find problem occurs when your program tries to move a data value, or to store a result in a memory that doesn't exist: Your PLC will probably fault (stop running) if this happens, and a fault code will be saved so that the programmer can find the problem. Even if the PLC does not stop, there will still probably be a fault code or fault bit in memory. (Chapters 11 and 15 cover fault interrupts and troubleshooting faults in more detail.) In some PLCs, memory must be assigned before it can be used. While memory is often assigned automatically as the programmer writes the program, memory addressed using indirect or indexed addressing has to be specifically assigned by the programmer, since programming software can't predict which memory locations will actually be required. Advanced addressing techniques can sometimes cause your PLC to try to write to addresses that are outside the limits of the memory areas the programmer intended to use.

Perhaps the problem is due to the way that your data is coded into binary. Binary coding schemes include binary-coded decimal (BCD), unsigned binary notation, signed binary notation, and floating-point notation. The math instructions preprogrammed into a PLC can get the correct answer only if they operate on binary numbers in the coding format for which they were designed. If you use a PLC instruction to add two BCD numbers and the instruction was designed for signed binary numbers, the result will be wrong. Most PLC math instructions assume signed binary numbers, but unsigned math and BCD math instructions are still common and floating-point math is becoming more common. Use the correct instructions for the number coding system. Check to ensure that the numbers are actually stored in memory in the notation you expect: Your programming unit should be able to display the memory contents in binary, or perhaps you can write an instruction to output the binary value to a digital output module that isn't connected to actuators, to see the pattern of on's and off's. Use data conversion instructions, if necessary, to match data coding to the instructions available.

Perhaps the problem is due to the **size of the data unit.** If the data values are stored as signed binary Words (16 bits) and the math or logic instruction is designed to operate on signed binary double words (32 bits), the low 16 bits of the result will be correct some of the time, but sometimes the answer will be wrong because the instruction will look for the code indicating a minus number in the *high* 16 bits. (Perhaps you didn't realize those high bits were being changed every time you execute the math instruction; now you know why the address above where the result is stored changes all by itself.) Check the size of the actual operand and result data elements against the data unit sizes the instruction is designed to use.

Remember that there is a limit on the range of values a fixed-size binary number can represent, and remember that the range limits also depend on the coding format. A 16-bit binary pattern (a word), for example, can only have 2^{16} (or 65,536) different patterns, so it can represent numbers between −32,768 and +32,767, or between 0 and 65,536 (depending on whether the binary pattern is treated as signed or unsigned binary notation). If you add 1 to the unsigned binary number representing 65,536, the computer will generate the result 0 (unless the PLC is preprogrammed to prevent this—check your manual). Similarly, adding 1 to the signed binary number representing 32,767 could generate the result −32,768! If these situations can occur, perhaps you should be using larger data element sizes and instructions that operate on those larger data elements.

Most PLCs (all the PLCs covered in this book) offer status bits (or flags) that are affected by math and sometimes by logic operations. The status bits can be examined by Boolean instructions in your program (permanently, or just during debugging) and can sometimes be displayed on a programming unit screen. Typical status bits include one that is turned on if a math operation results in a zero (which might represent an error if you were adding 1 to 65,536), a bit that goes on if the result is in a form that might be interpreted as a negative number (an error if you were adding 1 to 32,767 and expected +32,768), a carry bit that is turned on if the result is too big for its data element size, and an overflow bit that goes on to indicate that a result is too large to be correctly represented in signed binary notation.

Most PLC programming software includes a histogram feature, which can be used to record the historical size of a number while the PLC runs. The histogram feature can be used to track the size of a number that the programmer suspects might sometimes get close to its limiting size, so that the probability of future math errors can be evaluated and prevented.

Floating-point notation was designed to represent very large or very small numbers. It uses a binary version of scientific notation to represent real numbers. Remember, though, that there is still a limit on the size of the numbers that can be represented in the standard 32-bit floating-point notation[9] (a 64-bit version exists but is not yet in common use in PLCs). Perhaps even more important, remember that a 32-bit binary number can only be arranged into 2^{32} different patterns, which means that it can't possibly be used to represent every possible decimal number between its upper and lower limits. *Floating-point math instructions round off the results to the closest possible 32-bit pattern.* You may find that comparing the floating-point result of a math operation to see if it is equal to a number that it should be equal to actually results in a "false" result because of round-off error. Remember that converting a floating-point number to another binary number format always includes either rounding off or truncation of the decimal portion of the number.

[9] See Chapter 5 for the range limits.

QUESTIONS/PROBLEMS

1. What is the central component in a PLC's CPU module? What other types of components does it control?
2. Identify the purposes of the bits in typical status register and describe how each could be used in a PLC program.
3. Which MPU register is always used as a PLC executes a MOVE instruction?
4. Logic operations such as NOT, AND, OR, and EXCLUSIVE OR can be used to change bit values in whole data words. Which of these logic operations can be used to mask selected bits "on"?
5. A sequencer allows you to write programs that read a pattern of digital inputs, ignore some of the inputs, compare the rest of the pattern with an input mask, and if the pattern matched, output a pattern of digital outputs. Which ladder logic or STL instruction(s) can be used in a PLC for the same set of operations *instead* of programming a sequencer?
   (a) To read a pattern containing 8 input image bits.
   (b) To mask some of the bits.
   (c) To compare the result against an input mask.
   (d) To output a pattern to 8 bits of an output image.

Allen-Bradley PLCs

6. Fill in what would be in B3:1 after the PLC executes the following instruction:

```
                 B3:0: 1111 1111 0000 0000
B3:1 before MVM instruction executes:
                       0101 0101 0101 0101
```

```
 ——MVM————————————————————————————
   Masked Move
   Source :                  B3 : 0
   Mask :                 &H 0FF0
   Destination :             B3 : 1
```

```
B3:1 AFTER MVM instruction executes:
                            _ _ _ _
```

7. Fill in what would be in B3:1 after the PLC executes the following instruction:

```
                 B3:0: 1111 1111 0000 0000
B3:1 before BTD instruction executes:
                       0101 0101 0101 0101
```

```
 ——BTD————————————————————————————
   Bit Field Distribute
   Source :                  B3 : 0
   Source Bit :                   6
   Destination :             B3 : 1
   Destination Bit :              8
   Length :                       4
```

```
B3:1 AFTER BTD instruction executes:
                            _ _ _ _
```

8. What number would be in N7:1 after the PLC executes the following instruction (answer with a decimal number)?

```
 ——CPT————————————————————————————
   Compute
   Destination :             N7 : 1
   Expression :
    10 | 3
```

```
                    N7:1 contains _____
```

9. The COMPUTE instruction performs a mathematical operation and places the answer in a destination address. Depending on the result of the mathematical operation, 4 other bits may be affected. Identify what would set each of those 4 bits.
10. Some instructions require (or allow) you to use square brackets in an address, as in this example: **T4: [N7:5].DN.** How does the PLC know to which timer's "done" bit this address refers?

Siemens S5 PLCs

11. Assuming that all eight sensors at a Siemens S5 input module 3 are on and that no other inputs are on, what will the eight actuators at output module 4 be after each of the following programs?

Program 1:

```
L IB3
T QB4   will turn the actuators:___
```

Program 2:

```
L IW3
T QW4   will turn the actuators:___
```

Program 3:

```
L IB3
T QW4   will turn the actuators:___
```

Program 4:

```
L IW3
T QB4   will turn the actuators:___
```

Program 5:

```
L KH0000
T QW4
A I2.0
L IB3
T QB4   will turn the actuators:
        __ while I2.0 is on
        and turn the actuators:
        __ while I2.0 is off
```

Siemens S7 PLCs

12. Assuming that eight sensors are attached at the Siemens S7 input module that provides data for IB3, *and assuming that all of these sensors are on while sensors controlling all other input image bits are off,* will the *eight actuators controlled by QB4* be on or off after each of the following programs?

Program 1:

```
L IB3
T QB4   will turn the actuators:___
```

Program 2:

```
L IW3
T QW4   will turn the actuators:___
```

Program 3:

```
L IB3
T QW4   will turn the actuators:___
```

Program 4:

```
L ID3
T QB4   will turn the actuators:___
```

Program 5:

```
L B#16#00
T QB4
A I2.0
L IB3
T QB4   will turn the actuators:
        __ while I2.0 is on
        and turn the actuators:
        __ while I2.0 is off
```

13. The STEP 7 ladder logic instruction that adds two 16-bit numbers (ADD_I) includes an output contact labeled **ENO.** The ENO output should be on if the ADD_I instruction is "enabled and operating." Even if the ADD_I instruction is enabled, the ENO output might not be on. Why not? (Don't just say "it's not operating properly." Explain what might be wrong!)

OMRON CQM1 PLCs

14. The OMRON CQM1's COMPARE instruction does not affect the result of the logic operation (RLO) as COMPARE instructions in an Allen-Bradley or Siemens PLC would. What does OMRON's COMPARE instruction affect?

15. The following OMRON CQM1 rung contains an instruction that compares two data words. What would you add to the program if you wanted to turn an output on when the numbers in 020 and 024 are equal? (Use either specific bit addresses, the OMRON names that describe the bit functions, or just state what the bit represents.)

```
|
|-----------[ CMP (20)              ]
|           [   1st Word :      020 ]
|           [   2nd Word :      024 ]
|
```

SUGGESTED PLC LABORATORY EXERCISES

For a system with:

- Four control panel switches: inputs 0 to 3
- Four indicator lamps or visible output module LEDS: outputs A to D
- Two spring-return-valve-controlled cylinders: outputs E and F
- One detent-valve-controlled cylinder: outputs G and H
- Three sensors to detect extension of each of the cylinders

For the PLC that you are learning to use:

1. Write a program that:
   (a) Copies a constant to an output image to extend all the cylinders while switch 0 is on.
   (b) Includes a counter and instructions that set the counter's accumulated value to 5 each time switch 0 goes from off to on.
   (c) Flashes indicator lamp A on and off at a 1/4-s cycle speed while the counter value's accumulated value is between 10 and 15. Test your program using the data screen to modify the contents of the counter's accumulated value.
   (d) Uses logical operators to mask all of the even-numbered contacts (including zero) of the output images for the indicator lamps OFF for 2 s after the sensor at cylinder E is actuated.
   (e) Doubles the size of a signed binary number in memory each time switch 2 goes from off to on (use addition if your PLC does not offer a multiplication instruction) and:
      - latches an output on after the signed binary number exceeds the allowable size of a 16-bit word, and
      - clears the number and the latched output each time switch 3 goes from off to on.
2. Write a program that:
   (a) Copies the four input image bits to a word of data area of memory in such a way that one of the switches controls the data word's high bit.
   (b) Uses COMPARE instructions, so that when the data word is:
      - Greater than 0, extend all cylinders by moving another data word value (constant or from memory) to the output image.
      - Less than zero, retract all cylinders using a single data word move.
      - Equal to zero, add the values from two unused memory words, placing the answer in a third. (Use the data monitor screen to enter and change data values and to observe the result value.)
3. Write a program that:
   (a) Contains a counter to count up whenever switch 0 goes on, and down whenever switch 1 goes on.
   (b) Uses another counter that counts the number of times the first counter's accumulated value changes, but only while its accumulated value is over 5. *Hint:* Compare the accumulated value against its value from the previous scan cycle.

7

MANIPULATING DATA IN FILES, BLOCKS, ARRAYS, AND STRUCTURES

OBJECTIVES

In this chapter you are introduced to:

- Definitions of arrays, structures, data blocks, and data files.
- Allen-Bradley, Siemens, and OMRON instructions (or function blocks) to:
  - Shift or roll bit arrays.
  - Shift arrays of data words:
    - FIFO and FILO shifts.
    - Sequencer instructions.
  - Move sets of data words within CPU memory.
  - Exchange sets of data words with I/O modules:
    - Block transfers of data with I/O modules (Allen-Bradley).
    - Sending and receiving of data via shared memory (Siemens).
    - Reading and writing serial data words (OMRON).
  - Comparing sets of data words and using status bits in Boolean logic.
  - Mathematical, logical, and conversion operations on sets of data words.
- Instructions for detecting undesirable values in memory (Allen-Bradley).

If you are using a Siemens PLC, you should read Chapter 8 before reading Chapters 6 and 7.

If a control system needs to perform data manipulation operations on data words, it is often necessary to perform the same operations on series of data words. It may be necessary to sum *all* the accumulated values for *each pair* of counters in two identical manufacturing systems. The same statistical process control (SPC) calculations might need to be performed on data from several independent sensors. Perhaps you have written a control program that doesn't waste time performing an SPC calculation every time a sample is taken but just saves a series of sampled results. Your PLC program should use the file of saved samples to do the calculation when time permits. Files of data are used in control of sequenced operations and in exchanges of blocks of data within memory or between the CPU and other controllers. Configuration data for intelligent I/O modules is often saved in a data file in the CPU's memory so that it can be transferred as a single block of data when required.

Data files, data blocks, arrays, and *structures* are all names used to describe sets of simple data values stored together in sequential memory locations. Unfortunately, every PLC manufacturer has its own unique set of instructions and terminology for manipulating data in files, blocks, arrays, and structures.

FILES, BLOCKS, ARRAYS, AND STRUCTURES DEFINED

The names **data file** and **data block** do not imply any standard definition of the *type* of simple data elements in the data set. Proprietary PLCs have their own definitions of what constitutes a *file* or a *block.* The meanings of the terms *array* and *structure* are better accepted, but even the IEC has avoided defining them as standard data units in the IEC 1131-3 standard, to avoid antagonizing supporters among PLC manufacturers.

An **array** is a set of *identical data units.* The data units are usually simple data elements such as bits, bytes, integers, or floating-point numbers. An array of data bits might occupy several data words. (A data word is sometimes treated as an array of bits without ever referring to it as an array.) Arrays of ASCII-coded bytes are usually called *strings.* An array can be *one-dimensional* (i.e., data elements all in a row), *two-dimensional* (i.e., a matrix of data elements), or may have more dimensions. A one-dimensional array would require one **indices** number to identify a member of the array (e.g., STUFF [4] would indicate data element number 4 in the array STUFF). A two-dimensional array would require two indices numbers (e.g., MORE [2,5] means the data element in row 2, column 5 of the array MORE). Additional dimensions would require additional indices numbers.

Structures are arrays of data *elements that do not have to be identical* and don't even have to be simple elements. A structure can, for example, be a set consisting of several bits, integers, arrays, and even other structures. Structures are usually considered single dimensional, so a single indices number can be used to address one element of the structure, even if that element is an array or structure. Additional indices numbers might be required to address a simple data element *within a subarray or substructure* (e.g., MYSTRUCT.4[2] refers to data element 2 in the one-dimensional array that is itself the fourth data element in the structure MYSTRUCT). PLCs sophisticated enough to offer structure data types usually also allow *symbolic addressing,* so data element names can be used instead of some indices numbers (e.g., MYSTRUCT.STUFF[4] is an alternative addressing method for data element MYSTRUCT.2[4]).

PLC-5

ALLEN-BRADLEY DATA FILES

SLC 500

Each data file in an Allen-Bradley PLC is intended to hold data elements of a common type. Some data types are structures or arrays, as described in Chapter 5. Addressing of data in data files is also described in Chapter 5.

S5

SIEMENS STEP 5 DATA BLOCKS

S5 data blocks are intended for the storage of 16-bit data words. Data blocks, and the data words within the data block, must be created *before* a program that reads or writes the data words can be executed. Up to 256 data blocks can be created, each with up to 256 data words.

Only one data block can be active at a time! The program must execute the **DATA BLOCK CALL** (C) instruction before using the data words in that data block. When a new data block is called, the previously active data block ceases to be active. If a program calls a data block, then executes a jump to another program, the data block remains active during execution of the subprogram. If the subprogram calls a new data block, the new data block remains active only until the subprogram ends. When the subprogram ends and the calling program resumes, the data block that was active before the jump becomes active again.

S7

SIEMENS STEP 7 DATA BLOCKS, ARRAYS, AND STRUCTURES

An S7 PLC can store arrays of similar data items, structures of dissimilar data items, and data blocks. Data blocks are like structures, in that they can contain data elements of assorted sizes. Arrays, structures, and data blocks can be defined in the variable declaration tables of STEP 7 programs or can be created using the data block creation features in the programming software. A user data type (UDT) template can be defined and used for creating structures and/or data blocks. How to create and address elements within data sets is described in Chapter 5.

Only two S7 data blocks can be opened (active) at a time: one shared data block and one instance data block. A shared data block cannot be used until an **OPEN A DATA BLOCK** (OPN) instruction is executed within the program to make that data block active. An instance data block is opened automatically when the function block program with which it is associated is called, and is closed automatically when that function block program ends. Any program can open a new data block to replace the one currently active (shared *or* instance). When a program is interrupted by a higher-priority class program, or when a program calls a subprogram, the S7 PLC records the numbers of the data blocks that were active before the interrupt or the call, in a B stack area of memory. When the subprogram or the interrupt service routine ends, the data blocks that were active before become active again. STEP 7 has an **EXCHANGE SHARED DB AND INSTANCE DB** (CDB) instruction, which turns the active shared data block into an active instance data block, and vice versa.

If a program tries to use a data memory that doesn't exist, the PLC will execute a synchronous fault routine (if it exists), then will halt. Chapters 11 and 15 provide more detail.

CQMI

OMRON CQM1 DATA SETS

OMRON doesn't use words like *file, data block, array,* or *structure,* but the CQM1 comes preconfigured with sequential memory locations that can be used as if they were arrays of data units. The memory areas are described in Chapter 5. Because the programmer doesn't create these memory areas, he or she must be careful that the memory that file-manipulation instructions use actually exists.

If an instruction tries to manipulate an array of memories that extends past the end of a memory area (e.g., using 20 holding register addresses starting from HR90 goes past the last holding register memory at HR99), the error status bit (SR 25506) will be set. The error status bit will also be set if the program tries to manipulate the portion of read-only DM memory that is reserved for configuration and status.

BIT ARRAYS AND BIT SHIFT INSTRUCTIONS

Most PLC programs have instructions to shift the bits of a single data word or of a set of data words to the right or left, as shown in Figure 7.1. Since computers can only manipulate whole data units (typically 16-bit data words), an array of data bits being shifted must always be a multiple of the size of the data unit. Some PLC instructions make it appear that bit array sizes are not multiples of whole data words, but they really are.

Before the shift:

word 1	word 2
1111_0000_0000_0000	0000_1111_1111_1111

After shifting one place to the right:

word 1	word 2
?111_1000_0000_0000	0000_0111_1111_1111

(a)

Before the roll that includes a carry bit:

word 1	word 2
1111_0000_0000_0000	0000_1111_1111_1111

After rolling one place to the right:

word 1	word 2
0111_1000_0000_0000	0000_0111_1111_1111

(b)

FIGURE 7.1
Two-word (32-bit) bit array (a) shifted right; (b) rolled right.

Notice that in the bit shift example of Figure 7.1a, the leftmost bit is shown as a "?" after the shift. It will actually be a 0 or a 1, of course, but its value depends on how the bit shift instruction operates. Some PLC bit shift instructions always shift 0's in, while other PLCs allow the programmer to control the value of the bit shifted in. PLCs sometimes offer a variation on the shift instruction called a **roll** instruction. In a roll, the bit that is shifted out of one end of the bit array is put back in at the other end. Some PLC roll instructions include the carry bit of the CPU's status register in the array of bits that it rolls. Figure 7.1b shows a roll in which the 0 from the carry bit is rolled into the leftmost bit of the array, and the new carry bit is the bit rolled out of the right end of the bit array. (These two bits are shown in boldfaced type.)

Shifting of bit arrays might be used, for example, in an automated system to keep track of workpiece status bits to be used by each workstation along a conveyor. Whenever a workpiece on the conveyor is found to be faulty, the status bit corresponding to that part is reset to 0. Each time the conveyor belt moves, the PLC will perform a bit shift operation so that the status bit for each workpiece shifts one place in memory as the workpiece moves one position along the assembly line. The PLC program will check the status bit corresponding to each workpiece before telling the workstations to begin working. Faulty workpieces will not be worked on, of course. New bits that are shifted in will indicate the quality of new workpieces entering the conveyor system. Bits that are shifted out of the array, as workpieces leave, should be passed to the next system in the automated manufacturing system.

PLC-5

SLC 500

ALLEN-BRADLEY BIT SHIFT INSTRUCTIONS

The PLC-5 and the SLC 500 PLCs offer **BIT SHIFT LEFT** (BSL) and **BIT SHIFT RIGHT** (BSR) instructions that can treat a sequential series of data words as a single bit array. The bit shift instructions shift the bits of the array *one* space each time the control logic to the instruction is seen as going true. Allen-Bradley bit shift instructions make it look as if bit arrays can have sizes that aren't multiples of 16, suggesting that they can affect part of a data word and leave the rest of the data word unaffected, but that isn't the way it actually works. Whole 16-bit data words are shifted.

The BSL instructions in Figure 7.2 will execute when its control statement goes from false to true (when input image bit I:4/3 turns on). The BSL instruction data entered by a programmer includes:

1. A *file address* (#B3:2 in this example); any word address can be entered. The rightmost bit of that word will be included in the bit array to be shifted regardless of whether the instruction is a BSL or a BSR. The "#" is essential, to tell the PLC to use S:24 and indexed addressing to keep track of the data words that it must shift one word at a time. B3: [2 + S.24] is calculated as B3: (2 + 0), then B3: (2 + 1), then (if the bit array was longer than in this example) B3: (2 + 2), and so on.

2. The *length* indicates the number of bits that the programmer wants to be included in the shift. Bits are counted from the rightmost bit of the "file" data word toward the more significant bits, and into the next data word(s) if a length over 16 is specified. The PLC will store the length number in the ".LEN" data word of the control element. The PLC will round this number up to the next full increment of 16 data bits to determine how many bits to shift. 32 bits will actually be shifted to shift the 20 bits of Figure 7.2.

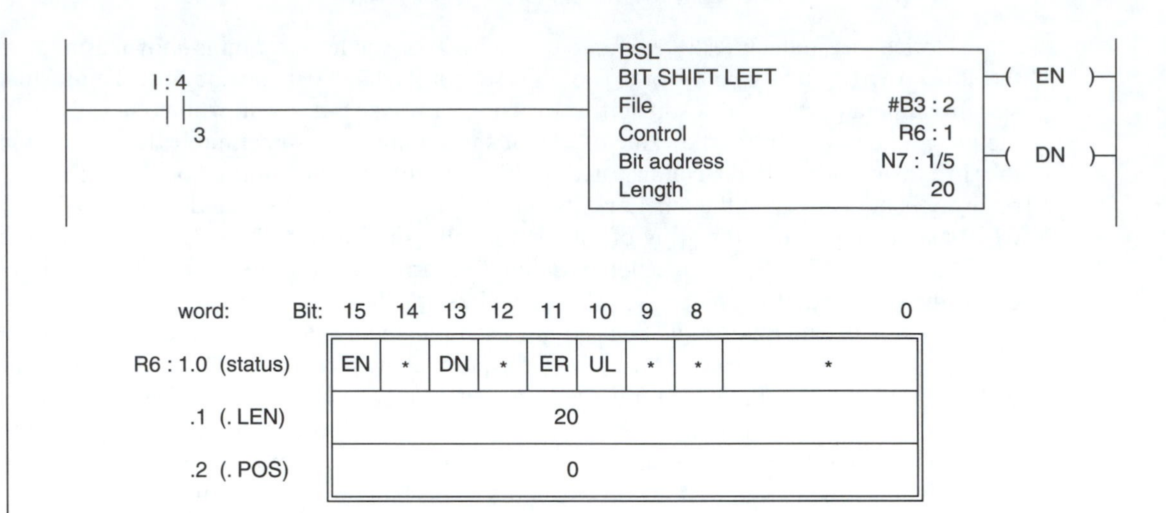

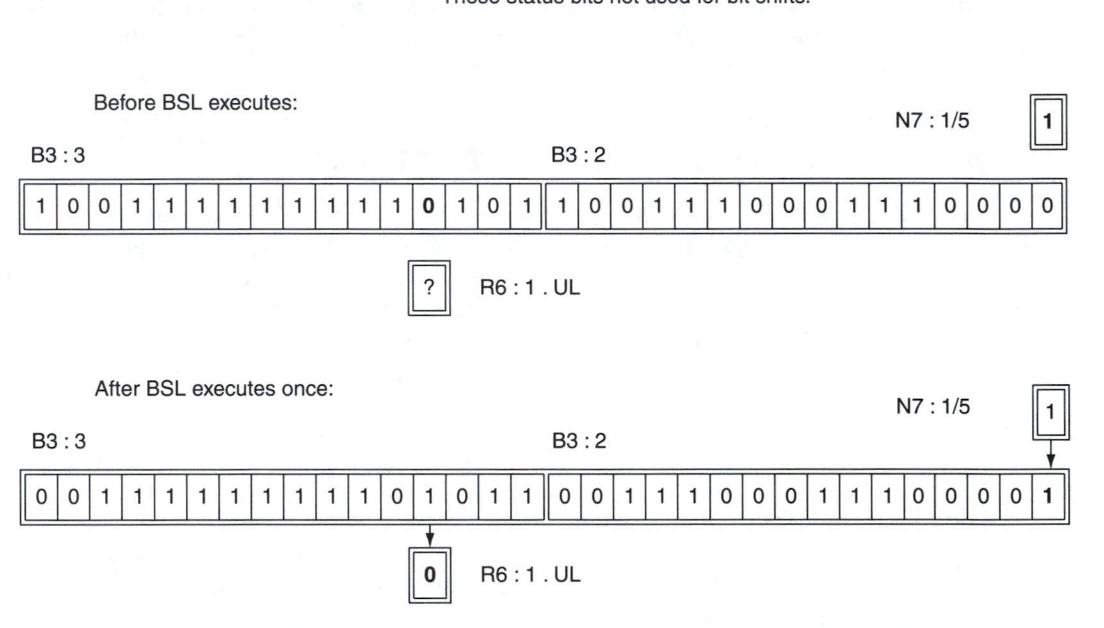

FIGURE 7.2
Allen-Bradley BIT SHIFT LEFT instruction, control block, and effect on memory contents.

3. A *control element* (Rf:x) is used by the PLC to store data that it needs to execute the file manipulation instruction. The ".POS" word of the control element's structure is used to keep track of how many bits it has already shifted. Four status bits are used by the BIT SHIFT instruction:

UL (*unload bit*) will contain the value (1 or 0) of the last bit that the programmer wanted shifted, after the shift executes. During a BSR it will receive its value from the low bit of the file address word. During a BSL it will receive its value from the data bit indicated by the control element's ".LEN" word (i.e., the twentieth bit in the example in Figure 7.2).

EN (*enable bit*) reflects the status of the instruction's control logic (which must go true to cause the instruction to execute).

DN (*done bit*) goes true after the instruction finishes executing, and stays true until the instruction's control logic goes false.

ER (*error bit*) will become true if the instruction tries to execute but fails, usually because of a programming error, possibly because the bit array includes bit memory that doesn't exist. This bit resets to zero when the control logic goes false.

4. A *bit address* is specified to indicate where the PLC should get a bit value to shift *into* the bit array. The value from this address (N7:1/5 in the example) will be shifted into the low bit of the file address word during a BSL, or will be shifted into the bit indicated by the .LEN data word during a BSR.

S5

SIEMENS STEP 5 BIT SHIFT INSTRUCTIONS

STEP 5 offers **SHIFT LEFT WORD** (SLW) and **SHIFT RIGHT WORD** (SRW) instructions, which unconditionally shift the 16-bit bit pattern in accumulator 1 each time they are encountered in the program. A number (0 to 15) following the instruction indicates how many places to shift the bits. Zeros are shifted in to fill the bit positions vacated. Bits shifted out are shifted out via the CC 1 bit, so the last bit shifted out can be examined if another instruction hasn't affected the CC 1 bit since the shift executed. For example,

```
                    Accumulator 1          CC1 bit
L KH F05F           1111 0000 1100 1111       ?
SRW 3               0001 1110 0001 1001       1
```

The SRW and SLW instructions can only be programmed in function blocks. Ladder logic versions are available.

S7

SIEMENS STEP 7 BIT SHIFT INSTRUCTIONS

STEP 7 offers the SHIFT LEFT WORD (SLW) and SHIFT RIGHT WORD (SRW) instructions that are offered in STEP 5 but they can now be programmed in any program block and they work slightly differently:

1. The **SHIFT LEFT WORD** (SLW) and **SHIFT RIGHT WORD** (SRW) instructions cause only the low word (16 bits) of accumulator 1 to be shifted, without affecting the high word.

STEP 7 adds some more shift and rotate (Siemens' word for roll) variations:

2. The whole 32-bit data value in accumulator 1 can be shifted using the **SHIFT LEFT DOUBLE WORD** (SLD) instruction or the **SHIFT RIGHT DOUBLE WORD** (SRD) instruction, followed by a number (0 to 32).

3. The low word, or the whole double word in accumulator 1, can be shifted right without changing the sign bit at the word's or double word's leftmost position, using the **SHIFT**

SIGNED INTEGER (SSI) or the **SHIFT SIGNED DOUBLE INTEGER** (SSD) instructions. A number (0 to 15 or 0 to 32) indicates how many bits to shift.

4. The 32-bit value in accumulator 1 can be rotated (rolled) right or left. When rolled, the bit values that exit at one end of the double word are immediately put back into the other end of the double word so that the bit pattern's sequence isn't lost. There is a **ROTATE LEFT DOUBLE WORD** (RLD) instruction and a **ROTATE RIGHT DOUBLE WORD** (RRD) instruction, which must be followed by a number (0 to 32) indicating how many bit positions to rotate.

5. Any of the above-listed shift and rotate instructions can be entered (in STL only) without a number to indicate how many bit positions to shift. If entered in this way, the S7 PLC reads the contents of accumulator 2 to determine how many bit positions to shift/rotate accumulator 1. For example,

```
                    Accumulator 1                      Accumulator 2
L +3                      3                                  ?
L MD4        11111111111110000000000000000000                3
RLD          11111111110000000000000000000111                3
```

6. A **33**-bit rotation is possible, but only if programmed in STL. The thirty-third bit is the CC 1 bit, which receives the bit rotated out and supplies a bit to be rotated into the other end of accumulator 1. These instructions don't require a number indicating how many bit positions to rotate, since they can only rotate the bit pattern one position per instruction execution. There is a **ROTATE LEFT VIA CC 1** (RLDA) instruction and a **ROTATE RIGHT VIA CC 1 (RRDA)** instruction.

There are ladder logic instructions for most of the shift and rotate instructions described above. A number (N) *must* be entered in ladder logic to indicate how many bit positions to shift/rotate, but the number can be any positive value up to hexadecimal FFFF. The ENO output of the ladder logic instruction is the same as the CC 1 bit, which means that it reflects the state of the most recent digit shifted or rolled out.

CQMI

OMRON CQM1 BIT SHIFT INSTRUCTIONS

The CQM1 offers four shift or rotate (roll) instructions that can alter a single data word, and two instructions for shifting larger bit arrays. The following instructions shift or rotate a single 16-bit data word in memory, one bit position every scan cycle while their control logic is true. If programmed in their differentiated form (with an @ prefix), they will cause a shift or rotate once each time the control logic goes from false to true. These instructions affect the carry bit (SR 25504), the equals zero bit (SR 25506 goes on if the 16-bit result is zero), and can affect the error bit (SR 25503):

1. ARITHMETIC SHIFT LEFT, **ASL (25),** and ARITHMETIC SHIFT RIGHT, **ASR (26).** ASL (25) causes all 16 bits to be shifted one place toward the most significant bit (bit 15). Bit 15 is shifted into the carry bit (SR 25504), and the previous contents of the carry bit are lost. A zero is shifted into the least significant bit (bit 0). ASR (26) causes all 16 bits to be shifted one place toward bit 0, bit 0 to be shifted into the carry bit, and a zero to be shifted into bit 15.

2. ROTATE LEFT, **ROL (27),** and ROTATE RIGHT, **ROR (28).** The ROL (27) instruction causes all 16 bits to be shifted one place toward the most significant bit (bit 15). Bit 15 is shifted into the carry bit (SR 25504) and the contents of the carry bit are shifted into the least significant bit (bit 0). ROR (28) causes all 16 bits to be shifted one place toward bit 0, bit 0 to be shifted into the carry bit, and the carry bit's value to be shifted into bit 15.

The following shift instructions can cause shifting of bits in *sets of multiple words.* The two instructions are programmed very differently.

1. The SHIFT REGISTER instruction, **SFT (10),** demonstrated in Figure 7.3, has two parameters. The first parameter is the *start word* (**St**) address (IR 130 in Figure 7.3), and the second is the *end word* (**E**) address parameter (IR 132 in Figure 7.3), so this instruction can cause all 48 bits of output words 130, 131, and 132 to be shifted together. The SFT (10) instruction can be programmed with as many as three Boolean logic control inputs. When the control logic at the *pulse (P)* input goes from false to true, all the bits in the shift register between the start word and the end word are moved one place toward the most significant bit of the highest address (toward bit 15 of word IR 132). The most significant bit of the shift register range (bit 15 of IR 132) is lost. The status (true = 1, false = 0) of the control logic on the *input bit (I)* input is shifted into the lowest bit of the lowest word (bit 0 of IR 130). If the control logic on the *reset (R)* input is true, *all* the bits in the shift register are reset to 0 and no shifts are allowed until this logic returns to false. The SFT (10) instruction does not have a differentiated form (can't be programmed with an @ prefix) because it inherently executes only once each time the pulse logic goes from false to true.

2. The REVERSIBLE SHIFT REGISTER instruction, **SFTR (84),** works similarly to the SFT (10) instruction but is programmed very differently, as the example in Figure 7.4 shows. As with the SFT (10) instruction, the start word and end word addresses are needed, but they are entered as the second and third parameters. A control word address (**C**) parameter is entered as a first parameter (address IR 012 in Figure 7.4 is a work area memory) and must be entered as an address, not as a constant. There is an optional differentiated form of the SFTR (84) instruction (which is used in this example), because in its undifferentiated form the SFTR (84) instruction causes a one-position shift every scan cycle that its control logic remains true. In the differentiated form (with the @ prefix), there will only be a single position shift each time the control logic goes from false to true.

The bits in the control word perform the functions of the control logic inputs of the SFT (10) instruction, and more. (The bits in the control address can be controlled by other rungs in

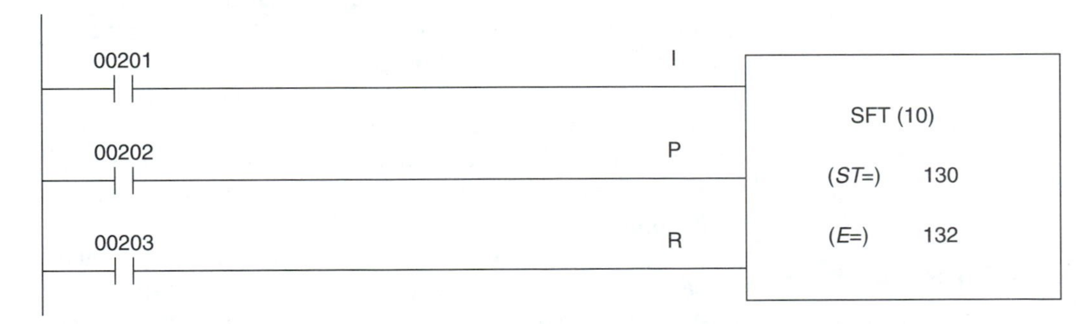

FIGURE 7.3
CQM1 shift register instruction in a program.

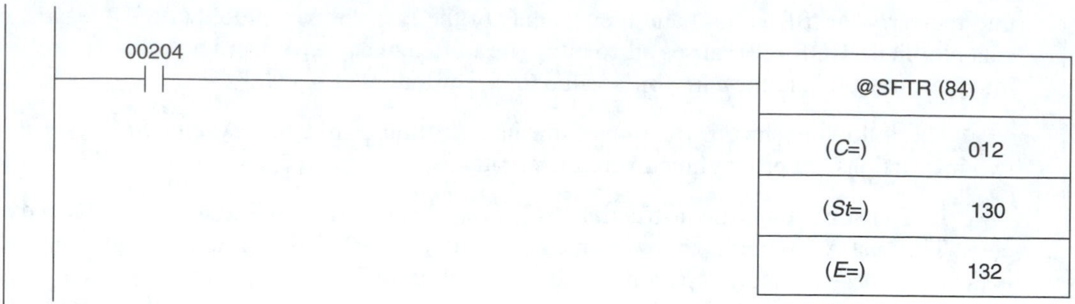

00204

@SFTR (84)	
(C=)	012
(St=)	130
(E=)	132

FIGURE 7.4
CQM1 reversible shift register in an example program.

the user-program). If the *reset bit* (bit 15) is off, the contents of the entire shift register (words IR 130 to 132) will be reset to zero and can't be shifted. The user-program must turn the reset bit (IR 01215 in the example) on so that the SFTR (84) instruction can work. The *input bit* (bit 13, IR 01213 in Figure 7.4) contains the value that will be shifted into the shift register at the right end if bits are shifted left, or at the left end if bits are shifted right. Bit 12, the *shift direction bit,* controls which way the bits will be shifted (1 for right, 0 for left). Bit 14 is called a *shift pulse bit* but should be considered an enable bit. The shift register will be frozen if the shift pulse bit is off, so the user-program must also turn this bit on. Bits 0 to 12 in the control address have no effect.

Other CQM1 shift instructions affect groups of bits (4-bit nibbles or 16-bit words) and are discussed in the following section describing array operations.

ARRAY SHIFT INSTRUCTIONS (INCLUDING FIFO AND LIFO)

Just as bits can be shifted, some PLCs allow *whole data words* (or *part words*) to be shifted up or down within sequential memory locations. A set of memory locations can, for example, be used to hold descriptive data for individual workpieces, and the descriptive data can be shifted up and down in a stack of descriptors as the workpieces move through an automated system.

Names like **first-in-first-out (FIFO)** and **last-in-first-out (LIFO)** describe two ways that data can be stored in and retrieved from a **stack** of memory locations. Data values that are put into the set of memory locations are said to be **pushed** onto the **top** of the stack. Data can be taken from the bottom of the stack, so that data that has been in the stack longest is removed first (FIFO), or it can be taken from the top, so that the data most recently put onto the stack is now removed (LIFO). The PLC needs to maintain stack-control information while pushing data onto the top of the stack and popping data off the stack.

Even if the PLC doesn't offer array shift instructions, a programmer can usually accomplish the same results through the use of indexed addressing or indirect addressing. An instruction can be made to operate on sequential data values if each time it operates, its indexing address offset or its indirect address pointer is incremented or decremented. Instead of actually shifting the data in memory, the address reference is thus shifted. FIFO and LIFO instructions often use indexed addressing internally in this way.

PLC-5

SLC 500

ALLEN-BRADLEY FIFO AND LIFO INSTRUCTIONS

The PLC-5 and SLC 500 both offer FIFO and LIFO instructions. Figure 7.5 demonstrates a program that uses a **FIFO LOAD** (FFL) instruction and a **FIFO UNLOAD** (FFU) instruction. The FIFO LOAD instruction places a 16-bit data value into a FIFO array each time its control logic goes from false to true (i.e., when B3/4 goes on). The FIFO UNLOAD instruction removes the oldest data item from the FIFO each time B3/5 goes on but only if the FIFO isn't empty (i.e., if R6:2.EM is not on). In entering FFL and FFU instructions, the programmer must enter parameters for:

1. A *source address* (FFL only) to indicate where the data word that is to be placed onto the stack is to be copied from. In the example, there are already three data words in the stack, so the next data word from the source address will be pushed into the fourth data position. The top of the stack is the highest-numbered address that contains a data word that has been pushed onto the stack, and the address of the top of the stack changes as data is pushed and popped from the stack.

2. A *destination address* (FFU only) to indicate where data that is popped (unloaded) from the stack is to be copied to. As shown in Figure 7.5, the oldest data in the stack is always at the lowest-numbered address, and when it is popped from the stack the PLC has to shift all the remaining data values one memory location so that the next-oldest data value is moved into the lowest-numbered address.

3. The *FIFO address,* complete with the essential "#", indicates where the stack's bottom address is (the lowest address number).

4. The *length* indicates how many memory locations are to be reserved for the stack, starting with the FIFO address. The maximum length is 64 words.

5. A *control element* must be entered for each FIFO stack. The FFL and FFU instructions must both use the same control element, which contains data indicating the maximum size of the FIFO (the length number) and the number of data words currently on the FIFO stack (which either the FFL or FFU can change). The control element also contains the following status bits:

- `EN` (*load enabled*), which will be on if the control logic for the FFL instruction is true.
- `EU` (*unload enabled*), which will be on if the control logic for the FFU instruction is true.
- `DN` (*done*), which will be true if the FIFO stack is full. When full, the PLC will not place new data words onto the stack.
- `EM` (*empty*), which will be true if there are no data words on the stack (all pushed words have been popped). It should be checked as part of the control logic for any FFU or LFU instruction to ensure that there is data in the stack before unloading a data item.

The LIFO instructions, **LIFO LOAD** (LFL) and **LIFO UNLOAD** (LFU), are similar to FFL and FFU. A LIFO differs from a FIFO in that data is unloaded (popped) from the top (highest address) of the LIFO stack, so there is never any need for the PLC to actually shift the data words up or down in the stack.

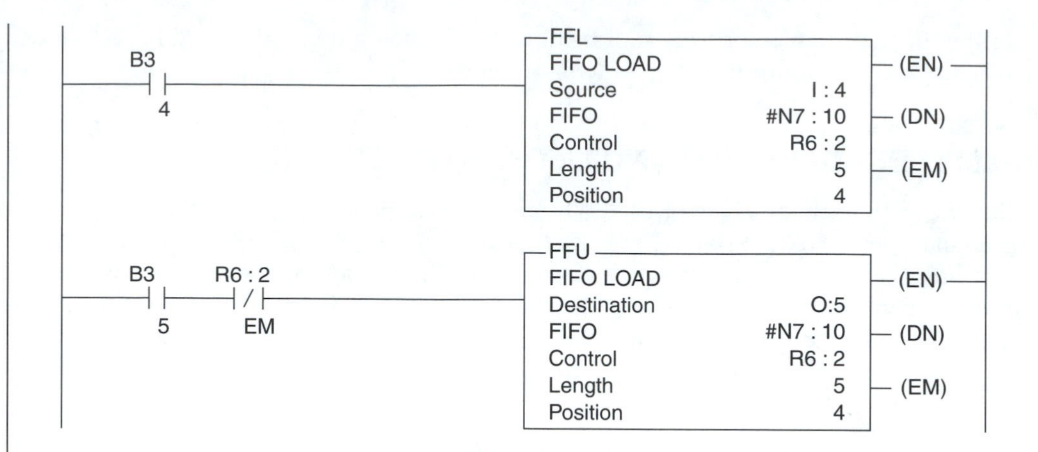

Word:	Bit: 15	14	13	12	11	10	9	8	0
R6:1.0 (Status)	EN	EU	DN	EM	*	*	*	*	*
1 (.LEN)	5								
2 (.POS)	4								

*These status bits not used for FIFO or LIFO control.

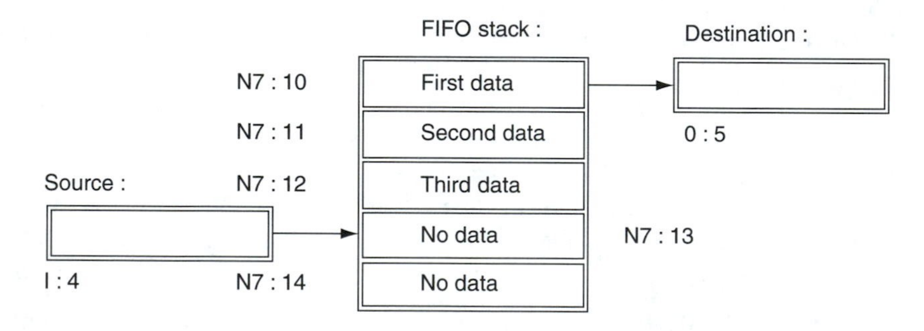

FIGURE 7.5
Allen-Bradley FIFO LOAD and FIFO UNLOAD instructions, control element, and the FIFO data stack.

S5

SIEMENS STEP 5 ARRAY SHIFT INSTRUCTIONS

The S5 line of PLCs do not offer any instructions that shift arrays of data words. The programmer must write his/her own program for shifting data in sequential memory locations. The indexed addressing **PROCESSING OPERATION** (DO) instruction and the **PROCESSING INDIRECT** (DI) instructions discussed in Chapter 5 can be used, and the structured programming techniques discussed in Chapter 8 also help.

S7

SIEMENS STEP 7 ARRAY SHIFT INSTRUCTIONS

STEP 7 does not offer any instructions specifically for shifting of data in structures and arrays. The programmer must write his or her own programs. Memory indirect addressing and register indirect addressing techniques, and the addressing techniques used for ARRAYS and STRUCTS (discussed in Chapter 5), can be used to make the job easier. The structured programming techniques discussed in Chapter 8 can also be used.

CQMI

OMRON CQM1 ARRAY SHIFT INSTRUCTIONS (INCLUDING FIFO AND LIFO)

Although OMRON's technical literature doesn't use the word *array,* the CQM1 can shift arrays of 16-bit data words, or arrays of 4-bit nibbles (which OMRON calls *digits*). OMRON's ASYNCHRONOUS SHIFT instruction is uniquely suited for shifting arrays of data words in the same way that products are conveyed through a production system with multiple storage buffers. FIFOs and LIFOs can be programmed using the SINGLE WORD DISTRIBUTE instruction in combination with the DATA COLLECT instruction. The CQM1 also has instructions designed to shift keyboard values into small arrays.

The **WORD SHIFT** instruction, **WSFT(16)**, can be used to shift 16-bit data words in an array up one memory location. The 16-bit value 0 is shifted into the lowest address as words are shifted up, and the contents of the highest memory address are lost as it is shifted out of the word array. The WSFT(16) instruction requires two parameters: the *start word* address (St) and the *end word* address (E). There is a differentiated form of the WSTF(16) instruction: @WSFT(16), which is needed because in its undifferentiated form, WSFT(16) shifts the array once every scan cycle while its control logic remains true. Additional instructions [e.g., MOV(21) instructions] can be included in the program to copy data words from the end word address before the WSFT(16) executes, and to copy new data words into the start word address after the WSFT(16) executes.

Two instructions can shift arrays of 4-bit digits: the **ONE DIGIT LEFT** instruction, **SLD(74)**, and the **ONE DIGIT RIGHT** instruction, **SRD(75).** (OMRON calls nibbles *digits* because the CQM1 often uses BCD coding, in which decimal digits are represented by binary nibbles.) The digit-shift instructions require a *start word* parameter (St) and an *end word* parameter (E), but St can be the same address as E to shift only the nibbles within a single word. When an SLD(74) instruction executes, it shifts 4 zero bits into the lowest nibble of the start word address. An SRD(75) shifts zeros into the highest nibble of the end word address. The other nibbles are all shifted left or right by four bit positions. Nibbles shifted left out of the high bits of one word are shifted into the low bits of the next-higher word, or vice versa. The nibble that is shifted out of the end of the series of nibbles is lost. Unless the differentiated form [@SLD(74) and @SLR(75)] is used, a shift occurs every scan cycle while the instruction's control logic remains true.

The **ASYNCHRONOUS SHIFT REGISTER** instruction, **ASFT(17)**, shifts data words up or down within an array of data words, but shifts a word only if there is an empty space in the array to shift it into. (A memory containing the value 0 is considered an empty space.) The memory that a data word is shifted out of then becomes an empty space (contains the value zero) and a data word can be shifted to that space the next time the ASFT(17) instruction executes. One data word is shifted for each empty space that exists in the array (leaving one empty space behind). ASFT(17) causes a new set of shifts each scan cycle while its control logic remains true, unless the differentiated form, @ASFT(17), is used. ASFT(17) requires a *control* parameter (C), followed by a *start word* parameter (St), then an *end word* parameter (E). The ASFT(17) instruction's control logic must be true for the instruction to execute, of course, but if the instruction executes, the control word determines what it will do. A 1 in the *direction control* bit (bit 13) of the control word makes ASFT(17) try to shift words down in memory, and 0 means up. The *shift enable* bit (bit 14) must be on; otherwise, ASFT(17) won't shift any data words. When the reset bit (bit 15) is on, the array is cleared (all words are changed to 0's). These control word bits can be manipulated by other instructions in the user-program. The other control word bits aren't used by the ASFT(17) instruction.

FIFO (first-in-first-out) stack arrays and LIFO (last-in-first-out) stack arrays can be implemented using the **SINGLE WORD DISTRIBUTE** instruction, **DIST(80)**, to put data words onto the stack and the **DATA COLLECT** instruction, **COLL(81)**, to remove data from the top or bottom of the stack. DIST(80) sets the error status bit (SR 25503) if the stack exceeds its assigned size. (These two instructions can also be used in a type of indexed addressing move, as we saw in Chapter 6.) The example in Figure 7.6 demonstrates how to use DIST(80) and COLL(81) to implement FIFO and LIFO stacks.

Prior to running the example program in Figure 7.6, the programmer has cleared the contents of data memories IR 040 and IR 060. Rung 1 contains a **@DIST(80)** instruction that has a control word (C) entered as a constant with the (BCD) value 9010. The control word's highest digit is 9, so @DIST(80) will perform a stack operation, and the following digits (010) indicate that the stack has a size of 10 words. Whenever bit 1 of input module 3 (IR 00301) goes from false to true, the differentiated @DIST(80) instruction increments the stack pointer and copies one data word from the *source* word (S = IR 001) onto the top of the 10-word stack and moves the other words in the stack down. A stack pointer is stored at the address specified by the *destination base* word parameter (DBs = IR 040), and the stack's top is actually one memory location higher (IR 041). Whenever new words are pushed onto the top of the stack (IR 041 is the top, even though it is the lowest-numbered address in the stack), the other words in the stack must be pushed to higher memory locations. The word that has been on the stack the longest is therefore at a memory location that can be found by adding the destination base parameter's address (IR 040) and the stack pointer (offset) value it contains. For example, if three words have been pushed onto the stack by the program in Figure 7.6, the contents of IR 040 will have been incremented to 0003 from its original value of 0000. The destination base parameter is IR 040, so the oldest value in the stack is at IR 043.

In rung 2, a differentiated form of the **COLL(81)** instruction executes once each time bit 2 of input module 3 goes from false to true. Because @COL(81)'s control word (C) has a **9** as its most significant digit, the @COL(81) instruction performs a **FIFO** removal of a data word from a stack, writing it to the *destination* word (D), which is IR 105 in this example. The length of the stack is 10, as indicated by the three low digits in the control word. The *source base* word (SBs) parameter is address IR 040, so this instruction uses the same stack as did the previous rung.

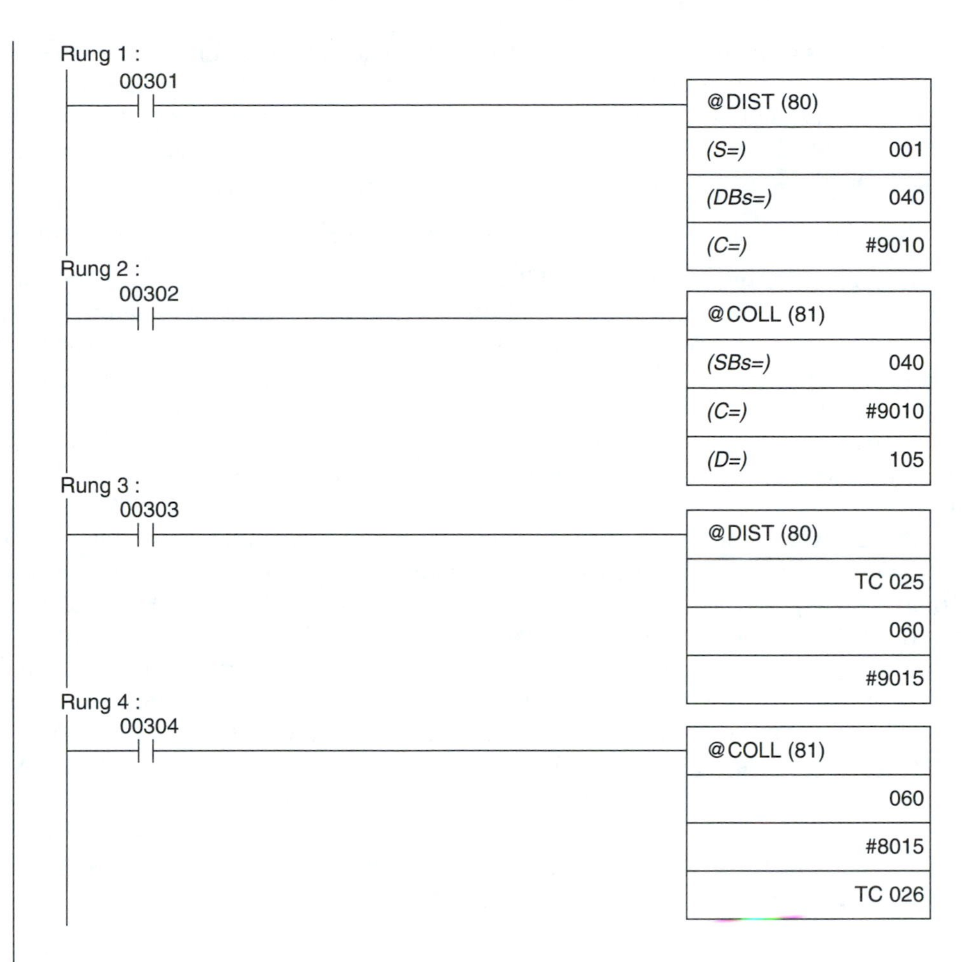

FIGURE 7.6
CQM1 FIFO and a LIFO in an example program.

When @COLL(81) performs a FIFO removal of a data word from the stack, it removes the oldest word from the stack (from IR 043 in the example so far), then decrements the stack pointer offset value at IR 040. Notice that IR 040 is used as a stack pointer by both the @DIST(80) and the @COLL(81) instructions, so rung 1 can continue to put data words onto the stack as long as rung 2 removes them fast enough that the stack size never exceeds 10.

In rungs 3 and 4, another @DIST(80) and @COLL(81) pair manipulate a 15-word stack that starts at IR 061 and has a stack pointer stored at IR 060. Present values from one counter (TC 025) are put onto the stack by rung 3 and copied to another counter (TC 026) by rung 4. This time, the @COLL(81) instruction's control word has **8** as its highest digit, so the @COLL(81) instruction performs a **LIFO** removal of data words from the stack. In a LIFO data removal, the data word at the top of the stack (IR 061) is always removed from the stack (if there is a data word there), the words in the stack are shifted toward the top, and the stack pointer is decremented.

Some CQM1 instructions are designed for capturing keyboard data into small arrays.

1. The TEN-KEY INPUT instruction, **TKY(18)**, can detect which of the 10 low bits of an input word is on (the IW parameter usually indicates the address of a digital input module), and will record that bit number (0 to 9) as the next BCD digit in a two-word array starting at the high nibble of the address specified by a first register address word (D1) parameter. The previous nibbles in the array are shifted right (toward the low nibble of the next address after D1), then out, so the most recent eight switch actuations can be saved in D1 and D1+1. Bits in an address specified as the second register address word (D2) parameter indicate which input bit was on last and whether that bit is still on.

2. The HEXADECIMAL KEY INPUT instruction, **HKY(−),** examines the low 4 bits of an input word (IW) address as it sequentially actuates the low 4 bits of an output word (OW) address. If a hexadecimal keypad is connected to those output bits and input bits, the result will indicate which of the 16 keys (0 to F) are being pressed. That hexadecimal value will be entered into the high nibble of the memory specified in the first register word (D) address parameter, the previous nibbles in that address and the next-higher address are all shifted right one place, and the lowest nibble of the higher word is lost. Eight keyboard entries are thus saved in the array. The bits of the second word above the first register word indicate the most recently actuated hex key (0 to 15 means 0 to F), and bit 4 of the output word address is on while the key is still being held on. Since execution takes multiple scan cycles to capture a single keystroke, the CQM1 turns a status bit (SR 25408) on while HKY(−) is executing. Turn HKY(−)'s control logic off to prevent it from continuously rereading the keypad.

MOVING FILES, ARRAYS, AND STRUCTURES

Some PLCs offer powerful MOVE instructions that can copy whole data sets or parts of data sets from one area of a PLC's memory to another area. Such an instruction might be used, for example, to copy the current input states of a whole rack of input modules to memory every hour for record keeping, or may be used to change a whole set of working data values whenever sensors detect that the type of product entering a workstation has changed. In some PLCs, block transfer instructions are used to copy arrays of data between the CPU's memory and an intelligent I/O module's memory.

Sequencers are ancestors to the modern PLC. A sequencer performs file move operations, copying data words from a series of memory locations, one at a time, to a single output port address to step a set of actuators through a predefined set of states. Some modern PLCs include **sequencer instructions.** Sequencer output instructions can be controlled by Boolean logic statements examining timer outputs (for timed intervals between output changes) or input bits (for event-triggered output changes). Logic statements controlling sequenced outputs may even contain sequencer compare instructions.

PLC-5
SLC 500

ALLEN-BRADLEY FILE MOVE, SEQUENCER, AND BLOCK TRANSFER INSTRUCTIONS

Both the PLC-5 and the SLC 500 offer a **FILE COPY** (COP) instruction that copies a whole series of data elements from one area of memory to another every scan cycle while the FILE COPY instruction's control logic remains true (not just once when the control logic goes true). Any type of data element can be copied, even structures such as the three-word structures used for counter control. The COP instruction actually copies data as 16-bit data words, so data can be copied between dissimilar data file types (even between integer files and counter element files), but remember that the COP instruction does not convert data as it copies it. In copying data between data files that have different sizes of data elements, the COP instruction will copy as many elements as required to fill the programmed length of destination elements. As shown in Figure 7.7, the instruction's source and destination parameters must both be preceded by the "#" symbol, because the PLC must use indexed addressing (automatically using status word S:24) to keep track of the addresses it is using in the source file and addresses it is using in the destination. In the example, there is no condition controlling the COP instruction, so every scan it copies 30 integer data words from addresses N7:4 to N7:33 into 10 three-word counter elements from C12:2 to C12:11.

The **FILE FILL** (FLL) instruction copies the contents of a *single* address (or constant) into all of the elements of the destination. The # symbol is not needed in the source address, of course, since only the single address is required.

Allen-Bradley PLCs offer a **SEQUENCER OUTPUT** (SQO) instruction, shown in Figure 7.8. Every time the instruction's control logic statement goes true, the SQO instruction copies the next data word from a series of memory locations to a destination address. The series of data words will have been entered into memory before the SQO instruction tries to use them. The programmer must enter:

1. The *file address* at which the series of data words starts. The # symbol is necessary because the PLC uses indexed addressing.

2. A *length* number, indicating how many data words there are in the sequenced series of output values. There must be one more data word than this number in the file of data words in memory, because the first data word (word 0) isn't considered part of the standard sequence. The SQO instruction outputs the value in word 0 only if the sequencer position is at 0 and the control logic statement is true when the PLC is switched into run mode. After the sequencer out-

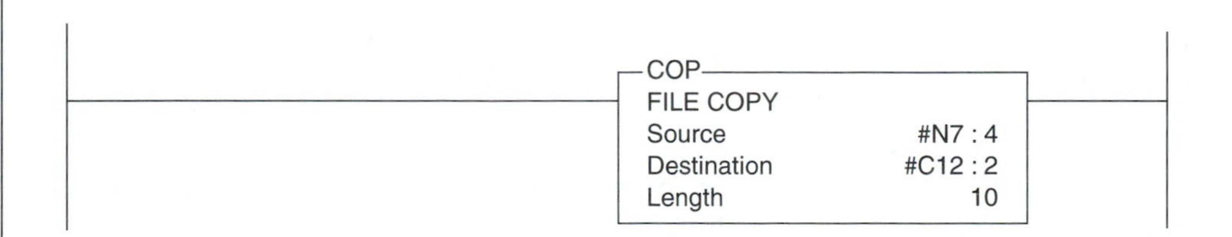

FIGURE 7.7
Allen-Bradley FILE COPY instruction.

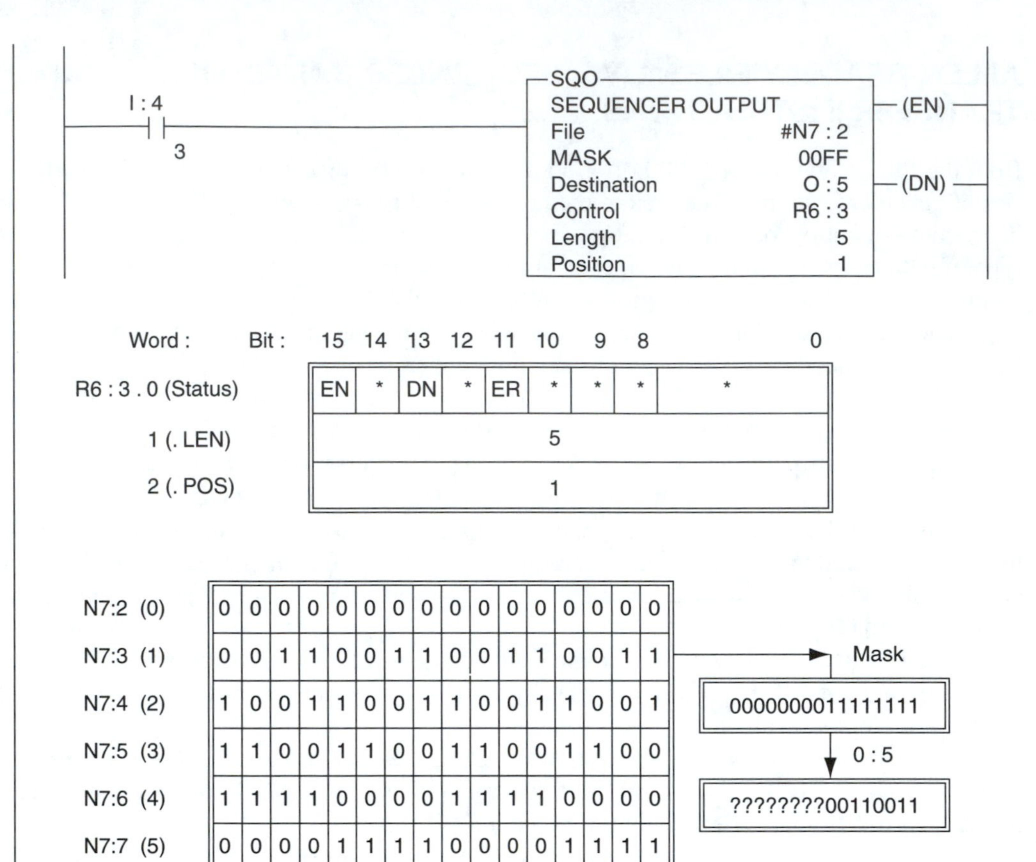

FIGURE 7.8
Allen-Bradley SEQUENCER OUTPUT instruction, control element, and a typical file of sequencer output words.

puts its last data word in the sequenced cycle, the next data word to be outputted will be word 1, not word 0.

3. A *mask* must be entered as a hexadecimal value. The mask indicates which of the 16 bits at the destination address are to be changed as the sequencer operates and which are to be unaffected by the sequencer. A binary 1 in the 16-bit binary mask means that the SQO instruction can overwrite the bit at that position in the destination address.

4. The *destination address* is where the sequencer's output values are to be written to. An output image address is generally used.

5. A *control element* is required so that the PLC can keep track of which position it is in the sequenced series of output words.

In the SLC 500 example in Figure 7.8, I:4/3 has just gone on for the first time, so the SQO instruction will copy the contents of data word 1 (N7:3, not N7:2) to the destination address (O:5) via the mask (hex 00FF), which only allows the SQO instruction to change the low 8 bits of the destination address. The high 8 bits at the destination address cannot be changed by the SQO instruction. As long as the input logic remains true, N7:3 will be copied to O:5 every scan cycle. When the SQO instruction's control logic goes false, the position value (at R6:3.POS) will change to 2 and the SQO instruction will stop writing to the output address. When the control logic goes true again, the contents of address 2 (N7:4) will be copied to the output address.

A SEQUENCER LOAD (SQL) instruction is also available, so a user-program can place data into a series of data words, one data word each time the SQL's control logic goes true. The SQL instruction includes specification of a source data address instead of a destination address, of course. The SQL instruction was originally designed to put data words into a file that would be used by another sequencer instruction, but the SQL instruction can be used to capture data from any source memory location into a sequenced series of memory locations.

The PLC-5 offers a SEQUENCER INPUT (SQI) instruction. The SLC 500 offers the same instruction but calls it SEQUENCER COMPARE (SQC). These instructions are intended to control SQO instructions, and they are covered later in the chapter when we examine file compare instructions.

SLC 500 SLC 500 Only: Sequencer Differences and the Swap Instruction

When programming sequencer instructions in an SLC 500 PLC, an address can be entered for the mask, and indexed addressing can be used for the mask and destination. Better SLC 500s (SLC 5/03 and above) offer a SWAP (SWP) instruction that can be used to exchange the high and low bytes of a series of 16-bit memory locations in bit, integer, ASCII, and string files. It is programmed with an indexed start address and the number of word elements to operate on.

PLC-5 PLC-5 Only: Block Transfer Instructions

In the PLC-5 system, intelligent I/O modules are called **block transfer modules,** because the CPU exchanges blocks of data with these I/O modules in response to BLOCK TRANSFER READ (BTR) instructions or BLOCK TRANSFER WRITE (BTW) instructions in the user-program. A BTW instruction is shown in Figure 7.9. A BLOCK TRANSFER instruction causes the PLC to pass a data exchange *request* to a communications handler in the CPU module. The PLC does *not* wait for the communication handler to finish exchanging the requested data with the I/O module before going on to the next instruction.[1] The communication handler handles data exchanges with I/O modules, one exchange at a time, in the order in which it receives requests from the program. Since the data exchanges requested by block transfer instructions are not necessarily performed within the same scan cycle as when they are requested, the communication handler must provide addressable status information to the main microprocessor so that the user-program can be programmed to react properly when the data exchange is over. Communication errors can occur if a program issues a second block transfer request for an I/O module

---

[1] The PLC does wait for BTR and BTW instructions to execute when they are programmed in an interrupt program, as we will see in Chapter 11.

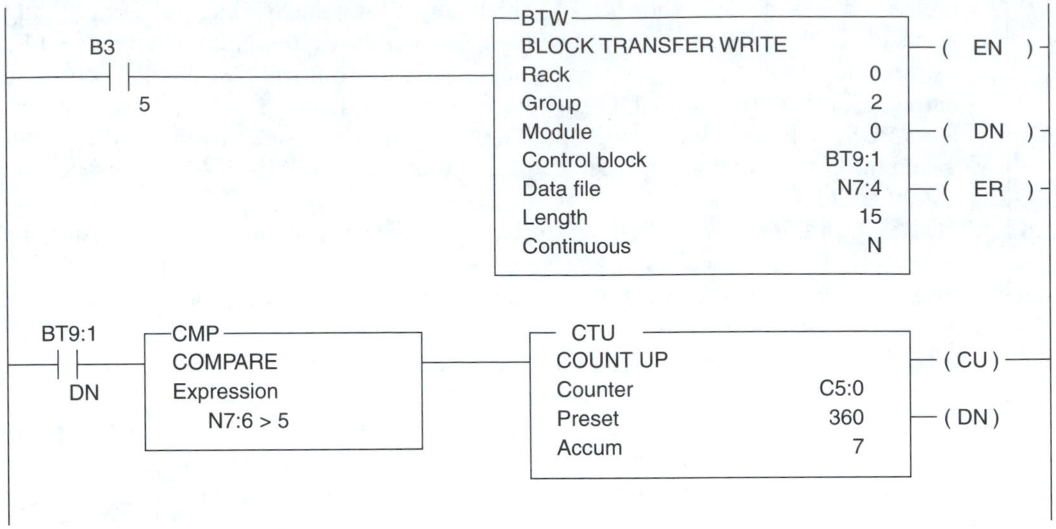

B3
5
BTW
BLOCK TRANSFER WRITE
Rack 0
Group 2
Module 0
Control block BT9:1
Data file N7:4
Length 15
Continuous N
(EN)
(DN)
(ER)

BT9:1
DN
CMP
COMPARE
Expression
N7:6 > 5
CTU
COUNT UP
Counter C5:0
Preset 360
Accum 7
(CU)
(DN)

FIGURE 7.9
PLC-5 program that reads data from an intelligent I/O module using a BLOCK TRANSFER READ instruction and counts the number of times one of the values read exceeds the value 5.

before an earlier block transfer is complete. The following explains the BLOCK TRANSFER READ instruction that is used in Figure 7.9, and the exchange of status information:

1. The BTR instruction will execute once (issue a request to the communication handler) every time the BTR's control logic changes from false to true, because the programmer selected N (no) for **continuous** operation when entering the BTR. [If the programmer had entered a Y (yes), the BTR instruction would automatically initiate a new read request each time a data set was read successfully from the I/O module, as long as the BTR instruction's control logic remained true.]

2. The location of the I/O module to read from is specified by **rack** (rack 0 in the example), **group** (2), and **module** (module is always 0 unless the rack is configured for two slots per group, when the right slot would be module 1). Rack/group/module addressing is sometimes referred to as an *RGM address.*

3. The **data file address** is the address of the first of the series of memory locations where the communication handler is to place the data it reads from the I/O module (N7:4 in Figure 7.9). The # symbol isn't needed. The programmer must also enter the **length** of the data block to read in 16-bit data words. (Fifteen words are specified in Figure 7.9, so the 15 data words will eventually be copied into addresses N7:4 to N7:18 by the communication handler.) If this were a BLOCK TRANSFER WRITE (BTW) instruction, the communication handler would copy data words from these addresses to the I/O module. The program would have to ensure that there was useful data in these addresses before initiating the BTW data transfer request, of course.

4. Instead of the simple three-word control element used for file manipulation inside a

CPU module's memory, a six-word block transfer control element structure (BT9:1 in Figure 7.9) is specified for a **control block.** The six words hold the I/O module address and the CPU memory addresses and are used by the communication handler to keep working data and status bits as the handler performs the exchange of data with the I/O module. The programmer has to use (at least) the **DN** *(done)* bit of the block transfer control element. (Other status bits are covered in Chapter 13.) A user-program can check the DN bit to determine when a communication handler has read data successfully in response to the BTR (or has written data in response to a BTW) instruction. If the DN bit isn't on, the data exchange hasn't been completed, and the user-program should not read data from the memories that could be overwritten by the communication handler at any time. (The DN bit goes back off if the BTR or BTW instruction requests a new data exchange.) Figure 7.9 includes a rung to count the number of times that a value obtained from an intelligent I/O module is over 5. It executes only after the block transfer control element's DN bit (BT9:1.DN) goes on, indicating that a block of data has been read from the I/O module.

S5

SIEMENS STEP 5 DATA SET MOVE INSTRUCTIONS (WITH SEND AND RECEIVE FUNCTION BLOCKS)

S5 PLCs do not offer instructions that move entire data arrays (or data blocks) within the PLC's standard data memory. The programmer must write multi-instruction programs to move sets of data words. The **PROCESSING OPERATION** (DO) and **PROCESSING INDIRECT** (DI) instructions and structured programming techniques can be used in such programs.

Preprogrammed function blocks are available in the larger S5 PLCs to move sets of data words (arrays, although Siemens doesn't call them arrays) between a CPU's memory and the memory of an intelligent I/O module. The **SEND** function **(FB244)** and the **RECEIVE** function **(FB245)** can be jumped to, to exchange data sets with intelligent I/O modules. Every time a PLC is put into run mode, the **SYNCHRON** function **(FB249)** may have to be executed once to synchronize the CPU and the intelligent I/O modules that exchange large amounts of data with the CPU. The **CONTROL** function **(FB247)** can be used to determine the status of the communication with an intelligent I/O module. The **RESET** function **(FB248)** can be used to terminate an intelligent I/O module's function and its associated exchanges of data.

To exchange data with an intelligent I/O module, the PLC copies a *frame* of data to or from an intelligent I/O module via a block of shared memory in the CPU module. *Data handler* chips in the CPU module copy data between the CPU's data memory and the CPU's shared memory and copy data between the shared CPU memory and memory in the I/O modules. The data handlers operate independent of the PLC's main microprocessor, except that they are assigned tasks by the above-listed function block programs. The SEND and RECEIVE function blocks tell the data handler *where* it is to copy data to/from in the CPU's data memory. Data frames that are too big for the shared memory space (or for the intelligent I/O module) can be broken into smaller sets of data by the data handler and transferred one set at a time. Alternatively, the user-program can use a series of sequentially initiated SEND or RECEIVE functions to transfer parts of the large data frame. *Status information* is maintained by the data handler for each data exchange task that it is assigned, and the status bits can be read by the user-program using the CONTROL function block.

The parameters that must be specified with a SYNCHRON (FB249) function call include:

1. SSNR *(interface number),* entered as a two-byte constant, usually in the form **KY 0, y.** "y", in the low byte, must be a number between 0 and 255 corresponding to the interface number assigned to an intelligent I/O module during system setup. Interface numbers are included for all data sets written to the shared memory block, so that the data handler can determine which intelligent I/O module it is for. If any digit other than 0 is entered as the high byte of the interface number, indirect addressing is indicated. Indirect addressing for communication with intelligent I/O modules is not covered in this book.

2. BLGR *(frame size),* a two-byte constant usually entered in the form **KY 0, y.** "y" is a code that specifies the maximum amount of data that can be exchanged with this I/O module per scan cycle. (Refer to your manual set for codes and for frame sizes allowable with specific intelligent I/O modules.)

3. PAFE *(error byte),* an address indicating where the function block should write a one-byte error code if the function block can't work as it is supposed to. A flag byte is usually specified.

The parameters that must be specified with a RESET (FB248) function call include:

1. SSNR and **PAFE,** as described above.

2. A-NR *(job number),* a two-byte constant usually entered in the form **KY 0, y.** "y" is a number between 1 and 233 that indicates which intelligent I/O module operation to reset or which communication job to cancel. Each intelligent I/O module comes with a manual identifying the jobs it can perform and their job numbers.

The parameters that must be specified with the CONTROL (FB247) function call include:

1. SSNR, A-NR, and **PAFE,** as described above.

2. ANZW *(job status word),* the address where the function block can write the first of *two data words (four bytes) of status data.* The job status word can be in flag memory or in a data block that is open when the function block is jumped to, so this parameter is entered as either **FWx** or **DWx.** "x" is the number of a flag word or a data word. The function block will report the status of any requested communications with the I/O module specified by the SSNR parameter by writing to the two flag or data words. The high word contains the number of data words exchanged so far, while the low word includes these important status bits:

0 and 1
- Indicate the status of communication processor I/O modules.

2
- Set when the data handler has successfully completed the job requested by the function block that specified this pair of job status words. The user-program should monitor this bit to ensure that valid data is being exchanged.
- Reset as soon as a function block with this job status word requests the data handler to do a new job.

3
- Set when the data handler has completed the job requested but an error was detected in the process.
- Reset as soon as a function block with this job status word requests the data handler to do a new job.

4
- Set while a data frame is being transferred between the CPU module and the intelligent I/O module in response to a SEND or RECEIVE function block request. It can take several scan cycles to send a large frame.
- Reset after the data frame has all been transferred.

5
- Set after a data transfer requested by a SEND function has completed. A user-program should not change the outgoing data or initiate a SEND to this intelligent I/O module until after this bit is set.
- Reset by the user-program after being read, or reset the next time a SEND is executed again with the same job status word.

6
- Set after a data transfer requested by a RECEIVE function has completed. A user-program should not evaluate the incoming data until after this bit is set.
- Reset by the user-program to acknowledge receipt of the data, or reset the next time a RECEIVE is executed with the same job status word.

7
- When set (by a user-program), SEND and RECEIVE requests are refused. Can be used to protect the data area of memory while the user-program reads or writes new data.

8-11
- Status bits will contain codes describing a communication error that caused bit 3 to be set.

12-15
- Not used.

The SEND and RECEIVE functions are the function blocks that actually cause arrays of data words to be copied between the CPU's memory and the memory of intelligent I/O modules. These function blocks should be initiated by JUMP CONDITIONAL (JC) instructions, as demonstrated in Figure 7.10, so that they initiate data transfer requests only when a data transfer is actually required. Whenever a conditional jump to a SEND or RECEIVE function is encountered in a program and the conditional jumps's control logic statement is false, the function is not completely ignored: The PLC updates the status words for that SEND or RECEIVE operation so that

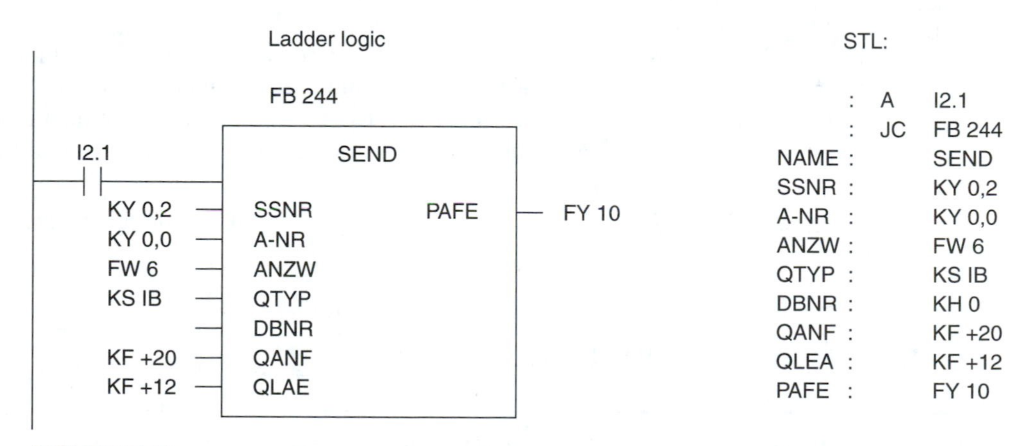

FIGURE 7.10
STEP 5 Instruction that conditionally jumps to the data SEND function block.

subsequent rungs which examine the status bits can work properly. The CONTROL function (FB247) therefore isn't needed to update status words if the SEND/RECEIVE functions are encountered in every scan cycle.[2]

The parameters that must be specified for a **SEND** (FB244) function block include:

1. SSNR, ANZW, and **PAFE,** as described above. In Figure 7.10, the data is being sent to the I/O module that is configured as interface 2; status data is kept in flag words 6 and 8 (flag bytes 6 to 9); and if there is an error in the data provided to the function block, an error code will be returned in flag byte 10.

2. A-NR (*job number*), which means something different here than it does in the RESET function. Still entered as a 2-byte constant in the form *FY 0, y,* but this time if "y" = 0 (as in Figure 7.10), the data handler is instructed to send the entire frame of data even if the data handler chip has to send it as a series of subsets. If "y" = a number between 1 and 255 (e.g., 15), it means that this SEND function is requesting the data handler to send only this part (part 15) of the larger frame of data.

3. The data source is specified using several parameters, including:

- **(a) QTYPE** (*type of data source*), which must be entered as a two-ASCII-code constant, usually in the form **KS xx.** "xx" must be one of the following alphanumeric codes: **DB, QB, IB, FB, TB, CB, or AS.**[3] The codes indicate that the data to be sent should be words copied from a data block (DB); or bytes copied from the output, input, or flag memory areas (QB, IB, or FB); or words copied from the timer or counter memory area (TB or CB) or from addresses that will be specified as absolute addresses (AS).
- **(b) DBNR** (*data block number*), entered as a 2-byte constant, usually in the form **KY 0, y.** "y" must be the data block number and is ignored unless the QTYPE parameter code was DB.
- **(c) QANF** (*start address of the source data frame*), entered as a 16-bit signed binary constant, usually in the form **KF x.** "x" is the numerical value part of the data word, or output byte, or other element that contains the first of the series of data values to be sent. For absolute addresses, see your manual.
- **(d) QLAE** (*length of the source data frame*), entered as a 16-bit signed binary constant, usually in the form **KF x.** "x" is the number of data units (QTYPE has defined data units as either words or bytes) to be sent, including the data unit from the start address and from sequential memory locations after that start address.

The program in Figure 7.10 specifies that 12 input bytes, IB20 through IB 31, are to be sent.

The parameters that must be specified for a **RECEIVE** (FB245) function block include:

1. SSNR, ANZW, and **PAFE,** as described above.

---

[2] If a SEND or RECEIVE is executed as part of an interrupt program or in an initialize program, or if it is in a part of a structured program that is executed only infrequently, the CONTROL function will be needed.

[3] Three other possible codes, which are not covered in this book, can be used to indicate that the data source will be indicated using indirect addressing.

2. A-NR, as described for the SEND function.

3. The data destination parameters that indicate where the RECEIVE function will put the data values that it receives. The parameters are similar to those described for the SEND function. The **ZTYPE, DBNR, ZANF,** and **ZLAE** parameters have the same meanings as the QTYPE, DBNR, ZANF, and ZLAE parameters described above, except that they have different names to reflect that they indicate destination addresses instead of source addresses.

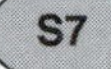

SIEMENS STEP 7 DATA SET MOVES (USING SYSTEM FUNCTIONS)

S7 PLCs do not offer individual instructions to move arrays of data items, although an ARRAY-type data item can be defined in STEP 7. User-written programs incorporating memory indirect addressing, register indirect addressing, and structured programming techniques must be used.

S7 PLCs include some preprogrammed system functions and system function blocks[4] that a user-program can call to move sets of data values or to execute a sequencer routine. Calls are executed unconditionally. (Structured programming techniques can cause the PLC to bypass a CALL instruction.) Some system functions that move sets of data within a single S7 PLC system include:

1. SFC 20 ("BLKMOV") to copy a block of data from any area of the CPU's data memory (except data blocks, timer data, or counter data) to another area of the CPU's memory (with the same exceptions). In STL, the complete call, with the required input and output parameters, would look like this:

```
CALL SFC 20
  SRCBLK  := x
  RET_VAL := y
  DSTBLK  := z
```

In this call example, "x" is an ANY pointer to identify the location and quantity of data elements to be moved (the source block) and "z" is another ANY pointer to indicate where the data is to be moved to (the destination block). "y" is a word address where "BLKMOV" can write an error code if it can't copy the data for any reason. Symbolic names can be entered. As we saw in Chapter 5, an ANY pointer constant can be entered in the following format:

P#Memory_prefix_Byte.Bit   Data_type   Number_of_items

For example:

P#M20.0 Word 5

which says: 5 words (= 10 bytes) starting with MB20

2. SFC 21 ("FILL") to copy the contents of a small block of data (the BVAL input parameter must be an ANY pointer to the small block), repeatedly until it fills a *larger block* of memory (pointed to by the BLK ANY input parameter).

---

[4] System functions are described in more detail in Chapter 8. A complete list of system functions is included in the Appendix.

3. SFC 79 ("SET") and **SFC 80 ("RSET")** to set or reset bits in the I/O area of memory. A pointer is entered for the SA parameter, indicating where to start setting or resetting bits, and an integer is entered for the N parameter to indicate how many bits to set or reset.

4. SFB 32 ("DRUM") is a sequencer. Because it is a function block program (not just an SFC), it uses an instance data block. **The following (static) VAR variables in each instance data block must be provided with data before SFB 32 can be used.**

- **(a) OUT_VAL, a two-dimensional array of bits (maximum 16 bits by 16 steps). Must contain the Boolean values for each of the (up to) 16 DRUM outputs for each of the (up to) 16 sequencer steps. Enter as a pointer to an array of Boolean elements.**
- **(b) S_MASK, another two-dimensional array of bits for a mask. A 1 in the mask allows the bit from OUT_VAL to be outputted. Enter as a pointer to the array.**
- **(c) DSP, the number of the OUT_VAL step to start at after being reset. Enter a number (1→16).**
- **(d)** Enter parameters for S_PRESET and DTBP if time-based sequencing will be used. S_SET is entered as the pointer to an array of time delays (up to 16 words) for each step and DTBP is entered as a word indicating the time base (in milliseconds) to use.

Other parameters can be passed to "DRUM," each time it is called. If these parameters aren't passed, they will have the values that they had the last time "DRUM" executed with this instance data block (perhaps the default values assigned with the static variables described above). The input parameters include:

- **(a)** DRUM_EN, a Boolean input to select between time-based sequencing (if DRUM_EN = 1) and event-based sequencing.
- **(b)** JOG, a Boolean input that causes the event-based sequencing to step when the Boolean control logic goes true.
- **(c)** EVENT1 to EVENT16, Boolean inputs that enable the timer to run for each of the 16 possible output steps, if DRUM_EN = 1. For purely time-based sequencing, these bits must all stay on. To use time delays in event-based sequencing, Boolean logic can turn these bits on to start the time delays at each step.
- **(d)** LST_STEP, a byte to indicate the number of the last allowable step before returning to step 1 (maximum: 16).
- **(e)** RESET, a Boolean input to reset the sequencer. See "Q" below.

The output parameters include:

- **(a)** OUT_WORD, enter an address where "DRUM" will write the 16-bit sequencer output.
- **(b)** OUT 1 to OUT16, enter bit addresses where "DRUM" can write the individual Boolean outputs, identical to the 16 bits of OUT_WORD
- **(c)** Q, a Boolean bit that goes on while the last step is being outputted, and indicates that "DRUM" must be reset before it can resume changing output values.
- **(d)** ERR_CODE, a word that DRUM can write an error code to if the DRUM SFB fails.

Various system functions exist to read and write the system area of memory or I/O module configuration data memory, so that the PLC program can respond to and change its own operating environment. These SFCs, which we will see again in Chapter 10, include:

1. SFC 51 (RDSYSST) to copy one of the sets of system status lists (containing status information) from the system data area of memory to data memory (M, L, D, I, or Q memory areas). In STL, a call to RDSYSST might look like the example shown in Figure 7.11.

The input parameters required with SFC 51 (RDSYSST) are:

(a) The REQ input parameter, which is of type BOOL. Although RDSYSST is called unconditionally, if the REQ input bit is false (e.g., if I2.1 in Figure 7.11 is off), the SFC will not copy any data.

(b) Two parameters, SZL_ID and INDEX, are 16-bit data words that must contain codes to tell the PLC what status list to copy from the system data memory. (In Figure 7.11, codes are used to read the 15 most recent records from the error diagnostic area of system memory.)

The output parameters include:

(a) RET_VAL, an address where RDSYSST can write error information if it can't work.

(b) BUSY, a Boolean bit the RDSYSST holds on while it is copying data.

(c) SZL_HEADER, a two-word structure where RDSYSST will report the number of bytes in the data records it has read from the system area of memory, and the number of data records it has read. After the program in Figure 7.11 has executed, MYSTRUCT.0 will contain 300, indicating that 300 bytes were read, and MYSTRUCT.1 will contain 15, because 15 records were read.

(d) DR, an ANY pointer telling the PLC where to put the data that it reads from the system area of memory. In Figure 7.11, 150 words of memory starting at MW40 are used, since each error record is 10 words in length.

2. SFC 58 (WR_REC) can be used to copy a block of data, other than configuration data, to an intelligent I/O module. An example STL program using "WR_REC" is shown in Figure 7.12.

The input parameters needed by SFC 58 (WR_REC) include:

(a) REQ, a Boolean input, which will allow WR_REC to send a block of data to the intelligent I/O module if it is true.

(b) IORD, a single-byte code that identifies the destination of the data as being either in the peripheral input area of memory (IORD = W#16#54) or the peripheral output area (IORD = W#16#55).

(c) LADR, an integer identifying the logical address of the intelligent I/O module. In Figure 7.12, the data will be written to the I/O module at address PQ 128.

(d) RECNUM, an integer identifying the record number for this set of data, which determines where the intelligent I/O module will place it in its memory.

FIGURE 7.11
STEP 7 call of SFC 51 (RDSYSST) to copy a block of status data from system memory to data memory.

```
CALL SFC 51
 REQ           := I 2.1
 SZL_ID        := W#16#01A0
 INDEX         := W#16#000F
 RET_VAL       := MW10
 BUSY          := Q 7.1
 SZL_HEADER    := MYSTRUCT
 DR            := P#M40.0 Word 150
```

FIGURE 7.12
STEP 7 call of SFC 58 (WR_REC) to copy a block of data to an intelligent I/O module.

```
CALL SFC 58
 REQ          := I 2.1
 IORD         := W#16#55
 LADDR        := 128
 RECNUM       := 4
 RECORD       := P#MW60.0 Word 5
 RET_VAL      := MW56
 BUSY         := M 55.0
```

(e) RECORD, an ANY pointer that describes the size and location of the data that is to be copied to the I/O module.

Two output parameters must also be specified. RET_VAL is the address where

(a) WR_REC can write a one-word error code if it fails, and BUSY is a bit that

(b) WR_REC will hold true while it is copying a data block to the destination intelligent I/O module.

3. SFC 59 (RD_REC) can be used to read blocks of data from an intelligent I/O module. Parameter assignment is the same as for SFC 58 (WR_REC). If the RECNUM parameter has a value over 1, it is user data that is being read from the I/O module, which may not be status information.

4. SFC 52 (WR_USMSG) is used to copy a user-defined one-word-long error event ID number (supplied as the EVENTN input parameter) and three words of user-defined error information (in the INFO1 and INFO2 input parameters) into a record in the diagnostic buffer area of system memory, as part of a standard 10-word diagnostic message; or to send the four words of input data to an external device such as an operator panel. This system function is discussed in Chapter 10.

5. SFC 55 (WR_PARM), SFC 56 (WR_DPARM), SFC 57 (PARM_MOD), and SFC 59 (RD_REC) to write blocks of data containing configuration data to intelligent I/O modules, or to read blocks of status information from I/O modules. These are discussed in Chapter 10.

CQMI

OMRON CQM1 FILE, ARRAY, AND STRUCTURE MOVE INSTRUCTIONS

Without ever using the word *array,* the CQM1 has an impressive assortment of instructions that can move sequential arrays of bits, nibbles (OMRON calls them *digits*), bytes, and words. Some instructions can convert between lines and columns in 16-word arrays. Other instructions are specifically designed to read values from input devices into byte arrays, or to copy values from byte arrays to display devices. Still other instructions [e.g., MSG(46), TXD(48), and RXD(49)] are intended to be used to move arrays of data units via a communications network, and are discussed in Chapter 13.

1. The BLOCK TRANSFER instruction, **XFER(70)**, copies *up to 9999*[5] *data words* (the first

---

[5] Of course, the entire range of source and destination addresses must exist. The error status bit (SR 25503) will be turned on if any instruction tries to write to a destination address that doesn't exist or is beyond the allowable range of an area of memory (e.g., HR 99 is the last address in the holding register area).

parameter, N, must be a BCD number between 0000 and 9999) from a memory area starting at a start source word address (S, the second parameter) to a memory area starting at a starting destination word address (D, the last parameter). Unless the differentiated form is used, the XFER(70) instruction executes every scan cycle while its control logic is true. An XFER(70) instruction is shown in Figure 7.13. In this example, 24 data words are copied from long-term storage memory locations HR10 through HR33 to memory locations DM1000 through DM1023 each time the PLC is put into run mode (when the first cycle flag bit, SR25315, is on).

2. The BLOCK SET instruction, **BSET(71),** causes the single 16-bit word in the source word address (S, the first parameter) to be copied to every memory location between the start word address (St, the second parameter) and the end word address (E, the last parameter).

3. The MOVE DIGIT instruction, **MOVD(83),** moves all or part of a 4-nibble array (one 16-bit data word) from a source word address (S, first parameter) to a destination word address (D, third parameter) and can roll the digits (nibbles) in the process. A digit designator (Di, second parameter) indicates in BCD which of the source digits to move first (Di's low nibble = 0 to 3), which nibble of the destination word to move it into (Di's nibble 3 = 0 to 3), and how many additional following digits to move (Di's second lowest nibble = 0 to 3). The fourth nibble of Di should be set to zero.

4. The TRANSFER BIT instruction, **XFRB(−)** (available in the CQM1-CPU4x only), can copy up to 255 bits of a bit array to another location in memory. The array can begin at any bit in the memory specified in the source word (S, second parameter) and can extend into higher addresses. It can be copied so that it starts at any bit location in the address specified in the destination word (D, third parameter) and can extend into higher addresses. A control word (C, first parameter) must contain a value in its low nibble to indicate the starting bit in the source word (hexadecimal values 0 to F indicate the source bit number, 0 to 15), a value in the second nibble to indicate the start bit in the destination word (0 to F again), and a value in the high byte to indicate how many bits to copy (hexadecimal 00 to FF means 0 to 255 bits).

5. Two instructions copy 16-bit values between columns in 16-word arrays and the bits of a single 16-bit word (called a *line*). The COLUMN-TO-LINE instruction, **LINE(−),** copies one bit from each of 16 data words starting at a source (S, first parameter) address and extending into higher-numbered addresses, to a single destination word address (D, second parameter). A column bit designator (C, third parameter) indicates which bit to copy from each word in the ar-

FIGURE 7.13
OMRON XFER (70) instruction for copying an array of data values.

ray of source words (enter a BCD value between 0000 and 0015). The lowest-numbered address provides bit 0 for the destination address. The LINE-TO-COLUMN instruction, **COLM(−),** performs the opposite operation: copying a 16-bit data word to a single column in a 16-word array.

6. A DIGITAL SWITCH INPUT instruction, **DSW(87),** reads the settings of either *four or eight standard digital switches*[6] into a 4-nibble array or into an 8-nibble array. An input word address (IW, first parameter) and an output word address (OW, second parameter) indicate which input and output modules to use in decoding the switch positions. The low 4 bits of the output module will be switched on sequentially in the reading process, to power the digital switches one at a time (or two at a time if eight switches are being read). When powered, a digital switch outputs a 4-bit BCD code that is read through the low 4 bits of the input module (and simultaneously through the next four input contacts if eight switches are being read). Each 4-bit result is saved in one nibble at the address specified as the first register word (R, third parameter) and simultaneously in the nibbles of the next address if eight switches are being read. Status bit SR 25410 is on while DSW(87) executes. One execution takes 12 scan cycles. Readings repeat as long as the control logic remains true.

7. The SEVEN-SEGMENT DISPLAY OUTPUT instruction, **7SEG(88),** copies nibble values (representing hexadecimal or BCD characters) from memory to an output module that can be connected to a set of seven-segment display chips. Four (or eight) separate display chips can be written to. 7SEG(88) copies nibble values, one (or two) at a time, to the low 4 (or the low 8) bits of the output module, then repeats until a total of 4 (or 8) nibbles have been written to the 4 (or 8) display chips. 7SEG(88) also switches the next-higher four contacts of the output module on sequentially as it outputs nibble values, to select the display chip (or pair of chips) that the current nibble value(s) is to affect. Nibble values are copied from the address specified in the first source word (S, first parameter) (and from the next address if eight chips are being written) to the output image address specified in the output word (O, second parameter). A control data value (C, third parameter) must contain a code indicating whether four or eight characters are to be outputted, and whether it is necessary to invert the 4-bit nibble values or the chip select states for the display hardware (see your manual for codes). Status bit SR 25409 and one output image bit are held on during the 12 scan cycles it takes to perform this operation. Writing repeats while the control logic remains true.

COMPARING FILES, ARRAYS, AND STRUCTURES

Some PLCs offer instructions that can compare entire arrays of data values against other arrays of values or against other individual values. The results of the individual comparisons can be used to control other instructions. Manufacturing processes controlled by sequenced output bit patterns (described earlier in the chapter) often require the sequencing to be controlled by waiting for the correct state of inputs. For each sequenced output step, there is a pattern of input states that must be matched before the next output step can be initiated.

---

[6] If the high byte of the CQM1's configuration word DM 6639 contains a 00, four digital switches are read through the low four contacts of an input module. If the DM 6639 high byte is 01, an additional four switches are read through the next-lowest 4 bits of the same input module.

PLC-5

SLC 500

ALLEN-BRADLEY FILE COMPARE INSTRUCTIONS

The PLC-5's SEQUENCER INPUT (SQI) and the SLC 500's SEQUENCER COMPARE (SQC) instructions are very similar, but not identical. Both compare a value from a source address (through a mask) to one of a series of data words in a sequencer file, and both perform the comparison only while their control logic is true. A mask indicates which bits of the source must match the sequencer file word for the data words to be considered equal. The presence of 0 bits in the 16-bit mask means that there does not have to be a match in this bit position.

In both the SQI and the SQC instructions, the first data word in the input file (word 0) is used as a startup pattern and is not included in the compare sequence the next time the sequence recycles. If the programmer wants to reuse the first data word, another program instruction (e.g., a MOV) must be used to write the value 0 to the .POS word of the SQI or SCC's control element. The differences are explained below.

PLC-5 SEQUENCER INPUT

The PLC-5 SQI instruction is used in a Boolean logic statement. The SQI instruction compares an input pattern against whichever word its control element points to. If they are equal, the SQI instruction is true; otherwise, the SQI instruction is evaluated as false. The SQI instruction does not increment the position (.POS) word in its control element. The .POS value in the control element must be changed to increment the pointer. It can be changed by another instruction (by a MOV or a CPT instruction, for example) to select a new input file value to compare against the value from a source address.

The SEQUENCER OUTPUT (SQO) instruction (discussed earlier) does automatically increment the .POS value in its control element each time its control logic goes true, so when an SQI instruction is being used to control an SQO instruction, they should have the same control element so that the SQO instruction will increment the SQI and the SQO steps together. Figure 7.14 shows how an SQI can control when the SQO steps through a sequence, which in turn increments the SQI sequence. Both instructions are at position 2 in the example (R6:3.POS = 2), so the SQO instruction is writing the low 8 bits of the value from word 2 of its output file (N7:2

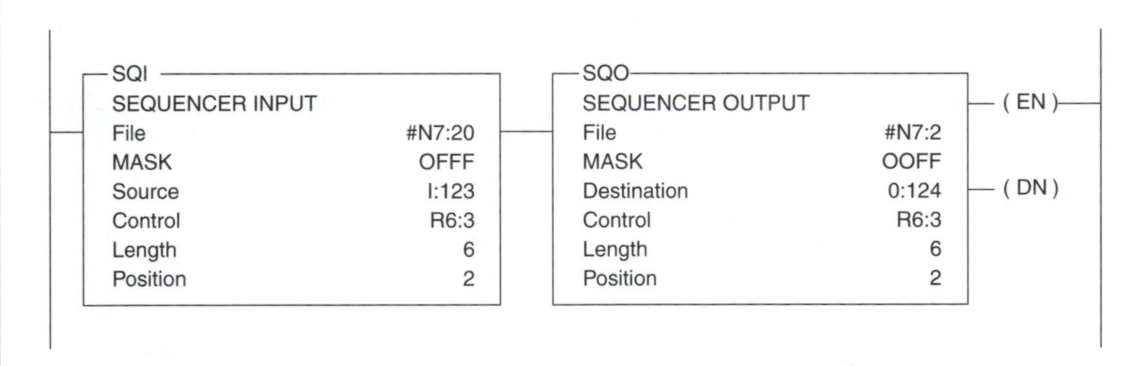

FIGURE 7.14
PLC-5 SEQUENCER INPUT and SEQUENCER OUTPUT pair.

+ 2 = N7:4, outputted through the hexadecimal mask "00FF") to the destination address (O:124). The SQI instruction is comparing the value from the low 12 bits (mask = hex 0FFF) from its source address (I:123) to word 2 of its output file (N7:20 + 2 + N7:22). As soon as the SQI generates a true (when I:123 = N7:22) the SQO will increment R6:3.POS and output the next word (N7:5) from its data file, and the SQI instruction will start comparing the source value to the next data word (N7:23) in its input file. This is one of the few places (other than with FIFO and LIFO instructions) where two instructions can share the same control element.

SLC 500 SEQUENCER COMPARE Instruction

The SQC instruction is considered an output element in an SLC 500 program. It does increment its own .POS word in its own control element, but only when its control logic goes from false to true. Since an output element cannot be used in a Boolean logic statement to control another output element (even to control a sequencer output instruction), an additional status bit has been added to the control element for an SQC. The **FD** (found) bit goes on when the source input pattern matches the sequencer file word pattern that it is being compared against, and off if the words don't match. The FD bit can be examined in the Boolean logic statements that increment the SQO and the SQC so that they increment together. To work in an SLC 500, the PLC-5 program from Figure 7.14 would have to be rewritten as shown in Figure 7.15.

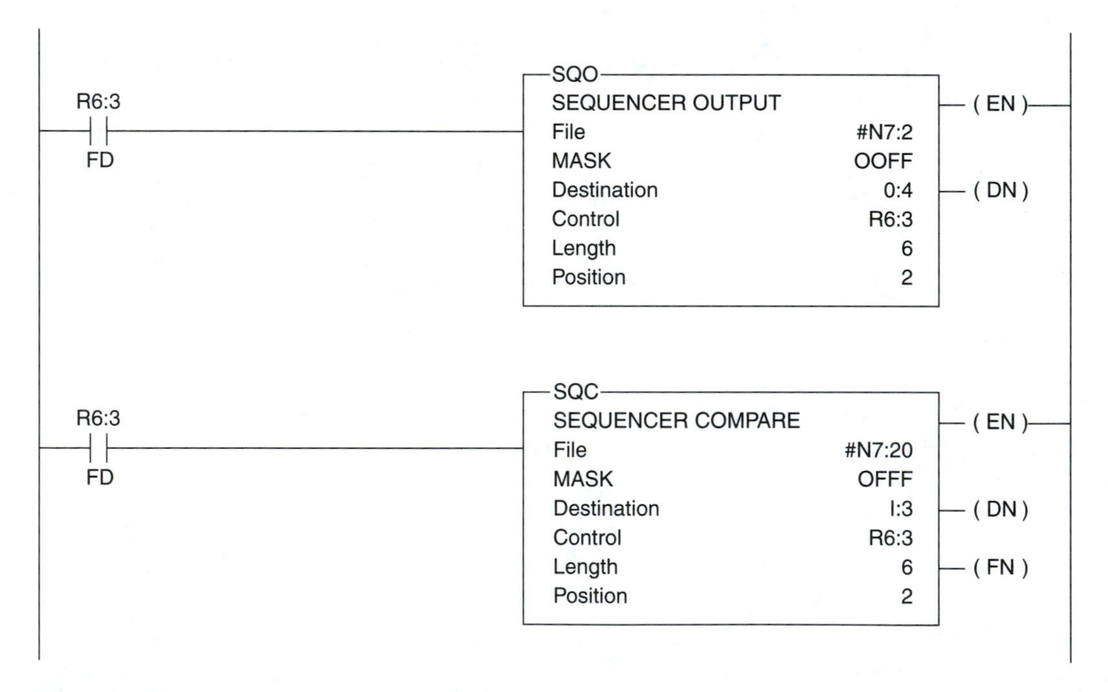

FIGURE 7.15
SLC 500 SEQUENCER COMPARE with SEQUENCER OUTPUT.

PLC-5 FILE SEARCH AND COMPARE (FSC), FILE BIT COMPARE (FBC), and DIAGNOSTIC DETECT (DDT)

The PLC-5 offers a **FILE SEARCH AND COMPARE** (FSC) instruction that can compare a series of data words in a file against corresponding data words in another file, or to compare each word in a file sequentially against a single data value entered as a constant or as an address. When a match is found, the comparing operation stops and a status bit is set in the instruction's control element indicating that a match was found. An FSC instruction in a program would look like Figure 7.16. Data that must be entered with an FSC instruction includes:

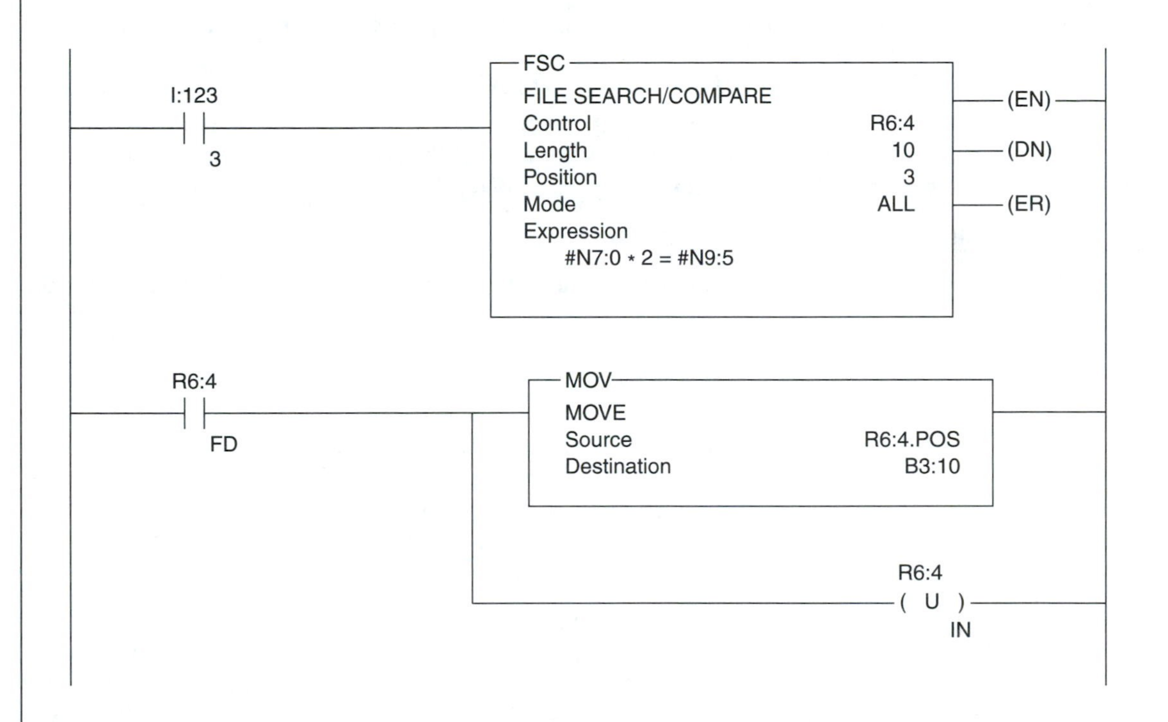

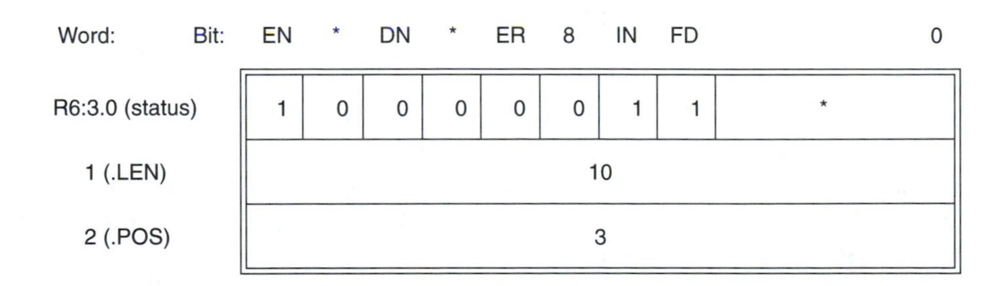

FIGURE 7.16
PLC-5 FILE SEARCH AND COMPARE instruction and its control element.

1. The **length,** to indicate the number of words that are being compared.

2. An **expression,** similar to the expression that would be entered in a simple COMPARE (CMP) instruction, except that addresses in the FSC expression can be preceded by the # symbol. This symbol indicates that this address is the first address of a file that should be examined one element at a time and allows the PLC-5 to use S:24 in indexed addressing. In the example in Figure 7.16, N7:0 will be multiplied by 2, then the result will be compared with N9:5, then N7:1 will be multiplied and compared with N9:6, and so on, until 10 comparisons have been performed.

3. A **control element,** which will be used to keep track of which data words of the file have been compared. In addition to the usual **EN, DN,** and **ER** bits, the FCS instruction uses:

FD *(found),* which is latched on each time the expression is found to be true, and unlatched if the expression is false.

IN *(inhibit),* which is latched on as the FD bit is turned on. As long as IN is on, the FSC instruction will stop executing, so the position of the word(s) in the files that made the expression true will be in the .POS word of the control element. Unlatching the .IN bit in a program allows the FSC instruction to resume comparing values.

4. A **mode** of one of the following types:

ALL to indicate that all comparisons are performed once each time the FSC's control logic goes from false to true. Comparisons will still stop if one match is found, but resetting the control element's .IN bit will allow the FSC instruction to resume comparing values in the rest of the files.

INC to cause the next comparison in the sequence to be performed each time the control logic goes true (will restart the sequence after comparing the last values in the files)

n where "n" is a number from 1 to the length value. To cause that number of comparisons to be performed each scan while the control logic remains true, until the last values in the files are compared. To restart comparing, the logical statement controlling the FSC instruction must go from false to true again.

In the program in Figure 7.16, the FSC instruction compares all 10 data values each time I:123/3 goes on. If no match is found, the FSC instruction is ready to repeat next time I:123/3 goes on again. If a match is found (as in this example), the FSC instruction will not continue comparing values but will turn the Found (.FD) and Inhibit (.IN) status bits on. The next rung of the sample program will execute because the FD bit (R6:4.FD) is on. This rung will store the position in the files at which the match was found (R6:4.POS contains 3) into a bit file data word, then reset the .IN bit so that the FSC will compare the remaining seven values in the next scan cycle.

The PLC-5's **FILE BIT COMPARE** (FBC) instruction was designed to assist in debugging a control program. It does so by comparing a file of input image data words against a second file of reference data words in memory (perhaps a pattern reflecting what the programmer expects the input images to look like) and records the locations of any mismatches in a third file. FILE BIT COMPARE is covered in Chapter 15.

The **DIAGNOSTIC DETECT** (DDT) instruction is identical to the FBC instruction except that whenever a mismatch is found, the reference file bit is changed to reflect the state of the input data bit from which it differed. The DDT instruction is, therefore, better suited to detect changes in input conditions and to record them in the order that they change.

S5

S7

SIEMENS FILE, ARRAY, AND STRUCTURE COMPARE INSTRUCTIONS

Siemens S5 and S7 PLCs do not offer any instructions or preprogrammed functions that compare sets of data values. The programmer must write programs to perform this function if it is required.

CQMI

OMRON CQM1 FILE, ARRAY, AND STRUCTURE COMPARE INSTRUCTIONS

The CQM1 has three instructions that can be used to compare a single data value against values in an array, and one instruction that can be used to compare values in one array against values in another array.

1. The TABLE COMPARE instruction, **TCMP (85),** is used to compare a single 16-bit compare data word (CD, the first parameter) against 16 words in sequential memory locations in what OMRON calls the *comparison table,* starting with the lowest-numbered address (TB, the second parameter). It performs the entire set of comparisons each scan cycle while its control logic remains true, writing 1's into a single 16-bit result word (R, the third parameter) to indicate the locations of words that are equal to the compare data word. A 1 in bit 5 of the result words, for example, would mean that word 5 in the comparison table was equal to the compare data word. Zeros are written to the result word if the values compared are not equal. There is a differentiated version, @TCMP (85), that will execute once to compare all 16 data words to the compare data word each time its control logic goes true.

2. The DATA SEARCH instruction, **SRCH (−),** also compares a single data word against an array of other data words, but the compare data value can be compared against any number of words. N, the first parameter, indicates how many data words to compare, and R_1, the second parameter, indicates the lowest address in the range of values to be compared against the compare data value. CD, the third parameter, must supply the *address* of the single value that is to be compared against the numbers in the range. The entire set of values is compared each scan while the control logic remains true, and if ANY matches are found, the EQ bit (SR 25506) is set AND the location of the *first (lowest addressed)* matching value is written into memory *in the address above the compare data address.* The location of this lowest-addressed matching value can be recorded in either of two ways: If the range of values was in the DM area of memory, the DM address number is saved; if the "range" is in any other area of memory, the number that is stored in the memory above the compare data value is the number of memory addresses offset between the start of the range and the memory where the match was found.

3. The BLOCK COMPARE instruction, **BCMP(68),** compares a single compare data value (CD, first parameter, can be a constant) to see if it is within one or more of 16 ranges of values. Each of the 16 ranges is identified by a lower limit and a higher limit, so 32 limit values must be entered into sequential memory locations before executing the BCMP(68) instruction, starting with the first range's low limit value in the lowest address and the mating high limit in the next, then the next range's low limit, and so on. Enter the address of the first (low) limit as the first comparison block word (CB, second parameter). The BCMP(68) instruction puts a 1 into one bit of a single 16-bit result word address (R, third parameter) if the compare data value is within the corresponding range, and a 0 if it isn't. If, for example, the seventh range is 40 to 200 and the compare data value is 200, bit 7 of the result word will be set to indicate that 200 is (just barely) within the seventh range. All 16 comparisons will be made each scan cycle if the BCMP(68) control logic remains true, but a differentiated version, @BCMP(68), is available.

4. The **MULTIWORD COMPARE** instruction, **MCMP(19),** compares 16 words starting from the first word of the table 1 address (TB1, the first parameter) against 16 words starting at the first word of the table 2 address (TB2, second parameter). If the values being compared are equal, a 0 (not a 1) is written into the corresponding bit position in a result word address (R, third parameter). If the values are unequal, a 1 is written instead. All 16 values are compared each scan while the control logic remains true, unless the differentiated form of the instruction, @MCMP(19), is used to force the 16 comparisons to be executed only when the control logic goes from false to true.

MATH AND LOGIC WITH FILES, ARRAYS, AND STRUCTURES

To perform mathematical or logical operations on data sets, a PLC has to repeat the same math/logic operation for each value in the set of data it is operating on. Instructions that do this have built-in looping operations, and they use indexed addressing so that they operate on data from a different address each time they repeat the loop. A good programmer could write functions that perform math/logic operations using looping, indexed addressing, and math/logic instructions that operated on one value at a time, but a few PLCs already offer instructions that make it unnecessary to write those looping programs.

PLC-5

SLC 500

ALLEN-BRADLEY FILE MATH AND LOGIC INSTRUCTIONS

Allen-Bradley's **FILE ARITHMETIC AND LOGIC** (FAL) instruction can perform the same mathematical and logical operations as Allen-Bradley's COMPUTE (CPT) instruction, except that it can perform those operations using data from files and can put results into files. As shown in Figure 7.17, the FAL instruction includes an expression describing the math/logic operation and specifies a destination, the same as a CPT instruction does. The FAL instruction, however, can include addresses preceded by the # symbol, which tells the PLC to use indexed addressing. The # symbol can be included in as many addresses as the programmer wants in the expression and/or in the destination address. The FAL instruction causes the PLC to execute once as a simple CPT instruction, evaluating the expression and placing the result in memory as if none of the addresses had a # symbol, but the FAL instruction then increments all the addresses with a # symbol by one and repeats the process, then increments and repeats as often as is necessary. The programmer should ensure that there is an array of operators in memory for each operator address that is preceded by a # and that there is enough memory for the result values in the destination array.

Figure 7.17 also shows the additional information the programmer must enter, similar to entries that we have seen in other file data manipulation instructions. The information includes:

1. A **control** element address and a **length** number indicating how many times the expression must be evaluated. The PLC uses the control element's .POS data word to keep track of the number of times the FAL instruction has looped as it performs the series of calculations.

2. A **mode** entry, as in the FBC instruction, which can be:

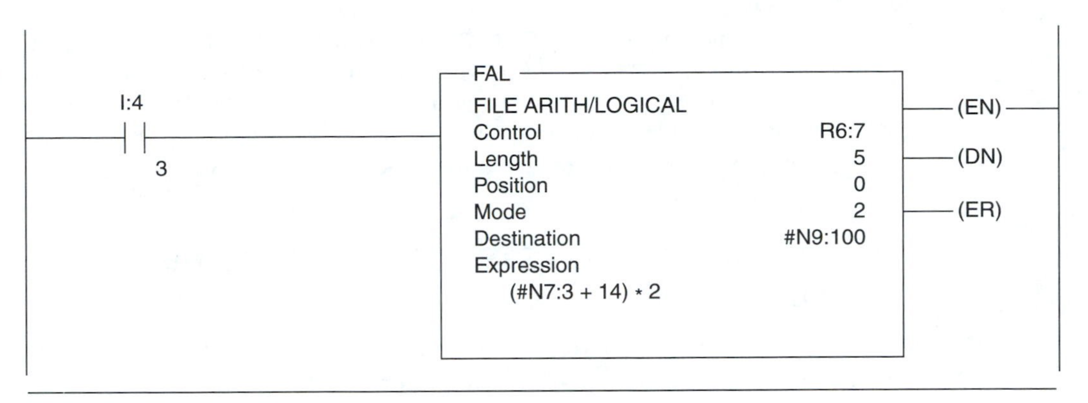

Scan number	I:4/3	These calculations are performed:
1	False	None
2	True	N9:100 = (N7:3 + 14) * 2 N9:101 = (N7:4 + 14) * 2
3	True	N9:102 = (N7:5 + 14) * 2 N9:103 = (N7:6 + 14) * 2
4	True	N9:104 = (N7:7 + 14) * 2
5	True	None
6	False	None
7	True	N9:100 = (N7:3 + 14) * 2 N9:101 = (N7:4 + 14) * 2
8	False	None
9	True	N9:100 = (N7:3 + 14) * 2 N9:101 = (N7:4 + 14) * 2

FIGURE 7.17
PLC-5 file arithmetic and logic instruction in which files are used for expression evaluation and result storage.

`ALL` so that the full series of calculations will be performed each time the FAL instruction's control logic goes true.

`INC` so that only one calculation will be done and the pointer will increment for the next calculation each time the instruction's control logic goes true.

n where *n* is a number between 1 and length, so that the instruction will perform that number of calculations each scan cycle while the instruction's control logic remains true, until the whole file has been operated on. To restart the FAL instruction, the instruction's control logic must go false, then true again. Figure 7.18 shows an FAL instruction programmed to execute in this mode.

S5
S7

SIEMENS FILE, ARRAY, AND STRUCTURE MATH AND LOGIC INSTRUCTIONS

Siemens S5 and S7 PLCs do not offer any instructions or preprogrammed functions that will perform math or logic operations on sets of data values. The programmer must write programs with loops and indexed addressing to perform these functions if they are required.

CQMI

OMRON CQM1 FILE, ARRAY, AND STRUCTURE MATH AND LOGIC INSTRUCTIONS

The CQM1 only offers a very limited set of instructions that operate on arrays of data values. The programmer must write programs with loops and indexed addressing to perform other functions if they are required.

Some instructions that examine multiple data values were covered earlier. As we saw in Chapter 6, **Max(−)** and **Min(−)** are used to find the single maximum or the minimum value from an array of values in memory. **AVG(−)** calculates the average of a set of data values, but the individual values are read from the same memory location once per scan, not from an array of data values already in memory, so it was covered in Chapter 6. AVG(−) does, however, create an array of data values as it reads them from the single memory location and saves them, because AVG(−) doesn't complete calculation of the average until the array is complete.

The CQM1's math/logic instructions that operate on arrays of data include:

1. The **SUM(−)** instruction, shown in Figure 7.18, which adds the words or bytes from

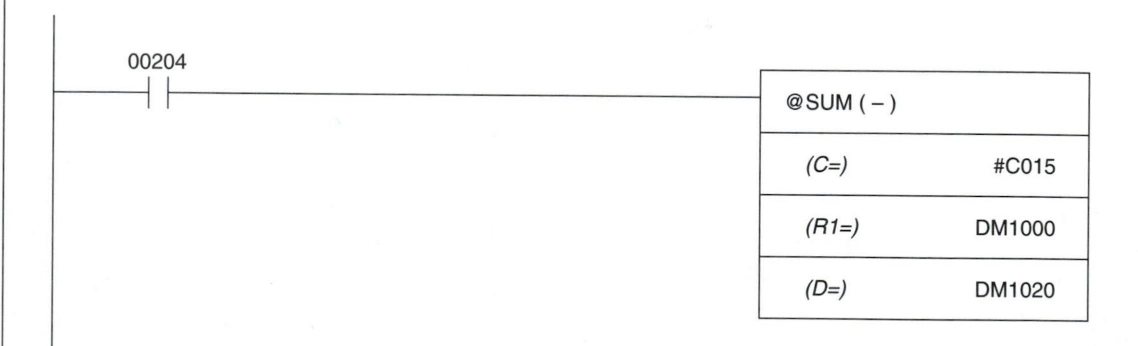

FIGURE 7.18
OMRON SUM(−) instruction.

a series of memory locations. It recalculates the sum every scan cycle while its control logic remains true unless the differentiated form, @SUM(−), is used, in which case it recalculates the sum only when the control logic goes from false to true. A control word (C, the first parameter) indicates the size, number, and type of data values to sum. The low 12 bits of C must be a BCD number between 000 and 999 to indicate the number of addresses containing data to be summed. The high two bits of C indicate that the data values are to be interpreted as BCD (if bits = 00) or as unsigned binary (if 01) or signed binary (if 11). The other two bits of C (bits 12 and 13) indicate to sum the entire 16-bit data word (if bit = 00), or to sum just the rightmost byte from each word of the array (if 11), or to sum all bytes (if 10). Next, a range word (R1, second parameter) indicates the start address of the values to be summed. Finally, a destination word (D, third parameter) indicates the first of the two addresses where the 32-bit result will be stored. The low 16 bits are stored in the specified address and the high 16 bits in the next-higher address.

In Figure 7.18 the differentiated form of SUM(−) is used, so a sum is calculated once each time IR 00104 goes from false to true. The control parameter is entered as a constant and indicates that 15 values are to be summed (from DM 1000 to DM 1014 since the R1 parameter says DM 1000). The values from those addresses are to be treated as 16-bit signed binary values (the high hex digit in the control word parameter is C, which is 1100 in binary). The 32-bit signed binary result is stored in addresses DM 1020 and DM 1021.

2. The FRAME CHECKSUM instruction, **FSC(−),** performs an Exclusive OR operation using all the words or bytes that are contained in a series of memory locations. It is entered with the same parameters as a SUM(−) instruction, except that the high bits of the control word parameter should always be 00. The result of the Exclusive Or is only a single byte or word, so it can be stored in a single destination memory location. A differentiated form, @FSC(−), is also available (used in serial communications; see Chapter 10).

Two instructions are available to convert arrays of hexadecimal values in nibbles to/from arrays of ASCII codes:

1. The ASCII CONVERT instruction, **ASC(86),** converts up to four hexadecimal values to ASCII codes each scan cycle while its control logic remains true. A differentiated version, @ASC(86), is available. A source word (S, first parameter) indicates the address that contains a 16-bit value to be interpreted as four hex digits. The digit designator word (Di, second parameter) indicates which digit of the source word to convert first (number 0 to 3 in Di's low nibble), how many additional digits from the source address to convert (0 to 3 in the next-lowest nibble), what parity to use in the ASCII codes (if Di's high nibble = 0, no parity; 1 means even; 2 means odd), and whether to write the first ASCII code to the low or the high byte of the destination address (if Di's second highest digit = 0, write into the low byte; if 1, into the high byte). The output array can be distributed into as many as three destination addresses. A destination word (D, third parameter) indicates the lowest address into which ASCII codes are to be placed.

2. The ASCII-TO-HEX instruction, **HEX(−),** does exactly the opposite of the ASC(86) instruction, converting ASCII codes from bytes starting in the address indicated by the source word (first parameter) to hex codes to be stored into a single destination address (third parameter). The digit indicator (second parameter) has the same meaning as in ASC(86). If the parity of an ASCII code does not match the parity indicated by Di, the error flag bit (SR 25503) is set.

TROUBLESHOOTING

One major difference between the instructions covered in Chapter 6, which generally operated only on data in one or two memory locations, and the instructions in this chapter, which operate on data in many more memory locations, is the time they take to execute. Programs with several conditional file/array/structure instructions can become very *nondeterministic* in nature. This means that their execution time can be highly variable, perhaps variable enough to cause control problems. To avoid high variability in program execution times, the programmer might wish to limit the number of file/array/structure instructions that execute per scan, or may want to program the instructions so that they operate on large arrays in several steps (so that the array operation is executed over several scans). If array operations are programmed so that they take several scans to complete, the program should use the status bits that some PLC instructions make available to detect when the operation is done, or should include other mechanisms to detect when an array operation is finished. Another way to make a control program operate in a more deterministic way is to configure your PLC to use timed interrupts for interval-critical operations or to execute with a fixed minimum scan time. These techniques are discussed in Chapter 11.

Instructions that exchange blocks of data with intelligent I/O modules also take long times to execute. Some PLCs have separate communication handler chips that accept data transfer requests from a user-program and which execute the data transfers one at a time in the order the requests are received, and sometimes without synchronizing the data transfers with the scan cycle. These instructions typically manipulate more complete sets of status words and bits, and the programmer should use the available status information to ensure that the incoming and outgoing data blocks aren't changed at the wrong times.

Most of the problems that can occur while executing programs with file/array/structure manipulation instructions are the same as the types of problems that can occur when using instructions that manipulate individual data words, but it is easier to cause those problems! When an instruction is supposed to make use of a series of 100 data words starting at an address that the programmer knows exists, it is easier to forget that there may not be 99 higher addressable memory locations free or available. Also, since the program listing contains only the starting address, it is easy to forget and to write another instruction that changes the contents of one of those 100 addresses when it shouldn't be changed. As the programmer develops a program, he/she should maintain a record of the memories used by the program so far and what type of data each memory contains. (Recording the data type will help the programmer avoid accidentally using an instruction that is meant to operate on a different data type.)

Remember that PLCs are computers, and computers are designed to manipulate whole data words at a time. If your manual suggests that an instruction manipulates part of a data word and leaves the rest of the word unchanged (as some bit-array shift instructions appear to do), *be suspicious.* Read the manual carefully, or better yet, try the instructions (manipulating unused data bits) to see what the instruction does to the other bits in the words being manipulated.

Usually, a PLC will fault when a nonexistent address is used, and will store fault bits and/or fault codes describing the error, but not always. For example, some OMRON CQM1 PLCs don't have data memory between addresses DM 1024 and DM 6143, but all CQM1 PLCs allow programs to try to read or write those nonexistent addresses! A CQM1 will also allow instructions to try to write to the read-only parts of data memory between DM 6144 and DM 6655.

In a structured Siemens program, remember that data blocks must be called before they are used, and remember that as control jumps between organization blocks, function blocks, and other programs, the PLC may automatically change the data blocks that are active. Although this is usually helpful in that it prevents the programmer from having to reopen a data block after each jump, it can lead to problems.

Find out if your programming software can detect improper memory references before the program is entered into the PLC's memory. When programming off-line, configure the programming software so that it prepares programs for exactly the right make and model of PLC that will run the program. This will enable the programming software to detect memory references that won't work with your model of PLC. Some programming software can be configured to watch for some types of programming errors and let other errors past.

Most programming software offers a cross-reference option, which will tell you every rung that contains a reference to each address. Use it to verify that addresses used by file, array, or structure manipulating instructions aren't used by other instructions, too.

File/array/structure manipulation instructions use indexed addressing internally. To do so, they must make automatic use of a memory location somewhere to hold an offset value. Two things can go wrong here: (1) If the same memory location is used to store the offset value for two different functions, the file/array/structure instruction may change the offset, causing the other instruction to use an incorrect memory reference; and (2) if a file/array/structure instruction is interrupted, another instruction can affect the offset value before the instruction resumes. For example, several Allen-Bradley instructions use the offset at S:24 automatically, and S:24 is also required for indexed addressing.

Some PLCs offer instructions that can be used to detect unexpected changes in the contents of an array or to detect differences between data in an array and the data that should be in the array. Allen-Bradley's FILE BIT COMPARE (FBC) and DIAGNOSTIC DETECT (DDT) instructions are examples. These instructions are covered in Chapter 15. Most PLCs offer instructions that can be used to *make* a PLC fault. If a program uses a simple COMPARE or a FILE COMPARE instruction as a condition for causing the programmable fault while a program is being debugged, undesirable memory content changes (or failures to change) can be trapped and the causes diagnosed. In Chapter 11 we describe user-generated faults and how to cause them.

QUESTIONS/PROBLEMS

1. What is the difference between an array and a structure?
2. What would be in B3:0 and B3:1 after an Allen-Bradley executes the following instruction?

```
Before BSL executes:
          B3:0 - 1111 1111 0000 0000
          B3:1 - 0101 0101 0101 0101
          N7:3 - 0000 0000 0000 0101
```

```
   BSL
   Bit Shift Left
   File :            #B3 : 0
   Control :          R6 : 3
   Bit Address :     N7 : 3.1
   Length :               18
```

```
After BSL executes once:
          B3:0 - __ __ __ __
          B3:1 - __ __ __ __
```

3. **(a)** What Allen-Bradley PLC-5 instruction is used to exchange data between the CPU and an I/O module, if the module needs more data than one 16-bit word (e.g., an analog I/O module)?
   (b) If the module address in the instruction identified in part (a) is 111, where is the I/O module? Use a sketch to show *clearly* and *exactly* the location of the module in the rack.
   (c) As part of the instruction, you must specify a block transfer control element, which includes a done bit. What component in the PLC-5 turns on the done bit?

4. Fill in the blanks for the following Allen-Bradley file arithmetic/logic (FAL) instruction so that it would perform all three of the following calculations each time its control logic goes true:

```
N7:10 = N7:0 * 3 + N9:0
N7:11 = N7:1 * 3 + N9:0
N7:12 = N7:2 * 3 + N9:0
```

```
FAL
File Arith/Logic
Control :       ________
Length :        ________
Position :
Mode :
Destination :   ________
Expression :    ________

________________________
```

5. An Allen-Bradley control element (e.g., R6:0) contains how many data words? What does the PLC keep in each of those data words?
6. Some PLC-5 instructions require (or allow) you to use a # in an address (e.g., #N7:1 instead of N7:1). What does the # mean? Answer in simple English. You will not receive full marks if you just repeat what the manual says about indexed addressing unless you make it clear that you know what *indexed addressing* means.
7. Allen-Bradley's FBC (FILE BIT COMPARE) instruction can be used during program debugging. How many files of data does the FBC instruction use? How many control elements does it need?

SUGGESTED PLC LABORATORY EXERCISES

For a system with:

- Four control panel switches: inputs 0 to 3
- Four indicator lamps or visible output module LEDS: outputs A to D
- Two spring-return-valve-controlled cylinders: outputs E and F
- One detent-valve-controlled cylinder: outputs G and H
- Three sensors to detect extension of each of the cylinders

1. Write a program with file manipulation instructions (or function calls) to:
   (a) Copy five data words from one area of data memory to another area whenever switch 0 is on. Use the data monitor and data write capabilities of your PLC to enter data into the first data area and to observe changes at the second data area.
   (b) Shift all the bits of the output image data word (or byte) that controls the indicator lamps. The shift should move all bits up one position each time switch 1 goes from off to on. Include logic to copy the state of the image bit that is shifted out of the word (or byte) into the lowest bit of the image word (or byte).
   (c) Shift an array of 16-bit data words one place in an array of five data words each time sensor 2 goes from off to on. New data words (shifted in) must be copied from data memory (which you will change using the data screen). After the array has had five data words shifted in, turn on an output bit and copy each data value that is subsequently shifted out to a memory location separate from the array.
2. Write a program with a four-step output sequence. (Use sequencer instructions if available.) The program should monitor the four switches and control the cylinders as follows:
   (a) Start by extending all cylinders if all switches are off,
   (b) Retract all cylinders when switches 0 and 1 are both on,
   (c) Extend cylinder G–H when all switches go off again,
   (d) Retract cylinder G–H and extend cylinders E and F when switches 1 and 2 go back on.
   (e) Repeat.

   Add timer(s) to the sequencer program so that there is at least a 3-s delay between output changes.
3. If programming a PLC-5, write a program with block transfers to copy the contents of N7:1 continuously to an analog output module while an input switch remains on.
4. If your PLC offers a compare instruction that can operate on files (Allen-Bradley and OMRON), write a program that examines a set of five data words each time an input bit goes on, and turns an output on if any of the data words are equal to 25.

8

PROGRAM STRUCTURE AND STRUCTURED PROGRAMMING

OBJECTIVES

In this chapter you are introduced to:

- Disabling user-program segments with master control relay/reset (MCR) and interlocks.
- Bypassing or repeating user-program segments by jumping forward or backward.
- Conditional execution of subprograms (subroutines, functions, function blocks, etc.).
- Parameter passing to make subprograms reusable.
- Effects of disabling or bypassing instructions such as timers and one-shots.
- Changing a PLC's scan-cycle characteristics with configuration of a CPU.

This chapter covers the structured programming capabilities offered by modern PLCs: techniques that allow portions of a user-program to be skipped past or to be repeated more than once during a single scan cycle. Sections of a user-program may be disabled in such a way that they turn all their outputs off or so that their output states are frozen in their last state. Portions that are conditionally executed may contain user-written control programs that will be called only when necessary. PLC programmers now demand this flexibility, even if it means that PLC scan cycle times may become more irregular.

In this book we have emphasized the importance of the PLC scan cycle, which limits the PLC to repeating these three steps:

1. Create a static image of the **input conditions** in the first scan cycle step, like taking a snapshot.

2. Execute a **user-written program** during the second step. The user-program builds an output image that the PLC saves until the user-program has completed executing. The PLC executes the user-program one instruction at a time, starting from the first instruction. Ladder logic rungs are executed starting at the top left and evaluating one branch at a time until the lowest rightmost output element is executed.

3. Copy output image data to the **output modules** during the third scan cycle step, like projecting an image.

The programming techniques covered in this book so far have been simple unstructured programs: The instructions were all executed, and they were executed in the order in which they were placed in the program. No subroutines or function calls or jump instructions were included to cause a deviation from the top-down execution order. In early PLCs, user-programs had to be written in this simple unstructured format to facilitate the simulation of real-time control. Unstructured programs make the intervals between the input image snapshots and output image projections as regular (as deterministic) as possible. *Beware:* Structured programming techniques, although powerful, allow the scan time to become variable. In Chapter 11 we explore interrupt programming, which can be used to restore deterministic behavior to highly structured programs.

We will examine structured programming techniques of three major types in the following order:

1. Instructions that allow a PLC to bypass or to repeat rungs in a program or to disable the outputs of selected rungs in the program under some conditions. (Professional programmers might argue that these features aren't "true" structured programming techniques, but they do change the structure of a PLC program.)

2. Instructions that make the PLC execute other programs called functions, subroutines, program blocks, or program files when conditions require their execution. We use the generic name *functions* in this book except in those sections dealing with specific PLCs. PLCs that offer user-written function capabilities usually also allow the use of variable names in functions instead of requiring absolute addressing of memory. (Professional programmers would be satisfied with these structured programming techniques.)

3. Configuration options that can cause the PLC to deviate from the normal scan cycle to execute more than one program per scan, to change which program is scanned, or to include extra I/O scan steps. (Some professional programmers wouldn't even know what this means.)

Note to Readers Learning How to Program Siemens PLCs

The Siemens PLC programming languages make extensive use of structured programming, so it is best to understand those techniques even before learning the simple instructions. That's why it is recommended that you read this chapter before reading Chapters 6 and 7.

The section of this chapter that deals with the use of functions is the most important, followed by the section about Jumps. You may wish to skim those sections now, return to Chapters 6 and 7, then study this chapter more completely after finishing Chapter 7.

INSTRUCTIONS THAT AFFECT EXECUTION IN A SINGLE PROGRAM

Master Control Reset/Relay

A **master control reset,** sometimes called a **master control relay,** is the software equivalent of a commonly used electrical lockout circuit used to switch part of an automated system off while the rest of the system continues to run. There are some significant differences between the software master control reset and a hardware master control reset, however, so the user is advised not to depend on the software MCR to perform the safety lockout operation. The software MCR should be used to enable/disable non-safety-related control functions.

A software master control reset (MCR) disables the outputs of a series of rungs. The MCR instruction must be entered on two rungs. The first rung has a conditional statement controlling an MCR start instruction and is sometimes called the MCR **startfence.** The second rung, entered immediately following the rungs that the MCR is intended to control, must have an MCR end instruction and is often called the MCR **endfence.**

When the conditions on the MCR startfence rung are:

TRUE — The MCR has no effect. The rungs between the MCR startfence and endfence are enabled to execute normally.

FALSE — The MCR becomes active. The output elements on the rungs between the MCR startfence and endfence act as if their control logic statements are false, regardless of the actual Boolean result of the control logic statements. For example,

- If a rung has a standard **-( )-** output instruction, then despite the truth of the control statement on that rung, the bit it controls is turned off.
- If a rung has an **-(L)- or -(U)-** output instruction, the bit it controls is unaffected by the truth of the logic on those rungs, so it remains frozen in the state it was in before the MCR zone became active.

In some PLCs, an MCR can only affect specific output instructions, while in other PLCs the MCR can affect any conditional output instructions. The programmer is advised that MCRs can cause some unusual effects if the PLC operating system allows them to affect functions that a hardware designer wouldn't normally build into a hardwired master control relay circuit. Here are some examples:

- An **on-delay timer** may reset when the MCR control logic goes false and the MCR zone becomes active, then restart when the MCR zone logic goes true again, even if the logic statement controlling the timer hasn't been false!
- An **off-delay timer** may start timing when the MCR zone control logic goes false (the condition that would disable most other functions) if its control logic was true before the MCR zone become active.
- A MOVE instruction may change an output value to zero, instead of freezing that value, when the MCR zone becomes active.

PLC-5

SLC 500

ALLEN-BRADLEY MASTER CONTROL RESET

Figure 8.1 is an example of an MCR as it might be programmed in an Allen-Bradley PLC-5 or SLC 500 (the example uses PLC-5 addresses). There are seven rungs between the startfence and the endfence, so the outputs of those rungs are affected by the condition statement controlling the MCR startfence. The MCR endfence must be programmed unconditional, as shown. If an endfence is not entered, the end of the user-program acts as the endfence.

If the logical statement controlling the MCR startfence is false (i.e., if I:040/1 is off), nor-

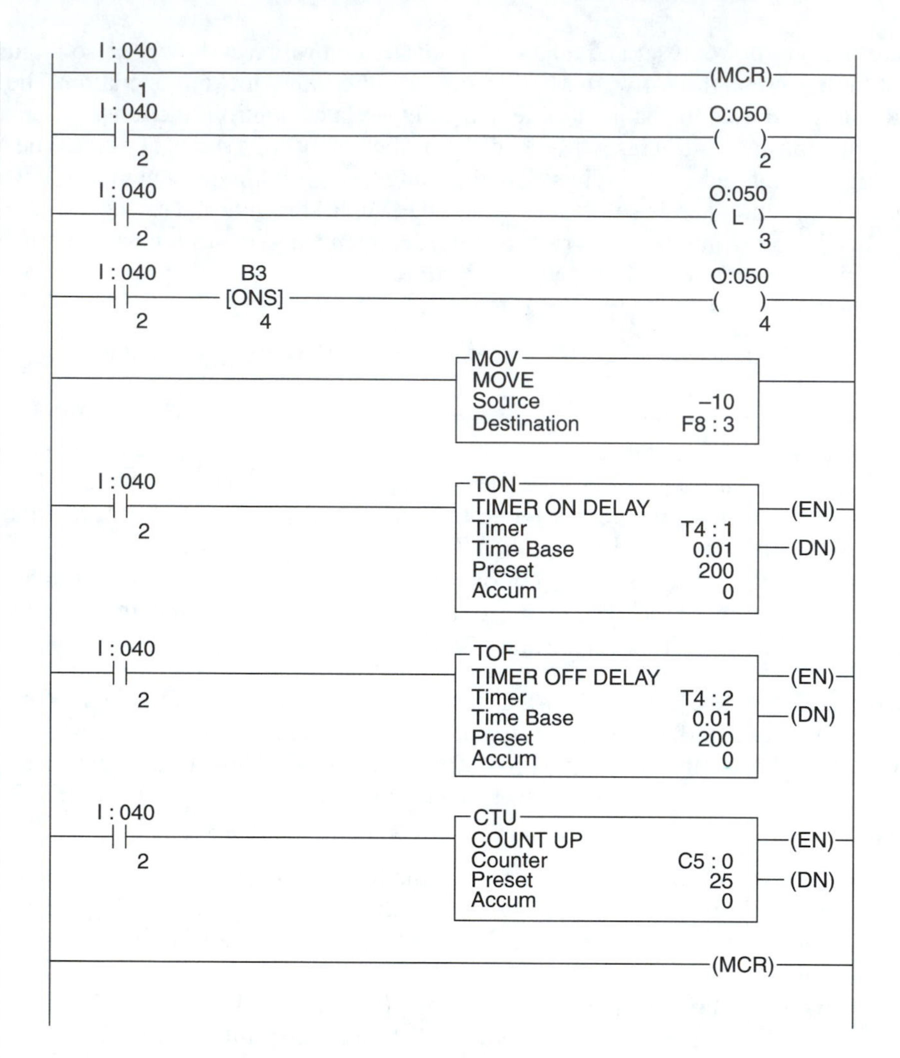

FIGURE 8.1
Allen-Bradley master control reset example.

mal execution of the rungs between the startfence and the endfence is suspended. Even if the logical statement controlling the affected rungs is true (i.e., if I:040/2 is on), then:

- Output O:050/2 will be off.
- O:050/3 cannot be latched on by I:040/2. If I:040/2 was on before the MCR became active, O:050/3 will have been latched on and will stay on. If a rung somewhere else in this program turns O:050/3 off, this rung within the MCR zone can't latch it back on.
- The one-shot bit, B3/4, and O:050/4, which the one-shot controls, will remain off, regardless of the state of I:040/2.

- The MOV instruction will not execute even though it is programmed as unconditional, so address F8.3 will not be affected by the MOV instruction.
- The on-delay timer's accumulated value (T4:1.ACC) will reset (T4:1.TT and T4:1.DN will go off).
- The off-delay timer (T4:2) will start, just as if I:040/2 has gone off (so T4:1.TT and T4:1.DN will go on for the next 2 s).
- The count-up element (C5:1) will turn its CU status bit (C5:1.CU) off.

When I:040/1 goes on, making the MCR inactive and reenabling the rungs in the MCR zone, then:

- O:050/2 and O:050/3 will go on if I:040/2 is on.
- If I:040/2 is on, the one-shot will turn the output bit, O:050/3, on for one cycle even if I:040/2 did not change state while the MCR was active, and will also turn bit B3/4 on. If I:040/2 is off, no one-shot will be generated even if I:040/2 did change state while the MCR was active.
- The on-delay timer will restart from zero if I:040/2 is on even if I:040/2 was on when the MCR became active and hasn't changed state.
- The off-delay timer will reset if I:040/2 is on but will not be affected if I:040/2 is off.
- The count-up element (C5:1) will turn its CU status bit (C5:1.CU) back on and will increase its accumulated value if I:040/2 is on even if it hasn't been off.

If an Allen-Bradley PLC program contains a statement to jump to a subroutine and that statement is within an active MCR zone, the subroutine will not be jumped to. (Subroutines are a structured programming feature that is covered later in this chapter.) The functions that are controlled by the subroutine that isn't allowed to execute will remain frozen until the subroutine is allowed to execute again. If an MCR startfence is in a subroutine, and the subroutine doesn't include an endfence (or if an instruction ends the subroutine before its endfence rung executes), the end of the subroutine will act as the missing endfence.

Overlapping MCR zones are not allowed in an Allen-Bradley PLC. Programs execute slightly faster when an MCR becomes active because the control logic can be assumed to be false.

S5

SIEMENS STEP 5 (NO MCR)

STEP 5 does not offer MCR instructions. To accomplish the same effect as an MCR, a STEP 5 programmer must specifically write programs to make control logic in individual segments go false, as it would if an MCR became active. For example, the logic that would control an MCR's startfence could be programmed in every segment that would otherwise be programmed between the startfence and the endfence. If the MCR control logic is complex, it could be used to control a flag bit that could be examined in several segments.

Other structured programming features that will be described later (such as conditional jumping to program blocks or to function blocks) can be used to conditionally execute logic to turn selected outputs off, overriding other logic that would otherwise control their state. Remember, though, that if more than one logical statement in a program tries to control the same output, the *last* statement that executes will actually control the output's state!

SIEMENS STEP 7 MASTER CONTROL RELAY

Siemens has added master control relay features to the STEP 7 operating system, largely to satisfy the U.S. market. While offering some advantages over the MCRs that U.S. PLCs have offered for years, STEP 7 MCRs are severely limited in the types of output instructions they can affect.

The only instructions that can be affected by a STEP 7 MCR are those listed below:

- The STL language OUTPUT instruction, =, and the equivalent ladder logic instructions: −( ) and **-(#)-.**
- The STL language SET and RESET instructions, **S** and **R,** and the equivalent ladder logic instructions **-(L)-, -(U)-, SET-RESET FLIP FLOP**, and **RESET-SET FLIP FLOP**.
- The STL TRANSFER instruction, **T,** and the ladder logic equivalent **MOVE** instruction. (Note that the STL LOAD instruction, L, executes despite an MCR.) If TRANSFER (or MOVE) is within an active MCR zone, it will write zero to the destination address.

As in earlier chapters, in this chapter we discuss primarily STL language instructions. There are ladder logic equivalents in STEP 7 for each STL instruction that is covered here. In STEP 7, an MCR zone must start with the conditional **BEGIN MCR** instruction, **MCR(,** and end with an unconditional **END MCR** instruction, **)MCR.** There must be an **)MCR** instruction for each **MCR(** instruction. Siemens also requires the use of two other MCR control instructions: the unconditional **MCR ACTIVATE** and **MCR DEACTIVATE** instructions, **MCRA** and **MCRD.** As shown in the following example program, an MCR only affects instructions that are between the MCR(and)MCR instruction pair and between the MCRA and MCRD instruction pair.

A single MCR zone in a single program:

```
A      M0.0       // Opens an MCR zone
MCR(
A      I4.0       // Statement in MCR zone
=      Q5.0
MCRA              // Activates any open MCR zone(s)
A      I4.1       // Statement in active MCR zone
=      Q5.1
S      M3.0
R      M3.1
L      +15
T      MW2
CU     C1
)MCR              // Closes the MCR zone
A      I4.2       // Statement outside MCR zone
=      Q5.2
MCRD              // Deactivates any open MCR zone(s)
```

In the program:

- Q5.0 is controlled only by I4.0, even if M0.0 is false, since = Q5.0 is programmed ahead of the MCRA instruction. Similarly, Q5.2 is controlled only by I4.2 because it is programmed after the)MCR instruction.
- Q5.1 is controlled by I4.1, unless M0.0 is false, in which case Q5.1 will always be off.

- M3.0 cannot be set, and M3.1 cannot be reset by I4.1 while M0.0 is false.
- The value +15 will be loaded into accumulator 1, regardless of the state of M0.0. If M0.0 is true, +15 will be transferred to MW2. If M0.0 is false, the value 0 will be transferred to MW2.
- The count-up (CU) instruction is unaffected by MCRs, so C1 will increment each time I4.1 goes true, despite the state of M0.0.

STEP 7 instructions that call other program blocks (functions and function block programs are discussed later in this chapter) are not affected by MCRs, so other program blocks can be called from within an active MCR zone. The MCR will not affect the instructions in that program block unless the program block executes an MCRA instruction. Use of MCRA and MCRD in a called program block does not affect whether the MCR is active in the calling program. The following example program shows a program block being called from within an MCR zone.

A single MCR zone in a structured program:

```
A      M0.0     // Opens and activates an MCR zone
MCR (
MCRA
A      I4.3     // A conditional call of function 10
CC     FC10
A      I4.5     // Statement in MCR zone after a call
=      Q5.5
)MCR            // Closes and deactivates MCR zone
MCRD

    FC10
      Function 10, called from within the MCR zone

      A      I4.3     // Statement preceding MCR activation
      =      Q5.3
      MCRA
      A      I4.4     // Statement following MCR activation
      =      Q5.4
      MCRD            // Deactivates MCR zone(s) in function 10

      BE              // Ends function 10
```

In the example above:

- Function 10 is called when I4.3 is true, even if M0.0 is false, because the CONDITIONAL CALL instruction (CC) is not affected by MCRs. In function 10:
  - Q5.3 is only affected by I4.3, even if M0.0 is false, because MCR zones are automatically deactivated at the time of a program block call.
  - Q5.4 will be false if M0.0 is false, despite the state of I4.4, because function 10 is called from within the MCR zone controlled by M0.0, and because Q5.4 is controlled in function 10 after an MCRA instruction is executed.
- Q5.5 will be false if M0.0 is false, despite the state of I4.5, and despite whether function 10 executed. Function 10 includes an MCRD instruction, but when a called program block ends the effect of MCRA and MCRD statements in the called program block ends and the pre-call MCR activity level is resumed.

STEP 7 allows **nesting of MCRs,** to a maximum depth of eight! As many as eight conditional MCR(instructions can be executed before the first)MCR instruction is needed. The first)MCR statement that is encountered ends the MCR zone started with the most recently executed MCR(instruction. Instructions inside more than one set of MCR(and)MCR pairs are dependent on the start conditions of *all* the MCR zones they are within. As shown in the following example, if even one of those MCR(control statements is false (and if an MCRA has been executed), all the instructions inside the MCR zone with the false control statement are affected by the MCR.

Nested MCR zones in a single program:

```
MCRA            // Activates all MCR zone(s)
A      M0.0     // Opens first MCR zone level
MCR(
A      I4.6     // Statement in first MCR zone level
=      Q5.6
A      M0.1     // Opens second MCR zone level inside first
MCR(
A      I4.7     // Statement in second nested MCR zone
=      Q5.7
)MCR            // Closes the second MCR zone
A      I4.6     // Statement in first MCR zone level
=      M3.6
)MCR            // Closes the first MCR zone
MCRD
```

In the example program shown above:

- Q5.6 and M3.6 will be off if M0.0 is false, despite the state of I4.6. They are not affected by M0.1 since Q5.6 is controlled before the second MCR zone is opened, and M3.6 is controlled after the second MCR zone is closed.
- Q5.7 will be off if either or both of M0.0 or M0.1 are false, despite the state of I4.7.

In a later section of this chapter we discuss instructions that can be used to bypass or to repeat parts of a STEP 7 program within a single scan cycle. The programmer is advised to avoid jumps, especially conditional jumps, which bypass or repeat MCR instructions. Even if you believe that you can handle the tangled web of nested MCR zones and activations that might occur, think about that poor programmer who might have to modify your program when you move into management.

OMRON CQM1 INTERLOCK (MASTER CONTROL RESET)

The CQM1 has a master control reset feature, but OMRON calls it an **interlock.** The startfence for an interlocked range of rungs must contain logic to control an **IL(02)** (**INTERLOCK**) instruction, and the endfence must contain an unconditional **ILC(03)** (**INTERLOCK CLEAR**) instruction. If the logic on the startfence is true, the rungs between the IL(02) and the ILC(03) operate normally, but if the logic is false, all the logical statements on those interlocked rungs are treated as if they are false. Output instructions will execute as if their logic is false or has just gone false, so bits controlled by nonlatching output instructions are turned off, timers are reset, and counters stop counting.

One-shots controlled by the DIFU(13) or DIFD(14) instructions (Omron calls one-shot instructions *differentiate instructions*) may execute after the IL(02) logic returns to true but only if their control logic is different when the interlock ceases to be active. If, for example, the logic controlling a DIFU(13) instruction is true when the interlock becomes active and is true when the interlock ceases to be active, the DIFU(13) instruction will *not* generate a one-shot even if the DIFU(13)'s control logic changed several times while the interlock was active. On the other hand, if the DIFU(13) control logic was false when the interlock became active and is true when the interlock ceases to be active, the DIFU(13) instruction *will* generate a (delayed) one-shot.

Overlapping interlocked zones are allowed (although the programming software may issue an error warning), but the first ILC(03) instruction that is encountered will terminate *all* active interlocked regions.

Jump Instructions

Jump instructions allow a user-program to bypass or repeat some instructions in a program during a scan cycle. Although this capability doesn't constitute what non-PLC programmers call true structured programming, it does allow a programmer to write programs that are structured to execute in an order other than just top-down.

Jump instructions are output instructions, so a Boolean logic statement on a rung with a jump instruction controls whether the PLC will immediately jump to a *labeled rung* elsewhere in the program. The label instructions do not affect program execution except when the jump is executed. Outputs on rungs that are bypassed by a forward jump are effectively *frozen* since the rungs are not executed. Program execution time will be reduced if a jump instruction causes the bypassing of part of the program.

Some PLCs do not allow backward jumps because they can cause endless loops, leading to a watchdog timer fault. Since a PLC inherently repeats the user-program when it returns to execute another scan cycle, backward jumps should not be necessary anyway. The entire program will execute again in a few milliseconds even without a backward jump.

PLC-5

SLC 500

ALLEN-BRADLEY JUMP INSTRUCTIONS

When the conditional statement controlling a PLC-5 or an SLC 500 **JUMP** (JMP) instruction is true, the PLC will skip forward or backward through the program file to the rung that begins with a **LABEL** (LBL) instruction that has the same number as the JMP instruction. There can be several JMP instructions with the same numbered destination label, but each LBL must have a unique number (see your manual for the allowed range of numbers), and the LBL instruction must be the first element on its rung. The LBL instruction must be on a rung that is completed with at least an output instruction, and that rung's operation will not be affected by the presence of the LBL.

The example in Figure 8.2 shows how a PLC-5 might be programmed to use JUMP instructions:

1. While I:040/1 is true, the rungs between the JMP instruction and the LBL instruction are bypassed, so:

(a) Output O:050/2 remains frozen in its last state even if I:040/2 changes.
(b) The MOV is not executed even though it is unconditional.

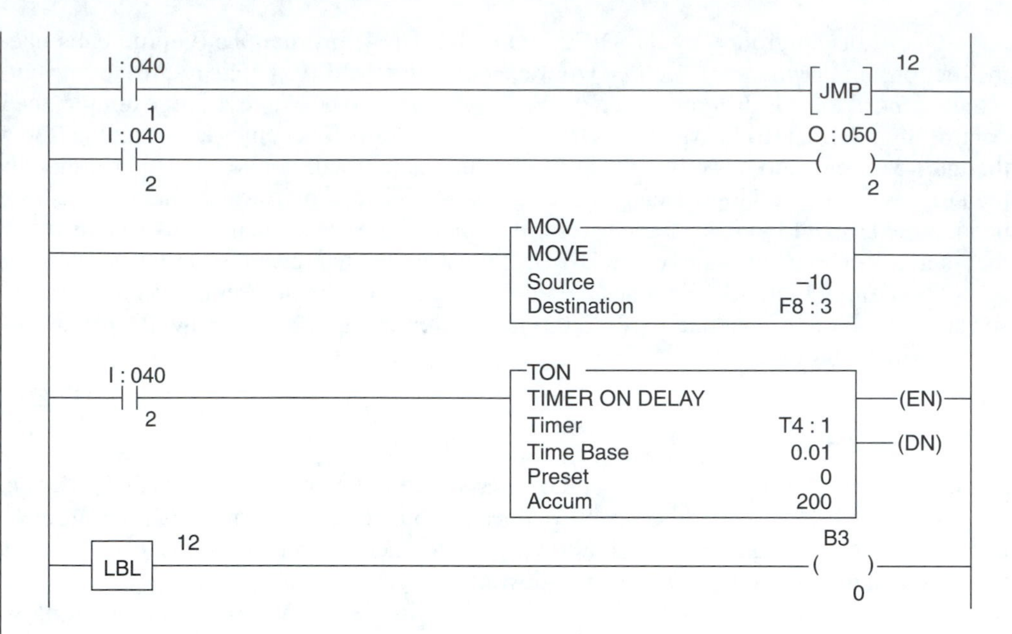

FIGURE 8.2
Allen-Bradley PLC-5 program with a forward jump.

(c) The on-delay timer's accumulated value (T4:1.ACC) freezes because the PLC doesn't update it as it would normally do every scan cycle as it executed the rung. The .TT and .DN bits remain in their last state. (.ACC will resume incrementing as if no time passed when the TON instruction resumes executing.)

2. While I:040/1 is false, the program executes as if there were no JMP or LBL instructions.

> Bit B3/0 will be on unconditionally whether or not the JUMP executes. It is only programmed here to show that the LBL instruction must be on a complete rung. Normally, the rung with the LBL instruction would contain a useful control operation.

The example in Figure 8.3 shows a JUMP that shouldn't be programmed in an Allen-Bradley program. Consider what happens in a scan cycle and see if you can figure out what is

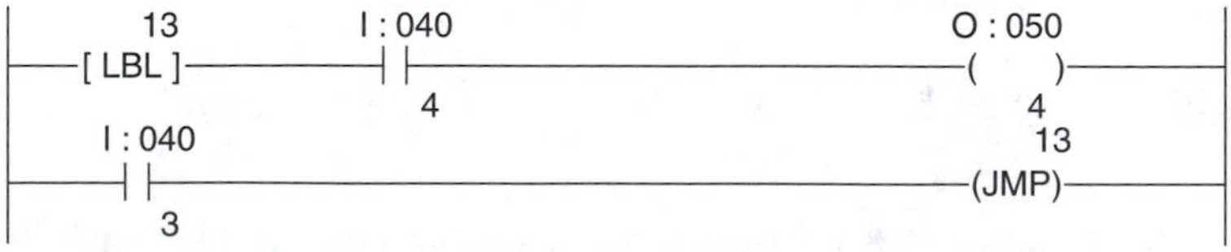

13
[LBL]
I:040
4
O:050
4
I:040
3
13
(JMP)

FIGURE 8.3
Allen-Bradley program with a jump that will cause the PLC to fault.

wrong with this program before you read the explanation telling you what is wrong. The program looks like it should repeatedly monitor the input associated with I:040/4 to control the output associated with O:050/4 as long as the input associated with I:040/3 is on. (My wording should give you a hint.)

Here's why the program in Figure 8.3 will cause a fault: The PLC won't fault while the input associated with input image bit I:040/3 remains off. The PLC will repeatedly execute its scan cycle, turning the output controlled by output image bit O:050/4 on or off as the input that controls input image bit I:040/4 goes on and off. The JUMP will not execute, so the normal scan cycle will be unaffected by the JMP instruction.

After the sensor associated with input image bit I:040/3 goes on, the input image bit I:040/3 will be made true during the next input scan step, and then the user-program will start to execute. The state of input image bit I:040/4 will be written to output image bit O:050/4, then the JMP instruction will execute. The unchanged state of I:040/4 will again be written to O:050/4, then the jump will execute yet again, since there is nothing in the program to change input image bit I:040/3. Even if the sensor that should control I:040/3 goes off, the PLC cannot get out of the user-program to execute the input scan, so input image bit I:040/3 will never go off. Notice that the output scan doesn't execute either, so the actuator controlled by 0:050/4 will remain in its last state (as it was before the loop started executing) despite the state of I:040/4.

PLCs are designed not to allow this type of endless loop to execute for too long. All PLCs run a watchdog timer while the scan cycle executes, and the PLC will *fault* (stop running) if the watchdog timer exceeds a preset time. Watchdog timer faults are examined in more detail in Chapter 11, and we will see then how to change the default watchdog timer preset value.

S5

SIEMENS STEP 5 JUMP INSTRUCTIONS

Jumps within single STEP 5 programs can be programmed only in function blocks[1] and can be programmed only in STL. Jumps can be forward or backward. The following example shows a typical STEP 5 CONDITIONAL JUMP (JC) instruction in an STL program:

```
     : L KH00    ;  Load 8 bits, all zeros, into the accumulator
     : A I4.0    ;  If input I4.0 is on, then
     : JC = past;      jump past the next instruction
     : L KHFF    ;  Change the accumulator's contents to all 1's
past: T QB5      ;  Send whatever is in the accumulator to an
                 ;     output module
```

The word *past* is an alphanumeric label that the programmer makes up, which cannot exceed four characters, of which the first must be a letter. Note that labels must be entered to the left of the colons. The JUMP command must indicate the label to jump to, and the label identifier must be preceded by an = sign to indicate that the jump is to a label within the function block. Jumps can be either forward or backward, and several jump instructions can cause jumps to the same label. There should, of course, not be two identical labels in the same program.

---

[1] By default, a Siemens PLC executes an organization block as part of its normal scan cycle, and STEP 5 does not allow the programming of jumps within an organization block. The information here on jumping within a program will be useful only after you learn how to program a jump to a function block, which we cover later in the chapter.

Several STL instructions (e.g., LOAD and TRANSFER) execute unconditionally but they can be made to execute conditionally by programming conditional jumps to bypass them. The example program above demonstrated how a JC instruction can be used to make a LOAD **(LKHFF)** conditional. Some ladder logic instructions can be programmed with conditions in function blocks but not in other STEP 5 programs (e.g., MOVE). STEP 5 automatically translates ladder logic programs into STL before sending them to the PLC, and can insert jump instructions (invisible in ladder logic) making them conditional if programmed in function blocks. Some ladder logic instructions are only available in function blocks, several of which automatically include jumps as part of how they operate.

There are several other STL jump instructions in addition to the CONDITIONAL JUMP (JC) instruction shown above. There is an **UNCONDITIONAL JUMP** (JU), and several instructions that will cause jumps dependent on the state of the status bits maintained by the operating system as it manipulates data words. The status word–dependent JUMP instructions include **JZ, JN, JP, JM,** and **JO,** which will cause the jump to occur if the last number manipulated was **Z**ero, **N**ot zero, a **P**ositive number, a **M**inus number, or **O**verflowed the accumulator's signed 16-bit capacity.

S7

SIEMENS STEP 7 JUMP INSTRUCTIONS

In STEP 7, jumps can be forward or backward and can be programmed into any type of program block. Each destination label must be a unique alphanumeric label of up to four characters, of which the first must be a letter. There can be several jump instructions to jump to the same destination label, but there can't be two identical destination labels in the same program.

Jump instructions look significantly different in ladder logic from how they look in STL, so a ladder logic example is shown. Figure 8.4 shows the ladder logic **JUMP** (JMP) instruction and shows a label in the destination network. The JMP instruction can be programmed conditionally (as shown) or unconditionally. The JMP executes if the preceding conditions (if any) are true. There is also a ladder logic **JUMP-IF-NOT** (JMPN) instruction that executes if its control statement is false and which must be programmed conditionally. Note that the destination

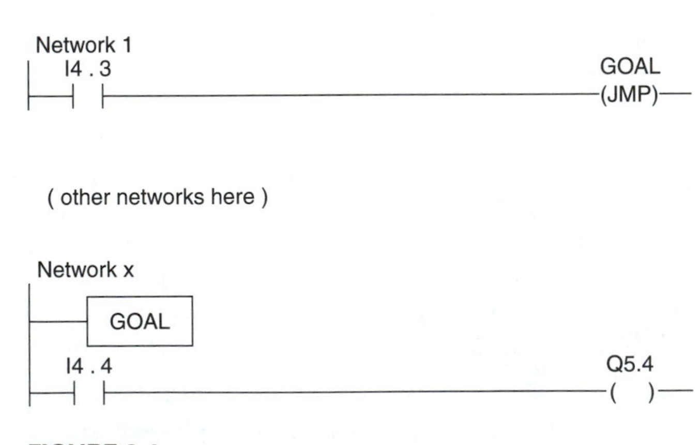

FIGURE 8.4
STEP 7 ladder logic JUMP instruction.

label must be entered as the first element in the destination network and appears to precede the network in ladder logic.

The STL equivalent of the ladder logic program of Figure 8.4 includes the STL CONDITIONAL JUMP (JC) instruction:

```
         A      I4.3
         JC     GOAL

                (other instructions here)

GOAL:    A   I4.4
         =   Q5.4
```

In the STL program above, the destination label (GOAL) must be entered with a colon (:) because the STEP 7 editor doesn't print colons on every line as the STEP 5 editor did, but the destination label must still be on the left of the colon and the program must still be on the right of the colon. In STEP 5 the JC command could be used to call other program blocks, but it can't in STEP 7, so the JC command doesn't need an = sign with the label (GOAL).

If you program in STL, you will note that a lot of the STL instructions execute unconditionally.[2] Even the LOAD (L) and TRANSFER (T) instructions that must be used to copy data values into and out of the accumulators execute unconditionally. You can make unconditional instructions execute conditionally by programming conditional jumps to bypass them. If you translate some ladder logic instructions into STL, you will see that they translate into a set of STL instructions, including conditional jumps past unconditional STL instructions.

STEP 7 only offers the two ladder logic JUMP instructions covered above, but offers an extensive range of JUMP instructions for programmers who like STL. The STL JUMP instructions that are equivalent to the ladder logic JUMP instructions include:

- **JC,** the CONDITIONAL JUMP instruction.
- **JU,** which causes an UNCONDITIONAL JUMP.
- **JCN,** which causes a jump if the RLO of the preceding logical statement is false.

STL JUMP instructions for which there are no ladder logic equivalents include:

- **JCB** and **JNB,** which do the same as JC and JCN, respectively, but also copy the RLO into the BR bit of the status word first.
- **JL,** which allows the PLC to select from a list of possible jump destination labels, depending on what number is in accumulator 1 when the JL instruction executes. (See your programming manual for more details.)
- **JBI** and **JNBI,** which cause jumps if the BR bit is set (JBI) or if it's clear (JNBI).
- Instructions that cause jumps dependent on the result of the most recently executed math operation: **JZ,** if a **z**ero resulted; **JN,** if a **n**onzero resulted; **JP,** if a **p**ositive value; **JM,** if negative; **JPZ,** if positive or zero; and **JMZ,** if negative or zero.
- **JOV** causes jumps if a signed binary number too big for the accumulator was generated

---

[2] You can write STL programs with conditional statements preceding unconditional instructions such as the LOAD instruction, but when you run the program you will find that the LOAD instruction executes regardless of the truth (the RLO) of the conditional statement.

by the most recent math operation, and **JOS** causes a jump if any math operation has done so since the last time the OS (overflow set) status bit was cleared.

- **JUO,** which causes a jump if the most recent math operation was a floating-point math operation, but generated a floating-point number that was not in valid floating-point format.
- **LOOP,** which is used to program repeated loops in a program and which is covered later in the chapter.

CQMI

OMRON CQM1 JUMP INSTRUCTIONS

Unlike in other PLCs, an OMRON JUMP executes if the control logic controlling the jump instruction is *false.* Figure 8.5 demonstrates use of OMRON's conditional JUMP instruction, **JMP(04),** and the unconditional **JUMP END** instruction, **JME(05).** The jump will not execute unless input image bit IR 00503 is off.

An OMRON JMP(04) instruction must include a numerical label (01 to 99) matching the number in the JME(05) instruction. The same label number cannot be used in another JMP(04) or JME(05) instruction in the same program. The JME(05) instruction is the jump destination but must be programmed as an unconditional output instruction. You cannot program a backward jump in an OMRON PLC.

Label number 00 *can* be used instead of a number between 01 and 99, but the rules are slightly different then. Any number of JMP(04) and JME(05) instructions can use label number 00. A JUMP instruction to label 00 will cause a jump forward to the next JME(05) with label 00.

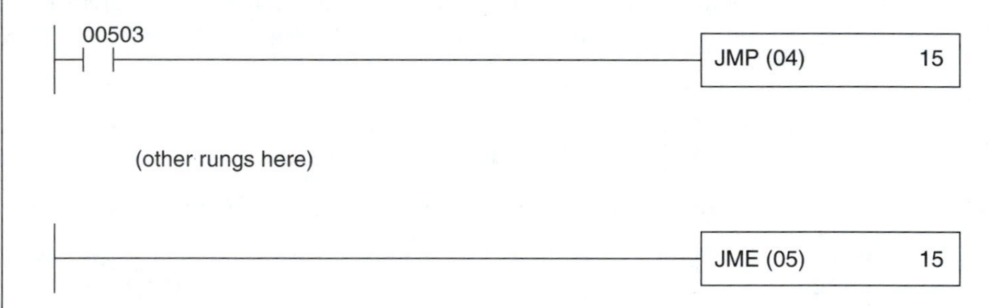

FIGURE 8.5
OMRON PLC program with a JUMP.

Loops

Loops are common elements in many structured programming languages but haven't been available in PLC programming languages until recently. In a structured-language loop, the computer jumps backward to execute a set of instructions and maintains a counter to ensure that the backward loop executes a predetermined number of times.

PLC instructions that operate on arrays of data elements (covered in Chapter 7) include built-in loops (invisible to the PLC programmer) to operate on the elements in the array one at a time. In this section we briefly examine the instructions that a PLC programmer can use to create his or her own loops. Loops are not covered in much detail in this book because they are somewhat redundant in a PLC. A PLC's operating system causes the entire program to scan (loop) repeatedly anyway! Loops can be useful in some applications, such as where multiple data words must be moved during a single scan, but misuse of loops can cause the watchdog timer to cause a PLC fault.

Only two of the PLCs covered in this book offer loop instructions, discussed below.

ALLEN-BRADLEY PLC-5 FOR-NEXT LOOPS

The PLC-5's **FOR** (FOR) and **NEXT** (NXT) instruction pair is used to cause a specific number of repetitions of the rungs between the conditional rung controlling the FOR instruction and the unconditional rung containing the NEXT rung. The set of loops is executed every scan, while the conditional statement controlling the FOR statement remains true.

In the FOR instruction, the programmer must enter a label number, matching the label number in the matching NXT instruction, and must enter an address for the index so that the PLC can keep track of the number of times it loops back. Also in the FOR instruction, the programmer must enter an initial value, a terminal value, and a step size, so the PLC knows how often to repeat the looping process (e.g., entering 0, 5, and 1 for initial value, terminal value, and step size would cause five loops; entering 100, 110, and 2 would also cause five loops).

An optional **BREAK** (BRK) instruction on a rung between the FOR and the NEXT instructions can be used to allow early (conditional) termination of looping.

SIEMENS STEP 7 LOOP

When an S7 PLC executes the **LOOP** instruction, it decrements the 16-bit number in the low word of accumulator 1 (treating the number as an unsigned binary number). If the resulting number is not 0, the PLC reads the label following the LOOP instruction and jumps within the program block to the instruction that is preceded by the same label. The jump would typically be backward, so that the LOOP instruction will be executed again in the same scan cycle.

To use the LOOP instruction, the program must first load a positive number indicating the number of times to loop (size limited by the 16-bit unsigned binary format maximum: 64,565) into accumulator 1. The program should save (TRANSFER) the loop counter from accumulator 1 immediately after the LOOP instruction (the instruction preceded by the label will be the next instruction executed in the loop) and restore (LOAD) the loop counter to accumulator 1 immediately preceding each execution of the LOOP instruction; otherwise, instructions inside the looping range may affect accumulator 1's contents. The following example program shows a LOOP instruction that will execute the instructions in the loop five times:

```
       L    +5
more:  T    MW10

       (instructions in the loop)

       L    MW10
       LOOP more
```

INSTRUCTIONS THAT AFFECT WHICH SUBPROGRAMS OR FUNCTIONS ARE EXECUTED DURING PROGRAM SCANS

Using a single PLC program to control several processes, or even to control one complex process, can get quite messy. A better approach is to write a main program that calls subprograms as they are required. The multiprogram structure makes it easier to understand which parts of the program affect which processes, and also makes it possible for teams of programmers to work together on the same control project. Each programmer can work on his or her own subprogram.

When a main program calls a subprogram, the main program temporarily stops executing while the subprogram runs from start to finish. The main program then resumes running, starting with the instruction following the subprogram call. Subprograms can call other subprograms. Figure 8.6 shows how the user-program step in the three-step scan cycle is affected by the use of subprogram calls.

If a multiprogram project is planned carefully enough, some subprograms can be made **reusable.** A reusable subprogram can be called more than once by the main program to control more than one process, with the main program providing different parameters to the reusable subprogram each time it is called. For example, Figure 8.7 includes a reusable subprogram to compare a sensed temperature against a desired temperature setpoint and to increase or decrease heat output to correct the difference. The same subprogram is called twice by the main program, and the main program passes different parameters to the subprogram each time it is called, to tell the subprogram which temperature sensor, which setpoint temperature, and which heater to use during each call.

A subprogram can share data areas used by other main programs and subprograms or may use its own private data area (as demanded by IEC 1131-3). Subprograms that have their own

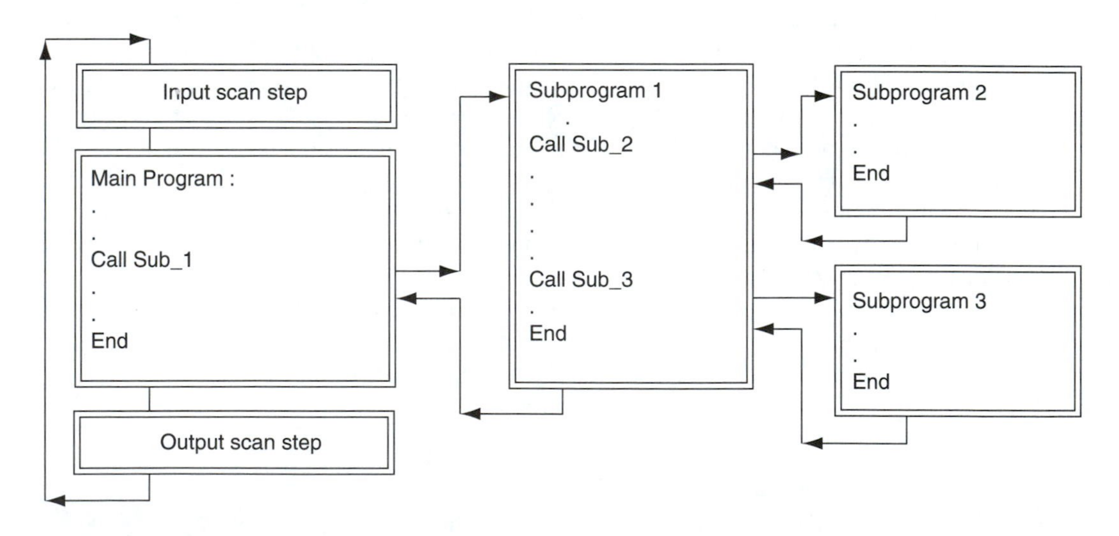

FIGURE 8.6
Main program with calling of subprograms.

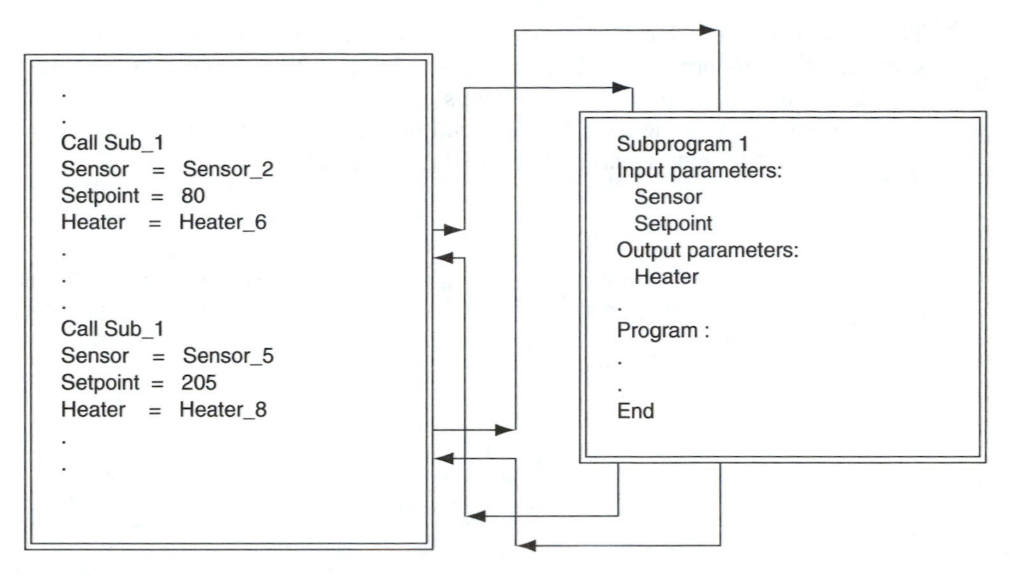

FIGURE 8.7
Multiple calls of a reusable subprogram.

data area usually need to receive data as **input parameters** from the main program when they are called and must pass **result parameters** back to the main program when they finish. In Figure 8.7, the main program provides two input parameters (a sensor address and a temperature setpoint) to the subprogram and provides a destination parameter (the heater address) for the subprogram's result.

PLC manufacturers all have their own names for subprograms. Some call them **subroutines,** some use the term **program files,** others use names such as **function blocks.** IEC 1131-3 (the International Standards Organization's guide for standardization of PLC programming, which is covered in Chapter 9) says that programs, function blocks, and functions can be called by any other program, function block, or function. (Programs, function blocks, and functions have slightly different characteristics, which IEC 1131-3 clearly defines.)

PLC-5

SLC 500

ALLEN-BRADLEY SUBROUTINE CALLS

A newly manufactured PLC-5 or SLC-500 PLC automatically executes the program in program file 2 as the user-program during its scan cycle. Programming software recognizes this, so program file 2 is the program that is created by default when the programmer enters program edit mode. (The defaults can be changed.)

Any Allen-Bradley program file (including file 2) can include a JUMP TO SUBROUTINE (JSR) instruction to call any other program file

which should start with a SUBROUTINE (SBR) instruction to identify the program file as a subroutine and to accept any input parameters that the calling program may try to pass. The SBR instruction must be the first element on the first rung in the subroutine program file.

The SBR instruction does not affect the logic on the rung on which it is programmed. A subroutine can in turn use another JSR instruction to call a second program file down to a maximum of eight levels of nested calls. Figure 8.8 demonstrates an unnested subroutine call. The first call doesn't count as a nested level. Each called program file should start with an SBR instruction.

This instruction (perhaps in MCP file 2) would call a subroutine program file (file 4) if input I : 040 / 2 is on. The programmer has opted to specify addresses for one input and one output parameter.

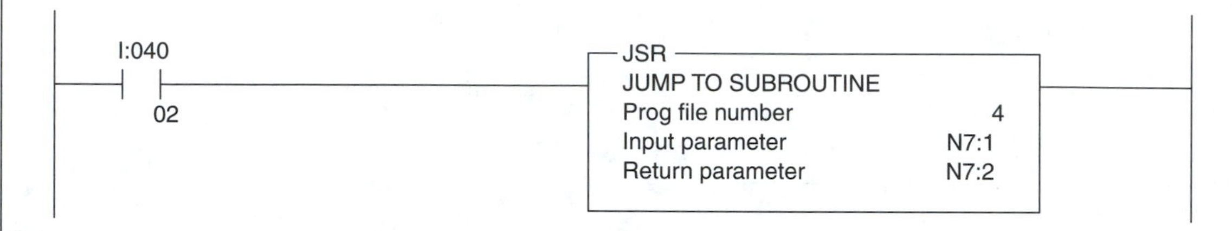

The SBR instruction in program file 4 says to accept an input parameter from the calling program. Here, it accepts a copy of the data from N7:1, and puts it into N9:0. File 4 then manipulates the *copy* of the data from N7:1 data (changing N9:0.)

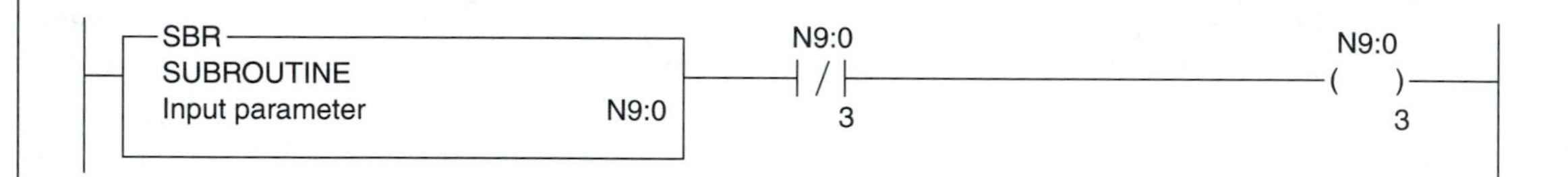

Subsequent rungs in file 4 (not shown) manipulate the data in N9:11.

The final rung in file 4 contains a RET instruction that passes control back to file 2, and gives file 2 a *copy* of the new data from N9:11.

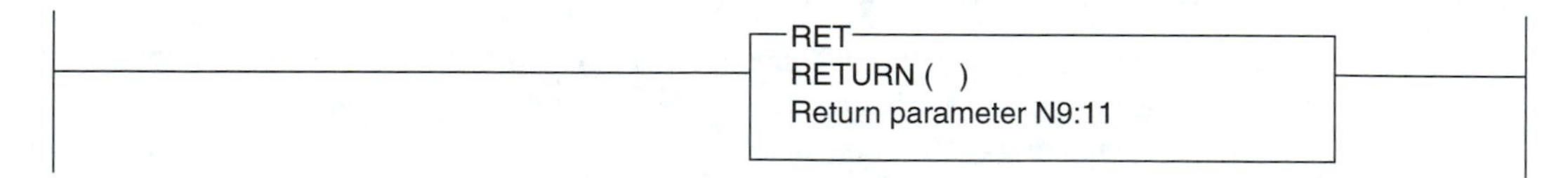

The JSR instruction in file 2 isn't finished until the PLC copies the data from N9:11 into N7:2, then the next rung of file 2 executes.

FIGURE 8.8
PLC-5 program that jumps to a subroutine in another program file.

Every subroutine should end with a RETURN (RET) instruction, which can specify return parameters, and which causes a return to the program file that called the subroutine. Return parameters (if any) are accepted from the subroutine; then the instruction following the JSR instruction is executed. It is important that all subroutines end with a RET instruction so that the first program that called the first subroutine can eventually resume executing.

Allen-Bradley PLCs only have one data memory area, which Allen-Bradley calls a **data table,** and which consists of several **data files.** Any program files can read or write data in any data table, so values changed by a subroutine can affect the operation of other program files that use the same data addresses. Because there is only one shared data table, it is not necessary to pass parameters between a subroutine and a program that calls the subroutine unless the subroutine is written to be reusable.

A calling program may (optionally) pass **input parameters** to a subroutine program file that is written to be reusable. Passing input parameter constants or addresses in a JSR instruction causes the creation of copies of the input data in the memory addresses specified in the subroutine's SBR instruction. The JSR instruction in a calling program can also (optionally) specify addresses as **return parameters.** When a subroutine ends, the PLC copies data from memory addresses specified in the subroutine's RET instruction into the memory addresses specified by the main program's JSR instruction. Figure 8.8 shows a PLC-5 program with a JUMP TO SUBROUTINE instruction with input and with return parameter specifications.

S5

SIEMENS STEP 5 FUNCTION CALLS

By default, organizational block 1 (OB001) is executed as the user-program during the scan cycle of a S5 PLC, and programming units default automatically to editing OB001. Any block (including OB001) can call any other block as a subprogram (except data blocks, of course). Blocks can be called to up to 16 nesting levels deep, using UNCONDITIONAL JUMP or CONDITIONAL JUMP instructions as in the ladder logic and STL examples in Figure 8.9.

In an S5 PLC, all programs and functions share the same flag memory area and the same input and output image memory. A STEP 5 program can have as many as 255 data blocks, only one of which can be open at a time, so programs can be written so that each subprogram has its own data block. Any main program or subprogram can call any data block, so STEP 5 data blocks do not provide foolproof protection of one subprogram's memory space from another subprogram. When a data block is active, any subprogram that is subsequently called uses that data block by default, unless the subprogram calls its own data block. When each subprogram ends, returning control to the program that called it, the data block that was active before the call automatically becomes active again.

Three types of main programs and subprograms are used in STEP 5:

1. Organizational blocks (OB 1 to 255) are intended to be used as main programs, to organize the PLC's operation by consisting only of conditional and unconditional calls of FBs, PBs, or of other OBs. Some OBs (e.g., OB 1, OB 2, OB 13) can be called automatically by the operating system when they are required.

2. Program blocks (PB 1 to 255) are intended to be used for simple control programs but don't offer any PLC programming capabilities not offered by OBs.

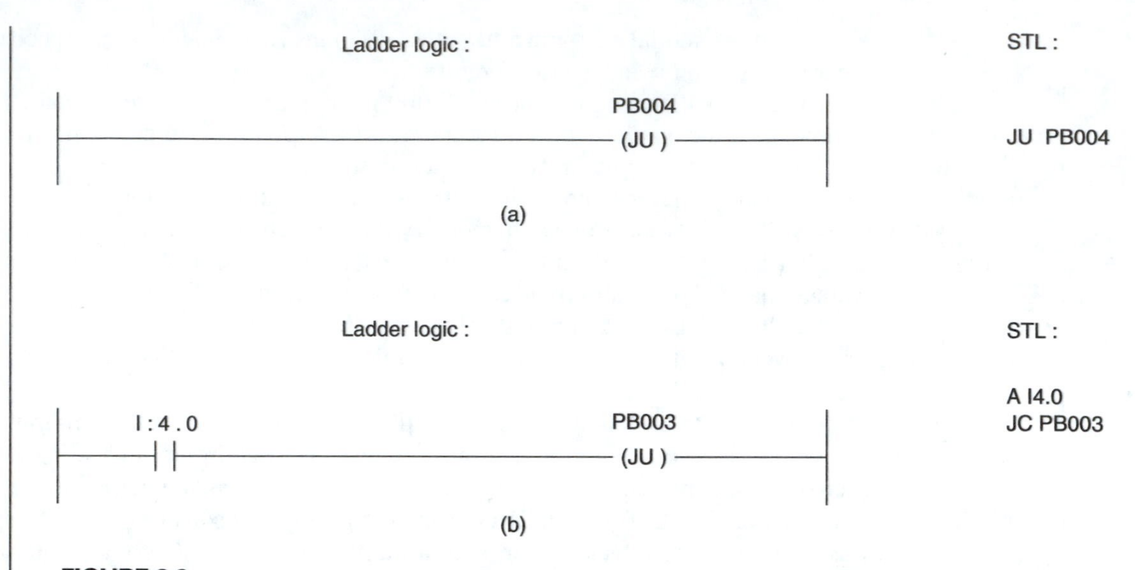

FIGURE 8.9
Jumping to program blocks in STEP 5: (a) unconditional jump; (b) conditional jump.

OBs and PBs that are called as subprograms do not need to contain labels identifying them as subprograms and do not need return instructions. When an OB or PB function finishes, the PLC automatically resumes running the program that called the OB or PB.

3. Function blocks (FB 1 to 255) allow more powerful instructions and memory-access capabilities than is possible in OBs and PBs. Only in a FB, for example, can a program read or write the RS memory areas to change the PLC's configuration, or use JUMP instructions to bypass LOAD and TRANSFER instructions to make them conditional. FBs are a little more difficult to use.

The following steps are necessary in order to include a function block in a program:

1. Create the function block (FB). This must be done before you try to program a call to the FB by another program. While you create a FB, the programming software will ask you to enter:

(a) A function name (up to eight characters). In the following example, the name *control* has been assigned. You can refuse to enter a name: Just press [Enter].

(b) Up to 40 declared parameters. Four parameters are declared in the following example program. You don't need to enter any parameters: Press [Enter] if you have no more parameters to declare. You declare **formal variable names** while declaring parameters. Your FB program can then use the variable names in place of addresses for the data it manipulates. At the DECL: prompt, enter a line consisting of three fields for each parameter:

formal-operand **parameter-type** **data-type**

where

- **formal-operand** is a variable name, up to four characters in length. Using a name to represent a data value or a data address will be discussed later in an example. The first formal variable name declared in the following example is "sw1".
- **parameter-type** can be:
  - **I,** to indicate an input parameter. The calling program must provide the *address* of an actual data value that the FB can use when the FB is called. (The address does *not* have to be an input image address.)
  - **Q,** to indicate an output parameter that this function will return, to the address specified by the calling program, when this FB ends. (Does not have to be an output image address.)
  - **D,** to indicate that the calling program will provide an input parameter consisting of a constant value (instead of the address for the value) when it calls this FB.
  - **B,** to indicate that the calling program will provide an input parameter consisting of a data block number, function block number, or program block number that the formal variable name will represent during execution of the FB.
  - **T** or **C,** to indicate that the calling program will provide an input parameter consisting of a timer or counter address that the formal variable name will represent.
- **data-type** is one of:
  - **BI, BY,** or **W** (if the parameter-type was I or Q), to indicate the size of the input or output parameter (either a **BI**t, a **BY**te, or a **W**ord). Data from input, output, flag, or data blocks can be passed to or returned by the FB.
  - **KF, KH, etc.** (if the parameter-type was D), to indicate the format of the constant data that the calling program will provide.
  - No data-type identifier is required if the parameter-type was T, C, or B. These data types imply that the data provided by the calling program will be a timer number, a counter number, or the number of a data, function, program, or organizational block.

After you have named the function and declared the variables (or declined to do so), you can enter a program into the function block. You can use more powerful instructions in FBs than in OBs or PBs. STEP 5 calls these extra instructions (some of which can only be programmed in STL) supplementary instructions and system operations, and they include instructions for:

- Jumping within the FB
- Additional timer and counter control operations
- Performing bit logic using timer or counter accumulated values and using data in data blocks
- Logic operations and shift operations that manipulate whole data words at a time
- Changing the PLC's **RS** data areas, to change the PLC's configuration, including how the PLC handles interrupts

If parameter passing is used, the function block must address the parameter data using its formal variable names. An = sign must precede any formal variable name (e.g., =sw1 in the following example). As shown in the example, using named parameters to represent data or data addresses in an FB allows the FB to be reused, with different input and output addresses, when called from more than one place in the structured PLC program. The FB will perform the same operations on whatever data it is passed.

2. In the program(s) that will call the FB, program JU or JC statement(s) to call the function block by its function block number. Your programming unit will immediately display the function name you entered when you created the FB and will prompt you to enter actual addresses or constant values for each formal variable that you declared in writing the FB.

The following example function block, FB001, declares four parameters with formal variable names. It identifies the first two (sw1 and mem) as input parameters (I) to be received from the main program, the third (act) as an output parameter (Q) that this FB will return to the main program, and the fourth (dat) as a block (B, a data block in this example). All data-types are identified as single bits (BI), except the data block parameter, for which a size specifier isn't required. The program in FB001 changes the output parameter (act), depending on the values for the two input parameters; uses the DO command[3] to call a new data block (dat); copies a data word (DW001) from that data block to an analog output module (QW072); then ends. Notice that you can use absolute addressing (DW001 and QW072) in a function block if you want. (Boldface indicates what the programmer had to enter.)

```
FB001
     NAME : control
     DECL : sw1   I  BI
     DECL : mem   I  BI
     DECL : act Q BI
     DECL : dat B

          : A   =sw1
          : AN =mem
          : =   =act
          :
          : DO =dat
          : L DW001
          : T QW072
          : BE
```

A program that calls FB001 might look like the following. During programming of the jump instruction calling FB001, the programming unit prompts the programmer to enter addresses for parameters declared in the FB. (Programmer entries are all shown in bold.)

```
          : C   DB002
          : JU FB001
     name : control
     sw1  : I4.2
```

[3] STEP 5 allows the DO command to be used only with formal block parameters: OB, PB, FB, SB, and DB. If the formal parameter represents an organization block, program block, function block, or sequence block, DO means JU (jump). If the formal parameter represents a data block, DO means C (call).

```
mem  : F3.2
act  : Q5.4
dat  : DB006
     :
     : L DW001
     : T QW74
     : BE
```

The example program calls DB002 (data block 2) before calling FB001. After FB001 has been called and FB001 ends, Q5.4 will have a new value provided by FB001, the value from DW001 of DB006 will have been copied to QW72 for outputting as an analog value, and the main program will then copy DW001 of DB002 to QW74. (Because DB002 was the active data block before FB001 was called, it becomes active again after FB001 ends.)

Another "JU FB001" instruction can be placed elsewhere in the structured program, where the programmer can respond to the prompts for parameters by entering a completely different set of addresses. FB001 will accept these different addresses; assign the names "sw1", "mem", and so on, to them; execute its program; and return the result. Performing the same set of operations using different data values and/or addresses is possible only if the FB is passed parameters and uses the formal variable names.

Some FBs (240 to 251) are preprogrammed (in ROM) for specific tasks that Siemens thought users might use. If your program calls these blocks, your program must provide the actual constants or actual addresses for the parameters declared in the preprogrammed function blocks. Some of the function blocks are described in this book, or you can see your manual for the parameter specifications that each preprogrammed FB requires.

Some OBs (31 and 251) are preprogrammed (in ROM) for specific tasks that Siemens thought users might use. If your program calls OB251, OB251 will read and write data words, so your program must provide data blocks in a specific form and open one of them before calling OB251. See Chapter 12 for an explanation of the data block structure required by OB251.

Some OBs (1, 2, 3, 13, 20, 21, and 34) are not preprogrammed, but if they exist, they will be called by the STEP 5 operating system when STEP 5 thinks they should be run, despite what the programmer may have intended when he or she programmed them. See Chapter 11 for an explanation of why these OBs execute.

S7

SIEMENS STEP 7 FUNCTION CALLS

OK, make sure that you are in a comfortable position and are prepared to spend a little quiet time reading. If you have been thinking you would like a cup of coffee soon, now is a good time to go get it. STEP 7 requires the programmer to use a program-preparation process that makes structured programming very easy and powerful but is very different from how most PLC programmers are accustomed to program. It will take a few minutes to describe the required structure before we get into how to actually write structured programs.

Projects and CPU Programs

Before any program can be entered, the programmer must first use the STEP 7 programming software to create a **project** and then to create (at the very least) one PLC **station** (sometimes called a CPU program in Siemens literature) within the project. The programmer must then enter configuration information for the PLC system in the station. Then the programmer can enter the program(s), which Siemens calls **logic blocks,** and can enter any **data blocks** that the logic blocks require. Before or during programming of logic blocks and data blocks, the programmer can make entries into a **symbol table** to define symbolic names that can be used in any program in this PLC. The configuration and the symbol table become a part of the station that is saved in this programming unit as well as being downloaded to the PLC.

The concept of having a project containing a station might seem unnecessary to a programmer whose system consists of a single PLC, although having the station made up of separate configuration, logic block, data block, and symbol table components makes sense even in simple systems. Most control systems now consist of at least one central PLC with intelligent I/O modules and remote I/O modules. The project structure starts to make more sense when you recognize that a lot of PLCs are now used in manufacturing systems consisting of multiple PLCs interconnected via multiple local area networks, all working together.

A single project (e.g., a manufacturing plant) might contain several stations, or may even consist of multiple **subprojects,** each containing several stations. The configuration data in each station in a system of PLCs interconnected by local area networks must include configuration information describing this PLC's communication connections and unique address in the local area network. A STEP 7 project can also include a **global data table** defining data items that a local area network must copy between this PLC's memory and the memory of other PLCs in the network. Using the project system to define the structure, interconnections, and data exchange requirements in the overall system means that the programmer does not have to program or configure each PLC separately to exchange data. We examine local area networks and global data in more detail in Chapter 13.

A simple complete project example (Project_A) and its components are shown in Figure 8.10.[4] Project_A includes a symbol table, two local area networks (MPI Network_X and Profibus-DP Network_Y), one of which contains a global data table, two stations (S7-300 Station_AA and S7-300 Station_AB), and one subproject (Sub_Proj_AC). The station called Station_AA contains system data blocks containing the configuration data describing the hardware under that PLC's control and the setup parameters for that hardware. Station_AA also contains two intelligent modules, including a communication processor to connect to a special network, and (of course) one CPU module. The CPU module contains two logic blocks (OB1 and FB4, named Program_1 and Program_2), one shared data block (DB2), and one instance data block (DB3).

Configuration Data for the Station

Project creation and station configuration are covered in Chapter 10, but we provide a brief overview in this section.

During creation of a project, the programmer must enter data describing the type and location of the CPU, power supply module, and all I/O modules, including signal modules (SMs),

---

[4] The STEP 7 programming software screens show the project tree a little differently. Some components (such as hardware configuration) appear in a secondary window only when the items that they are components of are selected in the main window.

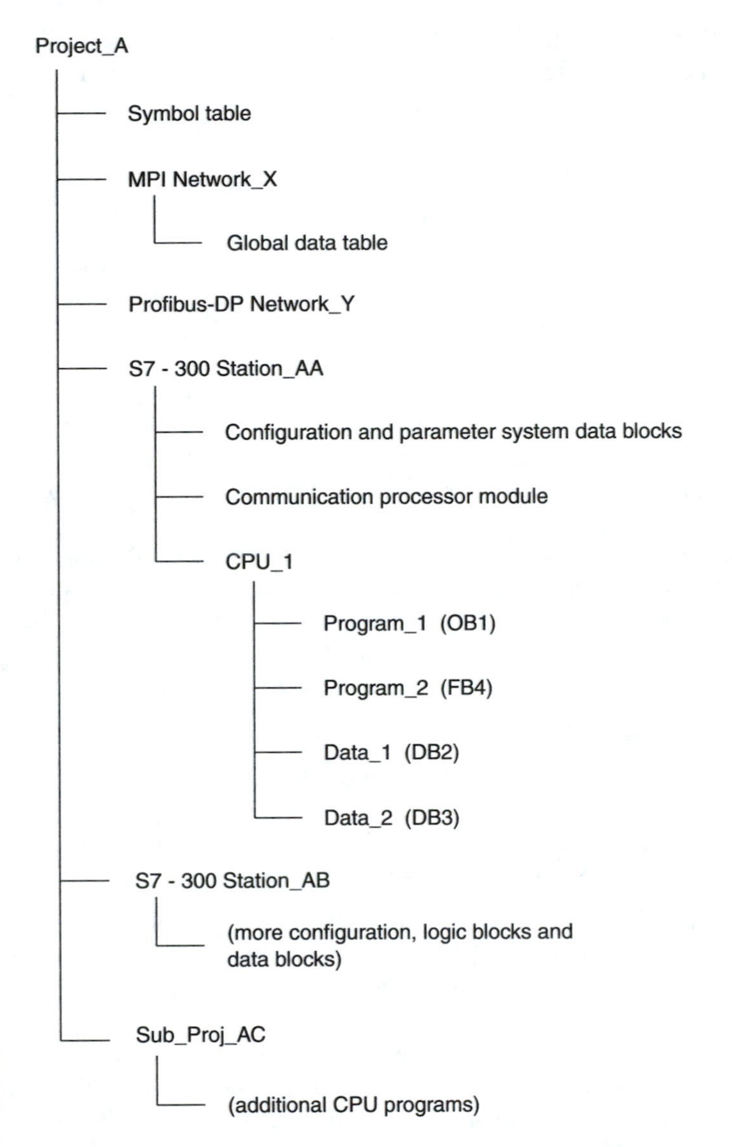

FIGURE 8.10
Small STEP 7 project, showing some of the components.

function modules (FMs), and communication processors (CPs). Parameters are assigned to modules that need further configuration, using STEP 7 dialog screens.[5] The STEP 7 programming unit assembles the configuration data into **system data blocks (SDBs),** then downloads

[5] If the user has purchased one of the special I/O modules that Siemens calls function modules (FMs) or communication processors (CPs), the modules come with software that becomes part of the STEP 7 programming software when installed, and which contains the necessary parameter configuration routines.

them to the CPU. The CPU stores system data blocks in its retentive memory and rewrites module configuration data to the appropriate modules every time the PLC is powered up, so once the configuration data has been entered, the programmer doesn't have to worry about it again until the PLC system needs to be changed.

STEP 7 Operating System Calls to Organization Blocks

By default, when an S7 PLC is in run mode, it executes **organization block 001 (OB1)** as the user-program every scan cycle. The programmer will usually use OB1 as the main program in a structured program and will program OB1 to include conditional and/or unconditional calls to subprograms.

Whenever an S7 PLC is switched into run mode, it always looks for **organization block 100 (OB100).** If OB100 exists, the STEP 7 operating system will execute the program it contains, once, before starting to execute the scan cycle. OB100 should be programmed to, for example, perform any one-time operations that a programmer wants the PLC to perform at the start of each shift.

The STEP 7 operating system also automatically tries to execute other organization blocks under certain conditions. The other organization blocks and the conditions under which they run are described in the section on configuration options later in this chapter and again in more detail in Chapter 11.

Structured Programming in STEP 7

The entire user-program for an S7 PLC *can be* written as a single program in organization block 1 (OB1),[6] but STEP 7 allows and encourages structured programming, in which a main program (e.g., OB1) calls subprograms that in turn can call other subprograms. Siemens refers to programs and subprograms as *logic blocks,* of which there are three types:

1. **Organization blocks (OBs)** are the main programs that can call other logic blocks. An OB executes in response to a call from the STEP 7 operating system. OB1, for example, is called once each scan cycle. The user cannot program a call to an OB but can enter programs into OBs.

2. **Function blocks (FBs)** can be called from any user-written logic block, even from another function block. Siemens S7 PLCs come with some preprogrammed function blocks,[7] which are called **system function blocks (SFBs).** Every call of an FB or an SFB *must* include the name of an instance data block to open with the FB/SFB. The following examples show STL instructions that call function block 3 with instance data block 25, and call system function block 4 with instance data block 125:

```
Call FB 3, DB25
Call SFB 4, DB125
```

---

[6] You can even write an entire program into a single network if you are programming in STL, but breaking a logic block into several networks improves the program's readability without affecting its performance.

[7] Check the specification sheet that comes with your CPU module; not all CPUs have SFBs.

3. Functions (FCs) can be called from any user-written logic block, even from another function. Siemens S7 PLCs come preprogrammed with some functions, called **standard functions (FCs)** and **system functions (SFCs).** The following examples show STL instructions to call function 20 and to call system function 46.

```
Call FC 20
Call SFC 46
```

There are only four standard functions: FC4, FC6, FC7, and FC8, all preprogrammed to convert dates and times to/from the standard constant format called DATE_AND_TIME format, which is described in Chapter 5. The standard functions require the program to pass parameters to the FC. See your manual set for parameter descriptions.

There are quite a few system functions (SFCs), which are preprogrammed to do things like moving data sets, working with the watchdog timer or the *system clock* (Siemens' name for the real-time clock), controlling interrupts, and so on. Most require the program to pass parameters. (SFC 46, used in the example above, just causes the PLC to go into stop mode and does not need parameters.) Some SFCs are covered in other chapters of this book. See your manuals for a complete list of SFCs and the parameters they need.

Parameter Passing in STEP 7 Structured Programs

A programmer can use shared memory (input/output images, bit memory, shared data blocks, etc.) throughout the logic blocks in a structured program so that all logic blocks share the same address space, but STEP 7 allows and encourages the passing of parametric data between logic blocks so that each logic block controls its own data in its own memory space and cannot accidentally change a data value that another logic block needs. It is easier to read and to modify a program that contains declarations of parameters with symbolic names, and which use symbolic names instead of absolute addresses in the program.

The key to parameter passing is the **variable declarations** that each logic block can include. When programming in ladder logic or in STL, the programmer enters the declarations into a **variable declaration table** that is an integral part of each logic block (the other part of a logic block is the program that manipulates the data). System function blocks (SFBs), system functions (SFCs), and the preprogrammed standard functions (FCs) come preprogrammed with variable declaration tables, of course. A variable declaration table in a logic block might look like the one shown in Figure 8.11 (which would normally display the logic block's name in the top right corner). The variable declaration table is explained below.

We saw a variable declaration table very similar to the one in Figure 8.11 in Chapter 5 when we covered STEP 7 instance data blocks, but this time we examine the contents of a variable declaration table from the point of view of passing parameters and reserving memory for a logic block's use. The variable declaration table in Figure 8.11 includes six records to declare six variables. Each record includes the following columns:

1. The **absolute address** (Addr.) of the memory that will be reserved for a data element when this logic block executes. The programmer doesn't enter the absolute address number:

Declaration Table [project/CPU Program/Logic Block name]					
Address	Declaration type	Name	Data type	Initial value	Comment
0.0	in	switch_1	BOOL		
0.1	out	cylinder	BOOL	False	
2.0	in_out	loop_cnt	INT	20	
4.0	stat	work_dat	INT	0	
0.0	temp	push_cnt	INT	0	
2.0	temp	my_pie	REAL	3.14	

FIGURE 8.11
STEP 7 variable declaration table (as it might look for a function block).

The STEP 7 programming software assigns them after the programmer finishes entering each table row. In this function block variable declaration table example, two bits are reserved in the first *instance data block* byte for two Boolean data units representing states of a switch and an actuator, then four more bytes in the instance data block (starting at byte 2) are reserved for two INTs (integer values). Notice that the address numbering restarts at 0.0 after the fourth record, because the following records reserve space in the *local stack memory* for an integer value (two bytes) and a real value (four bytes). In a function block, the variable declaration table must always declare instance data block variables first (the first four variables in Figure 8.11), then declare "temp" variables for the local stack area of memory (the last two variables in Figure 8.11). Organization blocks and functions can't have instance data blocks, so all their declared variables are given space in the local stack memory.

The program that contains the variable declaration table in Figure 8.11 can use absolute addressing to access the data in the instance data block or the local stack memory. The address **DIX 0.1,** for example, can be used in examining or outputting to the Boolean bit declared in the second record (with the name #cylinder). The address **DIW 2** can be used in manipulating the integer value declared in the third record (#loop_cnt). (I will explain the # characters later.) The local stack memory's real element (#my_pie) can be addressed using its absolute address: **LD 2.**

2. The **declaration type** column identifies the source, destination, and lifetime of each data value. Data elements must be declared in the following order in the variable declaration table:

(a) **In** parameters are values that are passed to this logic block when it is called by another logic block. The data values passed to a subprogram (as constants, addresses, or pointers) must match the data types of the variable declared here in the subprogram's variable declaration table. "in" parameters are also described as **VAR_INPUT** type in Siemens literature.

(b) **Out** parameters cause this logic block to return that variable's value to the calling logic block when this program ends. The calling logic block must indicate an (appropriately sized) address for the subprogram to write the output parameter to. "out" parameters are also described as **VAR_OUTPUT** type.

(c) **In-out** parameters receive an initial value from the calling logic block and return the variable's final value to the calling logic block. The calling program must supply an address or a pointer. Also described as **VAR_IN_OUTPUT** type.

Organization blocks cannot have any in, out, or in-out parameters. **Function blocks** copy in and in-out parameter values into the instance data block that is called with the function block, then copy in-out and out parameter values from the instance data block to the calling logic block as the function block ends. **Functions** copy *pointers* for in, out, and in-out parameters into the local stack memory, so that as the function runs it actually reads and writes values in memory locations specified by the calling logic block. If a calling logic block does not pass a parameter, the preexisting value from the instance data block (if any) is used, and if there isn't an instance data block, the initial value in the variable declaration table is used (initial values are covered below). If none of the above provide values, the variable is assigned the value zero.

- **stat** parameters are *not used in passing parameters* during logic block calls. "stat" parameters can be declared only in the variable declaration table of a function block. Declaring a stat variable reserves space in the instance data block so that the program can store a variable's value in the instance data block until the next time the function block is called. Also described simply as **VAR** type.
- **temp** parameters are *not used in passing parameters* during logic block calls. Declaring a temp parameter only reserves space in the local stack memory until the logic block ends, when the reserved local stack memory space is returned to the general pool and values stored there are lost. Also described as **VAR_TEMP** type.

3. A **block-local symbolic name** should be (but doesn't have to be) entered. If entered, this logic block program can address the data item using a # followed by the block-local symbolic name instead of having to address the data item using its absolute address. Complex data elements can only be addressed using symbolic names. Using symbolic names in a program allows the programmer to modify a program address by changing the actual address in one place in the variable declaration table instead of having to change it in many places in a program. The following examples show how data from the variable declaration table in Figure 8.11 could be addressed using their block-local symbolic names:

```
#switch_1      means the same as DIX 0.0
#my_pie        means the same as LD 2
```

If the # prefix is left off, the STEP 7 editor will add it to your program if the symbolic name matches a name declared in the variable declaration table.

STEP 5 also allows the programmer to use a symbol table to define (nonlocal) symbolic names for shared memory addresses. (Nonlocal) symbolic names can be used in any logic block running in the S7 PLC if they are included in quotation marks (e.g., "big_data"). If the quotation marks are left off, STEP 7 will look for a block-local declaration of that symbolic name in the variable declaration table before looking at the symbol table. Be sure to include the quotation marks when addressing nonlocal symbolic names or your program may accidentally address a block-local address with the same name.

4. The **data type** column defines how much instance data block or local stack memory must be reserved for a data item, and the **initial value** column allows the programmer to assign initial values in constant form for some types of data items. Allowable data types and constant formats include the types described in Chapter 5, including:

(a) Elementary data types:

BOOL
BYTE and **CHAR**
WORD and **DWORD**
INT and **DINT**
REAL
TIME and **DATE** and **TIME_OF_DAY (TOD)** and **S5TIME**

(b) Complex data types (initial values cannot be entered for complex data types):

DATE_AND_TIME (DT)
STRING
ARRAY
STRUCT

(c) Any complex data type that the user has defined by creating a **user data type (UDT)** data block. The data type column would contain the UDT data block with its data block number (e.g., **UDT3**).

The data type can also be one of the data types described by Siemens as *parameter data types,* including those listed below (for which initial values cannot be entered):

(a) **Timer:** causes the called logic block to reserve two bytes of memory and to accept parameter values consisting of the letter "T" and a number between 0 and 255, identifying a timer, from the calling logic block.
(b) **Counter:** like the timer parameter except that the calling logic block must pass the letter "C" and a number between 0 and 255, identifying a counter.
(c) **Block_FB, Block_FC, Block_DB, Block_SDB:** each of which causes the called logic block to reserve 2 bytes of memory and accept the number for a logic block (FB or FC) or a data block (DB or SDB) from the calling logic block. The calling logic block must pass the parameter by its full name: the block type identifier (FB, FC, DB, or SDB) followed by a number from 0 to 255, or the calling logic block can specify a block by its symbolic name.
(d) **Pointer:** causes the called logic block to reserve space to store an address specification in pointer constant format. Whenever the called logic block reads or writes

values to this variable, it will actually read or write to the address specified by the calling logic block. Pointers can always be used to pass values to functions and function blocks, but cannot necessarily be declared as an "out" data declaration type. (See your manuals.) Remember that the pointer parameter is stored in 6 bytes of memory in the format shown in Figure 8.12. The allowable codes for memory areas include a code that indicates the "previous" local stack area of memory. This means that a logic block can call a function (or function block) and can pass a pointer parameter that will allow the function (or function block) to access the local stack memory that belongs to the calling logic block.

(e) **ANY:** in which case the called logic block reserves 10 bytes of space for an ANY pointer. Since an ANY pointer includes a description of the type and number of data items to which it points, the calling logic block can use it to pass any type of elementary or complex data element to the called logic block (e.g., if this logic block is called from two places in a main program, the main program can pass an ANY pointer to an integer value during one call and an ANY pointer to an array in the next call). As with a standard pointer parameter, an ANY pointer parameter can give a function (or function block) access to various memory areas used by the calling logic block, including the calling logic block's local stack memory.

Using STEP 7 Call Instructions and Instructions That Terminate Logic Blocks

There are two ladder logic instructions to program a call to another logic block, but there are three STL instructions. Ladder logic also offers two instructions to end the logic block but a variety of ways of ending an STL logic block.

A ***ladder logic*** **CALL FC/SFC FROM COIL** instruction can be used conditionally or unconditionally to call a function (FC) that has no in, out, or in-out parameters, in which case it will appear like this in your program:

```
                           FC11
--------------------------( Call )
```

Besides calling the function, the rung shown above will also affect some of the status bits in the status word (see your manual).

Byte:																	Byte:
0	DB number (0 if not a DB)																1
2	Code for memory area								0	0	0	0	0	b	b	b	3
4	b	b	b	b	b	b	b	b	b	b	b	b	b	x	x	x	5

FIGURE 8.12
STEP 7 pointer format.

To program a ladder logic call of an FB, SFB, or SFC, or to an FC that has declared in, out, or in-out parameters, the programmer can use the STEP 7 browser to select the logic block as if the logic block were an instruction. (Obviously, the logic block that is being called must already be programmed.) The ladder logic element will look like the element shown in Figure 8.13. The calling program must provide data constants or addresses for the in, out, and in-out parameters required by the logic block being called. In Siemens' terminology, a variable declaration table creates **formal parameters,** and the calling program must provide **actual parameters.** The ladder logic call example in Figure 8.13 is to the function block with the variable declaration table shown in Figure 8.11. Figure 8.13a shows the ladder logic and STL equivalent before actual parameters (or instance data blocks) are entered by the programmer, and Figure 8.13b shows the same programs after the programmer has entered typical actual parameters.

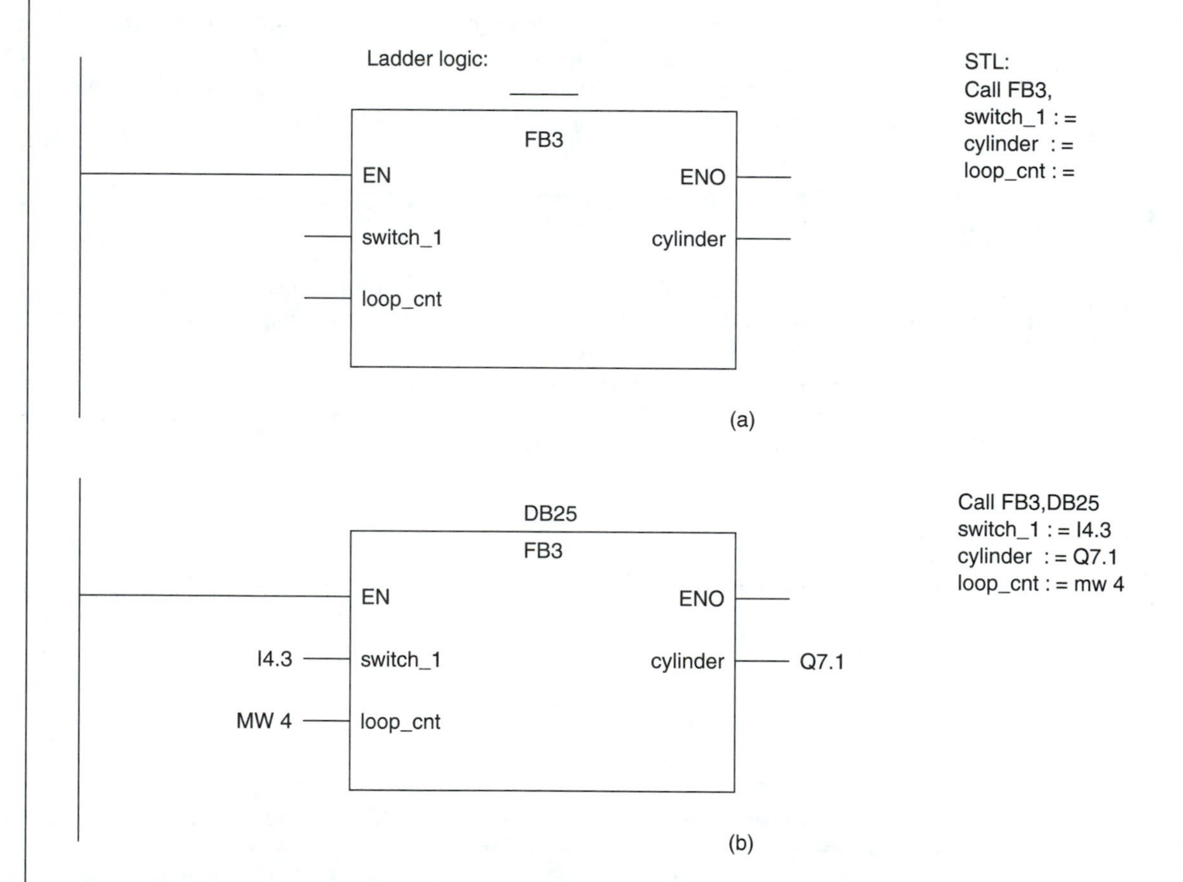

FIGURE 8.13
Calling a STEP 7 function block that contains the variable declaration table shown in Figure 8.11: (a) before entering actual parameters; (b) after entering actual parameters.

Notice that all the in and in-out parameters are shown on the left of the ladder logic element. The actual parameters for in parameters can be constants or addresses, but in-out parameters must be addresses. The out parameters are shown on the right of a ladder logic element, and the actual parameters must be entered as addresses. There are a few rules to keep in mind as you enter actual parameter addresses:

1. Addresses for *elemental data elements* can be entered in either absolute address, symbolic name, or pointer format.
2. Addresses for *complex data elements* must be entered as pointer constants or as symbolic names.
3. Symbolic names cannot be used in passing:
   (a) Complex data if the call is made from a function.
   (b) Pointer or ANY pointers unless the call is made from an organization block.
   (c) Timers, counters, or blocks if a function is being called from another function.

The ladder logic element in Figure 8.13 includes a Boolean input labeled EN, at which the programmer can attach logic to make the call execute conditionally. The ladder logic element also includes an output labeled **ENO** that provides a Boolean output that the programmer can use to control another operation, even though the ENO output isn't included in the variable declaration table for FB3 as shown in Figure 8.11. All ladder logic functions and function blocks have this ENO output, which outputs the state of the BR status bit of the status word. All SFC and SFB programs provided by Siemens include logic that resets the BR bit if the SFC or SFB fails to execute for any reason and sets the BR bit otherwise. If you are programming in ladder logic, it is up to you to include instructions that control the BR bit in the functions or function blocks that you write. If you do not do so and your program includes a ladder logic statement that is controlled by the logic state of the ENO output from the FC or FB, you could get a nasty surprise.

To control the BR bit (and therefore the ENO output), include a **SAVE** output element or a **RETURN** output element at the end of every FC or FB program you write. The SAVE instruction affects only the BR bit, while the RETURN also affects other status bits (see your manual). If the SAVE or RETURN instruction is programmed to execute unconditionally, the BR bit will always be set. If the SAVE or RETURN instruction executes conditionally in your program (as in Figure 8.14), the BR bit will only be set if the RLO of the conditional statement is true.

Siemens SFCs and SFBs often include an out parameter named **RET_VAL** that will contain a code to identify the error. The error codes are covered in Chapter 15. You can include an out variable named RET_VAL, and can program your logic blocks to write to RET_VAL if you want to.

FIGURE 8.14
Condition return in STEP 7.

The three **STL** instructions for calling other logic blocks include the UNCONDITIONAL CALL (UC), CONDITIONAL CALL (CC), and CALL instructions. The **CALL** instruction is executed unconditionally, so it is the equivalent of the unconditional ladder logic call shown in Figure 8.13. CALL must be used to call a logic block that has declared in, out, or in-out parameters. The CALL instruction executes unconditionally, so if you want to make the CALL conditional, you must program a conditional JUMP, as we saw earlier in this chapter, to jump past the CALL instruction.

The **UNCONDITIONAL CALL** (UC) instruction is similar to the CALL instruction except that UC can only be used to call an FC or an FB that does not need any in, out, or in-out parameters. One advantage in using UC is that the FC or FB number with the UC instruction can be indicated in memory indirect format (e.g., FC5 can be called with the instruction UC FC[MW4] if MW4 contains the number 5), or the UC instruction can call a function whose number has been passed to this logic block as a parameter of type BLOCK_FC (e.g., UC #my_prog calls FC5 if the variable name #my-prog is a BLOCK_FC type parameter with the actual value 5, as passed to this logic block).

The **CONDITIONAL CALL** (CC) instruction is similar to the UC instruction except that it can follow a conditional statement that will determine whether the CC instruction will cause a function to be called. If the CC instruction does not cause the FC to be called, the CC instruction does not affect the RLO.

It is possible to write an STL function or function block program so that it sets or resets the BR bit as it ends, as all Siemens SFC and SFB programs do. The STL **SAVE** instruction copies the RLO bit status to the BR bit, so if SAVE is controlled by a logical statement, the BR bit will store the result of the logical statement. Alternatively, the STL instructions **SET, NOT**, or **CLR** can be used to set, invert, or clear the RLO bit before unconditionally executing the SAVE instruction.

All STL logic blocks must have a **BLOCK END** (**BE**) instruction programmed as their last instruction. When the PLC executes this instruction, it always terminates the current logic block and returns control to the logic block that called this function or function block, or returns control to the STEP 7 operating system if it is an organization block that is ending. An STL program can also include a **BLOCK END UNCONDITIONALLY** (BEU) instruction, which has the same effect as a BE instruction. Unlike the BE, which can only be the last instruction, the BEU instruction can be programmed anywhere in a logic block. In STL, the programmer can also program a **BLOCK END CONDITIONAL** (BEC) instruction. If the RLO is true when BEC executes, no further instruction in this logic block will execute, just as if a BEU instruction had executed.

Your STL-language FC or FB can also declare a **RET_VAL** parameter and write error codes to it so that they will operate like the Siemens-written SFCs and SFBs. If you program a function using a text editor instead of using the STEP 7 ladder logic or STL editor, you can declare a data type for each function you write, and (unless the declared type is VOID) then STEP 7 will automatically add a record to the variable declaration table to create an out parameter named RET_VAL when it translates your text file for an S7 PLC.[8]

---

[8] We do not cover editing programs in text format in this book, but text format programming is based on STL, is easy to learn, and offers some advantages not otherwise possible, such as allowing you to create a program that STEP 7 will refuse to display later.

Effects of STEP 7 Structured Programming on Microprocessor Registers

Whenever one logic block calls another logic block, the STEP 7 operating system automatically stores information from some microprocessor registers (information regarding the logic block that is doing the calling) into an area of memory called the block stack **(B stack)** so that that logic block can be resumed when the called logic block ends. The stored information includes:

- The name of the calling logic block and the next instruction in that program
- The numbers of the shared and the instance data blocks active at the time of the call
- The address in the local stack where the calling logic block's data starts

  The B stack can contain eight sets of these data, which limits an S7 PLC to eight nested levels of logic blocks calling other logic blocks.[9]

If the logic block that is being called is a function block, the STEP 7 operating system then switches the contents of the microprocessor registers containing the numbers of the instance data block and shared data block (effectively turning the instance data block into a shared data block) before putting the number of the new instance data block into its register.

When a called logic block ends, the contents of the B stack are put back into the microprocessor registers they came from, so the calling logic block resumes, with access to the data blocks and local stack memory it was using before the call took place.

Calling and executing another logic block can affect some microprocessor registers in ways that the programmer should be aware of. If the called logic block manipulates complex data elements, or even if its variable declaration table declares complex data elements as in, out, or in-out parameters, the STEP 7 system has to use the shared data block register and address register 1. These register contents should always be restored in the called logic block before they are used. After returning from the called logic blocks, the microprocessor registers that were not saved on the B stack can have been changed by the called logic block. These registers include the accumulators, the address registers, and some bits of the status Word (including the RLO bit). Some status word bits are automatically changed by a call (see your manuals).

Symbols in STEP 7 Structured Programs

Symbolic names can be defined in a CPU program's symbol table, in a logic block's variable declaration table, or in a data block definition.[10] Symbolic names must begin with an alphabetical character and are case sensitive (i.e., "1spot" is not a valid name, and "onespot" is not the same symbolic name as "Onespot").

---

[9] The STEP 7 operating system can interrupt a structured program to cause another organization block to execute. Each organization block represents a priority level, and each priority level has its own B stack, so the new organization block will have eight new levels of nested calls available even if the interrupted organization block already has several calls active.

[10] You can't declare symbolic names for data blocks and user data types while creating data blocks. They must be declared in the symbol table.

An example **symbol table** is shown in Figure 8.15. A symbol table can contain records assigning symbolic names to I/O addresses (I and Q memory areas, but not P memory), to bit memory (M), to data blocks (DB and UDT), to timers and counters (T and C), and even to logic blocks (OB, FB, FC, SFB, and SFC). The data type column tells the PLC how big a data item each symbolic name represents. After a symbolic name has been defined in a symbol table, that symbolic name can be used anywhere in the STEP 7 programming software instead of using the absolute address it represents as long as you continue to use the same programming unit. Symbolic names are kept in the memory of the programming unit, not in the PLC's memory. In a user-program, symbolic names can be entered in quotation marks (e.g., "big_var"). The compiler will add the quotation marks if the compiler doesn't find a similarly named block local symbol.

Using symbolic names to represent I/O addresses allows I/O contacts or module locations to be changed without requiring major program changes. After an I/O contact or module is moved, the programmer only has to change the absolute address once in the symbol table. Any logic block using the symbolic name will then use the new absolute address. Assigning symbolic names to non-I/O addresses aids program readability and allows programs written for one type of PLC to be easily modified for another PLC with a different memory structure.

Symbolic names that are declared in variable declaration tables in logic blocks or are declared in data blocks are known as **block local symbols**. A block local symbol is valid only when the logic block or data block that defines it is open. You can declare the same symbolic name in different logic blocks or declare the same symbolic names already defined in the symbol table, but that is discouraged. The programmer can enter block local symbols preceded by a # character (e.g., #small_var) or let the compiler add the #. Symbolic names for variables declared in variable declaration tables and in data blocks are saved in the memory of the pro-

[Name of project and CPU program] \symbols			
Symbol	Memory address	Data type	Comment
Cylinder_1	Q 7.1	BOOL	Bad part ejector
Bad_parts	C 6	COUNTER	How many bad?
GoodBad_rat	MD 4	REAL	Ratio:bad/good
Template_1	UDT 5	UDT 5	A data structure
Program_AB	FB 5	FB 5	Check part quality
Data4_progAB	DB 34	DB 34	Instance for FB 5

FIGURE 8.15
Typical symbol table in a STEP 7 CPU project.

gramming unit and are not downloaded to the memory of the PLC. If you upload a program from a PLC to another programming unit and view the program, the program will not contain the symbolic names.

Sequence of Entering a Structured Program in STEP 7

It may already be obvious that STEP 7 demands that you must write structured programs in a specific order:

1. Projects and stations must have already been created, complete with all the configuration information needed for your PLC.

2. If you intend to use user data types as templates for other data blocks or for data structures in variable declaration tables, create them now. (See Chapter 5.)

3. If you want to use symbolic names that will be valid in all logic blocks (to represent inputs and outputs, for example), plan them now and enter them in the symbol table. You can, if you wish, access this table as you write programs later to make changes and additions. It isn't absolutely necessary to create symbolic names before using them in STEP 7 logic blocks, but you won't be able to compile a logic block until you have defined all its nonlocal symbols in the symbol table.

4. Create the functions (FCs) starting with the lowest-level functions first, then functions that call the functions that you have already created. If any functions call function blocks (FBs), don't create those functions until you have created the function blocks they need (after step 6). If you want, you can just create functions with just variable declaration tables now, and enter the programs later.

5. Write your function blocks (FBs), at least to the point of completing the variable declaration tables. You can declare data elements of type UDT, since you have created the UDTs in step 2. You do not need to write SFBprograms, nor do you have to download them to the PLC. Your programming unit software and your PCL both contain copies.

6. Create the instance data blocks (DBs). During creation, you have to indicate which function block (FB) each instance data block references. The variable declaration table in that function block will define the instance data block's contents automatically. If necessary, create several instance data blocks for a single function block. View the instance data blocks if you want to modify the initial values, but you won't be allowed to change anything else.

7. Create the organization blocks (OBs) that call the FBs and FCs you have already created.

8. Create the shared data blocks (DBs). (You could have done this at any time after step 2, but you must do it before trying to run your program.)

CQMI

OMRON CQM1 SUBROUTINE CALLS

Subprograms are called **subroutines** by OMRON and are programmed at the bottom of the main program but before the END (01) statement. Any program (including a subroutine) can contain a SUBROUTINE ENTER instruction, **SBS (91) N,** to call a subroutine numbered 0 to 255 (N) when the rung's logic is true, as shown in Figure 8.16. Nesting to 16 levels of subroutine calls is allowable.

A conditional subroutine enter instruction in the main program:

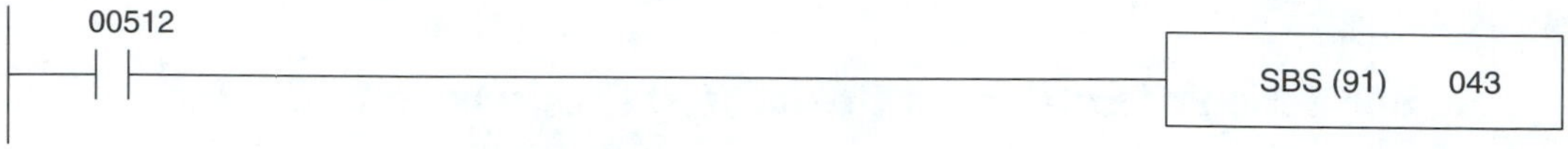

00512

SBS (91) 043

The SUBROUTINE DEFINE instruction identifying the start of the subroutine, after the last rung in the main program:

SBN (92) 043

The RETURN instruction on the last rung in subroutine 43 :

RET (93)

After the last RET (93) :

END (01)

FIGURE 8.16
Subroutine calls and subroutine definitions within an OMRON PLC program.

Each subroutine must begin with a **SUBROUTINE DEFINE** (output) instruction, **SBN (92) N,** identifying the subroutine number. Each subroutine must end with a **RETURN** instruction, **RET (93).** The first SBN (92) N instruction that is encountered tells the CQM1 that the main program is over, so do not enter an SBN (92) N before the end of your main program. After the last subroutine's RET (93) instruction, the **END (01)** instruction is required.

Parameter passing cannot be done with CQM1 subroutine calls. All CQM1 programs and subroutines share the same data space.

CQM1 Subroutine Notes

It is possible to configure and program an OMRON PLC so that the OMRON operating system calls subroutines at times outside the control of the main program. We examine those methods briefly in the next section of this chapter and in more detail in Chapter 11. To avoid possible conflicts between subroutines called by SBS (91) instructions and the subroutines that the operating system can call automatically, the programmer should avoid using SBS (91) to call subroutines numbered 000 to 003 and subroutines mentioned in STIM (69) instructions or in the data files used by CTBL (63) instructions.

CONFIGURATIONS THAT AFFECT PROGRAM EXECUTION

A PLC is expected to execute a three-step scan cycle. That's what makes a PLC different from other types of computers. PLCs use a scan cycle because it has been proven to allow relatively fast, predictable, and trouble-free control of automated processes. Now, with the powerful structured programming features we have seen in this chapter, it is possible for a programmer to write programs that make a PLC operate in a way that is neither fast, predictable, nor trouble-free.

As strange as it may sound, most PLCs now offer users the option to change the PLC's configuration so that the PLC will deviate from the standard scan cycle. The PLC can be set up to do things such as running more than one main program in the scan cycle or performing input and output scans at times other than immediately preceding and following the main program scan.

Most modern PLCs can be set up so that the PLC's operating system can interrupt the standard scan cycle at precisely timed intervals, or in response to a change at an input module, to execute a user-written program that isn't a standard part of the scan cycle. These options allow the programmer to force a PLC to execute parts of the structured user-program at predictable time intervals, or to respond to some input conditions quickly. By taking some user-program out of the standard scan cycle, the program can sometimes be simplified so that it will again run trouble-free. Chapter 11 covers interrupts in detail, so in this section we describe only *what* the programmer can make the PLC do with interrupts.

ALLEN-BRADLEY PLC-5 CONFIGURATION FOR STRUCTURED PROGRAMMING

By default, program file 2 is the only MCP (main control program) that the PLC-5 executes in its scan cycle, but the PLC-5's configuration can be changed to execute as many as 16 separate program files as **MCP files** (MCP A through MCP P) in every scan cycle, with or without I/O scans between each MCP execution.

Configuration to change the MCP program files to include in the scan cycle can be done using the **PROC CONFIG** screen. Enter program file numbers (1 to 999) for each of the (up to) 16 MCPs. For each MCP file in the list, the programmer can indicate whether the PLC-5 should perform an I/O scan after the MCP file executes (0 means include), and can disable individual MCPs temporarily without removing them from the list (1 means disable).

Whenever a program is saved to disk, the data table contents, which include configuration data in the status file, are also saved. When the program is restored, the status file including the PLC's configuration is restored, too.

Instead of using the configuration screen, MCP configuration can be changed by a program. Instructions to change PLC configuration can be included in a subroutine that is called each time the PLC is put into run mode, using the **first scan bit (S:1/15)** to call the subroutine.

This method ensures that configuration is reset once per shift when the PLC is started. Advanced PLC programming may make it desirable to dynamically change the PLC's configuration as the PLC runs, depending on sensed conditions in the workplace. To change MCP configuration, the program must include instructions to modify the following configuration data:

- **MCP file numbers** are stored in every third status data world location, starting at S:80 and extending to S:125 (the two status data words following each MCP file number are used by the PLC-5 to record the latest execution time of the MCP and the maximum scan time for that MCP).
- The 16 bits that control whether I/O scans follow each of the 16 MCPs from MCP A to MCP P are status file bits **S:78/0 to S:78/15.**
- Any of the MCPs can be disabled by setting one of the 16 bits in **S:79.**

The PLC-5 can be configured to execute a program written in **sequential flowchart (SFC)** language if the SFC program is included in the list of MCP programs (program file 1 is reserved for an SFC program, but any file can contain an SFC program). SFC programming is covered in Chapter 9.

The configurable **interrupt features** are listed briefly here (they are described in detail in Chapter 11):

- PLC-5 programs can include IMMEDIATE INPUT and IMMEDIATE OUTPUT instructions, which read or write I/O modules directly without waiting for an input scan or an output scan.
- A PLC-5 can be configured to interrupt the scan cycle to run a *selectable timed interrupt* (STI) program file at precisely timed intervals.
- The PLC-5 can be configured to interrupt the scan cycle to execute a *processor input interrupt* (PII) in response to changed input states at an input module.

SLC 500

ALLEN-BRADLEY SLC 500 CONFIGURATION FOR STRUCTURED PROGRAMMING

By default, program file 2 is the only MCP (main control program), and the user cannot change this. The programmer can only write programs in file 2 that conditionally call subroutines in other program files. The conditions can be written to simulate sequential flow chart (SFC) programming so that each subroutine file must complete a control function before the next subroutine can be called.

The SLC 500 has a first scan bit (S:1/15) that automatically goes true for only the first scan cycle after the PLC is switched into run mode. The programmer can use the first scan bit to call a subroutine, ensuring that some control functions (e.g., the count of parts made today) are reset once per shift when the PLC is started.

The following SLC 500 interrupt features are listed briefly here (they are described in detail in Chapter 11):

- IMMEDIATE INPUT and IMMEDIATE OUTPUT instructions can read or write I/O modules directly without waiting for an input scan or an output scan.
- SLC 500s can be configured to interrupt the scan cycle to run a *selectable timed interrupt* (STI) program file at precisely timed intervals.

- The SLC 500 can be configured to interrupt the scan cycle to execute a *discrete input interrupt* (DII) program file in response to changed input states at an input module. The SLC 500 also offers another variation called *I/O interrupts.*

S5

SIEMENS STEP 5 CONFIGURATION FOR STRUCTURED PROGRAMMING

By default, the STEP 5 operating system repeatedly scans organization block 1 (OB 1). This configuration cannot be changed. The programmer can, however, program OB001 to include jumps to other blocks (OBs, PBs, and FBs) effectively including those other blocks in the list of blocks that will be executed each scan cycle. By making the jumps conditional, the programmer can define conditions under which some blocks are disabled.

The S5 series of PLCs scans outputs and inputs only after the whole program has executed, and this configuration can't be changed. If the programmer wishes to cause an I/O scan between executions of blocks, he/she must get creative in the writing of conditions for executions of blocks. For example, OB001 can toggle a flag bit each scan, executing only PB001 if the flag is on and only PB002 if the flag is off. There will be a full scan cycle, complete with an I/O scan, between each execution of PB001 and PB002.

STEP 5 offers an optional **control system flowchart** (**CSF**) programming language, which can be used to program sequence blocks. CSF allows creation of flowchart-style programs in which step programs repeat each scan until a "transition" state becomes true, allowing the PLC to stop repeating the old step program and to start executing the next step program (which will also have a transition condition). If CSF programming software hasn't been purchased, a creative programmer can program conditional jumps in organization blocks, so that each control function must complete its work before it is disabled and the next control function is enabled. Only the pretty graphical display of the flowchart will be missing.

The following interrupt options are described in more detail in Chapter 11:

- In larger S5 PLCs, direct access addressing (addresses with the "P" prefix) causes the PLC to read or to write the I/O module, so the PLC does not have to wait for the input scan or the output scan to update data.
- OB 20 executes every time an S5 PLC is powered up, and OB21 runs every time it is switched from program mode to run mode. If OB 20 and OB 21 exist, they can be used to initialize the PLC before normal repetitive scanning of OB001 starts.
- OB 13 (and OBs 10 to 12 in some S5 PLCs) runs at timed intervals,
- OB 2 (and OBs 3 to 5 in some S5 PLCs) runs when an input module generates a hardware interrupt signal.

SIEMENS STEP 7 CONFIGURATION FOR STRUCTURED PROGRAMMING

An S7 PLC executes its scan cycle repeatedly while it is in run mode, reading data from input modules into input image memory, then executing the program in organization block 1 (OB 1) before copying the contents of the output image table to the output modules. The programmer can't change this cycle. A programmer can write programs so that OB1 enables logic blocks to run in a sequential manner such that each logic block is disabled until input conditions indicate that the previously active logic block's work is complete.

Siemens offers optional programming languages, called S7 Graph and S7 HiGraph, which aid in the creation of **sequenced program structure.** S7 Graph is similar to the standard IEC 1131-3 programming language called structured flowchart (SFC). Siemens also offers an optional programming language based on the IEC 1131-3 standard for structured text. Siemens structured text programs are actually written in C and/or Visual Basic, so they can be written using what non-PLC programmers recognize as structured programming languages.

There is a system function, **SFC 47 (Wait),** which a user-program can use to cause the PLC to stop executing the user-program for a short time, then to resume. The programmer enters a parameter indicating how long to wait. Wait extends the time of a scan cycle.

An S7 PLC does not scan all the I/O modules during the scan cycle. The S7 PLC only has I/O image memory for the digital I/O module range of addresses (up to byte 127; see Chapter 5). To program reading or writing of other I/O addresses, the programmer must use **peripheral input** (PI) or **peripheral output** (PQ) address prefixes, which force the PLC to read or write the I/O modules, not an image of the I/O data in memory. The programmer can even use peripheral input or peripheral output addressing to read or write digital I/O modules, bypassing the I/O image memory.

Whenever an S7 PLC is switched into run mode, the STEP 7 operating system will immediately execute the program in organization block 100 (OB100) if the programmer has created OB100, and will then start the scan cycle that includes OB1.

When a programmer creates a program for an organization block other than OB 1, the programmer is enabling the operating system to use that organization block as an interrupt service routine! Each organization block has its own **priority class,** and an organization block will not interrupt the program that is running unless it has a higher priority class.[11] OB 1 has the lowest priority class (1), so the STEP 7 operating system can interrupt OB 1 to execute any other organization block. OB 100 has almost the highest priority class (27), so it can't usually be interrupted. Chapter 11 covers interrupts in more detail, but a list of the interrupt OBs follows:

`OB 35` executes at precisely controlled **cyclic intervals.** The programmer can set the cycle time during configuration of the CPU program. OB 35 has priority level 12. Siemens has reserved OBs 30 to 38 for additional cyclic interrupts.

`OB 40` is called to execute as soon as an I/O module requests a **hardware interrupt.** The programmer can configure I/O modules to generate interrupt requests during parameter specification while entering configuration data for the CPU program, or the program can include calls to SFC 55, 56, or 57 to change an I/O module's configuration parameters as the PLC program runs. OB 40 has priority level 16. Siemens has reserved OBs 40 to 47 for additional hardware interrupts.

`OB 10` is called to execute at a specific **time and date,** either once or at the same time every hour (or week or month, etc.). The programmer can set the time and repetition of OB10 by calling SFC 28 ("SET_TINT") and can activate the interrupt with SFC 30 ("ACT_TINT"). Pending time and date interrupts can be examined with SFC 31

---

[11] It is possible to modify the priority classes in some S7 PLCs: See your manuals.

("QRY_TINT") and canceled with SFC 29 ("CAN_TINT"). OB 10 has priority level 2. Siemens has reserved OBs 10 to 17 for additional "time-of-day" interrupts.

OB 20 executes after a specific **delay time** after SFC 32 ("SRT_DINT") executes. Pending interrupts can be examined with SFC 34 ("QRY_DINT") and canceled with SFC 33 ("CAN_DINT"). OB 20 has priority level 3. Siemens has reserved OBs 20 to 23 for other time delay interrupts.

Other organization blocks respond to events that the programmer usually would prefer never to happen, such as when a program error or a hardware fault is detected. Organization blocks that run in response to errors and faults are described in Chapter 11.

Siemens also offers four system functions (described in Chapter 11) that can be called in a program to change how the PLC responds to interrupts:

- **SFC 39 ("DIS_IRT")** is used to disable one or more priority classes of interrupts, depending on the parameter values the program supplies with the call.
- **SFC 40 ("EN_IRT")** is used to reenable interrupts disabled with "DIS_IRT".
- **SFC 41 ("DIS_AIRT")** makes interrupts with a higher priority class than the one executing wait until this organization block ends. (Lower-priority-class interrupts always wait.)
- **SFC 42 ("EN_AIRT")** cancels one "DIS_AIRT". When all "DIS_AIRT" calls are canceled, any high-priority-class interrupts that are waiting can execute without waiting for this organization block to end.

CQMI

OMRON CQM1 CONFIGURATION FOR STRUCTURED PROGRAMMING

Only the main program of a CQM1 can be scanned, and it will execute from start to end each scan. Flowchart programming language can be simulated using the **STEP (08),** and the **SNXT (09)** and **SR 25402,** the "step start flag." (See your manual set.)

The CQM1 can be configured with a **minimum scan time** so that a new scan cycle will not start until at least that length of time after the start of the previous scan cycle. Specifying a minimum scan time in the PLC's configuration data forces the PLC to execute its scan cycle at repeatable intervals, making the PLC's response more predictable but predictably slower. The OMRON PLC can be configured so that the user-program always writes directly to output modules instead of only changing output states during the output scan step.

The following OMRON PLC *interrupt capabilities* are listed briefly here (they are described in more detail in Chapter 11):

- The program can include an **STIM (69)** instruction to set the operating system up to interrupt the main scan cycle and to execute a subroutine at precisely timed intervals.
- The OMRON PLC can also be configured so that whenever the input state changes at input address IR 00000 (or IR 00001, IR 00002, or IR 00003), the operating system will automatically interrupt the scan cycle to scan a few input modules, then execute subroutine 0 (or 1, or 2, or 3). Configuration data must be entered into the appropriate data

memory words to enable the input bits to initiate interrupt, to specify which input modules to scan, *and* the program must include the **INT (89)** instruction to enable (unmask) the interrupt capabilities.

- High-speed counters can **count the input state changes** at IR 00004 to IR 00006 (or at an optical encoder input port), even if the pulses change faster than the scan cycle would normally be able to detect. The PLC can be configured to interrupt its scan cycle to scan a few input modules and then run a subroutine when the count exceeds a preset range. Configuration data must be entered into data memory to enable the high-speed counting and to specify which input modules to read, and a **CTBL (63)** instruction and a data table must also be used to indicate the subroutine(s) to run and the count values at which to run the subroutine(s).

TROUBLESHOOTING

What can go wrong in a structured PLC program? Lots! Every structured programming procedure you use makes the PLC operate differently from the way PLCs were designed to operate. The PLC will no longer simply execute a three-step scan cycle, and it won't necessarily even run the user-program from top to bottom.

Troubleshooting Master Control Relays

Master control relays (under any name) work by making the logic on the affected rungs appear to be false. If the instruction is designed to perform some operation when its control logic changes, an MCR can cause the instruction to work even if its logic hasn't changed, or can make the instruction miss a logic change, or can delay the instruction's operation by making it appear that the logic state change was delayed. Be sure that you understand how MCR activity can affect other instructions in your PLC.

If your program has jump statements, it is possible that your program may execute a jump past an MCR endfence, so that the MCR-affected zone becomes larger than you intended. Beware of nested MCRs. Nesting of MCRs is allowed in some PLCs, but in most PLCs the first MCR endfence ends *all* nested MCR-affected zones.

Troubleshooting Jump and Loop Instructions

Backward jumps and loops can cause a PLC to fault if the backward jump is repeated often enough to cause the watchdog timer to time out. Remember that while your user-program is executing, your PLC won't perform output scans or input scans, so there isn't any point in using a backward loop while waiting to see an input change unless your program uses immediate input instructions, and even then you may encounter trouble with the watchdog timer.

The result of jumping forward is to cause the PLC not to work out the logic on the rungs bypassed and not to update the states of the output instructions controlled by those instructions. The effect is that the states of the outputs controlled by those instructions remain frozen. Instructions that operate when input logic states change (counters and one-shots, for example) can miss the actual changes in state. Instructions that operate over multiple scans (timers and some file-manipulation instructions) act as if those scans didn't happen. A timer that has begun to time a 5-s pulse of spray paint may time for 1 s, become frozen because a jump bypasses it, then finish timing the remaining 4 s three weeks later when the jump ceases to occur. That's a lot of wasted paint!

Be careful with jump labels. Most PLCs allow several jump instructions to cause jumps to a single label but won't allow multiple labels with the same number. Other PLCs (e.g., OMRON) require a unique label for each jump instruction, except label 00, and you can have several labels with the number 00.

Troubleshooting Subprogram Calls

Careless structuring of nested subprogram calls can lead to the same problems as backward jumps but will usually cause the PLC to fault sooner. If, for example, subprogram A calls subprogram B, which in turn calls subprogram A again, you have a continuous loop and the user-program part of the scan cycle can't end. Since a PLC needs to save information in stack memory to allow it to return from a call, and since a typical PLC has only enough stack memory to allow approximately 16 pending returns, the PLC will usually overfill the stack memory and fault long before the watchdog timer times out. If an otherwise well-designed structured program tries to call one too many nested subprograms, the PLC will fault. Your programming software will usually include program debugging tools that look at the subprogram stack to tell you if stack overflow was the cause of your problem, and may even let you see which programs were waiting for subprograms to end when the fault occurred.

Not jumping to a subprogram can lead to the same problems as when a jump instruction bypasses part of a program. The logic that doesn't execute can't update the outputs it controls, so they remain frozen in the state they were left in the last time the logic did execute.

Timers can be a big problem for inexperienced programmers as they write their first structured program. If a program uses the same logic to control a call of a subprogram and to control a timer in that subprogram, the timer instruction will never be executed when that logic is false, so it will never reset. (A new programmer may not realize that the same timer number can be used in another timer instruction elsewhere in the program, where the PLC can see its control logic as false, as long as the program is structured so that the PLC won't execute both timer instructions during the same scan cycle.)

A structured program will usually contain conditional calls to subprograms, so any combination of subprograms might execute during any scan cycle. If more than one subprogram uses the same memory address (even timer or counter addresses), one of the subprograms may make the other subprogram operate incorrectly. Use parameter passing so that subprograms only get a copy of the original data, allowing each subprogram to manipulate its own copy without affecting the original or affecting other subprograms. When parameter passing is used, the programmer must be careful that all the needed parameters are passed, that they are passed in the correct order, and that the data formats match in the calling and called programs. When passing large data elements and sets of data, and when using indirect addressing, indexed addressing, or pointers as addresses, the programmer must ensure that the PLC doesn't try to use more memory than is available, or the PLC will fault. (See the "Troubleshooting" section in Chapter 5.)

Siemens PLCs offer data blocks, which are a useful tool in keeping data sets separate from each other, but the programmer must be careful when using data blocks. The user-program must open a data block before it can be used, and the user-program can only read/write data elements that have been created or the PLC will fault. Faults may occur if a subprogram is used to open a data block for the main program. The PLC remembers which data block was open before the subprogram was called, and when the subprogram ends the PLC reopens the old data block. Faults can also occur in an S7 PLC if a data block is used in a subprogram after a complex data element is passed to it, because the PLC uses its shared data block pointer register during the call and the shared data block number will have been lost.

Especially in STL-like languages, math operations and compare operations that use the microprocessor registers can fail because subprograms can change the contents of a PLC's accumulators or other registers (including status bits). The call instruction itself may not affect the accumulators and other registers, but the instructions in the called program will! Ladder logic elements usually manage most of the microprocessor's registers well, but the Allen-Bradley offset register (S:24) is one that might change and affect the program when a subprogram ends.

Troubleshooting Programs Configured to Allow Deviations from the Scan Cycle

If you have configured your PLC so that the PLC's operating system allows or causes deviations from the standard scan cycle, remember that the scan cycle does still execute. Even if your PLC doesn't fault because of too many nested interrupt levels, it may fault because too many interruptions of the scan cycle allowed the watchdog timer to time out before the scan cycle finished. Your programming software should be able to identify this type of problem and may even be able to tell you which programs were waiting to resume executing at the time of the fault. Some PLCs maintain a status word where they save the longest scan time observed. The programmer can check this status word, even if the PLC hasn't faulted, to see how close the PLC has come to faulting.

Some PLCs (e.g., Siemens) automatically try to execute interrupt subprograms, so be sure that you don't write a program into an organization block that the operating system looks for unless you want interrupts to occur. Other PLCs (e.g., OMRON) automatically look for specific subroutines to execute, but only if interrupts are specifically enabled. If you think you might need to enable OMRON interrupts someday, don't use subroutines numbered 00 to 03.

QUESTIONS/PROBLEMS

1. PLCs are often described as deterministic controllers because their control interval is predictable and repeatable. Why is the main control program of a PLC without structured programming more deterministic than one with structured programming?
2. What happens to nonretentive relay outputs between the startfence and the endfence of a master control relay when the MCR's conditions are not true? When they are true?
3. What happens to retentive relay outputs between the fences of a master control relay when the MCR's conditions are not true? When they are true?
4. What is wrong with the following program?

IN_1 (MCR)

IN_1 OUT_2 ( )

(MCR)

6. What's wrong with this program? What will happen, and why?

5. When the logical statement controlling a JUMP instruction goes true (false in an OMRON PLC), how are the rungs between the JUMP and the destination LBL affected?

13 [Label] IN_1 OUT_2 ( )

IN-3 13 (Jump)

7. Why can the use of parameter passing make your subroutines/functions reusable?
8. What does your PLC do when it encounters a RETURN instruction in a subroutine or function?
9. How does a timed interrupt affect the standard scan cycle?

Allen-Bradley PLCs

10. For the following program:
    (a) What will happen when IN_1 GOES OFF while IN_2 stays on? Answer in the space below each rung. (Be specific; the timer consists of three data words.)
    (b) What will happen to the timer when IN_1 goes back on? (Be careful; explain clearly.)

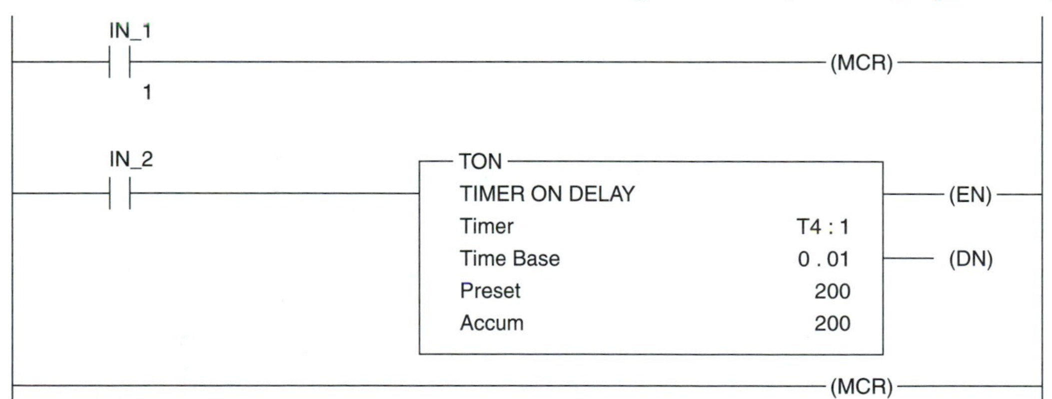

11. What will N7:1 and N7:2 contain after the following program executes?

    Before program executes:
    N7:1 – 0000 0000 0000 0000
    N7:2 – 0000 0000 0000 0000

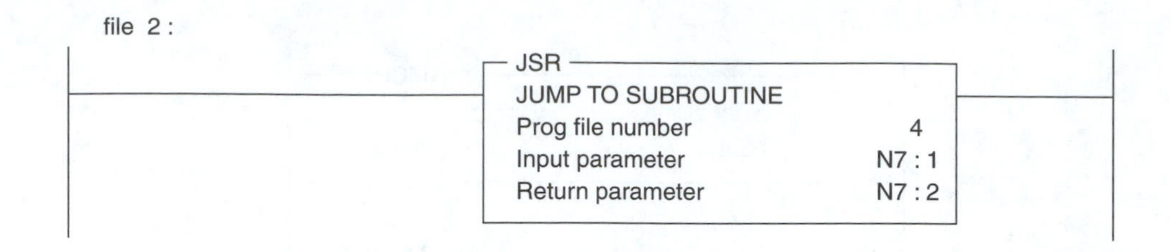

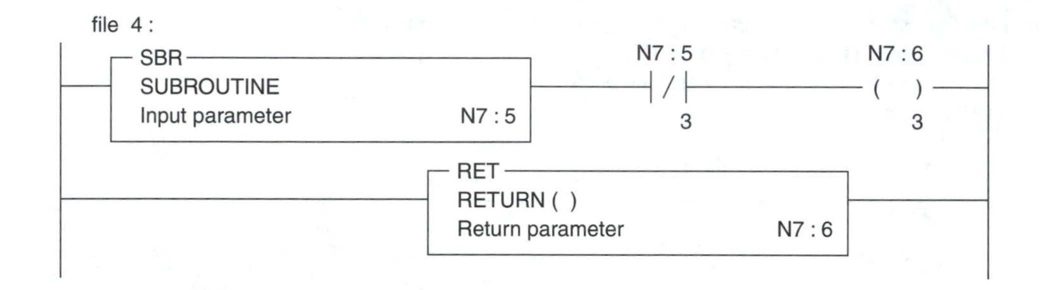

After program executes:
N7: 1 – ____
N7: 2 – ____

12. An Allen-Bradley PLC-5 executes ________ as the main control program every scan cycle (unless you change the configuration).
13. If you changed a PLC-5's configuration so that file 2 was no longer an MCP (main control program) but file 3 was, how would the PLC-5's scan cycle be changed?

Siemens PLCs

14. What is wrong with the following program? ("x" is a time and is not what is wrong.)

```
OB1
    A I4.0
    JC "Function_1"

"Function_1"
    A I4.0
    L x
    SP T1
    A T1
    = Q7.0
    BE
```

15. A Siemens PLC executes __________ as the main control program every scan cycle. A cyclic timed interrupt in an S __ PLC would be in ______.

Siemens S5 PLCs

16. What number would be in FY25 of the Siemens S5 PLC after it executed the following program?

    Before program executes:
    FW0 contains KH0008
    C0 contains KH0008

```
     L FW0
     L C0
     ==F
     JC +past
     L KHFF
past:T FY25
```

FY25 contains ___

17. There are five kinds of STEP 5 blocks for programming PLCs. What are these five block types, and how does each differ from the others?
18. In passing parameters to a function block in STEP 5, the parameter's formal operand, parameter type, and data type must be specified (declared) in the function block. If a parameter with the formal operand name "mine" is declared as

    DECL: **mine Q BY**

    (a) What *size* of data unit does the main program exchange with the function block?
    (b) Does the main program supply the data to the function block or receive it *from* the function block?

Siemens S7 PLCs

19. What number would be in MB25 of the Siemens S7 PLC after it executed the following program?

Before program executes:
MWO contains W#16#0005
C0 contains W#16#0008

```
      L MW0
      L C0
      <>0
      JC past
      L B#16#FF
past :T MB25
```

MB25 contains ______

20. What number would be in MB25 of the Siemens S7 PLC after it executed the following program?

Before program executes:
MWO contains W#16#0005
CO contains W#16#0008

```
      L MW0
      L C0
      == I
      JC past
      L B#16#FF
past: T MB25
```

MB25 contains ______

21. Which of the following declaration type(s) are possible in a Siemens function block (FB) but not in a function (FC)? Explain why not.

IN
OUT
IN-OUT
STAT
TEMP

SUGGESTED PLC LABORATORY EXERCISES

For a system with:

- Four control panel switches: inputs 0 to 3
- Four indicator lamps or visible output module LEDS: outputs A to D
- Two spring-return-valve-controlled cylinders: outputs E and F
- One detent-valve-controlled cylinder: outputs G and H
- Three sensors to detect extension of each of the cylinders

1. Write a program with MCRs, jumps, and subroutine/functions to:
   (a) Turn indicator lamp A on when switch 0 is on but jump past this rung when switch 1 is on.
   (b) When programming an Allen-Bradley or OMRON PLC:
   - Use an on-delay timer to turn indicator light B on 0.75 s after switch 0 goes on. Now add a master control reset, controlled by switch 2, so that the timer works if switch 2 is on. Hold 0 on and experiment with the MCR. If the timer requires its control logic to go from false to true to reset the timer, why is it restarting each time the MCR zone is enabled?

   When programming a Siemens S7:
   - Use a move instruction to copy the four switch bit values to the indicator lights. Add a master control reset, which allows the move only if the sensor at cylinder E is not actuated. Set some switches, then actuate the sensor at cylinder E.

   (c) Jump to a subroutine or function if sensor 3 is on. The subroutine or function should turn indicator lamp C on when switch 0 is on. (Absolutely essential only if passing parameters!)

2. Write a structured PLC program (in STL if programming a Siemens PLC) to:
   (a) Cause all three cylinders to extend as soon as the PLC goes into run mode.
   (b) In your main program, monitor the control switches and call other programs' files/functions/subroutines depending on the switch positions. These other programs can include jumps and calls to still other programs.
   - When switch 0 goes from off to on, retract all cylinders regardless of where they are when the switch goes on.
   - While switch 0 stays on, cycle the cylinders. When the sensors indicate that all cylinders have been retracted, extend them again.

When the sensors detect that all cylinders have been extended, retract them again. (Restrict airflow to slow the actuations.) When switch 0 goes off, the cylinders should stop in whichever position they were in when the switch went off.

- When switch 1 is on, retract all cylinders regardless of switch 0.
- When switch 2 is on, all valve solenoids should immediately go off, regardless of what other switch is on. Flash the output lights repeatedly. When switch 2 goes back off, all lights should go off.

3. Write a program that uses a (Siemens) function block or an (Allen-Bradley) subroutine, which accepts input parameters and generates output parameters (not possible with OMRON). Name the parameters appropriately if your PLC allows you to name parameters. The program should include:

(a) A function block/subroutine program should end without doing any work if an input parameter bit is off; otherwise, it should:

- Add an increment value passed from the calling program to a current value data word (which should also be passed from the main program), saving the result as the temporary parameter.
- If the result is higher than the upper limit value entered as a second input parameter, set it to the same value as the upper limit.
- If the result is lower than a lower limit contained within the function block/subroutine, copy that lower-limit value to the temporary word.
- Copy the temporary word value to the output parameter.

(b) A main program that includes two calls to the function block/subroutine, each providing separate parameters (via data blocks if programming a Siemens PLC).

- The first call should work only while switch 0 is on. Pass parameters containing the input image of switch 1, a positive increment value (as a constant), the current value, and pass an upper-limit value by address reference. Accept a new value for the current value from the function block/subroutine.
- The second call should work only while switch 0 is off. Pass the address of switch 2, a negative increment value as a constant, the same current value passed by the other call, and the address for an upper-limit value which is half as big as that passed with the other call. Accept a new value for current value.

(c) Use the data monitor and data write screen to enter upper-limit values and to observe the current value as you test your program. (If using a Siemens PLC, you may have to add lines to your main program to copy data between data blocks and memory areas that the data monitor can access.)

9

IEC 1131-3: THE COMMON PROGRAMMING LANGUAGE

OBJECTIVES

In this chapter you are introduced to:

- An overview of the IEC 1131 standards.
- The IEC 1131-3 requirement for a common program structure consisting of configurations, resources, tasks, and program organizational units (POUs), with examples.
- Using variable declarations to define:
  - Data unit sizes.
  - Variables that are local (accessible only to a single POU and its subprograms) or global (accessible to other POUs).
  - Variables that are temporary, are to be passed between this POU and the POU that calls it, or which are to be stored as pointers to memory shared by other POUs.
- Instances of function blocks, algorithms that contain their own data.
- The standard functions.
- Standard programming languages (LD, IL, ST, SFC, FBD, and perhaps CFC).
- Standard instructions.

INTRODUCTION TO IEC 1131

In 1979, the International Electrotechnical Commission (IEC), a sister group to the International Standards Organization (ISO), formed a committee to develop a common standard, IEC 1131, for programmable controllers. In 1997, renumbering of the IEC standards led to IEC 1131 being renumbered as **IEC 6-1131.** Users still refer to the standard by its old number, so we also do so in this book.

There are six sections to the IEC 1131 standard. This book is concerned primarily with the third part, IEC 1131-3, which defines the requirements for **programming language(s)** for use with PLCs. The standard allows programming in any combination of *five different pro-*

gramming languages, and pressure is mounting to allow a sixth programming language. Other sections include:

```
IEC 1131-1      standard terminology
IEC 1131-2      PLC hardware and testing requirements
IEC 1131-4      selecting and installing PLC components
IEC 1131-5      communications between PLCs
IEC 1131-6      fuzzy logic in PLC programs
```

The IEC 1131 committee includes representatives from most of the major PLC manufacturers. Rockwell Automation (which owns Allen-Bradley) and OMRON are members. Siemens is not, but it has so much influence through its support of standardization throughout Europe that it is probably best that it isn't a member directly. To get the agreement of so many manufacturers, the IEC 1131 standard includes many compromises and only calls for voluntary compliance. Even if the standard does not become universally accepted (and it wouldn't be the first important international standard to suffer that fate), it will still set the tone for future developments (like other "failed" international standards have done).

> In the 1970s the ISO developed a voluntary standard for intercomputer communications, called the *open systems interconnect* (OSI) standard. General Motors used it as the basis for a compulsory standard, which it tried to impose on all of its computer component suppliers. General Motors' Manufacturer's Automation Protocol (MAP) standard failed, perhaps because of the cost required to build all the compromise requirements into computers. Despite the failure of MAP, the OSI's seven-layer model for intercomputer communication services remains the basis for new computer communication networks, including the IEC 1131-5 standard for PLC communication requirements.

PLCopen is an organization that encourages acceptance of the IEC 1131-3 standard. PLCopen also includes representatives from the major PLC manufacturers, including Rockwell Automation, OMRON, and Siemens. PLCopen selects parts of the IEC 1131-3 standard to implement, develops acceptance tests, and certifies commercially available PLC programming languages as being IEC 1131-3 compatible. PLCopen also makes compromises; it has developed three levels of compliance: **base** level, **portability** level, and **full compliance** level. At each level there are compliance and testing requirements for each of the five allowable programming languages, so suppliers can seek certification of compliance for individual languages. Meeting the *base-level requirement* means only that the programming language includes the basic instruction set and allows programming using the structured programming techniques the IEC has laid out. The *portability-level certification* will require additional instructions, but more important, it will require that the programming software can convert programs to and from a neutral file format so that programs written for one PLC can be downloaded to another brand of PLC using a different manufacturer's programming software. The process of developing standards is slow. As of this writing PLCopen is almost finished writing the base-level compliance requirements and has begun working on the portability-level requirements. Inevitably, PLCopen is finding that its members want PLCopen to advocate changes to the IEC 1131 standard, so PLCopen is now involved in pressuring the IEC to accept the sixth programming language mentioned earlier.

IEC 1131-3 PROGRAMMING LANGUAGES

The ultimate goal in developing a standard for PLC programming languages is *to allow a programmer to write a PLC program without knowing (or caring) what PLC the program has to run on, in a language and format that all PLC programmers use.* The standard says that any single PLC program should work as long as it only contains components written in one or more of the five (or six) languages. Before we discuss the languages, we have to examine the required program structure.

The writers of the IEC 1131-3 standard foresee a day when a PLC programmer can purchase *standard program components* and assemble them into special-purpose control programs.[1] To prepare for that future day, the IEC 1131-3 standard requires that all programs be written in a common structured way to accommodate standard program components. The required structure is good, but the terminology that IEC 1131-3 uses to define the structural components is perhaps the biggest impediment to acceptance of the standard. Don't let the terminology in the following section distract you from recognizing the very useful structure it describes.

Those readers who have been learning Siemens' STEP 7 programming languages will find that STEP 7 is very close to the programming language described in the IEC 1131-3 standard, although there are some differences that might provide difficulties. The terminology used in STEP 7 is just close enough to IEC 1131-3 terminology so that the differences can be confusing. STEP 7 also provides some features not mentioned in the IEC 1131-3 standard (such as passing pointers as parameters) and accomplishes some of what IEC 1131-3 requires in unusual ways (such as when a function block is called with an instance data block). Remember that compliance with IEC 1131-3 is voluntary, that PLCopen hasn't yet decided which parts of IEC 1131-3 require certification beyond a base level, and that neither group wants to stifle growth. STEP 7 programmers will have to forgive me for using ladder logic exclusively in examples in the first parts of this chapter. The other readers of this book haven't yet been introduced to other programming languages.

COMMON ELEMENTS OF AN IEC 1131-3 STRUCTURED PROGRAM

Algorithm and Data Types

Even before writing a program, future programmers will have sets of standard algorithms and templates for data structures available. The programmer will declare (create) additional types (algorithm and data templates) as his or her career continues. A standard **algorithm type declaration** will consist of routines written using standard IEC 1131-3 instructions to perform common operations. For example, a standard algorithm might be written to drive an actuator to correct the difference between a desired system output and an observed system output, and the algorithm should work just as well to control motor speed as to control room temperature. Standard **data structure type declarations** will consist of a defined order of arrangement of standard IEC 1131-3 data elements. An example data structure might be two 16-bit signed binary numbers followed by two Boolean elements (bits). This structured data type could be used to

---

[1] Computer programmers who write in C do this now when they purchase "toolboxes" of standard C language routines and include them in the programs they write.

contain a speed setpoint from an operator station and an observed speed from a sensor in the two signed binary numbers, for use by the standard control algorithm we described earlier. The algorithm could set or reset the two data bits to drive a motor one way or the other. The same data structure could also contain a temperature setpoint and an observed actual temperature instead, so it could be used by the standard algorithm to control the two bits to indicate a need for temperature increase or decrease.

The IEC 1131-3 standard defines the formats to be used for standard type declarations of algorithms and data structures. We will examine these individual formats after we see the overall structure of the IEC 1131-3 system into which they must fit.

If standard algorithms and data structures are available, programming may consist entirely of assembling the standard components into a structure to perform the control operations that a specific manufacturing process requires. Writing a new program will consist of declaring (entering) a top-down description of the structure using a programming unit. The programming unit will compile the final program, translating it into the binary machine language understandable by a specific type of PLC, and will download it to the PLC (or to several PLCs).

Configuration

The top-down programming approach requires the programmer to start by declaring what the IEC standard calls a configuration. The configuration declaration statements describe the system of PLC "Resources" that will work together in this configuration, the data values they will share globally, and defines the data that this configuration must make available for computers outside the configuration to *access* (e.g., so that a plant supervisory computer system can monitor the process, or so that a supervisory controller can order changes).[2] The following is the global and access variable section of a configuration declaration example that is included in the IEC 1131-3 standard of 1993. In rewriting it I have left the standard IEC terminology in uppercase and entered the user-defined names in lowercase. Parts of the example won't make sense to you until you have read more of this chapter.

```
CONFIGURATION cell_1
    VAR_GLOBAL w: UINT;
    END_VAR
    .
    .
    .
    VAR_ACCESS
     able       : station_1.%IX1.1      :BOOL READ_ONLY;
     baker      : station_1.p1.x21      :UINT READ_WRITE;
     charlie    : station_1.z1          :BYTE;
     dog        : w                     :UINT READ_ONLY;
     alpha      : station_2.p1.y1       :BYTE READ_ONLY;
     beta       : station_2.p4.hout1    :INT READ_ONLY;
     gamma      : station_2.z2          :BOOL READ_WRITE;
    END_VAR
END_CONFIGURATION
```

[2] Siemens STEP 7 uses different terminology. In STEP 7 a configuration is called a *project,* a resource is called a *station,* and global data can be declared in a STEP 7 symbol table.

Resources

Within the configuration declaration, the programmer must enter a set of declaration statements for each of the **resources.** When the IEC first met to define a resource, they envisioned it as being a single computer, programmed to control a single interrelated set of actuators and sensors, although they permitted the definition to be interpreted so that an intelligent I/O module running as a slave of a CPU module could be called a resource separate from its CPU master. Since then, developments in computers have led to multiprocessor arrays in CPU modules, so a single CPU module may actually contain several computers. Other developments have given us fast multitasking operating systems, so a single computer can run several independent programs concurrently as if it were more than one computer. The definition of exactly what a resource is has become a little fuzzy (to say the least), but a programmer can still think of a resource as *something that acts as if it were a single computer programmed to control a single process.* Resource declaration statements identify the make and model of the microprocessor in this single computer, list the "programs" and/or "function blocks" that this resource must execute (from the available program templates), and set up the resource's operating system with "*tasks*" to define the conditions that must exist for the resource to execute each program. Other statements within the resource definition declare the data values that can be globally shared, but only by programs in this resource, and declare the data values within this resource that the configuration can allow other computers outside the configuration to access.

The configuration declaration example from above has been expanded below to include resource definitions with variable declarations for each resource. The added statements are shown in boldface.

```
CONFIGURATION cell_1
    VAR_GLOBAL w: UINT; END_VAR

    RESOURCE station_1 ON processor_type_1
        VAR_GLOBAL z1: BYTE;  END_VAR
        .
        .
    END_RESOURCE

    RESOURCE station_2 ON processor_type_2
        VAR_GLOBAL z2          :BOOL;
                   AT %QW5     :INT;
        END_VAR
        .
        .
        END_RESOURCE

        VAR_ACCESS
         able    : station_1.%IX1.1    :BOOL READ_ONLY;
         baker   : station_1.p1.x21    :UINT READ_WRITE;
         charlie : station_1.z1        :BYTE;
         dog     : w                   :UINT READ_ONLY;
         alpha   : station_2.p1.y1     :BYTE READ_ONLY;
         beta    : station_2.p4.hout1  :INT READ_ONLY;
         gamma   : station_2.z2        :BOOL READ_WRITE;
        END_VAR
END_CONFIGURATION
```

Tasks

The resource definition includes declaration statements defining **tasks.** The task declaration statements define the conditions under which the computer's operating system should initiate the execution of programs or function blocks, but does not specify which programs or function blocks to execute. Task declaration statements must specify a task name and whether the task's initiating condition will be a single event (a rising edge at an input sensor) or is to occur at precisely timed intervals. The task declaration statement can (optionally) indicate a priority level for this task. If a priority number is entered, a *preemptive* system of tasks is set up so that tasks with high priority numbers can interrupt the execution of programs or function blocks associated with tasks with lower priority numbers. (An interrupted task will resume when the task that interrupted it has run to completion.) If priority values are not entered, a *nonpreemptive* system of tasks is set up, in which case tasks will start and execute to completion, in the order in which their task definition statements become true, regardless of how long they have to wait to execute.

In the following example from the IEC 1131-3 standard, task declaration statements have been added in boldface.

```
CONFIGURATION cell_1
    VAR_GLOBAL w: UINT; END_VAR
    RESOURCE station_1 ON processor_type_1
        VAR_GLOBAL z1: BYTE; END_VAR

        TASK slow_1 (INTERVAL := t#20ms, PRIORITY := 2) ;
        TASK fast_1 (INTERVAL := t#10ms, PRIORITY := 1) ;
        .
        .
    END_RESOURCE
    RESOURCE station_2 ON processor_type_2
        VAR_GLOBAL z2       :BOOL;
                   AT %QW5  :INT;
        END_VAR

        TASK per_1 (INTERVAL := t#50ms, PRIORITY := 2);
        TASK int_1 (SINGLE := z2, PRIORITY := 1);
        .
        .
    END_RESOURCE
    VAR_ACCESS
     able    : station_1.%IX1.1     :BOOL READ_ONLY;
     baker   : station_1.p1.x21     :UINT READ_WRITE;
     charlie : station_1.z1         :BYTE;
     dog     : w                    :UINT READ_ONLY;
     alpha   : station_2.p1.y1      :BYTE READ_ONLY;
     beta    : station_2.p4.hout1   :INT READ_ONLY;
     gamma   : station_2.z2         :BOOL READ_WRITE;
    END_VAR
END_CONFIGURATION
```

Programs

The resource definition must include declaration statements defining the **programs** that are to be run by this resource. Each program declaration statement must indicate the name of the standard program template to use, a unique name for this instance of use of the program template, and the name of the task that initiates execution of this program. If no task is specified, the program will be executed as the main user-program upon each execution of the PLC's scan cycle, but will be assigned the lowest priority so that it can be interrupted by any preemptive task. If more than one program is declared for execution with any one task (including no task), the programs will be executed in the order in which they are declared for that task. As part of the program declaration, declaration statements must also identify the simple data element and the data structure elements that the program will use as it executes. These variable declaration statements define data values that must be passed between this program and other programs in the configuration. Variable declarations turn a standard program template into an instance of the program. Additional program definition statements can declare global data values to be shared among the function blocks and functions that make up this program, or can declare the values from this program that the resource can make available to the configuration and which the configuration can then allow other computers to access.

The example that we have been using from the IEC 1131-3 standard is developed further below, with program declarations added in boldface.

```
CONFIGURATION cell_1
    VAR_GLOBAL w: UINT; END_VAR
    RESOURCE station_1 ON processor_type_1
        VAR_GLOBAL z1: BYTE; END_VAR
        TASK slow_1 (INTERVAL := t#20ms, PRIORITY := 2);
        TASK fast_1 (INTERVAL := t#10ms, PRIORITY := 1);

        PROGRAM p1 WITH slow_1 :
                  f(x1 := %IX1.1);
        PROGRAM p2 :
                  g(out1 => w,
                       .
                       .        );

    END_RESOURCE
    RESOURCE station_2 ON processor_type_2
        VAR_GLOBAL z2       :BOOL;
                   AT %QW5  :INT;
        END_VAR
        TASK per_1 (INTERVAL := t#50ms, PRIORITY := 2);
        TASK int_1 (SINGLE := z2, PRIORITY := 1);

        PROGRAM p1 WITH per_1 :
                  f(x1 := z2,
                  x2 := w);
        PROGRAM p4 WITH int_2 :
                  h(hout1 => %QW5,
                       .
                       .        );

    END_RESOURCE
    VAR_ACCESS
     able       : station_1.%IX1.1    :BOOL READ_ONLY;
     baker      : station_1.p1.x21    :UINT READ_WRITE;
     charlie    : station_1.z1        :BYTE;
     dog        : w                   :UINT READ_ONLY;
```

```
    alpha      : station_2.p1.y1    :BYTE READ_ONLY;
    beta       : station_2.p4.hout1 :INT READ_ONLY;
    gamma      : station_2.z2       :BOOL READ_WRITE;
   END_VAR
END_CONFIGURATION
```

Function Blocks

It is possible to program a PLC to execute individual **function blocks** directly from the resource level instead of using programs to call the function blocks. This capability looks like it might have been a last-minute addition to the IEC 1131-3 standard, since it requires declaration statements within a program definition to define function block calls that have nothing to do with that program! The function block declaration statement is very similar to a program declaration statement. It identifies the function block template, a name for this instance of the function block, the task that is to initiate the function block (not necessarily the same task as the one that calls the program that contains the function block definition) and a list of variables to pass to this instance of the function block.

The final set of additions have been made to the IEC-supplied example below, assigning function blocks to tasks.

```
CONFIGURATION cell_1
    VAR_GLOBAL w: UINT; END_VAR
    RESOURCE station_1 ON processor_type_1
        VAR_GLOBAL z1: BYTE; END_VAR
        TASK slow_1 (INTERVAL := t#20ms, PRIORITY := 2);
        TASK fast_1 (INTERVAL := t#10ms, PRIORITY := 1);
        PROGRAM p1 WITH slow_1 :
                  f(x1 := %IX1.1);
        PROGRAM p2 :
                  g(out1 => w,

                    FB1 WITH slow_1,
                    FB2 WITH fast_1     );
    END_RESOURCE
    RESOURCE station_2 ON processor_type_2
        VAR_GLOBAL z2        :BOOL;
                   AT %QW5   :INT;
        END_VAR
        TASK per_1 (INTERVAL := t#50ms, PRIORITY := 2);
        TASK int_1 (SINGLE := z2,  PRIORITY := 1);
        PROGRAM p1 WITH per_1 :
                  f(x1 := z2,
                  x2 := w );
        PROGRAM p4 WITH int_2 :
                  h(hout1 => %QW5,

                    FB1 WITH per_2 );

    END_RESOURCE
    VAR_ACCESS
     able      : station_1.%IX1.1     :BOOL READ_ONLY;
     baker     : station_1.p1.x21     :UINT READ_WRITE;
     charlie   : station_1.z1         :BYTE;
```

```
    dog      : w                     :UINT READ_ONLY;
    alpha    : station_2.p1.y1       :BYTE READ_ONLY;
    beta     : station_2.p4.hout1    :INT READ_ONLY;
    gamma    : station_2.z2          :BOOL READ_WRITE;
    END_VAR
END_CONFIGURATION
```

PROGRAM ORGANIZATIONAL UNITS

IEC 1131-3 defines three types of **program organizational units (POUs)** that can contain instructions in ladder logic or in another allowable programming language. The program organization units include *programs, function blocks,* and *functions.*[3] Functions can also be called *procedures.* The IEC 1131-3 standard requires that if a POU is executing and contains calls to other POUs, the calls cannot be *recursive.* This means that a POU that has not finished executing (it might be interrupted) cannot be called again until it does finish executing. The standard also specifies that data sets that a POU is using can't be used by other POUs until the first POU has finished with them. These requirements will force PLC manufacturers to carefully plan the interaction between the execution of the user-program (including user-programs that interrupt other user-programs) and the I/O scans and serial communications that also read and write data memory. In essence, the operating system must ensure that each program organization unit has its own memory so that it can copy all the data that it needs into that memory before it starts executing, and can keep its working data where other POUs or communication handlers can't change it. Output data from the POU must be copied from the POU's memory when the POU finishes executing.

All three types of POU are similar in that they must consist of a section that **declares the variables** that the algorithm will use, followed by a section containing the **algorithm.** The variable declaration section assigns variable names for each data element the algorithm needs, indicates the type of data the name represents, indicates whether the actual data value must be provided to or outputted by this POU, can declare an initial value for the variable, and can indicate if the variable has some special characteristic, such as the ability to keep its value even after a failure of PLC power. We examine variable declarations in the next section after we have examined the three types of POUs.

Programs

Execution of a program can be initiated by a task defined in a resource, which in turn is part of a configuration. Programs are intended to be user-written, but to facilitate the use of reusable program organization unit templates the program should consist of little more than conditional and/or unconditional calls of function blocks or functions. A program cannot call another program.

Functions

A function can be called from any POU: a program, a function block, or even from another function. Functions can operate on data passed to them from the calling POU, and always gen-

---

[3] Siemens STEP 7 calls programs *organization blocks.*

erate a single result that is returned to the POU that calls the function. Writing a function, therefore, is a little like creating your own instruction! The IEC 1131-3 standard includes definitions of several functions, effectively extending the instruction sets of IEC 1131-3 languages. There is a function called COS, for example, which looks like Figure 9.1 when it is included in a ladder logic program. The programmer must enter a number, an address, or a variable name outside the box beside the input variable (IN). The COS function will calculate the cosine of that value and will output the result to the address (may be represented by a variable name) at the output (OUT). A function will execute only if the Boolean logic connected to its ENABLE (EN) input is true. Standard functions have an ENABLE OUTPUT (ENO) output that goes true after the function executes successfully. The ENO output can be used in following Boolean statements (until another function is called). In Figure 9.1, COS executes unconditionally. It gets a 32-bit number (the "D" in MD4 indicates a 32-bit double word) from memory address 4, treats it as a floating-point format real number, calculates its cosine, then places the 32-bit floating-point result into address MD8. The example does not show any connection at the ENO output, so this program cannot detect whether the COS function worked properly.

Standard Functions IEC 1131-3 defines standard functions[4] for many basic arithmetic and numerical operations:

ADD, MUL, SUB, DIV, MOD, EXPT, MOVE, ABS, SQRT, LN, LOG, EXP, SIN, COS, TAN, ASIN, ACOS, ATAN

for selecting one out of several input values:

SEL, MAX, MIN, LIMIT, MUX

for logical operations:

AND, OR, XOR, NOT

for comparing values:

GT, GE, EQ, LE, LT, NE

for shifting bit arrays:

SHL, SHR, ROR, ROL

for converting data types:

_TO_*

where "*" and "**" are data types defined in IEC 1131-3

and for manipulating character strings:

LEN, LEFT, RIGHT, MID, CONCAT, INSERT, DELETE, REPLACE, FIND

---

[4] You can probably still find the complete IEC 61131-3 standard on the Internet. At the time of this writing, it could be downloaded from Allen-Bradley's FTP site at FTP//FTP.CLE.AB.COM/STS/IEC/SC65BW67TC13

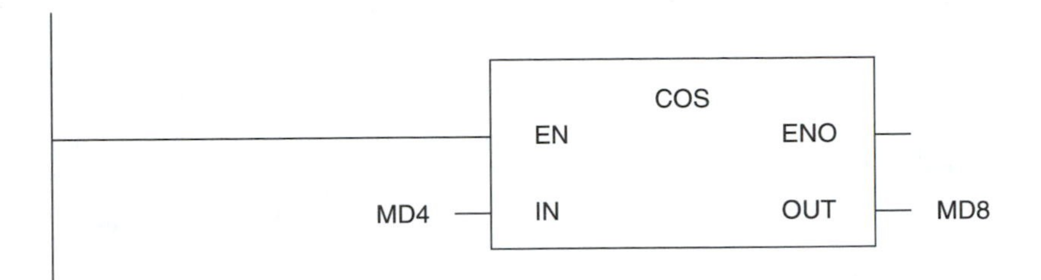

FIGURE 9.1
Function in ladder logic.

Most IEC 1131-3 standard functions are *overloaded functions,* which means that they are programmed to accept any format of binary data and to output a result that has the same format if possible.[5] The COS function is an example of a function that is only partially overloaded. If the input address in Figure 9.1 had been ML4, the COS function would have interpreted the 64-bit input (L means a 64-bit long number) as a 64-bit floating-point format number and would have generated a 64-bit floating-point result. If, on the other hand, the input variable is specified as MW4 (W means a 16-bit word size), the SIN function will fail to work because it would have to generate a 16-bit result, and since there is no standard for depicting a real number in 16 bits, the SIN function's overloading does not allow it to work with 16-bit numbers! Other IEC 1131-3-defined functions have more overloading capabilities. ADD, for example, can accept two or more input numbers coded in any binary format (all in the same format) and will generate a sum in the same format. Even specially coded Time_and_Date values can be added, because Time_and_Date is an IEC 1131-3 standard format. A MOV function's input doesn't even need to be a number. MOV can move any type of data element, even an array of ASCII characters or a user-defined data structure. Where overloading is allowed, the programmer, of course, has to ensure that the destination address size matches the input address size.

User-Written Functions A programmer can write his or her own functions. Figure 9.2 shows a function type declaration containing a ladder logic algorithm that uses standard IEC 1131-3 functions as building blocks. The example function calculates the sine of 16-bit signed binary values. The function accepts an input value that is 100 times the true radian value and outputs an integer value that is 100 times the actual cosine (so that it can include two decimal places in the result). Take a minute to read and understand Figure 9.2. The first line in the function type declaration declares the function's name (INT_COS_X_100) and the data unit size of the function's output (SINT means a 16-bit signed integer). *A function must always output its result by writing to a variable with the same name as the name of the function.* In creating a function, the programmer declares the function as having a data type, which also defines the data type of the function's result. A function definition also declares variables for values that must be provided by the program that calls this function (between VAR_IN and END_VAR) and for storage of temporary data as the function executes (between VAR and END_VAR). The IEC 1131-3 explicitly forbids programmers from writing their own functions with overloading

[5] The IEC 61131-3 standard defines exactly how much overloading each of its standard functions should have.

```
FUNCTION int_cos_X_100 : = SINT

VAR
      Temp1, Temp2, Temp3, Temp4  : REAL
END_VAR

VAR_IN
      Rad_X_100 : SINT
END_VAR
```

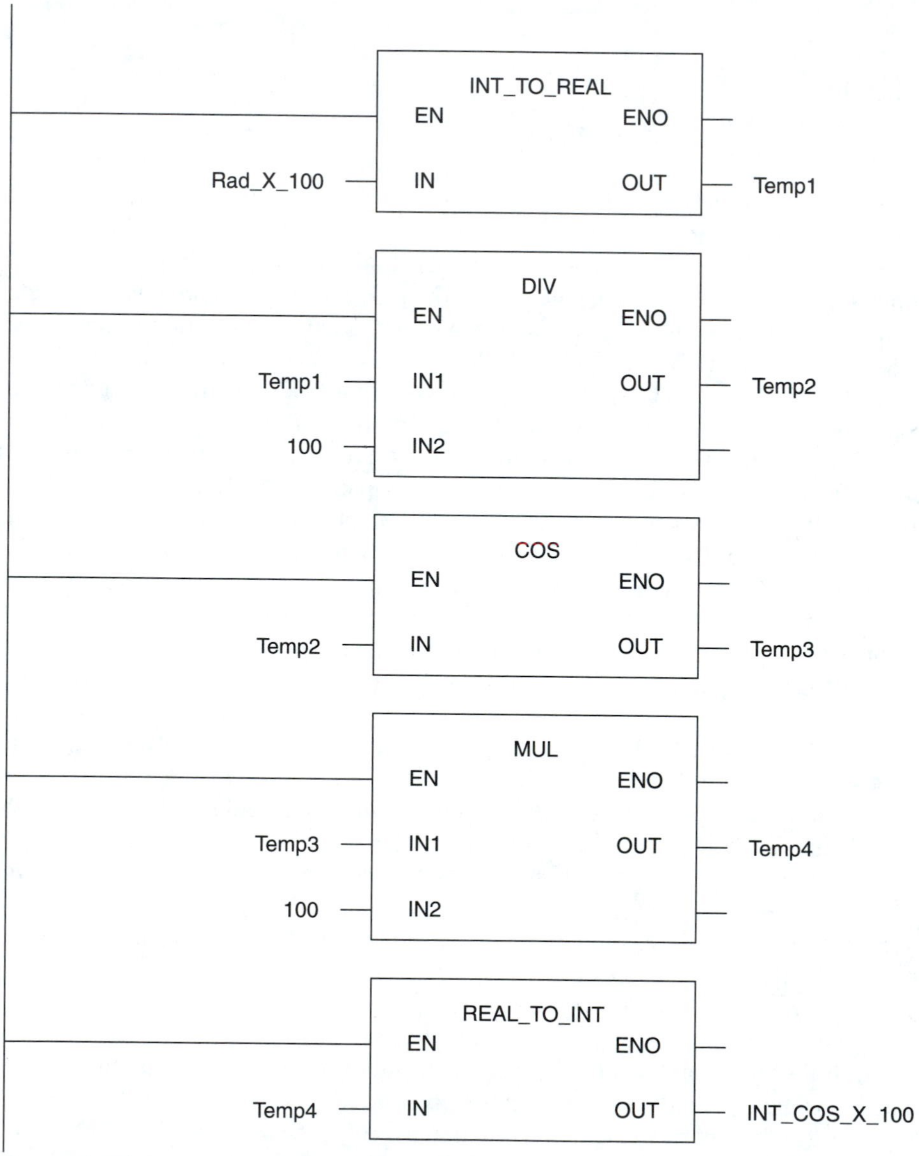

INT_TO_REAL
EN ENO
Rad_X_100 — IN OUT — Temp1

DIV
EN ENO
Temp1 — IN1 OUT — Temp2
100 — IN2

COS
EN ENO
Temp2 — IN OUT — Temp3

MUL
EN ENO
Temp3 — IN1 OUT — Temp4
100 — IN2

REAL_TO_INT
EN ENO
Temp4 — IN OUT — INT_COS_X_100

FIGURE 9.2
User-written function in ladder logic.

capabilities. Only the standard functions, which the programmer buys as part of the IEC 1131-3 compliant programming software package, can be overloaded.

To keep the example simple, the ENO outputs aren't used in Figure 9.2, and each component function is programmed to execute unconditionally (it would be better if each function executed conditionally upon successful execution of the preceding function).

When a function is called, the PLC's operating system sets aside memory to store the values for each variable declared in the variable declaration section. When the function finishes executing, the memory that was assigned for its use is released for other uses, so functions cannot set aside working data for later use. A function cannot, for example, be used for an operation in which a new average is calculated every time a new measurement is made, because the function can't save the old measurements it would need to calculate a new average.

Function Blocks

Function blocks are the most important program organization units in the IEC 1131-3 standard. A function block can be a reusable algorithm. Whenever a function block is called, it is called as an instance of a function block template. Function blocks are like functions, except in a couple of ways. Function block variable declaration sections can declare more than one output parameter. An even more important difference is that whenever an instance of a function block is executed, the PLC's operating system sets aside memory for *permanent* storage of data. The data becomes part of the instance of the function block and can be reused by the function block instance each time it executes.[6] A function block therefore could be used to recalculate a new average after making every measurement, because a function block instance can save old measurements.

Because a function block instance is part algorithm and part data, it is sometimes treated as an algorithm and sometimes as data.

1. A function block can be called from another POU (a program, another function block, or even a function) to execute its algorithm. A function block can even be called directly by a task if a resource declaration sets up the task-to-function block connection.

2. Before a function block can be called by a POU or a task, a function block instance must be declared as if the function block were a data element. The function block instance can be declared as either a global variable or a local variable for use inside a single POU. The declaration tells the operating system to reserve data memory for use by the function block instance. The declaration specifies which function block to use as the template and specifies a unique name for this instance of the function block. The function block's template, previously entered as a type declaration, describes what memory requirements each of its instances has. (Reserving memory for a function block instance is called *instantiation,* which is a great word to use during your next job interview. Your instructor will probably ask you to define *instantiation* on your next test to see if you actually read this far.)

3. Function block instances can be passed as parameters because they are part data. A PLC's operating system reserves memory for the exclusive use of each function block instance,

---

[6] Siemens STEP 7 function blocks accomplish the same effect in a different way. You must call a unique instance data block with a STEP 7 function block, and the data is saved in the instance data block, not in the function block instance.

so passing a function block instance to another POU is sometimes the only way that some of those data values can be shared.

Standard Function Blocks IEC 1131-3 defines some standard function blocks that perform operations similar to standard PLC programming instructions. The standard function blocks are:

```
TON       on-delay timer
TOF       off-delay timer
TP        timed pulse
RTC       real-time clock
CTU       up counter
CTD       down counter
CTUD      up-down counter
R_TRIG    one-shot, rising edge
F_TRIG    one-shot, falling edge
SR        bistable latch, SET input dominant
RS        bistable latch, RESET input dominant
SEMA      semaphore; useful in creating a busy signal
```

Figure 9.3 shows the CTUD function block (the most complex of the standard function blocks), as it would appear in an IEC 1131-3 ladder logic program. The instance name for the function block (my_counter) is shown in its usual position, above the box. Other counters, with other instance names, can be used elsewhere in the program, or the same counter instance can be called elsewhere. Boolean logic statements and parameters have been entered (in lowercase) for each input and output variable. The variable name "count_up" represents a Boolean element that can be manipulated elsewhere in the program. If count_up goes TRUE, the counter's cur-

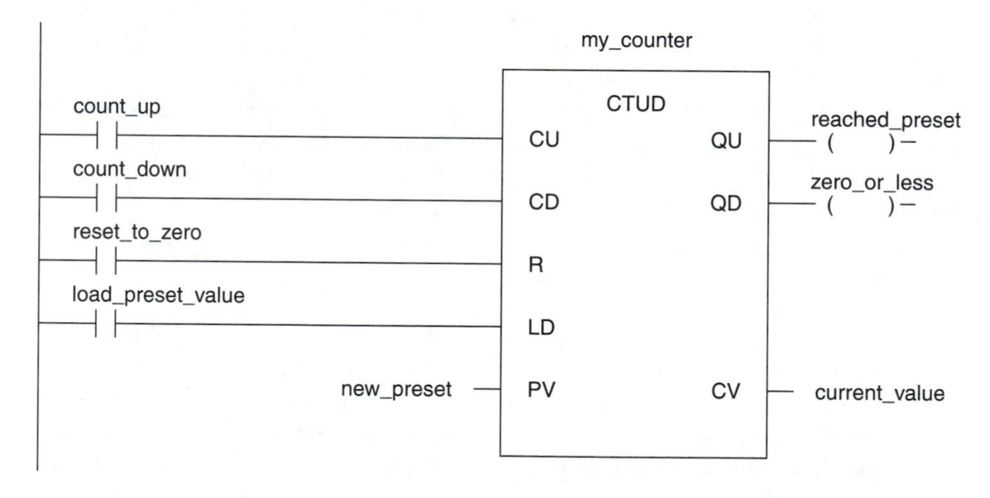

FIGURE 9.3
Standard function block for an up-down counter.

rent value (used internally as "CV" and outputted as "current-value") will increment. Similarly, if load_present_value goes true, the new_preset value, which is stored internally as "PV," will be copied to "CV" in this instance of CTUD.

The IEC 1131-3 standard provides some *examples* of function blocks for **PID servo control** and for **statistical process control** operations. By the time you read this, there should be standard function blocks for these and other functions.

VARIABLES AND VARIABLE DECLARATIONS

The use of variables and the passing of parameters between components *in* the configurations and computers *outside* the configuration are central to the IEC 1131-3 standard. Viewed from outside a PLC configuration, the configuration's important aspects include things like the actuators it controls and the control commands it can accept. Viewed from a function block instance, the configuration's most important aspects include the ability to provide working data, to keep old values, and to accept output values when the function block instance ends. The configuration must satisfy needs of all its *clients* (humans in the sales department as well as functions deep within the configuration) so that they all have access to the values they need, and only to the values they need. Variable declarations at the levels where the variables are needed satisfy this need.

Variable Declarations at the Configuration Level

Declarations of variables at the highest level define the connection of the PLC system to actuators, sensors, and to other computers. The very highest level is the configuration level, which begins immediately with the keyword CONFIGURATION at the start of a computer file defining the PLC system.[7] The programmer should declare variables at this level only for values that all or most of the resources need.

Global Variables at the Configuration Level The first set of variable declarations following the CONFIGURATION keyword must be prefixed with the keyword **VAR_GLOBAL** and must end with the keyword END_VAR. Between these keywords, individual declarations must be entered in this format (square brackets indicate optional entries, italics indicate user-selected entries):

var_name [AT %dir_rep_addr] : data_type [::5 init_value] ;

For example,

```
motor_1 AT %QW42 : INT ::= +35
```

[7] The IEC 61131-3 standard does not specify how keywords and other required file syntax elements are put into a file, so the programming software can assemble the file for the programmer, allowing the programmer to use a GUI interface to enter data into a menu in any order.

where

- *var_name* is any unique name containing standard alphanumeric characters.[8] Uppercase characters are treated as lowercase. The first character must be a letter. Underscore characters (_) are ignored, but you can enter them for readability. The IEC 1131-3 standard requires a PLC to examine only the first six characters (not counting underscores), so the programmer should ensure that all variable names have different patterns in those first six characters.
- AT *%dir_rep_addr* defines the variable as one that represents a **directly represented variable,** which means an actual PLC address. Actual PLC addresses must be entered as a % sign (as shown) followed by a two-letter code and then a number. The first letter in the code must be:

```
I     for an input memory address
Q     for an output memory address
M     for an internal data memory address
S     for PLC status (reserved for future standards)
```

  The second letter must be:

```
X     bit (may be assumed if no second letter entered)
B     byte (8 bits)
W     word (16 bits)
D     double word (32 bits)
L     long word (64 bits)
```

  The number's size is limited only by the size of the PLC's memory. It can be made up of several numbers separated by periods to represent a hierarchical memory structure. For example, %IX4.3 might represent bit 3 of byte 4 in the input image area of memory.
- *data_type* can be any of the IEC 1131-3 recognized data types or can be a user-defined data type. The IEC 1131-3 recognized data types include:
  - **INT** for a 16-bit signed integer value.
    - The prefix **S, D,** or **L** (e.g., SINT, DINT, LINT) changes the size to 8 bits, 32 bits, or 64 bits, respectively.
    - The prefix **U** (e.g., UINT, USINT, UDINT, ULINT) changes the data type to unsigned binary.
  - **BOOL** for bits.

    When used as a *data_type* in a nonglobal declaration in a program organization unit (which we examine later), **BOOL R_EDGE** and **BOOL F_EDGE** are allowable variations that define *var_name* as a bit that will be set to TRUE for one execution of the program organization unit after the Boolean state of the value at *dir_rep_addr* becomes true or becomes false, respectively.

  - **BYTE** for bytes (8 bits), **WORD** for 16-bit words, **DWORD** for 32-bit double words, or **LWORD** for 64-bit long words.
  - **REAL** for 32-bit floating-point notation, or **LREAL** for 64-bit floating-point notation.

---

[8] IEC 1131-3 specifies the ISO 646 basic code table characters. If you stick to standard alphanumeric characters, you won't have any trouble.

- **TIME, DATE, TIME_OF_DAY,** and **DATE_AND_TIME** are specially formatted data types to store data with the obvious uses. The short forms **TOD** and **DT** can be used in place of TIME_OF_DAY or DATE_AND_TIME.
- **STRING** for storing a string of ASCII characters.
- A data structure that is contained in a **function block instance.** Enter the name of the function block instance. The function block must have already been declared in a type declaration file in the programming unit. The function block type declaration definition, which is described later, defines what memory to reserve.
- A user-defined data type, called a **derived data type** in the IEC 1131-3 standard. Must have been defined previously in a type declaration file in the programming unit. A derived data type declaration file must start with the keyword TYPE followed by a unique name, include a list of variable names and data types, and must end with the keyword END_TYPE. The programmer would enter the derived data type's name for *data_type* in the declaration above. A derived data type template can contain:
  - A simple variable declaration (one of the types shown above), in which case the result is just like renaming a data type. For example,

    ```
    TYPE my_real : LREAL ;
    END_TYPE
    ```

  - A single variable declaration followed by two numbers defining the limits of this variable's allowed values. For example,

    ```
    TYPE my_real : LREAL (-4.3, +4.3) ;
    END_TYPE
    ```

  - A set of alphanumeric names that can be used as input/output values by program organization units instead of having to pass numerical values: for example,

    ```
    TYPE mode-switch (load, run, unload) ;
    END_TYPE
    ```

  - A data structure consisting of an array of similar data type elements in one or more dimensions. The keyword ARRAY must follow the TYPE (name) keyword, and the limits of the array and the data types of the array elements must be specified. The following example defines a 4 × 3 array of integers:

    ```
    TYPE my_array
       ARRAY [1..4,1..3] OF INT;
    END_TYPE
    ```

  - A data structure consisting of several elements of different data types. The declarations must be between the keywords STRUCT and END_STRUCT, which will be between the TYPE (name) and END_TYPE keywords: for example,

    ```
    TYPE my_struct
      STRUCT
        switch   : BOOL
        speed    : INT
        position : my_array
      END_STRUCT
    END_TYPE
    ```

Note that some derived data types require declarations of the data types of their components. You can use a derived data type as a data type in another derived data type declaration! You can, for example, create a data type that consists of a structure that contains an array (as in the structure example above). Variable names declared in derived data type declarations allow hierarchical addressing of individual components of derived data type structures. If, for example, a global variable was defined as

```
my_mess : my_struct
```

then individual components of "my_mess" could be addressed as follows:

```
my_mess.switch              for a single bit value
my_mess.position            for the entire 4 × 3 array
my_mess.position[2.2]       for one integer from the array
```

- *init_value* can be an initial value that will be put into the memory assigned for this variable when the configuration is initialized. Subsequent program execution can change the value. Initial values must be entered in the *literal* formats defined in the IEC 1131-3 standard. If initial values are not entered, the PLC's operating system will set all values to zero except date values, which will be set to Jan. 1, 0001.

Access Path Declarations at the Configuration Level Following the declaration of global variables at the configuration level, which ends with the END_VAR keyword, the IEC 1131-3 standard requires the programmer to enter declarations for resources. After the last resource declaration and the last END_RESOURCE keyword, the programmer can list the variables that the configuration will allow computers outside the configuration to access. The declaration list starts with the keyword **ACCESS_VARS** and ends with the keyword END_VARS just before the keyword END_CONFIGURATION ends the entire configuration file. Access path declarations are entered in the following format:

global_name : local_name : data_type [parameter]

where

- *global_name* is the name that a remote computer would use to read or write the variable. Subject to the same limitations described for global variables.
- *local_name* can be:
  - A global variable name defined at the configuration level.
  - A hierarchical variable name defining (as necessary) the resource, program, and function block and the name of an input, output, or global variable declared at that level.
  - A directly represented variable address.
- *data_type* is as described for global variables.
- An optional *parameter* can be added to limit access to this variable from outside the configuration. Parameter keywords include READ_WRITE or READ_ONLY, which shouldn't need explanations. READ_ONLY is the default, so does not need to be entered.

Variable Declarations at the Resource Level

Only global variables can be declared at the resource level, using the same GLOBAL_VAR/declarations/END_VAR structure as described above. Variable declarations at this level are made immediately following the declaration of RESOURCE..ON.., which assigns a name to the resource and defines the type of processor in the PLC. When declared at this level, the declared variable names and memory they represent are not accessible to other resources in the configuration. Use variable declarations here to assign variable names to I/O addresses controlled by the program organization units in this resource, and to assign memory for values that this resource's program organization units must share.

The global variables declared at this level can also be made available to a computer outside the configuration if the configuration's ACCESS_VAR section defines the *local_var* name using a hierarchical structure consisting of the resource name, a period, then the variable name as declared here.

Variable Declarations at the Program Level

The programmer can declare several types of variables at this level, which is the first level that can contain an algorithm to manipulate data. Some declarations will define variables that haven't been defined above this level, but other declarations are required to make the PLC set aside memory to contain copies of variables that have been declared at higher levels. There is a good reason for maintaining multiple copies of the same data element. Each individual program can manipulate its own copies of shared data entries and can take as long as necessary to execute, without running the risk that another program (even one that interrupts this program to execute at a higher priority level) will change the data it needs. If each program had to use data in shared memory locations, programs could change each other's working data.

Variable declarations at the program level are not made in the file that contains configuration, resource, and task declarations. Program-level variable declarations are made in the template for each program organization unit, which is contained in a **program type declaration file** in the programming unit, as shown in the example in Figure 9.4. The program type declaration file must begin with a declaration statement containing the keyword **PROGRAM** followed by a name for the program template, the variable declarations, then the program algorithm, and the file must end with the keyword END_PROGRAM. Several sets of variable declarations can be included, and each set must end with an END_VAR keyword. (The example program in Figure 9.4 includes one example of every type of variable that can be declared in a program.) The algorithm must be written in one of the five (or six) allowable programming languages. In the algorithm, all data must be referred to using the variable names declared in the variable declaration section.

The example in Figure 9.4 shows only enough of the algorithm to demonstrate how program-level variable names are used. Remember that a program shouldn't do much data manipulation; it should call function blocks and functions to manipulate data. When a program calls a function block or a function, the program must pass parameters to that lower-level program organization unit, as shown in the example, in which variables are provided at the IN contact for each function.

The resource definition section of a configuration file must call an instance of a program, as shown in Figure 9.5. A program instance must be assigned a unique name in the resource statement that declares the program. For each uniquely named program instance that is called, the PLC will set aside memory for each variable declared in the program's data type template. In Figure 9.5, "my_prog" (from Figure 9.4) is used as a program type template for the program instance named "run_it". run_it executes whenever task_2 is initiated. The resource must pass

```
PROGRAM my-prog

VAR
        one : INT;
END_VAR

VAR RETAIN
        two : REAL;
END_VAR

VAR CONSTANT
        three : INT := 100;
END_VAR

VAR_TEMP
        four : BOOL := FALSE;
END_VAR

VAR_INPUT
        five : INT := -10;
END_VAR

VAR_OUTPUT
        six : INT;
END_VAR

VAR_IN_OUT
        seven : REAL;
END_VAR

VAR_EXTERNAL
        eight : INT;
END_VAR

VAR_GLOBAL
        nine AT %IX20.2 : BOOL;
END_VAR
```

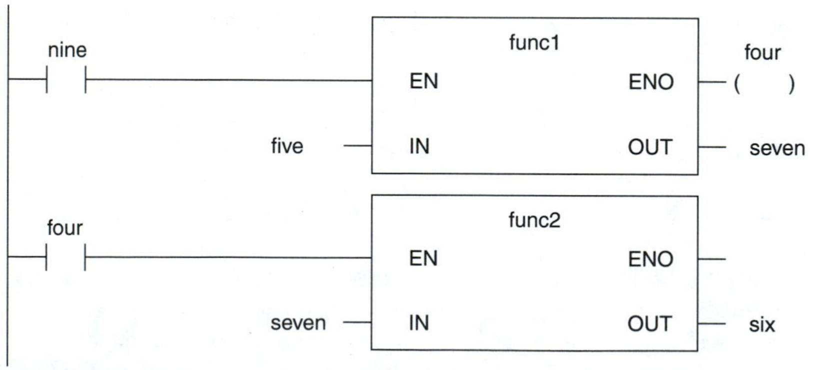

```
|   nine                          +-----------------+   four
|---| |---------------------------| EN    func1 ENO |---(   )
|                                 |                 |
|                         five ---| IN          OUT |--- seven
|                                 +-----------------+
|   four                          +-----------------+
|---| |---------------------------| EN    func2 ENO |---
|                                 |                 |
|                        seven ---| IN          OUT |--- six
|                                 +-----------------+
END_PROGRAM
```

FIGURE 9.4
Program type declaration.

parameters to a program instance when the instance is called, and must accept output parameters from the program instance as it terminates. The variable declarations in the program type template define the parameter passing requirements. In Figure 9.5, for example, the parameter my_int is passed for the program's input variable, "five".

The example in Figure 9.4 showed the types of variables that can be declared in a program (with only one example for each type). The individual variable declarations are no different from the declarations we examined for configuration-level global variables, but it does make a difference which variable section the declaration is in. Variables declared in the section begin with the following keywords:

1. **VAR** has space reserved in the PLC's memory while the program instance is executing. When initialized by any variable declaration, memory contents are set to zero (usually) unless the variable declaration defines an initial value. The calling program does not pass a parameter for a VAR variable.
2. **VAR_TEMP,** added in 1997, means the same as VAR.
3. **VAR_INPUT** reserves memory but also requires the resource to pass a parameter containing a value for "five". The example declaration includes an initial value (−10), which will be used only if the resource does not pass a parameter. Note that the resource example in Figure 9.5 passes one of its previously declared variables (my_int) for the instance program to copy into the memory reserved for "five".
4. **VAR_OUTPUT** reserves memory for a value to which the program should assign a value. When the program terminates, a value will be passed to the resource. Note that in this example the resource provides a directly represented variable address (%QW4) for the program instance to write to.
5. **VAR_IN_OUT** breaks some of the rules that IEC 1131-3 lays out for memory use by program organization units. The program instance saves a pointer to the address that the calling program passes as a parameter (my_real). As the program instance executes, it can read and write the value stored at that address. In this example, if program run_it contains a statement that writes a value to "seven", the PLC will actually write the value to my_real! VAR_IN_OUT variables have more value in function blocks than in programs.
6. **VAR_EXTERNAL,** like VAR_IN_OUT, gives the program instance access to a memory location outside its own memory areas. In this case the variable "eight" must be a variable name that has been declared as global at the resource or configuration level.
7. **VAR_GLOBAL** creates global parameters that can be shared with the function blocks and functions that execute under this program's control. In the example a directly referenced variable address (I × 20.2) is defined in the program type template as the data source for the bit that is assigned the name "nine". Declarations at the program level are the lowest level at which direct referenced variable addresses can be used.

FIGURE 9.5
Resource declaration that creates an instance of a program.

```
PROGRAM run_it WITH task_2
        my_prog ( five   := my int,
                  six    => %QW4,
                  seven := my_real)
END_PROGRAM
```

Any of the variable declaration sections, except VAR_GLOBAL, can include the keywords RETAIN or CONSTANT. The following two examples show the effect of adding these keywords in a VAR declaration section.

1. **VAR RETAIN** to reserved space in EEPROM memory for variables, so that their values are not lost if there is a power failure while the program instance executes. (Some PLCs save all data in EEPROM or in battery-backed RAM, so this keyword would have no effect.) Function block instances cannot be declared as variables that are retained.
2. **VAR CONSTANT** are standard VAR variables, but a variable declared as CONSTANT cannot be changed by the algorithm, so an initial value should be supplied with the declaration ("three"'s initial value is 100 in this example). CONSTANT should be used only for VAR or VAR_TEMP variables.

Variable Declarations at the Function Block Level

A function block template is defined by creating a type declaration file in the programming unit. A **function block type declaration file** starts with the keyword FUNCTION_BLOCK followed by the name for this function block template and must end with the keyword END_FUNCTION_BLOCK. Between these keywords, variables must be declared, as they are declared in a program type declaration file, but with differences as noted below. Below the variable declarations, the function block type declaration file must contain an algorithm. Function block algorithms should be written so that they can be reused.

When a function block is instantiated, the PLC's operating system sets aside memory for some of the variables declared in the function block's type declaration section. The PLC does not release this memory until the PLC is reset. Data in this memory is saved between the times when the function block instance executes. Because of the data *persistence,* variable declarations are slightly different:

1. **VAR** variables retain their values between executions of the function block instance algorithm. Initial values defined in the function block type declaration are not used after the first time the function block instance is executed.
2. **VAR_TEMP** variables are the only values not retained between execution of the function block instance.
3. **VAR_INPUT** values only receive new values when a function block instance is called and only if the POU passes new parameters.
4. **VAR_OUTPUT** values can only be changed by executing the function block instance again.
5. **VAR_IN_OUT** and **VAR_EXTERNAL** values can be changed by other POUs while the function block instance is inactive, since both refer to values held outside the function block instance's memory.
6. **VAR_GLOBAL** variables cannot be declared at this level.

Directly represented variable addresses cannot be used anywhere in a function block. If you want a function block instance to manipulate a directly represented variable, you must declare a variable for the directly represented variable at the program level or higher, and pass that variable to the function block instance as an input, output, In_Out, or external variable.

Variable Declarations at the Function Level

Function templates are written as **function type declaration files,** just like creating function block templates except for a few differences. The type declaration file must begin with the keyword FUNCTION, of course, and must include a name for the function. This first line must also declare a **data type** for the function. The declaration of a data type for the function automatically declares the function's only output parameter, which has the same name as the function. Other variable declarations can be made in variable declaration sections after this first line, then the algorithm must follow the last variable declaration. Variable declarations and allowable variable sets are as described for program type declarations, except that there cannot be any **VAR_OUTPUT, VAR_IN_OUT,** or **VAR_EXTERNAL** declarations. The only possible output variable has already been declared. The function should write a value to a variable with the name given to the function. Whenever a POU calls a function, the function will return this value. **VAR_GLOBAL** variables cannot be declared at this level.

Directly represented variables cannot be used in functions. You must declare a variable for the directly represented variable at the program level or higher, and pass that variable to the function as an input or as the function type.

PROGRAMMING LANGUAGES IN IEC 1131-3

The IEC 1131-3 defines five allowable languages, and pressure is mounting to force acceptance of a sixth. All six languages are variations on programming languages now used in control systems. The six languages include three text-based languages:

1. Ladder logic (LD), which is based on the ladder logic languages developed from relay logic standards. Ladder logic is the favorite PLC programming language in North America.

> The graphics you see in a ladder logic program all represent fairly simple text instructions. Some programming software actually allows you to enter ladder logic programs using a text keyboard.

2. Instruction list (IL), which is based on the assembler languages that are native to all microprocessors. Instruction list–like languages are commonly used to program PLCs in Europe.

3. Structured text (ST), which looks a lot like the C programming language that is the current favorite in the personal computer world, but is more closely related to Pascal.

Two graphics-based languages are now included in IEC 1131-3:

4. Sequential function chart (SFC), which is based on Telemechanique's GRAFCET language. (Telemechanique is now part of Groupe Schneider.) Rockwell Automation offers a similar language for programming Allen-Bradley PLC-5s. SFCs are a graphical way of programming sequenced calls of subprograms.

5. Function block diagram (FBD), which looks a little like electrical schematics with ladder logic–like programming elements thrown in.

The sixth is a graphics-based language, which may be included in the IEC 1131-3 standard if its supporters succeed:

6. Continuous function chart (CFC), which is similar in some ways to FBD but uses symbols and connections based on diagrams now in use in the process control and motion control industries. A separate standard, IEC 1499, is being prepared to standardize CFC language.

In this book we do not attempt to teach any of the six languages but describe each in enough detail for the reader to recognize the advantages of each. Knowing where to use each language is important because the IEC 1131-3 standard says that a single configuration can be made up of program organization units that are written in any of the five (so far) allowable languages.

Programming software is now available that claims to offer all five IEC 1131-3 languages, and most new programming software packages claim to offer at least one IEC 1131-3 language. Remember when you read those claims that compliance with the IEC 1131-3 standard is voluntary and is not verified by the IEC. PLCopen plans to certify compliance with IEC 1131-3 at a base level, a portability level, and at a full-compliance level, but has only started certifying compliance at the lower levels. PLCopen requires manufacturers to state which parts of the IEC 1131-3 standard they do not comply with if they claim partial compliance.

One consideration that the IEC 1131-3 standard does not make clear is where type Declarations for user-defined data elements and algorithm templates, variable names, and program comments *are* to be stored. They must be available at the programming unit when a PLC configuration is being entered, of course, but the standard does not say that they must be downloaded into a PLC's memory. A programming unit could remove all comments and variable names and could even change the program structure as it compiles an IEC 1131-3 compliant configuration into machine language. Unless standards are agreed on, it may be impossible to use another programming unit for control system monitoring or modification.

Ladder Logic

Ladder logic in IEC 1131-3 looks like other ladder logic programming. IEC 1131-3 calls rungs **networks,** and networks can have branches. Besides the standard EXAMINE ON and EXAMINE OFF Boolean instructions, elements have been standardized for examining a bit for positive and negative transitions:

—| |—	—| / |—	—| P |—	—| N |—
EXAMINE ON	EXAMINE OFF	POSITIVE transition	NEGATIVE transition

There are several standardized Boolean output instructions, four of which should need no description:

—( )—	—(/)—	—(S)—	—(R)—
COIL	NEGATED COIL	SET COIL	RESET COIL

There are some new Boolean output elements, two of which provide one-shots of true when their Boolean logic control statements change state:

—(P)—	—(N)—
POSITIVE transition	NEGATIVE transition

and three of which are used to store a Boolean output into Boolean variables that are declared with the keyword RETAIN, so that their values will be kept by the PLC even if power is lost:

—(M)—	—(SM)—	—(RM)—
RETENTIVE	SET	RESET
MEMORY	RETENTIVE	RETENTIVE
COIL	COIL	COIL

Other familiar ladder logic instructions, such as MOVE and TIMER, are invoked by calling standard functions or standard function blocks.

The standard functions defined within IEC 1131-3 all include EN input contacts, so that they can be programmed to execute dependent on a Boolean logic statement; and all include ENO outputs, so that the successful completion of the function can control other Boolean elements. User-written functions or function blocks can be used in ladder logic, regardless of the language they were written in originally, as long as they include at least one Boolean input so that they can be attached to a ladder logic network. A ladder logic network that includes calls of standard functions might look like Figure 9.6, in which "value" is increased by two each time "switch" goes from false to true, but is held to 100 or less. Once the programmer gets used to entering parameters as variable names (e.g., "value" and "temp") or as literals (e.g., +2 and +100) outside the graphical box, the transition to programming in IEC 1131-3 ladder logic is easy.

Figure 9.7 shows a standard on-delay timer function block as it might be used in a ladder logic network. The function block type declaration name is "TON". The instance of the function block is given the name "my_ton" by the programmer. The timer's preset time (PT) is entered as a literal (10 s in this example), and the Boolean bit represented by the variable name "light" will go on 10 s after the Boolean bit "switch" goes on. Other networks in the program can read the elapsed time (ET) that is outputted by the function block instance into a variable named "so_far".

While IEC 1131-3 includes a standard for JUMP instructions in ladder logic, it recommends that jumps should not be programmed inside a ladder logic program organization unit.

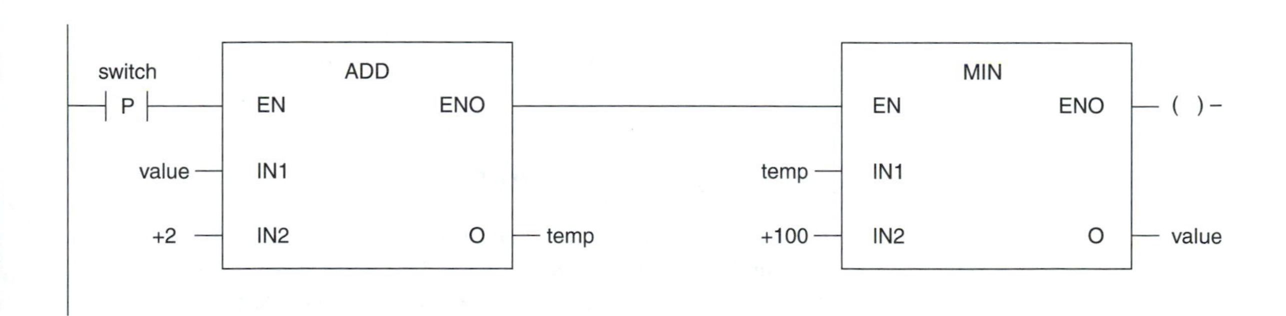

FIGURE 9.6
Ladder logic network with standard functions.

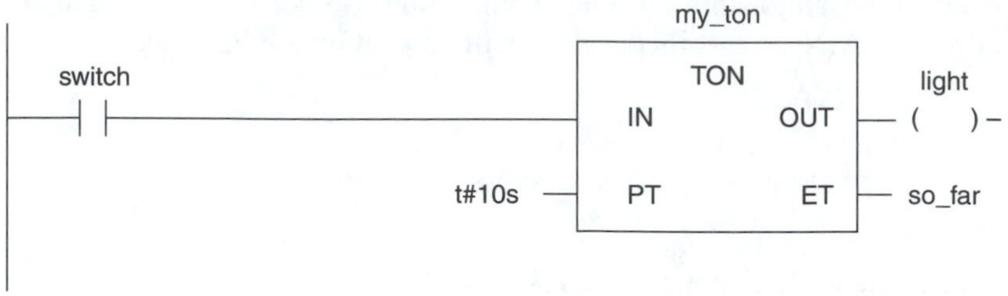

FIGURE 9.7
On-delay timer function block in a ladder logic network.

Instruction List

The instruction list programming language looks a lot like the Siemens' statement list (STL) programming language that we have already covered except that the instructions are a little different. Boolean logic instructions include

LD, AND (or &), OR, XOR, ST, S, R,)

where LD (load) can load a bit value to initiate a Boolean logic statement, and ST (store) can write the result of a Boolean operation to an output address.

Any of the Boolean operators except S, R, and) can have a modifier that **inverts** the state of the bit they examine or control. For example, LDN would load TRUE if the bit it was loading was false.

The operators that develop Boolean logic results (e.g., AND, OR, XOR) can also include a "(" modifier to begin developing another Boolean statement within the statement already begun. For example:

```
LD    switch1
XOR ( switch2
AND   switch3
)
ST    light
```

will turn "light" on if "switch1" is on, or if both "switch2" and "switch3" are on, but not if all three switches are on.

Instructions that operate on non-Boolean data values include

LD, ST, ADD, SUB, MUL, DIV

LD (load) and ST (store) can be used in Boolean logic, or they can be used to copy non-Boolean values into and out of the microprocessor's accumulators. The address that follows the LD or ST instruction determines whether a Boolean or non-Boolean operation is required.

Operators can manipulate data values of any standard IEC 1131-3 data type, returning a result with the same data type. All the operators except LD and ST can include the "(" modifier to force following data manipulation operations to be executed before the pending operation.

For example,

```
LD   value1
MUL( value2
ADD  value3
)
ST   result
```

will force the PLC to add "value2" to "value3" before multiplying the sum by "value1". The final result is saved as "result".

Data **comparison** instructions include

GT, GE, EQ, NE, LE, LT

which can accept any standard data type and will return a Boolean logic result. Any of these operators can include a "(" modifier.

Instructions for structured programming include

JMP, CAL, RET

any of which can include a "C" modifier to make it operate conditional upon the preceding Boolean logical statement, and/or an "N" modifier to invert how it conditionally executes. For example,

```
LD          switch
JMPCN       there
```

will cause a jump to the label "there" conditional upon "switch" not being true.

The CAL instruction must be followed by brackets enclosing a list of the parameters (if any) being passed to the function or function block that is being called. The standard on-delay timer function block (TON) shown in ladder logic format in Figure 9.7, for example, has input variables named IN and PT and output variables named OUT and ET. Figure 9.7 would look like this in instruction list language:

```
CAL my_ton (IN:switch, PT:t#10s, OUT:light, ET:so_far)
```

Unconditional calls that are not negated do not require use of the CAL instruction. The instruction above could also be written as

```
my_ton (IN:switch, PT:t#10s, OUT:light, ET:so_far)
```

IEC 1131-3 includes instructions specifically designed to write new values to the input variables of function block instances. The input variable name is used as the instruction, and the name of the instance function block is included with the instruction! For example, the on-delay timer function block (TON) has input variables named IN and PV, and these input variable names can be used as instruction list instructions, as in the following example:

```
LD      switch
IN      my_ton
LD      t#10s
PT      my_ton
my_ton  (OUT:light, ET:so_far)
```

Passing of variables in function block calls can be eliminated *completely* by reading and writing the function block's variables using standard instruction list instructions that address the variables in the form: instance name.variable name, as in the following example.

```
LD      switch
ST      my_ton.IN
LD      t#10s
ST      my_ton.PT
my_ton
LD      my_ton.OUT
ST      light
LD      my_ton.ET
ST      so_far
```

If IEC 1131-3 compliant programs can ever be made truly portable, it must be possible for a program written for any PLC to be loaded into any type of programming unit running programming software from any supplier and then to be downloaded to a PLC.[9] For this to be possible, a **neutral language format** is required for program files as they are being moved from one supplier's product to another's. Definition of a neutral file format for data definition has already been agreed upon, but a neutral format for algorithms is still being discussed. **Instruction list** has been proposed as the neutral language format for algorithms by instruction list's European supporters. The instruction list language is close to the native machine language that all microprocessors use, and all the other IEC 1131-3 language components can be expressed in instruction list (although the exact translations might be controversial). The North American suppliers aren't happy with the proposal to use instruction list.

Structured Text

Most non-PLC programmers program in languages such as C or Pascal. If PLCs are to be accepted as standard control computers by programmers who can't program in instruction list or in ladder logic, they must be programmable in a structured text language such as C or Pascal.

The concept of declaring variables, assigning variable names and data types, and passing variables is new to most PLC programmers, but it is very familiar to C and Pascal programmers. C and Pascal both allow the defining of functions or procedures that can be called from within another program, so the concept of reusable algorithms is familiar. Declaring a data structure as a data type, then passing data elements of that data type, is central to object-oriented programming (OOP) and to Plug_and_Play.

The language that structured text is based on (Pascal) includes instructions for logical operations, which work fine on data elements of type Boolean, and offer the other data types and data manipulation instructions that are used in PLCs. If you know Pascal or C, you can learn to use the IEC 1131-3 set of instructions defined for structured text.

---

[9] It is necessary for downloading software to be written specifically for an individual PLC type, since the microprocessor and memory requirements of each PLC will continue to differ.

Structured text offers some of the advanced programming features available in Pascal and C for structured programming. These features include:

1. The **IF..THEN..[ELSE..END_ELSE]..END_IF** construct, which can use a conditional statement to select an operation to execute from one or two options.

2. The **CASE..OF..[ELSE]..END_CASE** construct, which can be used to select a single operation to execute out of a long list of possible actions.

3. The **FOR..DO** construct, the **WHILE..DO** construct, and the **REPEAT..UNTIL** construct, which can be used to cause a program to loop repeatedly through a section of algorithm until a conditional statement becomes true (or until it becomes false).

Sequential Function Chart

PLCs have been used to control the sequencing of manufacturing processes for as long as they have existed. We examined how to write sequencer programs in Chapter 3. Sequential function chart (SFC) language offers the programmer a simplified way to write very complex sequencer programs by building graphical diagrams showing each **step** in the sequence, the **transition** event that causes one step to finish and the next to begin, and which shows alternative step sequences. In SFC, every step must be followed by a transition, and each transition can enable only *one* step.

Step actions and transition conditions can be programmed right on the SFC diagram if they are simple, and they can be programmed in any of the IEC 1131-3 languages. If the step action algorithms or the transition condition algorithms are complex, they can be written as functions or function blocks that are called by the SFC. The algorithms for transitions, of course, must return a Boolean result to enable or disable the transition. Algorithms for steps can be more complex and are limited only by the rules that all programs and function blocks are subject to, including the rules about variable declarations and parameter passing. An SFC can even call function blocks written in SFC language.

SFC Step Actions In an SFC algorithm, the SFC chart's steps must have **step names** and the steps must indicate which **actions** to perform when the sequence reaches this step. Actions are actually functions or function block instances that the SFC program calls, so the transitions are really conditional statements that control the calls. The SFC diagram includes the name of the function or function block instance beside the step. A step can have zero or more actions. (If there are no actions, the PLC does nothing every time the SFC executes, until the next transition becomes true.) Each action can be programmed with an optional **action qualifier** to specify when or for how long the step's action should continue to be executed after this step becomes active. An optional **indicator variable** can also be entered in an SFC program. The indicator variable is an output variable of the action's function or function block and is meant to be used to monitor that step's activity.

Figure 9.8 shows a transition–step–transition sequence, with transition controls programmed in ladder logic. When "tank_full" goes true, the previous step is deactivated and step "Warm" starts. The action qualifier is "N", which means that the action must execute every time the SFC executes until the next transition becomes true. The action performed is contained in a

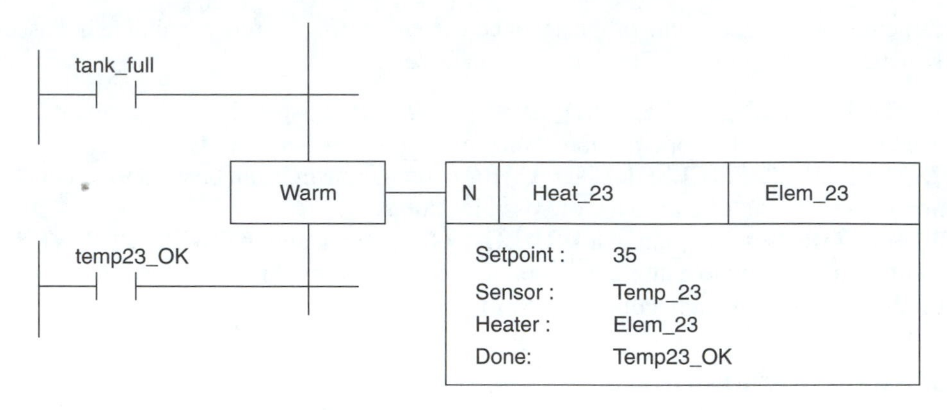

FIGURE 9.8
Transition–step–transition sequence.

function block "instance" named "Heat_23". Heat_23 isn't shown here, but the parameters that it requests give us an idea of what it does. It requires a temperature setpoint, the address of a sensor to read, the address of a heater to turn on, and a Boolean output for "done". While the value in variable "Temp_23" is lower than 35 degrees, Heat_23 will hold the heater (Elem_23) on. When the temperature reaches 35, Heat_23 will turn Elem_23 off and turn temp23_OK on. Turning temp23_OK on enables the next transition and disables this step.

Action qualifier "N" was used in the previous example. The allowable **action qualifiers** include:

N — to cause the action to execute every scan of the SFC program while this step is active (same effect as not entering an action qualifier).

P — to cause the associated action to execute just once when the step becomes active.

L time — where "time" indicates the maximum time that the action will execute unless the step ceases to be active sooner.

SL time — where "time" indicates the total time that the action will execute each time the step becomes active even if the step ceases to be active before time runs out.

D time — where "time" indicates a time delay before the action can start executing. The action will only continue to execute while the step is active.

S — to start the associate action executing and keep executing it even after this step ceases to be active.

SD time — where "time" indicates a time delay before the action starts executing. Execution will start and continue even after this step ceases to be active.

DS time — where "time" indicates a time delay before the action starts executing. Execution will not start unless this step is still active when time runs out, but once started it will continue to execute after the step ceases to be active.

R — to stop an action started with an S, SD, or DS action qualifier.

An SFC program can include instructions that inspect the status of steps in the program. For each step the PLC maintains a Boolean **step active flag,** addressable as ⟨*step_name*⟩.X, and

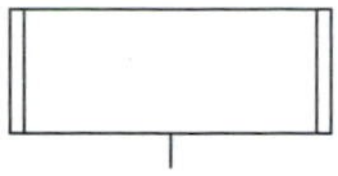

FIGURE 9.9
Initial step box.

an **elapsed time,** addressable as ⟨*step_name*⟩.T. The time is stored in IEC 1131-3 TIME format. Using a step's elapsed time to trigger the next transition is one way to keep a step active for a fixed time. For example,

- "Warm.X" will be true while the step in Figure 9.8 is active.
- "Warm.T" will have a value equal to the time that Warm has been active since the last time it became active.

SFC Execution Sequence Next we describe how SFC programs can be structured to control how they execute.

> From this point on, examples will show only the step blocks and transition marks, without showing the action identification blocks or transition conditions that they must include.

There must be an **initial step,** which is visibly different from other steps because it has double vertical lines, as shown in Figure 9.9.

The initial step will be the first step the PLC executes when the PLC is started.[10] The PLC will reexecute the action(s) associated with this step (subject to the effect of any action qualifiers) each time the SFC algorithm executes until the next transition becomes true.[11] After the transition following a step becomes true, the step ceases to be active and the next step will become active.

SFC charts can **diverge** into one or more of several alternative paths. There are two ways of programming divergence. In an **exclusive divergence,** the SFC algorithm will select one out of the several possible vertical paths to follow, and steps on the other paths will not be executed. The SFC flowchart must show flow from the preceding step to more than one path, each of which must start with a transition, as in Figure 9.10.

When programmed as shown above, the programmer must write the transition conditions (at T1 and T2, etc.) so that only one transition can be true at any time; otherwise, the actual path the PLC selects may be unpredictable. Exclusive divergence can be programmed to force the PLC to evaluate the leftmost transition condition first and then to examine the next leftmost, until a true condition is found. To force this, enter a star (*) at the point of divergence as in Figure 9.11.

The programmer can force another order of transition evaluation if the star is entered and numbers are entered indicating the order (1, then 2, then 3, etc.) to evaluate the transition conditions, as in Figure 9.12.

Divergent paths should converge, following a transition on each of the parallel paths, as shown in Figure 9.13. Note that transitions are required ahead of the convergence.

---

[10] Some PLC manufacturers allow the PLC to be configured to restart SFC execution from where the sequence stopped when the PLC was last powered-off.

[11] You may read elsewhere that a step should execute one more time after its transition goes true. That was an error in the standard that was corrected in 1997.

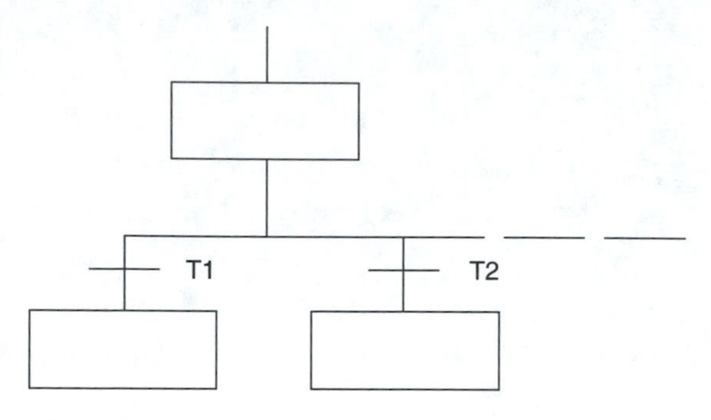

FIGURE 9.10
Exclusive divergence.

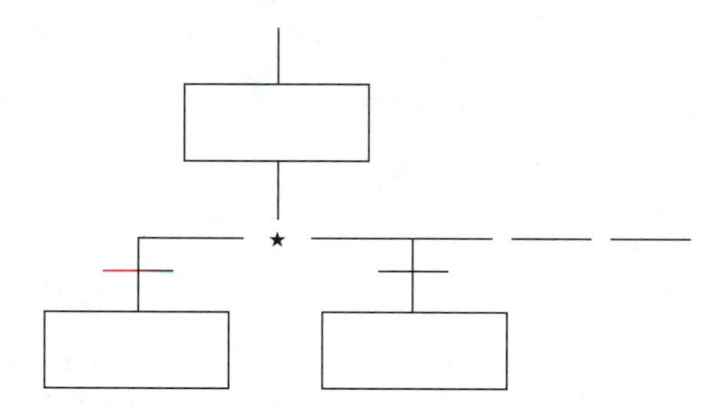

FIGURE 9.11
Exclusive divergence forcing left-to-right transition evaluation.

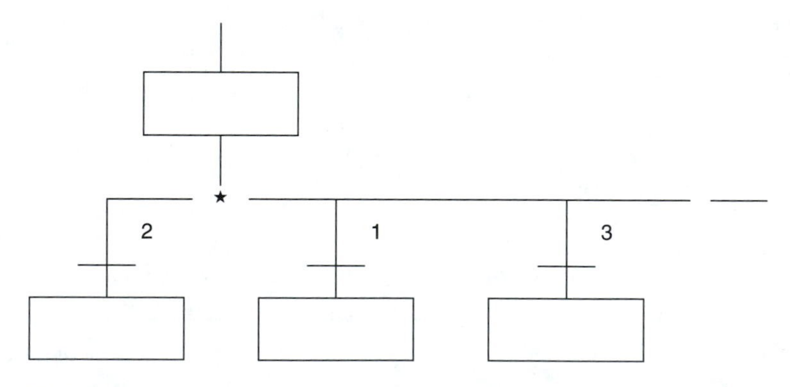

FIGURE 9.12
Exclusive divergence forcing programmer-selected transition evaluation order.

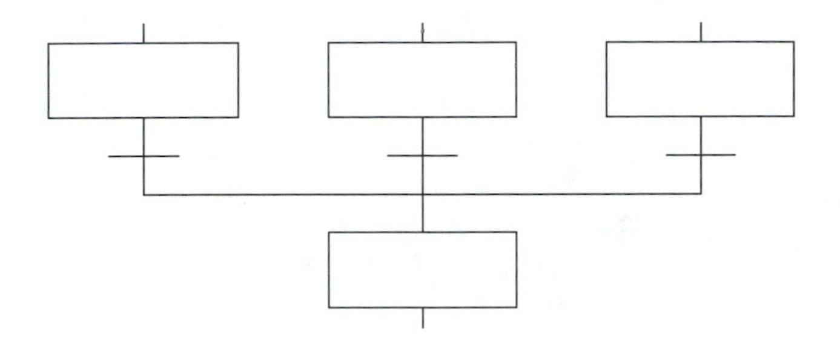

FIGURE 9.13
Convergence of exclusive divergent paths.

An exclusive divergence path does not have to include *any* steps, so it can be used as a way of bypassing or returning to repeat steps, as shown in Figure 9.14.

The other way of programming divergence is to use **simultaneous sequence divergence.** A transition must be programmed *before* the paths are shown to diverge. Double lines indicate the divergence and convergence. An SFC simultaneous sequence divergence and convergence is shown in Figure 9.15. After the transition ahead of the divergence becomes true, the PLC will start executing the steps on all the divergent paths. (In fact, a computer can do only one thing at a time, so it actually executes the step on the leftmost path first, then the step on the next path

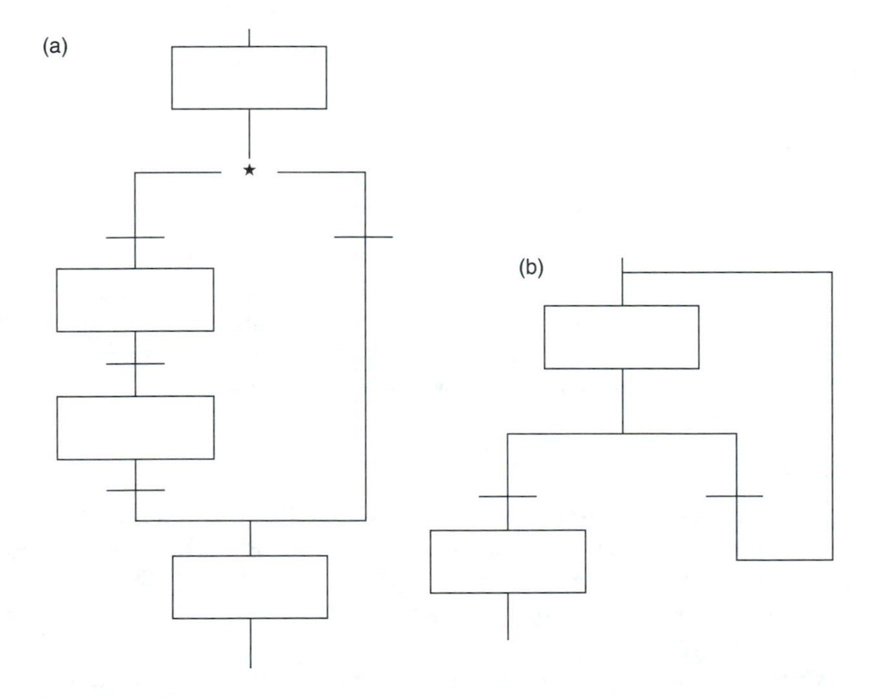

FIGURE 9.14
Exclusive divergence paths to (a) bypass selected steps; (b) repeat selected steps.

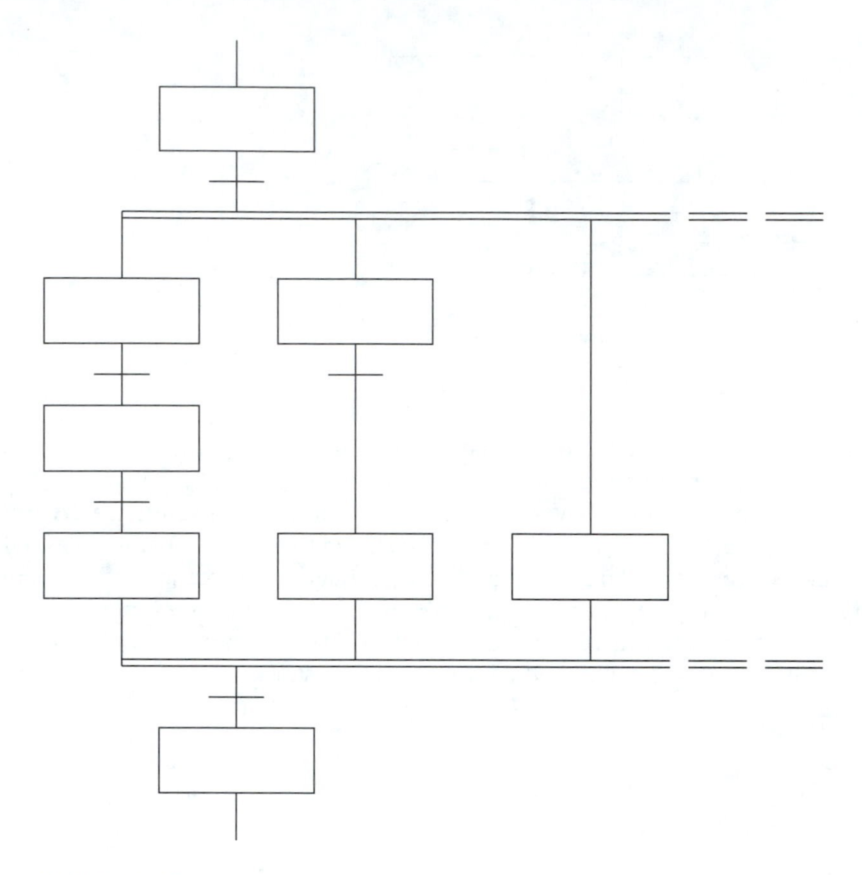

FIGURE 9.15
Simultaneous sequence divergence and convergence.

to the right, etc.) Each path can contain several steps and transitions (and even other divergences and convergences), but the paths must not contain separately programmed transitions after their last steps. Instead, a single transition is programmed below the convergence point. This transition is evaluated only if the last step on *all* the converging paths is active.

Function Block Diagram

Programming a function block diagram (FBD) is very much like programming ladder logic calls to functions and function blocks, but without the ladder logic Boolean logic graphics. For Boolean logic, the function block elements such as AND and OR must be used, with Boolean input parameters. Function blocks can also be programmed to include Boolean logic statements that affect their internal operation.

Each FBD element (representing a function or a function block) is depicted as a box, with input variables shown at the left and output variables on the right, and with the formal names of the parameters shown inside the box at the appropriate contact points. The function or function block name is shown inside the box, and the names of function block instances are shown above the box.

The difference between FBD algorithms and ladder logic's function block calls lies in how variables are shown to be passed between elements in the FBD network. In FBD, one element's output parameter contact can be drawn to connect to a second element's input parameter contact. These parameter-passing lines can even be drawn to show parameters being passed to elements that are above and/or to the left of the element that generates the parameter, as shown in Figure 9.16. The Torque output from my_speed is fed back to my_position so that my_position can detect a fault if a large requested torque isn't causing position error to decrease. Although the FBD looks like it has been programmed with an endless loop, the PLC will use the Torque output of my_speed as the Requested_Torque input for my-position during the *next* time this FBD diagram executes. IEC 1131-3 says that a programmer should be able to select the order that function blocks execute in an FBD. ("want", "posn", "max", "drive", and "stop_motor" are parameters that this FBD exchanges with the program that calls it.)

Continuous Function Chart

Continuous function chart (CFC) language is not currently an IEC 1131-3 approved language, and if it does become approved, it may be assigned a different name. Users of current process control and motion control software want to use the CFC language because it is similar

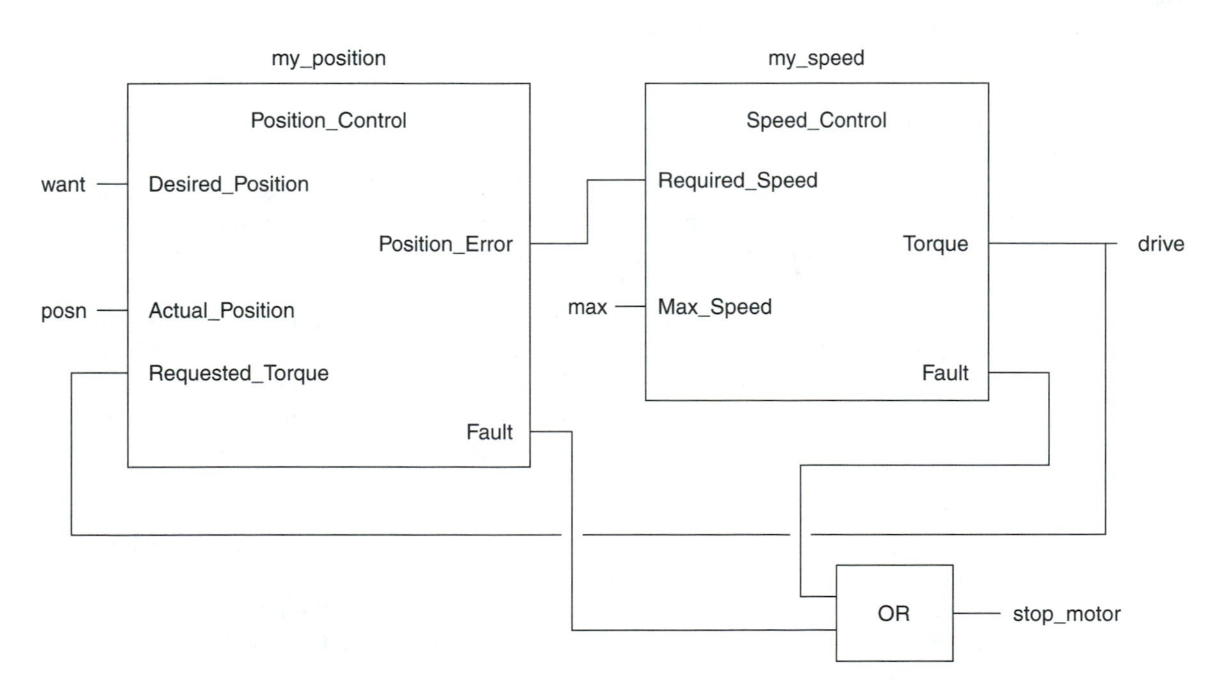

FIGURE 9.16
Function block diagram network with feedback.

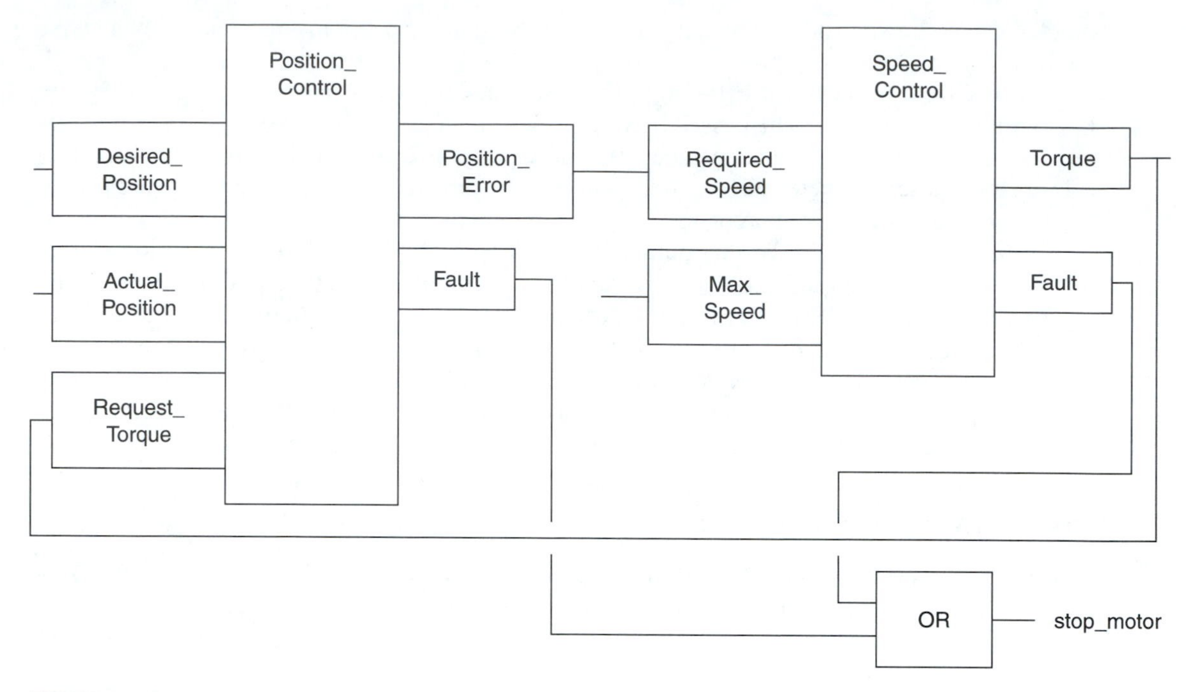

FIGURE 9.17
Continuous function chart network with feedback.

to some of the language they now use. The CFC language will include standard algorithms for process control and for motion control.

CFC algorithms are similar to FBD algorithms in that they consist of elements that can be shown with output variables from one connected to input variables in the other. Figure 9.17 shows how the FBD algorithm from Figure 9.16 might look in CFC (the input and output parameters for the CFC have been left off).

There will be a visible difference between FBD and CFC, because formal parameter names are shown in taglike attachments on the outside of the CFC element boxes instead of being shown inside the box. The external appearance of the variables reflects the one significant difference between the CFC language and the existing IEC 1131-3 languages. CFC programs will ignore the difference between functions and function blocks. The programmer won't have to declare an instance of a function block to get a program element that can use long-term memory to store parameters. The external tags on the graphical elements represent CFC program variables that can be retained as long as the control program continues to operate (or longer), which are not part of the function, but which the function can use.

It may be possible to develop CFC programming software that automatically creates a new instance for every CFC call of a standard function block, so that CFC will conform to the structure requirements of IEC 1131-3, but if the instances are invisible to the programmer, the ability to pass parameter values becomes severely limited.

SUMMARY

IEC 1131 has been renumbered as IEC 61131 but is still popularly known as IEC 1131. Part 3 (IEC 1131-3) defines a set of standards that will allow any PLC program to be downloaded into any PLC. PLCopen, an organization not officially associated with the IEC, is developing certification requirements that can be used to verify that programming software is compliant with IEC 1131-3 standards.

IEC 1131-3 describes five languages that can be used to program PLCs. Programming software does not have to offer all five to be IEC compliant. The five languages are allowed because of different historical PLC applications. Ladder logic (LD) is a standardized variation of the language that grew out of relay logic diagrams in North America. Instruction list (IL) is similar to assembler languages and is most popular in Europe. Structured text (ST) is similar to Pascal and C, so it will allow "mainstream" programmers (those who haven't worked with PLCs) to program in a language that is similar to those they use now. Sequential function chart (SFC) is a relatively recent graphical language for programming sequenced control programs. Function block diagram (FBD) is a graphical language favored by electrical engineers and the designers of process control systems. There is pressure to add a sixth language, continuous function chart (CFC), to the standard. CFC is common in the process control and motion control industries.

All five languages must facilitate the structure that IEC 1131-3 imposes on programs to make them portable between PLCs. Besides simply standardizing the instruction set, IEC 1131-3 requires variable declaration and parameter passing to allow better control over which program has the rights to modify data that may also be used by other programs.

IEC 1131-3 describes program organizational units (POUs). Every POU has an algorithm section and a variable declaration section. The variable declarations define formal parameter names for values that can be used in the algorithm, specify a data type for each variable, and indicate whether this POU requires a parameter to be passed for this variable when this POU is called from another POU. Variable declaration and parameter passing allow each POU to have exclusive rights to the data it needs, yet allows copies of data items to be passed between POUs. Some data elements can be declared at a higher level, which allows them to be used globally, by several POUs.

There are three types of program organization units. A function generates only one output parameter, and its output value is always the same when it is supplied with the same input parameter values. A function block needs permanent storage for some of its data because it can be programmed to use remembered data from previous executions. To indicate which set of data a function is to use, instance names are declared for each function block and its unique data set, and function blocks are called by the instance names. A program should contain calls of function blocks and functions, so programs should be used to organize the order and conditions under which functions and function blocks are called.

Function blocks and functions can call each other, but a program cannot be called by any POU. Instead, the programmer can declare a program as associated with a task and must declare the conditions that must exist for a task to call the programs. Tasks can call programs to execute in response to a sensed condition, or to execute at repeated timed intervals or, if the program is not declared with an initiating task, it will execute every scan cycle. Tasks can be assigned priority levels so that low-priority tasks will pause to allow higher-priority tasks to execute. A single PLC is called a resource and can be configured to execute multiple tasks. Resources are typically components in a larger control system, complete with interresource communications capabilities. The whole control system is called a configuration. Figure 9.18 is a copy of Figure

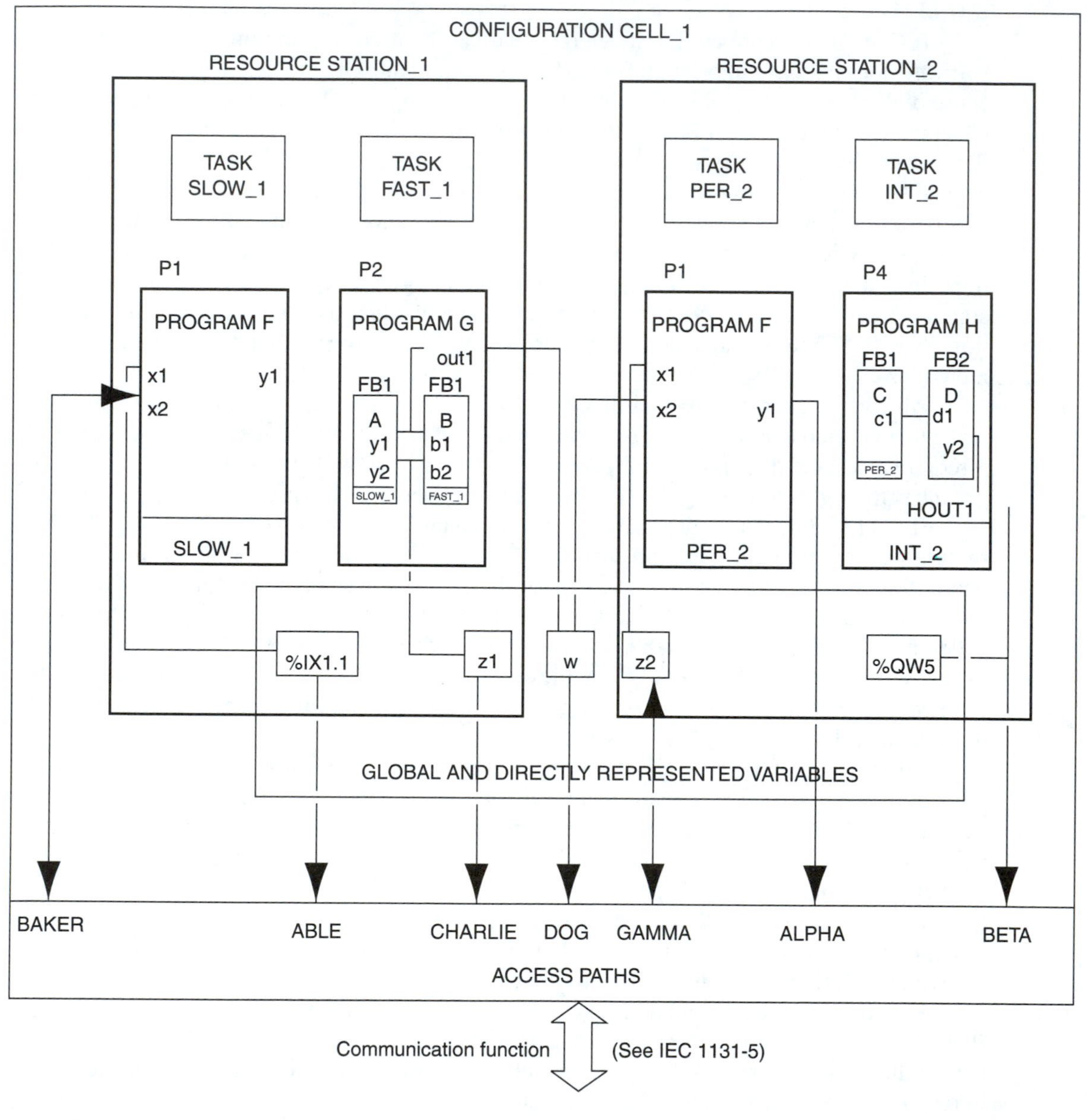

FIGURE 9.18
IEC 1131-3 software model.

19b from IEC DIS 1131-3, which shows the encapsulation of a control system into a sample configuration.

TROUBLESHOOTING

IEC 1131-3 does not define any required tools for determining why a program doesn't work. In fact, since functions and programs release their memory when they finish executing, the programmer can't always examine memory to find when and where an incorrect result was generated. When used carefully, the program structure that IEC 1131-3 requires should help avoid programming problems, because new programs should consist almost entirely of calls to old, tested functions and function blocks. IEC 1131-3 also requires that the programming unit should detect mismatches in data types (e.g., passing a real number to a function that needs an integer), so these types of problems should become a thing of the past.

Until the standard is fully understood, and until PLCopen (or someone else) can impose compatibility certification, the troubles that a user can expect are the troubles that IEC 1131-3 is attempting to stamp out: *incompatibility.* You may find that you have written a program in instruction list (conforming perfectly to the IEC 1131-3 standard) but when you try to download it to another PLC (which also conforms to the standard), it can't be downloaded because that other PLC can't translate your instruction list program into a language it recognizes! Other problems might be discovered when you try to upload a program from a PLC to an IEC 1131-3 programming unit, only to find that the compiled program in the PLC's memory can't be decompiled because not enough of the required information was downloaded by the programming unit on which the program was originally written. IEC 1131-3 does not define what information has to be compiled and downloaded into the PLC's memory beyond the understanding that the program must execute as programmed.

PLCopen's existing certification requirements include the requirement that manufacturers who claim that their products are in compliance with IEC 1131-3 must also explicitly state which sections of the standard they don't comply with. The IEC 1131-3 standard includes many tables with numbered sections. Each numbered section defines one component of the standard. Before purchasing a PLC or programming software, the purchaser should request a statement of compliance and then interpret what that statement says. PLCopen has an Internet Web site (http://www.plcopen.com) where they maintain information on the development of standards. The purchaser will also require a copy of the IEC 1131-3 standard, which can be purchased from IEC or from your national standards organization, or can be downloaded from Rockwell Automation's Allen-Bradley FTP site on the Internet (at ftp//ftp.cle.ab.com/sts/iec/sc65bw67 tc13). Remember that IEC 1131 became IEC G1131 after 1997, and it will be some time before the numbers are changed everywhere. (Marketing people have been advertising "IEC 1131-3" compliance for a couple of years now, and they may not want to change at all.[12])

---

[12] You know that RS-232C port that most computers have? Well, it is actually an EIA-232D port and has been since the early 1990s, but computer marketing departments haven't changed their literature yet.

BIBLIOGRAPHY

Brendel, W., and Tiegelkamp, M., *Uniform PLC Programming in Accordance with IEC 1131-3,* PLCopen, The Netherlands.

Lewis, R. W., *Programming Industrial Control Systems Using IEC 1131-4,* Institute of Electrical Engineers, London, 1995.

IEC Subcommittee 65B/WG7, *IEC DIS 1131-3,* IEC, Geneva, 1993.

IEC Subcommittee 65B/WG7/TF3, *Committee Draft: Amendments to IEC 1131-3,* IEC, Geneva, 1997.

QUESTIONS/PROBLEMS

1. Is compliance with IEC 1131-3 compulsory? Does the IEC test and certify compliance?
2. **Configuration** declaration statements describe the system of PLC __________ that will work together and the __________ they will share. **Resource** declaration statements identify __________, list the __________ and __________ that the resource must execute, set up __________, and declare __________ that can be within the resource. **Task** declaration statements define the conditions under which the operating system should ________________________________. **Program** declaration statements indicate a unique name for __________ of use of the program template and the name of the __________ that initiates execution of this program. **Function block** declaration statements identify a name for the __________ of the function block, the __________ that is to initiate the function block, and a list of __________ to pass to this instance of the function block. **Functions** always generate __________ that is returned to the program that calls the function.
3. What is an overloaded function?
4. How does a function differ from a function block in how it uses memory? Use the word *instantiation* correctly in your answer.
5. What are global variables, access variables, temporary variables, static variables, input variables, and in-out variables?
6. Identify the five current languages specified as IEC 1131-3 languages, with a sentence describing each.
7. What is an action qualifier as used in an SFC-language program? How is a step with an S action qualifier different from a step with a standard N action quantifier?
8. How does an SFC exclusive divergence differ from a standard (nonexclusive) divergence?

10

PLC SETUP AND CONFIGURATION

OBJECTIVES

In this chapter you are introduced to:

- How to assemble an Allen-Bradley, Siemens, or OMRON PLC system, complete with remote I/O, proprietary local area network, and programming unit. DIP switch and jumper setting are included.
- The configuration that is required when the programming unit, and then the PLC system, is set up for a user-selected application.
- How to program a PLC to initialize its configuration as it starts up and to modify its own configuration as it runs.

In this chapter we do not discuss connecting power supply lines to a PLC or between PLC components. For safety reasons you should follow the installation instructions supplied with your PLC.

INSTALLING AND CONFIGURING A NEW PLC

When you open the boxes containing your first newly purchased PLC system, you should find:

1. *Manuals* describing how to install, configure, program, and troubleshoot the equipment you have purchased. *Familiarize yourself with these manuals before interconnecting or providing power to the PLC system.* In this chapter we prepare you in general for installing a PLC system. Your manual set tells you exactly what you have to do.

2. A *CPU* module, sometimes called a processor or a PC.

3. A *programming unit.* This will probably consist of software from the PLC manufacturer, which you must install onto a personal computer. Some manufacturers sell complete programming units that are industrial-hardened personal computers with the software already installed. Some PLC manufacturers sell pendant programming units that are adequate for programming very small PLC controllers. In this chapter we assume that you have programming software that is to be installed on a personal computer.

4. *Interface hardware* to connect the CPU module to a *programming unit.* This may simply be a serial communications cable or may include an RS-232C/current-loop converter. The hardware may include a communication card for installation in your personal computer for high-speed programming and program monitoring, or for connecting to a local area network containing multiple CPU modules.

5. A *power supply* module (unless your PLC is an integrated type, with the power supply built in).

6. *I/O modules* specific to your application, unless your CPU module has enough built-in I/O capability to control the small system you need to control. In this chapter we assume that you have a modular PLC system with digital, analog, and intelligent I/O modules.

7. A *bus system* to connect I/O modules to the CPU module. In this chapter we assume that your PLC system has a central rack (containing the CPU and some I/O modules) and remote racks of I/O modules. Increasingly, small groups of sensors and actuators are being connected into PLC systems via sensor/actuator buses or fieldbus local area networks. Sensor/actuator buses and fieldbuses are discussed in Chapter 15.

INSTALLING THE HARDWARE

Some components may have DIP switches or jumpers that need to be set before assembling the components. If the PLC has remote racks, they must be connected to the main rack. Finally, the programming unit must be connected to the CPU or to a LAN containing several CPUs.

ALLEN-BRADLEY PLC-5 HARDWARE INSTALLATION

PLC-5 Chassis Settings

PLC-5 modules plug into a rack that Allen-Bradley calls a *chassis.* A jumper in the chassis must be placed to indicate whether the power supply module is mounted externally on the end of the chassis or is a module that will be plugged into the chassis. DIP switches in the chassis must be set to indicate:

1. The **type of addressing** that modules in this rack will use:
   (a) **Two-slot addressing.** A pair of adjacent slots is called a *group* and is represented by one 16-bit word in the input image table and one 16-bit word in the output image table. A 16-slot chassis is needed for one complete eight-group logical rack. Each group of two slots can contain:
   - Two 8-bit I/O modules, in which case the left-hand slot (slot 0) is represented by the low byte of the 16-bit words in the input image or output image table. The right slot (slot 1) is represented by the high byte of the I/O image word.
   - One 16-bit input module and one 16-bit output module.
   - One 16-bit input module and one 8-bit output module.
   - One 8-bit input module and one 16-bit output module.

- One single-slot block transfer module and one 8-bit input or output module.[1]
- Two single-slot block transfer modules.
- One two-slot block transfer module.

(b) Single-slot addressing. Each slot is represented by a 16-bit input image word and a 16-bit output image word. Every group has only slot 0; there is no slot 1. Only eight slots are required in a rack to be considered a full logical rack. A 16-slot chassis would be considered two logical racks. A single slot group can contain:

- Any 8- or 16-bit I/O module or a block transfer module (including a two-slot block transfer module, which is addressed as if it were in the leftmost of the two slots it will occupy).
- 32-bit I/O modules! Each pair of adjacent slots must contain no more than one 32-bit input module and one 32-bit output module. The PLC-5 will automatically assign the lower-numbered I/O image addresses for the low 16 bits of each module and the higher I/O image addresses for the high 16 bits of each module.

(c) 1/2-slot addressing. Each slot is represented by two 16-bit input image words and two 16-bit output image words. The low 16 bits of I/O from the module are represented by the word with the lowest address. Only four slots are required for a complete logical rack. A 16-slot chassis would be considered four logical racks. Each slot could contain any of the I/O modules listed for one-slot addressing. If I/O modules with 32 input contacts and 32 output contacts become available, one could be installed in each slot of a chassis configured for 1/2-slot addressing.

2. How output modules in this rack will affect their **outputs if a hardware fault is detected.** Options include to **reset all outputs** (usually the safest option) **or to retain their last state.** Even if the "retain last state" setting is used, there are some situations in which the outputs will be reset in the event of a hardware fault. See your manual.

3. Memory-control options, if this chassis will contain a CPU module. Options include:

(a) Allow or disallow clearing of CPU memory from a programming unit.

(b) Always transfer EEPROM data to memory at reset, or never do so, or do so only if the CPU's memory is detected as faulty (e.g., if battery backup has failed).

4. Restart options, if this chassis is to be a remote I/O rack complete with a remote I/O adapter module instead of a CPU. Options include allowing the CPU module to reset this rack after it faults or requiring a switch connected to this chassis to be used in resetting this chassis.

PLC-5 CPU Module Settings

DIP switches on the side of the PLC-5 CPU module are used to:

1. Set a **data highway plus address** if this CPU is to be included in a DH+ local area network.

2. Configure the **serial interface channel.**

3. Select **scanner mode** or **adapter mode** for this CPU module, and if adaptor mode is selected:

---

[1] The CPU and a block transfer module use 8 bits in the I/O image table during block transfers of data, so you can't install a 16-bit I/O module in the same group as a block transfer module.

(a) A rack number and the first I/O group number
(b) The size of the data unit to exchange with the scanner mode CPU

(See Chapter 13 for more details on the data highway, serial communications, and remote I/O scanning.)

PLC-5 Remote I/O Adapter Settings

Racks of remote I/O must have either a CPU module configured as an adapter module, or a 1771-ASB remote I/O adapter (RIO) module in the leftmost slot, where a CPU module would otherwise reside. DIP switches on the side of a 1771-ASB remote I/O adapter module are used to:

1. Assign a rack number and first I/O group number.
2. Select whether or not to ignore (not scan) the last four slots in this chassis.
3. Select a serial communication speed for exchanging data with the scanner mode CPU.
4. Select whether this chassis is to be treated as a primary chassis or as a complementary chassis. If another chassis contains an RIO module configured as a primary chassis, this RIO module can be configured as a complementary chassis with the same rack and I/O group numbers. This allows unused I/O image space to be used. If, for example, the primary chassis has an input module providing data for I:123 but not an output module to receive data from O:123, a complementary chassis can contain an output module to receive that data. (The complementary chassis can also have input modules to complement output modules in the primary chassis.)

PLC-5 Block Transfer and Intelligent I/O Modules

Some I/O modules have DIP switches and jumpers that have to be set before installing the module into a chassis.

- Analog input modules, for example, may have jumpers to configure the module for voltage or current input signals.
- Analog output modules may have jumpers that can be used to cause the outputs to go to minimum, maximum, or to the middle of its range, or to hold their last value when the PLC program stops running.

Block transfer modules and intelligent I/O modules come with their own manuals. See those manuals for specific requirements.

PLC-5 Programming Unit Communication Interface Card

For faster data exchange than is possible through a serial port, Allen-Bradley offers a selection of communication interface cards for personal computers. (Most have "KT" in their identification numbers.)

While "plug-n-play" technology is eliminating the need to perform hardware settings on interface cards for modern computers, most interface cards still have DIP switches that are used to:

1. Select an address for the communication card so that the personal computer can read and write data through the card to the PLC's CPU module. If you can use the default address setting on the card as shipped, you won't have to change the default configuration of your programming software to use the interface card.

2. Select which interrupt the card will generate to interrupt the personal computer. If the personal computer already has an interface device that generates the default interrupt, another interrupt must be selected. Two Allen-Bradley communications cards can share the same interrupt. "No interrupt" can be selected, although it isn't recommended because the user must then ensure that the personal computer is programmed to poll the interface card repeatedly to receive data from the PLC.

Connecting Remote I/O Racks to the PLC-5 Main Rack

Simply daisy-chain[2] channel 1B of the PLC-5 scanner-mode CPU to the remote I/O adapter modules or adapter-mode CPUs in the remote chassis of I/O modules. Allen-Bradley recommends Belden Blue cable. Connect an 82-ohm terminating resistor across contacts 1 and 2 at the last remote chassis along the daisy chain. (And you thought this was going to be hard. Wait until you see how easy it is to configure the scanner-mode CPU to use remote I/O.)

Channel 1B is the default scanner channel. You can use another channel if you want, but you will have to change the default CPU communication port configuration, too, as we will cover later.

Interconnecting PLC-5 CPUs Into a Data Highway Plus (DH+)

Simply daisy-chain channel 1A of the PLC-5 CPU module to the same channel of the other PLC-5 CPUs. Allen-Bradley recommends Belden Blue cable. Install an 82-ohm termination resistor across connections 1 and 2 at both extreme ends of the daisy-chained connection of the DH+.

Channel 1A is the default DH+ channel. You can use another channel if you want, but you will have to change the default CPU communication port configuration, too, as we will cover later.

Connecting the PLC-5 Programming Unit

Install the communication card into a slot in the personal computer. Plug the programming-unit interface cable into the card (or into an RS-232C connector if you didn't purchase a communication card). Connect the other end of the communication cable into the CPU module at the plug closest to the DH+ connector. (The DH+ connector is in channel 1A now; if you change the communication configuration to move the DH+ interface to channel 2, you must move the programming unit connection after making the DH+ change.)

---

[2] In a daisy chain the master module and the first remote module are connected together, then another identical cable is attached to the same contacts of the first remote module and attached to the second remote module. Then connect from the second to the third, and so on.

SLC 500

ALLEN-BRADLEY SLC 500 HARDWARE INSTALLATION

SLC 500 CPU Module Settings

Jumpers on an SLC 500 CPU module are placed to select either 120 or 240 V ac as the input power and to indicate the size of the EEPROM module installed in the CPU module.

Connecting SLC 500 Remote I/O Racks to the Main Rack

An SLC 500 modular PLC, with a CPU module, can still be seen as a remote I/O rack by a scanner-mode PLC if it has a 1747-DCM module. The scanner-mode PLC can be a PLC-5 or may be an SLC 5/02 or higher with a 1746-SN module. Remote I/O connections to SLC 500s are made by daisy-chaining 1747-DCM modules to a 1747-SN module.

Interconnecting SLC 500 CPUs into a Data Highway 485 (DH-485) or Data Highway Plus (DH+)

An SLC 5/04 CPU module can be connected directly to a data highway plus (DH+) network, with PLC-5 CPUs and other SLC 5/04 modules, by daisy-chaining the CPUs together using the three-wire DH+ connections. Channel 1 is the default DH+ port.

Any SLC 500 CPU can be connected into a data highway 485 (DH-485) by connecting each CPU module to its own 1747-AIC module, then daisy-chaining all the 1747-AIC modules together using the four-wire DH-485 connection. At each end of the daisy chain, terminals 5 and 6 in the connector must be connected to provide terminating resistance. At one end (not both) of the daisy chain, terminals 1 and 2 should be connected to connect the cable shield to ground. Channel 0 is the default DH-485 port in an SLC 5/04.

You can connect an SLC 5/04 to a DH+ via one port, and to a DH-485 via its other port, so that the SLC 5/04 can transfer messages from one network to the other.

Connecting the SLC 500 Programming Unit

There are several ways to connect a programming device to an SLC 500 or to a network containing SLC 500s. The most common method is to connect the SLC 500 to the serial port of a personal computer containing programming software using a 1747-PIC interface cable. For higher-speed exchanging of data, you can purchase a communication interface card from Allen-Bradley and install the card in your personal computer. Read the section on the PLC-5 programming unit communication interface card if you have a communication interface card to install. Your SLC 500 manual set includes details on other connection methods.

S5

SIEMENS S5 HARDWARE INSTALLATION

S5 Chassis Settings

In the S5-100U series PLCs, the power supply, CPU module, bus units, and (optionally) an interface unit are hung on a DIN rail. The power supply must be wired to the CPU and to the bus units, but the other components are connected by plugging ribbon cables from one module to the next. Before plugging an I/O module into a bus unit, a dial switch in the bus unit must be turned to the orientation matching a molded-in post on the back of the I/O module, so that you can't accidentally plug the wrong type of I/O module into a slot set for a different I/O module type. The switch setting also tells the CPU how much data to read/write at this slot.

In the S5-115U series of PLCs, all modules are plugged into a Siemens backplane. A mechanical **coding element** can be installed onto the backplane so that only one type of I/O module can be inserted into the backplane at that position, but the coding element serves no electrical function.

S5 I/O Modules with Interrupt Capability

Some S5-115U digital input modules with interrupt signal-generating capabilities can have their interrupt capability disabled by removing a jumper before inserting the modules into the backplane.

S5 Intelligent I/O Modules

Intelligent I/O modules have their own computational capabilities. The definition of intelligent I/O modules used in this book includes Siemens' intelligent I/O (IP) modules and Siemens' analog I/O modules. Communication processing modules and interface modules are also intelligent but are discussed in the next section. Switches on intelligent I/O modules and analog I/O modules are used to select such things as:

- Analog signal types and ranges (sometimes selected by inserting the appropriate measuring range module)
- Binary number format to assume in converting between binary and analog
- How many of the module's analog input channels are in use (to allow faster conversion if fewer signals are converted)
- Enable/disable checking for wire breaks in an analog signal circuit
- AC supply frequency for analog input modules (to allow ac noise at that frequency to be filtered out)
- Interfacing or compensation requirements for some sensor types
- Timer or counter module preset values
- Whether pulses from incremental encoders are to be counted as 1X (count positive-going signals at one channel only), 2X (count all signal changes at one channel), or 4X (count all signal changes at both channels)

Connecting S5 Expansion Unit Racks to the Main Rack

There are two ways to connect expansion unit racks of I/O modules to a central rack containing an S7 CPU module:

1. In a *centralized rack system,* total cable length must be 2.5 meters or less. Cables interconnect interface modules (IMs) in each rack, including the central rack. Each IM module must be assembled as the rightmost module on the rail or backplane. The central rack has to have a power supply, but the expansion unit racks don't.

2. In a *distributed rack system* (not available for the S5-100U), expansion unit racks and a central rack with an S5-115U CPU module can be in a system up to 3 kilometers long (depending on the IM module selected). Up to 64 racks can be interconnected. Each rack (including the central rack) must have an interface module (IM), usually as its rightmost module. Only centralized system IM modules can be mounted to the right of distributed system IM modules.

Each of the racks interconnected into a distributed system can also be part of a centralized rack system, as shown in Figure 10.1. The distributed system IM module can be in any of the

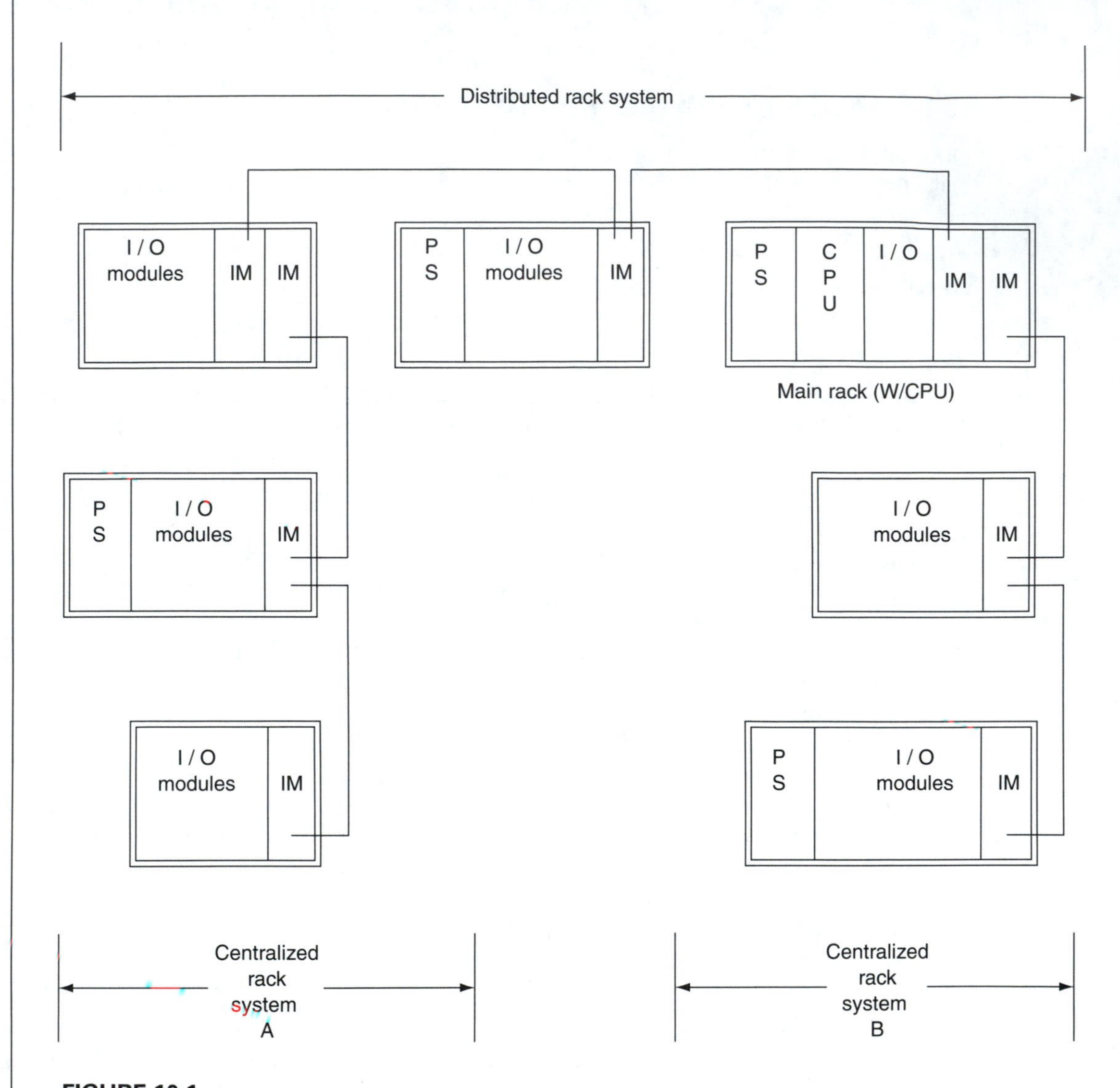

FIGURE 10.1
Combined S5 distributed and centralized system containing one central rack and several expansion unit racks.

racks of a centralized system. Only one rack of a combined distributed–centralized system needs a CPU module, of course.

If you need only one expansion unit, a preassembled component consisting of two IM modules and a cable can be used. If more expansion units are necessary, IM modules, cables, and terminators must be purchased separately, and the IM modules will have DIP switches and perhaps jumpers that must be set to:

1. Select a range of addresses for each expansion unit rack.
2. Indicate how much I/O each rack contains. Some modules require 32 bits of digital or 16 channels of analog I/O, while others are small enough for 16 digital bits or eight analog channels.
3. Enable one or two of the interface module's communications ports. (A terminator must be inserted in an unused port.)
4. Indicate the length of the longest connecting cables connected to a (distributed) IM module.
5. Indicate how to react if a "not-ready" state is detected at a (distributed) port.
6. Specify a unique interface number for each IM module (if the IM module has a 1-kilobyte memory to exchange data with the CPU and if the CPU is going to be configured for page frame communications).

Remote I/O racks can also be connected via a Profibus local area network (explained in a following section).

Interconnecting S5 CPUs into an L1 Local Area Network

An S5 CPU module can be connected to an L1 interface unit through the CPU serial port, also called the *programming port.* The interface module converts L1 electrical signals to programming port signals. L1 interface units are daisy-chained together using two twisted pairs of signal wire and a shield. The shield should be connected to ground at one end of the L1 network. An L1 network master is needed in the network, and it can be in a personal computer or an S5-115U PLC. S5-100U PLCs can only be slaves on an L1 network.

Interconnecting S5 CPUs into a Profibus Local Area Network

There are three types of Profibus network, so connecting to a Profibus network requires the correct interface. The original is often just called **Profibus,** but sometimes the letters "FDL" or the descriptor "part 1" are used to differentiate it from the others: **Profibus-DP** and **Profibus-FMS.** (The differences are explained in Chapter 13.)

An S5-115U CPU module can be connected into a Profibus-FDL or Profibus-DP network through an interface (IM 308-C) module. The S5-115U can be an active node, which means it can contain a program that commands data exchanges via the Profibus network. An S5-115U can also be the Profibus controller. A CP 5431 module acts as a master on either a Profibus-DP or a Profibus-FMS, performing control functions commanded by a user-program at the S5-115U CPU.

An S5-100U or an S5-115U CPU module can be connected into a Profibus-FDL or a

Profibus-DP network via a communication processor (CP 541) module. The CP 541 connects to the S5 CPU via the programming port, not via the PLC's bus system. S5-100U CPUs can only be passive components in a Profibus-DP network, which means they can only send/receive data in response to commands from active nodes on the Profibus-DP.

If you are using a Profibus-DP network so that a rack of S5-100U I/O modules can be used as slaves of another controller, then instead of paying for an S5-100U CPU and a CP 541, install the I/O modules into a distributed I/O module (e.g., an ET200U with a CP 318 Profibus-DP interface module). Distributed I/O modules are always passive nodes.

Other ET200 distributed I/O modules, with CP modules, can also be connected to a Profibus network. Interfaces are available to connect whole actuator-sensor interface (AS-I) sensor/actuator nets as Profibus-DP slaves. Personal computers with Siemens-supplied interface cards can also be connected and can be active components.

Communication processor (CP) modules have DIP switches and jumpers that can be set to:

1. Enable the CP module to exchange status and control information with the CPU module via communication flags in an area of the CPU's memory reserved for that purpose. The jumpers select which 32-byte area of the communication flag area the CP module will work with.

2. Specify a unique interface number for this IM module (if the IM module has a 1-kilobyte memory to exchange data with the CPU and if the CPU is going to be configured for page frame communications).

Connecting the S5 Programming Unit

A cable with built-in RS-232C to 4-to-20 milliampere (voltage to current) conversion is available to connect between a personal computer's RS-232C port and an S5 CPU module's programming port.

SIEMENS S7 HARDWARE INSTALLATION

Installing S7 CPU and I/O Modules

Other than mounting the S7 PLC into a control panel and wiring it to the ac supply and to sensors and actuators, there is very little hardware setup required for an S7 PLC system. Modules in the central rack simply plug together using connectors provided with the modules.

A memory module card can be plugged into some S7 CPU modules. A memory card contains flash EPROM memory and stores critical data while the power is off. (Battery-backed memory is also available, and some integrated function module CPUs contain EEPROM[3] mem-

---

[3] Siemens uses "EPROM" in its manuals when discussing the electrically erasable PROM type of memory that is more properly called EEPROM.

ory and therefore don't necessarily need a memory card.) Software configuration (covered later) tells the CPU what data is retentive and should therefore be stored on the memory card or in EEPROM. Part of a memory card can be programmed (using a separate programming device) with data that can't be changed by the CPU module. The operator can clear *almost* all the CPU's RAM memory and retentive memory contents[4] and reload the data from the memory card (or from EEPROM) by performing a **memory reset.** A memory reset is necessary if the CPU detects a problem with the contents of RAM or retentive memory.[5] To perform a complete memory reset, use the keyswitch on the CPU module as follows:

1. Switch the PLC to STOP, then
2. Switch to MRES and hold until the STOP LED flashes, then
3. Switch to STOP again, then
4. Switch to MRES again and hold until the STOP LED stops flashing.

Some analog I/O modules do need hardware configuration before being inserted into a PLC rack. These modules have a range selector module that may have to be removed, rotated, then reinserted to select a voltage or current analog signal range.

Wiring connectors are terminal-strip units that plug into the front of I/O modules. You must depress a connector lock on the I/O module to allow inserting the connector. Each I/O module comes with a coding key that is transferred to the first wiring connector that is inserted into the I/O module. The coding key then prevents that wiring connector from being inserted into a different type of I/O module. (You can remove the coding key from a wiring connector if necessary.)

Siemens numbers their S7 rack slots a little differently from how they are numbered in other PLCs.[6] S7 slot numbering starts at the very left with:

- Slot 1, which contains the power supply module.
- Slot 2 must contain a CPU, or it is considered empty.
- Slot 3 must contain an interface module (IM), or slot 3 is considered empty.
- Slots 4 to 11 can contain any arrangement of I/O modules.

Siemens offers three types of I/O modules:

1. **Signal modules (SMs)** for digital or analog I/O
2. **Function modules (FMs),** intelligent I/O modules with their own microprocessors for specialized process control
3. **Communication processors (CPs),** to handle connecting the PLC to communication channels other than those the CPU can connect to via its built-in serial ports

---

[4] Error messages in the diagnostic buffer, the MPI node address for this CPU, and the operating counter's contents cannot be cleared.

[5] A memory reset is also required before transferring a new program from a programming unit to a CPU. A programmer can initiate the memory reset from the programming unit.

[6] The slot that an I/O module is in affects its address, as we saw in Chapter 5. Some S7 PLCs allow the user to assign user-assigned addresses instead of using the slot-oriented addresses. We do not cover user-assigned addresses in this book.

S7 Distributed I/O System Installation

A distributed I/O system has a **central rack** that is connected to **expansion racks** over short distances so that a single CPU module can control more I/O modules. Each central rack and expansion rack must have an interface module (IM module). Local area networks such as Profibus can make more distant I/O modules controllable from a central CPU module but require more sophisticated communication processor (CP) module interfaces and are not discussed in this section.

Preassembled connecting cables are used to connect expansion racks (with IM modules) to a central rack (containing the CPU module and an IM module). Some S7-400 IM modules require setting a hardware switch to select a rack number for addressing purposes, but S7-300 IMs are self-configuring and don't even require terminators.

S7 Local Area Network Installation

Siemens offers three types of **local area network** interfaces:

1. Multipoint interface (MPI) networks, to which any S7-300 (or higher) CPU can be connected via its programming port. (Siemens calls a programming port an MPI port.) MPI connectors must be wired together in a daisy-chain configuration. Every MPI connector contains a termination resistor, but only the connectors at the ends of each segment of the daisy chain should have their termination resistors switched on. Before interconnecting CPUs via wired MPI connectors, each CPU and each programming unit must be configured with its own unique MPI address, as described in the section on first-time configuration of an S7 CPU.

One MPI segment can contain only 32 nodes, but several segments can be interconnected to allow as many as 126 nodes. RS-485 repeater units can be used to connect segments together. Adding an RS-485 repeater to a segment reduces the allowable number of nodes in the segment by one, but the repeater does not need an MPI address and does not reduce the number of addressable nodes in the whole MPI network.

To connect one S7 CPU into more than one MPI segment (or to other types of networks), the S7 PLC must have a communication processor (CP) module for each additional network connection. Other than installing the wiring, no hardware configuration is required.

2. Profibus networks, of which there are three varieties: flexible message service (FMS), distributed peripheral I/O (DP), and process automation (PA). Most S7-300 CPUs require Profibus CP modules to connect to Profibus networks, but S7-400 CPUs can be directly connected to a Profibus-DP network via their programming port, after they have been configured with unique Profibus addresses, as described later.

As with MPI networks, there can be only 32 nodes in a single Profibus segment, but RS-485 repeaters can be used to interconnect segments so that a total of 125 addressable nodes can be connected.

Two types of components can be connected to a Profibus network: **controllers** and **slaves.** A controller might be an S7-400 CPU or a CP module in any S7 PLC system. One controller in each segment must be designated as the **Profibus master** (node 1), to control use of the network by other controllers. Slaves are similar to expanded I/O racks in that they generally do not have CPU modules (although Profibus CP modules are available that allow an S5 PLC, complete with a CPU, to be a slave in a Profibus network). A slave can only respond to communication commands from controllers; it cannot initiate data exchanges. Some slave devices have DIP switches that are used to assign node numbers, but no other hardware configuration of a Profibus network is required.

3. Industrial Ethernet,[7] requiring CP modules. Industrial Ethernet is based on the IEEE 802.3 standard for interconnecting computers. Ethernet is the most common communication protocol for office networks and differs from Profibus and MPI (and from most PLC networks) in that it does not schedule each controller to have a turn (called a *token*) to communicate. Up to 2024 stations can be connected to a single Industrial Ethernet subnet, and each has the right to attempt to transmit whenever it finds the network is available (CSMA/CD protocol).

S7 Programming Unit Installation

Programming devices (including personal computers with STEP 7 software) can be connected directly via their RS-232C ports to an S7's MPI port using an MPI interface cable. Adapters are available so that the programming unit and an MPI network can share the single MPI port of an S7 CPU. A programming unit connected to a CPU that is connected into an MPI network can access the other nodes on the network.

For more permanent installations of a programming unit into an MPI, Profibus, or Industrial Ethernet local area network, the programming unit must have a *network interface card* installed. A Profibus network can only have one programming unit connected in this way. S7 network interface cards are "plug-n-play," so they do not require hardware configuration. New interrupt numbers and memory space allocations may have to be assigned for the card (via software, explained below) if the personal computer already uses the default interrupt and/or memory for other purposes.

OMRON CQM1 HARDWARE INSTALLATION

CQM1 modules plug into the sides of each other and are locked together by sliding locking tabs toward the back of the modules. The completed assembly is hung on a DIN rail by lowering the rear locking tabs on each module, hanging the assembly on the rail, then raising the tabs to lock each module to the rail.

The CQM1 is intended for use in small PLC systems. It does not allow adding racks of remote I/O, but CQM1s can be connected to communicate with each other.

CQM1 CPU Settings

The six DIP switches in the face of a CQM1 CPU module can be set before or after assembling the CPU into a PLC. The DIP switches allow you to select:

1. Pin 1, when off, protects program *memory* and the configuration memory in addresses DM 6144 to DM 6655 from being written to. If you want to modify the program or configura-

---

[7] Fast Ethernet (100 Mbps or better) has been selected as the backbone for the open fieldbus foundation standard controller network, to the chagrin of Profibus supporters.

tion, or if you want the PLC to load a program and configuration from a memory cassette automatically when it starts, you must turn this switch on. A backup battery must be installed to retain the data in these areas if protection is selected.

2. Pin 2, when on, causes the PLC to copy the contents of a *memory cassette* into RAM when it starts. Switch 1, of course, must be off and a cassette with program and configuration data must be inserted into the CPU module.

3. Pin 3, when on, causes the PLC to display an *English-language* message on the programming unit when it starts. If this pin is off, the message will be in Japanese.

4. Pin 4, when off, prevents the changing of the default expansion set *instruction* uses. Some expansion instructions are covered in the programming manual without function numbers [e.g., AVG(−)], while others do have function numbers [e.g., PULS(65)]. If you want to use an instruction without a default function number, you must have pin 4 on, and you must reassign one of the default numbers to that instruction before using the function number in a program [e.g., AVG(−) becomes AVG(65) and PULS(65) becomes PULSE(−)]. We will see how to reassign function numbers later.

5. Pin 5, when on, assigns a default protocol to the CPU's RS-232C port. The default is 7 data bits plus a bit for even parity, plus one start and two stop bits, at 9600 bps. When pin 5 is off, an alternative protocol, assigned during software configuration of the CPU, is used.

6. Pin 6, when on, holds AR 0712 on, and when off, holds AR 0712 off. AR 0712 is called the *DIP switch pin 6 flag* (for obvious reasons) but is not automatically used by the CQM1 for any purpose. The programmer can write a program to monitor this bit.

As the explanation of pins 1 and 2 imply, you can insert a memory cassette into a CPU module, and the memory cassette can contain a program and configuration data. Several types and sizes of memory chips are available for insertion into memory cassette. A pair of DIP switches on the memory cassette must be set to indicate the size of memory chip installed. A memory chip can either be an EPROM chip, programmed with a program and configuration from a CQM1 using a PROM burner, or an EEPROM chip. EEPROM chips can be written to by a CQM1 CPU module (setting AR 1400 causes the CPU to download its current program and configuration to the memory module). To prevent accidental overwriting of an EEPROM, the memory cassette has a *write-protect DIP switch,* which prevents overwriting when turned on. *Never remove or insert memory cassettes with EEPROM chips when the PLC's power is switched on.*

If a CQM1 is installed without a backup battery for retaining program and configuration data, program and configuration data should be on a memory cassette, CPU pins 1 and 2 should allow loading RAM from the memory cassette, and the programmer should be aware that the CQM1's other memory contents may not be dependable after restarting the PLC. The DM, HR, AR, and TC memory areas may not be cleared and **output OFF** bit (SR 25215) may not be off when it should be. The user-program should prepare these memories before using them, perhaps by using the BLOCK SET instruction, **BSET(72),** to clear the HR, AR, CNT, and the non-configuration portions of DM memory and by using the **always OFF flag** (SR 25314) to hold the output OFF bit off.

In some CPUs (CQM1-CPU42E) four *rotary pots* on the CPU module can be turned to adjust the contents of memories IR 220 to 223. The CPU will scan the pots on each scan cycle and put BCD numbers between 0000 and 0200 into memory to reflect their state. (The pot settings can be used for purposes such as manual settings for timer preset values.)

Connecting CQM1 I/O Modules

Only analog I/O modules require setting. *Analog input modules* need to have pairs of DIP switches set to select a voltage or a current range for each input channel. Another DIP switch selects whether two or four input image words will be reserved for analog input values, yet another selects whether input values will be averaged. *Analog output modules* have jumpers for each of their two channels. The jumpers enable or disable that channel's ability to output negative analog signals. Analog I/O modules are supplied with power from a power supply module that plugs into the PLC as if it were an I/O module and is then connected to the analog I/O module(s).

The CQM1 has an unusual I/O addressing system. Slots are numbered from the CPU module (which is slot 000). Input modules are assigned I/O image addresses 001 to 011, and output modules are assigned I/O image addresses 101 to 111. The first input image address assigned is always 001, the second is 002, and so on, no matter how many output modules are inserted between them. Output modules are similarly assigned addresses 101, 102, and so on, no matter how many input modules are inserted between them. Analog power supplies in the rack do not affect addressing.

Each digital I/O module is assigned 16 bits (one word) of I/O image memory, but analog I/O modules can have two or four words of I/O image, depending on how many channels they have.

Connecting the CQM1 Programming Unit

A connecting cable that includes a protocol converter is available to connect the CQM1's peripheral port to the RS-232C port of a personal computer running programming software. Alternatively, the user can build an RS-232C cable (specifications are given in the CQM1 manual set) to connect the CQM1's RS-232C port to the personal computer's RS-232C port.

Interconnecting CQM1 CPUs for Communication

CQM1 CPUs can be connected to peripherals (e.g., printers, bar code readers, etc.) via their RS-232C ports, or they can use the RS-232C ports to communicate with each other. Two CPUs can communicate in a point-to-point configuration if the three-wire RS-232C cable connects the signal ground/SG pins of both CQM1 RS-232C ports together and connects the RXD/RD[8] pin from each CQM1 to the TXD/TD pin of the other. Up to 32 CQM1s can be connected into a local area network, where they will act as slaves to a single host computer (not a CQM1) so that the host can perform communication control to allow the slaves to communicate. The signal ground/SG lines of the CQM1 ports and the host computer must all be connected together; the TXD/TD lines of the CQM1s must be connected together and to the host computer's TXD/TD line.

FIRST-TIME CONFIGURATION OF A PLC SYSTEM TO PREPARE IT FOR AN APPLICATION

If you have purchased a small PLC with a preconfigured programming unit or a hand-held pendant, you may simply be able to turn the power on and proceed immediately to program the

[8] The old RS-232C standard used such pin names as RXD and TXD. The RS-232C standard has been renamed IEA-232D, and new names (i.e., RD and TD) are now used for the same pins. Most users (but not OMRON) still use the old nomenclature.

PLC. More powerful and flexible systems require a few more steps before you are ready to program. You may have to:

1. Install the programming software onto your personal computer. In this book we say little about installing software.

2. Configure the software's offline programming characteristics. Offline programming allows a programmer to write programs and data files without being connected to a PLC. The programs and data can later be downloaded to a PLC.

3. Configure the programming software so that it can communicate with your PLC's CPU module and configure its online programming characteristics.

4. Customize the PLC's operating characteristics.

5. Configure the PLC CPU module to communicate with other controllers and remote I/O.

6. Configure intelligent I/O modules to work the way an application requires.

ALLEN-BRADLEY PLC-5 FIRST-TIME CONFIGURATION

Installing the PLC-5 Programming Software and Offline Programming

During software installation you will be asked if you want to enable or disable **privilege password protection.** If you answer yes, you will be able to set up a system of passwords to protect your programs, data, and configuration from being changed by unauthorized users. (If you answer yes, be sure to perform the configuration described in the section on configuring PLC-5 access privileges in this chapter before an unauthorized user does it and locks you out.) After running the installation program to load the programming software onto your personal computer's hard drive, start the programming software.

Allen-Bradley's 6200 programming software includes a *link mode* selection in its "display characteristics" configuration menu. Turning link mode on causes the personal computer to require an ACK signal from the PLC whenever the personal computer transmits a packet of data to the PLC. Turning link mode on will therefore slow the communications down, but will also make communications between the personal computer programming unit and the PLC more dependable and will help to avoid unsafe conditions that might exist if data packets are damaged during transmission.

Configuration of the programming software for *offline programming* involves selecting the type of PLC-5 CPU to which the program will be downloaded and selecting display characteristics such as how the programming unit will display the rungs, rung labels, symbols, and so on. Offline configuration is not discussed here.

Connecting to the PLC-5 CPU and Online Programming

For the programming software in the personal computer to communicate with a PLC, it's online programming characteristics must be set by selecting:

1. An alphanumeric *terminal name* for this programming terminal (optional).

2. The *communication port* of the personal computer to which the PLC is connected. This selection consists of:

(a) Identifying the type of communication card installed in the personal computer.

The programmer will also be asked to enter the communication card's address, as set on the card's DIP switches. (Allen-Bradley's ABHELP software can usually read the card's DIP switch setting if you haven't recorded it.) If you are using a serial port instead of a communication card, you will be asked to set the serial communication protocol of the serial port selected.

(b) Identifying the type(s) of data highway connecting this programming unit to the PLC-5 CPU that it connects to by default for online programming.

(1) Choose **DH+** to connect to a CPU that is connected to the same DH+ link as the programming terminal. (Although you have probably plugged the programming cable into a connector on a PLC-5 CPU module, it is actually connected to the DH+ that includes this PLC-5.) Figure 10.2 shows a personal computer connected to a DH+ link containing two PLC-5 CPUs, either one of which could be the default PLC for online programming.

As a *network access* option, select **Local** and enter:

- The PLC address (DH+ address) of the CPU module to connect to
- The terminal address (DH+ address) of this programming terminal

(2) Choosing **DH+** also allows you to connect to a CPU on another DH+ that is connected to this DH+ via a data highway. Figure 10.3 shows a personal computer connected to a DH+ link which is connected to another DH+ via a data highway. Any PLC-5 on either DH+ could be the default PLC for on-line programming. As a network access option, select **Remote (DH+),** then enter:

- The PLC address (DH+ address) of the CPU module to connect to

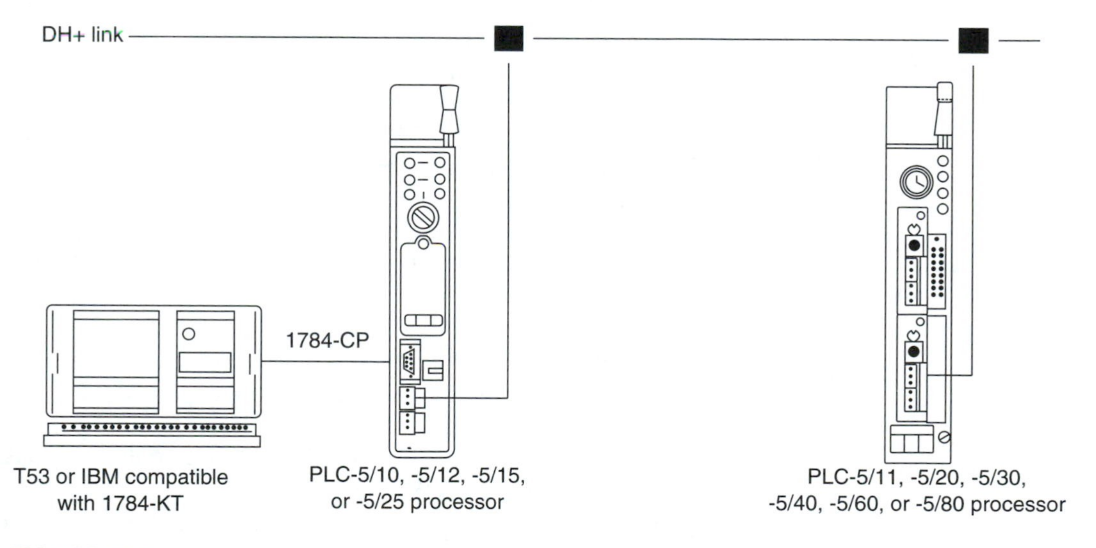

FIGURE 10.2
One programming unit and two PLC-5 CPUs on a local DH+.

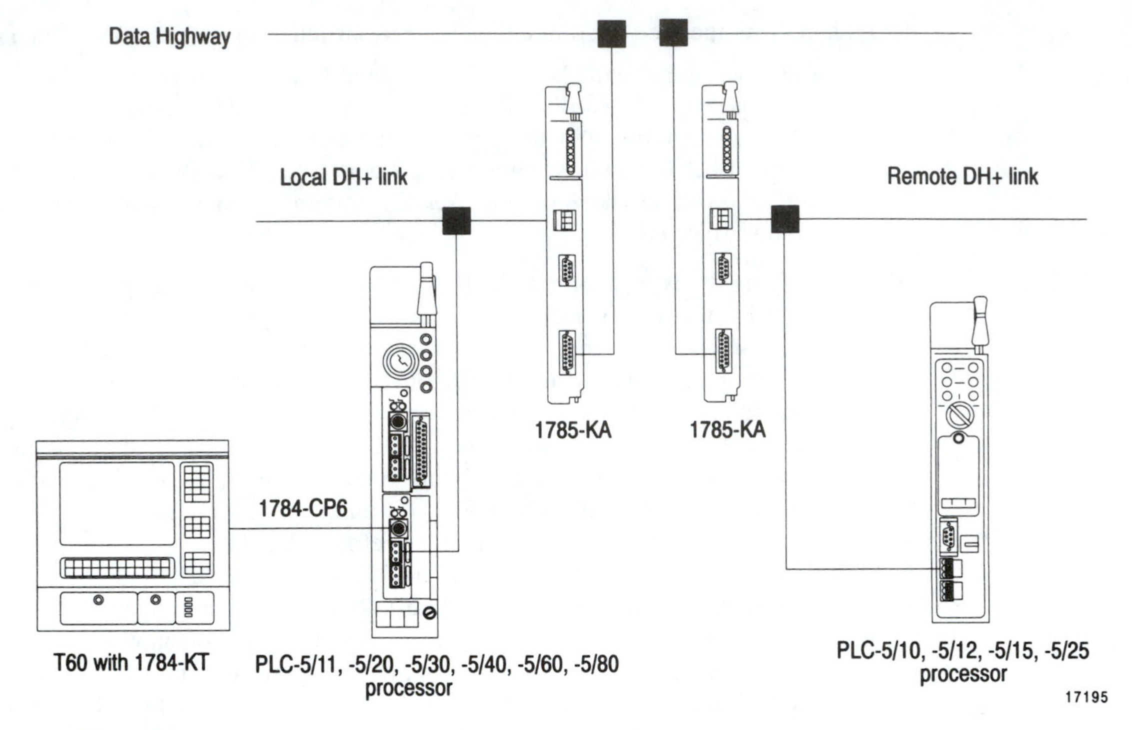

FIGURE 10.3
One PLC-5 programming unit, two data highway pluses, and a data highway.

- The terminal address (DH+ address) of this programming terminal
- The data highway (not the DH+) address of the 1785-KA/KE module in this network that acts as a local bridge link to the data highway system containing the remote DH+
- The data highway (not DH+) address of the 1785-KA/KE module in the remote DH+ that acts as a remote bridge to connect that DH+ to the data highway system

(3) **DH+ routing** allows connection to a PLC-5 if the electrical path to the other CPU includes an Allen-Bradley pyramid integrator backplane. You must also enter:
- The PLC address (DH+ address) of the CPU module to connect to
- The terminal address (DH+ address) of this programming terminal
- The local DH+ address of the bridge 5130-RM/KA module that links this DH+ network to the backplane
- The address of the destination link 5130-RM/KA module that links the remote DH+ network to the backplane

(4) **DH II** allows connection to another PLC-5 via a data highway II (DH II). You must also enter information identifying the:

- *Source (SRC) link,* including the address of the programming unit (**source user**) and the 1779-KP5 (**source node**) on the local DH+, and the address of the DH II (**source link**) through which the source node communicates to a destination node on the remote DH+. If the source node and destination node are connected via a single DH II, the source link address can be entered as 0. If other links are included between this DH II and a remote DH II, enter the actual address of the local DH II network.
- *Destination (DST) link,* including the address of the PLC-5 (**destination user**) and the 1779-KP5 (**destination node**) on the remote DH+, and the address of the DH II (**remote link**) through which 1779-KP5 communicates to the remote DH+. If the DH II links the destination node directly to the source node, its remote link address can be entered as 0. If other links are included between this DH II and a remote DH II, enter the actual address of the remote DH II network.

Configuring the PLC-5 CPU

Now that your programming unit has been told *how* to connect to a PLC-5, you can connect to the CPU module by selecting **online programming.** In online programming mode you can edit programs and change memory contents, but before doing so, you will want to customize the PLC's operating characteristics for an application by calling configuration screens and entering data.

CPU configuration data is stored in the *status file (S:)* area of memory as you enter configuration data using the configuration screens. (The structure of the status file area of a PLC-5's memory is shown in Appendix A. It also includes words reflecting the PLC's status.) PLC programs can read any of the data in the status file. Some configuration data is **dynamic,** which means that a PLC program can change it and the changes become effective immediately. Other configuration data is **static,** which means that changes become effective only when the PLC is switched into run mode, even if the changes are made while a user-program executes.

PLC-5s have backup batteries, so status file data including configuration changes will remain in effect (even if power is disconnected) until a programmer or a program changes the configuration again. When you save a program to a programming unit, you save the status file, and when you reload that program to a PLC, you also reload the configuration.

The **processor configure** screen for the enhanced PLC-5 CPU is shown in Figure 10.4. The classic PLC-5 status screen contains many of the same configuration options but looks different. As the programmer moves the cursor through the screen's fields, the actual status file addresses for the configuration option are displayed at the bottom of the screen and the programmer can enter new data.

The processor configuration screen gives the programmer access to configuration options, including the following (we do not attempt in this book to cover all configuration options).

Startup Configuration The low-order bits of the user control bits (in S:26) can be set or reset to control how your PLC will restart if power is restored with the PLC in run mode and each time the PLC is switched into run mode.

If you set bit S:26/1, then whenever the PLC restarts, it will set a major fault bit (S:11/5)

```
                              Processor Configuration

User control bits:        00000000 00000000    RESTART LAST ACTIVE STEP
Fault routine prog file no.:   0               Watchdog (ms):                   500
I/O status file:               N20             Communication time slice (ms):   100
VME Status File:               N34

Processor input interrupt    prog file no.:   0    module group:   0
                             down count:      0
                             bit mask:        00000000 00000000
                             compare value:   00000000 00000000

Selectable timed interrupt   prog file no.:   0    setpoint (ms): 0

Main control program A:      prog file no.:   0    disable: 0    I/O update: 0
                     B:      prog file no.:   0    disable: 0    I/O update: 0
                     C:      prog file no.:   0    disable: 0    I/O update: 0
                     D:      prog file no.:   0    disable: 0    I/O update: 0
                     E:      prog file no.:   0    disable: 0    I/O update: 0

Press a function key, page up or page down, or enter a value.
S:26/15 =
Rem Prog    Forces:None                          5/V40 File NP540V
         Proc       VME
        Status     Config
          F2         F4
```

FIGURE 10.4
Processor configure screen (enhanced PLC-5 CPU).

so that the PLC will immediately try to execute the fault routine. The fault routine's file number must also be entered (into S:29) using this configuration screen. If the fault routine clears the major fault bit, then when the fault routine finishes executing, the PLC will start executing standard scan cycles. If bit S:26/1 is not set when the PLC restarts, the PLC will start executing standard scan cycling immediately. Fault routines are described in Chapter 11.

Bit S:26/0 only affects the execution of sequential flow chart (SFC) programs. If S:26/0 is set when the PLC is restarted in run mode or when the PLC is switched into run mode, the PLC will start executing the scan cycle as usual but will resume executing sequenced operations programmed in SFC starting with the step that was active when the PLC stopped last. If S:26/0 is not set when the PLC restarts, the sequenced set of SFC steps will restart at the initial step. SFCs are described in Chapter 9.

Main Control Programs The PLC-5 can be configured to execute as many as 16 user-programs, called main control programs (MCPs), during each scan cycle, or to execute as few as zero. MCP file numbers entered using the processor configuration screen are stored in every third status file word, from the first MCP file (MCP "A"), whose number is in S:80, to the sixteenth MCP (MCP "P" is identified in S:125). After each MCP file number in the status file, the next word stores the duration of the most recent scan of the MCP, and the following word stores the longest scan time for the MCP.

The PLC-5 can be configured to execute, or not to execute, an output scan and an input scan after any or all of the MCPs. The 16 bits of word S:78 contain the I/O update bits for the 16 possible MCP files. Setting a bit in this word (e.g., bit S:78/0) disables the I/O update after the corresponding MCP (e.g., MCP "A").

The processor configure screen can be used to disable execution of any or all of the MCPs, without requiring their removal from the configuration. The 16 bits of word S:79 are used. If a bit is set, the corresponding MCP will not execute.

Watchdog Timer As each scan cycle starts, the PLC-5 starts a watchdog timer. If the timer runs out before the scan cycle ends, the PLC will fault (see Chapter 11 for more detail). A programmer can use the processor configure screen to enter a time, in milliseconds, for the watchdog timer. This value is placed into S:28. (As each scan cycle ends, the PLC-5 writes the time the scan took to complete into S:8.)

Interrupts Three types of interrupts can be configured using the processor configuration screen. All are described in detail in Chapter 11.

1. *Fault routine* **file numbers** can be entered (into S:29).
2. A *selectable timed interrupt (STI)* **file number** can be entered (into S:31), and an **interval time** for executing the STI file can be entered (into S:30).
3. A *processor input interrupt (PII)* **file number** can be entered (into S:46). The programmer can enter a **module group number** and a **bit mask** (S:47 and S:48, respectively) to configure which input module and which contacts can be interrupt sources, and can enter a **compare value** and a **down count value** (S:49 and S:50, respectively) to configure the type of bit transitions and to indicate how many are required before an interrupt occurs.

I/O Status An I/O status file number should be entered using this processor configuration screen if remote I/O racks will be scanned by this PLC-5 module. (The CPU will have been set up in scanner mode using the DIP switches.)

Entering a file number (e.g., 13) reserves an integer file (N13) for the PLC-5 to store I/O status pertaining to I/O racks. The I/O status file will contain **two 16-bit data words for each I/O rack** in the PLC-5 system (N:13.0 and N:13.1 for rack 0; N:13.2 and N:13.3 for rack 1; etc.). The first data word contains status information, and the second is used to configure and control the I/O rack. (Configuration and control data for I/O rack 1 will be placed into S:13.3.)

Within the data words, each data bit contains status or configuration information for **two I/O groups,** so 4 bits are needed to configure or to control a standard eight-group I/O rack:

1. Bits 0 to 3 of a rack's configuration word are used to inhibit I/O groups when the bit is set (e.g., if S:13.3/0 is set, groups 0 and 1 of rack 1 are inhibited and cannot provide input image data to the CPU or accept output image data from the CPU). Normally, entire racks would be inhibited at a time, not just two groups.
2. Bits 8 to 11 of a rack's configuration word are used to reset I/O groups if there has been a fault. As we discussed above, DIP switches at the rack can set to allow the CPU to provide the reset or to require a manual reset (e.g., setting bit S:13.3/9 resets I/O groups 2 and 3 of rack 1). Normally, an entire rack would be reset at a time.
3. Bits 4 to 7 and 12 to 15 are not used.

The status word for a rack contains bits indicating that I/O modules are present (bits 8 to 11) and that an I/O rack has faulted (bits 0 to 3) (e.g., if bit S:13.2/2 is set, there has been a fault in rack 1). Any fault in an I/O rack causes all 4 fault bits to be set.

Communication Time Slice The PLC-5 system performs serial communication with other PLC-5s via the DH1 system, with remote intelligent I/O modules via the remote I/O scanning system, and with other devices via other serial ports. Communication-handling chips do most of the work, but the CPU's main processor must "service" the communication handlers (issue communication requests and exchange data) during a housekeeping step in the I/O steps of the scan cycle. (See Chapter 13 for more detail.) Communication tasks may be ordered by commands in the user-program or may be in response to requests from another PLC-5, from an intelligent I/O module, or from another peripheral device, so the amount of housekeeping time needed per scan cycle may vary.

If a time of 0 is entered for the communication time slice, the PLC-5 will perform all the communication tasks required per scan, so scan time will be made more variable. If a time of other than 0 (milliseconds) is entered, the PLC-5 will use that amount of time (plus the basic 3.1 milliseconds always required for housekeeping) for communications servicing every scan cycle, regardless of how many (or how few) communications tasks are actually required.

Configuring PLC-5 Communication Channels

The PLC-5 has several communication ports with connectors on the faceplate. Communication port configuration is easier in online programming but is also possible in offline programming. To change configuration, the programming unit's channel overview screen, shown in Figure 10.5, must be used. It shows the default uses for each channel, allows the operator to change the defaults, and allows the programmer to select a channel for further configuration or for monitoring of its communication status.

Communication configuration data is saved when a program is saved. When the program is downloaded to a CPU module, the communication configuration data is also downloaded. Communication channels can be configured for:

1. *Data highway plus (DH+).* Channel 1A is the default DH+ channel. If you must change channel 1A's purpose, configure another channel for DH+ and connect the programming unit to the other channel first. More than one channel can be configured as DH+ channels, so that a PLC-5 can be used as a link between two DH+ networks.

Configuration of a channel for DH+ involves entering the addresses of two unused integer files, one to store a 40-word **diagnostic file** for DH+ communications at this node, and the other to store a 64-word **global status flag file** for the DH+ network. Channel 1A's DH+ **baud rate** must be 57.6 kilobaud, but 230 kilobaud is possible through channel 1B or 2B. The node address of a PLC-5 connected via channel 1A is fixed by the DIP switch settings, but a node address can be entered for a DH+ channel other than 1A. If a DH+ is going to be connected into a system with more than one DH+ networks, each DH+ must have a unique link ID number, and you must enter that **DH+ link ID** when configuring the channel. All nodes connected to the same DH+ must be configured with the same link ID number, of course.

2. *Remote I/O scanning.* Remote I/O scanning allows a PLC-5 that is set up as a scanner-mode CPU (using the DIP switches) to read input data from, and send output data to, remote I/O adapter modules or PLC-5 modules set up as adapter-mode CPUs.

Channel 1B is the default remote I/O scanning channel. Further configuration involves specifying an unused integer file to use as a **diagnostics file,** specifying a **baud rate** for the

```
                                Channel Overview


  Channel 0:     SYSTEM (POINT-TO-POINT)

  Channel 1A:                        DH+
  Channel 1B:               SCANNER MODE
  Channel 2:           EXTENDED LOCAL I/O


  Channel 3A:                        N/A







 Press a function key or enter a value.
 >
 Rem Prog    Forces:None                             5/40L File BATCH
 Accept   Channel   Node          Channel        Channel                Select
  Edits     Priv     Priv          Config         Status                 Option
   F1        F2       F3             F5             F7                     F10
```

FIGURE 10.5
PLC-5 communication channel overview screen.

channel (which must match the baud-rate setup using the remote I/O adapter module DIP switches), indicating whether **complementary I/O** racks exist (if a remote I/O adapter module is configured as a complementary rack) and establishing a **scan list.** A scan list identifies which I/O racks and part racks exist, so that the PLC-5 won't waste time reading or writing nonexisting I/O points. If the configuration is done in online programming and the CPU module is not in run mode, if an I/O status file has been specified using the processor configure screen, and if none of the rack inhibit bits of the I/O status file are on, the PLC-5 can create the scan list for you! Simply select **CLEAR LIST**, then select **AUTOMATIC CONFIGURATION,** then read the scan list to verify that it matches the racks you have connected.

3. If the PLC-5 being configured has been set up (through its DIP switches) as an adapter-mode CPU instead of as a scanner-mode CPU, one channel (usually channel 1B) must be set up as a *remote I/O adapter mode* channel. This PLC-5 will provide input data to, and will accept output data from, a scanner-mode CPU.

Further configuration involves entering the configuration data you would have set up using the DIP switches on a remote I/O adapter module (e.g., **rack number, baud rate, etc.**), but you must also enter the addresses of **two unused BT elements** (or two unused integer files) for this PLC-5 to use as **control elements** in exchanging data with the scanner-mode PLC-5.

4. Channel 0 has a standard 25-pin connector, and it is intended for connecting the PLC-5 to standard computers and peripherals. DIP switches select part of channel 0's serial communication protocol (RS-232C, or RS 422B, or RS 423), but the channel overview screen is where a programmer can configure other serial communication protocol, starting with selection of either:

- **DF1 point-to-point** protocol (the default, for connecting this PLC-5 to one other computer)
- **DF1 slave mode** protocol or **DF1 master mode** protocol (to allow interconnecting of multiple computers and PLC-5s)

- ASCII protocol, also called **user mode** (to connect to a peripheral such as a bar code reader or a serial printer)

These serial communications options are mentioned again in Chapter 13, but they are outside the scope of this book.

Still other channel configurations are offered (extended I/O rack scanning, Ethernet communications, specialized coprocessor interfacing) but are not discussed here.

Configuring PLC-5 Intelligent I/O and Analog I/O

PLC-5 intelligent I/O modules are called **block transfer (BT)** modules. **Analog input and output** modules are BT modules. Other BT modules include high-speed counter modules and power-monitoring modules.

A PLC-5 CPU must be configured to recognize the intelligent I/O modules with which it must communicate before block transfers of data can take place. When a PLC-5 program is saved to disk, any intelligent I/O configuration data that the CPU module contains is also saved. Uploading the program to a PLC-5 reloads the configuration data. There are several steps in the I/O configuration process for a block transfer I/O module:

1. See the manual that comes with your BT module to find how many block transfers (and therefore how many BT elements) are required and how many data words each data block must contain.

2. Write a program containing the BLOCK TRANSFER READ (BTR) and BLOCK TRANSFER WRITE (BTW) instructions needed to communicate with the BT module. Entering BTR and BTW instructions automatically creates the **BT elements** (or the integer file control elements) and the **data blocks** that you will need for the next step in I/O configuration. Data that you enter with the instructions (I/O module address, data block location and size) also gets put into the BT control element. (*Hint:* Keep a record of which BT control elements you have programmed and what each is used for.)

3. Select the **I/O overview** option or cursor to a BTW or BTR instruction and select **I/O edit,** then select **add new module** and use the following menus to select the exact type of I/O module that you are configuring the CPU to use. Once a module type is selected, a screen like the one in Figure 10.6 will be displayed. Enter the **BT control elements** that you created in step 2. (Don't you wish you had kept that record I recommended?) The PLC-5 will find the block transfer data file address and length and will display that data on the screen. When you accept the configuration so far, you will see the I/O overview screen again, and it should include the newly selected I/O module description, address, BT elements, and data files, as shown in Figure 10.7.

4. Select EDIT and you will see a module edit screen, specific to the type of BT I/O module, where you can enter further configuration data. One or more other screens of configuration requirements may also be available by selecting channel edit/setup, counter setup, output/device setup, or a similar option, depending on the type of BT I/O module you are configuring.

Module edit configuration data might include:

(a) *For analog input modules:* the signal **Sampling rate,** input **filtering,** the **binary number format** of the number to generate during analog-to-digital conversion, and selection of a **sensor type**

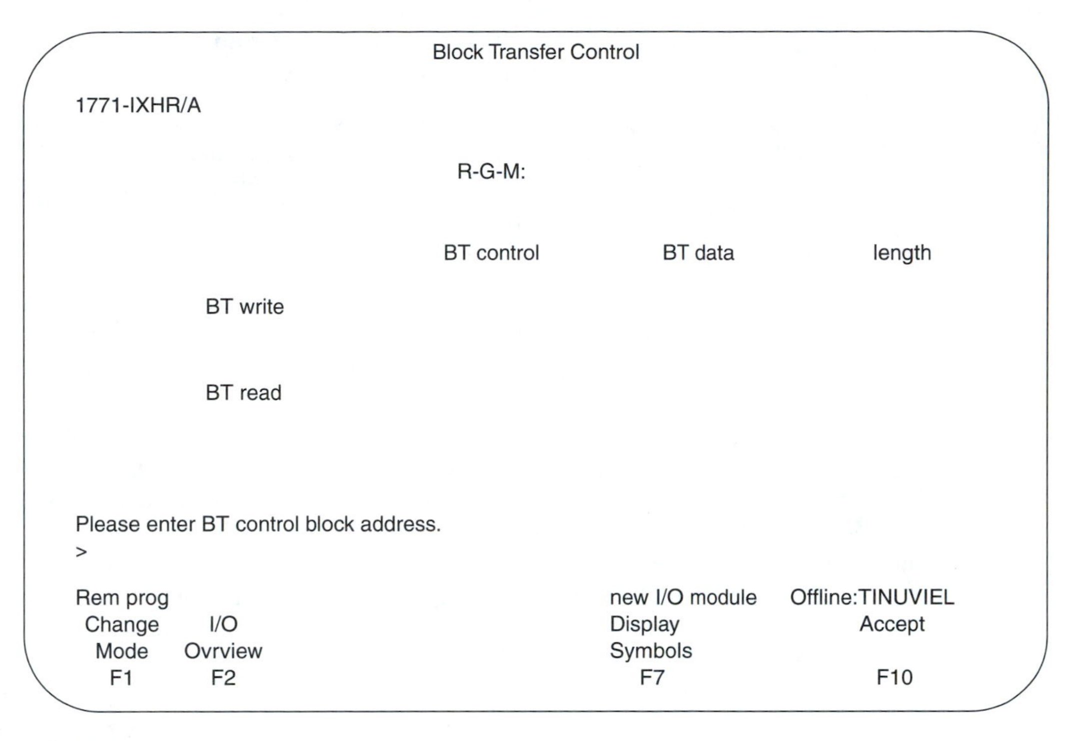

FIGURE 10.6
Entering BT elements for a PLC-5 BT module.

```
                        I/O Module System Overview

 R-G-M   Cat.No./Ser.    BTW data        BTR data        BTW ctrl    BTR ctrl

 1-0-0   1400-PD/A       N10:6    43     N16:6     43    N10:1       N16:1


Use keys to position cursor, then press a function key.

Rem Prog                                    mod 1 of 1    Offline:  START1
Change   Edit   Monitor Delete     Add New  Display  Copy  Change
 Mode                              Module   Symbols        Display
  F1      F2      F3      F4         F6       F7      F8     F9
```

FIGURE 10.7
PLC-5 I/O overview screen after adding one BT I/O module.

(b) *For analog output modules:* the **binary number format** of the binary number to be converted to an analog output signal

The **channel edit** screen is where you can select characteristics such as an **analog signal type and range.** You can often specify an upper and a lower number to represent the endpoints of an analog range, so that an analog input module will **scale** analog input signal values into that engineering unit range, or so that an analog output module will scale engineering units to analog output signals. You can also enter a new output value for an analog output channel.

During this step and the following step, the PLC-5 is accepting the entries you are making and writing them into the appropriate locations in the data file(s) that your BTW instruction(s) will eventually write to the BT module. When the BTW executes, this data will configure the intelligent BT I/O module.

If you are working in online edit mode, the I/O edit software will offer you the option of immediately transferring this new configuration data to the BT I/O module instead of waiting for the BTW instruction to execute. This is particularly useful if you want to change an analog output value.

5. Edit your PLC program to execute:

(a) BTW instruction(s) to write configuration data to BT modules at least once every time your PLC restarts executing its user-program. Other BTW or BTR instructions for the I/O module should not execute until after the configuration data has been transferred successfully. As a condition for execution of other BTW or BTR instructions, examine the "done" bit of the BT control element (BTx.y.DN) for the BTW instruction that writes configuration data to the BT module.

(b) BTR instruction(s) that read status information (including input data) from the BT I/O module as frequently as your application needs.

(c) BTR instruction(s) that write other control data (including analog output values) to the BT I/O module as frequently as required by your application.

(d) Instructions to use incoming data after a BTR instruction has read successfully from a BT I/O module. Your manual set and/or the I/O monitor screen are helpful in determining the actual addresses where the BTR instruction stores specific data items. You may want to write your program so that the incoming data is operated on only after each successful BTR execution, as indicated by the "done" (.DN) bit of the BTR's BT element.

(e) Instructions to write outgoing data to the data blocks that BTW instructions write to BT I/O modules. Use your manual set and/or I/O monitor/edit screens to determine which addresses of the data blocks you should write to. You may wish to write your program so that a BTW instruction executes only once after each change of the outgoing data, by toggling the logic controlling the BTW instruction, or by toggling the "enabled" (.EN) bit of the BTW's BT element off while holding the BTW's control logic true.

Configuring Access Privileges

Assigning privileges allows you to prevent unauthorized reading and/or modification of the program, data, or configuration of a PLC-5, while allowing authorized users to have access.

(After performing the configuration described so far in this chapter, you can see the value in preventing unauthorized modification!) You enable or disable this feature during software installation.

Even if you have not installed the passwords and privileges option, you can still assign (or remove) a password for a PLC-5's memory contents using the processor functions menu. The next time a programming unit tries to access this PLC-5's memory, the user will be asked for the password, and if he/she doesn't know it, monitoring and editing access will be denied.

If you have installed the passwords and privileges option, a more selective access system is possible. You can assign a system of privileges for accessing this PLC-5's memory. Each PLC-5 must be configured for access privilege separately. To make use of the selective privilege features, follow these steps:

1. Select PASSWORDS AND PRIVILEGES from the **processor functions** menu. A screen showing the four privilege classes and the privilege option will appear.

2. Select MODIFY PASSWORD. Enter a **privilege class name.** Four privilege class names (classes 1 to 4) are available. Enter a **password** that will apply to the privilege class. *This password will be saved in the PLC-5 processor and in the programs you download from the PLC. If you forget the password, you will have to remove all battery-backup and EEPROM data storage devices so that powering down will allow the password to be lost, with everything else in the PLC-5's memory. The saved files will be useless to you.* (*Hint:* Memorize the password now.)

3. Select TOGGLE PRIVILEGES. Toggle the rights that you want the users at each privilege level to have. Assign maximum rights to privilege class 1 for a supervisor, who must be able to change other classes' privileges if necessary. Privileges include the rights to create, delete, or edit programs and data files, to upload or download files from the CPU, to use I/O forcing, and even to change the processor mode (run or program mode) from a remote terminal.

4. Select CHANNEL PRIVILEGES from the **channel overview** in the **General Utilities** menu. Default class 1 privileges will be shown for channels to which a programming terminal can be connected. Since class 1 (should) have the rights to change privileges, your programming terminal can be used to assign default (lower) class levels to apply to future users accessing this PLC-5 via each channel.

5. Individual nodes on a communication channel can be assigned higher privilege classes than the (low) classes you set up in step 4. You may want a programming unit in the engineering department, or a PLC running a supervisory program, to have higher privilege levels than the default. Select NODE PRIVILEGES from the same menu where you found CHANNEL PRIVILEGES, toggle the channels using **select privileges,** enter the station address and (if on another DH+) the link ID of the node, then enter the (higher) privilege class.

6. Individual program files or data files in your PLC-5 can be configured to require a programming unit to have a specific privilege class before that programming unit is allowed to view and change the file. Use the Processor Function menu to select a program file, or the Memory Map menu to select a data file, then toggle the privilege class requirement for that file using the TOGGLE PRIVILEGES option.

Any user can get higher privilege class rights than the default by selecting **Enter Password** and entering the password for a higher class.

SLC 500

ALLEN-BRADLEY SLC 500 FIRST-TIME CONFIGURATION

SLC 500 Offline Programming Configuration

If your SLC 500 PLC system includes specialty I/O modules with M0/M1 memory, you will probably want to select **Enable M0/M1 Monitoring** during software configuration. To write programs offline, **Offline Configuration** must first be selected. In Offline Configure, you must select the type of CPU module for which the program is being written, and enter a file name for the program that the CPU module will contain.

You must also use the Offline Configure menu to enter I/O configuration data consisting of a description of the racks and I/O modules in each slot of the PLC system controlled by this CPU. Module descriptions can be selected from menus, or there is a **Read Configuration** option that causes the programming unit to connect to the PLC, examine the PLC system, and automatically generate the I/O configuration system description.

Because Read Configuration requires the programming unit to connect to the PLC, the I/O configuration menu includes an Online Configure option. Online configuration is described below.

SLC 500 Online Programming Configuration

You must perform **Online Configuration** before entering Online Programming. In online configuration, specify which communication interface hardware is installed between the programming unit and the PLC. Interface equipment specification is done by selecting from a menu of interface options.

You must also select **Driver Configuration** to do further configuration for the interface hardware. A driver interface configuration screen for the selected interface hardware appears. At this screen the programmer must configure items such as:

1. An address for the programming terminal (0 to 31), so that a networked PLC system can operate with multiple programming units.

2. The address of the SLC 500 CPU module to connect to when online programming is initiated (1 to 31). By default, node 1 is connected to, and is the default address of, a newly purchased SLC 500 CPU. If/when you change the address of the CPU during configuration (CPU configuration is described next), you must modify this online configuration parameter.

3. The address of the communication interface card in the personal computer (if used), so the programming software can communicate through it. This number should match the DIP switch setting on the card.

4. An interrupt request (IRQ) number for the communication interface card, so that it can notify the programming software when it receives data from the PLC (use the default IRQ number unless the default IRQ number is already being used by another device in your personal computer).

5. Communication baud rate.

6. If the interface hardware between the programming unit and the SLC 500 includes hardware connecting DH+, DH−485, modems, and so on, the addresses of the interface module(s) must be entered with descriptions of any other communications protocols that might be required. See your manual set for more information.

Configuring the SLC 500 CPU

CPU configuration data is stored in the status file (S:) area of memory as you configure the SLC 500 CPU via the processor status screens. The **processor status** screens for the SLC 500 CPU display the contents of the status file area of memory, with labels identifying the meaning of each data item. Processor status screens are available through the data monitor menu. When the cursor is on a configuration item, the programmer can enter new configuration data. As the programmer moves the cursor through the processor status screens' fields, the actual status file address for each configuration or status option is displayed at the bottom of the programming unit screen.

SLC 500s have backup batteries, so status file data, including configuration changes, will remain in effect (even if power is disconnected) until a programmer or a program changes the configuration again. Whenever you save a program to a programming unit, you save the status file, and when you reload that program to a PLC, you also reload the configuration.

PLC programs can read any of the data in the status file. Some configuration data is **dynamic,** which means that a PLC program can change it and the changes become effective immediately. Other configuration data is **static,** which means that changes become effective only when the PLC is switched into run mode. The status file also contains words that only reflect the PLC's **status.** The structure of the status file area of an SLC 500's memory is shown in Appendix B.

Configuration options that can be modified include the ones discussed below (only the better SLC 500 models offer all these options; Appendix B indicates which processor offers which option).

Startup Configuration Bits S:1/10 to S:1/12 are the **load memory module control bits.** They are used to select the conditions under which the PLC will copy the contents of its EEPROM module into RAM.

Bit S:1/9 is called the **startup protection fault bit.** If set, the PLC will try to run the fault routine program after a power failure while the PLC is in remote run mode. The **fault routine's file number** must also be entered (into S:29) using this status screen. If the fault routine clears the major error halted bit (S:1/13), then when the fault routine finishes executing, the PLC will start executing standard scan cycles. If bit S:1/9 is not set when the PLC restarts, restarting depends on bit S:1/8. Fault routines are described in Chapter 11.

Bit S:1/8, the **fault override at powerup bit,** will cause the PLC to clear its major error halted bit (S:1/13) whenever power is switched on, so that the PLC can try to run its program again. If S:1/8 is not set when power is restored, bit S:1/13 is not cleared, so an operator must intervene to clear bit S:1/13 after every major fault.

Math Control Set bit S:1/14, the **math overflow selection bit,** if you intend to use 32-bit addition and subtraction. This bit affects what result is stored in the event of a math overflow or underflow.

Set bit S:34/2 to **disable the math flags** from being set during math operations that use **floating-point numbers.** Bits S:0/0 to S:0/3 and S:5/0 will not be set in the event of a carry, overflow, zero, negative sign, or for a minor error overflow trap, respectively. (S:0/0 is used for other purposes during floating-point math anyway.)

Watchdog Timer As each scan cycle starts, the SLC 500 starts a watchdog timer. If the timer runs out before the scan cycle ends, the PLC will fault (see Chapter 11). A programmer can enter a time, in 10-ms units, for the watchdog timer. This value is placed into S:3H, the high byte of word S:3. (As each scan cycle ends, the SLC 500 writes the time it took to complete that scan into S:3L, the low byte.)

A *scan time timebase* is selected using S:33/13. If this bit is set, the timebase changes from its default 10-ms value to 1-ms units. The average and maximum scan time values that the PLC writes to S:22 and S:23 (respectively) are also affected.

Real-Time Clock S:37 to S:42 contain year, month, day, hour, minute, and second values (respectively) and are updated as the PLC runs. (Year numbers are OK until after the year 65535!) You can disable the real-time clock timekeeping by writing zeros to all these words.

Interrupts Four types of interrupts may be configurable (depending on the type of SLC 500) using the processor status screens. All are described in detail in Chapter 11.

1. A **fault routine** file number can be entered (into S:29).
2. A **selectable timed interrupt (STI) file number** can be entered (into S:31), an **interval time** for executing the STI file can be entered (into S:30), and a **time unit** (1 or 10 ms) for the STI timer can be selected using S:2/10 (set for 1 ms). STIs can be disabled by changing the **STI enabled bit** (S:2/1) from its default set state.
3. A **discrete input interrupt (DII) file number** can be entered (into S:46). The programmer can enter a **slot number** and a **bit mask** (S:47 and S:48, respectively) to configure which input module and which contacts are to be interrupt sources, and can enter a **compare value** and a **counter preset** value (S:49 and S:50, respectively) to configure the type of bit transitions and how many are required before an interrupt occurs. If S:2/12 is changed from its default set state, **DII interrupts will be held** until this bit is set again.
4. Specialty I/O modules can be configured to generate I/O interrupt requests. The program file number to execute in response to an I/O interrupt is entered during I/O configuration (discussed next), but the programmer can use the processor status screen to **disable I/O interrupts** from slots 1 to 31 by clearing bits S:27/1 to S:28/14 (respectively). These bits are set by default.

I/O Slot Enables The programmer (or the program) can disable I/O in any of its 31 I/O modules (slots 1 to 31 and the CPU module in slot 0) by resetting bits S:11/0 to S:12/14. Bit S:11/0 controls slot 0, bit S:11/1 controls slot 1, and so on. These bits are all set by default. When an I/O slot is disabled, the CPU's input image table is not affected by input module changes, and output module's states are not affected by changing the CPU's output image table.

Communication and Message Control In this section we describe the configuration options setup using status file (S:) words. Communication using an SLC 500 is covered in detail in Chapter 13.

Set S:1/14 to deny a programming device access to monitor the program and data at this PLC unless it already has a copy of the processor file.

You can limit the amount of time spent servicing communication channels during the I/O part of the scan cycle. The amount of housekeeping time to service communication channels per scan cycle may vary greatly unless it is limited.

1. Set S:2/15 to limit the CPU to one communication servicing task for channel 1 each time channel 1 is serviced.

2. Set S:33/7 to limit the CPU to servicing one channel 1 communication task initiated by a message (MSG) instruction even if S:2/15 is not set.

3. Set S:33/5 to limit the CPU to one communication servicing task for channel 0 each time channel 0 is serviced.

4. Set S:33/6 to limit the CPU to servicing one channel 0 communication task initiated by a message (MSG) instruction even if S33/5 is not set.

5. Set bit S:34/0 to prevent this CPU from passing messages between its DH+ connection at channel 1 and its DH-485 connection at channel 0.

6. Set bit S:34/5 to enable this CPU to pass messages between its DH+ connection at channel 1 and its DF1 connection at channel 0.

7. Set bit S:34/1 to cause this CPU to maintain a record of which stations are active on the DH+, using a "DH+ active" table.

Bits S:33/14 and S:33/15 configure and control the DTR signal at channel 0.

The size of the default data unit in the **common interface file (CIF)** can be changed from the default 16 bits to 8 bits, by setting S:2/8. Message (MSG) instructions programmed to use PLC-2 type read/write type automatically read/write the SLC 500's CIF (see Chapter 13 for more details on CIF files). Data file 9 in the SLC 500 is automatically used as the CIF file (so the programmer would be wise to reserve data file 9). Other PLC-2 compatible devices also automatically access the SLC 500's CIF file, and some, especially those that accessed the SLC 500 via a network bridge or gateway, need 8-bit data unit sizes.

Some configuration data that is stored in the status file is actually entered using the channel configure screens (explained next) but can be overwritten by a user-program. These status file values include:

1. S:15L (the low byte in word S:15) contains a **node address** (1 to 31 for DH-485, 1 to 63 for DH+).

2. S:15H (S:15's high byte) contains a code number for a **baud rate** for a DH-485 network.

3. S:34/3, if set, causes this CPU to pass a **global status word** (the contents of S:99) via a DH+ network. S:34/4, if set, causes this CPU to accept global status words from other CPUs on the DH+, copying them into S:100 to S:163 (S:100 from node 0, S:101 from node 1, etc.). A user-program can write data words to S:99 and read data from S:100 to S:163.

Configuring SLC 500 Communication Channels

Some communication channel configuration options must be selected using the **Channel Configure** screens. These configuration parameters are stored in the CPU module, but not necessarily in the status file, and may be saved into separate files when an SLC 500 program is downloaded to a personal computer. Whenever a saved program is uploaded to an SLC from a personal computer, the Channel Configure setup is also uploaded. Remember that when you change some configuration parameters (such as baud rate or node number) at one node, you may also have to make corresponding changes at other nodes or at a programming unit or at an operator interface panel that exchanges data with this node.

Two channels (0 and 1) must be configured using Channel Configure. Either channel can be configured as a:

1. *DH+ channel.* A **node address** (in octal, from 1 to 77) must be entered for this CPU, and the **baud rate** can be increased from 57.6 kilobaud. If there is more than one DH+ in the communication system, a **link number** must be entered to indicate to which DH+ this channel is connected. The **global status word** transmit-enable and receive-enable status bits (S:34/3 and S:34/4, discussed in the preceding section) can be changed from this screen.

2. *DH-485 channel.* A **node address** (1 to 31) must be entered for this CPU. The **maximum node address** can be reduced from 31, to reduce lost communication time, if there are fewer controllers connected to the DH-485. You can also increase the **token hold factor** up to 4, allowing this CPU to transmit that number of messages each time its turn to "talk" on the network comes around. The **baud rate** can be changed from 19200.

3. *RS-232C DF1 channel.* Allows two-way communications between this CPU and RS-232C serial ports at one or more other computers (including modems and including other SLC 500 components). The RS-232C serial communication protocol at this SLC 500 must be configured to match the other computer's protocol. Several DF1 options, with different configuration screens, are allowed. (See the SLC 500 reference manual and the RS-232C protocol of the other computer.)

4. *ASCII channel.* Allows this CPU to send and/or receive ASCII messages as controlled by ASCII commands in the user-program.

Configuring SLC 500 Intelligent I/O

Specialty I/O modules are intelligent I/O modules for the SLC 500. Each has its own memory and each has its own configuration requirements. To configure a specialty I/O module, use the **SPIO (special I/O) Configure** option from the **Configure I/O** menu that you see when you first configure Offline Programming or when you select Processor Change while in Offline Programming. After selecting a specialty I/O module for a PLC slot, the SPIO configure menu allows you to select the:

1. ISR program file to execute in response to an I/O interrupt from the module in this slot.

2. Number of **input and output image words** to reserve for the specialty I/O module in this slot, and how many of those words to include in each I/O scan.

3. Number of **M0 and/or M1 data words** that this module has, so that they can be monitored if M0/M1 monitoring was enabled (one of your earliest configuration choices).

4. Number of **words in the G data files** in this specialty I/O module, and the contents of those G data words. G data words are used to configure the I/O module. See the manual that comes with a specialty I/O module for the meanings of G data for that module.

S5

SIEMENS S5 FIRST-TIME CONFIGURATION

Connecting to the S5 CPU and Online Programming

Select Online Programming ("Edit PLC"). No configuration is required.

Configuring the S5 CPU

S5 CPU configuration data is stored in 16-bit words in the CPU's **system data (RS)** area of memory. Optional configuration software makes configuration easier, but isn't necessary. Data can be entered into system data memory using the data screen, or user-programs containing function blocks can write to the system data memory area. (Only STEP 5's function block programs can read or write the system data memory.)

The following system data memories contain modifiable configuration data:

1. **RS 96** contains the **watchdog timer** setting in its low byte. This number is the number of 10-ms time units before the PLC will fault in the event of an unusually long scan cycle. To change the setting to 200 ms, these instructions in a function block will work:

```
L KF +20 ; (20) 10-ms time units = 200 ms
T RS 96
```

2. **RS 97** controls how frequently OB 13 is called as a **timed interrupt** service routine. RS 97 contains "10" by default and the unit size is 10 ms, so the default setting is for OB 13 to execute every 100 ms. Timed interrupts are enabled by default. S7-115U CPUs have time values in **RS 98** to **RS 100** to control cyclic interrupt using OB 12 to OB 10, respectively. (See Chapter 11 for more detail.)

To **disable timed interrupts and process input interrupts,** the command **IA** can be executed in a function block program.

3. **RS 8** to **RS 10** to control a **real time clock (RTC).** The three words contain two 3-byte pointers: one to where the (up to) 44 bytes of clock data should be stored, and one to an RTC status (and control) word. RTC clock data does not have to be in the same areas of memory as the RTC status word

The 3-byte **RTC data pointer** includes:

- A code in the high byte of **RS 8,** identifying the **area of memory** in which to store real-time clock data. Hexadecimal 44 means a data block, 45 means flag memory, 49 means input image (PII) memory, and 51 means output image (PIQ) memory.
- A number in the low byte of **RS 8,** indicating the data block number (2 to 255) or the starting address in flag or PII or PIQ memory. Remember, you need 44 bytes of memory starting at this address.
- A number in the high byte of **RS 9,** indicating the starting data word address. (Ignored unless the RTC data is in a data block, of course.) This will be the low word of 22 data words containing the 44 bytes of clock data.

The 3-byte **RTC status word pointer** includes:

- A code in the low byte of **RS 9,** identifying the memory area containing the status word. (Same codes as for RTC data above.)
- A number in the high byte of **RS 10,** indicating the data block (2 to 255), or the flag/PII/PIQ address containing the high byte of the status word. (The next-higher flag/PII/PIQ address will contain the low byte of the status word.)
- A number in the low byte of **RS 10,** indicating the data word that contains the status word. (Ignored unless status word is in a data block.)

If you have entered pointer data for a real-time clock, you will want to set the clock. To set the clock, enter the following RTC data into the 44 bytes (or 22 words) of RTC clock data (the first byte is called byte 0 here):

- The **actual time,** entered into bytes 8 to 15. This data will be copied into the working real-time clock in bytes 0 to 7 when you toggle the status word bit that sets the RTC time. Enter the following BCD values (use KC constants) into byte:

```
 8    number of years until a leap year (KC 0 to 3) (This number is
      only used internally and is not copied to byte 0.)
 9    weekday number (KC 1 to 7)
10    day (KC 1 to 31)
11    month (KC 1 to 12)
12    year (KC 00 to 99)
13    hour (either KC 0 to 23 for 24-hour cycle, or KC 1 to 12 for AM
      and 81 to 92 for PM)
14    minute (KC 0 to 59)
15    second (KC 0 to 59)
```

- A **load prompt** interval time, entered into bytes 16 to 23. After you toggle the status word bit that causes the CPU to use the load prompt, the CPU will repeatedly set bit 5 of the high byte of the status word bit at this interval. (Enter an invalid number to disable a portion of the interval timer.) Enter the following interval times into bytes:

```
16    (not used)
17    weekday number (KC 1 to 7)
18    day (KC 1 to 31)
19    month (KC 1 to 12)
20    (not used)
21    hour (either KC 0 to 23 for 24-hour cycle, or KC 1 to 12 for AM
      and 81 to 92 for PM)
22    minute (KC 0 to 59)
23    second (KC 0 to 59)
```

- A new time for the **operating hours counter,** entered in bytes 30 to 35. This value is copied into the working operating hours counter in bytes 24 to 29 when you toggle the status word bit that sets the operating hour counter. The working operating hours counter keeps track of the time that the CPU is in run mode. Enter the following times into bytes:

```
30    (not used)
31    seconds (KC 1 to 59)
```

```
32      minutes (KC 1 to 59)
33      hours (KC 0 to 99)
34      100 hours (KC 0 to 99)
35      10,000 hours (KC 0 to 99)
```
[9]

You will also want to make use of the status word to which the second pointer in system data memory points. The bits that you will want to use in configuring (setting) the real-time clock include:

- The low-byte bits (next byte after the flag/PII/PIQ byte address pointed to or the low byte of the data word pointed to):
  - **(a) Bit 2,** when set, causes the CPU to copy the real-time clock setting from data bytes 8 to 15 to the working clock bytes 0 to 7. The S7 operating system clears this bit after successfully performing the copy. If copying was successful, but the operating system found faulty data, bit 0 will be set.
  - **(b) Bit 1,** when set, calculates hours in the 12-hour mode (1 to 12 and 81 to 92). When clear, uses the 24-hour cycle (0 to 23).
  - **(c) Bit 4,** when set, causes the PLC to update the real-time (in bytes 0 to 7) even while PLC is in stop mode.
  - **(d) Bit 5,** when set, causes the PLC to record the time that it was last switched out of run mode. The time will be stored in data bytes 36 to 43 in the following (familiar) format:

```
36      (not used)
37      weekday number (KC 1 to 7)
38      day (KC 1 to 31)
39      month (KC 1 to 12)
40      year (KC 00 to 99)
41      hour (KC 0 to 23 or KC 1 to 12/81 to 92)
42      minute (KC 0 to 59)
43      second (KC 0 to 59)
```

- The high-byte bits (the flag/PII/PIQ byte address pointed to or the high byte of the data word pointed to):
  - **(a) Bit 2,** when set, causes the CPU to copy a new operating hour counter time from data bytes 30 to 35 to the working operator hour counter in bytes 24 to 29. The S7 operating system clears this bit after successfully performing the copy. If copying was successful but the operating system found faulty data, bit 0 will be set.
  - **(b) Bit 1,** when set, enables the operating hour counter. When clear, operating hours are not counted.
  - **(c) Bit 6,** when set, causes the CPU to use the load prompt time setting in data bytes 16 to 23. The S7 operating system clears this bit after successfully performing the copy. If the operating system finds faulty data, bit 4 will be set.

  Each time the prompt time runs out, the STEP 5 operating system will set bit 5 of this same byte. Presumably, your program will include a response routine that executes when bit 5 goes on, and which will reset bit 5.

[9] Does Siemens really expect one of these PLCs to run for 115 years?

Configuring S5 Communication Channels

In this chapter we describe the configuration required to allow an S5 PLC to communicate via serial ports or via communication processor modules. Communication is described in Chapter 13.

S5 CPU **serial ports** are sometimes called *programmer ports,* because they are used to connect to a programming unit. Serial ports can also be used to connect two S5 PLCs in a **point-to-point** configuration, or to connect up to 30 L1 slaves to an L1 master in a single **L1 network.** In either case, you must enter network addresses and pointers into the CPU's system data area of memory[10]:

1. **RS 57's high byte** can contain a node number for a programming unit if you want the programming unit connected at this PLC to have access to other CPUs via the L1 network.
2. **RS 57's low byte** must contain a unique node number. Enter an L1 slave number (1 to 30) if an L1 network is being used. (S5 CPU modules can only be L1 slaves. L1 masters are explained below.) For point-to-point connections, one CPU must be node 0 and the other CPU must be node 1.
3. **RS 58 to RS 63** must contain data pointing to the location of a receive coordination byte, a send coordination byte, a send mailbox, and a receive mailbox. Coordination bytes and mailboxes can be in data blocks or in flag memory.

   (a) For the **receive coordination byte:**

   ```
   RS 58, high byte     data area ID
   RS 58, low byte      DB or flag byte
   RS 59, high byte     data word
   ```

   (b) For the **send coordination byte:**

   ```
   RS 59, low byte      data area ID
   RS 60, high byte     DB or flag byte
   RS 60, low byte      data word
   ```

   (c) For the **send mailbox:**

   ```
   RS 61, high byte     data area ID
   RS 61, low byte      DB or flag byte
   RS 62, high byte     data word
   ```

   (d) For the **receive mailbox:**

   ```
   RS 62, low byte      data area ID
   RS 63, high byte     DB or flag byte
   RS 63, low byte      data word
   ```

   where

   - **Data area ID** must be:
     - The ASCII code for D (hex: 44) if the reserved memory is in a data block
     - The ASCII code for F (hex: 4D) if the reserved memory is in the flag memory area
   - **DB or flag Byte** indicates which data block or flag memory is used. Enter:

---

[10] FB 239 is preprogrammed in some S5-115U CPUs. FB 239 copies configuration data, provided to FB 239 as parameters, into the appropriate system data memories.

  - A data block number (2 to 255), if data area ID is D
  - A flag byte number (1 to 255), indicating the start of the data, if data area ID is F
- **Data word** indicates:
  - The data word containing the coordination byte in its low byte, or where the mailbox data starts, if data area ID is D
  - Irrelevant, if data area ID is F

Coordination bytes and mailboxes are described in Chapter 13. They are used during the exchange of data between PLCs on the L1 network.

For S5-115U CPU modules that have ASCII drivers built into their CPU serial ports, configuration can include:

1. Writing the number 1 to the **high byte of RS 46,** to select the ASCII driver for the serial port.

2. **RS 48 to RS 55** must contain data in the following format (similar, but not identical, to the format for configuring an L1 network):

(a) For the **ASCII parameter set:**

```
RS 48, high byte      data area ID
RS 48, low byte       DB or flag byte
RS 49, high byte      data word
```

(b) For the **send mailbox:**

```
RS 49, low byte       data area ID
RS 50, high byte      DB or flag byte
RS 50, low byte       data word
```

(c) For the **receive mailbox:**

```
RS 51, high byte      data area ID
RS 51, low byte       DB or flag byte
RS 52, high byte      data word
```

(d) For the **send coordination byte:**

```
RS 52, low byte       data area ID
RS 53, high byte      DB or flag byte
RS 53, low byte       data word
```

(e) For the **receive coordination byte:**

```
RS 54, high byte      data area ID
RS 54, low byte       DB or flag byte
RS 55, high byte      data word
```

Where **data area ID, DB or flag byte,** and **data word** are as described for configuring an L1 network

The **ASCII parameter set** is a set of 12 bytes of ASCII communication protocol such as baud rate, bits used per ASCII character, printer page control, and so on. See your manual for more detail. The data in mailboxes and coordination bytes are used in a way similar, but not identical, to how data in L1 mailboxes or coordination bytes are used (see Chapter 13).

An *L1 network* must have a master node, which is often a CP 530 communication processor module in an S5-115U PLC's rack. S5-115U CPU modules have data-handling function blocks that can be used to write configuration parameters to CP modules as well as to send and receive data via CP modules. Configuration of a CP module must be redone every time the PLC goes into run mode unless the CP 530 module has its own nonvolatile memory such as an EEPROM module. In Chapter 7 we described how to program data handling function blocks but did not discuss how they were used to write configuration data to communication processor modules. They include:

1. **FB 246,** SYNCHRON, used before other data handling function blocks, to synchronize and to set a maximum data block size for data exchanges between the CPU module and a CP module.
2. **FB 244,** SEND, to write configuration data to a CP module before the L1 network will operate, to send control data and message data to a CP module after configuration, or to send data messages via the L1 network. The parameters that are important while using SEND to configure an L1 network include:
   - **A-NR,** the **job number.** Enter as KY 0, y, where "y" is a job number indicating the purpose of this data transmission to the CP module. SEND must be executed once for each of the following job numbers to perform specific configuration operations:

     222 to configure the CP 530 module by writing a new **L1 system ID parameter (SYSID) set** to the interface area of memory. The data set, entered into memory by the programmer, must consist of ASCII-coded data, in fields separated by ASCII codes for carriage return (CR). (The CR is necessary after each field even if no data is entered for that field.) The fields must include data in this order:
     - Four optional identification fields. The first two fields can contain up to eight ASCII characters, the third up to 19 characters, and the last up to eight characters.
     - Two blank fields (enter as CR character codes).
     - A seventh field *must* be entered. It can contain a two-digit programming unit node number (if a programming unit will be used to access slave nodes via the L1 network), then an **/** character. No entry after the **/** means that this node number will be the L1 master node, number 0. (If you are programming this CP 530 to act as an L1 slave, you would enter a slave number between 1 and 30 here.)
     - Five more blank fields.
     - The twelfth field *must* include the interface number. The default interface number is 1, but if you enter a different number here in the SYSID data, you are reprogramming the CP module's interface number. You must then enter the new interface number as the SSNR parameter the next time you use a function block to communicate with this module. If you copy the CP module's memory to its EEPROM, the interface number will be reconfigured permanently.
     - The thirteenth field may contain a "Y" for yes or a "N" for no, indicating if this CP should automatically resume L1 master operations immediately af-

ter a power resumption. (Default = Y.)

- The fourteenth field may contain a new baud rate for the L1 network. (Default = 9600.)

43 to configure the CP 530 module by writing a new **polling list** to the module via the interface memory. The polling list must consist of up to 64 node numbers (in sequential bytes of memory) identifying the order in which slaves will be asked if they want to transmit via the L1 network. (Node numbers can be entered more than once to give some nodes more turns per polling cycle.)

44 to configure the CP 530 module by writing a new **interrupt list** to the module via the interface memory. The interrupt list must consist of up to 30 node numbers (in sequential bytes of memory) identifying the nodes that will be allowed to send high-priority messages and identifying the priority level of each node (highest-priority nodes listed first) in case two modes request access simultaneously. (Each node number can only be entered once.)

42 to write a **mode control byte** to **start or stop** the L1 network. If bit 0 of the byte is set, the L1 network will stop. If bit 1 is set, the L1 network will start to run. Configuration data can only be written to a CP 530 module while the L1 network is in stop mode, but job 42 can be used to turn the L1 network into run mode after it is configured.

3. FB 247, CONTROL, to read the status of the CP 530 module.

4. FB 245, RECEIVE, to read CP 530 module status but can also read messages received by the CP 530 for this PLC.

5. FB 248, RESET, to cancel any outstanding data handling function block job requests.

Some CP modules use interprocessor communication flags to control communications between the CP module and the CPU module. During hardware configuration, DIP switches will have been set on the CP modules indicating which area(s) of the CPU's interprocessor communication flag memory they should use to exchange messages with the CPU module. (Interprocessor communication flags are kept in a special area of memory with bytes numbered 0 to 255.) *To configure your CPU to use interprocessor communication flags,* write configuration data to **data block 1** (which is why we have recommended only using data block addresses 2 to 255 for other purposes). The configuration data that is in data block 1 tells the CPU which bytes contain **input** bits (which the CP module can write to) and which contain **output** bits (which the CP module can only read). You can create and edit DB1 while the PLC is in run mode, *but the PLC does not recognize modifications until the next time it goes into run mode.* To configure the use of interprocessor communication flags, create a data block 1 (DB1) containing:

1. A header consisting of the ASCII characters:

KS = **'MA'** in DW0
KS = **'SK'** in DW1
KS = **'01'** in DW2

2. The hexadecimal **code CE00** to indicate the start of a list of byte numbers for input flags, followed by the byte addresses of the interprocessor flag memory to be used as input bytes through which the CPU can read CP module status: for example,

KH = CE00 in DW3
KF = +10 in DW4
KF = +11 in DW5
KF = +20 in DW6

3. The hexadecimal **code CA00** to indicate the start of a list of byte numbers for output flags, followed by the byte addresses of the interprocessor flag memory to be used as output bytes through which the CPU can control CP modules: for example,

KH = CA00 in DW7
KF = +15 in DW8
KF = +30 in DW9
KF = +31 in DW10

4. The hexadecimal **code EEEE** to indicate the end of the list of interpersonal flag memory byte assignments: for example,

KH = EEEE in DW11

Configuring S5 Intelligent I/O

In this section we briefly outline configuring of a digital input module for interrupt generating. For configuration requirements of more sophisticated modules, see the manuals provided with those modules.

A digital input module with eight interrupt-capable contacts can be configured by transferring a 16-bit number to the module. Program the transfer instruction in a startup block (OB 21 and/or OB 22) and use direct access addressing so that the module is enabled as soon as the PLC starts running. The high 8 bits are used to enable interrupts from one or more of the eight contacts (a 1 enables that contact), and the low 8 bits each configure one enabled contact to generate an interrupt if voltage rises (if configuration bit = 0) or when voltage drops. The following program would enable the high six contacts of the digital input module in slot 1 (contacts 1.2 to 1.7), of which five contacts will generate an interrupt signal when voltage rises, while one contact (1.2) will wait until voltage drops.

```
L KM 11111100 00000100
T PW 1
```

Chapter 11 covers interrupts in more detail.

S7

SIEMENS S7 FIRST-TIME CONFIGURATION

Installing the STEP 7 Programming Software

The STEP 7 programming software's install program will load programming software onto a personal computer without any help from the user (other than changing floppies), except that the user must install a piece of **authorization code** before or during installation. The easiest method is to have the disk with the authorization available during installation, so that you can insert it when the installation program asks for it.

The STEP 7 programming software stores the authorization as hidden files in a hidden directory (C:\AX NF ZZ) on your personal computer's hard drive, and won't work if it can't find the authorization later. A defective cluster is also created. *Do not delete the defective cluster.*

An authorization can be lost during formatting of the hard drive, during installation of a new operating system, during compressing of disk space, or while simply erasing files that have defective clusters. Before doing any of these actions, the user should reinsert the authorization disk and run the **Authors** program to move the authorization from the hard drive to the floppy until after the potentially hazardous activities are done. The authorization can then be reloaded to the hard drive using the Authors program again.

Setting Up the STEP 7 Programming Unit/PLC Interface

The Windows 95/NT control panel is automatically configured with default programming unit interface parameters during STEP 7 installation, but you must finish the process manually. If you are using the personal computer's serial port and an MPI interface cable to connect directly to an S7 CPU's MPI port, select Windows' MY COMPUTER, then CONTROL PANEL where you will find STEP 7's SETTING PG/PC INTERFACE.

1. Select SETTING PG/PC INTERFACE.

2. Set the access point of application to **S7ONLINE.**

3. Select INSTALL and select an interface (to the programming port via a PC/MPI adapter cable).

4. Select PROPERTIES, select a **local station address** for the programming unit on the MPI network (different from any other programming unit's local station address), and check the **Not the Only Active Master** box if other programming units will be attached. Check to ensure that the **Com Port** number is correct.

If you are connecting your personal computer to a PLC network via a network interface card in your computer (instead of via the personal computer's serial port), the default settings installed with STEP 7 may include interrupt (IRQ) numbers or addresses that are already used by other personal computer components. To find out, select Windows' CONTROL PANEL, then SYSTEM, then DEVICE MANAGER, then COMPUTER, then toggle the INTERRUPTS and MEMORY options on to display the interrupt (IRQ) and interface card address assignments. Look for conflicts between the interrupts and the memory address defaults that your interface card and other computer components use. If any are found, note the available interrupts and memories. Return to the control panel and perform the four-step sequence described above, and add this fifth step, to correct any conflicts that you may have found:

5. Select the interface card type and display its PROPERTIES. Modify any defaults that need modification here. You can use interrupts 5, 10, 11, 12, 15, or sometimes 9. If an address space is not assigned automatically, use F00000H to F4004FH or F80000H to FC004FH.

Creating a STEP 7 Project

You have to create a **project** with a **station** before you can configure or program a PLC.[11] In creating a project, a separate section of programming unit memory is assigned to contain the project data that you are about to create and transfer to the PLCs in your system. Project data includes data that a programmer usually thinks of as part of a program but aren't necessarily written to an S7 CPU. Symbolic names, for example, are created during the programming process and are saved in the project but aren't downloaded to current S7 CPUs. If you later reconnect to the CPU via the same programming unit, you can still see the symbolic names, but if you connect to the CPU program using a different programming unit, you will only be able to see absolute addresses.[12]

The project structure helps you to maintain a logically ordered database containing the configuration, programming, and data memory assignments for all the PLCs in a complete interconnected PLC system. Separate projects should be created for separate PLC systems. Figure 10.8 shows an example project (named Project_A) and how it could be structured to contain data relevant to all CPUs in the PLC system (e.g., symbol table), to selected networks of PLCs (e.g., global data table), or only to specific CPUs in the system (e.g., configuration and parameter system data for Station_AA). The tree structure in Figure 10.8 appears slightly different from the way it would appear on a STEP 7 programming unit, which uses a side window to show some components.

To create a project

1. Start the STEP 7 programming software by selecting **SIMATIC Manager** from the Windows desktop. The first screen you see tells you to select a CPU, an MPI address, and a logic block for a station in a project. You should answer these questions by selecting the CPU type that you have installed, selecting OB1 for a logic block, entering a name for the project, then selecting **Make.** (The next time you start the SIMATIC manager, you can cancel this opening menu, then access the project using File–**Open**–Project.)
2. If you need to create a project and you have quit from the dialog boxes in step 1 without **Making** the project, select **File,** then **New,** then **Project,** then enter an appropriate **name.**

After creating a project you can create and edit programs and configuration files for the network and/or for individual stations on the network, and you can assign symbolic names that can then be used in any of the programs in the project.

Configuring an S7 Station in a Networked PLC System

For first-time installation, you will probably start by creating the configuration and parameter system data blocks for a PLC system before you write any programs (although you can write

---

[11] You can create program templates (especially function blocks) and data block templates (user data types) in libraries before creating a project. STEP 7 offers a browse option when editing projects, allowing copying from libraries.

[12] Siemens says that future S7 PLCs will accept symbolic names as part of the downloaded program. PLCOpen has specified a limited neutral file exchange format that they hope can eventually be expanded to allow any PLC programming software to be used (offline, of course) to create programs for later use in any PLC. The members of PLCOpen are still debating whether things like symbolic names should be considered part of a PLC program.

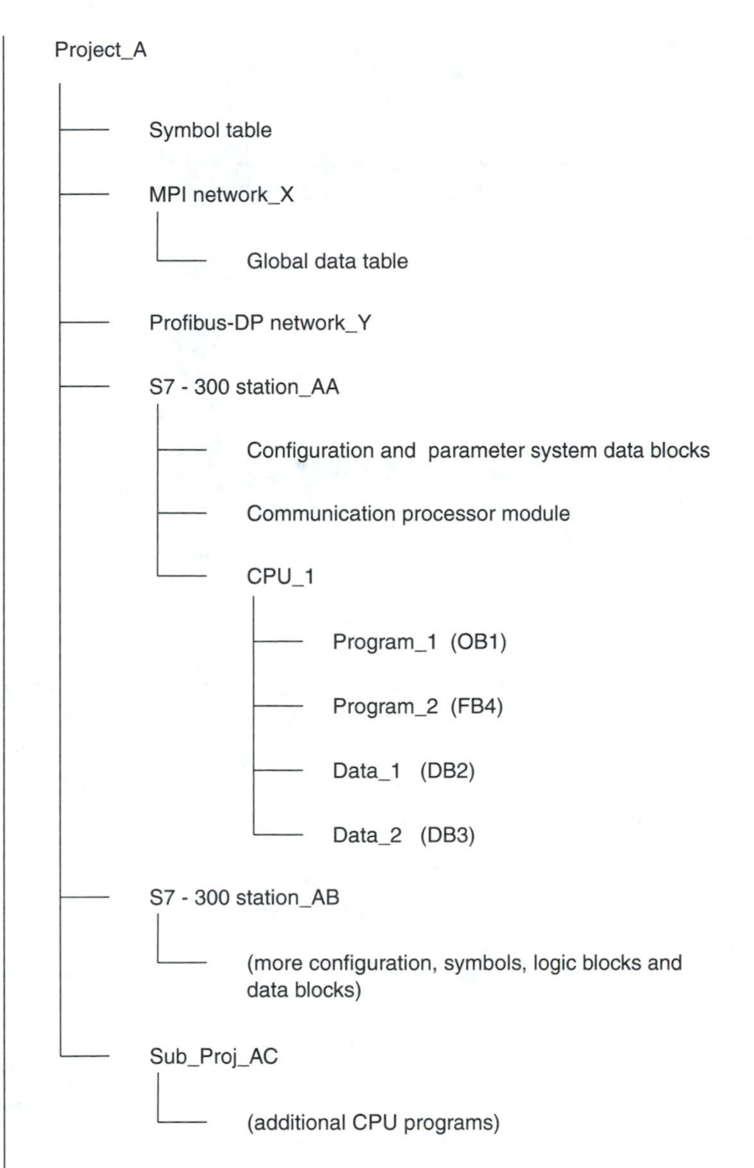

FIGURE 10.8
Small STEP 7 project, showing some of the components.

the programs first). The configuration data must include a description of the station(s) in the system and a description of the communication network(s) connecting stations. If you allowed SIMATIC Manager to create a project for you when the software started, skip to step 2 below. To create the configuration files:

1. From the **SIMATIC Manager** top menu, select **File–Open–Project,** then select the project that you created for this PLC system. A window split in half will appear, as in Figure

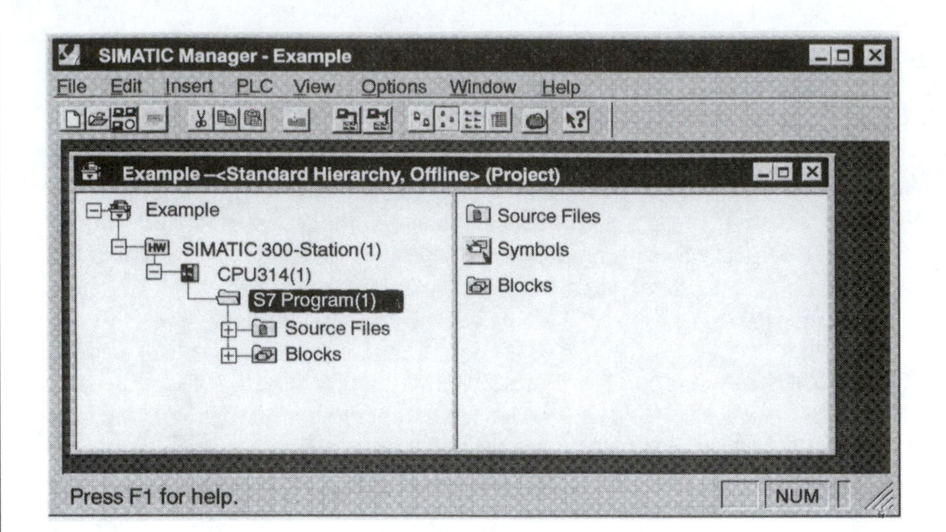

FIGURE 10.9
Typical STEP 7 project window.

10.9. The left side shows interconnected icons representing the data items stored in your project. The right side will show icons for components of items that you click on in the left window.

2. If you are going to interconnect PLCs for communication purposes, it is best to configure the PLC *subnets* now so that you can configure station communications during the next steps. One MPI subnet was inserted automatically when you created the project. Skip part (a) unless you only need more subnets.

(a) Click on the project name, then select **Insert–Subnet,** and select the type of interconnection: MPI, Profibus, Industrial Ethernet, or simply point-to-point, and select a highest address for your MPI or other network (for MPI subnets select either 15, 31, 63, or 126; default: 15). Repeat for each type of subnet that will interconnect components in the PLC project.

(b) Enter configuration data for each subnet (including the MPI subnet that was inserted automatically) by selecting the subnet, then selecting **Edit–Object Properties.** A dialog box will appear and you can enter a subnet name, a transmission rate, and a highest node address. Do this for each subnet.

3. Now enter configuration data describing each individual PLC in the system. If you have already selected a CPU type for one station and you don't need additional stations, skip to part (b) below.

(a) Select **Insert,** then **Station,** then select the type of station you are inserting (S7-300, S7-400, programmer, etc.) and enter an appropriate name for the station that you are configuring. There will usually be exactly one CPU module per station[13]

---

[13] S7-400 PLCs can have more than one CPU in a central rack. *Multicomputing,* as such a system is called, is not covered in this book.

in a project, although each station can include several intelligent I/O modules (IM, CP, or FM modules).

(b) Select **Hardware** by clicking on the new station icon in the left window and then double-clicking on Hardware in the right window. A hardware configuration table window will appear.

(c) Assemble components in the hardware configuration window to reflect the PLC system in the real world. Select **View,** then **Catalog** (if the catalog hasn't come up automatically). A new window appears listing standard Siemens PLC components.

(d) An S7-300 station hardware configuration will already have a rail item, but for an S7-400, find the central rack or one that corresponds to the one in your system, then drag it to the configuration table window.

(e) Find the power supply, CPU, and other modules that your real world system contains and drag them, one at a time, to the appropriate slots on the configuration table screen. (*Remember:* slot 2 only for a CPU, slot 3 only for an interface module)

(f) If your system includes **expansion racks:**

(1) Repeat rack and module selection for each additional rack. Each rack must have an interface module (IM), of course.

> If your station contains an S7-314 IFM module, you cannot insert an I/O module into slot 11 of rack 3 because the digital and analog I/O contacts that are "integrated" into the CPU are arranged addresses 124 to 129,[14] addresses that would normally apply to slot 11 in rack 3.

(2) (S7-400 only) Connect IM modules by double-clicking on the send IM, then selecting the **Connection Tab.** Select an expansion rack and click on the **connect button** of the send IM. Repeat for each expansion rack.

(g) If your system includes **Profibus-DP** master and slaves:

(1) A **Properties** dialog box will open when you insert the DP master into your configuration table. Select the subnet to which this module will be connected; enter a Profibus node address.

(2) Drag DP slave modules from the catalog to the connection icon in the hardware configuration window. A **Properties** dialog box will open, into which you can enter a Profibus node address and other Profibus parameters. (See Chapter 13 for more detail.)

(3) For DP slaves representing I/O rack interfaces, a new rack will be inserted into the hardware configuration window. For AS-i controllers as DP slaves, you will be allowed to select and insert AS-i slaves into the AS-i system.

4. Enter configuration parameters for each configurable module at this station. Select a **configurable module** (e.g., CPU, CP, etc.) by double-clicking on it in the hardware configuration table. A **Properties** dialog box will be displayed for that module. Enter the configuration

[14] Twenty digital outputs: Q124.0 to Q126.3; 16 digital inputs: I124.0 to I125.7; four analog input channels: IW128 to IW135; one analog output channel: QW128.

parameters you want. If the module has a clock, select **PLC–Set Time and Date** to set it. In following sections of this chapter we provide more detail for configuration of S7 CPUs and other modules.

5. Select **Station>Save and Compile.** The compile option causes the STEP 7 software to translate the configuration information into system data blocks (SDBs). If SDBs exist at the time that a program is downloaded to a CPU module, they are transferred to the CPU and change the configuration of the CPU and of its configurable I/O modules. There is a Station>Save option, which does not create system data blocks, that you should only use later when you are editing programs and don't want to modify the system's configuration.

6. Check to ensure that the SDBs have been created. (STEP 7 won't create them if you enter data with obvious errors.) Select **Station–Consistency Check.** STEP 7 will tell you why it fails to create SDBs.

7. Prepare your PLC to receive new configuration data: Perform a **Memory Reset** using the keyswitch method described in the S7 hardware installation section of this chapter, or through the software as follows:

- **(a)** Switch the PLC's keyswitch to STOP or to RUN-P, then
- **(b)** Select **File–Open–Accessible Nodes,** and enter the PLC's node number, which will be 2 for a CPU that hasn't been configured yet. (During configuration, you will assign a new node number. Use stickers on each CPU to record the node numbers you assign.)
- **(c)** Select **PLC–Operating Mode..,** and click on **STOP**.
- **(d)** Select **PLC–Clear/Reset.** (Note the STOP LED flashing as the PLC resets its memory.)

8. (While still in the Hardware Configuration screen) Select **PLC–Download–To Module..,** then let the dialog boxes help you verify your choices.

> Even if you were in online programming while you created or modified a module configuration, the new configuration data was not written to the CPU. It *must* be compiled, then downloaded. The new configuration parameters will come into effect the next time the PLC is put into run mode.

9. Exit from the Hardware Configure screen, and save your finalized project into the programming unit's memory, using **File–Save As.** The name you enter *must* have the extension S7P or the data won't be saved correctly. (Experienced personal computer users save their work at frequent intervals.)

The STEP 7 programming language includes several standard functions (SFCs) that can be used to modify the configuration of programmable I/O modules after this initial configuration has been completed. They do not cause the contents of the original system data blocks (SDBs) in the CPU's memory to be changed, so restarting the PLC will always restore the original configuration. SFCs for changing configuration are discussed in a later section.

Configuring the S7 CPU

When you double-click on the S7 CPU module in the hardware configuration table for a station (in step 4 above), a dialog box containing several tabbed pages appears. The tabs allow you to select register types, for which you can enter configuration parameters. The CPU's configurable registers are described below.

General Register Select the **MPI** button from this screen, then select a new **MPI address** for the CPU (2 to 126, default = 2), an MPI **subnet** via which the CPU will communicate, and a baud rate (the default: 17.5 kilobaud, cannot be changed).

> Siemens recommends that you change all CPU addresses to MPI addresses over 2. By not assigning address 0, 1, or 2, you are reserving them for when you have to connect replacement PG, OP, or CPU units (respectively) to the MPI network. Node numbers 0, 1, and 2 are the factory-set addresses of those devices. All nodes on your MPI network must have different MPI addresses, and all must have the same highest MPI address entry.

You must also activate the **This Node Is Connected to the Selected Subnet** box.

If the station is an S7-300 type and it contains additional communication processor (CP) modules or function (FM) modules that also need MPI addresses, you *must* do the station hardware selection before you configure the CPU module, so that MPI addresses sequentially following the CPU's MPI address can be automatically assigned to them as you enter the CPU's MPI address. For example, if a station contains a CPU and two CP modules requiring MPI addresses and you assign address 12 to the CPU, the CP module in the lower of the other two slots will automatically be assigned MPI address 13, and address 14 will be assigned to the remaining CP module. (So don't assign MPI address 13 or 14 to other CPUs.)

Startup Register Select whether the CPU must test its hardware each time it is completely restarted by being switched into run mode (default: "no"). S7-315-DP CPU modules can be configured to refuse to go into run mode if the detected Profibus-DP network configuration is different from the software configuration.

Time limits can also be placed on the amount of time a CPU module will spend waiting for a ready signal from a configurable module or trying to write parameters to a configurable module. (Station configuration parameters are retained *only* in CPU modules, so each CPU must automatically rewrite the necessary configuration parameters to all I/O modules each time the PLC is switched into run mode.)

Cycle/Clock Memory Register Select a **maximum cycle time** (called a watchdog timer time elsewhere in this book, default: 150 ms) and in some CPUs, a **minimum cycle time**. If any scan cycle exceeds the maximum cycle time, the PLC will fault. If any cycle is completed faster than the minimum cycle time, the PLC will delay before starting the next cycle. S7-400 CPUs allow programming of a background program in OB 90 that will execute whenever the CPU is idle while it waits for the minimum scan time to expire.

You can specify a memory address for a **clock memory** (default: none). If you want to use a clock memory, select a data memory byte (M0 to M255). The PLC will toggle the 8 bits of that address at the following rates:

Bit:	7	6	5	4	3	2	1	0
Rate(s):	2.0	1.6	1.0	0.9	0.5	0.4	0.2	0.1

You can also specify that time spent handling **communication via the MPI network** cannot exceed some percentage of the total scan time (default: 20 percent). A low percentage will ensure a more repeatable scan time but may slow the passing of data between stations in your PLC system.

You can limit the percentage of scan time that is spent in **testing the CPU memory** (default: 0 percent means don't test).

Retentive Memory Areas Register Select which addresses should be saved in nonvolatile RAM memory (NVRAM) and **not cleared** when the PLC is switched from stop to run mode, or when power is restored after being lost. If your PLC has a **backup battery**, all data blocks will also be retained, despite your choices in configuring this CPU register. Any data blocks on a **memory card or in EPROM** memory will be copied to RAM when the PLC restarts, despite the configuration entered here.

Depending on the CPU model, you may be able to configure up to the following amounts of retentive memory:

- **Bit memory** from M0 up to M255 (default: M0 to M15).
- **Timers** from T0 to T127 (default: none).
- **Counters** from C0 to C64 (default: C0 to C7).
- (S7-300 only) Up to eight **data block(s),** DB 1 to 127 (default: DB 1). For each, select the address of the first data byte to be saved (DBB 0 to 8191) and the number of data bytes to save (0 to 4096).

Interrupt Register You can change the **priority classes** of interrupts using this configuration option, but only in S7-400 CPUs.

Time-of-Day Register For S7 PLCs that offer time-of-day interrupts (explained in Chapter 11), you can **activate** the interrupt here and set its **start date and time** and its **interval time** (default: not enabled).

Cyclic Interrupts Registers For S7 PLCs that offer cyclic interrupts (explained in Chapter 11), you can set its **execution interval** (default: 100 ms). Change the execution interval to 0 ms to **disable** cyclic interrupts.

Diagnostics/Clock Register Select whether the CPU should record an extended history of PLC events (such as OB calls) in its **diagnostics buffer**, in addition to records of the other faults detected (default: no). When you configure I/O modules (described later), you will include specification of whether they can send fault event messages to the CPU. Select whether this CPU should send a copy of its last diagnostic buffer entry to another MPI mode each time it enters stop mode. The diagnostics buffer entry indicates why it has gone into stop mode (default: yes).

Select how to synchronize this CPU's clock with other clocks (default: no synchronization). You can select this CPU to be the synchronization master or a slave, and you can select the intervals at which synchronization is performed. Synchronization via the S7-400 backplane (also called the K-bus) is possible now (e.g., to a CP module connected to a Profibus or Ethernet network), and synchronization via the MPI port will probably be possible in future S7 PLCs. You can also enter a correction factor to help maintain synchronization between synchronization intervals, after you find how much error your CPU's clock typically exhibits.

Integrated Function Register If your PLC has an IFM-type CPU (I/O built into the CPU module), you can configure some digital inputs as **hardware interrupt** sources (explained in Chapter 11). The inputs that can be configured are:

- I 124.6 to I 125.1 in an S7-312 IFM
- I 126.0 to I 126.3 in an S7-314 IFM

Select whether the digital inputs are to be treated as sources for **interrupt inputs** (to call OB 40) or as high-speed **counter, frequency meter, parallel A/B counter,** or **positioning** sensor (e.g., encoder) input signal sources. If input interrupt use is selected, you can individually enable each of the four input bits (default: disabled) and select whether the interrupt is to be generated when the input signal rises or when it falls. (When OB 40 is called in response to an input interrupt, the STEP 7 operating system passes several parameters to OB 40, including OB40_POINT_ADDR, which is a 32-bit word, of which the low 4 bits indicate which input bit initiated the interrupt). Other interrupt options are discussed in Chapter 11.

Protection Register Eventually, Siemens plans to allow the user to set up a password, without which another user won't be able to change or perhaps won't even be able to monitor memory contents of a CPU.

Remember to return to step 5 of the configuration procedure to save, compile, and download these configuration parameters after you have finished entering them.

Configuring Other S7 Programmable Modules

During hardware configuration of a station, double-click on the hardware configuration table row that contains the specification for a programmable I/O module. A dialog box that pertains to that module will appear. When the user purchases a function module (FM) or a communication processor (CP), they come with software that becomes part of the STEP 7 programming software when installed and which contains the necessary configuration routines.

When configuration data for a programmable I/O module is changed from the default values and compiled, it is placed into two records (records 0 and 1) of a system data block (SDB). There is a separate set of SDB records for each programmable module.[15] Whenever the CPU restarts, it writes SDB configuration data to the programmable modules (if the programmer has changed the default configuration). The configuration in record 0 usually only enables or disables the module's ability to generate programmable error messages for sending to the CPU module's diagnostic buffer. Record 1 will contain additional configuration, including definitions of the types of detectable faults that will cause programmable messages and generation of

---

[15] Your manual set tells you which SDB is used for what purpose and describes the structure of the records. You only need to know these things if you want to write user-programs to modify the configuration. We discuss how to do so later.

diagnostic interrupts to cause the CPU module to call OB 82 immediately. (Chapter 11 covers interrupt response.) Even if record 0 disables sending of messages, some types of detectable hardware faults (nonprogrammable events) will still cause messages and diagnostic interrupts to be sent to the CPU.

User-programs can contain system functions (SFC 55, SFC 56, or SFC 57) to dynamically change a module's configuration, as we will see later in this chapter. Executing one of these SFCs does *not* change the SDBs in the CPU, so restarting the PLC restores the module configuration to its original state. Record 0 module configuration can only be changed by SFCs in an S7-400, but record 1 configuration can always be changed dynamically.

Besides the diagnostic response characteristics that have just been described, other configuration settings must be entered using dialog boxes for programmable I/O modules. Depending on the type of module, the following options may exist: (All configuration options described below become part of an I/O module's record 1 unless otherwise specified.)

1. *Digital input modules* can be configured to delay changing their input states for between 0.5 and 20 ms, to debounce inputs. This configuration becomes part of record 0. Digital input modules can be configured to generate input interrupt signals (causing the CPU to call OB 40) in response to digital input state changes. (Chapter 11 covers interrupt response.)

2. *Digital output modules* can be configured with "substitute" output values and can be configured to hold their current output states or to output the substitute states. It is unlikely that a first-time configuration would configure a module to hold or to substitute output values. These configuration parameters are intended to be changed by a user-program (using the SFCs for that purpose) in response to a detected environment condition.

3. *Analog input modules* can be configured for a selected type and range of analog input signal per channel (if hardware configuration isn't available on the module). Analog input modules can also be configured to generate an input interrupt signal (requiring the CPU to call OB 40) when the input signal exceeds an upper or a lower limit. The limit values must also be entered here.

4. *Analog output modules* can be configured to output a specific type and range of output signal per channel (if hardware configuration isn't built into the module). Analog output modules can be configured with a substitute output value (per channel) and can be configured to either hold the last output signal level or to output the substitute value (usually a zero value) when the PLC is put into stop mode.

Remember to return to step 5 of the configuration procedure to save, compile, and download these configuration parameters after you have finished entering them.

Configuring S7 Communications

In this section we present only a brief overview of the steps in the configuration process. Configuring an S7 PLC for communications is described in detail in Chapter 13. First, you must determine how you want each S7 CPU to communicate with other nodes and what types of network it is going to communicate via:

Global data circles can be set up so that data in one CPU's memory is automatically copied via an MPI network to the memory of a small number of other S7 nodes. To configure an S7 CPU to participate in global data circles in an MPI network:

1. Configure a project so that it contains at least two stations, both configured as being connected to an MPI network.

2. Select the MPI subnet, then select **Options,** then **Define Global Data.** A configuration table will open for configuration of global data for an MPI subnet.

3. Select the stations to participate in the global data circle.

4. Enter the address(es) at each station that you want to dedicate to global data circle communication.

5. Create additional global data circles for exchanging data with other sets of S7 CPUs.

6. Select **GD Table Compile** to create system data blocks (SDBs) from your global data configuration. If you have entered data correctly, the software will report when it has completed this phase 1 compiling step successfully.

7. Optionally, set a scan rate and/or create GD status words, then select compile again to perform a phase 2 compilation.

8. Save your work. The SDBs will be downloaded to a station in the global data circle the next time you download a program to a PLC.

Standard functions (SFCs) in user-programs can be used to send data to any other node on an MPI network or to get data from that node (see Chapter 13).

S7-400 PLCs can use **coordinated exchanges of data between nodes** on an MPI, Profibus-DP, or an Industrial Ethernet network by using system function blocks (SFBs) if the node that initiates the communication has been configured to have a connection to the other node, as described above. To configure an S7-400 PLC to have configured connections to other controllers:

1. Create and configure an MPI, Profibus, Industrial Ethernet, or a point-to-point subnet in the project. At least two modules capable of communications must be configured for the subnet. You have already done this if you have been following the instructions above.

2. Select a node you will want to connect to another node, select its **Connections** object, then select **Edit–Open Object.** A configuring connection window will open.

3. Select **Insert–Connection** and a new dialog box will open.

4. Select a *partner* **Station** and **Module** to establish a connection with, and select a **Type** of communication for, this connection. Select S7 single station if both the local and partner are S7 PLCs, or select another type depending on the connection between the local and partner.

5. Select **Edit–Object Properties,** and another dialog box will open. Select **Active** if you want this node to set up the communication connection when it starts.[16] Select **Send Operating Mode Message** if you want this PLC to notify the partner each time its mode changes (stop, run, fault). You can also select a direction for communication unidirectional communications (for bidirectional communication, both controllers must be configured with a connection).

---

[16] You can only configure dynamic communications in M7 PLCs. With dynamic connections, neither node will set up the connection automatically, but instructions in the user-program can do so.

6. Note the local ID numbers and perhaps partner ID numbers that are assigned. These ID numbers will be needed as parameters for the SFC or SFB calls for communication in the user-program.

7. Download the connection table to the nodes.

8. Connection tables cannot be uploaded from a CPU, so save the connection table on your programming unit after each time you make modifications or you won't be able to see the connections again!

Profibus-DP network masters maintain an area of memory for up to 4 input bytes and 4 output bytes for each DP slave. The Profibus-DP network automatically copies data between these memories and the appropriate DP slave's memory, as if the slave were in a slot in the DP master's rack system. To configure a Profibus-DP to communicate in this way:

1. Create and configure a Profibus subnet in the project.

2. While configuring each DP-capable module on the Hardware Configuration screen, select a Profibus subnet and activate the **"This Node Is Connected to the Selected Network"** box.

- **(a)** While configuring a DP master, use the dialog boxes to set up to 1024 bytes of the Profibus master's I/O memory through which the master can read or write I/O addresses in each slave. Optionally, also set up a diagnostics address for each slave, where a block of diagnostics data from the slave will be kept.
- **(b)** For DP slaves that do not have DIP switches to select their address ranges on the DP network, you must use their dialog boxes to configure specific slave memories to participate in the Profibus communications. A diagnostics address can be set up in intelligent slaves, where the slave will keep a block of data pertaining to the status of the Profibus connection with the Profibus master.

A Profibus-DP master can use SFC 14 "DPRD_DAT" or SFC 15 "DPWR_DAT" in user-programs to exchange data sets as large as 122 bytes with DP slaves. Configured connections must have been set up as described above.

Remember to return to step 5 of the configuration procedure to save, compile, and download configuration parameters after you have finished entering them.

S7 Symbol Table

A symbol table does not get written to a CPU when downloading other project components. The symbol table remains in the programming unit and allows the programmer/operator to use symbolic names rather than absolute addresses. In Chapter 5 we described how to create and edit a symbol table. During first-time configuration of a PLC system, the programmer may wish to import Siemens' standard symbol table into the project. The standard symbol file, SYMBOL.SDF, can be imported from C:\STEP7_V2\S7DATA\SYMBOL.

S7 Logic Blocks and Data Blocks

Organization blocks (OBs), functions (FC and SFC), and function blocks (FD and SFB) are logic blocks. A logic block contains a program algorithm and a variable declaration table. Logic blocks get downloaded to a CPU. Data blocks (DB, DI, and SDB) are also downloaded to the CPU. Variable declaration tables and data blocks are described in Chapters 5 and 9. The programming elements used in logic blocks are described in the first eight chapters.

Logic blocks and data blocks can be created either online or offline:

1. Select the CPU in a station in a project, then select **Insert,** then **S7 Block,** then select a block type (OB, FB, FC, or DB).

2. If you have selected a logic block type (OB, FC, or FC), select the programming language you wish to use (ladder logic, STL, FBD, or other languages if you have purchased them).

OMRON CQM1 FIRST-TIME CONFIGURATION

Configuring the CQM1 CPU

To configure a CQM1, you must write data to data memory addresses DM 6600 to DM 6655. These memories can only be written to using a programming unit, not by a program. The configuration data in these areas becomes effective under different circumstances:

1. DM 6645 to DM 6655 becomes effective immediately and can be changed while the PLC is in program or monitor mode. (Monitor mode is like run mode except that a programming unit can be used to view the memory contents and program status.)

2. DM 6615 to DM 6644 becomes effective when the PLC is switched into run or monitor mode and can only be changed while the PLC is in program mode.

3. DM 6600 to DM 6614 becomes effective when PLC power is switched off, then back on, and can only be changed while the PLC is in program mode.

The following descriptions of configuration options use “x” to indicate hexadecimal character locations that don’t affect this configuration option. It is a good idea to get into the habit of entering zeros for configuration options that aren’t used or for which the manuals imply that the value doesn’t matter. In fact, there are several places where configuration and/or instructions won’t work unless zeros are entered into addresses that seemingly aren’t used. OMRON has plans to expand the power of the CQM1 family of PLCs, and much of the functionality of the CQM1 is similar to other C series OMRON PLCs.

Although you must enter the appropriate configuration data into data memory (DM 6600 to DM 6655) to enable CQM1-configurable options, some of those features also require a user-program to execute instructions to activate those features, and the instructions often require the programmer to enter data into other memory areas before the features will work properly. In other sections of the book we describe the instructions and data requirements; in this section we cover only the essential data memory configuration for the following processes.

Startup Operations (DM 6600 to DM 6614, effective after cycling PLC power)

1. The data value in **DM 6600** controls which mode the CQM1 enters when power is restored. The options are:

`00xx` The switch on the programming console (pendant) controls the mode of the PLC if a programming console is connected; otherwise, the PLC starts in RUN mode.

01xx	The PLC resumes in the mode it was in when power was lost.
0200	Start in program mode.
0201	Start in monitor mode.
0202	Start in run mode.

2. Data in **DM 6601** affects the refreshing of I/O images, work memory, and communication messages in the IR and LR (link register) areas of memory when power is restored. Options are:

00xx	Reset all IR and LR memory. Also reset bits SR 25211 and SR 25212.
x1xx	If the **IOM hold bit** (SR 25212) was set when power failed, IR and LR data in memory was saved when power was lost, and the operating system will retain that data as it goes into run or monitor mode. If SR 25212 contained 0 when power failed, reset IR and LR memory.
1xxx	If the **forced status hold bit** (SR 25211) was set when power failed, any IR and LR data in memory that was being forced on or off was saved in that state when power was lost and the operating system will not clear the forces as it goes into run or monitor mode. If SR 25211 was cleared when power failed, all forced states are removed.

3. Data in **DM 6611** and in **DM6612** are used as compensation values to correct absolute encoder values read at ports 1 and 2 (respectively) of a CPU44-E. Do not enter values manually. Your manual describes how to make the CQM1 enter values into these addresses.

4. Data in **DM 6611** selects the use of ports 1 and 2 of a CPU43-E as either high-speed counter mode input ports (if **0000**) or as pulse output ports (if **0001**).

5. Other data words in this range are not used.

Pulse Output (DM 6615, effective upon switching into run or monitor mode) Enter xx**NN**, where "NN" is the number of the output channel to write pulses to in response to the SPED(64) instruction. The SPED(64) instruction (described in Chapter 11) sets up a pulse frequency of up to 1 kHz[17] and selects which output bit to pulse.

Cycle Time Settings (DM 6616 to DM 6619, effective upon switching into run or monitor mode)

1. The value in **DM 6618** sets the cycle monitor time (called a watchdog timer time elsewhere in the book):

0000	for the default 120-ms setting
0btt	to set the time:

- "b" = "1" for 10-ms units, "2" for 100-ms units, or "3" for 1-s units.
- "tt" indicates the number of time units (00 to 99).

[17] SPED (64) is also used to initiate faster pulses via port 1 or 2 of a CPU43-E, in which case the configuration in DM 6615 is irrelevant.

The CQM1 records the length of its longest scan cycle time in AR 26 and the length of its most recently completed scan cycle in AR 27.

2. DM 6619 is used to set a minimum cycle time. 0000 disables this option, but any other BCD number (0001 to 9999) sets a minimum scan time in milliseconds. If any scan cycle finishes in less than the minimum cycle time, the CQM1 waits until the minimum cycle time lapses before starting the next scan cycle. If a scan cycle is longer than the minimum cycle time, the CQM1 turns the **long cycle time flag** bit, **AR 2405,** on but executes the long scan cycle to completion before the next scan cycle starts.

3. DM 6615 can be used to set an upper limit on the percentage of the scan cycle time that is dedicated to servicing of the RS-232C port.

0000	Do not limit RS-232 port servicing time.
01nn	"nn" is the upper percentage limit setting (00 to 99 percent).

4. DM 6616 limits the servicing time for the peripheral port. Set as for DM 6615, above.

Interrupt Settings (DM 6620 to DM 6639, effective upon switching into run or monitor mode)

1. DM 6620 to DM 6625 are used to select the latency for the bits in individual bytes of input image, to debounce them. Bit values aren't changed unless the input voltage remains changed for the entire latency period.

```
xxxN        of DM 6620 affects bits IR 00000 to IR 00007
xxNx        of DM 6620 affects bits IR 00008 to IR 00015
xNxx        of DM 6620 affects bits IR 00100 to IR 00107
Nxxx        of DM 6620 affects bits IR 00108 to IR 00115

xxxN        of DM 6621 affects bits IR 00200 to IR 00207
xxN etc, x of DM 6621 affects bits IR 00208 to IR 00215
```

where "N" is

```
"0" for 8 ms (the default)
"1" for 1 ms
"2" for 2 ms
"3" for 4 ms
"4" for 8 ms
"5" for 16 ms
"6" for 32 ms
"7" for 64 ms
"8" for 128 ms
```

2. The four digits of **DM 6628** can each enable one *I/O interrupt* or one high-speed counter interrupt. An I/O interrupt can cause the main program to be interrupted to execute a subroutine. A high-speed counter can count input pulses and interrupt the main program to execute a subroutine when a setpoint value (SV) is reached. (For more detail, see the discussion of I/O interrupts in Chapter 11.) If DM 6628 contains:

```
0000    No input interrupts are enabled.
xxx1    IR 00000 can cause I/O interrupt 0 (calling subroutine 0) or
        can change high-speed counter 0's present value (PV).
```

```
xx1x    IR 00001 causes I/O interrupt 1 (subroutine 1) or HSC 1.
x1xx    IR 00002 causes I/O interrupt 2 (subroutine 2) or HSC 2.
1xxx    IR 00003 causes I/O interrupt 3 (subroutine 3) or HSC 3.
```

Even if DM 6628 enables interrupts, the interrupts are still masked (blocked) until an INT (89) instruction unmasks them. Another INT (89) instruction selects whether the input bit will cause an I/O interrupt or will cause a high-speed counter's present value (PV) to change.

3. DM 6629 can be used to set up as many as 16 high-speed timers. Unlike standard timers, the present value (PV) of a high-speed timer is updated as soon as each time unit expires, not just when the timer instruction is executed. (See the discussion of timed interrupts in Chapter 11 for more detail.) If DM 6629 contains:

```
0000    16 high-speed timers (TIM 000 to TIM 015) are enabled.
0100    No high-speed timers (other timed interrupts will execute
        more dependably).
01NN    NN timers (01 to 15) are enabled, starting with TIM 000.
```

4. DM 6630 to DM 6638 indicate which input image words to *refresh* (by reading the input modules) immediately before an interrupt service subroutine executes. DM 6630, for example, indicates which input image words must be refreshed before the subroutine for I/O interrupt 0 executes. If DM 6630 contains:

```
0000    (default) Do not refresh any input images.
NNPP    Refresh NN (01 to 12) words of input image starting at
        address PP (000 to 011).
```

In Chapter 11 we describe which data memory (DM) words to use to configure refreshing before I/O interrupts, high-speed counters, or timed interrupts.

Output Refresh Method (DM 6639, effective upon switching into run or monitor mode) If DM 6639 is:

```
xx00    Output image values are written to output modules after each
        scan cycle (as usual).
xx01    Output image values are written immediately to output modules
        as the program changes them.
```

Digital Switch Input (DM 6639, effective upon switching into run or monitor mode) If you have a digital switch connected at an input module and you intend to use the DSW(−) instruction to read the switch, DM 6639 should be configured with:

```
00xx    to read a 4-bit number
01xx    to read an 8-bit number
```

High Speed Counting of Encoder Pulses (DM 6642 to DM 6644, effective upon switching into run or monitor mode)

1. DM 6642 is used to configure any CQM1 to count input signal changes at the input contacts associated with input image bits IR 00004 to 00006 as pulses from an incremental encoder. The CQM1 will interrupt its processing to change high-speed counter 0's present value whenever it receives a pulse at one of these inputs. If DM 6642 is:

```
0000    Do not count (no encoder input).
01x0    Count between −32,767 and +32,767.
01x4    Count between 0 and 65,535.
010x    Reset count after the program turns reset bit (SR 25200) on,
        then input contact 00006 goes on.
011x    Reset count when the program turns reset bit (SR 25200) on.
```

For more detail, see the discussion of I/O interrupts in Chapter 11.

2. DM 6643 and DM 6644 are used to configure the additional input ports (DM 6643 for port 1, DM 6644 for port 2) of a CQM1-CPU43-E for incremental encoder input or of a CQM1-CPU44-E for absolute encoder input.

RS-232C Port and Peripheral Port Communications Settings (DM 6645 to DM 6654, effective as soon as changed) Use these data memories to configure a communication protocol for each of the two communications ports on the CPU module. The port can be configured for:

- Standard RS-232C communication protocol, to transmit and receive data messages with peripherals or other computers.
- One-to-one communications, to share memory with one other computer.
- Host link communications with several other CQM1s (via a host computer) in a local area network.

In Chapter 13 we describe how to configure communications and describe the program instructions to use the communications capabilities of the CQM1. Until you have read Chapter 13, leave all configuration parameters in this range zeros and turn the CPU DIP switch 5 on. While switch 5 is on, the CQM1's RS-232C port is hardware-configured to a default configuration and the settings in DM 6645 to DM 6649 have no effect.

Error Log Settings (DM 6655, effective as soon as changed) Controls how PLC errors are recorded.

```
xxxN    where "N" affects how to store PLC fault records in the error log.
        If N =
        1       Only save the first 10 error records.
        0       Save all error records. Older error records are lost be-
                cause the error log can hold only 10 records.
        2 to F  Do not store error records.
x0xx    If minimum cycle time is exceeded, set AR 2405.
x1xx    Don't set AR 2405 if minimum cycle time is exceeded.
0xxx    If low battery power is detected, set SR 25308.
1xxx    Don't set SR 25308 if low battery power is detected.
```

Other CQM1 CPU Configuration

An additional area of data memory, from **DM 6144 to DM 6568,** is also protected from being overwritten by a program, so the programmer can use a programming unit to store read-only data in this area. **DM 6569 to DM 6599** is also read-only, in that it can't be changed from a program, but the CQM1's operating system automatically uses it to store an error history (which is described in the section on fault interrupts in Chapter 11 and again in Chapter 15).

Configuring CQM1 Expansion Instructions

Eighteen function codes are reserved for expansion instructions. OMRON has preassigned 18 of the expansion instructions to the function codes, but the programmer can change the assignments. Instructions without numbers inside brackets are the unassigned expansion instructions, which can be assigned function codes as follows:

1. Select the **UTILITIES** menu, then select **SET INSTRUCTIONS** to get the Set Instruction menu.

2. Select **EDIT INSTRUCTIONS** to get a list of the available function codes, complete with the default expansion instructions that OMRON has selected for each.

3. Select **WRITE,** then cursor to the function code you want to assign to a different expansion instruction (changing the default instruction for that code).

4. Press [**F2**] to get a list of the expansion instructions, and cursor to the one you want to assign to the function code you have selected, then press [**Enter**].

5. Press [**F10**] to record the changes in the programming unit's memory. You can now use the newly assigned function code(s) in programs.

6. Select **SAVE INSTRUCTIONS** if you want to save the currently assigned expansion instructions for future use.

Saving CQM1 Configuration to an EEPROM Memory Cassette

To save the new configuration to a memory cassette, you must insert a memory cassette with its write-protection turned off. *Remember that you must never insert or remove a memory cassette while PLC power is on.* Set AR 1400, and the configuration and any program in memory will be copied to the memory cassette.

RECONFIGURATION DURING RESTART OF A PLC PROGRAM

Since users can change a PLC's configuration, it may be desirable for the user-program to reinitialize the configuration each time the PLC goes into run mode. Some configuration options can be changed by PLC programs as they execute, so a user-program can be written to modify the PLC's configuration as environmental conditions change.

PLC-5

SLC 500

ALLEN-BRADLEY RESTART CONFIGURATION

Allen-Bradley PLC-5s and SLC 500s use backup batteries to retain almost all the data in their memory while the PLC is switched off or in program mode. When the PLC restarts, some status data is lost from the status file, nonretentive timers are reset, input images are refreshed, and output image data is cleared, but all configuration data in the status file and even things such as counter accumulated values are saved. The user may want to program the PLC so that startup resets the contents of some memory to default values. In Chapter 11 we describe how to use a fault routine for restart purposes, or the programmer can use the first scan bit feature.

Whenever a PLC-5 or an SLC 500 is switched into run mode, it automatically sets a **first scan bit (S:1/15)** for the duration of the first scan. A PLC program can include a rung that examines the first scan bit to control resetting a configuration option (by writing a new value into the appropriate status file word). The rung can even control a jump to a subroutine that contains a full set of reconfiguration statements.

Dynamic configuration options change as soon as a program changes the status word. Static configuration options become effective only when the PLC is restarted. If it is necessary to change a static configuration word, the user-program should also set a major fault bit so that the PLC will stop and an operator will be forced to switch it back into run mode. (The program should set the major fault bit only if it detects the need to change a static status word, of course.) Your manuals tell you which status words are dynamic. Instead of using the first scan bit, an Allen-Bradley PLC can be configured to execute its fault routine each time it is restarted. Reconfiguration instructions can be in the fault routine.

All intelligent I/O modules, including analog I/O modules, lose their configuration when the PLC is put into program mode or is powered down. The user-program should ensure that configuration data is rewritten to intelligent I/O modules each time the PLC goes into run mode.

S5

SIEMENS S5 RESTART CONFIGURATION

Siemens S5 CPU modules can store CPU configuration in their battery-backed RAM, or in FLASH memory, or in EEPROM memory (depending on the CPU make and model), so CPU configuration can be saved while the CPU is in stop mode or is powered down. Some configuration parameters can be overwritten by a user-program, but if the CPU is provided with EEPROM or flash memory containing default startup parameter values, reloading the default values will be automatic every time that power is switched on. It is only necessary to program startup blocks (OB 21 and/or OB 22) to reload values that are excluded from EEPROM or flash memory, or if the startup configuration is conditional upon other PLC-detectable environmental conditions.

Except for a few intelligent I/O modules and communication processor modules, I/O module configuration is lost when the PLC is not in run mode. S5 PLCs should be programmed with startup blocks (OB 21 and/or OB 22) to write configuration parameters to I/O modules each time the PLC restarts unless the I/O module has an EEPROM module that stores the configuration.

S7

SIEMENS S7 RESTART CONFIGURATION

S7 PLCs store configuration data in system data blocks (SDBs), which are saved even when the CPU is not powered. System data blocks containing I/O module configuration are automatically written to I/O modules whenever the PLC is restarted. There are two ways that an S7 CPU might start:

1. If an S7-400 PLC was running without problems when power was interrupted and its mode or its memory contents haven't been changed (e.g., by switching a memory card), it performs a **restart** when power is restored. During a restart it rewrites I/O module configuration parameters to the I/O modules, executes the **restart organization block (OB100),** then resumes the program execution at the point at which it was interrupted. Nonretentive timer, counter, or bit memory, and I/O images are *not* reset during this type of restart.

2. If the S7-400 memory contents have changed, if a fault is detected, or if the PLC has to be switched back into run mode, a **complete restart** is executed. Other S7 CPUs can only restart using the complete restart method. During a complete restart, I-stack and B-stack contents are cleared, I/O images and nonretentive memory are reset, the **complete restart organization block (OB101)** is executed, then the scan cycle restarts from the beginning.

To change a PLC's configuration as part of day-to-day operations:

1. The operator can initiate a memory reset. A memory reset can be caused by a manual operation using the mode switch (see Siemens S7 hardware installation) or through a programming terminal. When a memory reset is executed, the CPU deletes almost everything in its RAM and NVRAM memory, including configuration data in SDBs. The only configuration parameters not deleted are the MPI address and baud rate, so the CPU can still receive data from a programming unit on an MPI network. If the CPU has a memory card installed, the contents of the memory card (typically, a default configuration and program) are automatically reloaded into RAM. If the CPU doesn't have a memory card, a new CPU program must be downloaded to the CPU from a programming unit.

2. Include calls to SFCs to change I/O configuration as the program executes. The SFCs can be called in OB100 and/or OB101 to establish a default configuration for each time the PLC is restarted or can be elsewhere in the user-program to change the PLC's configuration as required during normal program execution.

You can assemble records containing alternative configurations in the CPU's memory, for writing to an I/O module as required. Your manual set explains the structure of SDB data records for each type of I/O module. There are three system functions that can be called to initiate parameter exchanges with I/O modules:

SFC 55 "WR_PARAM" can be used to modify part of an I/O module's configuration (the part in record 1), to change the settings written to the module when the CPU wrote SDB data to it. Figure 10.10 contains an example program that uses "WR_PARAM" to enable an analog input module to generate hardware interrupts if an analog input value exceeds an alarm input value.

SFC 56 (WR_DPARAM) will restore one record (you specify which) of an I/O module's configuration to its original configuration as saved in the SDB by writing one record from the SDB to the module.

SFC 57 (PARM_MOD) sends a complete set of SDB parameters to the module, restoring its whole configuration and other programmable characteristics.

Other SFCs can be used to read a module's status (SFC 59, RD_REC) or to write records (SFC 58, WR_REC) other than those containing configuration data to an I/O module.

```
        A       I  4.0          // Whenever input I 4.0 goes from off to on,
        FP      M  4.0
        S       M  2.0          // set "WR_PARAM" 's REQuest bit M 2.0,
        R       M  2.1          // and reset "WR_PARAM" 's BUSY bit m 2.1,
        JC      init            // then jump ahead to the label "init"

        A       I  4.0          // But while input I 4.0 stays on
        A       M  2.1          // and "WR_PARAM" hasn't completed the transfer,
        BEC                     // skip execution of the rest of this block

        JU more                 // If neither condition is true, bypass this example

 init : L B#16#80               // Binary pattern which enables limit interrupts
        T DB2.DBB0              // to first byte of data block 2

        L B#16#10               // Hex value 10 put into data block 2, byte 6
        T DB2.DB6               //      as an upper limit value for analog

        L B#16#00               // Channel group 0
        T DB2.DBB7              // No other configuration bytes in DB2 are changed

  CALL "WR_PARAM"               // SFC 55, which can change record 0 or 1
   REQ      : =M 2.0            // Write to module while M 2.0 is true
   IORD     : =B#16#54          // Hex 54 indicates an input module
   LADDR    : =256              // Module is at first slot in rack 0
   RECNUM   : =B#16#1           // Hex 1 indicates changing record 1
   RECORD   : =P#DB2.DBX0.0 Byte 14 // Address of the 14-byte
                                //      data record to be sent to the module
   RET_VAL  : =MW4              // MW4 will contain the failure reason if
                                //      "WR_PARAM" fails
   BUSY     : = M 2.1           // Will stay set until "WR_PARAM" finishes

more :                          // The rest of this program. . .
```

FIGURE 10.10
STEP 7 program using WR_PARAM to configure an analog input module to generate hardware interrupts, even though it was not originally configured to do so.

CQMI

OMRON CQM1 RESTART CONFIGURATION

In a CQM1, configuration parameters in the CQM1's data memory addresses DM 6144 to 6655 cannot be changed while the PLC is in run or monitor mode. Since the configuration parameters discussed in this chapter are all in that range, it may look like you can't write user-programs that dynamically change the CQM1's configuration. Dynamic configuration changes are sometimes impossible. You can't, for example, change any of the scan time settings, or change which input words to refresh when an interrupt occurs, or change the numbers of high-speed timers or counters configured, or the setting of the additional ports (for CQM1s that have additional ports).

On the other hand, some DM parameters only enable CQM1 capabilities that the user-program must turn on (or off) by executing the appropriate instructions. To turn input interrupts on (or off), the user-program must execute the MODE CONTROL instruction: **INI (61)**, and sometimes must enter setpoint values into status register (SR) memories. To use high-speed counting, the REGISTER COMPARISON TABLE instruction, **CTBL (63),** must be executed, and data sets must be entered into memory. High-speed pulse generation is set up and started by executing a variety of instructions, including the SPEED OUTPUT instruction: **SPED (64).** (These and other examples of dynamic reconfiguration are explained in Chapter 11.) The user-program can monitor input conditions for situations requiring the use of enabled configurable features and can turn those features on or off as required. The programmer should latch status bits and/or use the differentiated form of the initiation instructions to prevent wasting time reinitiating a feature that has already been initiated.

Yet other configurable features that are enabled by DM parameters work only if user-program-controllable bits are on when the configurable option is called. These bits can be turned on by a programmer through a programming terminal, or the user-program can turn them on. The I/O memory hold configuration in DM 6601, for example, enables retaining of I/O memory at startup, but I/O memory will be retained only if SR 25212 is on when the PLC stops running. Several other dynamic reconfiguration options are controllable by writing to SR memory, some of which are described elsewhere in this book (in sections dealing with the features they affect).

There are two other status register bits that can be used to dynamically reconfigure the CQM1:

1. The **first cycle flag bit, SR 25315,** goes on for one cycle each time the PLC restarts its scan cycling. The programmer can write lines in a user-program to examine this bit and to execute the instructions needed to initiate the features needed each time the PLC goes into run mode.
2. The **output off bit, SR 25215,** holds all outputs off when set to 1.

TROUBLESHOOTING

Difficulties during setup and configuration may result in:

- Not being able to connect online to the PLC from your programming unit, or no longer being able to read or change selected memory areas
- PLC faulting, or otherwise behaving incorrectly
- Failure to communicate with other CPUs via a local area network, or communicating slowly
- Configured features not working

If you find that you cannot connect to the PLC for online programming:

1. Verify that your programming unit is attempting to communicate. If communication is via an RS-232C port, use a protocol checking device in the RS-232C line and look for activity on the transmit (TXD/TD) and receive (RXD/RD) lines when you attempt to use the software to send data to the PLC.

- If there is no activity, the programming unit's communication software is set up incorrectly. Try changing the COM port setting, or plug the interface cable into another serial port at the programming unit.
- If there is activity on lines other than the transmit and receive lines, check that your connecting cable is correct, especially if it wasn't made by the PLC manufacturer. Use a continuity checker to verify that the correct connections (and no other connections) are made from one end of the cable to the other.

2. Check that your programming unit and PLC are speaking the same protocol language.

- Have you changed any part of the PLC's default communication protocol? You may have changed the PLC's node address or baud rate using DIP switches, for example, so the programming unit is trying to connect to a node that doesn't exist. Change your programming unit's communication parameters to match those of the PLC.
- If configuration parameters have been written to the PLC in the past or if a memory cassette is installed in the PLC's CPU module, the PLC may now be using a communication protocol different from the default. If you can't determine its current protocol setting, remove any memory cassettes or backup batteries and reset the PLC's memory. This will delete user-programs and data memory contents as well as resetting the communication parameters. In the S7 series of PLCs, a memory reset does not cause the CPU to revert to a default node number, so you may still have to try connecting to nondefault nodes if the node number has been changed.

3. If communication is via an interface card in the programming unit, verify that the software's interrupt setting (IRQ number) and personal computer address setting match those set up on the interface card, and do not conflict with those already used by the personal computer for other purposes. Change the interface card's settings, if necessary. This may require changing DIP switch or jumper settings on the card and will require changing the communication setup in the software.

If you can no longer access part of the PLC's memory to monitor and/or change it:

1. Does your PLC have memory protection features controlled by DIP switches (as a PLC-5 has)? Check those DIP switches.

2. Does your PLC offer password protection of selected memory areas? Did someone else set up a password before you got around to it? If you can't find the password, you may be able to reload a PLC program with a configuration that doesn't require a password (if you thought to save a copy immediately after configuring the PLC). In the worst case you will have to clear the PLC's memory. If you aren't allowed to clear the memory through the programming unit, there may be a memory reset hardware procedure or you may have to remove the memory backup battery to dump RAM memory.

3. Some program information may not have been downloaded to your PLC from the programming unit. If you try to monitor that program from another programming unit, you can't use the information that is only saved in the original programming unit. Siemens PLCs do not contain symbolic names or program labels used in jumps, for example. Copy the missing files from the original programming unit to be able to use the missing data at a new programming unit.

If the PLC repeatedly faults or behaves incorrectly:

1. Use your programming unit to check the history of fault codes in the PLC's memory. (Saving of fault messages in the CQM1 can be turned off, so you might have to change the CQM1's configuration to save them first.) The codes will tell you what type of problem exists. During first-time PLC setup and configuration, the problem could be things such as an electrical fault at an I/O module, a cycle time fault (if the watchdog timer is set too short), or an attempt to use a program (e.g., an interrupt file) or a memory that doesn't exist yet.

2. Have you configured communications between the CPU and remote I/O properly? Check the LEDs at the remote rack to ensure the remote racks are running normally. Check any status bits your PLC may maintain for remote racks and/or for input states from remote racks.

3. Are address-control DIP switches set correctly? PLC-5 racks must be set for two-slot, single-slot, or 1/2-slot addressing, depending on the I/O modules you use. Some remote I/O interface modules and some intelligent I/O modules have DIP switch settings that affect the addresses they use in the CPU's memory.

4. Check to ensure that you haven't accidentally entered configuration data you didn't intend to enter. Disabling main control programs in a PLC-5, for example, or turning the output off bit on in a CQM1 will prevent the PLC from controlling outputs as it should.

If the PLC is not communicating via a local area network, or if communication is slow:

1. Check to see if your PLC can find all of the other PLCs that are active on the network. Some PLCs maintain a list of active nodes that you can read through the programming terminal, or the programming terminal itself may maintain a list of active nodes that your PLC should also be able to find. If even one node on the network is failing or is configured differently from the others, it may be preventing the network from working.

2. Check to see that this PLC has been configured to allow enough time for communications in its scan cycle. Test by disabling controlling of communication time so that all the required message activity occurs. Too little communication handling time can really be a problem in PLCs that are links between PLC networks, where the PLC has to handle message transfers that aren't requested in its own user-program.

3. Are your user-programs simply trying to send too much data via the network?

If the PLC fails to use a feature you have attempted to configure it to use, or if the PLC won't allow you to enter configuration data for a feature:

> "RTM" stands for **read the manual.** (Sometimes frustrated support personnel add another word, which I won't use in this book.) While manuals for industrial equipment and software are rarely written to be read by mere mortals, they usually contain the information you need, somewhere! Read everything you can find about the feature you want to use. Check the indexes for references to the feature in an obscure section of the manual. Follow other leads by reading the sections describing related features.

1. Verify that your model of PLC offers the feature you are trying to use! In this book we describe what top-end PLCs of each type can do.

2. Verify that you have entered the additional programs or data required for the feature to execute. Especially in OMRON PLCs, you often need to execute instructions to turn a feature on even after enabling the feature through configuration. (Do the instructions that turn the feature on ever actually execute?)

3. Some configuration changes aren't accepted by the PLC until the PLC is switched out of run mode and then back into run mode. Other configuration changes aren't accepted until the PLC's power is switched off, then back on.

4. Is your configuration saved when the PLC's power is switched off? Most PLCs have a backup battery that retains memory. Check the battery. *Always remove the battery while the PLC's power is on, when the battery isn't needed to retain RAM memory.* (On the other hand, removing the battery while power is off is a good way to clear the contents of RAM memory.)

5. You may have to set DIP switches or change the CPU configuration parameters to allow saving of configuration to a memory cassette or to allow the memory cassette's contents to be copied to the CPU's RAM memory when the PLC starts. If, on the other hand, a blank memory cassette or one with incorrect configuration data is installed, you may have to set DIP switches so that the memory cassette's data isn't copied over your good configuration data when the PLC is turned on. Have you damaged your EEPROM memory chips by inserting them or removing them while the PLC's power was on?

6. Configuration of Allen-Bradley block transfer I/O modules must be done carefully, or the PLC-5 will recognize an inconsistency and won't let you proceed. Save your program before starting to edit the I/O configuration so that you can clear the PLC's memory and reload the pre-I/O configuration program if necessary.

QUESTIONS/PROBLEMS

1. Where would you use a termination resistor?

2. What are some typical options selectable on analog input modules using hardware configuration?

3. How many CPU modules would exist in a system of three distributed racks?

4. How many CPU modules would exist in a local area network with three nodes?

Allen-Bradley PLC-5s

5. In configuring programming software for online programming of the PLC-5, you must enter two DH+ addresses. What are each of these addresses for?

6. **(a)** During online programming, you can configure your PLC's communication time slice by entering a time or by selecting automatic operation. What are some types of communication that happen during the communication time slice?

(b) What would happen if you entered a time for the communication time slice instead of allowing it to remain automatic?

7. Identify three examples of items you might want to change in the processor configuration of an Allen-Bradley PLC.

8. **(a)** How does a PLC-5's two-slot addressing system differ from the $\frac{1}{2}$-slot addressing system?

(b) If a PLC-5 is set up with one-slot addressing, every input module can provide data for ________ word(s) of input image data.

9. What are the DIP switches on a PLC-5's CPU module used for?

10. What can the DIP switches settings on a remote I/O module be used to indicate?

11. Why is it necessary to set up a scan list of remote I/O racks during configuring of a PLC-5?
12. If you want to prevent unauthorized users from changing a program, what will you do while setting up your PLC-5?
13. Intelligent I/O modules have built-in microprocessors and programming to perform specific functions under control of the PLC's CPU module as the programs you write are executing. Each time a PLC-5 goes into run mode, it must reconfigure its intelligent I/O modules by executing ________ instructions.

Allen-Bradley SLC 500s

14. When do you configure an SLC 500 to enable M0/M1 monitoring?
15. Which data word contains the I/O slot disable bits?
16. What types of communication channels does the SLC 500 have?

Siemens S5 PLCs

17. How does an S5 PLC know what type of I/O module is present in a bus unit slot?
18. How is a centralized rack system different from a distributed rack system?
19. How is an interface module (IM) different from a communication processor (CP)?
20. How do you configure cyclic interrupts?
21. How do L1 slaves exchange data with each other?

Siemens S7 PLCs

22. What types of local area network interfaces are available for S7 communications?
23. What configuration is required before S7 PLCs can exchange global data via the default MPI network?
24. Where is configuration data for intelligent I/O modules kept while power is off?
25. How do you configure cyclic interrupts?

OMRON CQM1 PLCs

26. How can you configure a CQM1 so that further configuration changes to DM 6144 to DM 6655 are disallowed?
27. **(a)** Can a user-program change the configuration data in DM memory?
    (b) Which DM memories contain configuration data that becomes effective immediately upon being changed?
28. How would you set up a CQM1 so that all output images are sent to output modules as soon as a user-program changes them?
29. The CQM1 does not have timed interrupts to force consistent intervals between critical control operations. How would you configure the CQM1 to cycle at consistent interval times?

11

INTERRUPTS

OBJECTIVES

In this chapter you are introduced to:

- What happens when a program is interrupted, and how it resumes after an interrupt service routine (ISR) finishes.
- Immediate I/O instructions and/or addresses to make a PLC read or write an I/O module during execution of a user-program.
- Configuring a PLC to execute an ISR immediately upon an input signal change.
  - High-speed counter applications.
  - Frequency monitoring applications.
- Configuring a PLC to execute an ISR at timed intervals or at a preset time or after a preset delay.
  - Pulse generation and pulse width modulation.
- Configuring a PLC to execute a fault routine to try to recover from a fault detected during program operation or after a power failure.
- Initialization routines.
- Communication functions and interrupts.
- Instructions and/or functions that modify a PLC's interrupt response.

Interrupt-handling features, now offered in many PLCs, can be used to make a PLC respond faster to changes in the workplace or to make the PLC respond at more consistent intervals. Some control applications, such as position control, need consistent control intervals. Other control applications, such as responding to an intruder in a robotic workcell, require immediate response.

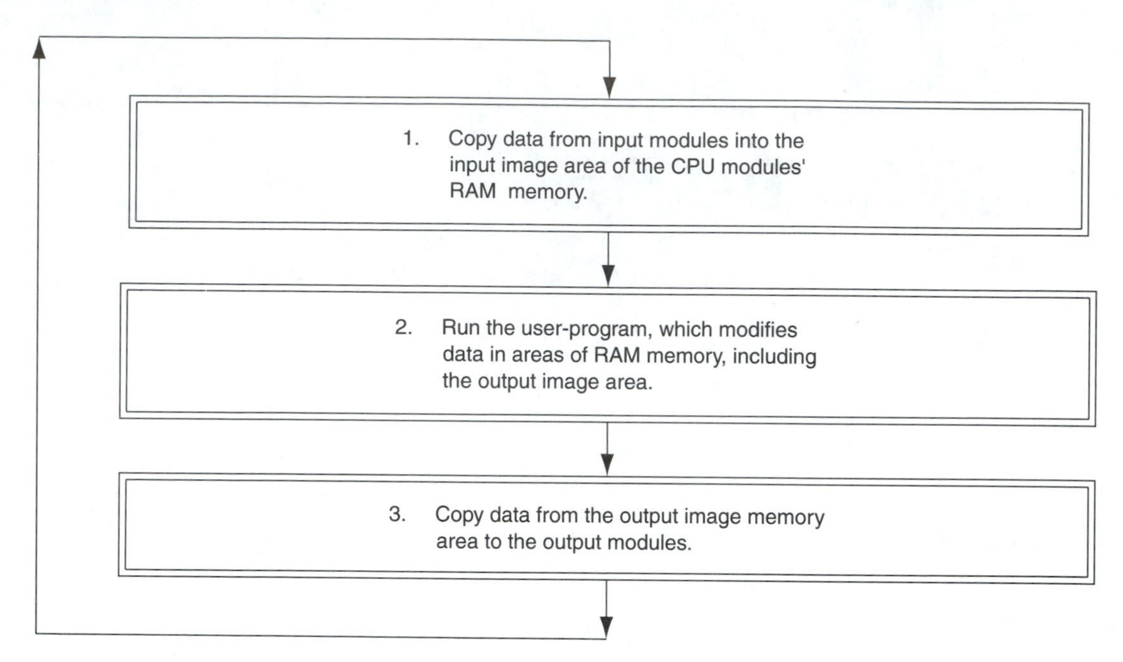

FIGURE 11.1
Standard PLC scan cycle.

THE PROBLEM

The standard PLC operating system causes the PLC to execute a three-step scan cycle repeatedly, as shown in Figure 11.1, when the PLC is running.[1] This scan cycle was originally devised so that PLCs could execute their control programs quickly and at consistently repeated interval times. Reading all the inputs and writing all the outputs together once per program execution is more efficient than reading/writing them singly as a control program executes. Consistent program repetition times makes it easier to predict how a controlled system will respond (so PLC control is more deterministic than other types of computer control). The scan cycle is still considered desirable, although it does have some undesirable side effects:

1. It inherently causes a response delay. The total time to complete the three steps is short, typically a few milliseconds, but even this delay can be objectionable if the system being controlled requires rapid response. If, for example, a workpiece enters a paint booth on a high-speed conveyor and moves past a spray paint nozzle for 15 ms before the paint starts, the workpiece is going to be left partly unpainted!

2. *Response time is variable.* Depending on whether an input state changes just before or just after the input scan step starts, the response delay may be one or two scan cycles in dura-

---

[1] Modern PLC scan cycles may also include steps for handling serial data communications, and several PLCs allow modification of the step sequence. For this chapter, assume that a three-step scan cycle is used.

tion. In one actual example, a PLC was programmed to extend a pneumatic cylinder as soon as an approaching pallet was detected, to lift the pallet off the conveyor, and hold it so that a robot could work on the workpiece it carried. Since the pallet was moving at 3 m/s, variations in the PLC response delay meant that the cylinder-actuated positioning mechanism usually missed the pallet, sometimes by as much as 3 cm.

3. Structured user-programs can vary in execution time, making the response time unpredictable (less deterministic). Programs can now contain conditional jumps to subroutines or conditional executions of instructions with long execution times (e.g., calculations). If conditions are such that several subroutine and long-execution-time instructions execute in one scan cycle but not in the next, execution time will vary quite a bit.

THE INTERRUPT SOLUTION

Most new PLCs offer **interrupts.** When the condition that the programmer has defined as an interrupt-condition occurs, the PLC is supposed to respond *immediately.* Whatever the PLC is doing at the time the interrupt condition is detected is stopped (interrupted), and data that the PLC was using is put aside for later. The interrupt-response program then starts. When the interrupt-response program finishes, the interrupted program's data is retrieved and the interrupted program resumes. Interrupt-response programs are properly called **interrupt service routines (ISRs),** which is what they are called in this chapter. PLC manufacturers prefer to make up their own terminology, so you will find ISRs called *subroutines* by OMRON, *blocks* or *functions* by Siemens, or simply *program files* by Allen-Bradley. Figure 11.2 shows what happens when a scan cycle is interrupted.

Six types of interrupt features are available with PLCs. The six interrupt types that are introduced here are described in more detail in the following sections. The terminology that is used is terminology used in the computer industry. PLC manufacturers each tend to invent their terminology for interrupt types, terminology that is often contradictory even between PLC models offered by a single manufacturer. (Terminology used by specific PLC suppliers is used in the sections of this chapter dealing with specific PLCs.)

1. The IMMEDIATE INPUT or IMMEDIATE OUTPUT instruction causes the PLC temporarily to suspend execution of the user-program, copy one data word from an input module to the input image area of memory or from the output image area of memory to an output module, then execute the next instruction in the user-program. Some PLCs offer immediate I/O features through the use of special addresses, so that standard PLC instructions can trigger an immediate input or an immediate output.

This first type of interrupt can be classed as a software interrupt since it occurs only when the program executes an instruction initiating the interrupt routine. The other five interrupt types can be classed as hardware interrupts. They are initiated by devices (hardware) that signal the PLC's microprocessor, indicating that they require immediate attention.

2. The **I/O interrupt** causes a PLC to respond to a change at an input contact. The input signal will be read by the CPU module without waiting for the input scan. The PLC immediately suspends what it is doing and executes a user-written interrupt service routine (ISR). Some PLCs offer an I/O interrupt variation called a **high-speed counter,** in which the built-in ISR only increments a counter.

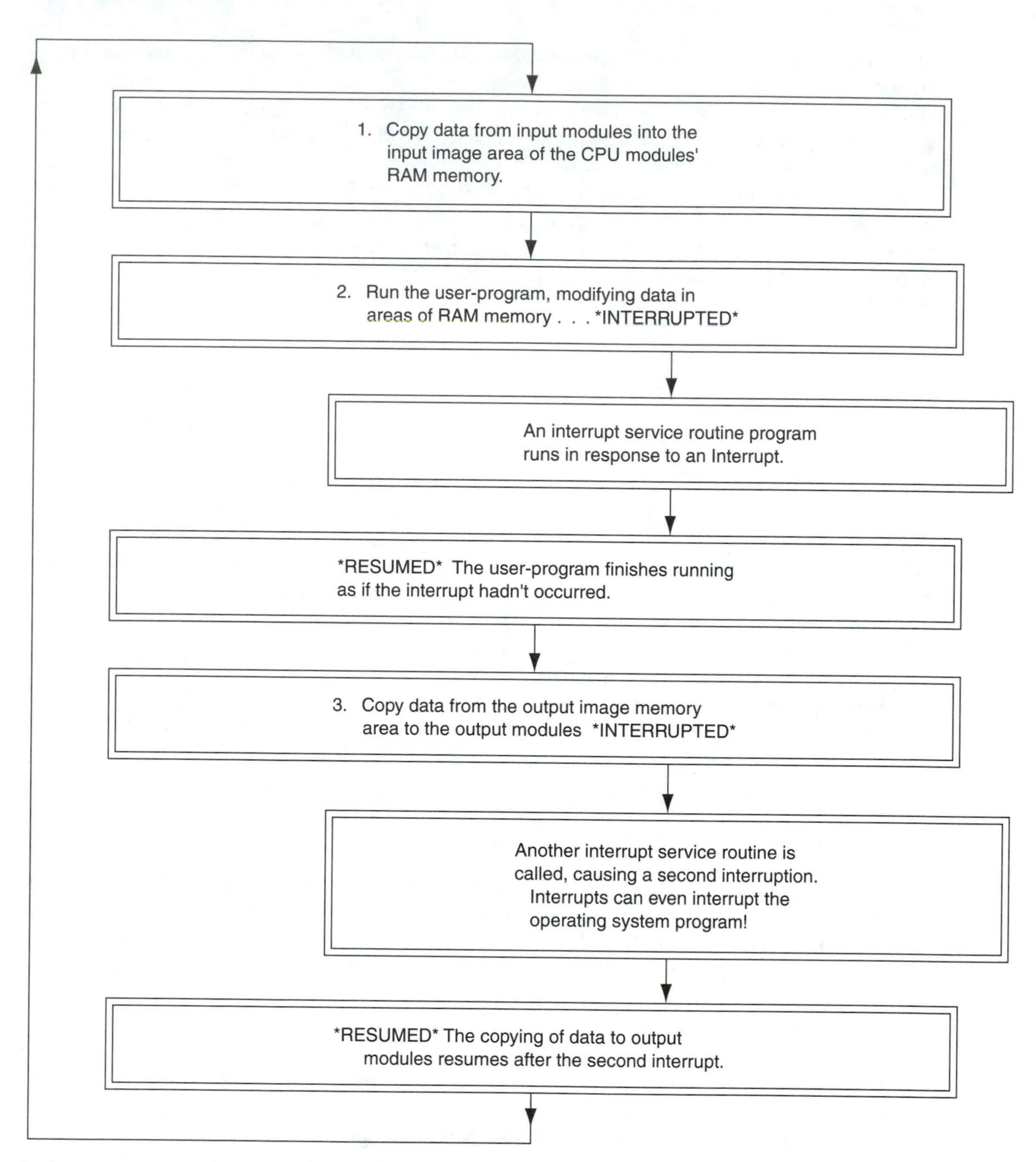

FIGURE 11.2
Interrupted PLC scan cycle.

3. The **timed interrupt** ensures consistent intervals between executions of a user-written ISR, or ensures that an ISR will be executed at a specific time. Some PLCs offer a timed interrupt variation called a **high-speed pulse output,** or a **pulse width modulation** feature, in which a built-in ISR turns a digital output on and off at precisely timed intervals.

4. The **fault routine interrupt** causes a fault routine ISR to execute if there is a PLC program error or hardware failure. Program errors and hardware failures aren't things that a PLC user would cause intentionally, but having user-written fault routine capability allows the programmer to control how the PLC responds to such undesirable conditions.

5. The **initialization interrupt** gives the programmer some control over how a PLC will restart after it has been stopped. PLC operating system programs often contain built-in initialization ISRs that are executed as soon as the PLC program starts to run. Initialization routines do things like setting or clearing selected data values in the PLC's memory (output image data is cleared by most PLCs) and/or automatically executing a user-programmed initialization routine. Some PLCs allow the user to write their own initialization routines.

6. The **communications interrupt** optimizes the use of serial communication links included in the PLC. Most communication control ISRs are built into the operating system program provided with the PLC and cannot be changed by the PLC user. With some PLCs, the programmer can select PLC configuration options to optimize communications and/or write programs that examine the status of serial communication channels to trigger communication routines.

Interrupt service routine (ISR) programs are not controlled by the scan cycle (except the immediate I/O type). ISRs execute only when specific conditions exist, conditions that may rarely occur, and the PLC saves time by not checking for those conditions. Alternatively, the conditions may occur several times during a single scan cycle, so the ISR will execute several times during the scan cycle. Scan cycles that are interrupted will, of course, take longer to complete than those that aren't.

All PLCs with interrupt capabilities define interrupt priority. When an ISR is running, it can only be interrupted by an interrupt of higher priority, not by an interrupt request with the same or lower priority. All ISRs have enough priority to interrupt a main scan cycle program. Fault routines generally have the highest priority, so they can cause interrupts of any other type of ISR. Timed interrupts usually have the lowest priority. Most PLCs queue interrupt requests if they aren't high priority enough to interrupt the program that is running. A queued interrupt request is responded to after the higher-priority interrupt's ISR finishes. Most PLCs also provide status bits that indicate if interrupt requests were delayed or were lost. The user-program can check the status bits to detect a delay in executing an ISR needed for a critical application.

Programmers can effectively change interrupt priority by careful use of **INTERRUPT ENABLE** and **INTERRUPT DISABLE** instructions. In most PLCs, interrupts are *enabled by default* until an INTERRUPT DISABLE instruction is executed. A programmer may want to disable interrupts during execution of a critical part of a main program (perhaps to prevent an ISR from changing memory that the main program is using) or may want interrupts to stay disabled until a specific input condition exists. The programmer may include an INTERRUPT DISABLE instruction in an ISR so that it won't be interrupted by another ISR (an INTERRUPT ENABLE instruction should execute before the ISR ends).

MORE DETAILED DESCRIPTIONS OF INTERRUPT RESPONSE

Readers with a good understanding of microprocessors may be interested in what actually happens in a PLC as the interrupt occurs. Other readers can skip ahead to the section describing immediate I/O interrupts.

During an I/O interrupt or a timed interrupt, the following interrupt sequence occurs:

1. An input port controller chip or a timer chip pulls the voltage down on an interrupt line connected to the microprocessor.

2. The microprocessor finishes the instruction or rung that it was executing when the voltage dropped on the interrupt line (some PLCs allow the user to change how fast the PLC responds). If interrupts are enabled, the microprocessor will store the contents of its internal registers in the stack memory portion of RAM. A stack pointer register in the microprocessor keeps track of the amount of data stored in the stack so that the data can be retrieved later.

3. The microprocessor retrieves the starting address of the appropriate interrupt service routine (ISR) from an area of memory set aside for this purpose. The address is put into the microprocessor's program counter register, so that the next instruction that the microprocessor runs will be the first instruction of the ISR.

Response to interrupts beyond this point varies from PLC to PLC. Sometimes, the address that the PLC has just retrieved is the address of a user-written interrupt service routine, so the user-written ISR starts immediately. In other PLCs, the address that the PLC retrieves is a ROM memory address, and the PLC performs some additional steps before and/or after calling the user-written ISR. These steps may even include a shortened version of a typical three-step scan cycle:

(a) A (usually reduced) input scan is executed.
(b) A jump to a user-written ISR is executed.
(c) After the user-written ISR ends, a (usually reduced) output scan is executed.

4. The final instruction in the interrupt service routine is an END or RETURN instruction. It causes the microprocessor to retrieve the data it stored on the stack at the time of the interrupt and to put it back into the microprocessor registers from which it came. The interrupted program then resumes automatically, exactly as if it hadn't been interrupted. (One of the data items retrieved is the contents of the program counter before the interrupt.)

Response to an immediate I/O instruction is similar to the description above, except that it is caused by the program instruction, not by a voltage change, and its ISR is completely programmed in ROM. Response to a fault interrupt is slightly different. If the fault ISR doesn't exist or if the fault ISR finishes without fixing the problem that caused the fault ISR to run, the PLC will not resume running the interrupted program. Instead, the PLC will go into stop mode, and the operator must find and repair the problem. After repairing the problem and switching the PLC back into run, the PLC usually restarts running its user-program from the beginning, not from the point where it was interrupted.

Immediate Input and Immediate Output Instructions

An **IMMEDIATE INPUT** instruction causes a PLC to interrupt user-program execution to read input data directly from an input module instead of using the slightly older data in the input image memory area. An **IMMEDIATE OUTPUT** instruction causes the PLC to interrupt user-program execution to write data directly to an output module, avoiding the wait for the PLC scan cycle to copy output data to an output module. Inclusion of immediate I/O instructions in the program does not prevent the PLC from reading or writing to the I/O module during the scan cycle, so inputs and outputs may be read and written more than once per scan cycle.

IMMEDIATE INPUT and IMMEDIATE OUTPUT instructions generally take longer to execute than instructions that read or write data in the I/O image areas of memory, so the programmer should use them only where necessary. Faster PLCs are reducing the execution time for immediate I/O instructions to the point that some PLCs must use immediate I/O instructions to read or write certain I/O modules! They don't have I/O image memory for those addresses and (obviously) don't include them in the input scan or output scan during the scan cycle.

Immediate input and immediate output interrupts are software-initiated interrupts. They are caused by instructions the programmer includes in a program, and they only happen when the PLC executes a rung that contains the instruction. The other interrupts discussed in this chapter are caused by events that are not under program control and can occur at any time during the PLC cycle.

IMMEDIATE I/O INSTRUCTIONS AND THE ALLEN-BRADLEY PLC-5

When the PLC-5 encounters an **IMMEDIATE INPUT** (IIN) instruction with the address of an input module that is in the same (local) rack as the CPU, the PLC copies a full data word from that input module into the input image file. If the input module is in a remote I/O rack or in an extended local rack, the PLC copies the most recently received data from that input module (from the communications data buffer[2]) and puts that data into the input image file. Subsequent instructions can then use the newly updated data in the input image file.

If the PLC-5 encounters an **IMMEDIATE OUTPUT** (IOT) instruction, it copies a full data word from the output image table to that output module if the module is in the CPU's local rack, or to the communication buffer for sending to the module if the module is in a remote rack or in an extended local rack.

Immediate I/O instructions take much longer to execute than most other PLC instructions, but they can be made conditional so that they do not execute every scan. In specifying input and output module addresses for these instructions, the "I:" or "O:" prefixes should never be used.[3]

Figure 11.3 shows a PLC-5 program that uses an IMMEDIATE INPUT instruction to read a 16-bit data word from a local input module. The program then changes output image data de-

---

[2] The PLC-5 stores incoming serial communications, including input image data from remote I/O racks, in a buffer area of RAM memory as it is received. During the I/O scan part of the PLC scan cycle, the CPU copies the buffered data into the input image file. Similarly, output image file data intended for remote I/O racks is copied from the output image file to the buffer during the I/O scan and is stored there until it can be sent via the serial communications channel. The process is described in more detail in Chapter 13.

[3] If an address with a prefix is specified, the PLC-5 reads the lowest 9 binary bits from that address, interpreting it as a three-digit octal address of the I/O module that the immediate I/O instruction will read from or write to. For example, if IOT N9:27 is programmed and if N9:27 contains the binary value 1010 101**0 1011 0010,** the PLC-5 will read the low 9 bits (010 110 010) from N9:27, interpret that as an octal address (262), and will write the contents of output image address O:262 to the digital output module in rack 26, group 2.

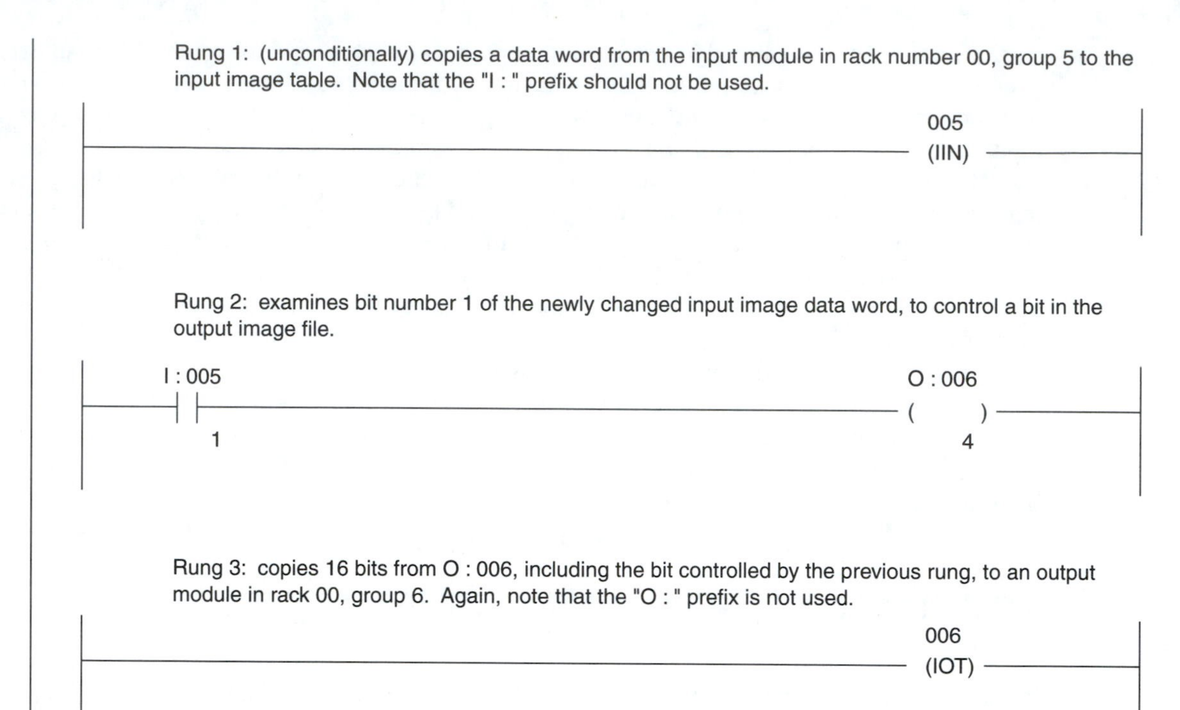
Rung 1: (unconditionally) copies a data word from the input module in rack number 00, group 5 to the input image table. Note that the "I : " prefix should not be used.
005
(IIN)
Rung 2: examines bit number 1 of the newly changed input image data word, to control a bit in the output image file.
I : 005
1
O : 006
( )
4
Rung 3: copies 16 bits from O : 006, including the bit controlled by the previous rung, to an output module in rack 00, group 6. Again, note that the "O : " prefix is not used.
006
(IOT)

FIGURE 11.3
PLC-5 program with IMMEDIATE INPUT and IMMEDIATE OUTPUT instructions.

pending on the new input image data and uses an IMMEDIATE OUTPUT instruction to send the modified output image word to a local output module.

BLOCK TRANSFER READ (BTR) and **BLOCK TRANSFER WRITE (BTW)** instructions are treated as immediate I/O instructions, but only if they are executed in an interrupt service routine (ISR) that runs in response to a timed interrupt, I/O interrupt, or a fault. (These interrupts are described later in the chapter.) When a BTR or BTW instruction is encountered in an ISR program, the PLC stops executing the ISR until after the block of data has been read from or written to the addressed I/O module. If the I/O module is in a remote rack, the block transfer takes significantly longer to complete, so enhanced PLC-5 models are designed to make use of the waiting time. Enhanced PLC-5s run the program that the ISR interrupted while the ISR itself is waiting for the remote block transfer to complete (unless the ISR is a fault routine).

SLC 500

IMMEDIATE I/O INSTRUCTIONS AND THE ALLEN-BRADLEY SLC 500

Allen-Bradley SLC 500 PLC's immediate I/O instructions offer a couple of improvements over those of the PLC-5. The SLC 500's instructions, which are called **IMMEDIATE INPUT WITH MASK (IIM)** and **IMMEDIATE OUTPUT WITH MASK (IOM),** allow the programmer to specify which of

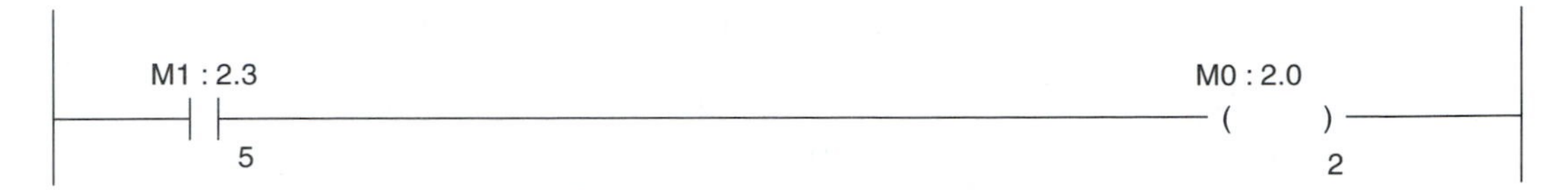

FIGURE 11.4
M0/M1 addressing for immediate I/O with SLC 500 specialty I/O modules.

the 16 bits of a data word are to be copied from an input module to the input image data table (or from the output image table to an output module). The other bits in the input image table or output module are not affected by these instructions. In addition, the SLC 500's IIM and IOM instructions allow the programmer to input or output a series of data words from a single input module or output a series of data words to an output module.

The SLC 500 also offers a **REFRESH (REF)** instruction that causes the PLC to interrupt its program execution to perform a full input scan, communications service scan,[4] and output scan before resuming the program execution. A **SERVICE COMMUNICATION (SVC)** instruction interrupts the program scan to service a specified communication channel before resuming the program scan.

Some specialty SLC 500 I/O modules *must* be read from or written to using a type of immediate I/O. A SLC 500 can be configured so that none, or only some, of the data words used by specialty modules are read and written by the CPU during its input and output scan steps, so the CPU doesn't maintain a complete copy of I/O data for these modules. The I/O modules contain the data in their own memory files. Instead of requiring the use of immediate I/O instructions that we have examined in this section, specialty I/O modules are read and written using standard instructions but with **M0** or **M1** addressing to tell the PLC to access data files in the module's memory rather than in the CPU's memory.[5] Figure 11.4 shows a program that reads and writes data from a specialty I/O module. M0 and M1 addressing looks like a cross between Allen-Bradley I/O image table addressing and data table addressing. M1:2.3/5, for example, means bit 5 of word 3 in data file 1 of the specialty module in slot 2. The prefix "M0" was originally intended to mean **m**odule **o**utput files and "M1" meant **m**odule **i**nput files, but the "0" and "1" now just indicate one of the two data files in the specialty module.

S5

IMMEDIATE I/O INSTRUCTIONS IN SIEMENS STEP 5

Siemens STEP 5 does not offer special instructions for immediate input and immediate output. Instead, S5 PLCs use standard input and output instructions with direct access addressing. Direct access addressing is similar to normal process input image (PII) and process output image (PIQ) addressing, except that the address prefix letter "P" is used instead of "I" or "Q". To read

---

[4] The communications service step in the scan cycle is when the PLC controls and exchanges data with the communication handling processors as they send and receive data via serial communications links. For further information, see Chapter 13.

[5] The ONE-SHOT RISING (OSR) instruction can't use M0 or M1 data file addresses. Bit shift, sequencer, and stack (LIFO and FILO) instructions cannot manipulate data in M0 or M1 data files.

the 16-bit value directly from the input module in slot 4, for example, the instruction L **PW**004 is used instead of L **IW**004. To write a 16-bit value directly to the module in slot 6, use T **PW**006 instead of T **QW**006. Output values written directly to an output module are automatically copied to the PIQ at the same time.

Some of the larger S5 PLCs don't have any input image (PII) or output image (PIQ) memory for analog I/O modules. In these PLC models, analog I/O values *must* be read from input modules or written to output modules using direct access addressing. When there is a corresponding PIQ address, output instructions that write data directly to a module also copy the data to the PIQ table simultaneously, so that when the PIQ table's contents are written to the output module during the scan cycle, it doesn't change the data written by the direct access addressed instruction. When data is read from an input module using direct addressing, the S5 PLC does *not* automatically overwrite the PII table with the new value. If the programmer wishes the PII to contain the new value, the program must write the value to the PII address, as shown in the following example:

```
L PW005     ; Read a word directly from the input module
            ;    in slot 5
T IW005     ; Update the value in the PII table for slot
            ;    5
```

Some midsized S5 PLCs (e.g., S5-103U) can't actually perform immediate I/O operations, although they allow the programmer to write programs with direct access addresses in them. In these PLCs, direct access addressing can only be used in an interrupt service routine (ISR) running in response to an I/O interrupt, a timed interrupt, or an initialization block (all covered later in the chapter). When an I/O interrupt or a timed interrupt occurs, these PLCs automatically copy data from input modules to a special **interrupt process image input (interrupt PII) table,** which differs from the normal PII table. Instructions in the ISR can use direct addressing to read data from the interrupt PII or to write data to an **interrupt process image output (interrupt PIQ) table.** After the ISR or initialization block ends, values in the interrupt PIQ table are automatically copied to the output modules, but only if the interrupt PIQ value was changed during execution of the ISR or initialization block. As in larger S5 PLCs, using direct access addressing to output a value also caused the value in the normal PIQ table to be updated. Since the interrupt PII/PIQ table is only updated by interrupting or restarting the PLC, use of direct access addressing in a standard scan cycle program will cause the PLC to fault.

One difference between using direct access addressing and addressing of data in the PII and PIQ tables is that the programmer cannot use direct access addressing for individual input or output bits. The instructions A P4.4 and = P5.3 don't work. To use direct access addressing to examine individual input bits requires three steps: A whole word or byte must be read using direct access addressing, then the word or byte must be stored in flag, data, PII, or PIQ memory, and then individual bits of the flag/data/PII/PIQ address can be examined. To use direct access addressing to write individual output bits: The individual bit values must be written into data/flag/PII/PIQ memory, then the flag/data/PII/PIQ word or byte must be copied to the output module (or to the interrupt PIQ) using direct access addressing. Figure 11.5 shows an STL program block that uses direct access addressing as described above to read one bit and control one bit. The direct access output will, unavoidably, copy the other seven flag data bits to the output module at the same time, so the example uses word logic instructions (covered in Chapter 6) to ensure that only one bit at the output module is actually changed.

```
PB002
Contains immediate I/0 statements

L PY002   ;Read the low 8 bits directly from the input module in slot 2, and
T FY002   ;Copy the value into flag byte 2 so that individual bits can be examined
A F2.0    ;IF the sensor attached at bit 0 of the input module in slot 2 is on, then
= F6.0    ;Turn on a flag bit in flag byte 6 (the other bits of flag byte 6 are always zero
           in this program)
L QB004   ;Get the data byte that reflects the current status of the low 8 bits at the
          output module in slot 4,and
L KH FE   ;Get a constant byte value containing "1" in every bit except the lowest, then
AW        ;Logical AND the two values. The lowest bit of the result will become 0, but
          the other bits remain the same as in QB004
L FY006   ;Get the 8-bit value from flag byte 6, which is all zeros except possibly
          F6.0' and
OW        ;Logical OR it with the result of the logical AND, so that the lowest
          bit is changed to the value of F6.0, but the other bits remain
          unchanged, then
T PY004   ;Transfer the result directly to the output module in slot 4
BE        ;Return to the block that called PB002
```

FIGURE 11.5
Immediate I/O instruction use in STEP 5.

S7

IMMEDIATE I/O INSTRUCTIONS IN SIEMENS STEP 7

Siemens STEP 7 PLC programs can read data immediately from input modules and write immediately to output modules. In fact, process-image input table and process-image output table memory exists *only* for the lowest 128 bytes of addressable inputs and outputs (bytes 0 to 127). Reading data from, or writing data to, addresses above 127 requires the programmer to use immediate I/O. Siemens calls immediate I/O access **direct access** in their STEP 7 literature. STEP 7 also offers four system functions to force immediate input, output, setting, or resetting of groups of I/O module addresses.

As with the STEP 5 language, STEP 7 doesn't include immediate I/O instructions. The programmer uses the standard instruction set but enters an address for what the STEP 7 literature calls **peripheral input memory** (e.g., "L **PI**W 003" instead of "L **I**W 003") to force the PLC to read values from the input modules instead of from the process-image input table, or enters an address for the **peripheral output memory** (e.g., "T **PQ**W 007" instead of "L **QW**

007") to force writing to the output modules. Any data written to peripheral output memory addresses for which there are process-image output addresses (bytes 0 to 127) are also automatically written to the process-image output table.

STEP 7 allows instructions to address bytes (e.g., L **PIB** 003) and 16-bit words (e.g., L **PIW** 003) of peripheral input and peripheral output memory, and adds the ability to address 32-bit double words (e.g., L **PID** 003) but still does not allow addressing of individual bits! You cannot, for example, enter the instruction "A PI 003.1". To examine input bits immediately, your program must copy whole peripheral input bytes, words, or double words into process-image memory, bit memory, local memory, or into a data block before bits can be examined. To write to an individual output bit immediately, a process image, bit memory, local, or data bit must be changed by the user-program and then the whole byte, word, or double word containing that bit can be transferred to the peripheral output memory. The process would be similar to that shown in Figure 11.5.

To copy multiple data values immediately from input modules to the PII table, the program can execute **SFC26 ("UPDAT_PI").** The program must provide a number (0 to 8) for the **PART** parameter to specify which part of the PII table to update. "0" means to update the entire PII, while other numbers indicate sections of the PII to update. The PII table can be configured in sections using the programming software. **SFC 27 ("UPDAT_PO")** can be used to copy all or part of the PIQ table to output modules.

SFC79 ("SET") is used to set a series of bits in output modules. If the bits are in addresses for which the PIQ contains data, the PIQ bits will also be set. With "SET", the programmer must provide a pointer (pointers were covered in Chapter 5) as the **SA** parameter to identify a starting address, and a number to identify how many bits to set as the **N** parameter. **SFC80 ("RESET")** can be called (with the same parameters) to reset a range of bits in output modules.

CQMI

IMMEDIATE I/O INSTRUCTIONS AND THE OMRON CQM1

The CQM1 offers only one immediate I/O instruction, the I/O refresh instruction: **IORF(97).** With this instruction, the programmer specifies the first, or start, I/O word and the last, or end, I/O word. The IORF(97) instruction causes the PLC to copy data immediately from input modules to input image data words in the range specified and to copy data from output image area data words to output modules in the range specified. Figure 11.6 shows this instruction, which will be executed every scan while input bit 12 from input module 5 is on. The instruction copies all the input module data (modules in slots 000 to 011) into the input area of memory (IR 000 to IR 011) and copies only two data words (IR 100 and IR 101) to output modules (module slots 100 and 101) because the specified range end ends at IR 101.

The CQM1 also offers a feature not found in Allen-Bradley and Siemens PLCs. The CQM1 can be configured for **direct refresh.**[6] When configured for direct refresh, IR data is copied to the appropriate output module as soon as it is changed by the program! You cannot limit the range of direct refresh outputs. Direct refresh does not affect inputs and there is no con-

---

[6] For direct refresh, write the data word xx**01** to address DM 6639. When configured for direct refresh, the CQM1 will also continue to write outputs to output modules at times when it would otherwise do so, including during the I/O scan cycle. DM 6639 cannot be changed while the PLC is in run mode.

FIGURE 11.6
Immediate I/O refresh during a CQM1 program.

figuration to force automatic immediate input. (We will see in the following sections on I/O Interrupts and Timed Interrupts that the CQM1 can be configured to perform input refreshing automatically whenever an interrupt program is executed, and there are lots of ways to cause interrupt programs to execute.)

The I/O Interrupt

An I/O interrupt is initiated by a signal from an input module, usually because of a change at a sensor attached to the input module. When the signal from the input module is received, the CPU module sets aside what it is doing and executes the interrupt service routine (ISR) assigned for that input condition. Once the ISR finishes executing, the PLC resumes what it was doing at the time that it received the interrupt request signal. I/O interrupts are always considered to be among the highest-priority interrupt levels. I/O interrupts can sometimes even interrupt an I/O scan!

An I/O interrupt is classified as a hardware interrupt because it is initiated by a signal from equipment (hardware) rather than by an instruction in the program (software). Some PLCs require the use of special input modules, capable of originating hardware interrupt signals. Other PLCs can be configured so that any input module connected by the CPU module can originate an interrupt. Still other PLCs only allow I/O interrupts in response to sensors attached directly to the CPU module.

Some PLCs can be set up to count changes at input modules and to initiate the ISR only when the count reaches a preset value. PLCs with built-in inputs in the CPU module may be capable of counting input changes very quickly and may offer a **high-speed counter interrupt** feature. If the user-program includes an instruction enabling this feature, the CPU interrupts to execute a high-speed counter ISR whenever the input signal changes. High-speed counter ISRs are preprogrammed in ROM as part of the PLC's operating system. The ISR simply increments or decrements a stored count value. Instructions in the user-program can examine and respond when the accumulated count value exceeds a user-selected preset. High-speed counter inputs are typically connected to sensors such as optical encoders, which generate pulses more quickly than most PLCs execute their scan cycle but require that no pulses be missed in the count. (Some PLCs without a high-speed counter feature offer separate counter modules, complete with their own microprocessors, to count changes at sensors such as optical encoders.)

I/O INTERRUPTS AND THE ALLEN-BRADLEY PLC-5

For their enhanced PLC-5 line of Allen-Bradley PLCs, Rockwell Automation calls I/O interrupts **processor input interrupts (PIIs).** (Yes, the same letters as used in Siemens' STEP 5 for *process image input.* Wouldn't it be nice to have a little standardization instead of this confusion?)

The PLC-5 can be configured to respond to a processor input interrupt (PII) request from *any* digital input module in a module group[7] if the input group is in the same rack as the CPU module. Any one or more of the 16 bits of input image from the group can be included in the set of bits to monitor. The other bits will be ignored. The PLC can be configured to react to either false-to-true or true-to-false transitions at each of the bits being monitored (some bits can be false-to-true while others are true-to-false). The PLC can be configured to run its PII response program every time one of the bits changes or to count the number of times changes occur and run the PII response program every time the count reaches a preset number up to 32,767. (The count resets to zero as the PLC goes from program to run mode.) The PLC-5 can be configured to run any program file as a PII response program. Although the PLC-5 does not do an input scan as it jumps to the PII program, it does save a record of the bit(s) that changed last at the input module group being monitored, causing the PII.[8] The PII program can examine this record to determine exactly which sensor(s) changed state last, causing the interrupt.

If a new PII interrupt occurs before the previous PII program has finished running, the new PII waits until the previous PII finishes. The PLC-5 will set bit S:10.12, a minor fault bit indicating that a PII overlap has occurred.

The programmer uses the processor configuration screen, shown in Figure 11.7, during program editing to specify the program file number of the program to be executed as the PII's

[7] Some restrictions: The I/O module must be in the local rack, not in a remote or extended local rack. The PLC-5 rack cannot be configured for two-slot addressing. A 32-bit module cannot be monitored if it is in a rack configured for one-slot addressing. If there is a block transfer module in the rack, then some transitions at the bits being monitored may be missed while block transfers are being performed.

[8] This "return mask" record is stored in data word S:51. S:51 is retentive, which means that bits which are on at the time a PII is called will still be set when the PII finishes unless instructions in the PII program explicitly clear them. If S:51 isn't zero when the PII finishes, the next return mask record will not be useful again and the PLC-5 will set a PII overlap bit (minor fault bit S:10.12).

```
                              Processor Configuration

User Control Bits        00000000 00000000  RESTART LAST ACTIVE STEP
Fault routine prog file no.:      0         Watchdog (ms):                  500
I/O status file:                  0         Communication time slice (ms):    3
VME Status File:                N34

Processor input interrupt   prog file no.:   0    module group:  0
                            down count:      0
                            bit mask:       00000000 00000000
                            compare value:  00000000 00000000
```

FIGURE 11.7
Defining PLC-5 processor input interrupts: the processor configuration screen.

interrupt service routine, the module group address indicating the rack and group number of the input module(s) that should be monitored for bit changes, a down count number indicating how often the bits must change before the ISR will be executed, a 16-bit binary bit mask value with a "1" in the position of each bit that should be monitored, and a 16-bit binary compare value with a "1" in each bit position where a false-to-true transition is to be counted (a "0" means that a true-to-false is to be counted).

Example

If the following data is entered at the processor configuration screen:

Processor input interrupt:

```
program file no.:  12        module group:  05
down count:         6
bit mask:          00000000  00000011
compare value:     00000000  00000010
                                   ||
                                   |└─bit number 0
                                   └──bit number 1
```

then the lowest 2 bits from the input module or modules in group 5 of rack 0 or 1 (whichever one contains the CPU module) will be monitored for interrupt requests. Each time bit 0 goes false or bit 1 goes true, the event is counted. After 6 of these events have been counted, the PLC will interrupt whatever it is doing, execute the program in program file 12, and then reset the count so that the interrupt will not occur until six more similar events are counted.

While entering data using the processor configuration screen, the programmer is actually entering configuration words into the status file 2.[9] Status data words can be changed by a PLC program. They do not have to be entered using the processor configuration screen. Instead of using the configuration screen, the programmer can include instructions in a program to put program file number, module group, down count, bit mask, and compare value data into the appropriate status data words. Some programmers include such instructions in an initialization program file, which runs every time the PLC goes into run mode, to ensure that PII configuration of a PLC remains the same even if an operator inadvertently modifies the configuration screen parameters. Except for the PII file number, changes to a PLC-5's PII configuration become effective only when the PLC is switched from program to run (or to test) mode, so initialization programs that write to the PII configuration status words should compare the new configuration against the old and should force the user to switch the PLC into run mode again if a difference is found.

The programmer can write a PLC program that changes which PII response program file to call. The PLC program shown in Figure 11.8 contains a MOVE instruction to change the data stored in the status file for the PII program file.

---

[9] The program file number is stored in S:46, the module group in S:47, the bit mask in S:48, the compare value in S:49, and the down count in S:50. S:51 indicates which of the monitored input module bits changed most recently to cause the count to increment. S:52 contains the accumulated count of the number of times the monitored bits have changed since the last time the PII ran (or since the PLC was put into run mode).

This rung sets the processor up to use program file 15 as an ISR in response to a processor input interrupt, if input switch I : 040/1 is on.

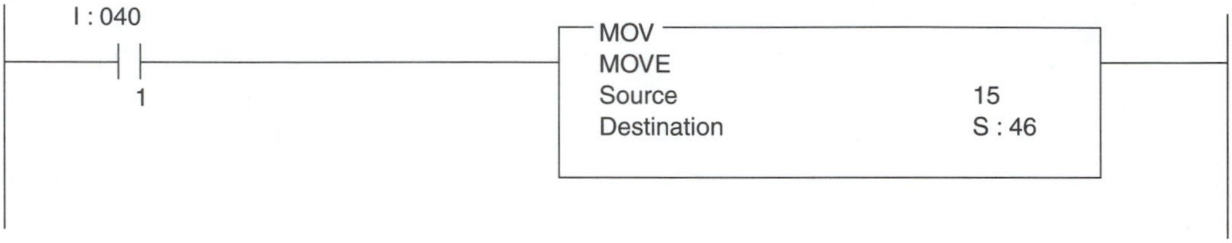

```
 I : 040                                    MOV
---| |--------------------------------------MOVE
    1                                        Source          15
                                             Destination     S : 46
```

Otherwise, if input I : 040/1 isn't on, this rung sets the processor up to use program file 16 to respond to the processor input interrupt.

```
 I : 040                                    MOV
---|\|--------------------------------------MOVE
     1                                       Source          16
                                             Destination     S : 46
```

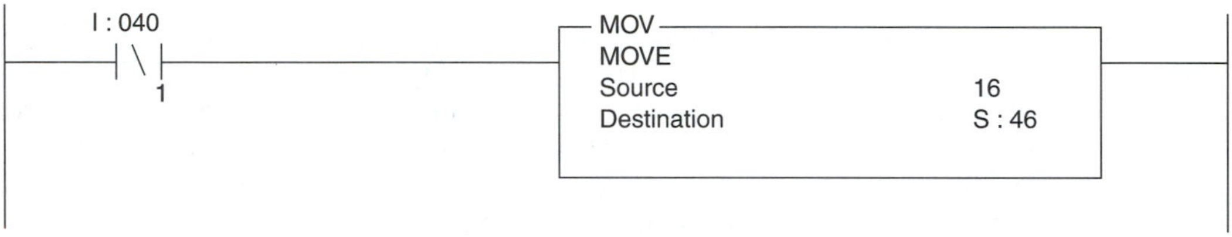

FIGURE 11.8
Allen-Bradley PLC instruction that changes the PLC's interrupt configuration.

SLC 500 I/O INTERRUPTS AND THE ALLEN-BRADLEY SLC 500

Instead of calling I/O interrupts processor input interrupts (PII), as they are called with the PLC-5, better Allen-Bradley SLC 500s offer exactly the same interrupts but call them **discrete input interrupts (DIIs)** (just to confuse programmers). Most SLC 500s also offer an additional, lower-priority I/O interrupt, actually called an **I/O interrupt,** which can be initiated only by a specialty I/O module.

For a description of discrete input interrupts (DIIs), the reader is advised to read the earlier section on processor input interrupts for the PLC-5. Even the configuration and status word addresses are the same, although the SLC 500 documentation sometimes uses different terminology.

The SLC 500's **I/O interrupt** feature is similar to the DII feature in that when the I/O module generates an interrupt request, the PLC sets aside what it is doing, executes an interrupt service routine (ISR), then resumes the task that was interrupted. There are some differences in priority, configuration, and interrupt signal initiation, however. Only a specialty I/O module can cause an I/O interrupt. The file number of the interrupt program used by a specialty I/O module is configured when configuring that specialty I/O module, not when configuring the PLC's processor. Each specialty I/O module can, therefore, have its own ISR. (Allen-Bradley's literature actually uses the letters "ISR" in referring to an interrupt service routine called by an I/O interrupt, but says that the letters mean **i**nterrupt **s**ub**r**outine—oh well, any standardization is better than none.) An ISR called by an SLC 500 I/O interrupt must have an INTERRUPT SUBROUTINE **(INT)** instruction as its first instruction, and a RETURN **(RET)** instruction at its end. An I/O interrupt has lower priority than a discrete input interrupt. In fact, the I/O interrupt has the lowest priority of the four hardware interrupts defined for the SLC 500 PLC. (Fault has highest priority, DII is next, followed by the timed interrupt called an STI, then finally the I/O interrupt.) I/O interrupts require quite a few status and control bits to prevent them from sometimes being ignored completely! Unlike DII interrupts, the PLC 500 includes the following status bits and words[10] used for I/O interrupts:

1. An **I/O slot enable** bit for each slot (despite what type of module is in the slot), which must be set if the slot contains a module that is to be included in the input or output scan steps. If the slot enable bit is zero, the CPU will not send any data to, nor accept any data (even an interrupt request) from, the module in this slot. If the slot contains a specialty I/O module, the CPU signals the module when its I/O slot enable bit value changes, because some specialty I/O modules change the way they operate when disabled.
2. An **I/O interrupt enable** bit for each slot, which controls whether an interrupt from a specialty module in the slot can cause an ISR to run at this time. If cleared by an instruction in the user-program or by a programmer at a PLC configuration screen, the slot cannot cause an I/O interrupt. The SLC 500 offers I/O INTERRUPT DISABLE **(IID)** and I/O INTERRUPT ENABLE] instructions that can be used to clear or set individual I/O interrupt enable bits.
3. An **I/O interrupt pending** bit for each slot, which is set whenever a specialty I/O module generates an interrupt request while the I/O interrupt enable bit for that module is not set, or if an I/O interrupt request signal from this slot hasn't caused an interrupt because a higher-priority ISR is already running. The I/O interrupt pending bits are used by the SLC 500 to allow the I/O interrupt ISR to run later, when conditions allow. A RESET PENDING INTERRUPT **(RPI)** instruction is available to clear I/O interrupt pending bits for selected slots, canceling interrupt requests that haven't been serviced yet.
4. An **I/O interrupt executing** word, which contains the slot address of the specialty I/O module that initiated the I/O interrupt ISR that is currently running, if any.

The Allen-Bradley *Micrologix 1000* PLCs have a **high-speed counter interrupt** that makes use of up to 4 input bits. The counter can be configured in a number of ways, including for use with optical encoders. Future SLC 500 PLCs will probably offer this more advanced high-speed counter, perhaps by the time you read this, so it is described here. After enabling the high-speed counter, the PLC monitors input contacts I:0/0 to I:0/3 and interrupts to update the counter when a signal change occurs. The high-speed counter can be set up so that when the high-speed counter's accumulated value reaches its preset, the PLC will interrupt to immediately write one word of data to O:0 AND/OR to execute program file 4. At least two instructions are needed in a user-program to use a high-speed counter.

[10] I/O slot enable bits are in S:11/1 through S:12/14 for slots 1 through 30. I/O interrupt enable bits are in S:27/1 through S:28/14, and I/O interrupt pending bits are in S:25/1 through S:26/14 for the same slot range. The interrupt executing word is in S:32.

1. The second instruction is a **high-speed counter (HSC)** instruction. There can be only one HSC instruction in your program. C5:0 is used automatically by the high-speed counter, and it includes 16 status bits! The HSC instruction must execute with a true conditional statement so that the PLC can set up a hardware high-speed counter. In the HSC instruction, enter:

(a) A counter **type** to tell the PLC how many of the four inputs to monitor and how to use the signals. Eight counter types can be selected, from a simple up counter that only counts rising signals at I:0/0, to a (bidirectional) encoder counter that monitors I:0/0 and I:0/1 as A and B encoder signals, I:0/2 as a Z (initialize) signal, and I:0/3, which causes the count to hold.

(b) A **high preset** value (C5:0.PRE), which is copied to the hardware high-speed counter when the HSC instruction first executes and at other times depends on the type of counter selected.

(c) An **accumulated value** (C5:0.ACC). The hardware accumulated value is used by the high-speed counter and is copied to C5:0.ACC only when the HSC instruction executes and/or when the program turns the update accumulated value bit C5:0.UA on.

Besides the usual status bits (DN, CU, CD, OV, and UN) and the UA bit mentioned above, C5:0 will contain bits that indicate when the accumulated value is above the high preset (HP) or below the low preset (LP), when interrupt file 4 has been called due to one of those conditions (IH or IL) or due to overflowing of the 16-bit signed binary numbering system (IV and IN), and when the (low-priority) interrupt file 4 is waiting its turn to execute (PE) or has been called again before it could execute (LS).

2. The first instruction in a user-program should be a **HIGH-SPEED COUNTER LOAD (HSL)** instruction, which provides more configuration for the hardware counter than is possible using the C5:0 counter structure called using the HSC instruction. With the HSC instruction, the user specifies a source address, indicating the first of five data words containing additional configuration data that will be copied to the hardware high-speed counter each time HSL executes with true control logic (so write the logic so that it is true only when reconfiguration is necessary). The five configuration words, in order, include:

(a) An **output mask.** Each "1" in this 16-bit value indicates a contact in O:0 that can be written to when the high-speed counter reaches its preset. (The default zeros in the hardware configuration disable writing to O:0.)

(b) An **output high source.** The 16-bit value to be written (through the mask) to O:0 when the high preset is reached.

(c) A **high preset.** Overwrites the high preset entered with the HSC instruction if HSL executes after HSC. In some types of timers, the high preset from HSC is reinstated each time the hardware high preset is reached (see your manuals).

(d) An **output low source.** The 16-bit value to write (through the mask) to O:0 when the low preset is reached *if* a bidirectional counter type is selected in the HSC instruction.

(e) A **low preset.** Relevant only if a bidirectional counter type is being used.

The Micrologix also offers **HIGH-SPEED COUNTER INTERRUPT ENABLE (HSE)** and **DISABLE (HSD)** instructions that can be used to enable or disable the calling of file 4 as an ISR. By default, file 4 can be called immediately after HSC executes. A **HIGH-SPEED COUNTER RESET AC-**

This first rung sets up and enables the high-speed counter to monitor input I : 0/0 and counts its transitions until there have been 360 transitions, at which time the counter's done bit will latch on, and the counter will restart automatically.

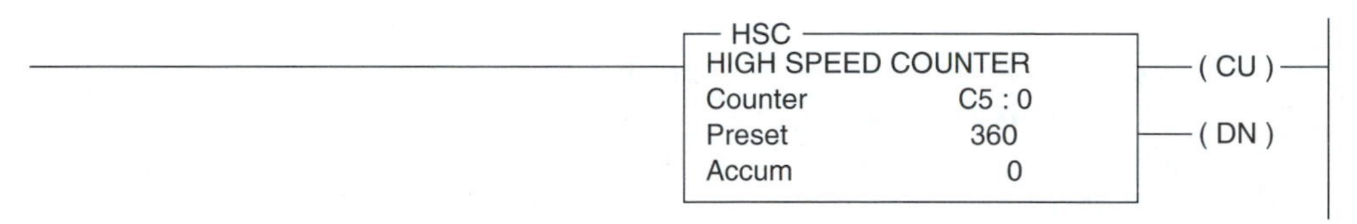

When the high-speed counter reaches 360 and turns its done bit on, output O : 0/4 is turned on by this rung. O : 0/4 will stay on while the done bit stays on.

C5 : 0
DN
O : 0
( )
4

Having responded to the high-speed counter reaching its preset, turn the counter's done bit back off. (Note that O : 0/4 will therefore only stay on for one cycle in this program.)

O : 0
4
C5 : 0
(OTU)
DN

This optional rung unconditionally copies the accumulated value from the high-speed counter to the counter element's accumulated value storage word, each time the rung is scanned.

C5 : 0
( )
AU

The high-speed counter's accumulated value (or actually, its value at the time of execution of the previous rung) can be compared against a value other than the high-speed counter's preset, to turn on another output.

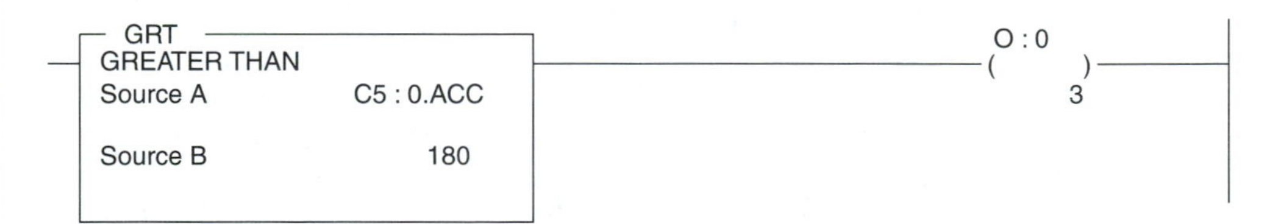

FIGURE 11.9
SLC 500 high-speed counter.

CUMULATOR **(RAC)** instruction can be used to change the contents of the hardware accumulator and C5:0.ACC simultaneously, and the RESET **(RES)** instruction resets the hardware counter simultaneously as it resets C5:0.

The SLC 5/01, with input contacts in its CPU, is the only SLC 500 that offers high-speed counting at the time of this writing. It has a more limited high-speed counter, which can only count changes at I:0/0, and only if a jumper has been cut in the CPU module. The SLC 500's high-speed counter does *not* initiate an interrupt service routine and does *not* write to an output when it reaches its preset. The slightly different version of the high-speed counter (HSC) is shown in the example in Figure 11.9. The HSC instruction is the only high-speed counter instruction that the SLC 500 can use.

S5

I/O INTERRUPTS IN SIEMENS STEP 5

A Siemens S5 PLC responds to what Siemens calls a **process interrupt** when an input signal either goes high or goes low. Its response to this type of interrupt occurs as follows:

1. The process interrupt has a reasonably high priority level. The S5's scan cycle or timed interrupt (explained later) is set aside when the MPU completes the instruction it is working on. Even I/O scan operations will be interrupted for a process interrupt. Initialization routines and fault routines (covered later) will not be interrupted.

2. In midsized S5 PLCs, which can't use direct access addressing to read or write I/O modules, a reduced input scan is executed. Input data is read from the few addresses where modules capable of generating hardware interrupts can be installed.[11] The input data is *not* copied to the standard process image input (PII) table but to a special interrupt PII table. Direct access addressing must be used to read/write the interrupt PII.

3. The interrupt service routine program that the programmer has entered in organizational block 2 (OB002) is run. No other hardware interrupts will be responded to until OB002 ends. OB002 may contain instructions to jump to other blocks, thus including those other blocks in the ISR. Function blocks that have been interrupted cannot be called during the ISR. If the programmer hasn't entered a program for OB002, interrupt signals from input modules will be ignored.

The larger S5-115U models offer more process interrupts, identified as interrupts A, B, C, and D. Interrupt A causes OB002 to execute; interrupt B causes OB003 to execute; C causes OB004; and D causes OB005. All have the same priority level, so if a process interrupt ISR is already running, none of the other three process interrupts can interrupt it. Direct access addressing reads/writes I/O modules in these PLCs.

4. When the block end instruction is encountered in OB002, those PLCs that have interrupt PII and PIQ tables execute a reduced output scan, copying only the interrupt PIQ table data that was *changed* during the execution of OB002, to the affected output modules.

---

[11] In the S5-103U, process interrupts can only be generated by 4-bit digital input modules or by a comparator module. These modules must be in slots 0 or 1, and these slots must be in an interrupt-type bus unit. When a process interrupt occurs, the S5-103U only copies one byte of data from each of these two slots into the lowest two bytes of the interrupt PII.

```
OB002
 This program will run every time a hardware interrupt signal is generated by an input module capable of
doing so.

L IB000        ; Loads the OLD (pre-interrupt) PII data for the input module in slot 0
L PY000        ; Loads CURRENT data directly from the input module in slot 0 (or from the interrupt PII)
><F            ; Compares the two values, and if they are different (indicating this module generated the
               ;   interrupt),then
JC PB010       ; Jump to program block 010

L IB001        ; Similar to above, but causes a jump to
L PY001        ; PB011 if the interrupt is from the
><F            ; input module in slot 1
JC PB011
BE             ; Return to the interrupted program (after PB010 or PB011 directly writes
               ; new output data to the output modules or to the interrupt PIQ)


PB010
Runs if called by OB002

T FY005        ; Copies the value in accumulator 1 (which is the value copied from PY000 in OB002) into flag
               ;   byte 5 so bits can be examined
A F5.0         ; If the sensor attached at bit 0 of input module 0 is on, then
= F6.0         ; Turn on a flag bit in flag byte 6

.              ; Other logic to respond if other sensors at input module 0 were responsible for the interrupt

L FY006        ; Copies the 8-bit value from flag byte 6 (including F6.0)
T PY004        ;   Directly to the output module in slot 4 (or to the interrupt PIQ table)
               ;   and also automatically to QB004, the PIQ table address for the same module
BE             ; Return to OB002
```

FIGURE 11.10
STEP 5 ISR.

5. The interrupted scan cycle resumes where it was interrupted. If other hardware interrupt or timed interrupt signals have been received during execution of OB002, only the most recently received interrupt signal will be responded to, even if the condition that caused the interrupt request no longer exists.

In the sample STEP 5 program in Figure 11.10, OB002 determines whether the interrupt was generated by the input module in slot 0 or the module in slot 1, and jumps to program block 10 (PB010) or to program block 11 (PB011) accordingly. PB010 examines individual bits (controlled by individual sensors) to determine the appropriate response. Program use of direct access addressed data at the bit level is accomplished by copying data bytes directly between the input/output modules and flag memory. Flag bytes 5 (FY005) and 6 (FY006) are used in the bit-level instructions for this purpose. The contents of the program PB011 are not shown in the example.

Process interrupts can be disabled in either of two ways. If there is no program in OB002, the PLC will ignore interrupt signals. If the user-program has executed the IA (interrupt disable) instructions, interrupt response will be delayed until the program executes an RA (reenable interrupts) instruction, at which time the most recently delayed ISR will execute.

S7

I/O INTERRUPTS IN SIEMENS STEP 7

STEP 7 calls I/O interrupts *hardware interrupts* in some places in their literature, and *process interrupts* elsewhere. When a hardware interrupt signal is received, the PLC interrupts whatever it is doing and calls OB 40. The user must have programmed OB 40, or the PLC will fault. Hardware interrupts are used in three ways in S7 PLCs:

1. **Hardware interrupts** can be initiated by input changes detected by digital input modules, by analog modules if the analog input value exceeds an alarm limit value, or by a CP (communication processor) or an SP (special) I/O module.

2. Some PLCs (e.g., S7-315-DP) can send a signal to a DP master PLC via a Profibus-DP local area network, to cause the DP master CPU to execute the hardware interrupt OB 40.

3. The smaller **integrated function module** (IFM) S7 PLCs have digital input capabilities built into the CPU module. These CPU-resident inputs can be used to trigger **(high-speed) counter** and **frequency monitoring** interrupts. (These are not called hardware interrupts in STEP 7 and do not trigger execution of OB 40.)

When the PLC's operating system program calls OB 40, it automatically provides 20 local variable parameters that the program in OB 40 can examine. These parameters include the address of the module that initiated the interrupt request signal, a number indicating the digital input number (if a digital module initiated the signal), or the current state of the analog module (if an analog module initiated the signal). Other parameters that are automatically passed to OB 40 (and which the program in OB 40 can examine) include the data and time of the call, the name of the organization block being called and its priority class, and sometimes some additional information describing why the call is being made. For a complete list of the 20 parameters and the variable names they are assigned, see your manual set.

A hardware interrupt is the second-highest-level interrupt in the STEP 7 system, so it will interrupt any program or other ISR that might be running, except for a fault ISR program. A new hardware interrupt signal will not cause OB 40 to be interrupted if it is already running.

> STEP 7 reserves OB 41 through OB 47 for future additional hardware interrupt ISRs. All hardware interrupts will have the same priority level, so they will not be able to interrupt each other.

If the programmer wants to disable hardware interrupts, or to disable other interrupts during execution of OB 40, there are two options:

1. Program a call to **SFC 39 ("DIS_INT")** to tell the PLC to disable interrupt requests. Interrupt requests received after DIS_INT has disabled them will be ignored completely (although information pertaining to each interrupt request will be stored in the diagnostics buffer at the time of the interrupt request). Parameters that the programmer provides with the call to DIS_INT indicate whether the PLC should ignore just a specific type of interrupt request, all interrupt requests of a specified priority class, or all interrupts except fault routines responding to programming errors (called synchronous interrupts in STEP 5). "DIS_INT" can be called by OB 40 to disable other interrupts while OB 40 runs. Although only a fault interrupt is high enough in priority to interrupt OB 40, disabling interrupts while OB 40 runs prevents those interrupts from being serviced after OB 40 finishes.

To enable interrupts again, the program would have to call **SFC 40 ("EN_INT")** and specify exactly which interrupt type or interrupt priority class(es) are to be reenabled. EN_INT doesn't have to be called in the same scan as DIS_INT. If interrupts are still disabled when a program scan ends, they will stay disabled during the input and output scan steps as well.

Figure 11.11 shows, in STL language, how a program would disable all hardware interrupts and reenable them later. In the SFC 39 (DIS_INT) call, mode parameter "1" (a hexadecimal digit, as indicated by the prefix "B#16#") indicates that an entire priority class is to be disabled. OB_NR parameter 40 is the OB number that hardware interrupts can call, so DIS_INT will disable all future hardware interrupt calls. RET_VAL is a 32-bit number that DIS_INT generates, and this call to DIS_INT indicates to store this number into bit memory at address MW 100. If DIS_INT is successful at disabling interrupts, this value will be zero; otherwise, an error code will be placed into MW 100 by DIS_INT. The program should include instructions to examine MW 100 after calling DIS_INT. The call to SFC 41 EN_INT includes a similar set of parameter values because it is intended to reenable the hardware interrupts. Only the RET_VAL parameter return address is different, because the programmer wants to maintain separate addresses for information on success or failure of DIS_INT and EN_INT.

2. Program a call to **SFC 41 ("DIS_AINT")** to tell the PLC to *delay response* to all interrupt requests. Interrupt requests received after DIS_AINT has been called will be held until **SFC 42 ("EN_AINT")** is called, or until the organizational block that first called "DIS_AINT"

```
CALL "DIS_INT"                  // Call SFC 39, to disable interrupts
  MODE      : =B#16#1           // Select a whole priority class
  OB_NR     : =40               // Specify hardware interrupt class
  RET_VAL   : =MD 100           // Accept SFC 39's 32-bit reply
.
.                               // Instructions here can't be
.                               // interrupted by hardware interrupts
.
CALL "EN_INT"                   // Call SFC 40, to reenable interrupts
  MODE      : =B#16#1           // Select a whole priority class
  OB_NR     : =40               // Specify hardware interrupt class
  RET_VAL   : =MD 104           // Accept SFC 40's 32-bit reply
.
.                               // Instructions here can be
.                               // interrupted by hardware interrupts
.
```

FIGURE 11.11
STEP 7 program that temporarily disables hardware interrupts.

ends. If DIS_AINT is called more than once before EN_AINT is called, then EN_AINT must be called an equal number of times before the organizational block can be interrupted to execute ISRs in response to interrupts. No parameters are required by DIS_AINT or by EN_AINT, but both output a RET_VAL parameter indicating the number of times that "EN_AINT" must be called before interrupts will again be allowed. Figure 11.12 contains an STL program that delays interrupt response.

If DIS_AINT is called in an ISR, it only affects interrupts that are of higher priority than the ISR organization block. Lower-priority interrupts would be delayed until the higher-priority organizational block ended anyway.

A Profibus-DP slave PLC (e.g., an S7-513-DP) can call **SFC 7 ("DP_PRAL")** to send a signal via the Profibus-DP local area network to cause the DP master PLC to interrupt and to run OB 40. When the slave calls DP_PRAL, it provides parameters indicating an imaginary peripheral input or peripheral output address. The slave PLC passes those parameters via the network so that they can be used as input parameters for the DP master's OB 40. OB 40 should be programmed to respond appropriately to hardware interrupt signals initiated from the imaginary peripheral I/O module.

Integrated function modules (IFMs) are S7 CPU modules with digital I/O built right into the CPU module. These IFM modules can be configured to monitor selected inputs and to interrupt to update a high-speed **counter** whenever the state of one of those bits changes. An IFM module can even be configured to compare the count value against limit values and to perform control operations before returning to the interrupted program. The STEP 7 documentation does not indicate a priority level for these high-speed counting interrupts but warns that even hardware interrupts can be delayed as they execute! There are four types of integrated high-speed interrupt functions available, but they all use the same built-in digital inputs, so the user can only select one:

```
CALL "DIS_AINT"                  // Call SFC 41, to delay interrupts
 RET_VAL   := MD 108             // Accept SFC 41's 32-bit reply, which
.                                // indicates the number of times
.                                // SFC 41 has been called
.
.                                // Instructions here can't be
.                                // interrupted by any interrupts

.
CALL "EN_AINT"                   // Call SFC 42, to reenable interrupts
 RET_VAL   := MD 112             // Accept SFC 42's 32-bit reply, which
                                 // indicates how many more times
.                                // SFC 42 must be called before
.                                // interrupts will be allowed
.
.                                // If the return value was 0, then
.                                // instructions here can be interrupted
.                                // by interrupts
```

FIGURE 11.12
STEP 7 program that temporarily delays interrupts.

1. A **frequency monitor** function can be used to count leading edges at an input contact[12] over a period of time and to calculate a frequency (in megahertz to 10,000,000) at the end of the period. The CPU will repeatedly recalculate the frequency every time the period interval elapses. The STEP 7 programming software must be used to configure the CPU to use either 0.1-, 1-, or 10-s intervals and to specify an instance data block to use (default = DB 62). In the user-program, the frequency can be read by executing **SFC 30 ("FREQ_MES")** and providing an address for the 32-bit result, **FREQ.** Other optional parameters that can be supplied with a call of SFC 30 include 32-bit signed binary upper- and lower-frequency limits (**PRES_U_LIMIT** and **PRES_L_LIMIT,** in MHz) and Boolean bits (**SET_U_LIMIT** and **SET_L_LIMIT**) to force SFC 30 to start using the limits. If limits have been defined, subsequent calls of SFC 30 can pass output parameter addresses to read the current limits (**U_LIMIT** and **L_LIMIT**) and status bits that will be on if the actual frequency is above the upper limit (**STATUS_U**) or below the low limit (**STATUS_L**).

2. A **counter** function can count high-speed pulses (to 10 kHz) at two inputs[13] and can immediately provide outputs and/or initiate a hardware interrupt (calling OB 40) if the count reaches either of two limit values. Siemens says a "counter event" occurs when the count reaches a value where a reaction is required. The STEP 7 programming software must be used to configure the CPU to count leading or trailing edges at (or to ignore) each of two inputs (a count-up input and a count-down input) and to configure whether a counter event is to occur when the count increases to or drops below either of the two compare values. Configuration also specifies what control operation to perform at a counter event: initiate a hardware interrupt and/or set (or reset) two output contacts. Configuration can also force resetting of the counter and/or setting new limits whenever a counter event occurs. DB 63 will be used as the instance data block, unless configuration changes this default.

The counter won't count until the user-program calls **SFC 29 ("HS_COUNT")** (as shown in Figure 11.13) at least twice, once with the Boolean parameters **SET_COMP_A** and **SET_COMP_B** FALSE, then again with them TRUE. The second call also should provide 32-bit signed values for the counting limits (**PRES_COMP_A** and **PRES_COMP_B**), should enable or disable control of the output contacts by passing a true/false Boolean value to **EN_DO,** and *must* enable counting by passing a true Boolean value to **EN_COUNT.** The PLC will then count up-count and down-count pulses at the two main contacts while the direction control input contact and the start/stop contact remain powered. Counting direction can be reversed if the direction contact voltage drops or will freeze if the start/stop contact voltage drops. The counter event actions that are configured will execute automatically. The user-program can execute SFB 29 again to read the 32-bit signed binary current count parameter (**COUNT**), to read the current limit values (**COMP_A** and **COMP_B**), or to check two status bits to see if the current count is above either of the limits (**STATUS_A** and **STATUS_B**).

3. The S7 314 IFM allows configuration of the CPU to run two high-speed counters, each with one limit value and a configured counter event response. Two inputs are used for counter A and another two for counter B.[14] Configuration can set up the two inputs as an up and a down

---

[12] The S7 312 CPU monitors I 124.6 for frequency monitoring. The S7 314 monitors I 126.0.

[13] The S7 312 CPU monitors I 124.6 to I 125.1 for up, down, direction, and start/stop (respectively). The S7 314 monitors I 126.0 to I 126.3 for the same purposes. Both CPUs control Q 124.0 and Q 124.1 as outputs A and B.

[14] I 126.0 and I 126.1 for counter A, I 126.2 and I 126.3 for counter B.

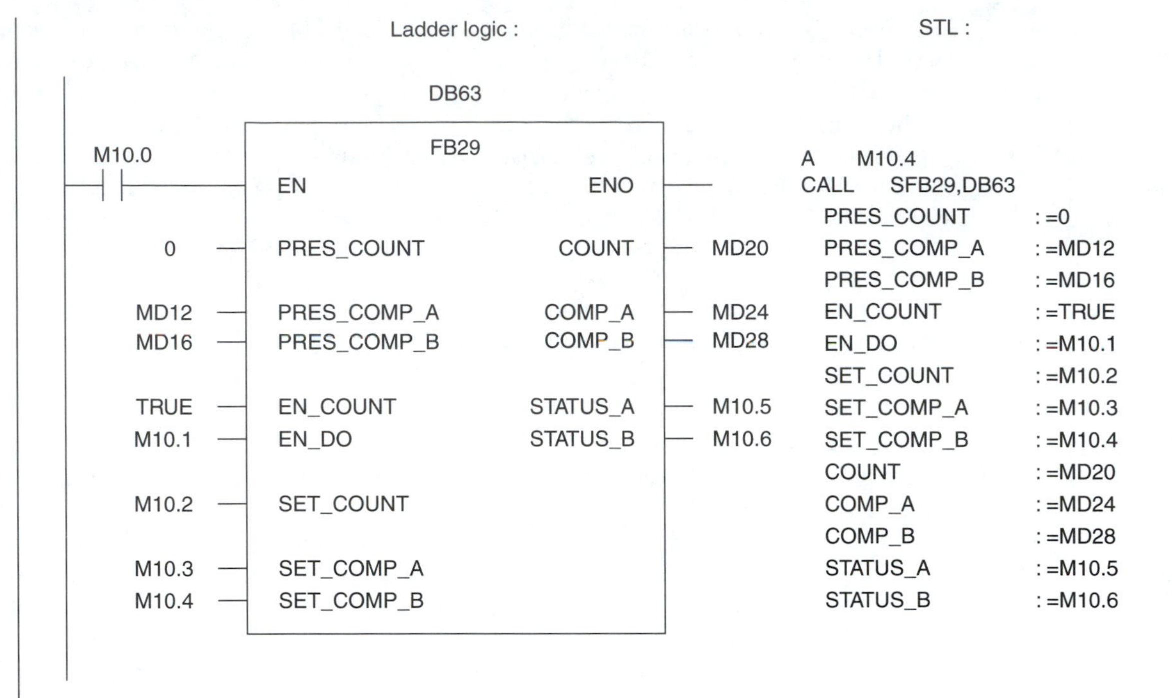

(Remember, since there is an instance data block that stores these parameters between calls of SFB 29, it isn't necessary to declare actual values for every parameter during every call as in this example.)

FIGURE 11.13
SFB 29 (HS_COUNT) as it would appear in a STEP 7 ladder logic network and as it would appear in an SFC program.

counter, or as one counter input plus one count-direction-control input. Other configuration options are as described for the high-speed counter in (2), except that there must be two configurations. Counter A uses DB 60 as an instance data block and counter B uses DB 61 unless configuration changes these defaults. The user-program must include a set of calls to **SFB 38 ("HSC_A_B")** with each of the instance data blocks, to set each counter's limit and to enable each counter. SFB 38 also allows a Boolean input to reset the count (to any value entered during configuration). SFC 38 does not provide a Boolean status output.

4. The S7 314 IFM also offers a motor-control function block, **SFB 39 ("POS"),** that includes counting of high-speed pulses from an incremental encoder. POS compares the current count (actual motor position) against a required count (desired motor position) and generates outputs to drive a motor to correct any error. The size of the positional error affects how strongly POS drives the motor to correct the error. POS uses instance data block 59, unless configuration changes this default.

In the user-program, POS can be called to "jog" the motor forward or backward (by setting the **POS_MODE_1** or the **POS_MODE_2** Boolean parameter). Switching **REF_EN-**

ABLE on while the count is changing in the right direction enables initializing the count. The count will then be initialized (to the signed 32-bit value supplied as input parameter **REF_VAL**) when a leading edge is observed at input contact I 126.2 or at parameter **SET_POS** when POS is called again. POS sets output parameter **REF_VALID** the first time it is executed after initialization has taken place.

After initialization (sometimes called *synchronization* or *homing*), POS can be called again with a position setpoint as input parameter **DEST_VAL,** a leading edge at input **SET_POS** to cause POS to accept the position value, and a leading edge at input **POS_STRT** to tell POS to start driving the motor. An unsigned 16-bit word value for parameter **SWITCH_OFF_DIFF** gives POS a tolerance range around the setpoint position, inside which the motor won't be driven. POS's output parameters include **ACTUAL_POS,** the current signed 32-bit count, and **POS_READY,** which means that the most recently desired position has been reached or the most recent jog motion has been turned off.

CQMI

I/O INTERRUPTS AND THE OMRON CQM1 PLC

OMRON CQM1 PLCs offer three variations of I/O interrupts. One variation is called **input interrupts.** Another variation is used as a **high-speed counter interrupt** and can only count input signal changes. The third variation of what OMRON calls I/O interrupts, called the **high-speed pulse output,** generates precisely timed output signals and is discussed in the section on timed interrupts.

Input interrupts, which execute in response to an input signal going on, are defined by OMRON as the highest-priority CQM1 interrupts, so they can interrupt any other ISR that might be running (but not the preprogrammed fault routines included with the CQM1). The input signal(s) must be provided to the PLC at one or more of the four lowest-numbered input terminals (00 to 03) on the CQM1's CPU module (module 000). OMRON has even defined priority among the four input interrupts, with interrupt 0 (from input terminal 00) having the highest priority, to interrupt 3 (from terminal 03), which has the lowest. The user must enter the ISR programs into specific subroutines: Subroutine 000 runs in response to an interrupt signal at input 00000, subroutine 001 for input 00001, SBR 002 for 00002, and SBR 003 for 00003.

OMRON's CQM1 must be configured to respond to input interrupt signals from the four input contacts.[15] Configuration data must be entered while the PLC is still in program mode. (Input interrupt configuration cannot be changed while a PLC program is running.) In the CQM1, interrupts are masked (disabled) by default, so after configuring the PLC for input interrupts the user-written program must execute an INTERRUPT CONTROL instruction, **INT(89),** to unmask input interrupt(s). The INT(89) instruction, as shown in Figure 11.14, must be executed with three parameters. A control code (CC) parameter must be provided as the first pa-

---

[15] Write a 1 into selected positions of the four hex characters of DM 6628, indicating which input bits will be able to cause input interrupts. A 1 in the least significant position enables input bit IR 00000 to trigger execution of the ISR in subroutine 000. The second least significant bit is for IR 00001, third for IR 00002, and leftmost for IR 00003 (e.g., 0101 in DM 6628 will enable bits IR 00000 and IR 00002 to trigger input interrupt subroutines 000 and 002, but IR 00001 and IR 00003 won't trigger input interrupts).

rameter. The second parameter is always 000. The third, control data (D), parameter's use depends on the control code. Control codes include:

1. CC = 000 is used to mask or unmask some or all input interrupts. The D parameter must then be a four-digit BCD value, of which the high three digits are always 000. The low digit represents a 4-bit code indicating whether to mask or unmask each of the four input interrupts. If bit 3 = 1, interrupt 3 (from contact 03) will be masked (disabled). If bit 3 = 0, interrupt 3 will be unmasked. Bits 2, 1, and 0 of parameter D control interrupts 2, 1, and 0 similarly.

2. CC = 002 is used to read which of the input interrupts are currently masked. The D parameter must be an address. INT(89) will write to the low 4 bits of the value at address D to reflect the masked states of the four interrupts. If bit 3 is on, that means the interrupt from contact 03 is masked, and so on.

3. CC = 100 masks all interrupts (not just input interrupts) and CC = 200 returns them to the state they were in before INT(89) with CC = 100 was executed. Interrupt signals received while masked by INT(89) with CC = 100 are recorded and will be responded to after INT(89) with CC = 200 executes. For some reason, the CQM1 will not recognize INT(89) with CC = 100 if it is executed inside an interrupt ISR! The D parameter isn't used but should be set to 0000.

4. CC = 001 is used to clear input interrupt requests that are waiting to execute. If an input interrupt has been prevented because a mask is set or because a higher-priority ISR is running, it will execute (once) when the input interrupt is unmasked or the other ISR finishes, unless INT(89) is used with CC = 001 to clear the waiting input interrupt. Again, the low 4 bits of D correspond to the four interrupts, and a 1 in one of those bit positions clears any waiting interrupt for the corresponding interrupt type.

5. CC = 003 is used to set up the input in **counter mode.** In counter mode the interrupt does not execute every time its associated input signal goes on. Instead, the CQM1 uses a counter to count input signals and causes the ISR subroutine to run only when the counter reaches its setpoint value. When the count reaches its preset, the count is also automatically reset to zero so that counting of input signals can resume. When CC = 003 is used, parameter D indicates which of the four interrupts will be set up in counter mode. A 1 in bit positions 0 to 3 of parameter D means that the corresponding input interrupt is *not* affected by this instruction, but a 0 will cause the corresponding input interrupt to be unmasked and set up in counter mode (and will cause the count to be refreshed back to a value of 0000, so the programmer will probably want to use the differentiated form of the instruction so that it only executes once each time its input logic goes true). If, for example, the low 4 bits of the 16-bit D value was 0111, input interrupts 0, 1, and 2 are unaffected, but input interrupt 3 is set up in counter mode with the count reset to 0000. Hexadecimal counter setpoints of between 0000 and FFFF (0 to 65,535) must be stored in SR 244 to SR 247 for interrupts 0 to 3, respectively, and can be changed during program execution, but changes will not be effective until INT(89) is executed again with CC = 003. Setting a setpoint to 0000 disables the interrupt. Counter accumulated values are stored in SR 248 through SR 251, for interrupt 0 through 3, respectively.

Only false-to-true transitions at the CPU's four lowest-numbered input terminals, 00000 to 00003, can initiate input interrupts (or be counted in counter mode). If one of these input signals turns on, and if that input has been configured as an input interrupt source and has been unmasked (see the sample program in Figure 11.7), the CQM1 will interrupt what it is doing and:

1. Perform a reduced input scan, reading only the input modules that it has been configured to read.[16]

2. Run the interrupt service routine corresponding to that input interrupt signal. (OMRON calls ISRs subroutines, and these subroutines are entered by the programmer just like normal subroutines except that JUMP TO SUBROUTINE instructions aren't necessary.) Input terminal 00 causes subroutine 000, input 01 causes subroutine 001, and so on.

3. Return to the interrupted program without sending new output data to any output modules (unless the CQM1 has been configured for direct refresh). The programmer must include I/O refresh instruction(s) in the ISR subroutine if immediate output is required.

In the example program in Figure 11.14:

1. The first rung enables two interrupts. It unmasks interrupt 0, moves the value 5 to data word SR 246 as the setpoint for interrupt 2, then sets interrupt 2 up in counter mode. (Remember, in setting counter mode using INT (89) with CC-003, a 0 in the low 4 bits of parameter D sets and enables the counter, so 1's must be used to indicate which of the interrupts are *not* to be affected by the instruction. The hexadecimal digit B is 1011 in binary.) It will now take five actuations of input 00002 to cause interrupt 2. The differentiated forms of both instructions are used (the @ does this) so that they execute only once whenever input IR 00203 goes on. Using the differentiated mode ensures that the counter will reset only when input IR 00203 goes from false to true.

2. The second rung disables all interrupts if input IR 00204 is on.

3. The third rung reenables all interrupts after clearing any pending calls to interrupt 0 if bit IR 00204 is off.

Interrupts 0 and 2 require subroutines 000 and 002 (the same numbers as the interrupts). Before putting the PLC into run mode, the programmer must have entered 0101 into DM 6628, enabling inputs 00000 and 00002 as input interrupt sources, and for this example has entered 0305 into both DM 6630 and DM 6631, indicating that three data words starting with IR 005 must be refreshed as either interrupt starts.

1. SBR 000 just copies the newly refreshed data word from input module 005 immediately to output module 104, using a MOV(21) followed by an IORF (97), controlled by the "always on" bit.

2. The program for SBR 002 is not shown in this example.

OMRON's **high-speed counters (HSCs)** make use of an I/O interrupt to initiate a preprogrammed ISR to count input signals received at the input contacts in the CQM1's CPU module. Most CQM1 models can only use HSC0 to count signals connected at contacts 00004 through 00006, but some models have two additional connector ports in the CPU and can also count input signals received at those ports using HSC1 and HSC2. In this section we describe

---

[16] For interrupt 0 (initiated by IR 000 00), DM 6630's most-significant BCD digits must contain the number of input words to read (00 to 08), the low digits must indicate the first module address to read from (00 to 07). DM 6631, DM 6632, and DM 6633 are similarly used for interrupts initiated by bits IR 00001, IR 00002, and IR 00003, respectively (e.g., if DM 6632 contains 0305, when IR 00002 causes interrupt 2 the CQM1 will copy input data from input modules 005, 006, and 007 into IR 005 to IR 007 before running the ISR subroutine 002).

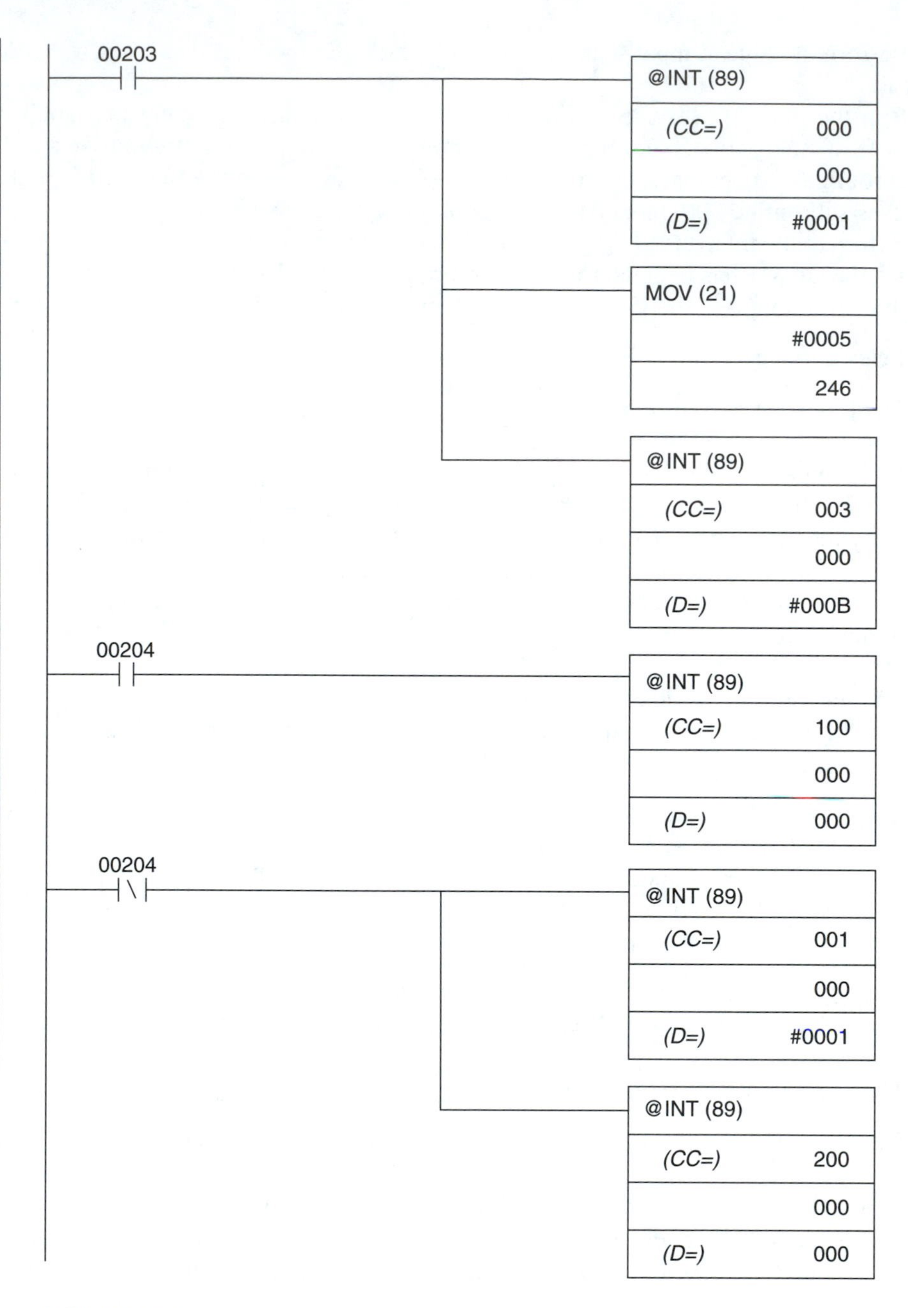

FIGURE 11.14
Program to use CQM1 input interrupts.

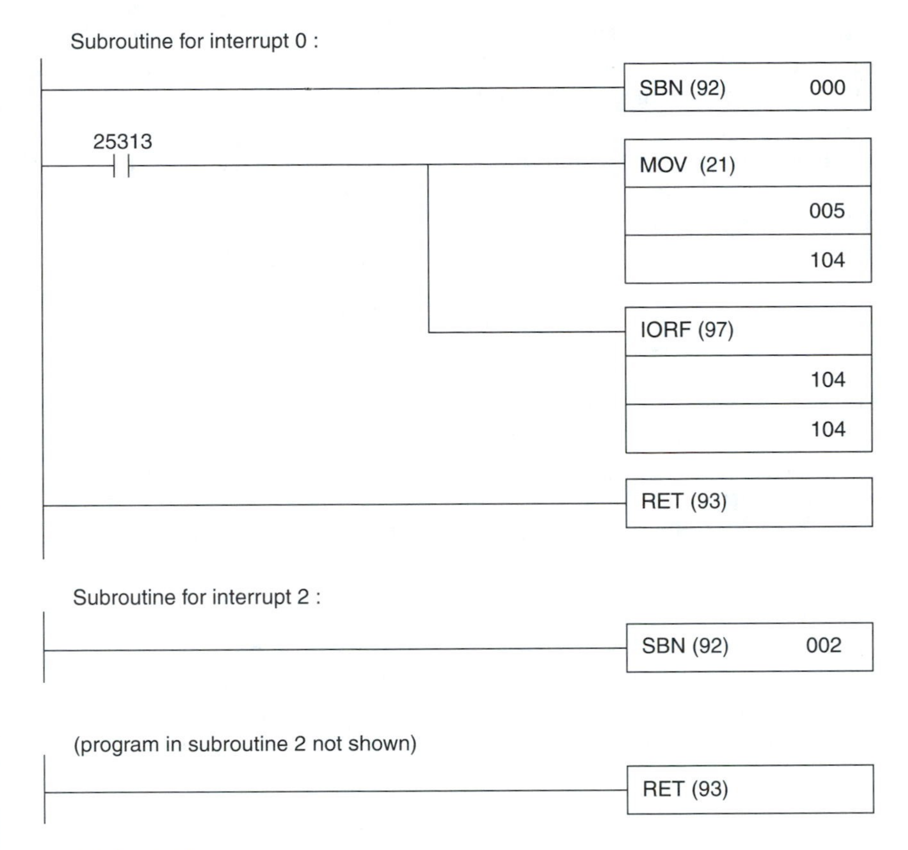

FIGURE 11.14
Continued

HSC0 first, then discuss HSC1 and HSC2. High-speed counting is the lowest-priority interrupt in the CQM1, so a high-speed counter interrupt service routine (ISR) may be delayed if another ISR is already running or if INT(89) has masked all interrupts.

Before any program can use HSC0, the CQM1 must be configured to use HSC0 by having the hexadecimal value **01**xx in DM 6642.[17] (The last two digits also affect the counter and are discussed later.) Configuration may also include entering a data word in DM 6638 to indicate that the CQM1 should start by reading data from up to 12 input modules whenever an HSC0 interrupt service routine (ISR) is executed.[18] These configuration values can be changed only while the CQM1 is in program mode. If configured to use HSC0, the CQM1 will reset HSC0's present value every time it is put into run mode and will count input signals arriving at input contacts 00004 (and possibly 00005) while in run mode. High-speed counter interrupts

---

[17] If DM 6642 sets the CQM1 up to use HSC0, the CQM1 cannot use interval timer 2.

[18] The most-significant two BCD characters of DM 6638 indicate how many input modules to read from (00 to 12), and the least-significant characters indicate which input address to start at (000 to 011).

are masked (disabled) by default, so the user-program must include a **REGISTER COMPARISON TABLE** instruction, **CTBL (63),** to "register" (provide) a *table* of setpoint values for the counter's present value to be compared against and a list of ISRs to run for each match found in the table. The CTBL(63) instruction (described below) can also be used to initiate comparisons, or comparisons can be started and stopped by the **MODE CONTROL** instruction, **INI(61),** after CTBL(63) has been executed to register a setpoint table. The counter's present value is copied into data memory addresses SR 230 and SR 231 automatically once per scan cycle, but the **HIGH-SPEED COUNTER PV READ** instruction, **PRV(62),** can update SR 230 and SR 231 during a program execution.

The HSC0 can be configured to count using either of two methods, depending on the configuration value in the lowest hex-character position of DM 6642.

1. If DM 6642 contains 01x**4,** HSC0 is in incrementing mode. It will count the times the input signal at 00004 goes on, will only count up, and will count from 0 to 65,535.
2. If DM 6642 contains 01x**0,** HSC0 is in up/down mode. This mode should be selected if an incremental encoder is being used as a position sensor. The CQM1 will increment or decrement the count whenever either the signal at 00004 (the encoder's A-phase output) or at 00005 (the encoder's B-phase output) changes. The counter's present value range will be −32,767 to +32,767.[19]

To reset HSC0's present value, one of three methods can be used. If DM 6642 contained 01**1**x when the PLC went into run mode, a program can set bit SR 25200 to cause HSC0's present value to reset to zero, after which the CQM1 will reset SR 25200. If DM 6642 contained 01**0**x, then HSC0's present value will reset only after the program sets bit SR 25200 and then the signal at input 00006 goes on. This signal could be from an incremental encoder's Z phase output. Despite the value in DM 6642, the **MODE CONTROL** instruction, **INI(61),** can be used to change HSC0's present value to zero or to any other value. The MODE CONTROL instruction is described later.

The REGISTER COMPARISON TABLE instruction, **CTBL(63),** indicates a table of eight-character BCD values that HSC0's present value is to be compared against and provides the subroutine numbers that will be called if matches are found. The programmer must have enterred the table of values into sequential memory locations before CTBL(63) is executed. There are two types of comparison tables that may be used:

1. A **target table,** which can contain up to 16 target values and associated subroutine numbers in the form shown in Figure 11.15. Normally, the subroutine runs when the counter's present value has been incremented to the target value, but if the first character of the subroutine number is F, the subroutine will execute after HSC0 has been decremented to the target value. After CTBL(63) has registered the example target table in Figure 11.15, and comparison has been initiated, subroutine 15 will be called when HSC0's present value increments up to +10,300. The counting will continue and subroutine 16 will be called if the count exceeds

---

[19] The CQM1 stores the present value in BCD format. SR 230 holds the four lowest BCD characters of the present value. SR 231 contains the high BCD character value in its low 4 bits. The other BCD characters of SR 231 will be 000 unless the present value is negative, in which case the most significant character will be F. If, for example, HSC0's present value was −12,345, then SR 230 would contain the value 2345 and SR 231 would contain F001.

	EG:
Number of target values in this table (BCD)	0002
First target: low 4 BCD digits	0300
First target: high 4 BCD digits	0001
First subroutine, if PV reached first target	0015
Second target: low 4 BCD digits	1111
Second target: high 4 BCD digits	0002
Second subroutine, if PV reached first target	F016
following values irrelevant	

FIGURE 11.15
Target table, as required for the CQM1 CTBL(63) instruction.

+21,111 and then decrements back down to +21,111. (If the first character of the high four BCD digits for a target value is F, the target value is negative.)

2. A **range table,** which must contain eight ranges, can be entered in the form shown in Figure 11.16. The subroutine will execute as soon as HSC0's present value enters a range, either by incrementing or decrementing. The data table must contain data for each of the eight possible ranges, even if all eight aren't required. (Subroutine FFFF must be entered for each range not used.) As in target tables, if the first character of a limit's high four BCD digits is F, the limit value is negative. Eight status bits in AR 11 reflect whether the present value is within the range table's ranges. If the example values in Fig. 11.16 are entered and CTBL(63) has registered them and comparison has been initiated, subroutine 17 will be called when HSC0's present value is found to have entered the range +10,333 to +20,444, and subroutine 18 will be called when HSC0's present value enters the range −20,666 to −30,555.

CTBL (63) should be used in its differentiated form, @CTDB (63), because it has a long execution time and should therefore be executed only when the data in the comparison table needs to be initiated (e.g., every time the PLC goes into run mode) or changed. Three parameters must be provided with CTBL (63):

1. The first is the P (port specifier) parameter. P must be 000 for high-speed counter 0, which uses port 0 (CPU input contacts 00004 through 00006).

2. The second parameter is the C (control) parameter. C indicates what type of comparison table is being entered (000 or 002 means target table; 001 or 003 means comparison table) and whether use of the table is to be initiated immediately (000 or 001 initiates comparisons and enables interrupts; 002 or 003 means wait until an INI(61) instruction initiates comparisons and interrupts.)

	EG:
First range: low limit: low 4 BCD digits	0333
First range: low limit: high 4 BCD digits	0001
First range: high limit: low 4 BCD digits	0444
First range: high limit: high 4 BCD digits	0002
First subroutine, if PV is in first range	0017
Second range: low limit: low 4 BCD digits	0555
Second range: low limit: high 4 BCD digits	F003
Second range: high limit: low 4 BCD digits	0666
Second range: high limit: high 4 BCD digits	F002
Second subroutine, if PV is in first range	0018
etc.	

FIGURE 11.16
Range table, as required with a CQM1 CTBL(63) instruction.

3. The "TB" (table) parameter is last and must contain the address of the first data word in the comparison table.

High-speed counter interrupts can be started and stopped by masking or unmasking all interrupts using INT(89) or by using the mode control instruction **INI(61).** If masked, the count will continue but no ISR will run, and status bits in AR 11 will continue to reflect whether the present value is within the ranges that would normally cause an ISR to execute. INI(61) must include three parameters:

1. The first, P (port specifier), parameter must be 000 to control HSC0.

2. The second, C (control), parameter can be 000 to unmask (initiate) use of the comparison table to cause interrupts, or 001 to mask (stop) its use. 002 means the following parameter provides a new present value for the counter. 004 is used for controlling pulse output and is not used for high-speed counters.

3. The third, P1 (first PV), parameter will be 000 unless the counter's present value is to be changed. If so, P1 must indicate the first address of the two addresses containing a new eight-BCD-character value for the present value. (As in the comparison tables, the first address must contain the low four BCD digits, and the next address must contain the high four BCD digits.)

@INI(61), the differentiated form, is recommended; otherwise, the present value will be changed every scan cycle while this instruction's control logic is true.

HSC0 will stop if it counts outside the allowable range (0 to 65,535 for incrementing mode, −32,767 to +32,767 for up/down mode). The eight-digit BCD present value will be set to 0FFF FFFF after counting above the range, or to FFFF FFFF after counting below the range, so the sixth and seventh digits, which should always be 0s, can be monitored to detect this condition. At the very least, the counter must be reset and the comparisons restarted before high-speed counting interruptions can resume.

Figure 11.17 shows a program using the high-speed counter interrupt feature. Before putting the PLC into run mode, DM 6642 contains 0104, to configure the CQM1 to use HSC0, to require a signal from input 00006 to reset the counter's PV and to use up/down counting. DM 6638 contains 0203, configuring the CQM1 to read input modules 003 and 004 into IR 003 and IR 004 each time HSC0 calls an interrupt subroutine. Two tables of range data have been prepared; the first at DM 0000 to DM 0039 and the second at DM 0040 to DM 0079. The program should also contain subroutines that the HSC0 can cause to execute, but the subroutines are not included in this example.

In the program in Figure 11.17:

1. Rung 1 uses the first scan bit (25315) to register a range data table starting at address DM 0000 and to initiate comparisons of HSC0's present value with the table. The differentiated version of CTBL (63) isn't needed since the first scan bit goes on only for the first scan.

2. When input IR 00102 goes on, rung 2 executes INI(61) to stop use of comparison tables for HSC0, executes CTBL(63) to change the range table but not restart use of comparisons to initiate interrupts, and finally, turns HSC0's software reset bit on so that the count will reset to zero when a signal is received at input 00006.

3. When input IR 00102 goes back off, rung 3 initiates the use of the new comparison table.

The **CQM1-CPU43-E** offers two more high-speed counters, **HSC1** and **HSC2,** that operate very similarly to the standard HSC0, except that HSC1 counts the input signals that are received at input port 1, and HSC2 counts signals at port 2, so the P parameter of CTBL(63), INI(61), and PRV(62) must be 001 or 002. The configuration requirements and the use of data words by HSC1 and HSC2 differ slightly from the description of HSC0 above:

1. DM 6611 selects use of ports 1 and 2 either as high-speed counter inputs or as pulse output ports. (Pulse output is described later in this chapter.)

2. DM 6634 is used to configure which inputs to refresh when HSC1 initiates an interrupt. DM6635 is used for HSC2.

3. DM 6643 is used to configure the way that HSC1's present value is reset. DM6644 is used for HSC2.

4. Present values are copied to SR 232 and SR 233 for HSC1 each scan. HSC2 uses SR 234 and SR 235.

5. AR 05 contains status flag bits for HSC1. HSC2 uses AR06.

There are a few differences between high-speed counter 0 and high-speed counters 1 and 2:

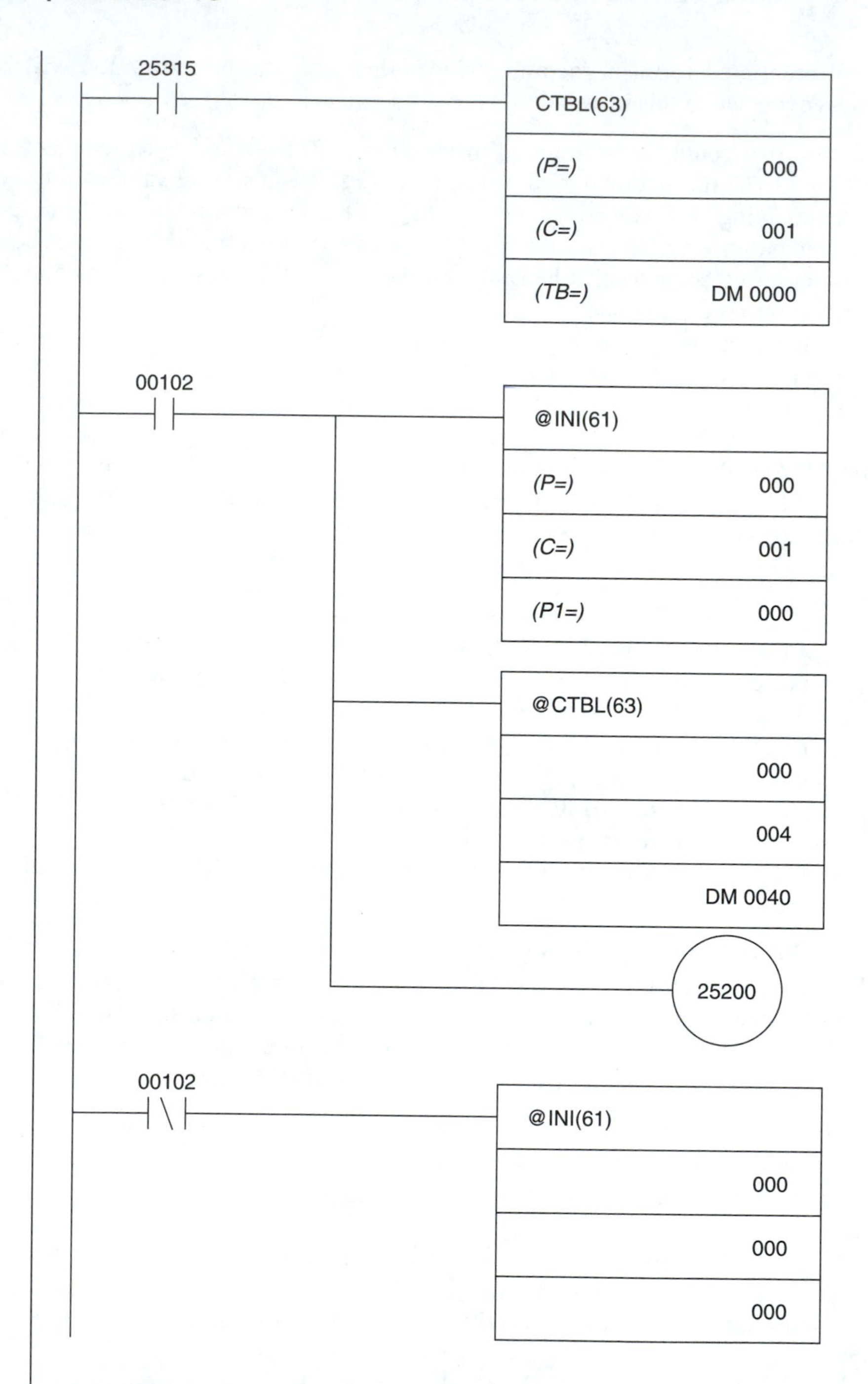

FIGURE 11.17
OMRON CQM1's high-speed counter interrupt.

1. Counting modes, configured in DM 6643 and DM 6644, are different. What is called up/down counting for HSC0 is called differential phase mode for HSC1 and HSC2. For HSC1 and HSC2, *up/down mode* means that a signal at the A-phase input causes the counter to increment, while B-phase inputs cause it to decrement. An additional pulse/direction mode is added, in which the signal at the A phase indicates whether signals at the B phase should cause incrementing or decrementing of the count.

2. DM 6643 and DM 6644 select linear or ring mode counting. In linear mode, HSC1 or HSC2 counts between −8,388,607 and +8,388,607. The counter stops if it counts outside the range. In ring mode the count continues in either direction from 0 as high as 64,999. The number of points in the ring (1 to 65,000) must be provided in the first two data words of the comparison table, before the target or range values. If the count exceeds the maximum value, it simply starts again at zero (and starts from the maximum if it counts below zero).

3. HSC1 has higher priority than HSC2. Both have higher priority than interval timer interrupts or HSC0, but lower priority than input interrupts.

For more detail, see your CQM1 manual set.

The **CQM1-CPU44-E** also offers HSC1 and HSC2, but they are designed to count changes in signals from 8-, 10-, or 12-bit *absolute encoders.* Unlike in the CQM1-CPU43-E described above, DM 6611 and DM 6612 aren't used to select between counter input or pulse output at ports 1 and 2 (the CPU44-E can't output pulses through ports 1 or 2), but instead can contain a compensation value that will be subtracted from HSC1's present value. Another difference is that DM 6643 and DM 6644 are used to configure HSC1 and HSC2 to monitor 8-, 10-, or 12-bit absolute encoders and indicate whether to convert the present value to degrees or leave it in BCD form. If conversion to degrees is selected, the maximum PV value will be 359, and comparison tables must be prepared with single data words for target or range limits.

Timed Interrupts

With **timed interrupts,** an interrupt service routine (ISR) runs at precisely timed intervals. An interrupt timer generates an interrupt request and the PLC's microprocessor interrupts what it is doing, runs the ISR, and then resumes the interrupted process. Intervals, of course, must be long enough that each timed interrupt ISR can finish before the next has to start.

Timed interrupts are used for control of processes that require consistent control intervals and for control processes that need only infrequent adjustment. Putting infrequently needed control routines into timed interrupt ISRs means they do not have to be executed every scan cycle, so the main program will execute faster (except, of course, during those cycles that are interrupted to execute the ISR).

Sometimes the entire user-program is written as a timed interrupt ISR instead of as the user-program for the scan cycle. Putting the entire program into a timed interrupt (and setting the interrupt time to a value at least as large as the longest possible scan interval) forces the PLC's control interval to be consistent, but consistently longer than otherwise possible. Some PLCs allow the programmer to configure the PLC with a minimum time between PLC scan cycles, effectively doing the same thing: making the main program a timed interrupt ISR.

Timed interrupts usually have lower priority than I/O interrupts. This means that a timed interrupt will not interrupt an ISR responding to an I/O interrupt, but an I/O interrupt can interrupt a timed interrupt ISR. Some PLCs execute an input scan cycle before executing a timed interrupt ISR and some don't. The benefit of using a timed interrupt can be lost if the ISR uses obsolete input data, so if the PLC doesn't automatically execute an input scan, the ISR should include immediate input instructions to read in and use the most recent possible input data. Similarly, some PLCs execute an output scan cycle to write modified output image table data to output modules immediately after the timed interrupt ISR completes, and some don't. The programmer may wish to use immediate output instructions in the ISR to force writing of output data to the output modules. Remember that immediate I/O instructions take longer to execute than most PLC instructions.

Some PLCs, if they have clocks that continue to run while the PLC is turned off, can be programmed to execute a timed interrupt ISR at a user-programmed date and time, or at the same time every day/week/month/etc. So far, PLCs all require that they be turned on and in run mode, or they won't run the timed ISR. That may have already changed by the time you read this. Even VCRs can turn themselves on to record programs at a preset time!

Several recent PLCs offer programmable high-speed pulse outputs. Some even offer pulse-width-modulated pulse outputs for analog output. These PLCs use timed interrupts to call pulse-control ISRs to turn the output bits on and off at precisely controlled intervals. The pulse-control algorithms are included in the PLC's operating system programs in ROM. The programmer needs to enter instructions and/or configuration data specifying the address of the output bit to be pulsed on and off, instructions to start and stop the pulses, and instructions to specify the frequency of the pulses and perhaps the percentage of time the output should be on (for pulse width modulation).

Some PLCs offer a programmable delay feature, by which a user-program can include an instruction to cause an interrupt at a precisely controlled time after the instruction executes. The program might, for example, execute an immediate output instruction, then initiate the time-delayed execution of an ISR that will perform an immediate input and examine the effect of the immediate output. During the delay time, the PLC executes its scan cycle normally.

Every PLC has a watchdog timer, which is a type of timed interrupt. Whenever a new scan starts, the watchdog timer restarts. If a scan time ever exceeds the watchdog timer setting, the scan will be interrupted and the PLC will go into fault mode. The scan time can include the main program's scan time plus the time required to run any ISR(s) that might also execute. Most PLCs allow modification of the watchdog timer's setpoint time, by changing the default configuration data for the PLC. Most PLCs also allow the user to enter a fault routine program that will execute when the watchdog timer reaches its preset. You can use the watchdog timer and fault routine to accomplish a time-delayed response (if your PLC doesn't offer a time-delay interrupt). A program can contain instructions to write a small time value into the watchdog timer setpoint. This becomes the delay time before the fault routine executes. The delay, of course, must not extend past the end of the current scan cycle because the watchdog timer is reset each cycle. The fault routine must include resetting of the fault and the watchdog timer so that the interrupted program can be returned to; otherwise, the PLC will stop running.

PLC-5

TIMED INTERRUPTS AND THE ALLEN-BRADLEY PLC-5

The PLC-5 can be configured to execute a selectable timed interrupt (STI) program file at timed intervals. Configuration data for the STI is entered using the same processor configuration screen as used to configure processor input interrupts (PII). The programmer only needs to enter the program file number containing the ISR and the setpoint interval time in milliseconds. The PLC-5's allowable interval time is from 1 to 32,767 ms. Entering a time of 0 ms disables the STI.

Example

If the following data is entered at the PLC-5 processor configuration screen:

```
Selectable Timed Interrupt: prog file: 11 setpoint: 1500 ms
```

then the PLC will start a timer as soon as it is put into run mode. When the timer reaches 1.5 s, the PLC will interrupt what it is doing and execute program file 11. When program file 11 is finished, the PLC will resume the program or I/O scan that was interrupted. The timer resets as an STI program file starts to execute, so this process will repeat every 1.5 s.

An STI interrupt is the lowest-priority PLC-5 interrupt. An STI ISR program file will not interrupt a fault routine, a processor input interrupt, or a previously begun STI program file. The STI will wait and will start when the higher-priority ISR finishes. If an STI is interrupted by a higher-priority interrupt, it will resume executing after the higher-priority interrupt ISR completes. The PLC-5 sets an "STI overlap" minor fault bit (S:10/2) if an STI cannot execute because another STI interrupt service routine is already running.

Block transfer requests in an STI program file are not queued. Instead, they are executed immediately after the block transfer instruction. The PLC waits until the block transfer is completed before the next instruction in the STI program is executed. (Continuous block transfers can't be programmed in STIs; the PLC-5 changes them to noncontinuous.)

Other characteristics of the PLC-5 are identical with those of the Allen-Bradley SLC 500 PLC and are described in the section of this chapter in which we discuss timed interrupts and the Allen-Bradley PLC-5 and SLC 500.

SLC 500

TIMED INTERRUPTS AND THE ALLEN-BRADLEY SLC 500

The SLC 500 can be configured to execute **selectable timed interrupt (STI)** program files at timed intervals, very similar to how a PLC-5 is configured. Configuration data for the STI is entered using the same processor configuration screen as is used to configure discrete input interrupts (DII). The programmer only needs to enter the program file number containing the ISR and the setpoint interval time. The SLC 500's STI interval can be from 1 to 255 increments of 10 ms (or 1-ms increments in some SLC 500 models). Entering a time of 0 disables the STI.

Example

If the following data is entered at the SLC 500's processor configuration screen:

```
Selectable Timed Interrupt: prog file: 11 setpoint: 150 10ms
```

the PLC will start a timer as soon as it is put into run mode. When the timer reaches 1.5 s, the PLC will interrupt what it is doing and execute program file 11. When program file 11 is finished, the PLC will resume the program or I/O scan that was interrupted. The timer resets when an STI program file starts to execute, so this process will repeat every 1.5 s.

An STI interrupt is the lowest-priority interrupt except for the type that is actually called an I/O interrupt in the Allen-Bradley SLC 500 literature. An STI program file will therefore not interrupt a fault routine, a discrete input interrupt, or an STI program file that is already executing, but will interrupt an I/O interrupt ISR. STI ISRs will, on the other hand, be interrupted if a higher-priority fault or a discrete input interrupt occurs. An STI that is delayed or interrupted by a higher-priority level interrupt will start (or resume) executing after the higher-priority interrupt ISR completes. (If a fault routine doesn't clear whatever fault caused it to run, the PLC will exit run mode and the STI will be canceled. STIs are held only during fault routines in better SLC 500s.)

SLC 500 PLCs have a more complete set of STI control and status bits than the PLC-5 has. Some bits can be changed by the user-program (e.g., in a fault routine to cancel any delayed STI interrupts). Other bits are only made available so that the user-program can monitor them (e.g., so that an STI can check to see if it was delayed and can be programmed to operate differently if it was). The bits, which aren't all offered by lower-level SLC 500 models, include:

1. An **STI enable bit** (S:2/1), which prevents STI interrupts if cleared but does not prevent the timer from running.
2. The **STI executing bit** (S:2/2), which is set automatically while an STI's ISR is running.
3. An **STI pending bit** (S:2/0), which is set automatically while an STI is delayed because it isn't high enough priority to interrupt an ISR that is already running, or because the STI enable bit isn't set.
4. The **STI lost bit** (S:36/9) and the minor fault **overrun bit** (S:5/10), which are both set automatically if STIs are enabled and the timer times out while there is still an STI pending. Only the overrun bit is set if interrupts aren't enabled at that time.
5. The **STI resolution bit** (S:2/10), which the programmer sets to cause the timer to count 1-ms increments rather than the default 10-ms increments.
6. The **interrupt latency bit** (S:33/8), to make interrupt response time shorter. If this bit is set when the STI timer expires, the PLC will interrupt when it finishes manipulating the data word that it is manipulating. If the bit is not set, the PLC will finish the program rung it is executing, or finish reading or writing an entire slot's input/output data, or finish the entire serial data packet it is sending/receiving before interrupting.

In the following section we describe STI characteristics common to the SLC 500 and the PLC-5.

PLC-5 SLC 500

TIMED INTERRUPTS AND THE ALLEN-BRADLEY PLC-5 AND SLC 500

Since the Allen-Bradley PLCs don't perform an input scan before running an STI program file and don't perform an output scan when the STI program file finishes, the use of immediate I/O instructions in the STI program file is strongly recommended. This will ensure that the intervals between the sampling of inputs and the control of outputs is as consistent as the timing of the STI program file execution. Other software interrupt instructions [TEMPORARY END (TND), REFRESH (REF), and SERVICE COMMUNICATIONS (SVC)] cannot be executed inside an STI program file.

STIs can be disabled or reenabled in any of several ways:

1. The interval time can be changed to 0 to disable STI interrupts, or set to a positive value (over 10 ms for the PLC-5 or over 2 ms for the SLC 500) to reenable them. When disabled, then reenabled in this way, the first interval time may not be correct.

2. Setting the STI file number to 0 disables STI interrupts, and setting it to the number of an existing program file reenables STI interrupts.

3. Using interrupt disable instructions and interrupt enable instructions. The PLC-5's **USER INTERRUPT DISABLE (UID)** instruction disables all interrupts, except the Fault routine, and the **USER INTERRUPT ENABLE (UIE)** instruction reenables all interrupts. The SLC 500's **SELECTABLE TIMED DISABLE (STD)** instruction disables only timed interrupts by latching the STI enable bit, and the **SELECTABLE TIMED ENABLE (STE)** instruction reenables them by latching the same bit. The SLC 500's STI timer continues to run, so an STI interrupt may execute as soon as an STE instruction is executed.

To force all program scans to execute at consistent time intervals, a PLC-5 or an SLC 500 can be configured so that all main control programs (MCPs) are disabled and an STI program file is configured to run at frequent intervals. (You can't configure these PLCs not to have MCP files. If you try, the PLC will automatically insert an MCP into the configuration when the PLC goes into run mode.) Unlike main control programs, only one selectable timed interrupt program can be executed per scan, but that one STI program can be written to call other program files. Input and output scans will continue to be executed every scan cycle even if the scan cycle doesn't include any MCP files.

Configuration data words entered at the configuration screen for STIs are stored in the status file.[20] They can also be entered and/or changed by a user-written program with MOVE instructions to change the appropriate status words. The SLC 500 also offers a **SELECTABLE TIMED START (STS)** instruction specifically for changing STI configuration. The user might write a PLC-5 or SLC 500 program that will set up STI configuration data every time the PLC is switched into run mode to avoid situations, for example, where a user might accidentally disable STI interrupts. A program can also be written to change its own STI configuration as it runs (although the first STI interval immediately following a changed interval time will not be correct). A program that changes its own STI interrupt interval and STI response program depending on external conditions is shown in Figure 11.18.

The PLC-5 and SLC 500 PLCs can also be programmed to change their watchdog timer interval as they execute.[21] If the watchdog timer time is intentionally set shorter than a PLC scan cy-

---

[20] In both the PLC-5 and SLC 500, the program file number is stored in S:31 and the setpoint in S:30.

[21] In the SLC 500, the watchdog timer interval is set by writing a value between 2 and 250 to the high byte of status word S:3 (S:3H), indicating the number of 10-ms intervals (from 20 to 2500 ms). The low byte of the same status word (S:3L) contains the elapsed time since the beginning of the current scan of the user-program. In the PLC-5, the timer interval, in milliseconds, is set by writing a value between 10 and 32,767 to S:28. S:8 contains the elapsed time in units of 1 or 10 ms, depending on the PLC-5 model, since this scan of the user-program started.

The first rung configures the PLC to execute program file 15 at 200-ms intervals if input I:022/3 is on.

I:022
03

MOV
MOVE
Source 15
Destination S:31

MOV
MOVE
Source 200
Destination S:30

But if I:022/3 isn't on, this next rung configures the PLC to execute program file 16 at 1000-ms intervals.

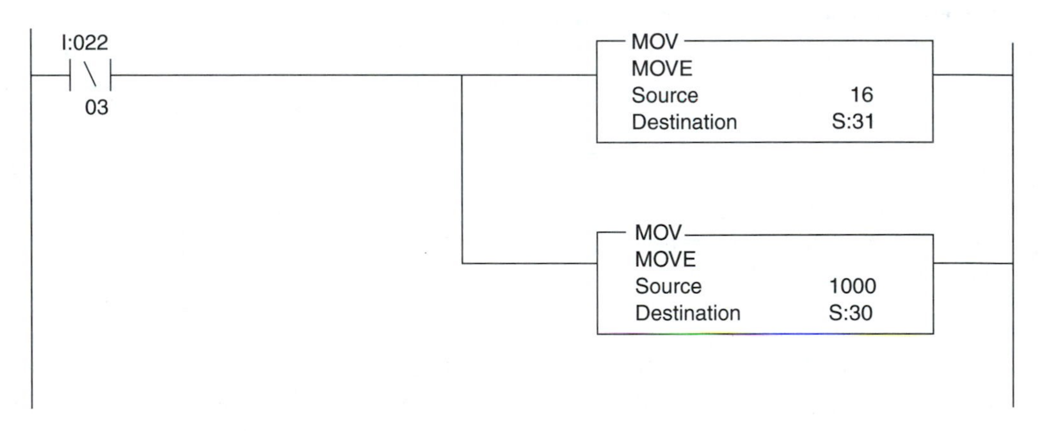

FIGURE 11.18
PLC-5 program that changes its own STI configuration.

cle (consisting of program scan time, I/O scan time, and communications time), the fault routine becomes, in effect, a timed interrupt routine that executes at a fixed time after the program scan starts. (Fault routines and fault codes are covered in more detail in a later section in this chapter.)

S5

TIMED INTERRUPTS IN SIEMENS STEP 5

Not all S5 PLCs offer timed interrupts. In those that do, the response to a timed interrupt differs from the response to an I/O interrupt in several ways:

1. The timed interrupt is the lowest-priority interrupt, so it will not interrupt an interrupt service routine (ISR) running in response to a process interrupt (STEP 5's name for an I/O interrupt), nor will it interrupt a timed interrupt ISR that is already running. Timed interrupt ISRs will be delayed until other ISRs end. A process interrupt, on the other hand, will interrupt an ISR that is running in response to a timed interrupt.

2. A timed interrupt is of lower priority than a standard I/O scan, so an I/O scan will finish execution before a timed interrupt is responded to.

3. When a timed interrupt is responded to in an S5-103U, it does a full input scan, copying the contents of the input modules into the interrupt process image input table (interrupt PII). Direct access addressing will cause these PLCs to read and write the data in the interrupt PII. (Larger S5 PLCs do not do these extra input and output scans. Direct access addressing causes larger PLCs to perform immediate input or output of data to/from I/O modules.)

4. A timed interrupt results in **organizational block 13 (OB013)** being run (like OB002, which is called in the event of an I/O interrupt). OB013 can contain jumps to other organization, program, or function blocks.

> Some larger PLCs offer additional timed interrupts that call OB010, OB011, and OB012.

5. In the S5-103U, direct access addressing causes output data to be written to the interrupt PIQ table (and to the normal PIQ table simultaneously). A short output scan is executed in these PLCs after OB013 finishes, copying all changed interrupt PIQ data to be copied to output modules.

6. The programmer must define the timed interrupt interval by writing a value to **reserved system word 97 (RS097).** This can be done directly by the programmer, using the programming software's data screen, or can be done in the user-program. Values of between 0 and 255 can be entered. The timed interrupt interval will be the number entered times 10 ms. If no value is entered, RS097 will contain the default value 10, so timed calls of OB013 will occur every 100 ms.

> Larger PLCs that offer additional timed interrupts require values in RS098 to set the interval time for OB012, RS099 for OB011, and RS100 for OB010.

Timed interrupts can be disabled by entering the value 0 into RS097, by not entering a program for OB013, or by executing the **IA** (**INTERRUPT DISABLE**) instruction in the program. If a timed interrupt would have occurred but has been disabled with the IA instruction, it will be executed immediately after execution of the **RA** (**REENABLE INTERRUPTS**) instruction.

One good place in the user-program to set up a timed interrupt is in an initialization program that runs once whenever the PLC goes into run mode (OB021 is reserved for use as an initialization program, as discussed later in this chapter.) The program in Figure 11.19 will set up the PLC to run the program in OB013 at precise 5-s intervals. This example does not show what OB013 does.

To force an S5 PLC to execute standard program scan cycles at consistent intervals, the programmer can set the time interval by writing to RS097, put the main program into OB013, and not enter any program in either of the default main program blocks, OB001 or PB001.

The **scan time monitor** (STEP 5's name for the watchdog timer) cannot be used to trig-

OB021
This organizational block runs once each time the PLC is switched into run mode.

```
JU FB001          ; Writing to system words can only be
   Time_Int       ; done in a function block
BE
```

FB001
Set up the timed interrupt to occur every 5 s by specifying how many 10-ms intervals between interrupts.

Name: Time_Int

```
L KF+500          ; For 500 X .01 sec = 5.00 seconds
T RS097           ; Put into the system word location
BE
```

OB013
Any program entered as OB013 will now run at precise 5-s intervals.

FIGURE 11.19
STL program to configure an S5 PLC to execute an ISR every 5 s.

ger a user-written fault routine program in an S5 PLC. When the scan time monitor times out, the PLC goes into stop mode. The programmer has some access to the scan time monitor in that the scan time setting can be changed by writing a value of between 1 and 255 to RS096, corresponding to scan time settings of between 0.01 and 2.55 s. During a program scan, if OB031 is called and if it contains at least a BE (BLOCK END) instruction, the accumulated time for the current will be reset back to zero, so that exceeding the scan time monitor time can be avoided.

S7

TIMED INTERRUPTS IN SIEMENS STEP 7

Siemens S7 PLCs above the S7-312 IFM level offer a greatly expanded range of timed interrupt options. Besides the standard timed interrupt, which is called a **cyclic interrupt** in STEP 7, other timed interrupts include the **time-of-day interrupt,** which can be set to occur once at a preset future day and time, or at the same time every day, or even at the same time during every minute, hour, week, month, or year; and the **time-delay interrupt,** which can be set up to occur a preset time after the program instruction that triggers it. There is also a **cycle time fault** interrupt, which is triggered for a variety of reasons, including when the PLC scan cycle exceeds the cycle time (watchdog timer) setting of between 1 ms and 6 s. S7 PLCs at the S7-312 IFM level and lower do not offer timed interrupts, but they can be configured to execute their main organization block program at a specific cyclic rate.

Timed interrupts are the lowest-priority interrupts in the STEP 7 system, so execution of ISRs in response to a timed interrupt will be delayed if a higher (or equal)-priority level ISR is already running. Some STEP 7 timed interrupts have higher-priority classes than other timed interrupts. Individual priority class levels are discussed later. An S7 PLC stores data on the I stack as it responds to a timed (or any other type of) interrupt, so that the interrupted program can be resumed when the ISR ends.

The programmer must enter interrupt service routines (ISRs) in specific organizational blocks if timed interrupts are to be used. If any timed interrupt organization block (other than for a cyclic interrupt) is called and its organization block hasn't been programmed (with at least a block termination instruction such as BEU or a CALL "STP"), a fault routine (OB 85) will run if it exists, or the PLC will immediately go into stop mode if OB 85 doesn't exist. In the following sections we tell you how to configure and enable timed interrupts and which organizational blocks you need to program for each type of timed interrupt.

The "**DIS_INT**" and "**EN_INT**" system function calls can be used anywhere in a user-program to disable and reenable some or all interrupt types, or the "**DIS_AINT**" and "**EN_AINT**" system functions can be called to delay or stop delaying higher-level interrupts from interrupting lower-level ISRs during a single program scan. (These system functions were discussed in the section on I/O interrupts in STEP 7.)

An S7 PLC can be configured to execute cyclic interrupts at fixed intervals of between 1 ms and 60 s, starting from when the initialization program (OB 100) finishes executing after the PLC is switched into run mode. In some S7 PLCs, the fixed intervals can be configured with a phase displacement of up to 60 s, so that the start of the first timing interval will be delayed by that amount of time after OB 100 finishes. Cyclic interrupts have a default priority class of 12, the highest-priority class of STEP 7 timed interrupts, but in some S7 PLCs the priority class can be changed to 0 (to disable cyclic interrupts) or to a priority class of from 2 up to 24.

Configuring cyclic interrupts will cause an interval timer to work while the PLC is in run mode, issuing interrupt request signals, so the programmer should enter an ISR program before a PLC program activating cyclic interrupts is started (although the PLC won't fault if a cyclic interrupt ISR is missing). The ISR must be entered as OB 35.[22] When the operating system calls OB 35, it passes several parameters to OB 35, including those that indicate that the interrupt is active and OB 35 has been called, the current configuration data for the cyclic interrupt (priority class, frequency setting, phase offset), and the date and time when OB 35 started. If the cyclic interrupt timer expires again while OB 35 is still running from a previous cyclic interrupt, the PLC will execute the fault routine (OB 80) if it exists, or will immediately fault and go into stop mode if OB 80 doesn't exist.

Up to eight independent **time-of-day interrupts** can be configured using the STEP 7 programming software or through calling **SFC 28** ("**SET_TINT**") in the user-program. Configuration options enable selection of the organization block number that is being configured as a time-of-day interrupt (OB 10 through OB 17), the date and time at which the first interrupt is to occur, and the interval periods at which the time-of-day interrupts are to run after they start. The interval options include interrupting once only at the specified time and date, or repeated interrupts every minute, hour, day, week, month, or year after starting.

Once configured, a time-of-day interrupt can be activated using the S7 programming software or through calling **SFC 30** ("**ACT_TINT**") in the user-program, specifying which of the configured time-of-day interrupts to activate (OB 10 through OB 17). After being activated, the operating system will call time-of-day ISR(s) at the specified time(s). Time-of-day interrupts have a default priority class of 2, which is just barely higher than the standard scan cycle's priority class of 1, so it will not interrupt any other ISR but will be held until no higher-priority interrupt requests remain. If a time-of-day interrupt organization block is called but hasn't been

[22] Siemens has reserved OB 30 through OB 37 for future cyclic interrupt capabilities.

programmed, the PLC will execute the fault routine OB 85, or the PLC will immediately fault and go into stop mode if OB 85 doesn't exist. If a time-of-day interrupt is still running when it is called again, or if the system clock value is advanced past a time-of-day interrupt's start time, the fault routine OB 80 is called, or the PLC will immediately go into stop mode if OB 80 doesn't exist. If the system clock is moved forward enough to bypass several time-of-day interrupts and OB 80 runs as a result, the STEP 7 operating system will cause the time-of-day interrupt(s) that were skipped to execute, as many times as they missed executing, after OB 80 finishes. The programmer may wish to cancel those outstanding interrupt calls by calling SET_TINT again.

Figure 11.20 shows the use of SET_TINT and ACT_TINT to configure and initiate time-of-day interrupt calls to OB 11 every hour, at 10 minutes after the hour, starting at 8:10 AM Friday, December 1, 1997. The variable #start has been declared as a variable of type

```
Network 1:

    Call "D_TOD_DT"                    // Function FC 3 defines the variable
                                       // "start"
      IN1      :=D#1997-1-12           // The birth date of HAL in the movie
      IN2      :=TOD#8:10:0.0          // "2001"
      RET_VAL:=#Start

Network 2:

    Call SFC "SET_TINT"                // System function SFC 28, used to
                                       // configure a time-of-day interrupt.
                                       // Parameters passed to SET-TINT include
                                       // specifications that:
      OB_NR   :=11                     // OB 11 is being configured
      STD     :=#Start                 // The first interval's starting time
      PERIOD  :=W#16#0401              // code 0401 indicates hourly intervals
      RET_VAL :=MW 100                 // return value goes to MW 100

Network 3:

    Call "ACT_TINT"                    // System funtion SFC 30 activates the
                                       // interrupt configured for:
      OB_NR   :=11                     // OB 11
      RET_VAL :=MW102                  // ACT_TINT's return value goes to MW 102
```

FIGURE 11.20
Configuring and starting time-of-day interrupts in a STEP 7 program written in STL.

DATE_AND_TIME in this example. The example program assigns a value to #start by calling system function **FC 3** (**D_TOD_DT**), which converts a date value constant (D#) and a time value constant (TOD#) to DATE_AND_TIME format. The RET_VAL parameter returned by system functions SET_TINT and ACT_TINT will contain bit patterns indicating reasons for their failure to configure or activate the time-of-day interrupt if they fail. Your program should include instructions to watch for these error codes.

SFC 31 (“**QRY_TINT**”) can be called to determine the status of a time-of-day interrupt, with a parameter specifying which interrupt's status is being queried (OB 10 through OB 17). The status word that QRY_TINT returns contains bit patterns that indicate whether the time-of-day interrupt is enabled, if the time setting is valid, if the interrupt is activated, if the interrupt has occurred, if the required organization block exists, and if a test function is active[23] (time-of-day interrupts are disabled while a program is in test mode). Refer to your STEP 7 programming manuals for the status word codes and their meanings.

The system function **SFC 29** (“**CAN_TINT**”) can be called in the user-program to cancel all future calls to any of the eight time-of-day interrupts specified with the call. Time-of-day interrupts can also be disabled by using the S7 programming software to change the priority class of time-of-day interrupts to 0. If SET_TINT is called to initiate a time-of-day interrupt a second time after that time-of-day ISR has already started or is waiting for a higher-priority class ISR to end so it can run, the waiting interrupt (or the rest of a time-of-day interrupt ISR that is running) will be canceled. Any ISR can therefore include SET_TINT instructions to cancel outstanding time-of-day interrupts without preventing future occurrences of the time-of-day interrupts from running.

A **time-delay interrupt** is initiated in a STEP 7 program if the conditional statement controlling a call to **SFC 32** (“**SRT_DINT**”) allows SRT_DINT to be called. After SRT_DINT has been called to start the time-delay interrupt, the PLC continues to run its user-program just as if SRT_DINT hadn't been called, but the PLC also starts running a timer. If SRT_DINT is called to restart the same time-delay interrupt again before the delay has expired, the delay time is reset and the delay restarts. When the timer reaches its preset value, even if that happens several scan cycles later, the PLC interrupts the scan cycle to run the time-delay ISR. If the PLC isn't in run mode when the timer expires, the time-delay ISR will execute as soon as the PLC is put into run mode. The delay time is accurate to within 1 ms.

The time-delay interrupt has the relatively low priority class 3, which means that it can only interrupt a time-of-day ISR but can be interrupted by any other interrupt type. If a higher-priority ISR is already running when the time-delay interrupt timer reaches its preset value, the time-delay ISR waits for the higher-priority interrupt(s) to finish before it starts. If another time-delay ISR is running when a time-delay timer reaches its preset, OB 80 is called or, if OB 80 doesn't exist, the PLC goes into stop mode. If the timer reaches its preset and the time-delay ISR doesn't exist, the PLC calls OB 85, or goes into stop mode if OB 85 doesn't exist.

Up to four different time-delay interrupt ISRs (OB 20 to OB 23) can exist. When SRT_DINT is called, the call must identify which organization block it is calling, the delay time in milliseconds (1 to 60,000 ms), and must provide an identifier that will be passed to the organization block when it is called after the delay time. Siemens calls this identifier parameter word

[23] A test function is active if a programmer is using a programming unit to debug a program online. While debugging, the PLC takes longer to execute each scan cycle because it must send status information to the programming unit.

a *sign*. The delay timer is reset each time SRT_DINT is called. If SRT_DINT is called a second time before the delay timer expires after the first time it is called, the time-delay interrupt occurrence will be timed from the last time SRT_DINT was called.

Time-delay interrupts can be canceled with the system function **SFC 33** (“**CAN_DINT**”) and interrupt status can be determined using the system function **SFC 34** (“**QRY_DINT**”). These calls must identify which of the ISRs (OB 20 to OB 23) they are canceling or querying.

```
Network 1:

      A   I  4.2                      // If input I4.2 goes on,
      FP M  1.0
      JNB m001
      CALL "SRT_DINT"                 // call SFC_32 start to an interrupt timer
       OB_NR    : =20                 // for time-delay OB 20
       SDT      : =T#100MS            // with a delay of 100 ms
       SIGN     : =W#16#1111          // and set the hex identifier to "1111"
       RET_VAL: =MW 100               // and store SRT_TINTs reply in MW 100
  m001:NOP 0

Network 2 :

      A   I  4.3                      // Like network 1, but triggered by I4.3
      FP M  1.1
      JNB m002
      CALL "SRT_DINT"
       OB_NR    : =20
       SDT      : =T#100MS            // Delay time is longer at 200 ms
       SIGN     : =W#16#2222          // and the identifier is hex "2222"
       RET_VAL: =MW 100
  m002:NOP 0

Network 3:

      A   I  4.4                      // But if input I4.4 is on,
      JNB m003
      CALL "CAN_DINT"                 // call SFC_33, canceling interrupts
       OB_NR    : =20                 // for time-delay OB 20
       RET_VAL: =MW 102               // and store CAN_TINTs reply
  m003:NOP 0
```

FIGURE 11.21
A STEP 7 program that may initiate a time-delay response when either of two conditions is detected.

The status word that is returned by QRY_DINT contains the same type of information that the QRY_TINT function returns (explained above in the description of the time-of-day interrupt).

Figure 11.21 shows how a program might have two different calls of the same time-delay interrupt ISR (OB 20). Since SRT_DINT resets the delay time each time it is called, only the last call received will be responded to if the conditions for both calls become true before OB 20 is started. Each call is jumped past except for the first scan after an input signal goes on. Each call initiates the same time-delay interrupt (OB 20), but uses different delay times and provides different 16-bit sign parameters as identifiers that will be passed to OB 20 when OB 20 starts. The program in OB 20 (not shown) can include logic to respond differently depending on which of the two sign parameters it receives. While I 4.4 remains on, CAN_TINT is called every scan cycle, canceling any delay-time interrupts before they occur.

TIMED INTERRUPTS AND THE OMRON CQM1

Omron's designers seem to like using timed interrupts, because they are used in several ways:

1. Interval timer interrupts can be used to run ISRs repeatedly at timed intervals, or once after a delay time.

2. The CQM1 can be configured to execute its scan cycle within a standardized minimum cycle time, or to stop if scan time exceeds a cycle monitor time.

3. Timed interrupts are used in pulse output.

4. High-speed timers can be configured to interrupt normal processing to update timer present values at precise intervals.

5. The PID instruction sets up timed interrupts to read input values. The PID instruction is described in Chapter 12.

The CQM1 can be configured to execute as many as three interval timer interrupts. As with other CQM1 interrupts, the programmer must enter configuration data into memory before running a program, and the program must include instructions enabling timed interrupts before they can interrupt the main program. Input interrupts and high-speed counter interrupts HSC1 and HSC2 have the highest priority of the timed interrupts. Interval timer interrupts are next, and interval timer 0 has higher priority than interval timer 1, which has higher priority than interval timer 2. Interval timer 0 is used automatically by the CQM1 to time pulse outputs through output modules (discussed later in this section) and can't be used for anything else while pulses are being timed. Pulse output therefore has the same priority as interval timer 0. Similarly, high-speed counter 0 (HSCO) uses interrupt timer 2, and the CQM1 must be configured to enable either HSCO or interrupt timer 2. If HSCO is selected, it replaces interrupt timer 2 as the lowest-priority interrupt in the CQM1 system. DM 6642 must contain **00**xx to enable interval timer 2 or **01**xx to enable high-speed counter HSCO.

DM 6636 through DM 6638 can contain configuration values that will cause the CQM1 to perform a partial input scan[24] whenever ISRs are called by interval timer 0, 1, or 2, respectively. The program must execute INTERVAL TIMER instructions, **STIM(69)**, to start or stop interval timer interrupts, to set timer setpoints, to indicate which subroutine to run as the ISR, and

[24] Up to 12 input modules can be scanned, starting at any module from 000 to 011.

to select the interrupt mode (one-time-only or repeating). STIM(69) is usually used in differentiated mode, @STIM(69), so that it doesn't continue to execute every scan while its input logic remains true. Three control parameters, C1, C2, and C3, are required with the STIM(69) instruction.

1. If C1 is 000, 001, or 002, interval timer interrupt 0, 1, or 2 is started in one-shot mode, which means that the ISR will execute only once, after a delay from the time the STIM(69) instruction executes. A C1 value of 003, 004, or 005 will start interval timer interrupt 0, 1, or 2 in scheduled interrupt mode, which means that the ISR will execute repeatedly at intervals starting from when the STIM(69) instruction executes. C1 = 006, 007, or 008 copies the present value of interval timer 0, 1, or 2 to data memory. C1 = 009, 010, or 011 stops interval timer 0, 1, or 2 and cancels any outstanding interrupts.

2. C2 is for the timer setpoint if C1 causes an interval timer to start. C2 may contain a constant (#0000 to #9999) indicating the setpoint in milliseconds or can contain an address where the first of two data words containing the setpoint is stored.[25] If C1 causes STIM(69) to read an interval timer's present value, C2 indicates the address to which the first of the two-data-word present value will be written.

3. C3 must contain the number of the subroutine to execute if C1 causes this instruction to start a scheduled timed interrupt. If C1 causes reading of an interval timer's present value, C3 contains the address where the CQM1 will store a number indicating the time, converted into tenths of milliseconds, since the present value was last updated (the timer time units are all larger than tenths of milliseconds).

Interval timer interrupts can be masked by executing the **INT(89)** instruction with a CC parameter of 100, or unmasked by executing INT(89) with CC = 200. Masking does not stop the timer from running, and ISRs requested by interval timer interrupts while they are masked will execute after they are unmasked.

Figure 11.22 shows STIM(69) being used to start two timed interval interrupts. The ISR subroutines are not shown.

1. In rung 1, the first scan bit (SR 25215) causes STIM(69) to run once when the PLC is put into run mode. Interval timer 1 is set up to cause scheduled interval timer interrupts, calling subroutine 015 every 25 s, because DM 0500 contains 0025, the number of time units in the setpoint, and the next memory, DM 0501, contains 1000, indicating time units of 1000 ms (1 s).

2. Rung 2 uses the differentiated form of STIM(69) so it will execute once each time input IR 00305 goes on. This rung starts interval timer 2 in one-shot mode so that subroutine 022 will be called once, 6000 ms after the STIM(69) instruction executes.

The CQM1 can be configured with a standard minimum cycle time, so that it starts each scan cycle a fixed time after starting the previous scan cycle. If the CQM1 has been configured with a nonzero time value (0001 to 9999 ms) in DM 6619, then after each scan of the user-program, the CQM1 will wait until the cycle time monitor reaches that time value before writing

---

[25] The address specified in C2 must contain a BCD number (0000 to 9999) indicating the number of time units in the setpoint, and the next address must contain a BCD number (0005 to 0320) specifying the size of each time unit in *tenths* of milliseconds. The possible interval time range is 0.5 ms to just under 320 s.

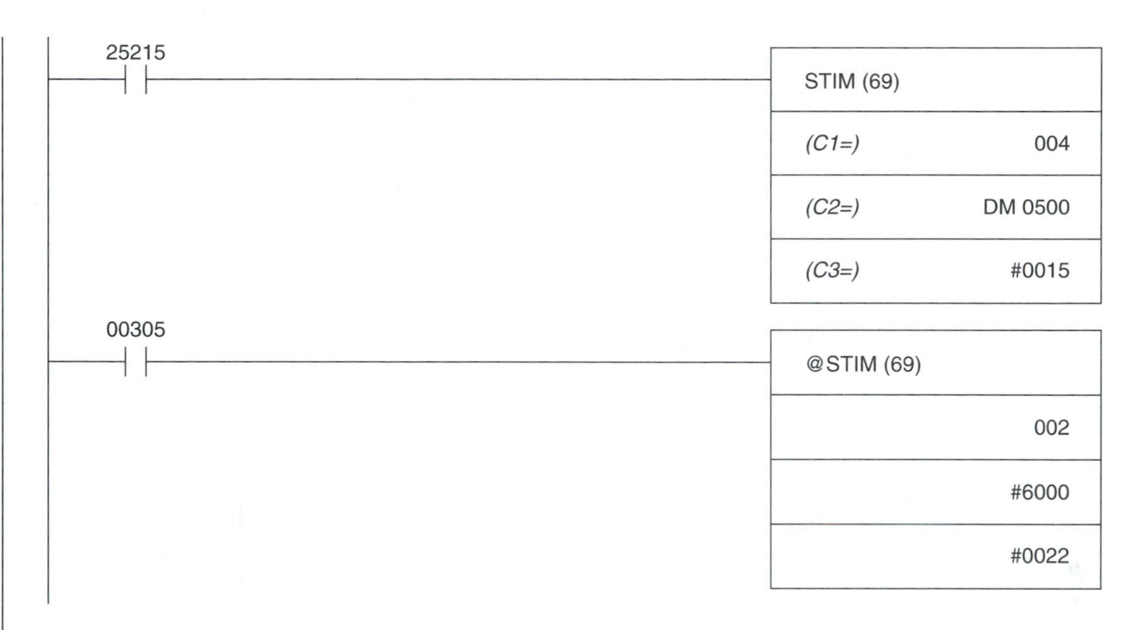

FIGURE 11.22
CQM1's interval timer interrupts.

to output modules and continuing with the next scan cycle. This value is a minimum scan time value. Scan cycles that take longer to execute will always finish before the next scan cycle starts, so the programmer should be confident that the time in DM 6619 is longer than the longest scan cycle. (Remember, scan cycles that are interrupted will take longer to execute than those that aren't.)

If any scan cycle exceeds the cycle monitor time setting in DM 6618, the CQM1 will immediately stop scanning, turn all outputs off, put an error code into the error log, and turn on the ERR/ALM and POWER indicator on the CPU faceplate. Cycle monitor time interrupts have no value as intentionally programmed timed interrupts, so will be left to be described in "fault routines and the OMRON CQM1."

The CQM1 offers programmable high-speed pulse output, which causes the CQM1 to interrupt its scan cycle to perform an immediate output turning a single output contact on or off. All CQM1 PLCs allow the programmer to select one bit in a single output module to be controlled, but the CQM1-CPU34-E also allows additional pulse outputs via its 15-pin connector ports in the CPU module.

Any CQM1 PLC can use pulse output to output 50 percent duty cycle[26] pulses via any digital output module contact. The programmer must specify the output module by entering a value (00**00** to 00**11**) into DM 6615, indicating which output module (1**00** to 1**11**) contains the controlled contact, and must configure the CQM1 to use direct output by entering the value xx**01** into DM 6639. Direct output means that all output values are written to the output modules as

[26] Fifty percent duty cycle means that the output will be on for half of each on/off cycle.

the program changes them. (Pulse output uses interval timer interrupt 0 to time changing of an output bit value.)

The user-program must execute the SPEED OUTPUT instruction, **SPED(64),** to start pulse output. SPED(64) must include a P parameter value from 010 to 150 to select bit 01 to 15, respectively, an M parameter containing 000 to select independent mode or 001 for continuous mode, and an F parameter containing a value of between 0002 to 0100 for frequencies from 20 to 1000 Hz. If **continuous** mode is selected, pulse output continues until another SPED(64) instruction changes the frequency to 0 Hz or until an INI(16) instruction (covered in "I/O Interrupts and the OMRON CQM1") is executed. If **independent** mode is selected, a SET PULSE instruction, **PULS(65),** must precede the SPED(64) instruction to indicate how many pulses to output before stopping.[27] Both instructions should be used in their differentiated forms, @SPED(6) and @PULS(65), so as not to slow the scan cycle down when they don't need to be executed, and to avoid restarting the pulse output timer and counter each scan cycle.

Figure 11.23 shows a CQM1 program that outputs pulses via contact 2 (P = 020) of output module 103 (DM 6615 previously set to 0003 by the programmer) continuously (M = 001) at 60 Hz (F = #0006) while input IR 00405 is off. (Before running the program, 0001 was entered into DM 6639 to indicate direct output.) When input IR 00405 goes on, INI(61) executes to stop the 60-Hz output, and then 75,000 pulses are output at 250 Hz via contact 11 of the same output module. (DM 0200 contains 5000, and DM 0201 contains 0007, before executing rung 2.)

In the CPU43-E version of the CQM1, pulses can be outputted via CPU ports 1 and 2 while pulses are being outputted via an output module. More options are available and are discussed here only in general terms. See your CQM1 programming manual for specifics. Some differences include:

1. Pulse output to ports 1 or 2 requires high-speed counter 1 or 2 (respectively), so HSC1 and HSC2 can't be used as counters while they are in use for pulse output.

2. SPED(64) and PULS(65) parameters indicate via which CQM1-CPU43-E port to output pulses. Three pulse outputs can be output independently and simultaneously. SPED(64) defaults to continuous pulse output unless PULS(65) has been used to indicate a number of pulses to output.

3. The value in DM 6643 before the PLC goes into run mode indicates whether ports 1 and 2 will output standard 50 percent duty cycle pulses or variable duty (pulse-width-modulated) pulses.

4. If standard pulse output is selected, the port selected outputs identical pulses at two contacts, with the pulses at one contact leading the other by 90 degrees (suitable for driving a stepper motor). PULS(65) is used to specify which output signal leads the other (to control stepper motor rotation direction) and can also specify that the pulse frequency should decelerate to (almost) 0 Hz before stopping. **PLS2(−)** can be used instead of PULS(65) and SPED(64), to ramp pulse frequency at a controlled rate up from 0 Hz to a specified maximum when they

---

[27] PULS(65) has a P and a C parameter, both of which must be 000 for pulses via an output module; and an N parameter, which must specify the first of the two data memory addresses containing the eight-digit BCD number of pulses. The first address contains the four low BCD digits, and the next address contains the four high BCD digits. A maximum of 16,777,215 pulses are allowed.

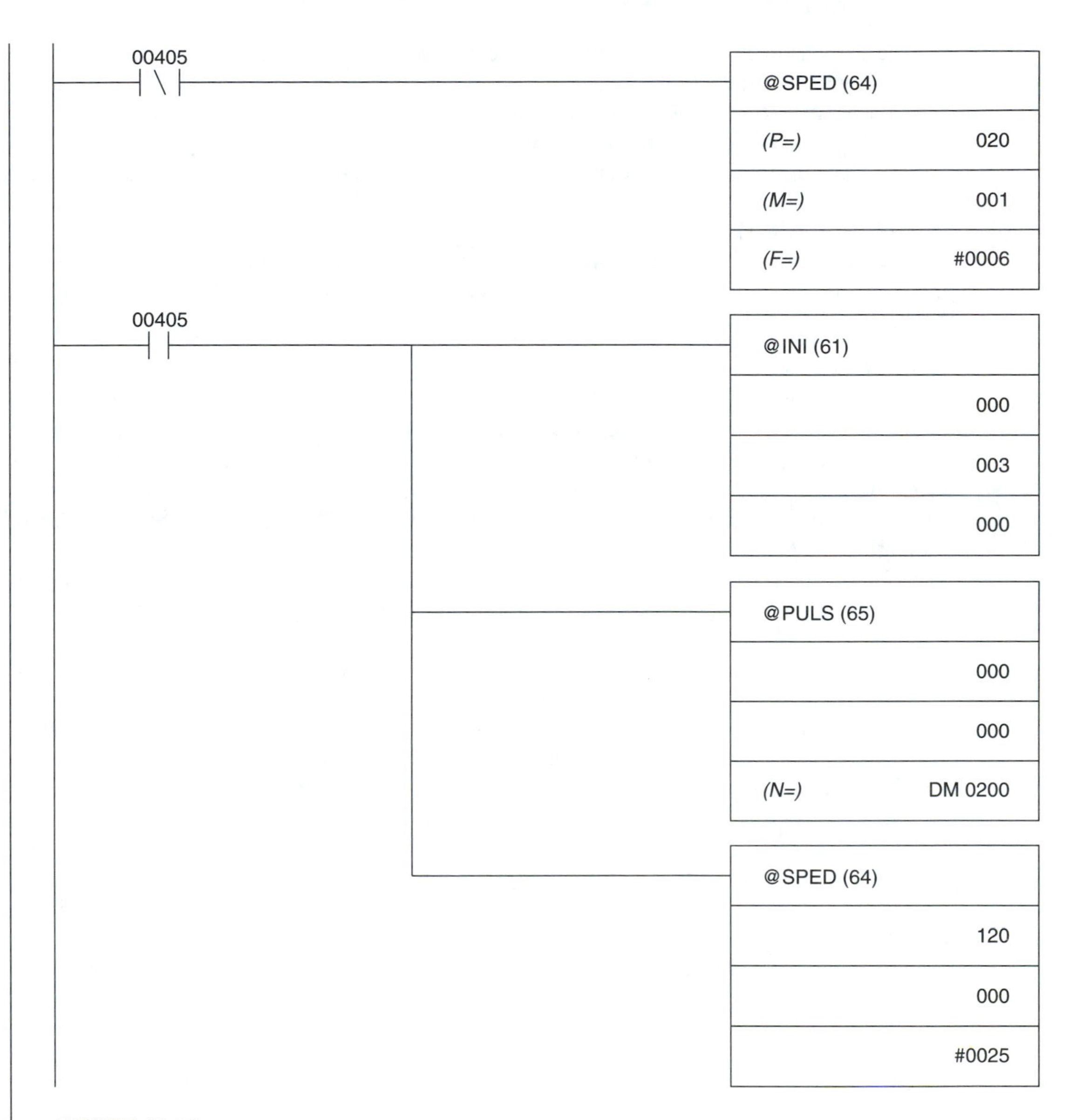

FIGURE 11.23
CQM1 program using pulse output.

start, and back (almost) to 0 Hz when they stop. **ACC(−)** can be used to ramp pulse frequency up or down after it has been started by another command. These features are ideally suited to control stepper motor speed and acceleration rates. PLS2(−) and ACC(−) require a high-speed counter and can be used only if DM 6611 is set to 0001, reserving HSC0 for use in pulse output.

5. If variable duty pulse output is selected, pulses are only outputted via one contact per port. They are started with a **PWM(−)** instruction, with parameters indicating the port, selecting one of three frequencies, and indicating a duty cycle between 1 and 99 percent. Variable duty pulse output is used to output analog dc.

6. The count of the number of pulses already outputted is updated each scan cycle in SR 236 and 237 for port 1 and in SR 238 and 239 for port 2. Several status bits are updated each scan cycle in AR 04 and 05 for port 1 and in AR 04 and 06 for port 2.

7. INI(61) can be used to stop pulse output at port 1 or 2. PRV(62) can be used to read more immediately updated status bits for pulse output than those in AR 04 to 06.

The **high-speed timers** (TC 000 to TC 015) interrupt the scan cycle when updating of their present value (PV) is required instead of waiting until the end of each scan cycle. Since OMRON doesn't describe high-speed timers as interrupts, they also don't tell us what priority-level interrupt they are, but the manuals warn that high-speed timers can degrade the operation of other interrupts, so they must have fairly high priority. When DM 6629 contains the value **00**xx, all 16 high-speed timers are enabled. The user can specify how many timers use interrupts, by entering the value **01XX** into DM 6629, where **XX** is a number between 00 and 15, indicating the highest timer number of the high-speed timers for which interrupt PV updating is enabled. The programmer may wish to enter a value here to limit the number of high-speed timers that are running so they don't degrade other interrupts (especially if pulse output at frequencies of over 500 Hz are required).

High-speed timers differ from standard timers in that the **HIGH-SPEED TIMER** instruction, **TIMH(15),** is used rather than the standard TIMER instruction, TIM. TIMH(15)'s setpoint value (SV) is entered in 0.01-s units (from 00.00 to 99.99), not in 0.1-s units. Using a high-speed timer with interrupt updating of its PV is useful only if scan cycle times exceed 0.01 s, but using the TIMH(15) instruction for a timer without interrupt updating of its PV will result in the timer having an inaccurate PV if scan times do exceed 0.01 s. Like other timers, the high-speed timer's present value (PV) decrements from its SV value to zero, at which time its completion flag goes on, but unlike normal timers, the completion flag can go on at any time during a program scan. Like other timers, high-speed timing will stop if the TIMH(15) instruction is between an active jump (JMP(04)) to a jump end (JME(05)) with jump 00. Like other timers, a high-speed timer will reset if it is within an active interlocked zone or if power to the PLC is interrupted.

Fault Routine Interrupts

A fault routine is executed automatically when a PLC detects that a serious problem exists. Fault routines are the highest priority level of interrupts, so anything the PLC is doing will be interrupted to allow the fault routine to execute. Early PLCs simply went into **fault mode** when they detected a serious problem. The PLC stopped running, turning a fault LED on the CPU's faceplate on, and turning all outputs off (or perhaps freezing them in their last state). Most PLCs saved (and most PLCs still save) a **fault code** in memory upon entering fault mode. In Chapter 15 we describe how to use fault codes to find why a PLC has faulted.

Better PLCs can be configured to execute a **user-written fault routine** when a detectable fault condition is recognized. Fault routines may be written to correct some types of problems, to send a message to an operator, or just to shut the PLC down in an orderly manner. After the fault is cleared by a fault routine, the PLC will resume executing the program that was interrupted when the fault was detected. All PLCs go into fault mode, as described above, if the error can't be identified or corrected.

Typical faults that can cause a fault routine to run include things such as the watchdog timer reaching its preset (indicating that a program scan cycle took too long to execute), an attempt to run a program file that doesn't exist, a failure detected while communicating with an I/O module or other CPU in the system, or simply a low charge in the backup battery. Some PLCs also allow the programmer to define and use their own user-defined faults. The user-written fault routine can then be used as a type of software interrupt, running only when the program contains an instruction causing the user-defined fault.

PLC-5

FAULT ROUTINES AND THE ALLEN-BRADLEY PLCS

SLC 500

The PLC-5 and SLC 500 have been preprogrammed to detect several types of **major** and **minor** faults. (Appendix H contains a list of these faults.) When a fault is detected:

1. The PLC interrupts whatever program it is running. Fault interrupts are the highest-priority level, so even other ISRs will be interrupted if a fault occurs.

2. One (or more) **fault bits** are set in the appropriate status words, and a **fault code** is put into another status word. The program file number and the rung number of the program that was running when the fault occurred are recorded into other status words. The status words can be displayed on the programming software's processor status screen.

> The PLC-5 sets fault bits in S:10 or S:17 if the fault is of the minor fault type, or in S:11 if the fault is a major type. It stores a fault code for major faults in S:12. The program file number and the rung number of the program that was running when the fault occurred is put into status words S:13 and S:14, respectively.
>
> The SLC 500 sets bits in S:5 for minor faults, or sets a single bit, S:1/13, for any major fault. It stores the major fault code in S:6. The program file number and the rung number of the program that was running when the fault occurred is put into status words S:20 and S:21, respectively.

3. If the fault was a **minor fault,** the interrupted program is resumed immediately, without running the fault routine. In Chapter 15 we describe the minor fault bits that will have been set, and how the programmer can use them.

4. If the fault was a **recoverable major fault** (described below), the PLC tries to run the fault routine.

- **(a)** If the programmer has entered a fault routine file number into address S:29, either from the configuration screen or through an instruction in the program, that program file will execute. Calls to other program files can be made from this fault routine.
- **(b)** If the major fault bit(s) are cleared by the fault routine before it finishes executing, the PLC will resume running the program file that was interrupted at the instruction following the instruction that was running at the time of the fault.

5. If:
   - The major fault is a **nonrecoverable fault** type, or
   - No fault routine file number is found, or
   - Another major fault is caused by the fault routine as it executes, or
   - A fault routine finishes executing without clearing all the major fault bits, then the PLC will go directly into fault mode.

Most of the recoverable major faults that can lead to execution of a fault routine are caused by **programming errors.** Some recoverable faults that can be caused by programming errors include the fact that the watchdog timer times out before a program scan finishes; the program tries to execute a jump to a label that doesn't exist or to a ladder logic program file that doesn't exist (including missing ISR files); an invalid value is used in indirect addressing, as a counter preset, or in certain other applications; or too many levels of nested subroutines are used. A recoverable major fault may also be caused by nonprogramming errors such as corrupted memory, or sometimes by faults at peripheral devices.

The PLC-5 and SLC 500 allow the program to generate a **user-defined recoverable major fault** complete with a fault code.

> In the PLC-5, the fault routine can be forced to execute if the user-program includes a JUMP TO SUBROUTINE (JSR) instruction to jump to the fault routine file. The fault routine should, of course, be programmed to respond to the conditions that caused the JSR to execute, but the fault routine will be called as an interrupt, not as a subroutine! One difference is that the PLC cannot pass normal parameters to the fault routine. The JSR instruction must, however, be programmed with a single parameter: a number between 0 and 9, which the PLC's operating system will write to status word S:12 as an error code as it calls the fault routine. (The fault routine does not need an SBR instruction to receive this value into S:12.) Allen-Bradley has intentionally not assigned meanings to the PLC-5 fault codes 0 to 9, so that the programmer can use these numbers as his/her own fault-identification codes. When the JSR instruction initiates the fault routine interrupt, the PLC-5 also sets a major fault bit (S:11/7), indicating that the recoverable major fault was caused by the JSR instruction. As with any other major fault bits, the fault routine must clear this bit before finishing or the PLC will go into fault mode.
>
> In an SLC 500 program, to force the fault routine to run the user-program must set bit S:1/13, the recoverable user fault bit. If the programmer wants to use a self-generated fault code, the program must write that fault code to S:6 before setting bit S:1/13. Allen-Bradley recommends that users use (hexadecimal) codes FF00 or FF0F, which means that Allen-Bradley intends to avoid using those fault codes in future SLC 500 models. The fault routine should reset bit S:1/13 before ending, or the PLC will go into fault mode.

Figure 11.24 shows a PLC-5 program that includes a JSR instruction to cause the fault routine to run when a situation that isn't normally a detectable fault situation (a rack malfunction, in this example) occurs. The fault routine in this example examines the fault code to determine why the program caused it to execute. (It might have been caused by an Allen-Bradley-defined fault.) The fault routine then responds by disabling the faulty rack and clearing the data

File : 2
Part of an MCP file.

Rung : 17
This rung calls file 7, the fault routine, if integer bit N15 : 0/2 (an I/O rack status bit indicating a fault in remote rack 2) is on while S : 27/2 is off (the rack control bit which would be on if rack 2 was disabled).

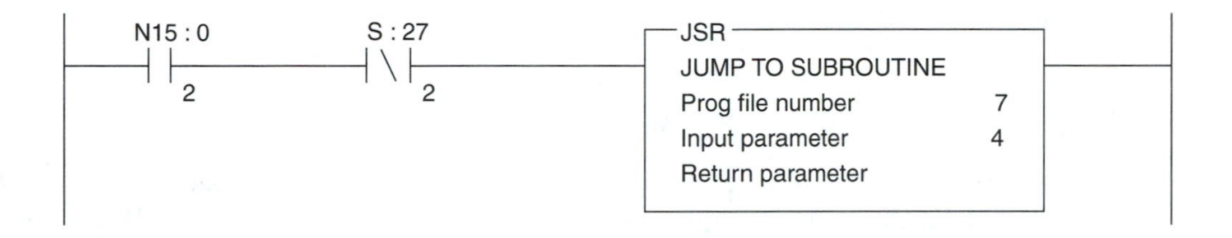

File : 7
The fault routine. This PLC's fault file status word, S : 29, contains the number 7, configuring program file 7 as the fault routine.

Rung : 1
This rung (one rung in a longer fault routine) checks to see if the fault code is user-defined fault 4. If it is, the rung clears the major fault status word (S : 11) and turns the I/O inhibit bit for rack 2 on, so that the PLC will stop scanning the faulty remote I/O rack.

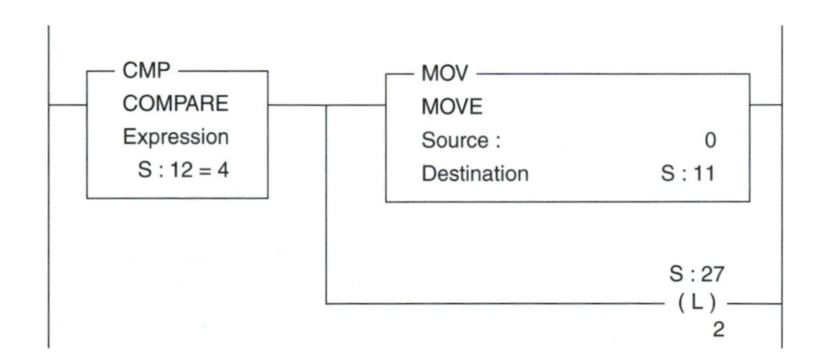

FIGURE 11.24
User-defined Allen-Bradley fault application.

word containing major fault bits. When the fault routine finishes, the PLC will resume running the MCP file 2 at rung 18, the next rung after the one that caused the fault interrupt. (Remember, this is only an example. In a real-world application, just disabling a faulty rack wouldn't be sufficient. The fault routine would also at least notify maintenance to repair and reenable the rack.)

PLC-5s and SLC 500s can also be configured to run a fault routine once each time the PLC goes into run mode after power has been switched off and back on. Use of the fault routine in this way, as an initialization routine, is discussed in the section on initialization interrupts later in the chapter. In the troubleshooting section of this chapter we describe how to avoid interrupts caused by timing out of the watchdog timer.

S5

FAULT ROUTINES IN SIEMENS STEP 5

The operating system in S5 PLCs automatically interrupts the program that is running, sets fault bits in the **ISTACK** area of memory, then may call an **organization block (OB)** as a fault routine when a fault is discovered. S5 PLCs offer only a few OBs that can act as fault routines, so for most faults, the PLC will go into stop mode. The operator must then find and correct the fault, as described in Chapter 15. Organization blocks that do exist can be programmed to correct faults and allow resumption of the interrupted program, or simply to affect a controlled shutdown of the PLC.

STEP 5 offers a STOP instruction **(STS)** that causes the PLC to react as if it faulted, and another STOP instruction, **STD,** that allows the current scan cycle to finish before causing the PLC to fault. STS and STD are useful during debugging of programs, and are discussed in Chapter 15.

The programs in the organization blocks that are called in response to faults may include instructions that examine the state of the fault-identification bits in the ISTACK and try to correct the fault, although not all of the ISTACK data is accessible in a program. The OB may even include the STS or STP instruction to put the PLC into stop mode if the problem can't be fixed and/or after performing a shut-down sequence. If a fault occurs and the response organization block hasn't been programmed, the PLC (usually) goes into stop mode. The OBs offered as user-programmable fault routines include:

- **OB 34,** which is called in response to a battery failure. If it doesn't exist, the PLC does not fault.
- **OB 23,** which is called if a direct access address is used in a program and the addressed module doesn't respond.
- **OB 24,** which is called if a module doesn't respond during the I/O scan or when communications between processors fails.
- **OB 27,** which is called if there is a mismatch between the formal parameters required by a function block and the actual parameters in a call of the function block.
- **OB 32,** which is called if a program tries to access a data block that hasn't been called, a data word that doesn't exist, or if the program tries to create new data words in a data block and there isn't enough PLC memory.

In the troubleshooting section of the chapter we describe how to avoid interrupts caused by timing out of the watchdog timer.

S7

FAULT ROUTINES IN SIEMENS STEP 7

Siemens S7 PLCs above the S7 312 IFM level offer several organization blocks (OBs) that act as fault routines, each triggered by a different type of PLC fault. Siemens defines the faults as belonging to one of two different classes. There are **synchronous error faults,** which are caused by errors in the program (so their timing is synchronized with the program scan time), and **asynchronous error faults,** which are caused by equipment failures.[28] In either case, when the error occurs the PLC stores information about the fault into a diagnostic buffer, and if the error occurred in an I/O module, into a diagnostics status list. The stored information includes an "event ID" error code describing the fault, the date and time the fault occurred, and (if the PLC has been configured to include extended diagnostics entries) the number of the organization block that was called to respond to the fault. (See your manual for event ID code meanings.)

S7 PLCs at the S7 312 IFM level and lower do not have organization blocks that execute in response to faults. These PLCs just store diagnostic data in the diagnostic buffer, then go into fault mode if a fatal fault occurs. If a fault-response organization block does not exist when it is called after fault information is saved, the PLC (in most cases) goes directly into stop mode. If the organization block (OB) *has* been programmed, the OB can end with a BEC or a BEU instruction to cause the interrupted program to resume (after taking some action to correct the problem, of course), or may call **SFC 46 ("STP")** to put the PLC into stop mode (perhaps after performing a shutdown sequence.) The STEP 7 system offers several options for writing fault-response organization blocks to determine the cause of the fault so that it can be corrected before the interrupted program is resumed:

1. When the operating system calls the organization block, it passes several parameters to the OB, including the same fault event ID code, time and date of the interrupt call, and other information that is written into the diagnostic buffer. The OB can include instructions to evaluate this information so that it can respond appropriately.

2. System status list information, including much more than just the information in the diagnostic buffer, can be retrieved by the OB by calling **SFC 51 ("RDSYSST")** with a parameter indicating what sublist of data is required. See Chapter 15 for more information on the system status list.

Figure 11.25 shows a simple STEP 7 program in STL which could cause a timer addressing error and a fault-response program in OB 121 which is called in the event of a synchronous error such as a timer addressing error. The program in OB 121 examines the event ID code that it receives as a parameter from the operating system when OB 121 is called, corrects the error if it is a timer addressing error, then returns to the interrupted program. If the error is any other type of error, OB 121 causes the PLC to go into stop mode.

Synchronous errors cause the PLC's operating system to call OB 121 or OB 122, depending on the type of error encountered as the program runs:

---

[28] In some places in the STEP 7 documentation, Siemens describes I/O interrupts (which they call *process interrupts*) as types of fault routines. They justify this unusual classification by saying that I/O interrupts often are used to detect and respond to alarm signals indicating that the controlled process has a fault. Siemens' process interrupt ISRs are not classified as fault routines in this book.

```
OB 1
  This program includes rung 25, which occasionally causes a synchronous error.


Network 25:

      A I 4.2                   // Causes  a pulse timer to run. The timer
      SP T [MW 10]              // number is obtained from memory address
                                // MW 10, which may contain a number
                                // that can't be used for timers

OB 121
  This organization block is called when a synchronous fault occurs.


Network 1:

      L "OB_121_SW_FLT"         // Read the event ID low byte that the
      L b#16#27                 // operating system provided, and compare
      <> I                      // it to the known event ID for a bad
                                 // counter address
      JC m001                   // If different, jump to the call STOP

      L w#16#5                  // Otherwise, put a valid timer address
      T MW 10                   // into MW 10, so the timer will work next
                                // time

      BEU                       // Then return to the interrupted program

m001:CALL "STP"                 // SFC 46, STP, puts the PLC into STOP mode
      BE
```

FIGURE 11.25
STEP 7 fault routine to correct a specific type of error, or to stop the PLC.

- **OB 121,** the "**programming error**" OB, is called in response to errors such as calling a function, function block, or data block that doesn't exist; attempting to use data memory that hasn't been assigned or is assigned for a different data type; or attempting to use invalid data values.
- **OB 122,** the "**module access error**" OB, called if an error occurs as the PLC is trying to read or write an I/O module.

OB 121 and OB 122 do not have their own priority class! They take on the same priority as the program they are interrupting. (This is the only circumstance in which an S7 PLC allows a program to be interrupted to run an OB of the same priority class.) Since they have the same priority class as the program they interrupt, OB 121 and OB 122 can be considered part of the program they interrupt. In fact, when OB 121 and OB 122 are called, they do not cause the PLC to save the values that were in the microprocessor's registers before the interruption. If OB 121 or OB 122 changes the contents of a microprocessor register, the original contents are lost.[29] This makes it possible for OB 121 and OB 122 to correct faulty working data, since the data was probably in a microprocessor register to cause an interrupt. The programmer should be careful that good data that the interrupted program will need when it resumes isn't changed, perhaps by saving the original values from the microprocessor registers that it affects, and restoring them before ending. STEP 7 offers a system function, **SFC 44 ("REPL_VAL"),** which allows the contents of accumulator 1 (where bad data values causing synchronous errors is often found) to be changed without affecting the contents of accumulator 2. If REPL_VAL is called, a data word included as a parameter with the call is put into accumulator 1.

Unlike other types of interrupts, OB 121 and OB 122 cannot be disabled or reenabled by calling the system functions DIS_INT, EN_INT, DIS_AINT, or EN_AINT (described in "I/O Interrupts in Siemens STEP 7"). They can, however, be disabled or reenabled by a different set of system functions:

SFC 36 ("MSK_FLT") can be called to cause a program to start masking (ignoring) selected types of synchronous errors. The programmer must provide two parameters with the call to MSK_FLT. One parameter is a 32-bit programming error filter and the other is a 32-bit access error filter. Setting one (or more) of those 64 bits will prevent one (or more) specific types of synchronous error from causing the programming error routine (OB 121) or the access error routine (OB 122) to run. After MSK_FLT sets a filter bit on, it stays on. (Calling MSK_FLT again with a new parameter with a 0 in the same bit position does not reset that bit to 0 and does not reenable that type of synchronous error!) MSK_FLT returns two 32-bit parameters that indicate which bits are still set. (There aren't 64 different causes for synchronous errors, so some bits aren't used. Those bits will always be 1's in the return parameters.) See a STEP 7 manual set for specific error filter bit purposes. When masked, synchronous errors will still cause the PLC to record an event ID error code in an error buffer. Diagnostics buffer entries are still made, too.

SFC 37 ("DMSK_FLT") must be called to allow a program to unmask (resume responding to) synchronous errors that MSK_FLT has masked, so that OB 121 or OB 122 will run when those errors occur again. As with MSK_FLT, a programming error filter parameter and an access error filter parameter must be included with the CALL DMSK_FLT instruction. If a bit is 1 in one of these masks, it reenables the synchronous fault type that was disabled by placing a 1 in the same parameter bit position when MSK_FLT was called. (Yes, use a 1 in MSK_FLT to mask, then use a 1 in DMSK_FLT to unmask!) The same two parameters are returned by DMSK_FLT as are returned with MSK_FLT. After using a 1 to unmask a synchronous error type, the returned parameter should have a 0 in the same bit position to indicate that it is now unmasked.

SFC 38 ("READ_ERR") can be called to examine the event ID error codes that were placed in the error buffer while errors couldn't be responded to because they were masked. With the call to READ_ERR, the program must provide programming error filter and access er-

---

[29] Microprocessor registers that OB 121 or OB 122 can change include the accumulators, where the microprocessor keeps values that load instructions have retrieved; the status register, containing bits which are set and reset as the microprocessor executes instructions; the address registers, containing pointers to data the microprocessor is using; and data block registers, containing the addresses of the data blocks in use. Some registers, including those that point to the tops of the local data stack, the block stack, and the interrupt stack, are not affected by calling OB 121 or OB 122.

ror filter parameters, as when calling MSK_FLT or DMSK_FLT, except that 1's in individual parameter bit positions now indicate that the programmer wants READ_ERR to return event ID codes that are in the error buffer because of those specific types of synchronous errors. READ_ERR returns one 32-bit parameter for program error event ID codes and another for access error event ID codes. Three of the 16-bit words in the two 32-bit return parameters are reserved for specific types of event ID codes, and the fourth word (half the access error return parameter) is not used at all. (See your manual set for additional details.) READ_ERR can return as many as three 16-bit event ID codes each time it is called, or it might take several calls of READ_ERR to return all the error codes. For example, if five access error event ID codes were in the error buffer, it would require five calls of READ_ERR to return them all. Each time READ_ERR returns an event ID code, it deletes that code from the error buffer, so when READ_ERR returns all zeros, it has returned all the event ID codes of the type(s) requested.

Figure 11.26 contains an example program that conditionally masks two types of synchronous errors (bad timer numbers and bad BCD numbers) so that they can't cause interrupts. Network 2 conditionally unmasks one error type (bad timer numbers). Network 3 contains a loop that reads error codes for each occurrence of a bad timer number while the interrupts were disabled, and runs FB 13 once for each time it occurred.

An asynchronous error causes an S7 PLC to call an organization block between OB 80 and OB 87, depending on the type of error:

- **OB 80,** the **"cycle time fault"** OB, is called in response to a timing error, such as when a timed interrupt can't run because the previous timed interrupt hasn't finished, or if the watchdog timer runs out.
- **OB 81,** the **"power supply fault"** OB, responds to a power supply failure. This is the only fault routine that won't cause the PLC to go into stop mode if it doesn't exist when it is called.
- **OB 82,** the **"diagnostic interrupt"** OB, responds when an I/O module fails if that I/O module is configured to enable its diagnostic interrupt capabilities.

    Diagnostic interrupts can be enabled using the STEP 7 programming software, or by executing a CALL WR_PARAM instruction in a user-program to send new configuration data to the I/O module. I/O modules capable of generating diagnostic interrupts also maintain a set of parameter values describing the fault. When OB 82 is called, it is provided with a parameter identifying the failed I/O module. OB 82 can include a CALL RD_PARAM instruction to read fault-identification parameters from the failed module to identify why it faulted. (See a STEP 7 manual set for the list of parameters. Diagnostic interrupt descriptions may be included with hardware interrupts.)
- **OB 83** runs if an I/O module is removed or the incorrect module has been inserted.
- **OB 84** runs in the event of an MPI[30] error or if a memory card is removed or replaced.
- **OB 85,** the **"OB-not loaded fault"** OB, is called if the program tries to run an organization block that doesn't exist or if there is an error accessing a peripheral module when updating the process image input (PII) or process image output (PIQ) tables.

---

[30] MPI is a local area network used to allow communications between PLCs and other computers.

```
Network 1:

      A I 4.2                                   // If input I4.2 is on,
      FP M 1.0
      JNB m001
      CALL "MSK_FLT"                            // call SFC 36 to disable errors
       PRGFLT_SET_MSK :=DW#16#42                // when bad timer
                                                // numbers are used and when non-BCD
                                                // numbers are treated as BCD numbers
       ACCFLT_SET_MSK :=DW#16#0                 // But not access errors
       RET_VAL          :=MW100                 // Store MSK_FLT's error information
                                                // (which this example doesn't
                                                // evaluate, but a real program should)
       PRGFLT_MASKED  :=MD102                   // Store return values showing which
       ACCFLT_MASKED  :=MD106                   // errors are now masked
m001 : NOP 0

Network 2:

      A I4.2                                    // If input I4.2 goes back off,
      FN M 1.1
      JNB m002
      CALL "DMSK_FLT"                           // call SFC 37 to reenable
       PRGFLT_RESET_MSK :=DW#16#40              // only the bad timer number synchronous error type

       ACCFLT_RESET_MSK :=DW#16#0
       RET_VAL            :=MW98                // Store DMSK_FLT's error information
                                                // (which this example
                                                // doesn't evaluate, but a real program
                                                // should)
       PRGFLT_MASKED      :=MD102               // Store return values showing which
       ACCFLT_MASKED      :=MD106               // errors are now masked

m002 : NOP 0

Network 3:

      A I 4.2                                   // If input I4.2 has gone off
      Fn M 1.2
      JNB m003
m005 : Call "READ_ERR"                          // retrieve codes from the error buffer

       PRGFLT_QUERY      :=DW#16#40             // But only for bad timer number
       ACCFLT_QUERY      :=DW#16#0
       RET_VAL           :=MW96                 // Store READ_ERR's error information
       PRGFLT_CLR        :=MD110                // Store return values containing
       ACCFLT_CLR        :=MD104                // event ID codes from error buffer

      L  MD110                                  // Compare the returned value against
      L  W#16#15260000                          // the event ID for a bad timer number
      <> I                                      // If not equal
      JC m004                                   // jump out of this loop.
      CALL FB15, DB10                           // Otherwise, call a function block once
      JU m005                                   // and loop back to read from the error
                                                // buffer again

m004 : NOP 0
```

FIGURE 11.26
STEP 7 program that masks and unmasks selected synchronous interrupts, and reads the error buffer.

- **OB 86** runs if there is a fault detected in the distributed power system or with communications with a remote I/O rack.
- **OB 87,** the **"communications fault,"** responds to an error detected in the data being shared with other PLCs in the local area network.

Asynchronous errors are the highest-priority class in the STEP 7 system. OB 80 to OB 87 have priority class 26. If OB 100, the initialization program, is running when the asynchronous error is detected, the asynchronous error OB is temporarily assigned priority class 28 so that it can interrupt OB 100's priority class 27. Asynchronous error OBs cannot be interrupted. If other asynchronous errors occur while OB 80 to OB 87 is running, their response OBs will be called sequentially after the currently running OB ends.

Asynchronous error fault routines can be disabled by calling DIS_INT, after which interrupt requests are ignored, or can be reenabled by calling EN_INT. Fault routines can also be delayed until the end of the program scan if DIS_AINT has been called, then allowed to start again if EN_AINT is called before the program scan ends. Use of these standard functions was explained in the section of this chapter where we discuss I/O interrupts in Siemens STEP 7.

FAULT ROUTINES AND THE OMRON CQM1

OMRON's CQM1 does not execute a user-written fault routine when it detects an error. When a fault occurs, the CQM1 stores an error code into an FAL area of memory, stores the error code and time of occurrence into an error log area of memory (if the error log feature is enabled[31]), and may turn on status bits in memory to describe the error further. Only three two-digit BCD error codes can be stored in the FAL area of memory. The error code most recently stored into the FAL area of memory is also stored in the low byte of SR 253. The error log starts at DM 6569 and extends to DM 6599. DM 6569 contains a pointer to where the next entry must be made into the log, followed by one three-word record for each of up to 10 errors. Each record contains an error classification (fatal or nonfatal), the error code, and the time and day of occurrence. CQM1 error codes and status bits are listed in Figures 15.7 and 15.8.

In the event of a **fatal error,** after storing error information the CQM1 stops immediately, turns all outputs off, and turns on an ERR/ALM indicator on the CPU module. In Chapter 15 we describe how an operator can find and correct major faults.

In the event of a **nonfatal error,** the CQM1 does not stop running. It stores the error code in the FAL area and in the error log, sets specific status bits, starts the ERR/ALM indicator on the CPU module flashing, then goes on with the scan cycle. Nonfatal errors include errors in transferring data between the CPU and a memory module, a problem with the PLC setup data words in the DM area of memory, cycle time exceeding 100 ms, or a bad backup battery. Problems communicating via a peripheral port are also considered nonfatal errors, but no error codes or status bits are stored when this occurs; instead, the indicator light at the port connector just *stops* flashing. A user-program can include instructions to look for error codes or status bits caused by nonfatal errors and can cause the CQM1 to execute response subroutines. The re-

---

[31] The lowest digit in DM 6655 at the time that the CQM1 goes into run mode determines error log configuration. xxx**0** causes the CQM1 to record the most recent 10 error codes and times; xxx**1** indicates that only the first 10 error codes and times are to be recorded; and xxx**2** indicates that any errors on the error log are not to be recorded.

sponse subroutines can send messages or even clear fault codes from the FAL memory area, but they execute under main program control, so they aren't interrupts and are not discussed here.

The CQM1 allows the user to define his/her own **user fault codes.** The CQM1 will store user fault codes in the FAL area of memory and on the error log when a program executes a failure alarm, the **FAL (06)** instruction, or a severe failure alarm, the **FALS(07)** instruction. If the FALS(07) instruction executes, the CQM1 will react as if a fatal error occurred, whereas the FAL(06) instruction has the same effect as a nonfatal error. In either case, the instruction requires one parameter: the user fault code that the CQM1 will store in the FAL area of memory and in the error log. The CQM1 system reserves the use of codes 01 to 99 for user fault codes. The differentiated form of FAL(06), @FAL(06), is recommended so that only one code is stored each time the instruction's conditions go true. Since they are under program control, the FAL(06) and FALS(07) can be considered to be software interrupts. In the troubleshooting section of the chapter we discuss how to avoid interrupts caused by scan cycles taking too long to execute.

Initialization Interrupts

When a PLC is switched into run mode or if power is restored after being interrupted while the PLC was running, the PLC will execute an initialization sequence before restarting execution of the scan cycle. The initialization sequence can be considered to be an interrupt service routine, although much of its execution may lag the actual source of interrupt (being switched *out* of run mode or having power turned *off*) by quite some time. What the PLC does during this initialization interrupt is often affected by how the user has configured the PLC and by the program(s) the user has written. First-time configuration and restart initialization are described in Chapter 10, but we discuss the interrupt characteristics of initialization routines here.

The actual interruption of the program that is running occurs when the PLC is switched into program mode or when power is interrupted. The PLC immediately[32] stops the scan cycle and turns all outputs off (or may freeze some of them in their last state if power is still available). Later, when the PLC is switched back into run mode or when power is restored, the PLC completes the interrupt process by executing an initialization ISR, then restarts execution of the scan cycle from the cycle's beginning (*not* from the point at which the cycle was interrupted by being switched out of run mode[33]). The restart sequence will be slightly different depending on whether the PLC stopped because of power interruption or because it was switched into program mode, and of course varies from one PLC model to the next.

If the PLC stopped because it was switched into program mode, and if power has not been lost while it was in program mode, restarting is relatively simple. When it went into program mode it will have (probably) turned its outputs off but will not have lost any data from memory, not even the output image data. While the PLC was in program mode, an operator may have

---

[32] Some PLCs offer a backup battery feature that allows the PLC to continue to run for a few milliseconds after the power has been interrupted, but this feature only delays the interrupt or allows the PLC to ignore short power interruptions.

[33] The PLC won't necessarily perform the same operation sequence each time it restarts because the data in its memory may be different. Sequential function chart programs don't necessarily restart their programmed sequence, for example.

used a programming terminal to change some data memory, including the program, working data, or even the configuration data. Some PLCs continue to scan inputs while they are in program mode, so the input image data may change. When the PLC is switched back into run mode, the PLC will start up using the latest configuration data, will clear nonretentive data memory (minimizing the danger that might result when the last program terminated before its end), will perform some **first scan operation,** and finally will begin executing the standard scan cycle. In some PLCs, the first scan operation consists only of setting a first scan bit in the status memory area so that a user-program can include instructions that monitor this bit to cause the PLC to perform specific functions each time it is switched into run mode. Other PLCs look for a user-written initialization routine which, if found, is executed before the scan cycle restarts. Some PLCs read new input values before running any user-programs, and others execute the user-written initialization routine before reading new input values. If the PLC runs a user-program before reading the latest input values, initialization routines should not be programmed to respond to data in the input image memory.

If the PLC was in program mode when power was interrupted, the PLC will have lost user-written programs, working data, and configuration data in all RAM memory except those RAM memory areas maintained in nonvolatile memory. Most PLCs offer nonvolatile RAM options such as battery-backed RAM, electronically erasable read-only memory (EEPROM) modules, or flash memory modules that can be used to store programs, data, and configuration while power is off. When power is restored, these PLCs look for an EEPROM or flash memory device as part of their power-up initialization routine. If found, the PLC will automatically copy EEPROM or flash memory contents into RAM memory. Battery backup of RAM contents is necessary only if EEPROM or flash memory isn't available. EEPROM memory is usually written to once, by the programmer, and will contain a complete user-program and startup data set so that the PLC will start up the same way every time it is turned on. Flash memory and battery-backed RAM memory can be changed by the PLC as it runs a program, so it may also contain working data reflecting what the PLC was doing last time it was running.

If power fails while the PLC is in run mode, *most* PLCs automatically switch themselves into program mode as part of the power-up initialization routine when power is restored. An operator must then switch them to run mode. The PLC then restarts as if power were interrupted while the PLC was in program mode. Other PLCs can be configured so that they will restart in run mode automatically after a temporarily power interruption. The user-program, essential data, and configuration data, obviously, must survive a power interruption by being maintained in battery-backed RAM or stored in an EEPROM or flash memory device that is copied to RAM at power restoration. These PLCs often look for and execute an initialization routine before restarting the standard scan cycle. The initialization routine is different from the one the PLC would execute when it is switched into run mode.

PLC-5 SLC 500

INITIALIZATION INTERRUPTS AND THE ALLEN-BRADLEY PLCS

The PLC-5 and SLC 500 do not clear any data memory, even input and output image file data, when they are started or stopped. While in program mode, they continue to scan inputs, so input image data will be up to date when the PLC returns to run mode. Before one of these PLCs starts executing its first scan cycle after being switched into run mode, it automatically sets a **first scan status bit (S:1.15).** The PLC turns this bit back off after it finishes the first scan cycle. The programmer can include instructions in the main program to make use of this first scan bit to cause initialization procedures to be executed. Any instructions (including JUMP TO SUBROUTINE) that are conditional upon the first scan bit will be executed as part of the first scan cycle but will not execute again until the next time PLC is stopped and restarted.

By default, a PLC-5 or an SLC 500 will restart in run automatically if it was in run mode at the time of a loss of power. The scan cycle will be restarted from the beginning, not from the point at which it was interrupted. These PLCs can be configured to execute a fault routine automatically after every power failure. If configured to do this, they set a major fault status bit every time they recognize that power has been turned on.[34] Since major faults are the highest interrupt priority class, the fault routine will execute as soon as the PLC tries to resume running the scan cycle, immediately if the PLC was in run mode at the time of the power interruption or later if it was not in run mode at the time of the interruption. If the PLC hasn't been configured with a fault routine, or if the fault routine finishes without clearing the major fault bit, the PLC will go directly into fault mode, requiring operator intervention. If the fault routine resets the major fault bit before it ends, the PLC will restart the standard scan cycle when the fault routine ends, without requiring operator intervention.

A PLC-5 can be configured so that whenever it restarts, it resumes running SFC files starting after the last sequence step that was completed the last time the PLC-5 was in run mode instead of restarting at the first SFC step. The PLC-5 can be configured this way by setting status bit S:26.0. (The SLC 500 does not offer SFC programming.) This configuration bit has no effect on ladder logic programs, which always restart from the beginning.

S5 INITIALIZATION INTERRUPTS IN SIEMENS STEP 5

When an S5 PLC is switched from program mode to run mode, it clears the input and output process images (PII and PIQ) and nonretentive memory, executes organization block 21, then enables outputs and restarts the standard scan cycle. If the programmer hasn't entered a program for OB 21, the scan cycle will be restarted without any other initialization other than the clearing of I/O images and nonretentive memory. Note that OB 21 should not be programmed to respond to input conditions by examining PII table values, because the input images remain cleared until after OB 21 finishes. Larger S5 PLCs can use direct accessing addressing in OB 21 because direct addressing causes the PLC to read input data from the input module, bypassing the PII table. Note also that outputs will remain off until the first standard scan cycle ends. They will not change immediately after OB 21 finishes.

If power is interrupted, then restored, while an S5 PLC is in run mode, the PLC executes the same procedure as described above, except that **OB 22** is executed instead of OB 21. Initialization interrupt routines OB 21 and OB 22 can be interrupted only by fault interrupts.

---

[34] To configure a PLC-5 to execute the fault routine whenever power is restored, set bit S:26.1. In an SLC 500, set bit S:1/9. The PLC-5 will trigger the fault routine by setting major fault bit S:11.5 when it detects power going on. The SLC 500 sets bit S:1/13.

S7

INITIALIZATION INTERRUPTS IN SIEMENS STEP 7

The S7-400 series of PLCs offer two ways of starting up, whereas lower-level S7 PLCs offer only one way. The startup method offered by both is called a *complete restart.* The method offered by S7-400 PLCs is called a *restart.* An S7-400 will execute a **restart** if power is restored after being lost while the PLC was executing a program without errors, but only if the CPU's restart selector switch is on WRST (not on CRST), and if there have been no changes to the PLC's memory or mode switches. An S7-400 CPU module can also be configured to execute a restart after being switched to program mode, then back to run mode. In this type of restart, the PLC rewrites configuration parameters to intelligent I/O modules, executes **OB 101,** enables the outputs, and then resumes executing the user-program starting with the step where it was interrupted when power was lost. The PLC does not change I/O images unless the PLC has been configured to clear them on a restart.

A **complete restart** is executed every time an S7-300 or lower PLC is switched into run mode or is powered back on if it was running a program when power failed or when an S7-400 PLC is started under conditions that do not allow a restart. In a complete restart, the PLC clears nonretentive data memory, clears the B stack and I stack, clears alarms, and loads configuration parameters that a programmer may have changed to the I/O modules attached to the PLC. If there has been a power loss since the PLC was last in run mode, it also writes parameters to intelligent I/O modules. The PLC then executes OB 100, the startup routine, before it enables the outputs and begins executing the main program cycle.

OB 100 and OB 101 can include user-written initialization programs. Both have priority class 27, so they can only be interrupted by the diagnostic interrupt OB 82, by a programming error OB 121, or module access error OB 122.[35] When the operating system calls OB 100 or OB 101, it passes several parameters that the user-written initialization routine can use. These parameters include:

1. A code identifying whether the PLC is starting because it has just been switched into run mode, or if power has just been restored after being interrupted while the PLC was in run mode.

2. A code identifying why the PLC last went into stop mode. It may have been switched into stop by the operator, or a fault may have caused the stop. The program can examine this code and the previously mentioned code so that the initialization program can be programmed to operate differently depending on why the PLC last stopped running.

3. A 32-bit data word that contains status bits indicating whether the configuration data has been changed while the PLC was in stop mode, what initiated this transition back into run mode, and whether the startup sequence ahead of OB 100 or OB 101 has been successful in restarting the PLC so far.

Direct access addressing can be used to perform immediate reads from input modules or to perform immediate outputs to output modules as OB 100 or OB 101 executes. When an S7

---

[35] OB 82 usually has priority class 26 but is assigned class 28 if it is caused by a hardware fault detected during execution of OB 100 or OB 101. OB 121 and OB 122 are assigned the same priority class as that of the organization block that they interrupt.

PLC is put into stop mode, it puts its outputs into a predefined state (as selected during configuration) rather than clearing them. While in stop mode, a programmer can write configuration changes to the CPU's memory.

CQMI

INITIALIZATION INTERRUPTS AND THE OMRON CQM1

By default, input images and output images are cleared when the CQM1 goes into run mode. The programmer can configure the CQM1 to skip clearing the I/O images when it goes into run mode, by setting the I/O memory hold bit (SR 25212) and by setting the second-most-significant BCD digit in DM 6601 to 1 (i.e., x**1**00). If DM 6601 isn't configured in this way, SR 25212 will be cleared automatically during the initialization routine, and then the I/O images will also be cleared.

The CQM1 has a **startup mode** status configuration word (DM 6600) that indicates how the PLC is to respond when power is restored after being interrupted. By default, DM 6600 contains **00**00, which configures the PLC to start up in run mode. If DM 6600 is changed to **01**00, the CQM1 will restart in the mode it was in when power was interrupted. If DM 6600 is set to **02**xx, the last two digits dictate the mode the CQM1 will be in after power is restored: **0200** for program mode, **0201** for monitor mode, and **0202** for run mode. If the CQM1 has a programming unit attached, it will start up in the mode set at the programming unit, despite the configuration value in DM6600.

The CQM1 has a **first cycle flag bit** (**SR 253 15**) that goes on for the first scan cycle after the PLC is switched into run mode or after the PLC returns to run mode after a power disruption. The main program can include instructions to examine this bit to cause execution of initialization operations during the first scan cycle.

Communication Interrupts

Modern PLCs include **communication handlers,** separate microprocessors that are dedicated to handling serial communications. The main microprocessor services communication handlers by monitoring their status and issuing them communication requests. The communication handlers often have direct access to the main memory. Because servicing can be accomplished during the I/O steps in a scan cycle, interruptions to service communication handlers are not as essential in a PLC as in most other computers! In this section we discuss the few communications interrupt features, and in Chapter 13 cover other programmable communications capabilities.

Some PLC programming languages include instructions to cause the PLC to interrupt user-program execution to execute servicing immediately instead of waiting until the I/O scan. These PLCs typically offer status bits controlled by what the communication handler is doing, so that the PLC can be programmed to execute communication servicing instructions when they are required to improve communication efficiency.

Data exchange tasks assigned to a communication handler are normally added to the end of a communication handler's queue of assigned tasks, but some PLCs offer instructions to cause a task to be placed at the beginning of the queue and/or instructions that cause the user-program to interrupt until after the communication handler has completed the data exchange task.

COMMUNICATION INTERRUPTS AND THE ALLEN-BRADLEY PLC-5

The PLC-5 offers few user-accessible instructions or configuration features to affect communication interrupts, probably in part because the PLC-5 has more communications capabilities than an average user would normally take the time to learn about. It is safest to prevent the user from changing the way the PLC-5 operating system handles communications. A PLC-5 CPU module contains communication handlers, and the main microprocessor controls them and exchanges data with them during a communication time slice part of a housekeeping scan step. These are not interrupts and are explained in Chapter 13.

There is one area where a user-program can force a communications interrupt. If a block transfer request is encountered in an ISR running in response to a selectable timed interrupt (STI), a process input interrupt (PII), or in a fault routine, the PLC-5 automatically interrupts the execution of the ISR and performs the block transfer. The ISR does not resume executing until the communications handler has completed the transfer of data. The programmer can *prevent* the ISR from being interrupted for a block transfer if the **USER INTERRUPT DISABLE (UID)** instruction is executed before the BLOCK TRANSFER READ or BLOCK TRANSFER WRITE (BTR or BTW) instruction. The **USER INTERRUPT ENABLE (UIE)** instruction should, of course, be executed after the BTR or BTW, to reenable interrupts. Block transfer instructions between a UID and a UIE place the requested block transfer in the communications queue, so the data transfer will execute eventually.

Next, we describe some characteristics of how the PLC-5 shares access to its communication channels between the communication handlers and the main program. While the programmer can't change these characteristics, it can be important for the PLC-5 programmer to know how the PLC-5 handles communications:

1. Immediate I/O instructions (IIN and IOT) conflict with block transfers in the local or extended rack because they make use of the same local communication bus while the user-program is executing, resulting in possible corruption of data on the data bus.[36] If the programmer must program block transfers and immediate I/O instructions for the local rack, the immediate I/O instruction should be conditional upon there being no active block transfers executing, which can be checked by examining the state of the .ST (start) bit for all block transfer instructions accessing all block transfer modules in the local rack.

2. Block transfers with modules in the CPU-resident rack interfere with monitoring of PII signals from the CPU-resident rack.

3. Block transfers to or from remote I/O racks cause the PLC-5 to interrupt remote I/O scanning of racks connected to the serial communication link so that the block transfer can execute, so remote digital I/O status updating becomes slower. A single block transfer can easily double the remote I/O scan time (e.g., it takes 2.7 ms to transfer a 10-data-word block on a serial link operating at 230.4 kbps, and the remote I/O scan time is normally only 3 ms). The PLC-5 limits the delay by executing only one block transfer per remote rack per remote I/O scan cycle (not counting block transfers requested in ISRs, of which any number can occur).

---

[36] Immediate I/O instructions for remote racks only read or write data from the remote I/O scanner buffer memory, so they don't interfere with block transfers to a remote rack, but they don't actually perform immediate I/O either.

4. A communication request made by a BLOCK TRANSFER READ **(BTR)** or BLOCK TRANSFER WRITE **(BTW)** in an ISR becomes higher priority than any other request in the communication queue, except for a request that is already active, so the interruption time is minimized. If the request is for a block transfer of data with a module in a remote I/O rack, the newer enhanced PLC-5 models execute the MCP program that the ISR interrupted while the ISR is waiting for the block transfer to complete.

SLC 500

COMMUNICATION INTERRUPTS AND THE ALLEN-BRADLEY SLC 500

The SLC 500's communications servicing functions are performed during a housekeeping step in the scan cycle. No data is sent or received until the housekeeping portion of the scan cycle is executed, so communications are usually performed without requiring any communications interrupts. Communication without interrupts is discussed in more detail in Chapter 13.

SLC 500 models starting with the 5/02 offer two instructions that can be used to cause a user-program to interrupt service communications immediately. The SERVICE COMMUNICATIONS **(SVC)** instruction causes immediate servicing of the serial communications channels, and the I/O REFRESH **(REF)** instruction (discussed earlier in the section on immediate I/O interrupts) causes a complete I/O scan, including servicing of serial communication channels. Newer SLC 500 models (the 5/03 and higher) allow the programmer to specify which of the two serial communications channels to service in response to the SVC or REF instruction. SVC and REF instructions cannot be programmed into other interrupt service routines, which implies that communications interrupts have been assigned relatively low priority. The SLC 500 offers several communications status bits that the SLC 500 operating system updates as serial communications execute, which can be used in a user-program to control whether SVC or REF instructions should execute. These bits include:

1. **Incoming command pending,** which is set if a communication request has been received from another controller. Servicing the communication clears this bit. Bit S:2/5 indicates that the request has been received via serial channel 1. S:33/0 indicates that a request was received via the serial channel 0 in an SLC 5/03 or higher.

2. **Message reply pending,** which is set if another controller has sent information that this controller requested in an earlier message and is cleared when this controller services the incoming communication. S:2/6 is used for channel 1 and S:33/1 indicates channel 0.

3. **Outgoing message command pending,** which is set when it is possible to send a message from this controller's queue of messages and is cleared when servicing the communication channel causes the message transition to begin. Bit S:2/7 indicates that channel 1 is ready, and S:33/2 indicates that channel 0 is ready.

S5

COMMUNICATION INTERRUPTS IN SIEMENS STEP 5

The STEP 5 manual set does not describe any communication interrupt control instructions or configuration settings that are available to the user.

COMMUNICATION INTERRUPTS IN SIEMENS STEP 7

The S7 PLC must be configured and programmed to exchange data with other computers in local area networks. The highly efficient network software accomplishes most data exchange operations without requiring interrupting of the PLCs, but there are some system functions that use interrupts with network communications. We examine those functions here.

The system function **SFC 35 ("MP_ALM"),** makes use of the MPI or Profibus network to interrupt all the other PLCs on the network to begin execution of **OB 60,** the "**multicomputing interrupt**" OB, in all PLCs simultaneously. Each PLC can have a different program in its OB 60, and a PLC will not fault if it lacks a program in OB 60. OB 60 has a priority class of 25, so if it exists, it will interrupt almost any other OB when a signal arrives via the local area network to make it run. The network signal can be initiated by any PLC that executes SFC 35 MP_ALM. To prevent conflicts that might occur if two PLCs both try to execute MP_ALM, none of the PLCs will recognize a signal to start executing OB 60 if OB 60 is executing on any PLC on the network. The PLC program that calls MP_ALM must provide a **JOB** parameter consisting of a number from 1 to 15. This value is sent to the other PLCs with the interrupt signal, where it is provided to OB 60 as the **OB60_JOB** parameter. The program in OB 60 can examine this value to know why the interrupt was called.

During configuration, S7 PLCs can be set up to participate in global data circles. Normally, the network copies data between PLC in the circle cyclically, but there are two system functions that can be invoked to interrupt the network to send or to receive global data. **SFC 60 ("GD_SND")** can be called to immediately send a global data packet. The parameters required include **CIRCLE_ID,** the number (1 to 16) of the subnet by which to send data, and **BLOCK_ID,** the number (1 to 3) of the specific PLC to which to send data. **SFC 61 ("GD_RCV")** is used in a similar manner but causes global data to be read immediately from a specific PLC on a specific subnet.

COMMUNICATION INTERRUPTS AND THE OMRON CQM1

The CQM1 can communicate in one of three modes, none of which uses any interrupt capabilities that can be controlled by a user-program. The CQM1's communications capabilities are covered in more detail in Chapter 13.

SUMMARY

Interrupts cause the scan cycle to be suspended while an interrupt service routine (ISR) executes, then (usually) to resume as if there had been no interrupt. High-priority interrupts can interrupt lower-priority interrupts.

Program instructions can cause the immediate input of data from an input module or the immediate output of data to an output module. Special addressing may be used instead of special instructions. Some PLCs allow multiple words of I/O image data to be sent or received with a single instruction, and the OMRON CQM1 can be configured to output all program-generated output data directly to output modules.

Timed interrupts are used to force the PLC to execute ISR programs at precisely timed intervals, thus making control more deterministic. The programmer will probably want to use immediate I/O instructions in timed interrupt ISRs, so that sampling of inputs and changing of outputs are performed at the same constant intervals as the timed interrupt ISR executes. Timed interrupts can also be used to make a PLC execute an ISR at the same time of every month, day, hour, or even minute. Timed interrupts can be used to start a delay timer so that an ISR executes a precisely defined time after the timer is started. Timed interrupts are generally the lowest priority of the PLC interrupt types, which means that they will not start if a higher-priority interrupt is already executing but will wait for that ISR to end before starting. Status bits are often set if this happens, so that the program can respond to any difficulties this may cause.

An I/O interrupt causes an ISR to be executed in response to a signal generated by a sensor connected at an input module. Some PLCs can be configured to monitor multiple signals and sometimes to count rapidly changing input signals and to execute an ISR when the count reaches a preset number. Some PLCs can only monitor input contacts built into the CPU module. I/O interrupts are higher priority than timed interrupts, but not as high as fault interrupts.

A fault interrupt, the highest-priority interrupt type, occurs when the PLC detects that something has gone wrong. If the problem detected is minor, the PLC will only interrupt long enough to store a fault code and perhaps set some fault bits in memory so that the user-program, or perhaps the human operator, can find them. If the problem is more significant, the PLC may look for and execute a user-written fault routine ISR after storing fault codes and then leave run mode. Fault routines can sometimes be written so that they correct the fault and clear the fault bits so that the PLC can resume executing the interrupted program, but severe faults usually cause the PLC to stop, after saving fault identification data. Once a PLC goes into fault mode in this way, the scan cycle interrupted cannot be resumed. It can only be restarted from the beginning once the cause of the fault is corrected. Most PLCs offer instructions that cause fault interrupts, perhaps only to allow a program to stop the PLC from running, but sometimes to allow triggering of the high-priority fault routine.

Initialization interrupts, almost as high in priority as a fault routine, respond to a PLC leaving run mode. When the PLC is switched back on or is put back into run mode, the interrupts finish executing by causing an initialization ISR to execute before restarting the usual scan cycle. Some PLCs clear part of the memory at this time, often including I/O image data. The rest of the initialization ISR usually executes before new input values are read in. Allen-Bradley and OMRON PLCs set a first scan bit, and it is up to the programmer to include instructions in the main user-program to monitor this bit and to perform any initialization that might be necessary. Siemens PLCs automatically execute a user-written initialization program before restarting the scan cycle. The PLC may start differently if power has been interrupted since the last time it executed, perhaps even executing a fault routine, or perhaps forcing the operator to switch the PLC manually into run mode.

Now that PLCs are being built with additional built-in microprocessors that handle only serial communications, some PLCs offer a small number of communication interrupts. Communication interrupts are generally initiated by instructions in the user-program. These (usually low-priority) instructions cause the PLC to interrupt the scan cycle immediately to provide the communication handler with the help it requires.

TROUBLESHOOTING

The order of execution of a program that includes interrupt capabilities is not as predictable as one that doesn't. A subroutine will execute only when it is called, but an interrupt service routine (ISR) can execute at any time while another program is executing. If you find that your PLC program is performing in ways that it shouldn't, suspect the ISRs! Some PLCs maintain a record of the most recent interrupts that occurred (e.g., Siemens S7 PLCs include interrupt events in their diagnostic buffers). Determining whether an ISR has executed is easier if your PLC maintains a history of ISR calls, because an ISR may be called and finish executing before you can see it work from a program monitoring screen. If the PLC does not maintain an interrupt history, you can use a simple trick such as programming a toggle and a toggle counter into an ISR, as shown in Figure 11.27, then monitor the counter's accumulated value to see if the ISR program is being called. (Use a data bit that retains its value between executions of the ISR!)

The recently introduced voluntary IEC 1131-3 standard requires a strict separation of data memory so that an ISR can't modify the contents of memory that the interrupted program is using, and also requires the operating system to ensure that interrupts can't occur until all the data that an ISR needs is available, but most PLC operating systems were designed before this standard was released. If you find that data values are changing when they shouldn't be changing, suspect your ISR programs again. Most programming software packages offer a cross-reference feature that can report all the places that an address is used in a program. Use cross-referencing to see if an ISR is capable of changing the data value. Careful use of interrupt disable and reenable instructions may be required to protect data values for those intervals in which they shouldn't be modified by ISRs.

Some interrupts are enabled by default (e.g., Siemens timed interrupts or cyclic interrupts), while other interrupts are disabled by default (e.g., Allen-Bradley STI timed interrupts). Some interrupts need to be configured *and* unmasked to be able to execute (e.g., OMRON interval timer interrupts). If ISRs aren't executing, or if they are executing unexpectedly, check what is required to start/stop them in this PLC.

Response to an interrupt signal is immediate, in theory, but there are several reasons why there might be a delay before an ISR executes. Some PLCs allow the programmer to adjust this delay time by configuring when the PLC can respond to an interrupt request: at any time during the scan cycle or only during execution of the user-program, after every rung, after every

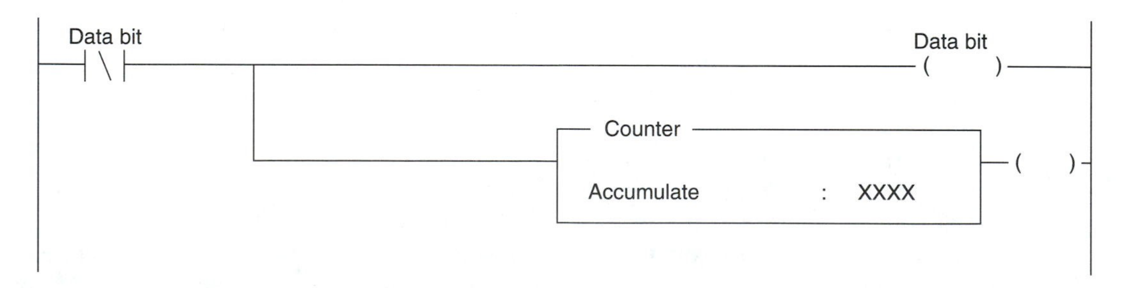

FIGURE 11.27
Simple ladder logic rung in an ISR to count every second time the ISR executes.

instruction, or partway through a long instruction. If the PLC allows interrupts only after a rung finishes, the program shouldn't contain rungs with exceptionally long execution times. Timed interrupts may not execute exactly when they should because they don't have high enough priority to interrupt other ISRs, even other timed interrupts. Most PLCs contain status bits that indicate whether an ISR had to wait before it executed. Monitor those status bits with the programming unit or in the user-program ISRs themselves. Use the status bits to drive counters if you suspect they are changing faster than you can see.

Even if an ISR is executing when it should, it may not be using the most recently available data. Most PLCs do not update the input image table as they call an ISR, so input image data may be out of date by as much as the longest scan cycle time (most PLCs maintain a status word showing their longest scan time). Similarly, data that an ISR writes to the output image table may not be written to output modules for as long as one full scan cycle. If these delays are not acceptable, use immediate I/O instructions or addressing in the ISR.

High-priority ISR programs can interrupt lower-priority ISRs. Whenever any program is interrupted, the PLC pushes data for the interrupted program onto the top of stack memory so that it can be retrieved when the interrupt ends. The size of the memory area reserved for the interrupt stack is limited, so there is a limit to the number of interrupted programs that can be held waiting to be resumed. If too many levels of interrupts are allowed, the stack overflows and loses some of the data required to be able to resume the interrupted programs. Generally, PLC manufacturers assign only a small number of interrupt priority levels so that their PLC won't run into problems by overflowing the stack. Modern PLCs, however, offer programmers increasingly more flexibility, so future programmers will have to be even more careful that they don't configure their PLCs so that stack memory can overflow; otherwise, the PLC won't be able to resume interrupted programs.

Stack memory is also used to store return addresses every time one program executes a normal call of another program, and some PLCs also use stack memory to store working data. The PLC may or may not use the same stack for all three purposes, so stack memory may overflow due to program calls and/or storage of working data. There were no universally agreed upon standards for stack memory before IEC 1131-3.[37]

If your program is failing to respond to critical high-speed counter accumulated values, it may be because the program is examining the counter's accumulated value by checking to see if it is equal to a target value. Remember that high-speed counters use interrupting to enable counting of input signals that change faster than the PLC scans, so a high-speed counter can count right *past* a limit value between scans. Instructions monitoring a high-speed counter's accumulated value should compare to see if it exceeds a limiting value rather than examining to see if it is equal to the limiting value. Some PLCs can be configured to interrupt their scan cycle immediately when a high-speed counter reaches its preset value (e.g., the SLC 500).

If your program is changing interrupt configuration as it executes but interrupts aren't working as they should, remember that some PLCs only accept certain types of configuration

---

[37] Of the PLCs explicitly covered in this book, only the Siemens S7 family of PLCs adhere to the IEC 61131-3 standard in terms of separate memory for each program block. It requires three types of stack memory to maintain that separation: For each of the STEP 7 priority classes there is an L stack that can contain up to 256 bytes of local data, and a B stack to hold data enabling eight levels of nested program calls. An S7 PLC also has one I stack to hold data for nested interrupts.

changes when they go into run mode! It may be impossible to change the interrupt response of your PLC unless you make the changes while the PLC is in program mode. If you want to write an initialization routine to set a default interrupt configuration, the initialization routine might have to compare the new configuration against the old and if they are different, execute an instruction to cause a user-initiated fault after writing the new values. The PLC will go into program mode and the operator will have to switch it back into run mode. The PLC-5 must be programmed this way to allow an initialization routine to change any PII configuration except for the PII file number. OMRON PLCs' configuration data memory values can be changed only while the PLC is in program mode, but interrupts can be turned on and off by instructions in the program.

Interrupts are intended to make a PLC respond more quickly or at more repeatable time intervals, but they may make the overall scan time of programs longer and more unpredictable. It may even happen that a PLC will unexpectedly go into fault mode after running flawlessly for a long time because an unusually large number of ISRs were all executed during the same scan and the watchdog timer timed out. PLCs offer several features to help a programmer anticipate, avoid, or recover from scan time irregularities and watchdog timer problems. Programming manuals often contain formulas that can be used to calculate how long a scan cycle can take in the worst case. Most PLCs also maintain a status word showing the longest scan time so the programmer can compare the calculated result against the actual scan times. As purchased, PLCs typically come configured with long watchdog timer settings but allow the programmer to change the watchdog timer time.

QUESTIONS/PROBLEMS

1. Name the step(s) in the three-step PLC scan cycle that can be interrupted by an I/O interrupt.
2. Describe, briefly, how a PLC handles a two-level interrupt request, one received during execution of its scan cycle and a higher-level request received during execution of the first interrupt's interrupt service routine.
3. Which is the one type of interrupt for which the interrupted program may not resume?
4. In terms of the data in an MPU's registers (e.g., accumulators), how does responding to an interrupt differ from responding to a subroutine call?
5. If your PLC program had to provide servo control of a process that changed too rapidly to be controlled properly with the delays imposed by a normal PLC scan cycle, what type of instructions would you use to read sensors and to output to actuators?
6. Would an immediate input/output instruction be considered a hardware interrupt or a software interrupt? A timed interrupt?
7. What does configuring a PLC to use timed interrupts make the PLC do?
8. What type of event would cause an I/O interrupt (sometimes called a hardware interrupt)?
9. Which typically has a higher priority level: a timed interrupt or an I/O interrupt?
10. Why do high-speed counters need to use interrupts?
11. Explain why using interrupts can cause control actions in a main program to become less deterministic. What type of interrupt can you use so that control actions can be forced to become highly deterministic again?
12. There are some practical restraints on how short a time interval should be used for a timed interrupt. Describe some of those constraints.
13. What types of events might trigger a fault interrupt?
14. Fault interrupt routines can sometimes be written so that they allow the interrupted program to resume after the fault routine finishes executing. What would the fault routine have to do?
15. If the PLC you are learning allows the use of user-

initiated fault interrupts, describe how you would set up and program a nonfatal interrupt.

16. What initiates an initialization interrupt? How would you configure and/or program the PLC you are learning to use to reset a counter as part of an initialization interrupt?

Allen-Bradley PLCs

17. Write the PLC-5 rung that would unconditionally read the status of 16 input bits from module 6 in rack 12. Where would the PLC put the 16 bits of data?
18. If N7:2 contains the value 0, what will the following instruction do?

```
|                     N7 : 2                    |
|------------------   (IIN)   ------------------|
|                                               |
```

19. Your PLC-5's watchdog timer is set too short, and the PLC faults occasionally. You want to write a temporary program that will increase the watchdog timer slightly each time the watchdog timer causes a fault. The PLC should resume operating all by itself after each time it faults, and eventually it won't fault at all. (After an hour you will delete this temporary program.) Describe the program carefully, but you don't need to actually write it or give actual status word/bit addresses.
20. To cause a PLC-5 to fault, you could program the PLC to set a __________ __________ bit.

Siemens PLCs

21. **(a)** Whenever a 55-100U PLC executes a hardware interrupt (also called an I/O interrupt), OB002 executes. What does the PLC do in addition to executing the organization block? Explain clearly what happens in the additional step(s), and indicate when the additional step(s) execute.

    (b) In the hardware interrupt organization block, you want a program to use the most recent information from input module 1, so you would program a load instruction as follows: **L ____B1** (fill in the blank).
22. If you wanted a Siemens PLC to fault immediately during program execution (to stop the program), you could program the __________ instruction in.
23. Write a Siemens STEP 5 or STEP 7 program to read the contents of one byte of data from an input module (input byte 25), and to examine one bit (bit 4) of the immediately input data.

SUGGESTED PLC LABORATORY EXERCISE

For a system with:

- Four control panel switches: inputs 0 to 3
- Four indicator lamps or visible output module LEDS: outputs A to D
- Two spring-return-valve-controlled cylinders: outputs E and F
- One detent-valve-controlled cylinder: outputs G and H
- Three sensors to detect extension of each of the cylinders

1. Write a program and/or configure your PLC so that it executes:

    (a) A timed interrupt program at 3-s (or 3000-ms) intervals. The program should turn indicator lamp A on when switch 0 is on.

    (b) Several (at least 10) immediate input and immediate output instructions, but only while switch 1 is on. With the PLC in run mode, monitor the scan time on the processor status screen as you actuate switch 0.

    (c) Reduce the watchdog timer time to a value that causes the processor to fault, but *only if* switch 0 is on (when the immediate inputs and outputs execute). (Test the program now.)

    (d) Write a fault routine that will run whenever the watchdog timer causes a fault. The fault routine must clear the appropriate fault bits (only the bits the watchdog timer sets) and must increase the watchdog time setting by one time interval each time it executes. (This would not be a good program in a production environment.)

12

PROCESS CONTROL

OBJECTIVES

In this chapter you are introduced to:

- Process control terms.
- Programming a PLC for process control without using any built-in PID instructions, using:
  - Setpoint, process variable, and control variable handling.
  - Scaling and limiting of variables.
  - Methods of making the control interval rapid and deterministic.
  - Inclusion of bias, disturbance, and feedforward values.
- Allen-Bradley, Siemens, and OMRON's PID process control instructions or function blocks:
  - Programming and data entry.
  - The equations: independent gains, dependent gains, and variations.
  - Explanation of proportional, integral, and derivative term effects and how they control steady-state error and overshoot error.
  - Using antiwindup and smoothing.
- Fuzzy logic as an alternative to PID control.
- Programming to allow manual control as the PID instruction executes, and automatic control that self-adapts to environmental constraints as the constraints change.

INTRODUCTION TO PROCESS CONTROL

Whenever you program a PLC to monitor inputs and to change outputs, you are using a PLC for process control. In this chapter we use the term *process control* to mean something slightly more complex: In process control, a **setpoint** command signal tells the controller exactly where the controlled process should be driven to, a sensor measures the processes's actual output and provides **feedback** to the controller, and the controller provides a **control signal** to an actuator telling it how hard it has to drive to reach the commanded state.

This definition implies that the PLC must perform calculations to determine how hard the actuator must drive. Several PLCs include functions to perform that calculation. *PID calcula-*

tions and *fuzzy logic algorithms* are the two most common process control functions. Figure 12.1 shows a typical process control application. In this example, a PLC is used to control the position of a motor-driven carriage moved by a ballscrew feed. A setpoint signal comes from a rotary potentiometer dial that is set by a human operator. The setpoint voltage is proportional to the potentiometer's rotational position, which is set to indicate the distance that the carriage should be from the motor. A linear potentiometer provides another voltage signal, which is proportional to the carriage's actual distance from the motor. The upper and lower position limits are fixed by the length of the feedscrew.

When the operator changes the position of the potentiometer dial, the PLC must cause the motor to turn until the linear potentiometer's voltage signal becomes the same as the voltage signal from the potentiometer dial. A process control **block diagram** of this system might look as shown in Figure 12.2. This block diagram shows one of the many possible process control configurations, a version that is adequate for the vast majority of PLC-controlled processes.

1. The **setpoint** in the block diagram is the voltage command signal from the rotary potentiometer. It is converted to a digital number by the PLC's analog input module.

2. The **error amplifier,** part of the PLC program, is examined in more detail later.

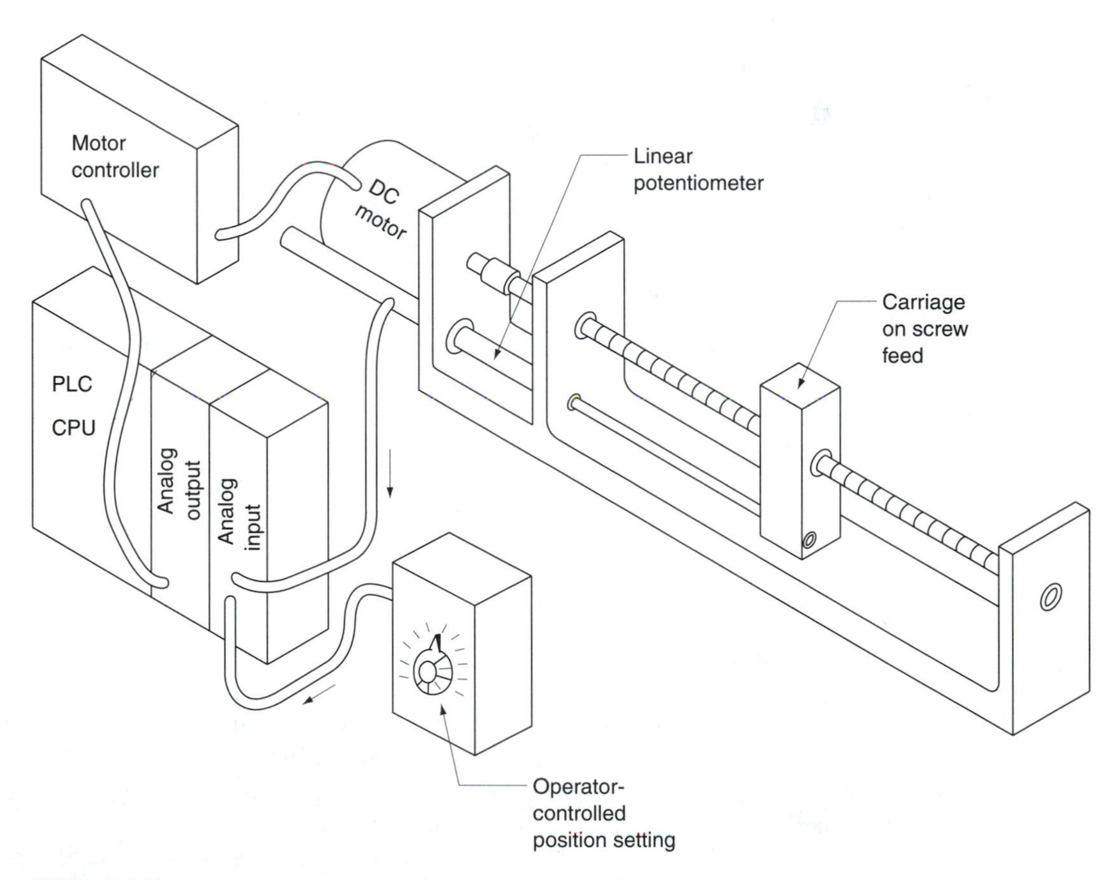

FIGURE 12.1
Process (position) control.

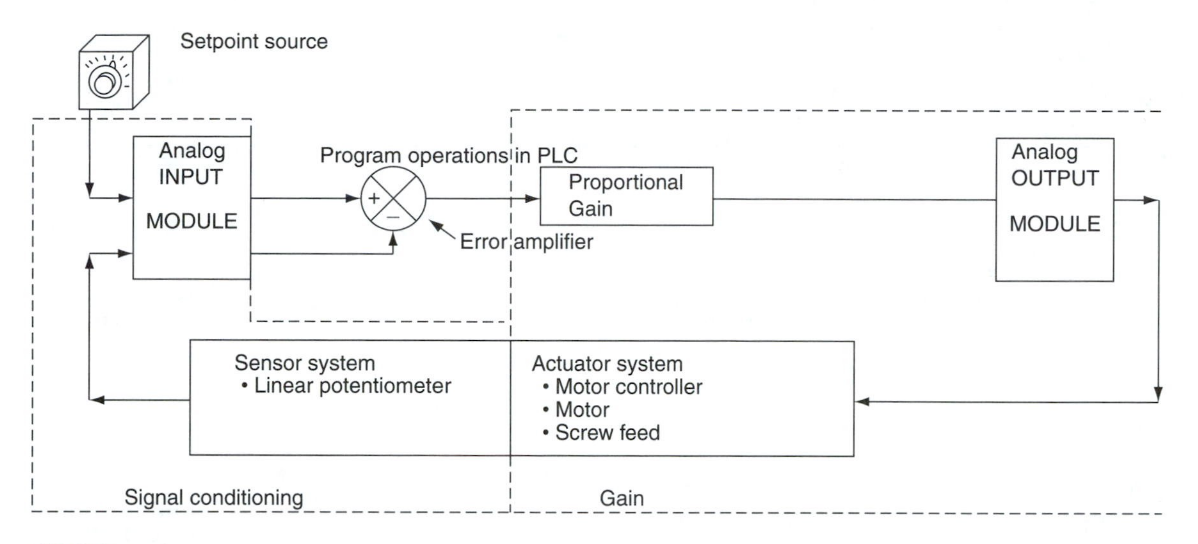

FIGURE 12.2
Process control block diagram.

3. The **gain** of the system includes calculations performed in the PLC program plus the effects of the following components in this example:
   (a) An analog output module connected to the CPU module. This module outputs an analog signal proportional to the digital value that the CPU module provides. Intelligent analog modules may adjust the value.
   (b) A motor controller, which amplifies the small analog voltage signal from the PLC to a higher-power dc supply to the motor.
   (c) A dc motor, which converts dc current to torque to change the screw feed's rotation speed.
   (d) A ballscrew and ballnut feed, which converts the rotational velocity of the screw into the linear velocity of the carriage along the screw.

   Controller manufacturers often ignore actuator components in the block diagrams in their literature, because users provide the actuator components. Despite what such a block diagram may suggest, a controlled system's gain behavior cannot be fully explained by what happens in the PLC. The characteristics of the actuator system and its environment must be considered.

4. The **signal conditioner** in this system consists of two components:
   (a) A linear potentiometer, which converts the position of the potentiometer's wiper into a voltage signal. (Sensor behavior is often omitted from block diagrams prepared by controller manufacturers.)
   (b) An analog input module, which converts the feedback voltage signal to a digital number and perhaps adjusts the result before providing it for the CPU module to read.

The signals in a block diagram of a process control system also have names, as shown in Figure 12.3:

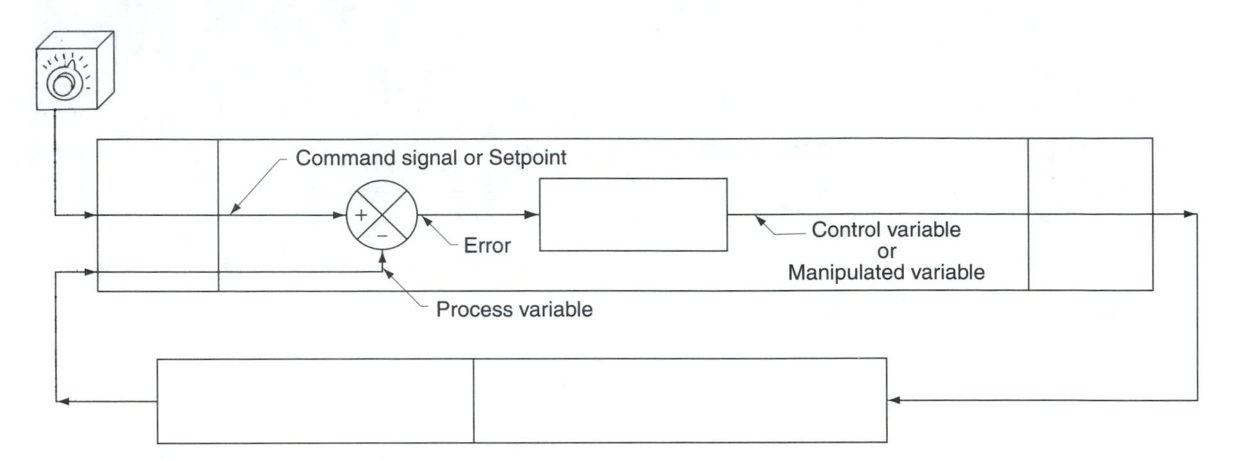

FIGURE 12.3
Signal names.

1. The **command signal** is sometimes called the **setpoint.**

2. The signal from the error amplifier to the gain block is usually called the **error signal.**

3. The final system **output** is often called the **control variable** or the **manipulated variable,** but it can be called by other names. In this example we would call it *carriage position* and measure it in length units. Control variables may be speed, temperature, pressure, or any other measurable quantity. If the block diagram does not include the actuator components, the signal that is provided by the controller to the unidentified actuator may be called the **control output.**

4. The **feedback signal** from the sensor is sometimes called the **process variable** because it varies proportionally with the measured state of the process being controlled.

THE PLC IN PROCESS CONTROL

In traditional process control theory, the error amplifier simply subtracts the process variable from the setpoint signal and outputs an error signal. All other functions of the traditional process controller are described in the gain box, which will include formulas describing the controller circuit's behavior and its digital computations.

A PLC programmer should think of the PLC as both an **error amplifier** (to calculate the error between the setpoint and the process variable) and as the **controller gain** (to perform additional mathematical operations on that error value). The PLC controller will provide a control signal to an actuator or to an actuator system (as in our motor and motor controller example). The actuator (part of the gain box in traditional process control theory) will never respond exactly proportionally to a PLC's output signal, but the additional functions are outside the control of the PLC program and outside the scope of this book.

A PLC program that will result in very simple **proportional** process control might look like the program shown in Figure 12.4. In this Allen-Bradley ladder logic program, the PLC program subtracts the process variable in address N7:2 from the setpoint in N7:1 to perform the error amplifier function, then performs a controller gain function by multiplying the error by a number from N7:50. Changing the value of the multiplier in N7:50 changes the proportional response of the control system. This example program puts the result in N7:100.

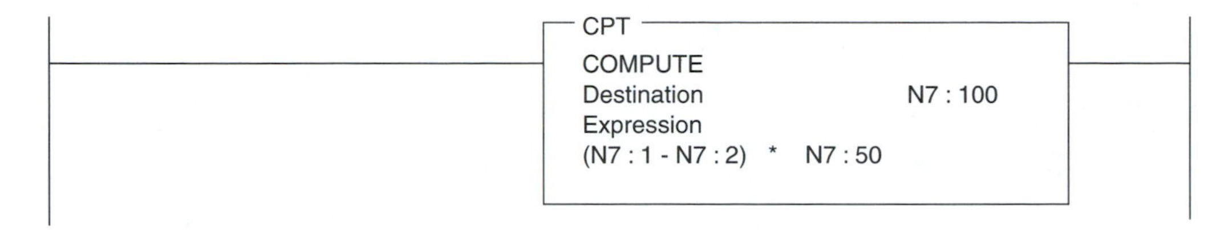

FIGURE 12.4
Very simple PLC-5 process control program.

The process variable value and (sometimes) the setpoint must be read from sensors via analog input modules, and the control variable output value must be written to an actuator system via an analog output module, so the program must include reading and writing of analog I/O modules. If analog I/O modules are directly addressable in your PLC, the process is simple. A STEP 5 STL-language program that executes the simple process control program, and which reads and writes analog I/O module addresses, is shown in Figure 12.5. The analog I/O mod-

```
          L IW072      ; Gets the setpoint value
          L IW074      ; Gets the feedback value
          - F          ; Subtracts feedback from setpoint
          T DW0        ; and stores the error value
          L KY + 0     ; Clear a working data word
          T DWI

          L DW2        ; Get the multiplier
          T DW3        ; and copy it to another working data word
          JZ =out      ; Bypass multiplication if multiplier = 0

add   :   L DW0        ; Add the calculated error
          L DW1        ; to the contents of working data word 1
          + F
          T DW1

          L DW3        ; Get the multiplier working data word
          L KF + 1     ; and decrement it
          - F
          JZ =out      ; Jump out of loop if result is zero
          T DW3
          JU =add      ; Otherwise, return for another loop iteration

out   :   L DW1        ; Get the result of the multiplication loop
          T QW112      ; and write it to an output module
          BE
```

FIGURE 12.5
STEP 5 STL program for simple process control.

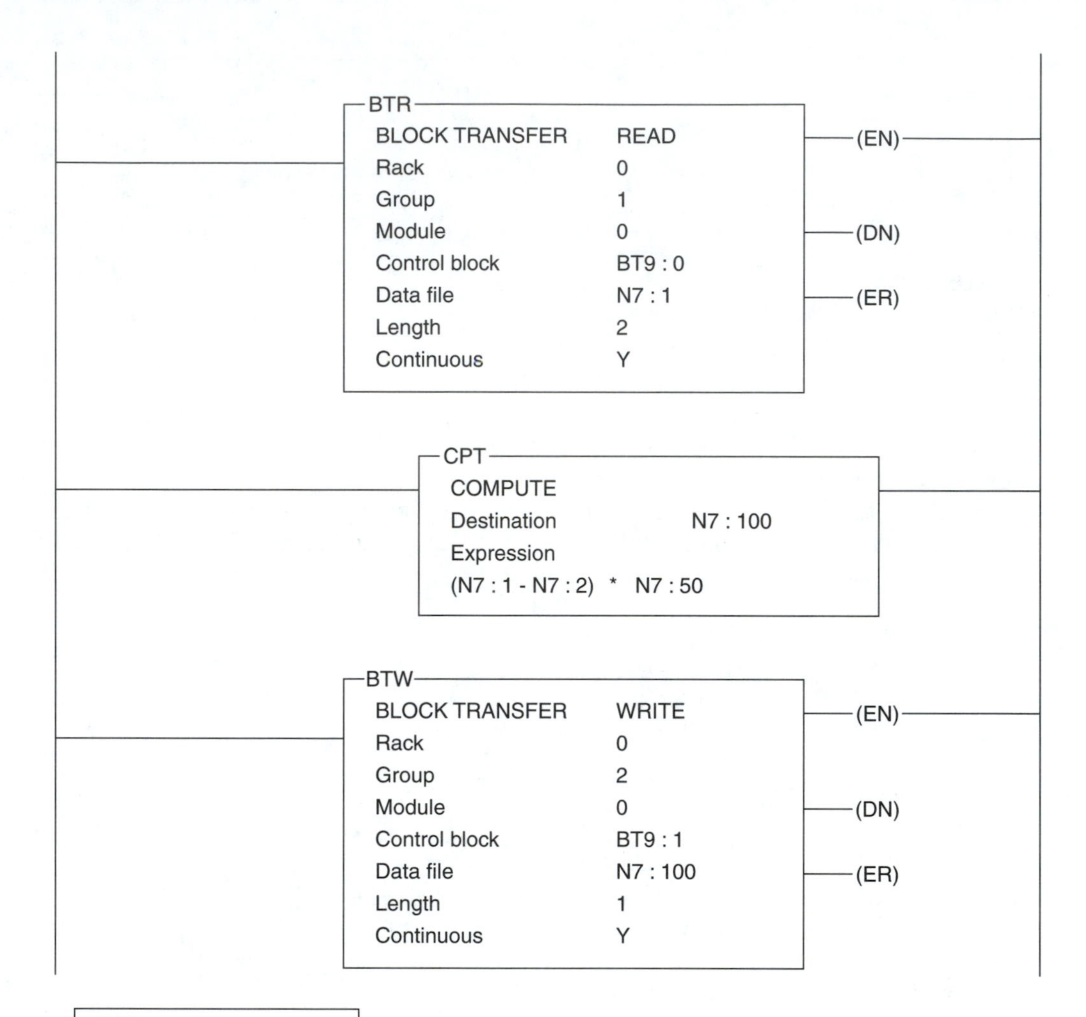

NOTE:
This and other BTR and BTW instructions in this chapter are written as if only analog data values had to be read and written. You will have to adjust programs for your PLC system.

FIGURE 12.6
PLC-5 program for simple process control.

ules for use in an Allen-Bradley PLC-5 system are not directly addressable, so block transfer instructions are required, as shown in Figure 12.6, to read analog values prior to performing the control calculation and to write the analog output value after the calculation.

The STEP 5 program in Figure 12.5:

1. Calculates the error.

2. Multiplies the error by a multiplier. STEP 5 doesn't have a multiplication instruction, so the program uses a loop to add the error value to itself as often as the multiplication requires. (*Note:* To make it more readable, this example program does not include error checking after math operations so it could generate a false result if the limits of the 16-bit number range are exceeded.)

3. Writes the multiplication result to the address of an analog output module.

In the Allen-Bradley PLC-5 program of Figure 12.6:

1. The first rung reads two values from an analog input module in rack 0, group 1, left-hand slot (slot 0). The first of the two values is the digital word representing the analog setpoint value, and it is put into the PLC's integer file 7 as word 1. The second value is the digitized analog value from a sensor, and it is put into the same integer file, but as word 2. The BLOCK TRANSFER READ instruction repeats the reading-in of digitized values as quickly as the analog input module can provide them.

2. The next rung performs the simple process control calculation.

3. The last rung executes a BLOCK TRANSFER WRITE instruction to copy the result of the subtraction from N7:100 to the analog output module that is in rack 0, group 2, in the right-hand slot (slot 0). The BLOCK TRANSFER WRITE repeats as quickly as data can be written to the analog output module.

IMPROVING THE PERFORMANCE OF PLC PROGRAMS FOR PROCESS CONTROL

There are several reasons why a PLC programmer might need to use a more sophisticated PLC program than the simple proportional process control programs shown above:

- The input and/or output values might require scaling.
- The output value might have to be limited to between maximum and minimum values.
- The control program's execution interval may have to be forced to be regular.
- Additional variables might need to be included in the calculations (e.g., to compensate for environmental conditions).
- A more sophisticated algorithm than the simple subtraction and multiplication may be needed (e.g., a PID algorithm or a fuzzy logic algorithm).
- The operator may need to take manual control of the system sometimes.
- The control output may have to be calculated differently under different conditions.

Scaling of the Process Variable and the Control Variable

A process variable value may have to be scaled before it can be compared to the setpoint value. The operation of a controlled system is easier to understand if setpoint values are in **engineer-**

ing units. Position-control engineering units may be inches, for example, so that when the PLC program provides the number 15 as a setpoint, it means that the position controller should drive the load to the 15-in. position. If the process variable number read from the position sensor via the analog input module isn't 15 when the load is at 15 in., scaling is required. Even if engineering units aren't used, scaling may still be required. A setpoint potentiometer, for example, might supply a dc voltage between 0 and 10 V as position setpoints, while the position sensor provides between 0 and 24 V over the same range. Unless process variable values are scaled so that they match setpoint values, the error calculation will be wrong.

Control variable value scaling is done to adjust the output values from the control algorithm so that they match the range of values required at the actuator system. If, for example, an analog output module converts signed 13-bit numbers into the ±10 V dc signal an actuator needs, scaling may be required to convert 16-bit numbers from the control algorithm into 13-bit numbers. A separate scaling calculation may be required to convert the control variable to engineering units for display purposes. Scaling may require two calculations, a multiplication and an addition. **Multiplication** is required to adjust the *span* of the unscaled values, and **addition** is required to offset the adjusted span so that it has the required *range limits.*

Figure 12.7 shows how a process variable might be scaled to match a set of position control setpoints in thousandths-of-an-inch engineering units. The position control system drives the load to positions between −5.000 and +5.000 in. from the center (setpoint range −5000 to +5000). The sensor is a linear potentiometer with a 12-in. stroke, and it outputs a dc voltage of

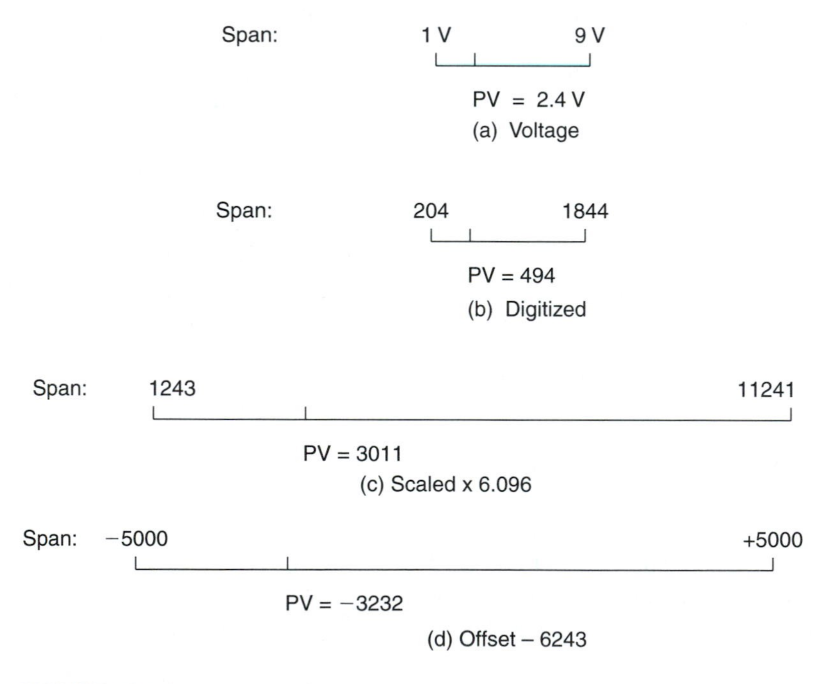

FIGURE 12.7
Process variable, before and after scaling: (a) voltage from position sensor; (b) unscaled values from the analog input module; (c) partially scaled by multiplying by 6.096; (d) scaling completed by adding an offset of −6243.

1 to 9 V dc over the 10-in. movement span of the load. The analog input module converts the 1 to 9 V dc signal to unsigned binary values of between 204 and 1844. Scaling must convert process variable (PV) values in the number range 204 to 1844 to numbers in the number range −5000 to +5000:

1. Multiplying:

$$\begin{aligned} PV \times \frac{\text{output span size}}{\text{input span size}} &= PV \times \frac{-5000 - +5000}{204 - 1844} \\ &= PV \times \frac{-10{,}000}{-1640} \\ &= PV \times 6.096 \end{aligned}$$

Note: These minus signs canceled. They don't always. Be careful not to reverse either span direction during the calculation.

If your PLC can't perform floating-point math, you may have to adjust the output voltage from the sensor so that an integer multiplier can be used.

2. Adding an offset:

$$\begin{aligned} &PV + [(\text{top of input span}) - (\text{top of multiplied output span})] \\ &= PV + [-5000 - (204 \times 6.096)] \\ &= PV + (-5000 - 1243) \\ &= PV + (-6243) \end{aligned}$$

Figure 12.8 shows a rung that could be added to the Allen-Bradley PLC-5 program of Figure 12.6 to scale the process variable value in N7:2 before using it in the process control algorithm. Note that the answer is stored back into N7:2, which is in a memory area where 16-bit signed integer values are stored, after multiplication by a floating-point number. The PLC-5 will perform the compute calculation in floating-point format and will then convert the answer into signed integer format for storage. There will be a small amount of round-off error.

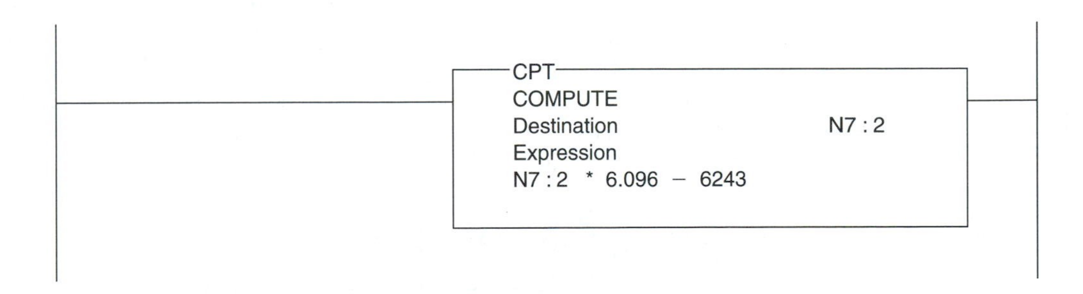

FIGURE 12.8
PLC-5 Ladder Logic rung to scale a process variable value prior to using it in a process control calculation

Some intelligent I/O modules can be configured to perform the scaling of analog values. If you are using a preprogrammed process control instruction that includes scaling (such as the PID instructions that we cover later in this chapter), don't do scaling in the I/O module. Every additional calculation consumes PLC execution time and has the potential to increase error in the result.

Limiting the Control Variable

If a process control calculation's range of control variable output values is too large for the analog output module and/or produces too large a range of analog output signals for the actuator system, scaling could be used to reduce control variables before they are written to the analog output module, but scaling would also reduce the strength with which the actuator system responds to any error. Limiting a control variable's maximum and minimum output value may be a better way to ensure that an actuator system isn't driven harder than it should be, while still allowing the system to respond strongly to small errors. In fact, if limits are imposed after the process control calculation, a large proportional multiplier can be used in the process control calculation to increase the response to small errors, while output limiting would prevent over-driving of the actuator system while error is large.

For example, assume that the 10-in. position control system that we have been using had a motor controller control signal that had to be limited to between +8 and −8 V dc. We could scale the control variable so that the analog output reaches only +8 V dc when the load is exactly −10 in. from where it should be (and −8 V dc at +10-in. error), but that would mean that when the load is 0.010 in. short of its setpoint position, the control system would supply only +0.008 V dc to correct the positional error. Instead, we select a large proportional factor (i.e., 80, which will generate the maximum output when positional error is 0.10 in.) and program the control system to limit the output to the range +8 to −8 V. At 0.010-in. error, the actuator system would receive a 0.8 V dc signal—100 times the strength we calculated if scaling alone was used, yet output will be only 8 V at a 10-in. error.

If in the Allen-Bradley PLC-5 program of Figures 12.4 and 12.6, the result of the process control calculation was being outputted via an analog output module that converted a 13-bit signed number (−2048 to +2047) to a +10 to −10 V dc signal, we would have to limit the size of the number to ±1638 to prevent the analog module from generating voltages over ±8 V dc. The rungs shown in Figure 12.9, inserted after the process control calculation but before the instruction writing to the analog output module, would do the job. (Since it isn't obvious that "1638" represents 8 V dc, this is one place where scaling the calculated control variable into engineering units of volts for display purposes could be used to make a program more readable.)

Preprogrammed PID instructions, offered by most PLCs, usually include output limiting as part of their preprogrammed functionality, as we will see.

Reducing Scan-Time Delay in Process Control

Process control routines need to execute at consistent intervals. Some PLCs offer preprogrammed process control instructions (e.g., PID instructions) that contain built-in timing routines to control their own execution rate. If necessary, the programmer can write a program with a timer to cause the process control calculation to be executed at measured intervals.

Changes values that are below -8 vdc to -8 vdc:

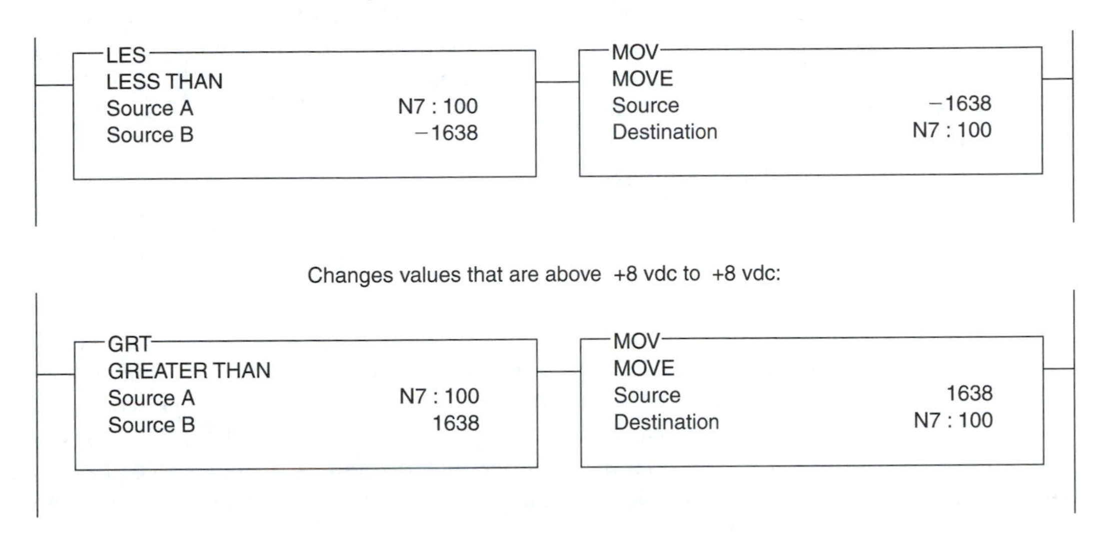

FIGURE 12.9
Limiting the control variable output after calculating the response required.

In some applications, variability in scan-time delay will still be a problem. A PLC executes its user-program one instruction at a time, and there is always a delay before a PLC can respond to a condition that requires control. The delay exists because:

1. The scan cycle creates delays. A PLC reads data from its input modules into input image memory before running the user-written program. After the program is finished, the PLC copies the program results from output image memory to output modules.[1] A sensed condition is always several milliseconds old before the PLC responds, and the response might be wrong by the time it is initiated.

2. PLC programs have variable execution times. Operations controlled by Boolean logic take longer to operate if the logic is true, especially instructions such as jumps to subroutines and file manipulation instructions.

3. Some PLCs scan some I/O modules asynchronously to the PLC scan cycle. I/O scanning is done by a separate microprocessor while the PLC's main processor executes the scan cycle. It is impossible to know how old input data is when the process control calculation is performed, and impossible to know how long before the result will get to the analog output module. The Allen-Bradley PLC-5 uses block transfers to exchange data with analog input and analog output modules. Communication requests initiated by block transfer instruction are queued and executed one at a time in the order in which they are commanded by the program.[2]

---

[1] Newer PLCs can read input data directly from analog input modules during program execution, and write output values directly to analog output modules, eliminating the scan cycle delay in response.

[2] Except for block transfer instructions in selectable timed interrupt (STI) programs or in processor input interrupt (PII) programs, which execute in synchronization with the program because the STI program (or the PII program) halts and waits for the block transfer to be completed before the PLC executes the next instruction.

4. Conversion between analog and digital values is not instantaneous. While an analog I/O module is performing the conversion, the process cannot be controlled. In the program example in Figure 12.10, an Allen-Bradley PLC-5 is programmed to request new sets of input values from an analog input module every time the previous read has completed. The analog input module will refuse to transfer a new set of data to the PLC-5 CPU until it has completed an *entirely new* analog-to-digital conversion of all analog input channels. There will be a delay that is not apparent to the programmer.

A control system that is undependable because of uncontrolled variability in its control interval is said to be **nondeterministic.** Here are some things a PLC user can do to reduce (but not eliminate) the delay and variability:

1. Select a PLC and I/O modules that work at a rate that is appropriate to the processes being controlled. (Not every process needs a fast controller.) PLCs are available that can execute entire programs in under 5 ms, while a more typical PLC's scan time would be about 25 ms. These values vary, of course, with the length of the program and the instructions used in the program. Instruction execution times are often included in PLC manuals so that a user can calculate (or estimate) scan times for programs. Figure 12.11 is an example of timing information for an Allen-Bradley PLC-5.

2. Use intelligent process control I/O modules. Some processes require control that is more real-time than a PLC's CPU can offer, so PLC manufacturers often offer intelligent I/O modules for process control. The modules plug into a standard PLC I/O rack, accept commands from the user-program, provide status information that can be read by a user-program, but perform process control functions using their own microprocessors to control directly attached actuators and sensors. PLC manufacturers are preparing a new generation of PLC CPU modules with built-in coprocessors for process control applications such as motor control.

3. The programmer may be able to write more efficient programs. Programs that use structured programming techniques such as jumps, subroutine calls, and interrupts (all covered elsewhere in this book) allow the programmer to write short programs if the programs are measured in terms of their memory storage requirements, but these programs may take longer to ex-

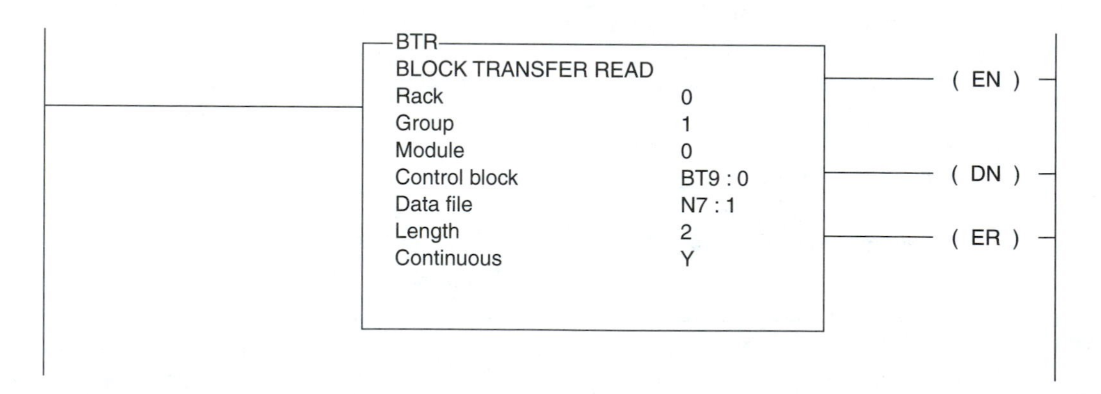

FIGURE 12.10
Allen-Bradley PLC-5 block transfer read from an analog input module.

Category	Code	Title	Execution Time (μs) Integer		Execution Time (μs) Floating Point	
			True	False	True	False
Logic	AND	and	5.9	1.4		
	OR	or				
	XOR	exclusive or				
	NOT	not	4.6	1.3		
Move	MOV	move	4.5	1.3	5.6	1.3
	MVM	masked move	6.2	1.4		
	BTD	bit distributor	10.0	1.7		
Comparison	EQU	equal	3.8	1.0	4.6	1.0
	NEQ	not equal	3.8	1.0	4.5	
	LES	less than	4.0	1.0	5.1	
	LEQ	less than or equal				
	GRT	greater than				
	GEQ	greater than or equal				
	LIM	limit test	6.1	1.1	8.4	1.1
	MEQ	mask compare if equal	5.1	1.1		
Compare	CMP	all	2.48 + (Σ[0.8 + i])	2.16 + Wi[0.56]	2.48 + (Σ[0.8 + i])	2.16 + Wi[0.56]
Compute	CPT					

[1] Use the larger number for addresses beyond 2048 words in the processor's data table.
i = execution time of each instruction (operation, e.g. ADD, SUB, etc.) used within the CMP or the CPT expression
Wi = number of words of memory used by the instruction (operation, e.g. ADD, SUB, etc.) within the CMP or CPT expression
Note: CMP or CPT instructions are calculated with short direct addressing

FIGURE 12.11
PLC-5 instruction execution times.

ecute than physically longer programs that allow the program to execute without jumping. Figure 12.12 shows two STEP 5 STL programs that do the same thing: add the data words of two data blocks and turn an output on if either sum is over 15. (STL has been chosen for this example because STL is very close to the final machine language program.) One of the programs is written in only 15 instructions, while the other needs 20, yet the 20-instruction program will execute faster because it does not include instructions to call other blocks. Compilers for some (non-PLC) programming languages allow the programmer to choose whether programs will be compiled into compact programs or into fast programs. PLC programming software may offer this option in the future.

4. Some PLC programming languages inherently produce more efficient machine language code. High-level graphic language elements, such as ladder logic, must be compiled

(a)

```
OB1
  Main program calls other
  programs:
        R Q7.0 ; Initializes flag
        C DB1  ; First data block
        JU PB1 ; First jump
        C DB2  ; Second data block
        JU PB1 ; Second jump
        BE
  PB1
  Subroutine in short program:
        L DW1
        L DW2 ; Processes data
        +F    ;   from either data
        L DW3 ;   block
        +F
        L  KY + 15
        >F
        S  Q7.0
        BE
```

(b)

```
OB1
  Whole program in one main
  program:
        R Q7.0 ; Initializes flag
        C DB1
        L DW1  ; processes data from
        L DW2  ;   first data block
        +F
        L  DW3
        +F
        L  KY+15
        >F
        S  Q7.0
        C  DB1
        L  DW1 ; Processes data from
        L  DW2 ;   second data block
        +F
        L  DW3
        +F
        L  KY+15
        >F
        S  Q7.0
        BE
```

FIGURE 12.12
(a) Short versus (b) fast programs.

(translated) into sequences of machine language instructions before the PLC can execute them. The translations usually include instructions that aren't necessary. Figure 12.13a shows a ladder logic element for a Siemens S5 on-delay timer with three contacts that the programmer has elected not to use. Figure 12.13b shows the STL equivalent, demonstrating that it requires one extra STL instruction[3] for each of the unused features. If the programmer writes the program directly in Siemens' STL programming language, the three NOP 0 instructions can be left out, as in Figure 12.13c.

5. Immediate I/O instructions (also called *direct access I/O*) can reduce the delay between when a process variable's current value is read and when the PLC outputs a new control variable. As we saw in Chapter 11, immediate I/O instructions cause the PLC to read input values directly from an input module and to write output data directly to an output module. Using immediate I/O instructions doesn't totally eliminate the response delay, because the PLC still

[3] "NOP 0" means "no operation." Siemens' compiler is "smart" enough to minimize, but not eliminate, the unnecessary machine language code for the unused functions.

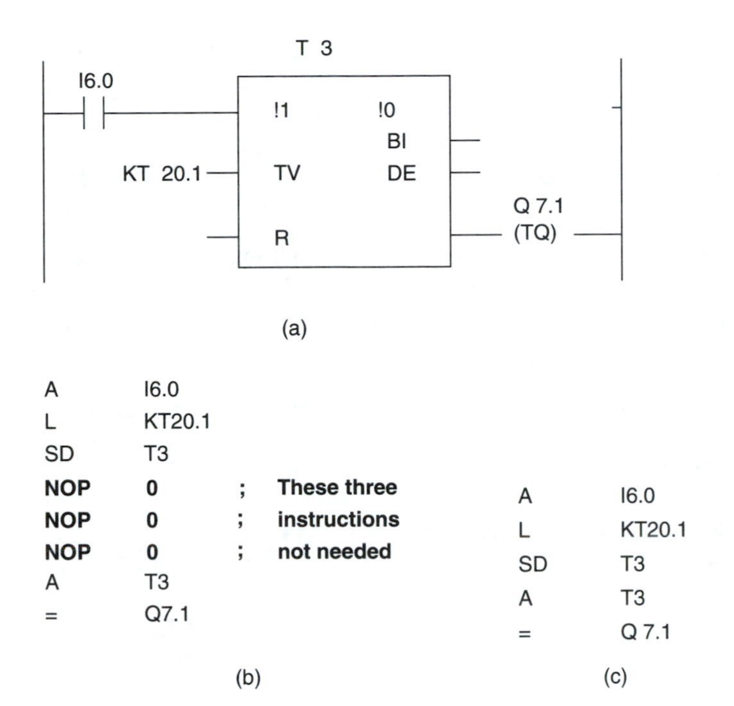

FIGURE 12.13
STEP 5 timer instruction: (a) programmed in Ladder Logic; (b) STL equivalent; (c) programmed efficiently in STL.

has to execute the process control calculations (including operations such as scaling and limiting) to calculate the control variable value, and analog I/O modules still require time to convert between analog and digital values.

IMMEDIATE I/O FOR PROCESS CONTROL IN THE ALLEN-BRADLEY PLC-5

The PLC-5's immediate I/O instructions aren't useful for analog process control. They can only read or write digital I/O modules, and they only read or write I/O modules immediately in the same rack as the processor. The BLOCK TRANSFER READ and BLOCK TRANSFER WRITE instructions (BTR and BTW) that read/write analog I/O modules can be made to execute immediately if they are programmed into interrupt service routines such as timed interrupt (STI) routines.

IMMEDIATE I/O FOR PROCESS CONTROL IN THE ALLEN-BRADLEY SLC-500

In the SLC 500 series of Allen-Bradley PLCs, analog I/O modules can be read from or written to using the IMMEDIATE INPUT WITH MASK (IIM) and the IMMEDIATE OUTPUT WITH MASK (IOM) instructions. The specialty I/O module can be addressed, if necessary, using M0 and M1 module addresses in standard instructions.

S5

IMMEDIATE I/O FOR PROCESS CONTROL IN THE SIEMENS S5

Siemens S5 PLCs allow direct access addressing as a way of forcing immediate I/O.[4] Programs are written the same as when using data from the input and output image tables, except that instead of addresses having the prefix I or Q, they have the prefix P (e.g., PW072 instead of QW072). The program in Figure 12.15 (in the next section) includes direct access addressing.

S7

IMMEDIATE I/O FOR PROCESS CONTROL IN THE SIEMENS S7

Siemens S7 PLCs allow peripheral I/O addressing as a way of forcing immediate I/O. Analog modules can only be addressed using peripheral I/O addressing. Programs are written the same as when using data from the input and output image tables, except that instead of addresses having the prefix I or Q, they have the prefix PI or PQ (e.g., PQW072 instead of QW072).

CQMI

IMMEDIATE I/O FOR PROCESS CONTROL IN THE OMRON CQM1

In a CQM1 program, the I/O REFRESH instruction **IORF (97)** can be used to immediately read from or write to I/O modules, including analog I/O modules, within the range of addresses specified with the instruction. If the PLC is configured for direct refresh (see Chapter 11), all output area memory words are copied to output modules as soon as they are changed by the program.

Timed Interrupts

As we saw in Chapter 11, some PLCs can be configured to execute interrupt service routines at timed intervals. Process control programs programmed in timed interrupt service routines provide very deterministic control of processes. Timed interrupt programs using immediate I/O instructions are both deterministic and quick responding.

PLC-5

TIMED INTERRUPTS FOR PROCESS CONTROL IN THE ALLEN-BRADLEY PLC-5

An Allen-Bradley PLC-5 can be configured to run one selectable timed interrupt (STI) program file at timed intervals. Block transfers (BTR and BTW) of data between a PLC-5 CPU and analog input/output modules usually occur asynchronously with the scan cycle after execution of a

---

[4] In the S5-103U PLC, direct access addresses can only be used in timed interrupt service routines (OB13) and in hardware interrupt service routines (OB2), or in program blocks called during the execution of those interrupt service routines.

block transfer instruction, but BTR and BTW instructions in timed interrupt programs take precedence over other block transfer requests that might still be waiting in the PLC's queue, and an STI program stops and waits for the block transfer to finish before the next instruction in the STI program is executed. The programmer should ensure that the STI program isn't called at a faster rate than the analog input module can convert analog input values into binary, of course. Figure 12.14 shows a BLOCK TRANSFER READ instruction in an STI program file. The instructions following the BTR only execute when the BTR completes.

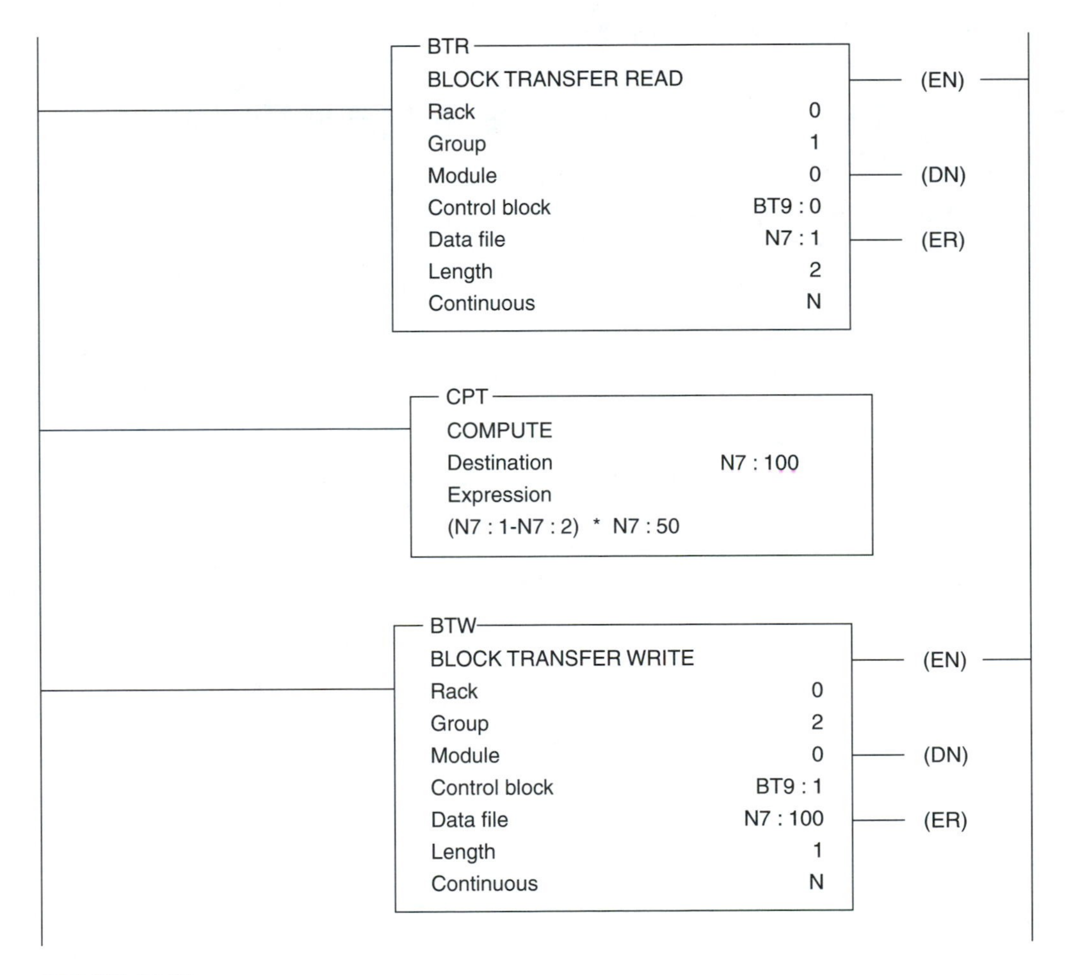

FIGURE 12.14
PLC-5 selectable timed interrupt program file that performs a process control function at timed intervals.

Block transfers between the PLC-5 CPU and I/O modules in remote I/O racks are significantly slower than when the I/O module is in the same rack as the CPU, so process control applications will work best if analog I/O modules are installed in the CPU-resident rack.

SLC 500

TIMED INTERRUPTS FOR PROCESS CONTROL IN THE ALLEN-BRADLEY SLC 500

An Allen-Bradley SLC 500 can be configured to run one selectable timed interrupt (STI) program file at timed intervals. Process control programs requiring consistent intervals between executions should be written into an STI file. The SLC's preprogrammed PID instruction can be configured so that it will execute properly when programmed into an STI program file, or it can be configured to use its own built-in timer to control its execution interval, in which case it should not be programmed into an STI.

S5

TIMED INTERRUPTS FOR PROCESS CONTROL IN THE SIEMENS S5

Siemens S5 PLCs automatically call the program in organizational block 13 (OB13) at timed intervals, with the interval defined by the value stored in reserved system word 97 (RS97), as we saw in Chapter 11. A process control program using OB13 would look like the example in Figure 12.15.

Refer to Chapter 11 for a detailed explanation of how S5 PLCs perform timed interrupts and what they do when a timed interrupt program contains direct access addressing ("PW" in

```
FB1
        Reserved system words can only be written to in function blocks, so this function block
        should be called by the user-program as the PLC starts.

        L KF 40      ; Puts the decimal valve (KF) 40 into accumulator 1
        T RS 97      ; Transfers this valve to reserved system word 97, setting up timed interrupts to occur
                     ;    every 40 * 10 ms intervals; if no value is written to RS97, its default value is 10, so timed
                          interrupts occur every (10 * 10 ms = ) 100 ms
        BE           ; Terminates FB1

OB13
        Any program in this block will automatically execute at timed intervals, with the intervals defined
        by the value in RS97 (every 400 ms now).

        L PW072      ; Reads two values from analog input modules,
        L PW074      ;    using direct access addressing
        JU PB1       ; Assume PB1 is the process control calculation
        T PW112      ; Writes an output valve to an output module
        BE
```

FIGURE 12.15
STEP 5 program using timed interrupts.

the addresses instead of "IW" and "QW"). Large S5 PLCs work differently from medium-sized S5 PLCs such as the S7 103U.

TIMED INTERRUPTS FOR PROCESS CONTROL IN THE SIEMENS S7

Most S7 PLCs offer cyclic interrupts that can be configured to interrupt the main program scan at timed intervals. The interval times are defined during configuration of the S7 CPU module (see Chapter 10). The program in OB 30 is the cyclic interrupt routine.

CQMI

TIMED INTERRUPTS FOR PROCESS CONTROL IN THE OMRON CQM1

The CQM1 offers both cyclic interrupt capability and the possibility of configuring the main scan cycle to execute at a minimum scan time, so the programmer can put a process control program either into a cyclic interrupt subroutine or in the main program. OMRON's preprogrammed PID instruction contains a built-in timer that conflicts with timed interrupt use, so do not program an OMRON PID instruction into an OMRON timed interrupt.

With some PLC systems, it is possible to configure I/O modules to allow or initiate data exchanges with the CPU module only at timed intervals or only under specific sensed conditions. I/O interrupts can often be configured so they will execute to run a process control program whenever an I/O module detects an event that requires immediate execution of the control program.

PLC-5

ALLEN-BRADLEY PLC-5 I/O MODULE-CONTROLLED DATA EXCHANGES

In the Allen-Bradley PLC-5 system, some analog input modules can be configured to convert analog input values into digital numbers at specified timed intervals and to ignore BLOCK TRANSFER READ (BTR) requests from the CPU module until the data has changed. If the PLC's main control program (MCP) contains a continuous BTR instruction, the input module will control the timing of the passing of data to the CPU. If the MCP only executes the process control calculations each time new data is received from the input module, the input module also controls the interval of execution of the control program. Since the execution intervals of MCPs are not controlled, and since block transfer requests from MCP programs must share the queue with other block transfer requests, the control program won't be as deterministic as an STI program, but for some control applications the degree of deterministic behavior should be adequate.

Figure 12.16 shows the configuration screen the programmer will see while configuring an analog input module to allow block transfer instructions at timed intervals. Entering a real-time sample rate of 0.5 would cause the module to hold each BTR request until 0.5 s have passed since the preceding block transfer. The instruction that is needed to initiate a BTR is also shown. The continuous BLOCK TRANSFER READ instruction initiates a new request as soon as the

1771 - IFE/A Module Edit R-G-M: 0-0-0

input type : single-ended

data format : BCD

digital filter : 0.00 seconds

real time sample rate : 0.0 second(s) (0 = disabled)

Press F9 (Toggle) to change selection.

Rem prog mod 1 of 4 Addr#12:TEST

Change Mode	I/O Overview	Channel Edit	Monitor	BT Data Tables	Default Config	Toggle	Accept
F1	F2	F3	F5	F6	F8	F9	F10

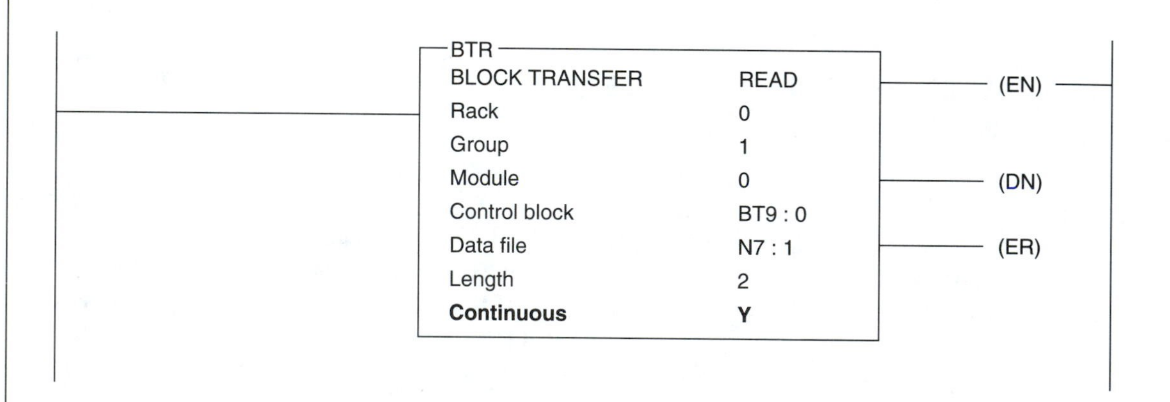

FIGURE 12.16
Configuration screen for a PLC-5 analog input module for timed release of data for block transfers, and the BTR instruction requesting continuous block transfers.

```
                         Processor Configuration

User Control Bits          00000000 00000000   RESTART LAST ACTIVE STEP
Fault routine prog file no.:      0            Watchdog (ms):                    500
I/O status file:                  0            Communication time slice (ms):      3
VME Status File:                 N34

Processor input interrupt    prog file no.:      0     module group:   0
                             down count:         0
                             bit mask:         00000000 00000000
                             compare value:    00000000 00000000
```

FIGURE 12.17
Event-driven interrupt configuration for process control using PIIs in the PLC-5.

previous block transfer read has been completed, but the module waits until the 0.5 s is up before providing the data.

Continuous block transfer instructions can't be programmed in interrupt service routines, but if an STI routine contains a (noncontinuous) BTR instruction that tries to read input values from an analog input module that has been configured to generate digitized values at intervals that are longer than the STI interval, the analog input module will force the STI program to wait for data each time the STI executes. Although there will be some wastage of time every time the process control program executes, the process control program will have maximum deterministic behavior since even the input sampling interval is tightly controlled.

In the PLC-5, event-driven interrupting of the program can be used to force the program to synchronize with external events, as detected by input modules. In the example in Figure 12.17, the PLC-5 has been configured to interrupt its scan cycle immediately to run program file 6 as a process input interrupt (PII) program whenever bit 5 of module 4 in rack 1 goes true twice. Rockwell automation recommends against using immediate I/O or block transfer instructions in PII programs, so using a PII to contain an event-driven control program has very limited value.

SLC 500

ALLEN-BRADLEY SLC 500 I/O MODULE-CONTROLLED DATA EXCHANGES

An SLC 500 PLC can be configured to respond to one or both of two types of interrupts from I/O modules. As described in Chapter 11, there are a discrete input interrupt (DII) and an I/O interrupt that can be initiated by specialty I/O modules. Either type of interrupt can be used to initiate a process control program.

S5

SIEMENS I/O MODULE-CONTROLLED DATA EXCHANGES

S7

Event-driven interrupt programming is available in Siemens S5 and S7 PLCs. Input modules capable of generating interrupt signals must be installed. Once installed and configured, I/O modules can be configured to initiate hardware interrupt signals under selected conditions (depending on the module). The S5 PLC executes the program in organizational block 2 (OB2) when a hardware interrupt signal occurs (if the PLC program hasn't given the command to disable interrupts), and S7 PLCs execute the program in OB 40. The programmer can write the same types of programs in an event-driven interrupt service routine as can be written in a timed interrupt service routine, including using immediate I/O addressing.[5]

Inclusion of Other Values in the Calculation of Output

The process control calculation examined so far in this chapter provides control output values proportional to the difference between the setpoint and the process variable. When error is zero, the control variable result is also zero. There are situations where control can be improved by including an offset. With an offset, the output does not go to zero when the calculated error is zero.

A control output signal can be offset by adding a *bias value,* a *disturbance value,* or a *feedforward value* during calculation of the control variable. Bias, disturbance, and feedforward values are used in similar ways, and the names are often used interchangeably, but each implies a slightly different reason why the offset is required.

Bias Value A bias value is a fixed value that is added during calculation of the control variable output value in a control system, usually after all other calculations have been performed. Figure 12.18 shows a situation requiring bias. A dc motor's speed is being controlled, and a speed sensor provides a process variable proportional to the motor's actual speed. In Figure 12.18a, without bias, the PLC controls the speed of the motor by providing a control variable output signal calculated as the speed error times a proportional gain (G). The control variable output controls motor torque via a motor controller. Control will be faulty because when the motor is running at exactly the right speed, the control variable will become zero, the motor's torque will become zero, and the motor's load will make it slow down.

To allow the speed controller in Figure 12.18a to work correctly, a speed bias value is needed. In Figure 12.18b, a bias value is provided by an operator-set potentiometer. The operator adjusts the pot when the motor is running, to compensate for energy losses, then leaves the pot alone. While the motor continues to run at the right speed, the error value will be zero and the control variable outputted by the PLC is equal to the bias signal's value, just large enough to hold the motor speed. If the speed sensor indicates that the motor's speed has deviated from the setpoint, the control variable signal will deviate from the bias level by an amount proportional to the speed error. A Siemens STEP 7 STL language example is shown in Figure 12.19 to demonstrate how a PLC program can add bias.

Disturbance Value A disturbance value, like a bias value, is used to modify the calculated control variable output value in a control program, often after all other calculations have been performed. A disturbance value typically comes from a sensor in the controlled system's environment. (Disturbance variables are sometimes called *environmental variables.*) The temperature control application in Figure 12.20a has a temperature sensor in the tank with the

---

[5] When a hardware interrupt occurs in an S5-103 PLC, the PLC only reads data from the input modules in slots 0 and 1 into the process interrupt image table.

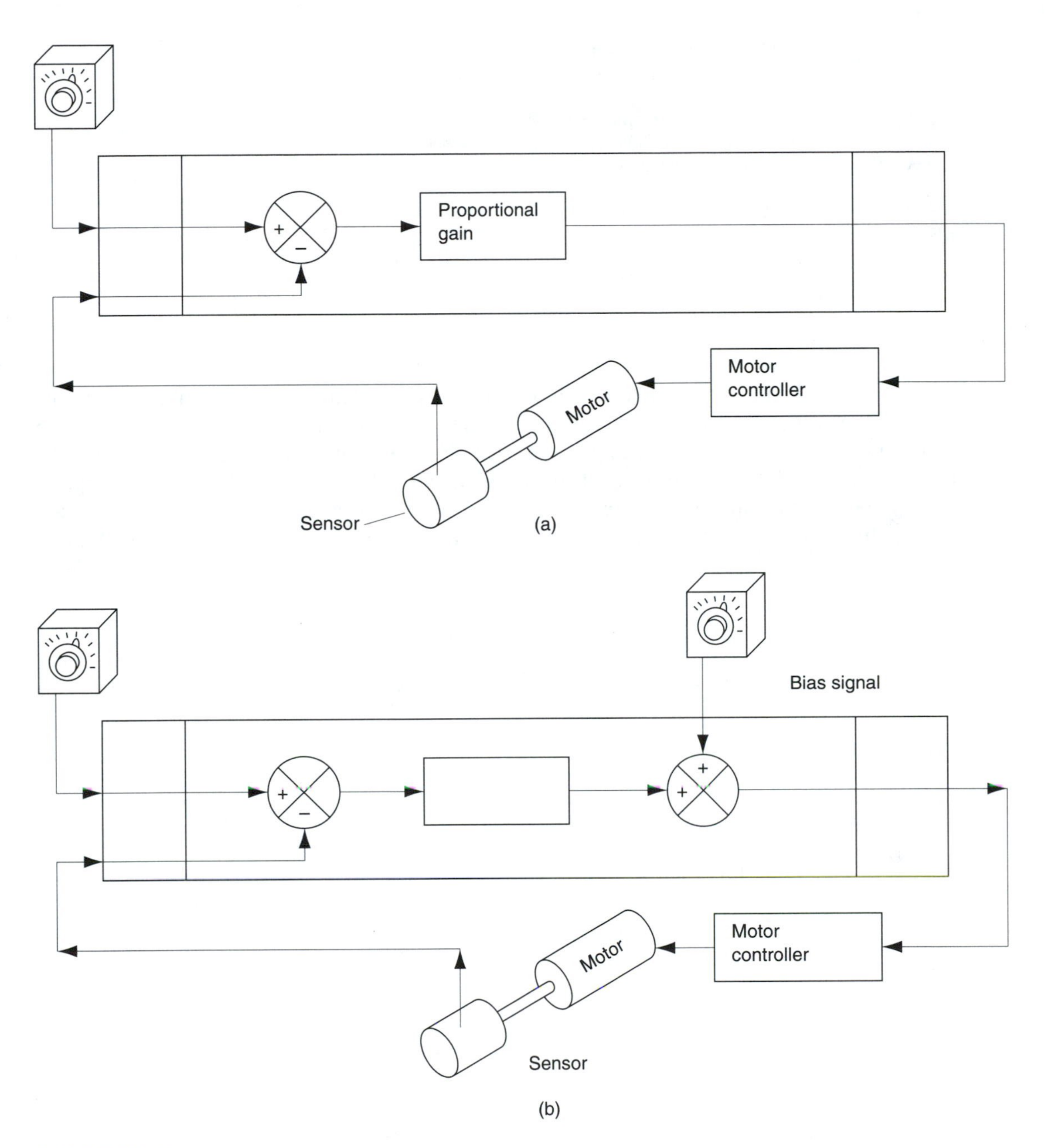

FIGURE 12.18
Motor speed control: (a) without bias; (b) with bias.

```
L PIW072        // Read integer setpoint
L PIW074        // Read integer process variable feedback
-I              // and calculate the error
L DW4           // Load integer proportional constant
*I              // and calculate the proportional control variable
L PIW076        // Read a bias value
+I              // and  add it to  the control variable
T PQW112        // Write final result to actuator
BE
```

FIGURE 12.19
STEP 7 STL program using bias.

liquid being heated and an additional temperature sensor outside the tank. Heat loss from the tank will be proportional to the square of the difference between the room temperature and the tank temperature, so the heater must generate extra heat to compensate for the heat that will be lost. Here a disturbance value is calculated as the square of the difference between the setpoint and the room temperature (the energy needed to hold the tank temperature while it is at its setpoint value). An Allen-Bradley PLC-5 program using the extra sensor in calculation of a dis-

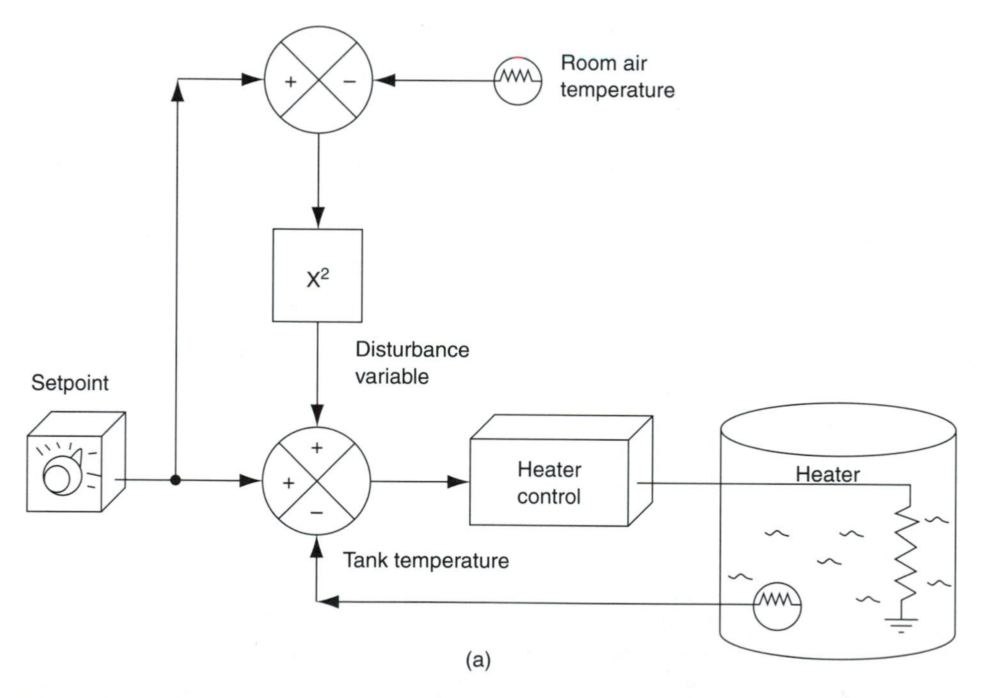

FIGURE 12.20
(a) Temperature controller, with compensation for environmental temperature. *(continued)*

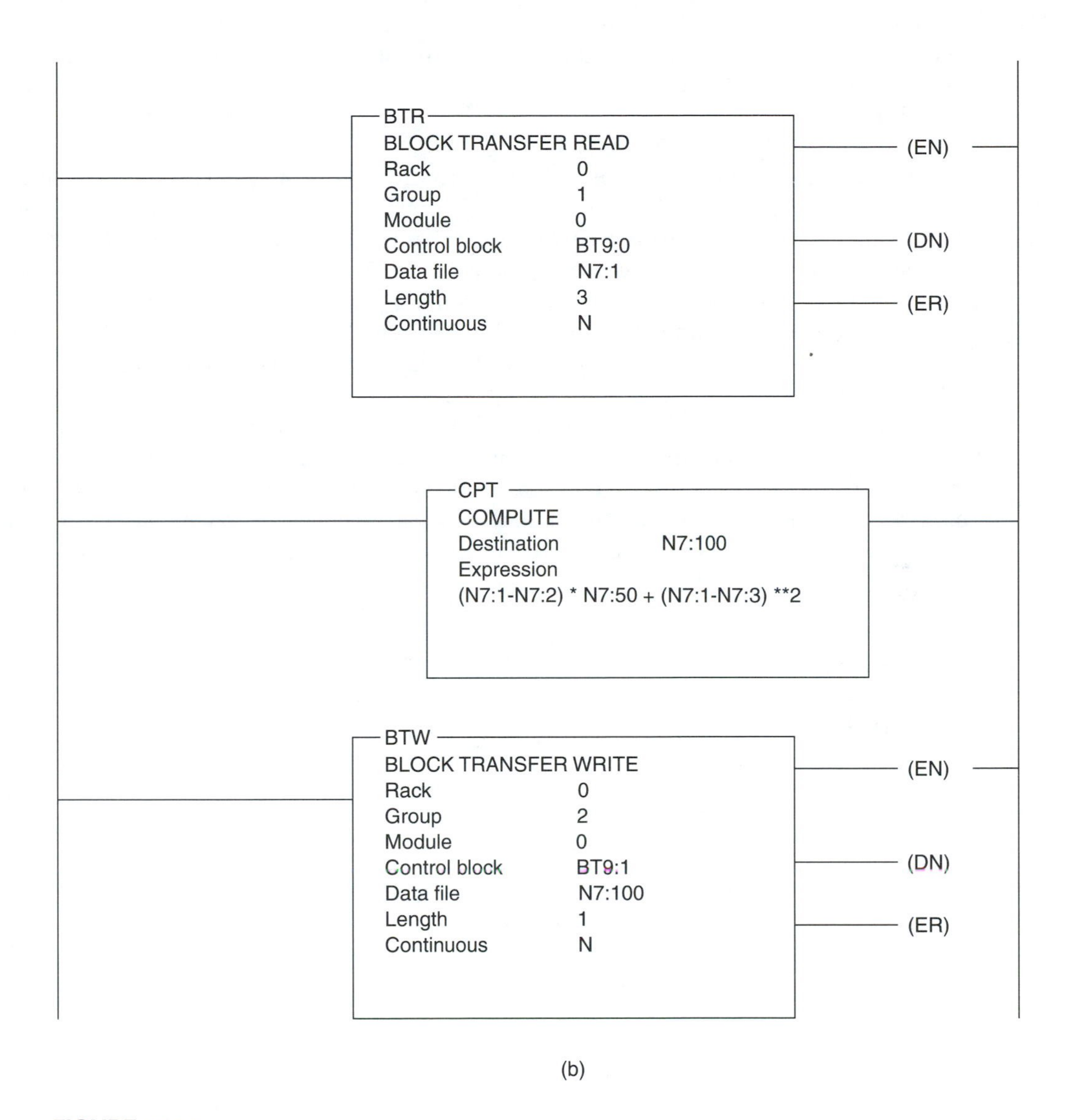

FIGURE 12.20
(continued) (b) PLC-5 program with disturbance offset calculated using environmental temperature input.

turbance offset value might look like the one in Figure 12.20b. A third analog input value is being read from an outside temperature sensor, and the process control calculation now includes

$$+(N7{:}12 - N7{:}3)^{**}2$$

This new formula addition adds a disturbance value that is calculated as the square of the amount the setpoint temperature (N7:1) exceeds the temperature of the air (N7:3) surrounding the tank. For simplicity, we assume that the air temperature never exceeds the setpoint temperature. (Another multiplier will actually be needed to adjust for insulation and heating system strength.)

Feedforward Value A definition for feedforward is not universally agreed on. In some literature (e.g., Allen-Bradley manuals) *feedforward* is used interchangeably with *bias* and actually means bias as described earlier. Other controller manufacturers define feedforward as something similar to an environmental variable, except that the feedforward sensor detects changes in incoming material. If changes in incoming material are sensed, the control system can anticipate error and correct for it before the error actually exists. When used this way, feedforward values are included in the error calculation before other process control calculations are performed. Figure 12.21a shows an example where feedforward might be used. The temperature control system has to control water temperature and must adapt to variations in incoming water temperature, so it is necessary to have a sensor detecting the temperature of the incoming water. Adjusting the heater output to match the incoming fluid temperature is crucial in this example because once water is past the heater, it is too late to change its temperature! (In fact, this example doesn't even include a sensor to detect final water temperature.) The control variable

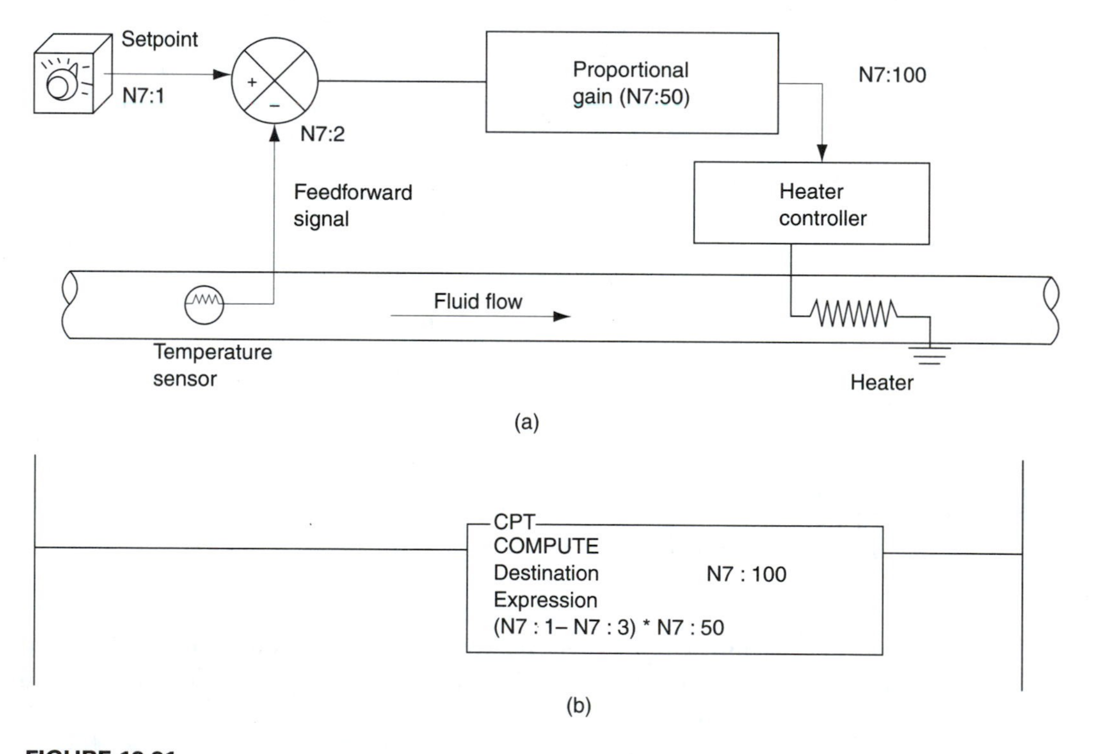

FIGURE 12.21
(a) Temperature control system with feedforward offset. (b) PLC-5 instruction calculating the output using a feedforward value.

signal sent to the heater is proportional to the difference between the setpoint and the temperature of the incoming water. The process control formula used in a PLC-5 program might look as shown in Figure 12.21b. The (N7:1–N7:3) feedforward term in the PLC-5 instruction increases as the temperature of incoming water decreases. (Control systems with feedforward can also include feedback.)

Sophisticated Process Control Routines

PID Control Sometimes, the inclusion of offset values in calculating the output value still won't yield satisfactory results. The controlled system may exhibit undesirable quantities of steady-state error or overshoot error. Using a PID equation to calculate the control variable can remove or reduce these error types.

Steady-state error, sometimes called **residual error** or **following error** (when the controlled system has a changing setpoint), shown in Figure 12.22, exists if the control system can't eliminate small differences between the setpoint and the actual system's measured output. A proportional control system will always exhibit some steady-state error because there will always be a range of error values where the control variable output is too small to cause the ac-

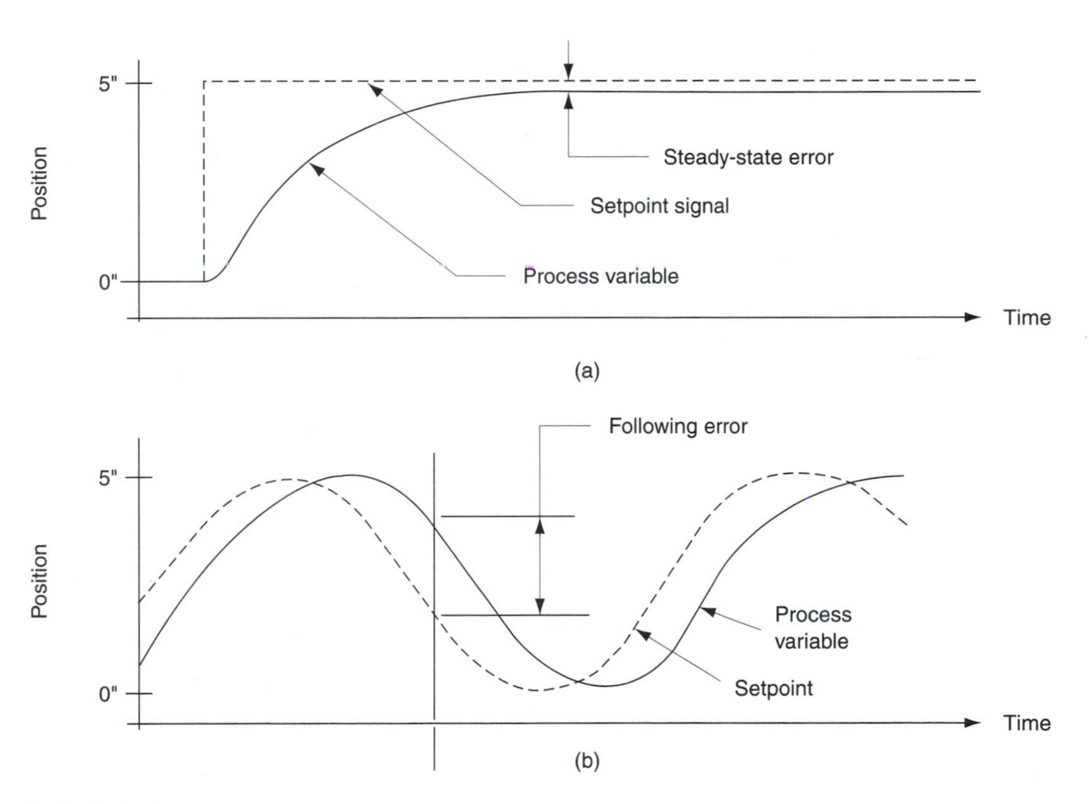

FIGURE 12.22
(a) Steady-state Error; (b) following error.

tuator to generate enough power to overcome resistance. A motor, for example, might not generate enough torque to overcome friction after it has moved a carriage to within 0.002 in. of the setpoint. The 0.002-in. error, which the control system can't correct, is steady-state error.

Overshoot, also called **dynamic error,** sometimes responsible for **oscillation** and sometimes leading to **instability,** is shown in Figure 12.23. Overshoot occurs if a system has too much response power and/or if capacitance and inertia in the controlled system is so large that the controlled system's output keeps changing even after the desired output is reached. If the reader were to swing a full wine bottle toward a brick wall at full speed, and began to stop the bottle at 2 in. from the wall, the undesirability of overshoot would become obvious. Trying the same maneuver at a lower speed, or with a nearly empty bottle, or underwater where friction helps slow the motion, will reduce overshoot and the resultant mess. If a control system receives a setpoint signal, disturbance signal, or feedforward signal that is oscillating at the controlled system's resonant frequency, overshoot might actually increase with each cycle, making the system unstable.

Many modern PLCs offer **preprogrammed PID process control routines** that can be called by the main program. **PID** stands for **proportional–integral–derivative,** which are the three types of calculations they can include in generating a control variable output value based on an error calculation's result.

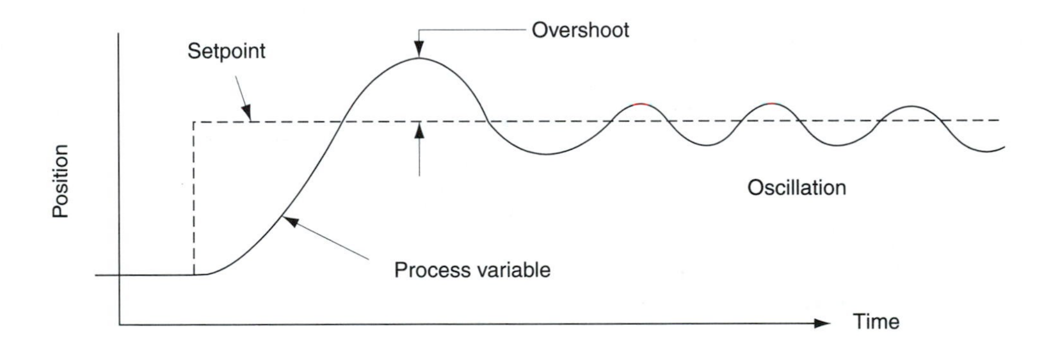

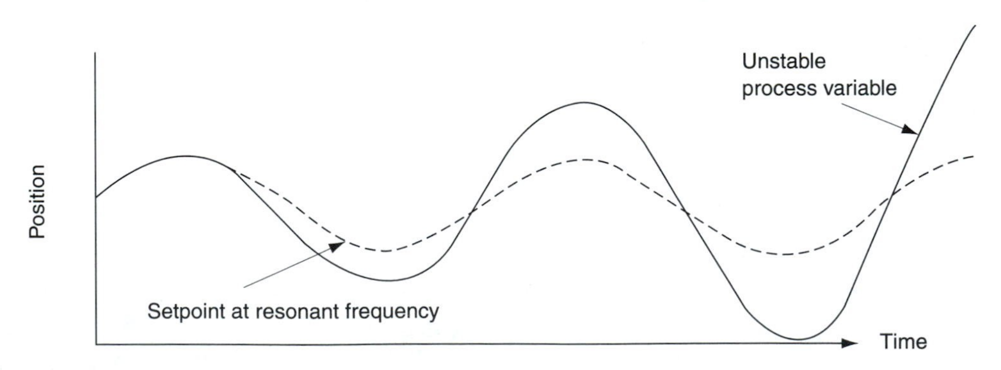

FIGURE 12.23
Overshoot, oscillation during settling, and oscillating instability.

Figure 12.24 shows the PLC-5 control program example that we have seen above but with the COMPUTE instruction replaced by a PID instruction. (We use the PLC-5's PID instruction in this introduction to PID instructions because it is visually uncomplicated, although making it work is a little more difficult than the instruction makes it look.) The PLC-5's PID instruction includes the source address for a process variable that is a value read from a sensor via an analog input module and generates a control variable that is the value that will be written to an actuator via an analog output module. We discuss the control block and tieback values later.

The PLC-5 program shown in Figure 12.24 should be programmed into a selectable timed interrupt (STI) so that it repeats at regular intervals. Like the earlier programs we have exam-

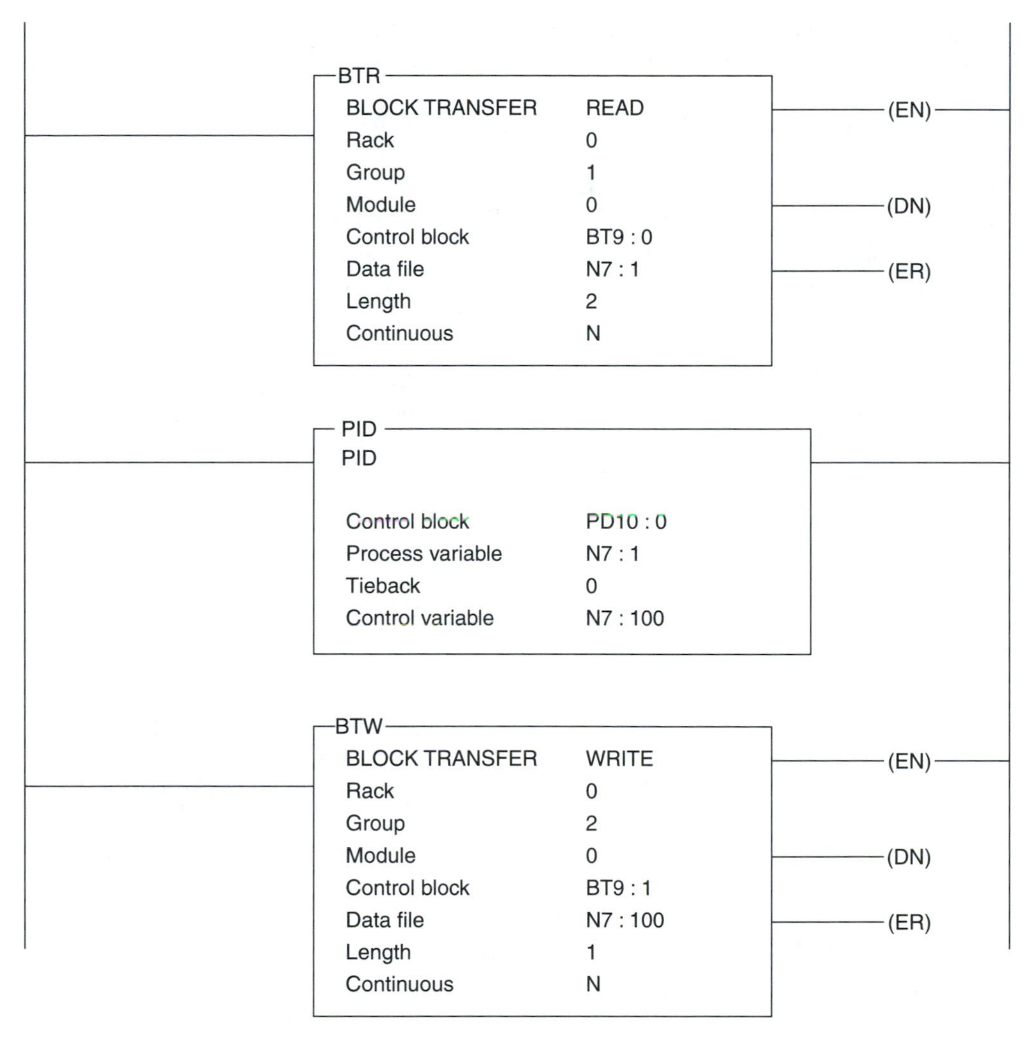

FIGURE 12.24
PLC-5 program with a PID instruction.

ined, the program in Figure 12.24 includes a BTR instruction that reads the process variable from an analog input module, then the PID instruction calculates the control variable, and finally, a BTR instruction writes the result to an analog output module.

At the opening to this chapter we examined a block diagram (Figure 12.2) for a closed-loop control system in which a PLC calculated error and multiplied the error by a proportional constant. PLC programs that call preprogrammed PID routines can do significantly more, as shown in Figure 12.25. Besides adding integral and derivative calculations to the proportional calculation, PID routines often allow offset values to be added, output values to be limited, and sometimes do scaling of process variable and/or control variable values. We will also see that the integral and the derivative calculations make it advisable to include other refinements.

What the PID Instruction Does The proportional control calculation examined so far performs the following operation:

$$\text{CV} = K_p \times \text{error}$$

where CV is the control variable (from the PLC), K_p is the gain in controller (in the PLC program), and error = setpoint − feedback (in the PLC program). The control variable is converted to an analog signal, then supplied to an actuator system, which may amplify the signal and which converts the signal to a mechanical force. Physical characteristics, such as friction and inertia, can affect the control system's final output. Figure 12.26 shows a system in which a dc motor is driven to control the position of a carriage on a screw feed. The setpoint signal and the process variable signals indicate the desired and actual carriage positions. After a large change in the position setpoint, the positional error will be large, so the motor will develop a high torque to drive the carriage toward the desired position. As the carriage nears its destination po-

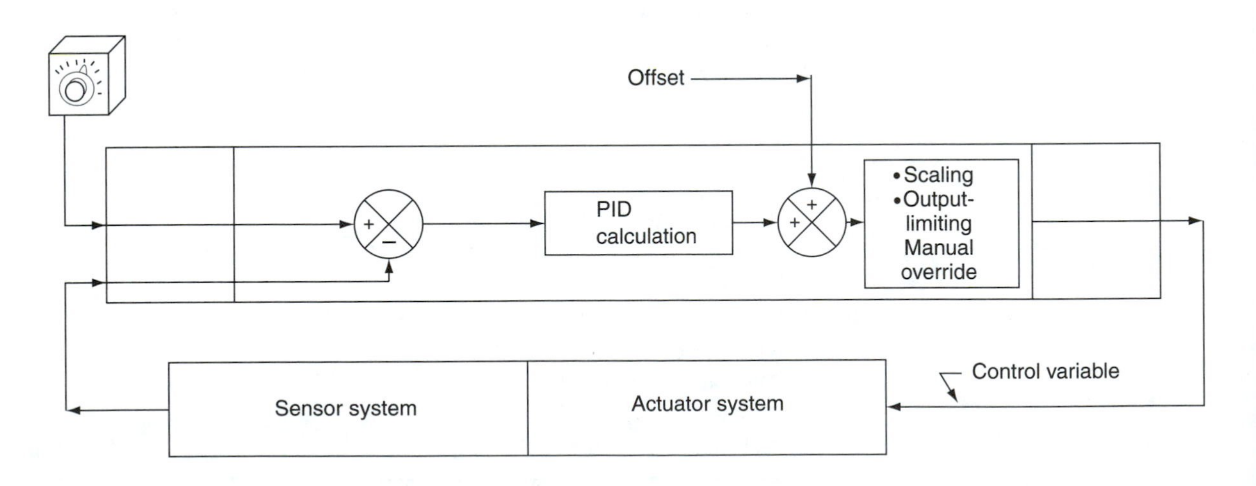

FIGURE 12.25
Control system block diagram, including a PLC running a PID routine.

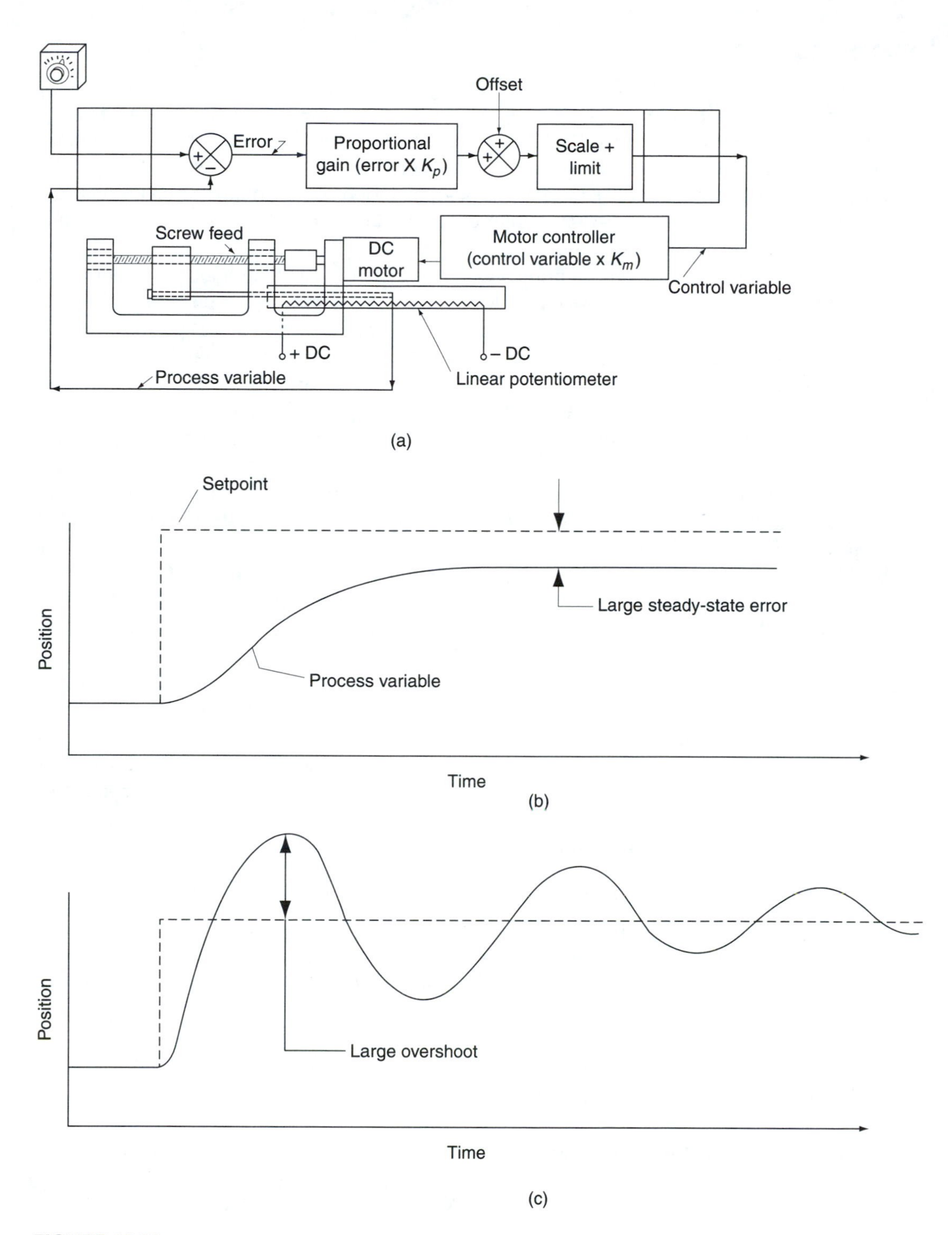

FIGURE 12.26
Response of a proportionally controlled system after a unit step change in setpoint: (a) controlled system; (b) response with low $K_p + K_m$; (c) response with large $K_p + K_m$.

sition, error reduces, so motor torque will be decreased and the carriage will slow as it approaches the setpoint position. Eventually, the carriage will stop, just short of the setpoint, because there won't be sufficient motor torque to overcome friction losses. Steady-state error remains. Steady-state error can be reduced by turning the proportional amplification up. This can be done by increasing the power at the motor controller or by installing a more powerful motor, or it can be done in the PLC program.

Unfortunately, increasing proportional gain also increases the likelihood of the other type of error that proportional control systems often exhibit: **overshoot.** In our position control example, if K_p were increased, the motor would generate more torque and more speed in response to error (within the saturation limits of the components and within the programmed limits of the PLC, of course). The rotor of the motor, the screw feed, and the carriage would all contain more kinetic energy as the carriage is driven harder toward its setpoint position. After the error goes to zero, inertia may cause the carriage to coast past (overshoot) the setpoint position. Overshoot is reduced by reducing K_p.

A controlled system's characteristics can be described in terms of capacitance and resistance. **Capacitance** is a measure of the energy-absorbing ability of the system being controlled, which makes it difficult for the controller to change the system's state (e.g., inertia due to kinetic energy in our carriage positioner, tank size in a water heating system). **Resistance** is a measure of the energy-loss rate in the controlled system (e.g., friction absorbs kinetic energy from the carriage positioner, heat radiates to the environment from a water heater tank).

In some controlled systems, the balance of capacitance and resistance is such that a K_p value can be found so that the system will have acceptably small values of both steady-state error and overshoot. In other systems, there may not be one acceptable K_p value to reduce both types of error, or the system's characteristics may change as the system runs so that the K_p that is required also changes. When proportional control isn't good enough, it may be necessary to add integral and/or derivative control in the calculation of output values. Integral control reduces steady-state error. Derivative control reduces overshoot. A formula for proportional plus integral plus derivative control would look like this:

$$\text{CV} = K_p \times \text{error} + K_i \times (\text{sum of error}) + K_d \times (\text{rate of change of error})$$

where CV is the control variable output value provided to an actuator by the PLC, K_p is the proportional gain (a unitless multiplier), K_i is the integral gain (seconds^{-1}), and K_d is the derivative gain (seconds). (A typical PLC's PID formula will look a little more complex, but we examine PID basics before exploring the other features.)

There are two common forms of the PID calculation (and variations of each, as we will see later). In some PLCs the programmer can select which form of the equation the PID instruction will use. The two PID equation forms include:

1. An **independent gains equation,** which looks like the one we have examined above:

$$\text{CV} = K_p(\text{E}) + K_i \int (E)\ dt + K_d \frac{d(E)}{dt}$$

where CV is the control variable output, E is the error, $\int (E)\ dt$ is the sum of error, and $d(E)/dt$ is the rate of change of error.

2. The ISA[6] standard **dependent gains equation:**

$$\text{CV} = K_c\left[E + \frac{1}{T_i}\int (E)\,dt + T_d\frac{d(E)}{dt}\right]$$

in which most of the variables are as in the independent gains equation; in addition, K_c is a proportional multiplier affecting the entire PID calculation result, $1/T_i$ is the reset gain (repeats/minute; T_i is sometimes called integral time), and T_d is the rate gain (minutes), sometimes called derivative time. Note that there is no K_p variable in the dependent equation, so the proportional component in the output cannot be calculated independently. After calculation of the PID value, the entire result is scaled by the single proportional multiplier K_c.

How the PID Instruction Corrects Error The integral term increases as long as error exists. It causes the control system's output to increase steadily as long as steady-state error exists, until the control variable is large enough to overcome the cause of steady-state error. The sum, or integral, of error is calculated by the PID instruction by performing the following calculation each time the PID result is recalculated:

new sum of error = old sum of error + (current error × time interval since last calculation)

Integral gain (K_p) and reset gain ($1/T_i$) are values that scale the integral output.

Here is how you might use integral control. Assume that you have a proportional control positioning system like that shown in Figure 12.26. You have observed that with a K_p value of 1, the carriage typically stops about 0.25 in. from the desired position (because of friction), and the analog output signal from the PLC is 0.25 V at this position. Since K_p is 1, then 0.25 V is the steady-state error output of your controller with unamplified error. Figure 12.27a shows the analog output of the PLC proportional controller after the carriage stops.

If, while the positional error remained 0.25 in., you replaced the proportional control formula with an independent gains PID formula with K_p and K_d set to zero, and entered an integral gain (K_i) of 1 s^{-1}, the PLC's analog output value would start at zero but would increase by 0.25 V (the equivalent of the output value due to error alone) every second that the steady-state error remained at 0.25 in. With $K_i = 2\ \text{s}^{-1}$, the analog output would rise at the rate of 0.5 V every second (twice the output due to error alone) while steady-state error remained at 0.25 in. Obviously, the larger the integral gain value, the faster the control variable output value will grow. Figure 12.27b shows the control variable's change over time if it acted alone, without proportional control, and Figure 12.27c shows the effect of proportional control (with $K_p = 1$) and integral control together. This formula is called an independent gains formula because changing one multiplier (e.g., K_p) has no effect on the output due to another term with its own independent multiplier (e.g., K_i). Figure 12.27d shows what the output of the proportional-plus-integral system would be if K_p were set to 3 (and positional error remained 0.25 in.).

---

[6] The ISA is a large, well-respected organization that provides standards for the process control industry in North America.

FIGURE 12.27
Effect of integral gain (K_i in seconds^{-1}): (a) error signal; (b) integral term alone; (c) integral plus small proportional; (d) integral plus large proportional.

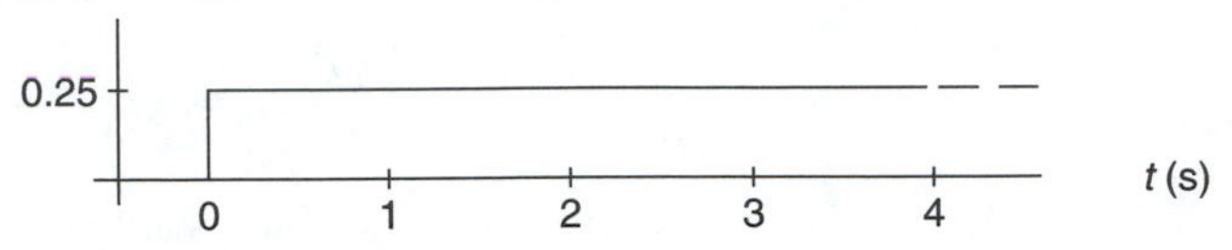

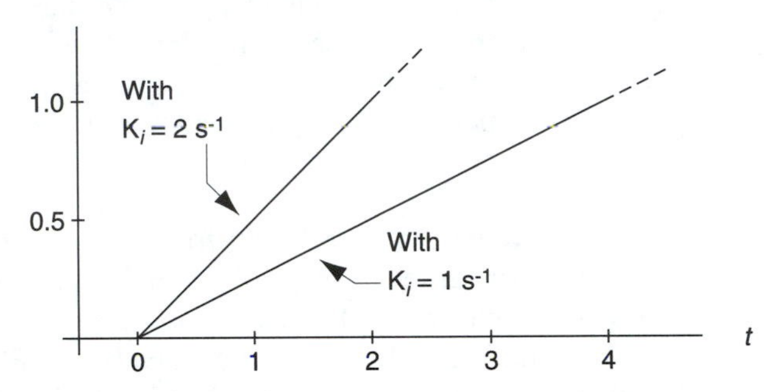

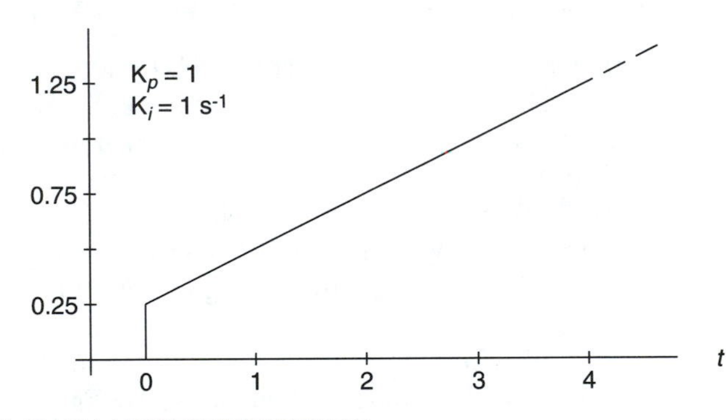

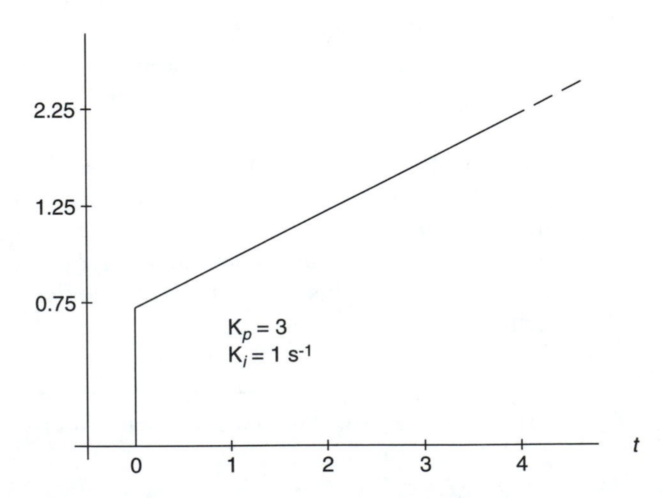

The dependent gain formula's integral calculation works in a similar manner, except in two ways: (1) The proportional multiplier (K_c) amplifies the sum of the error, integral, and derivative terms, so because there is no multiplier for the error term alone, the control variable output will always contain a component that is proportional to error; and (2) the integral term is calculated such that its value increases by an amount equal to the unmultiplied error term value every minute that steady-state error exists, and the reset gain ($1/T_i$) is given in units of repeats per minute. The component of the control variable that is due to the integral term *repeats* (becomes equal to) the component due to the error value after the number of minutes indicated by $1/T_i$. This definition is shown in Figure 12.28. Part (a) shows a steady state error that remains constant. The output due to error remains constant too. Part (b) shows the output due to the integral term alone. It rises to match the output due to error in one minute if $1/T_i = 1$, or if $1/T_i = 2$ it will rise to twice the output due to error in that minute. Part (c) shows the output due to error and integral combined, if Kc = 1. Part (d) shows that increasing kc to 3 increases the sum of error and integral terms.

Eventually, of course, with an integral term in the output, friction will be overcome and steady-state error will decrease. If the carriage returns and goes past the setpoint in the other direction, error would become negative, and the integrator would decrease output voltage at a rate controlled by the amount of error and the integral gain value until the control variable output was a large enough negative value to cause the carriage to come back to the setpoint position. The output due to the proportional term will be zero when the error is zero, but control variable output due to the integral term remains large when the error goes to zero.

The derivative term in a PID formula is intended to be used to prevent overshoot. The **derivative of error,** $d\,(E)/dt$, is a value equal to the rate of change of error. If a PID-controlled system is approaching the setpoint position, error is decreasing, so the derivative term will be negative. If the system is reducing error quickly, the derivative will not only be negative but will also be large. The independent gains formula version calculates the derivative value based on error change per *second,* and the dependent gains formula calculates it based on error change per *minute.* The multipliers **derivative gain** (K_d) and **rate gain** (T_d) are used to scale the effect that the rate of change of error has on the PLC's output value.

If, for example, the carriage in our position-control system is approaching the setpoint position at 3 in./s, the rate of change of error is −3 in./s. If the feedback signal from the position sensor varies by 10 V/in., the derivative term would be equal to (10 V/in. × −3 in./s) −30 V/s. If the derivative multiplier K_d is set to 0.01 s, the derivative term in the independent gains equation will reduce the control variable by 0.3 V, causing the speed of the motor to reduce. The faster error is reducing, the more the derivative term will reduce the speed of the system being controlled, reducing its momentum and reducing the chance of overshooting the setpoint. If the (negative) output from the derivative term becomes larger than the (positive) output from the proportional term (when position error is small), the motor will even be driven to decelerate, as demonstrated in the following calculation. In this example the carriage is within 1/2 in. of the setpoint but is still moving at 3 in./s. The carriage needs to be actively decelerated or its momentum will take it past the setpoint.

$$\text{CV} = K_p \times \text{error} + K_d \times \frac{d(\text{error})}{dt}$$

a) ERROR SIGNAL

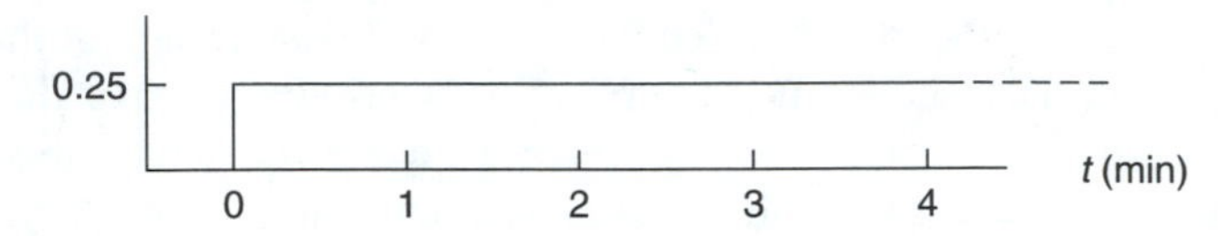

b) INTEGRAL TERM ALONE

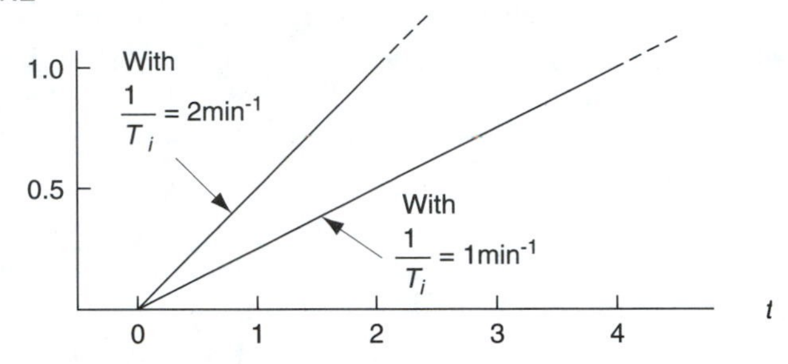

c) INTEGRAL PLUS SMALL PROPORTIONAL

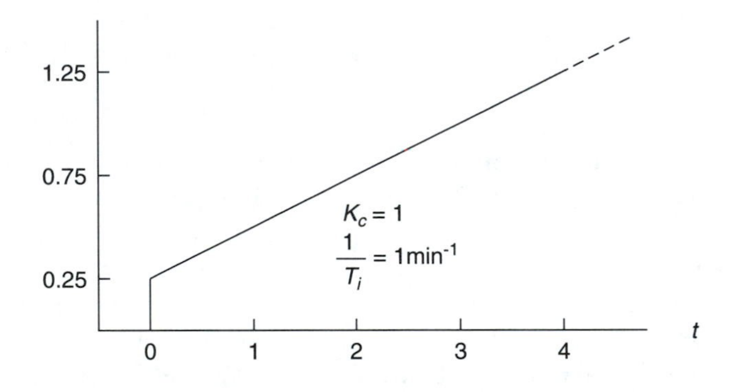

FIGURE 12.28
Effect of reset gain ($1/T_i$ in repeats per minute): (a) error signal; (b) integral term alone; (c) integral plus small proportional; *(continued)*

where $K_p = 1$, $K_d = 1$ s, error = 1/2 in., and $d(\text{error})/dt = -3$ in./s. Scaling of distance to voltage in this system is 10 V/in. Then

$$\begin{aligned}
\text{CV} &= K_p \times \text{error} + K_d \times \frac{d(\text{error})}{dt} \\
&= (1 \times \tfrac{1}{2}\text{ in.}) + (1\text{ s} \times \ -3\text{ in./s}) \\
&= (0.5\text{ in.}) + (-3\text{ in.}) \\
&= -2.5\text{ in.}
\end{aligned}$$

$$\begin{aligned}
\text{CV (in volts)} &= -2.5\text{ in.} \times 10\text{ V/in.} \\
&= -25.5\text{ V}
\end{aligned}$$

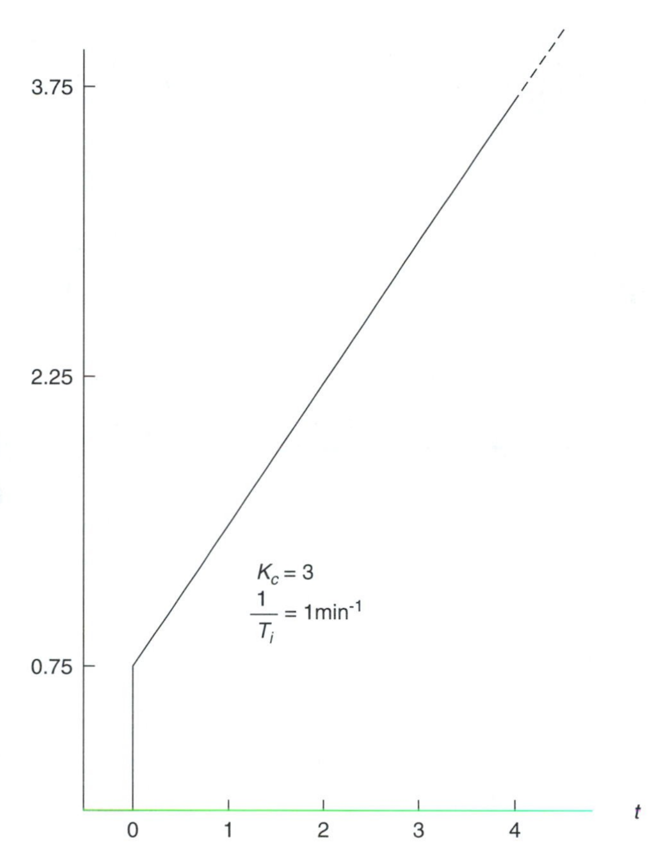

FIGURE 12.28
(continued) (d) integral plus large proportional.

The derivative term also helps prevent overshoot by its action when error is increasing. Error increases if the setpoint starts changing and the carriage hasn't yet accelerated to keep up with the changing setpoint. In this situation, the derivative term would be positive and would increase the motor torque, causing it to work harder to accelerate. Having a derivative component reduces the need to have a large proportional component. If the proportional multiplier can be reduced, the proportional term will be less likely to cause overshoot.

In the dependent gain formula, the derivative term works as in the independent gain formula except that the PLC calculates error change per minute, and rate gain (T_d) has units of min-

a) SETPOINT CHANGING

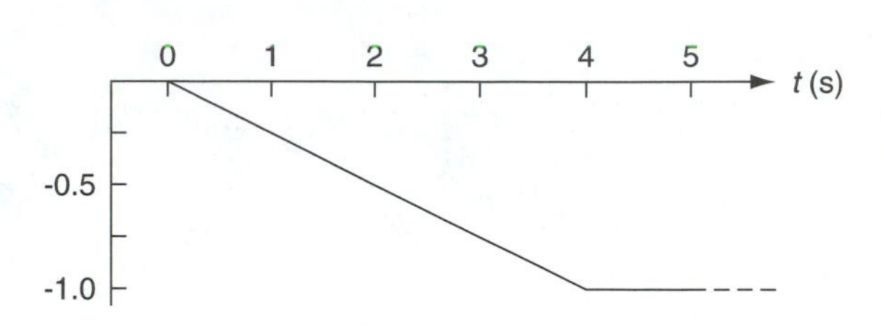

b) PROPORTIONAL TERM ONLY

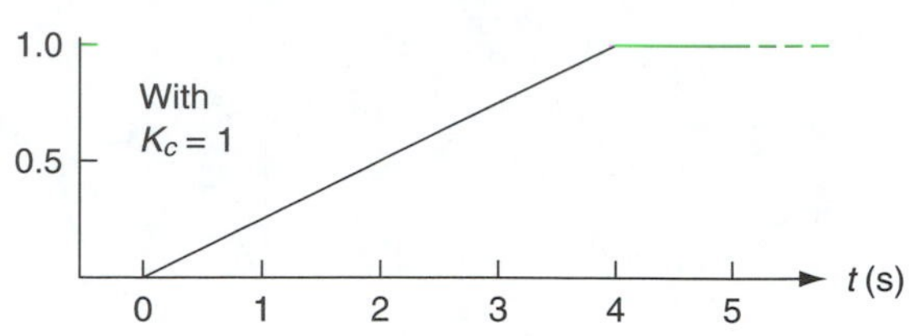

c) RATE OF CHANGE OF ERROR

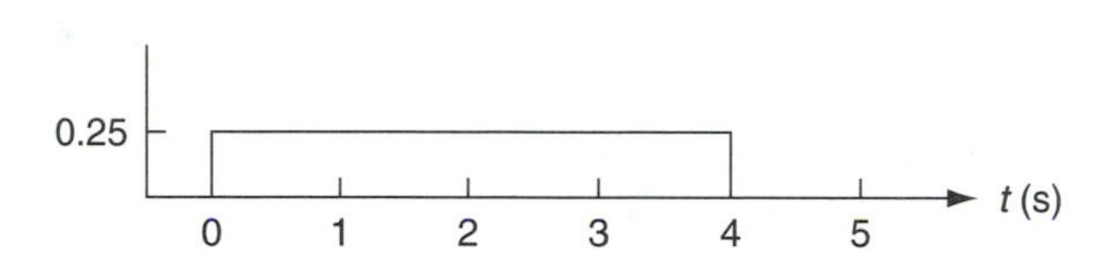

d) DERIVATIVE TERM ONLY

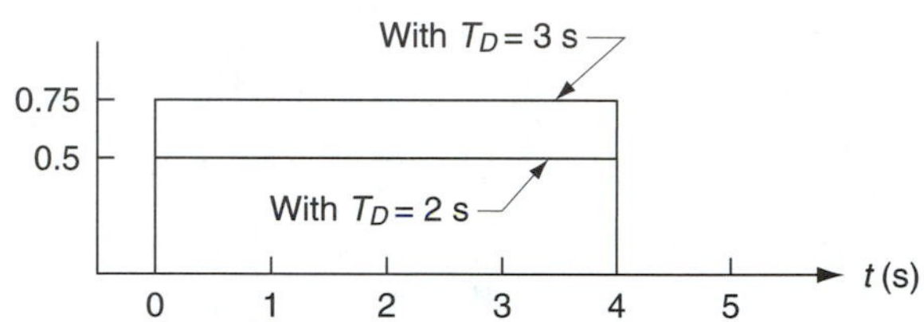

e) DERIVATIVE PLUS PROPORTIONAL (DEPENDENT GAIN)

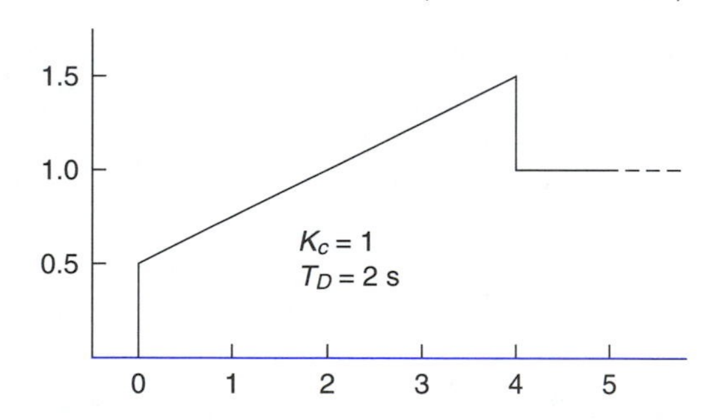

FIGURE 12.29
Derivative term, showing how it saves time in increasing control variable output in response to changing error: (a) setpoint changing; (b) proportional term only; (c) rate of change of error; (d) derivative term only; (e) derivative plus proportional (dependent gain).

utes. The rate gain multiplier in a dependent gain formula is often defined in terms of the amount of time the inclusion of the derivative term saves in raising the output value to some level, versus the amount of time it would take the (unmultiplied) error term alone to reach that level while error was changing at a constant rate.[7] Figure 12.29 explains this rather confusing

---

[7] The same definition applies to the K_d term in the independent gains formula, but only if the proportional constant, K_p, is equal to 1.

statement. Figure 12.29a shows a setpoint that is changing to cause the carriage to move backward at a constant rate. Assume (to keep this example simple) that something is preventing the carriage from moving. Position error is therefore growing at a fixed rate and an error-only controller's output with $K_c = 1$ would grow at the same rate, as shown in Figure 12.29b. The derivative of error is constant, as shown in Figure 12.29c, since the error is growing at a constant rate. If a T_d multiplier of 2 was used (as shown in Figure 12.29d), when the setpoint started changing, the derivative component of the control variable would immediately go to the value that the error-alone controller would take 2 min to reach, so using a T_d value of 2 saves 2 min in terms of how long it takes the output to reach that level. If T_d is 3, the output immediately goes to the value it would otherwise reach after 3 min of error-only control. In a dependent gain formula, the derivative component is always added to the unmultiplied error before the sum is multiplied by the proportional multiplier K_c, as shown in 12.29e.

Selecting the correct proportional, integral, and derivative gains can be difficult. A method for tuning PID control is discussed after we have covered how to program PID control instructions.

PLC-5

ALLEN-BRADLEY PID INSTRUCTIONS

SLC 500

There are four versions of the Allen-Bradley PID instruction. One version is available in the SLC 5/02, and another, very similar version in the SLC 5/03 and up (low-level SLC 500s do not have PID instructions). A third version, very similar to the SLC 500 versions, is offered in the PLC-5. The enhanced series of PLC-5s offers an improved fourth version. The following sections therefore contain a lot of duplicate information. *Italic type is used to indicate that a significant difference exists between the version being described and other PID versions.*

PLC-5

ALLEN-BRADLEY PLC-5 PID INSTRUCTIONS

With an Allen-Bradley PLC-5 PID instruction, as shown in Figure 12.30, the user must enter:

1. An address for the process variable. Each time the PID instruction executes, it will read the feedback value from this address and will calculate error, sum of error, and rate of change of error based on this value. *The PID instruction reads only the low 12 bits from this address, ignoring the contents of the high 4 bits. Integer values of from 0 to 4096 are thus required, and*

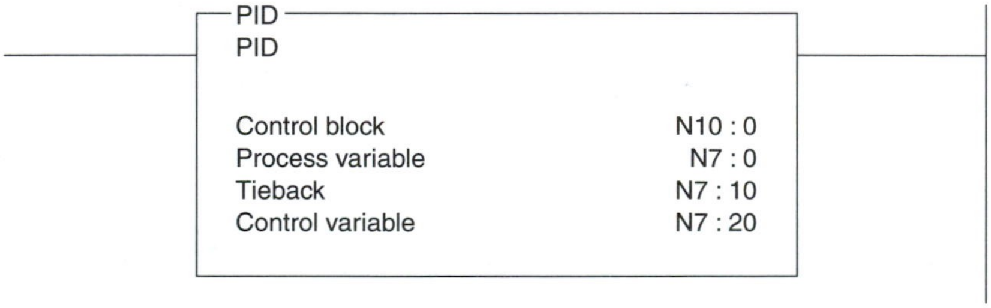

```
PID
PID

Control block        N10 : 0
Process variable      N7 : 0
Tieback              N7 : 10
Control variable     N7 : 20
```

FIGURE 12.30
Allen-Bradley PLC-5 PID instruction.

the program or the analog input module may have to scale process variable values into this range. The PID instruction can scale process variable values into engineering units for display purposes (see below), but the engineering unit values are not *used in the PID calculation.*

2. An address for a control variable. When the PID instruction finishes executing it will place its result, *a 12-bit unsigned value between 0 and 4095,* into this address. The PID instruction can be configured to limit the output so that it doesn't use the full range. If the analog output module requires numbers in another format, the program or the analog output module must perform scaling. If the contents of the control variable address ever exceeds 4095, and if the programmer has selected an integer control block for the PID instruction (see below), the control variable's value will be permanently changed (so don't allow any other instructions to write to this address).

3. *An address for a tieback value. If the system being controlled is switched into manual mode (see below), the control variable value will come from this address rather than from the result of the PID calculation. Figure 12.31 shows how a PLC programmed with a PID instruction can use either the PID formula or the tieback value to generate a control variable value. The value in the tieback address might come from an operator-controlled potentiometer attached to an analog input module, could be entered by an operator at a computer terminal, or could even be provided by another PLC on the data highway plus network. The tieback value must be a 12-bit value between 0 and 4095, which the PID equation may convert to a percentage value for internal use. Enter a "0" instead of an address for the tieback value if you don't intend to use manual control.*

The tieback value is also used in Allen-Bradley's bumpless transfer feature. When the controlled process is in manual mode, the value for the integral term is copied from the tieback address each time the PID instruction executes, so that when the control system is switched back into automatic, the output due to the integrator starts at the value that was being used in manual mode, and there won't suddenly be a large change in the control variable value.

4. The address of a control block. The control block is a set of data words that contain PID configuration parameters and working values that the PID calculation needs. Values in the

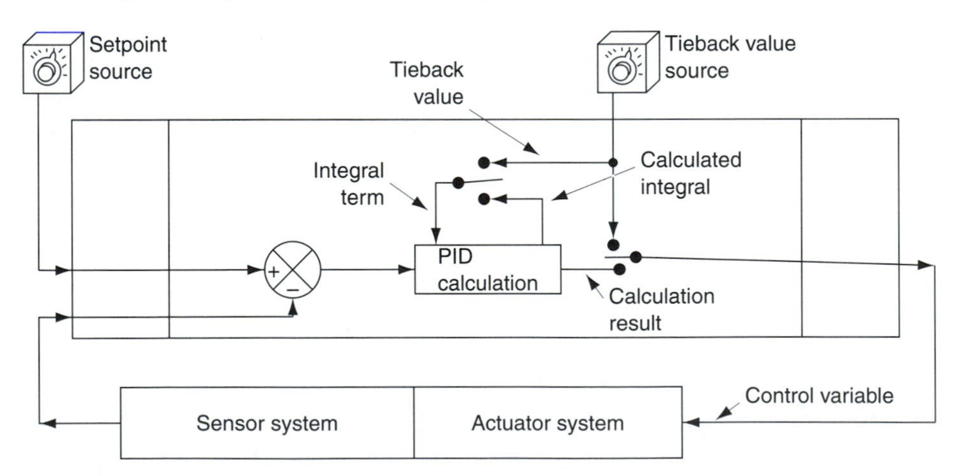

FIGURE 12.31
Selection of tieback value or PID formula result for the PID instruction's control variable output.

```
             equation: 1  (0:AB/1:ISA)              feed forward:        7
                 mode: 0  (0:auto/1:manual)     max scaled input:      300
                error: 1  (0:SP-PV/1:PV-SP)     min scaled input:        0
      output Limiting: 1  (0:NO/1:YES)                 deadband:        15
      set output mode: 0  (0:NO/1:YES)         set output value %:       0
    setpoint scaling: 0  (0:YES/1:NO)            upper CV limit %:     100
    derivative input: 1  (0:PV/1:error)          lower CV limit %:       0
   last state resume: 0  (0:NO/1:YES)             scaled PV value:      33
     deadband status: 1                              scaled error:       8
 upper CV Limit alarm: 0                              current CV %:       7
 lower CV Limit alarm: 0
 setpoint out of range: 0
             PID done: 0                                 setpoint:      25
          PID enabled: 0      proportional gain (Kc)      [.01]:        10
                              res. time (Ti) [.01 mins/repeat]:          1
                              derivative rate (Td)  [.01 mins]:          2
                              loop update time       [.01 secs]:        10


Enter value or press <ESCAPE> to exit monitor.
N10:0/0 =
Program     Forces:None       Data:Decimal           PLC-5/15 File DRILL1
```

FIGURE 12.32
Data screen for a PLC-5 PID instruction with an integer control block.

control block can be changed by other PLC program rungs that read or write the appropriate data word. A control block contains data such as:

- The setpoint value
- The multiplier constants (K_p, K_i, etc.)
- Interim calculated values the PID equation needs to store (e.g., the sum of error)
- Configuration bits for selection of the controller mode (manual/automatic) and selection of other optional features to be used in this PID calculation
- PID calculation result and status bits

Two types of control blocks can be used: an integer control block, *or a PD control block.* The type you select affects how the PID instruction works. Classic PLC-5s can only use integer control blocks, but if you have an enhanced PLC-5, you can program PID instructions with either integer or PID control blocks.

If an integer control block address (e.g., N10:0) is entered:

1. *The PID instruction will be executed once each time its control logic goes from false to true.*

2. *Calculations will be performed using 12-bit integer precision.*

3. An integer control block containing 23 integer (16-bit) words will be created by the PLC-5. The programmer can change values into the integer control block by placing the cursor on the PID instruction, then entering the data screen, causing *the screen in Figure 12.32* to be displayed. Cursor to the entry you wish to change (the actual addresses will be displayed at the bottom of the screen) and type in a new value. This data screen displays many (but not all) of the values a programmer may wish to monitor and change.[8]

---

[8] Some control block values should not be changed in run mode. Doing so could make the equipment that is being controlled perform in an unpredictable and even dangerous manner. See an Allen-Bradley instruction set manual.

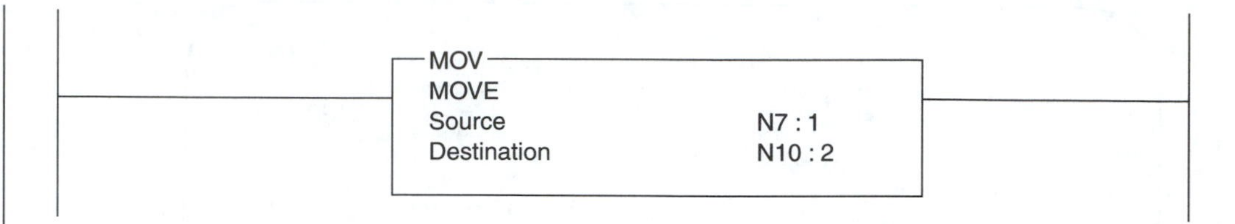

FIGURE 12.33
Using a MOVE instruction to write the setpoint into a PLC-5 integer control block.

The programmer can see and change the contents of an integer control block without identifying labels beside each entry if the data screen is invoked while the cursor is not on the PID instruction.

4. *Control block words and bits must be addressed numerically in a user-program. If, for example, the program needed to change the setpoint for the process, the programmer would have to know that the setpoint is stored in the control block's third data word (N10:2 if the control block starts at N10:0). Figure 12.33 shows how to write a new setpoint value into an integer control block before executing a PID instruction. The structure of a PLC-5 PID's integer control block is shown in Figure 12.34.*

Word:	Contains:		Term:	Entry Range:
0	Bit 15 Bit 13 Bit 11 Bit 10 Bit 9 Bit 8 Bit 7 Bit 6 Bit 5 Bit 4 Bit 3 Bit 2 Bit 1 Bit 0	Enabled (EN) Done (DN) Set point out of range Output alarm, lower limit Output alarm, upper limit DB, set when error is in deadband Resume last state (0=yes; 1=hold last state) Derivative action (0=PV, 1=error) Setpoint descaling (0=no, 1=yes) Set output (0=no, 1=yes) Output limiting (0=no, 1=yes) Control (0=reverse, 1=direct) Mode (0=automatic, 1=manual) Equation (0=independent, 1=ISA)		
1	reserved			
2	Setpoint		SP	0-4095 (unscaled) S_{min}-S_{max} (scaled)
3	**Independent:**	Proportional gain x 100 (unitless)	K_p*	0-32,767
	ISA:	Controller gain x 100 (unitless)	K_c*	0-32,767
4	**Independent:**	Integral gain x 1000 (1/sec)	K_i**	0-32,767
	ISA:	Reset term x 100 (minutes per repeat)	T_i*	0-32,767

FIGURE 12.34
Structure of a PLC-5 integer control block for a PID instruction. (*continued*)

Word:	Contains:		Term:	Entry Range:
5	**Independent:**	Derivative gain x 100 (seconds)	K_d*	0-32,767
	ISA:	Rate term x100 (minutes)	T_d*	0-32,767
6	Feedforward or bias		FF/Bias	0-4095
7	Maximum scaling		S_{max}	-32,768-+32,767
8	Minimum scaling		S_{min}	-32,768-+32,767
9	Dead band		DB	0-4095 (unscaled) S_{min}-S_{max} (scaled)
10	Set output		SETOUT	0-100%
11	Maximum output limit (% of output)		L_{max}	0-100%
12	Minimum output limit (% of output)		L_{min}	0-100%
13	Loop update time x 100 (seconds)		dt	0-32,767
14	Scaled PV value (displayed)			S_{min}-S_{max}
15	Scaled error value (displayed)			S_{min}-S_{max}
16	Output (% of 4095)		CV	0-100%
17 - 22	internal storage; do not use			

Important: Terms marked with an asterisk (*) are entered as Y_y x 100. The term itself is Y_y. The term marked with a double asterisk (**) is entered as Y_y x 1000. The term itself is Y_y.

FIGURE 12.34
(continued)

Enhanced PLC-5s allow selection of a PID control block. Enter a PD file address (e.g., PD9:0) to select the PID control block:

1. *The PID instruction will repeatedly execute every program scan while its control logic remains true.*

2. *PID calculations are performed using 32-bit floating-point math, resulting in more precision. (The control variable output will still be a 12-bit unsigned binary number.)*

3. *A PD data file containing 80 (16-bit) data words will be created. The programmer can view the data in the PID control block by placing the cursor on the PID instruction and entering the data screen. One of the data screens shown in Figure 12.35 will be displayed, and the programmer can toggle to the other data screen. (The same set of screens is displayed if the PD data file is accessed while the cursor is not on the PID instruction.) More PID operations and values can be monitored and changed[9] than is possible with an integer control block.*

4. *Every data word in a PID control block has an address mnemonic which must be used in user-programs. Writing the setpoint to a PD file requires the programmer to know the PD file number (e.g., PD9.0 in this example), and the setpoint address's mnemonic (.SP). The* MOVE *instruction could be written as in Figure 12.36. Address mnemonics, which are discussed later*

[9] See the warning in footnote 8 about changes while in run mode.

Setpoint:	0.000000	Proportional Gain (Kp):	0.000000
Process Var.:	0.000000	Integral Gain (Ki) [/secs]:	0.000000
Error:	0.000000	Derivative Time (Kd) [secs]:	0.000000
Output %:	0.000000		
Mode:	AUTO	Deadband:	0.000000
PV Alarm:	NONE	Output Bias %:	0.000000
Deviation Alarm:	NONE		
Output Limiting:	NONE	Tieback %:	0.000000
SP Out Of Range:	NO	Set Output %:	0.000000
Error Within Deadband:	NO		
PID Initialized:	NO		
A/M Station Mode:	AUTO		
Software A/M Mode:	AUTO		
Status Enable (EN):	0		

PID Equation:	INDEPENDENT	Engineering Unit Maximum:	0.000000
Derivative Of:	PV	Engineering Unit Minimum:	0.000000
Control Action	SP - PV		
PV Tracking:	NO	Input Range Maximum:	0.000000
		Input Range Minimum:	0.000000
Update Time (secs):	0.000000	Output Limit High % :	0.000000
		Output Limit Low % :	0.000000
Cascaded Loop:	NO	PV Alarm High:	0.000000
Cascaded Type:	-	PV Alarm Low:	0.000000
Master To This Slave:	-	PV Alarm Deadband:	0.000000
		(+) Deviation Alarm:	0.000000
		(–) Deviation Alarm:	0.000000
		Deviation Alarm Deadband:	0.000000

Press a function key or enter a value.
PD0 : 1.PE =
Rem Prog Forces: Disabled Data : Float Addr : Decimal PLC-5/40 Addr 4

Toggle	Specify Address	PID : 1: Monitor	Next File	Prev File	Next Element	Prev Element
F3	F5	F6	F7	F8	F9	F10

FIGURE 12.35
Data screens for a PLC-5 PID control block.

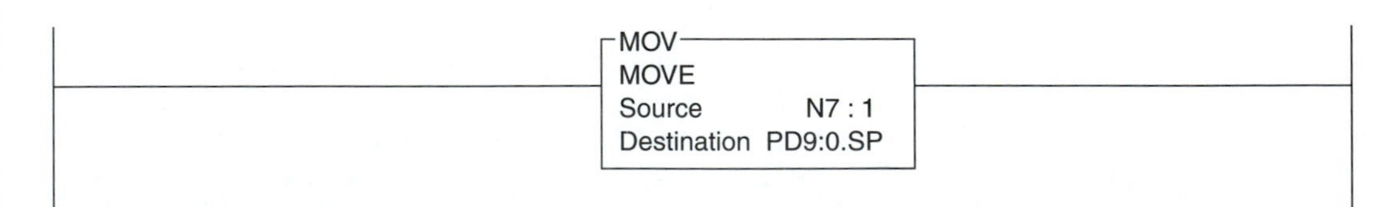

FIGURE 12.36
Using a MOVE instruction to write the setpoint into a PLC-5 PID control block.

Word:	Contains:			Range:
0	Control/Status Bits			
	Bit 15	Enabled (EN)	.EN	
	Bit 9	Cascade selection (master, slave)	.CT	
	Bit 8	Cascade loop (0=no, 1=yes)	.CL	
	Bit 7	Process variable tracking (0=no, 1=yes)	.PVT	
	Bit 6	Derivative action (0=pv, 1=error)	.DO	
	Bit 4	Set output (0=no, 1=yes)	.SWM	
	Bit 2	Control action (0=SP-PV, 1=PV-SP)	.CA	
	Bit 1	Mode (0=automatic, 1=manual)	.MO	
	Bit 0	Equation (0=independent, 1=ISA)	.PE	
1	Status Bits			
	Bit 12	PID initialized (0=no, 1=yes)	.INI	
	Bit 11	Set point out of range	.SPOR	
	Bit 10	Output alarm, lower limit	.OHL	
	Bit 9	Output alarm, upper limit	.OLL	
	Bit 8	DB, set when error is in DB	.EWD	
	Bit 3	Error is alarmed low	.DVNA	
	Bit 2	Error is alarmed high	.DVPA	
	Bit 1	Process variable (PV) is alarmed low	.PVLA	
	Bit 0	Process variable (PV) is alarmed high	.PVHA	
2,3	Setpoint		.SP	$-3.4\ E^{+38}$ to $+3.4\ E^{+38}$
4,5	Independent:	Proportional gain (unitless)	.KP	0 to $+3.4\ E^{+38}$
	ISA:	Controller gain (unitless)		0 to $+3.4\ E^{+38}$
6,7	Independent:	Internal gain (1/sec)	.KI	0 to $+3.4\ E^{+38}$
	ISA:	Reset term (minutes per repeat)		0 to $+3.4\ E^{+38}$
8,9	Independent:	Derivative gain (seconds)	.KD	0 to $+3.4\ E^{+38}$
	ISA:	Rate term (minutes)		0 to $+3.4\ E^{+38}$
10, 11	Feedforward or bias		.BIAS	–100 to +100%
12, 13	Maximum scaling		.MAXS	$-3.4\ E^{+38}$ to $+3.4\ E^{+38}$
14, 15	Minimum scaling		.MINS	$-3.4\ E^{+38}$ to $+3.4\ E^{+38}$
16, 17	Dead band		.DB	0 to $+3.4\ E^{+38}$
18, 19	Set output		.SO	0 - 100%
20, 21	Maximum output limit (% of output)		.MAXO	0 - 100%
22, 23	Minimum output limit (% of output)		.MINO	0 - 100%
24, 25	Loop update time (seconds)		.UPO	
26, 27	Scaled PV value (displayed)		.PV	
28, 29	Scaled error value (displayed)		.ERR	
30, 31	Output (% of 4095)		.OUT	0 - 100%

FIGURE 12.37
Structure of a PLC-5 PID control block, with mnemonics. (*continued*)

32, 33	Process variable high alarm value	.PVH	-3.4 E^{+38} to $+3.4$ E^{+38}
34, 35	Process variable low alarm value	.PVL	-3.4 E^{+38} to $+3.4$ E^{+38}
36, 37	Error high alarm value	. DVP	0 to $+3.4$ E^{+38}
38, 39	Error low alarm value	. DVN	-3.4 E^{+38} to 0
40, 41	Process variable alarm deadband	. PVDB	0 to $+3.4$ E^{+38}
42, 43	Error alarm deadband	. DVDB	0 to $+3.4$ E^{+38}
44, 45	Maximum input value	. MAXI	-3.4 E^{+38} to $+3.4$ E^{+38}
46, 47	Minimum input value	. MINI	-3.4 E^{+38} to $+3.4$ E^{+38}
48, 49	Tieback value for manual control (0-4095)	. TIE	0 - 100%
51	Master PID file number		0-999; 0-9999 for Enhanced PLC-5 processors only
52	Master PID element number		0-999; 0-9999 for Enhanced PLC-5 processors only
54-80	internal storage; do not use		

FIGURE 12.37
(continued)

in this chapter, are included in Figure 12.37. (A more detailed explanation of the data in a PD data file is included in Appendix J.)

Figure 12.38 shows a PLC-5 program with a PID instruction that uses an integer control block. The example is written as if it were in an STI (timed interrupt) routine. Block transfer instructions are unconditional in STI routines, so the example unlatches their enabled bits after they execute so that they can execute again next time the STI is called. Since a PID instruction using an integer control block executes only when its control logic goes true, the same trick has been used to force the PID instruction to execute again next time the STI executes (bit 15 of the first data word in the control block is the PID enabled bit).

Figure 12.39 shows a similar program, except that the PID instruction uses a PID control block. PID instructions using PID control blocks execute every scan cycle while their control logic is true, so the program in Figure 12.39 doesn't have to unlatch the control block's enabled bit.

Figures 12.40 and 12.41 are flowcharts copied from Allen-Bradley manuals. They show how a PID instruction executes when using an integer control block (Figure 12.40) and when using a PID control block (Figure 12.41). Notice that there are several differences, even in the terminology used. Refer to these flowcharts as you read the following section, which describes the contents of the two types of control blocks.

PLC-5 Control Block Data for PID Instructions

In this section we explain the integer and the PID control blocks used by PLC-5 PID instructions. It starts by covering data that is similar in both types of control blocks, then describes the

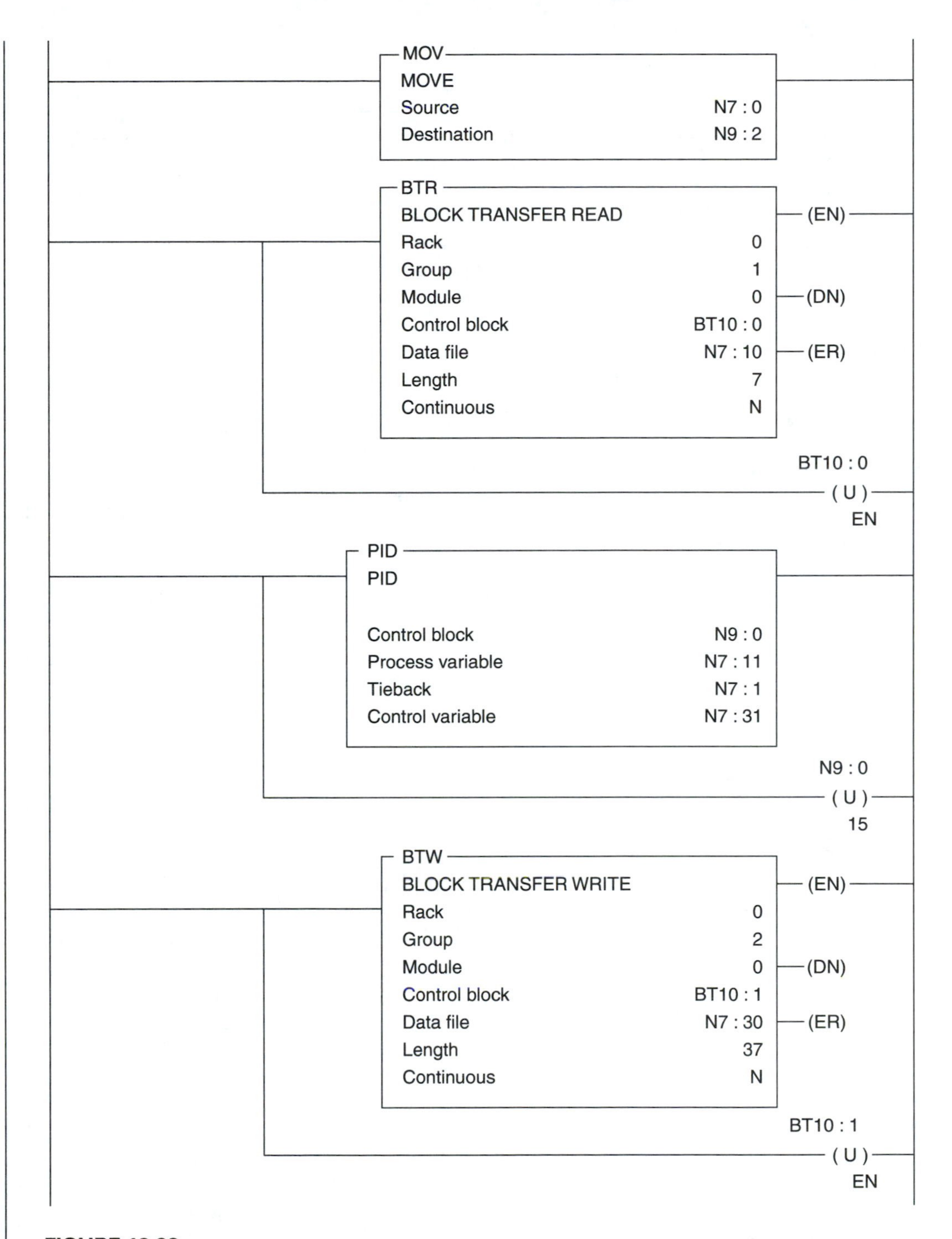

FIGURE 12.38
PLC-5 ladder logic program with a PID instruction that uses an integer control block.

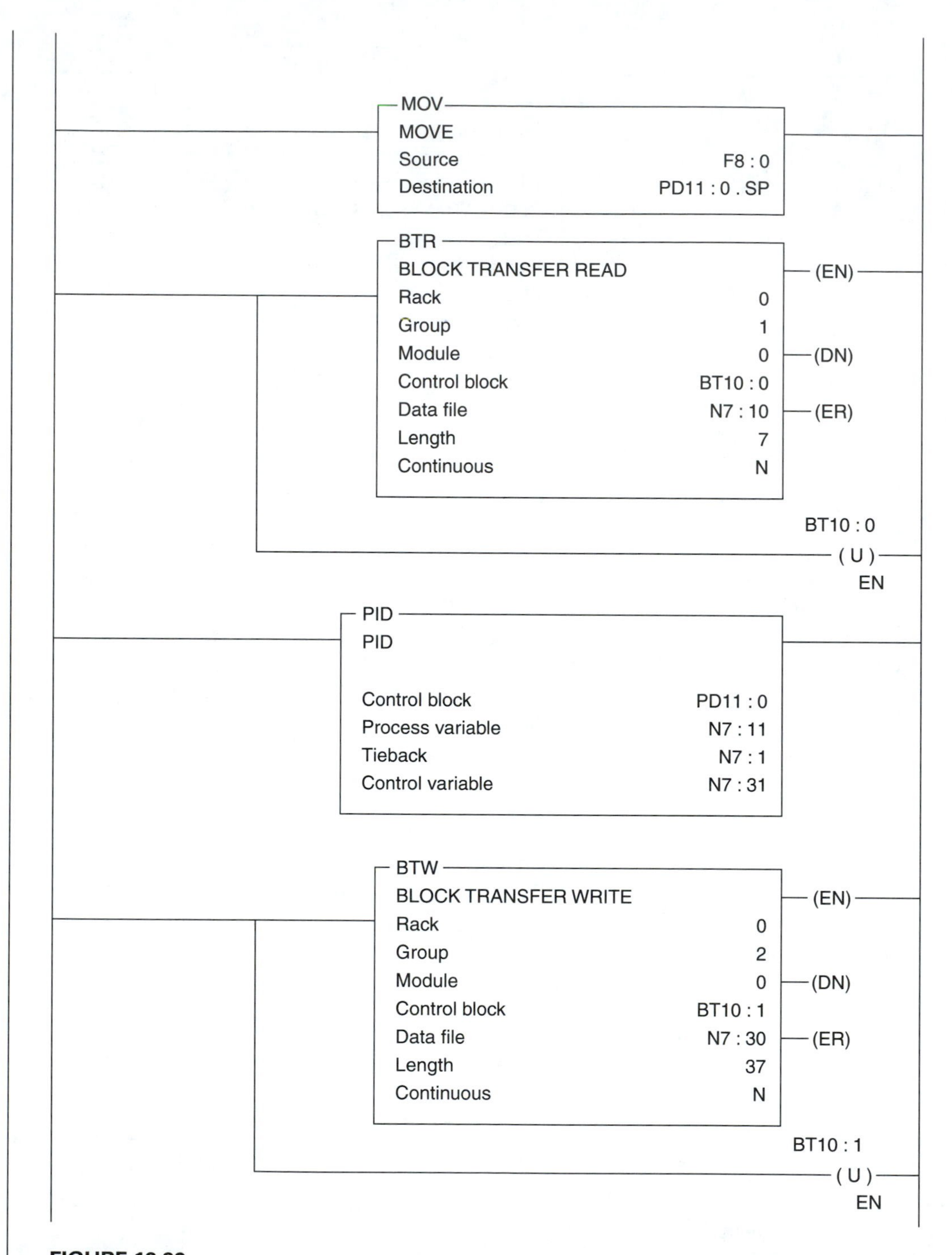

FIGURE 12.39
PLC-5 ladder logic program with a PID instruction that uses a PID control block structure.

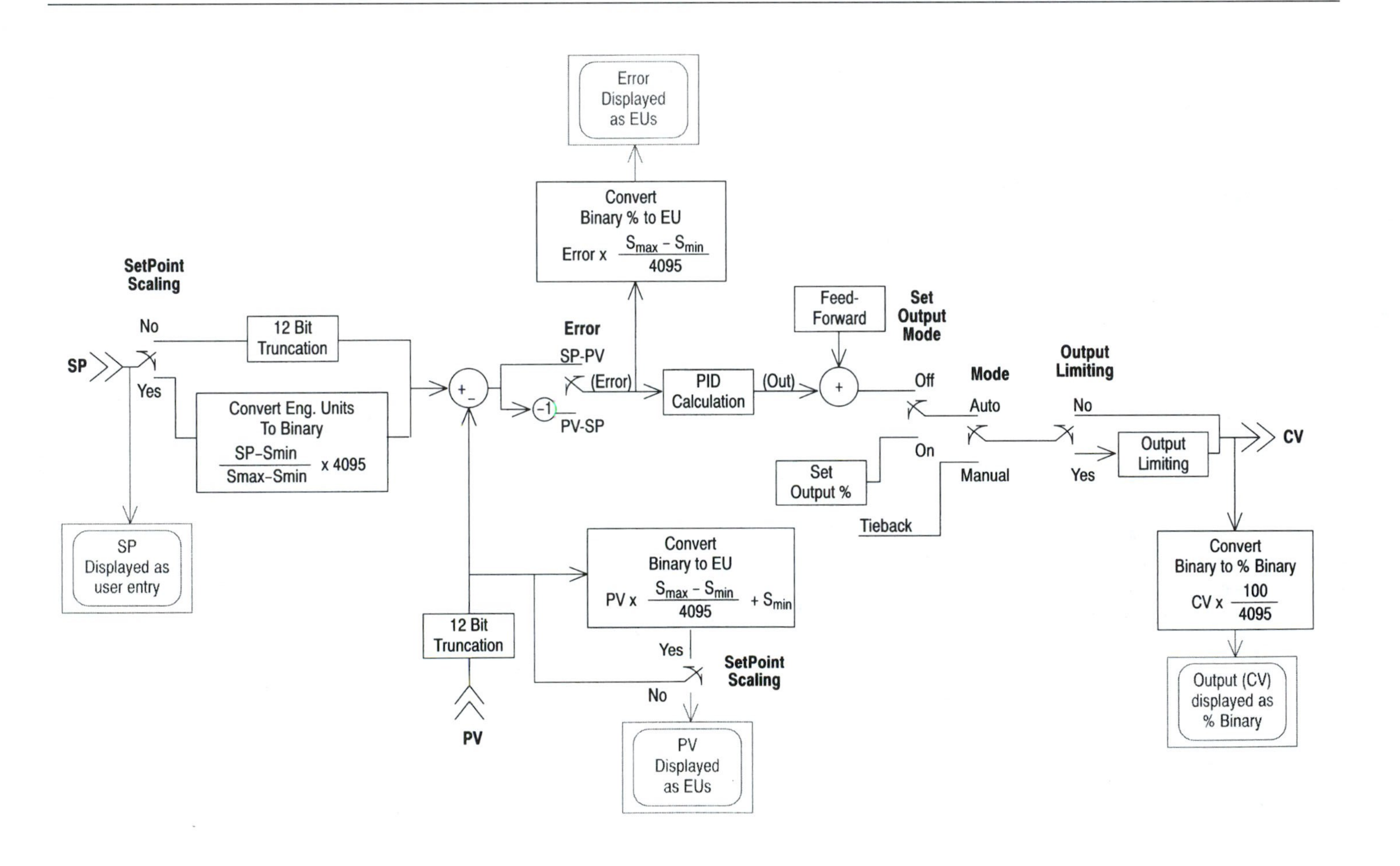

FIGURE 12.40
Operation of a PLC-5 PID instruction that is programmed to use an integer control block.

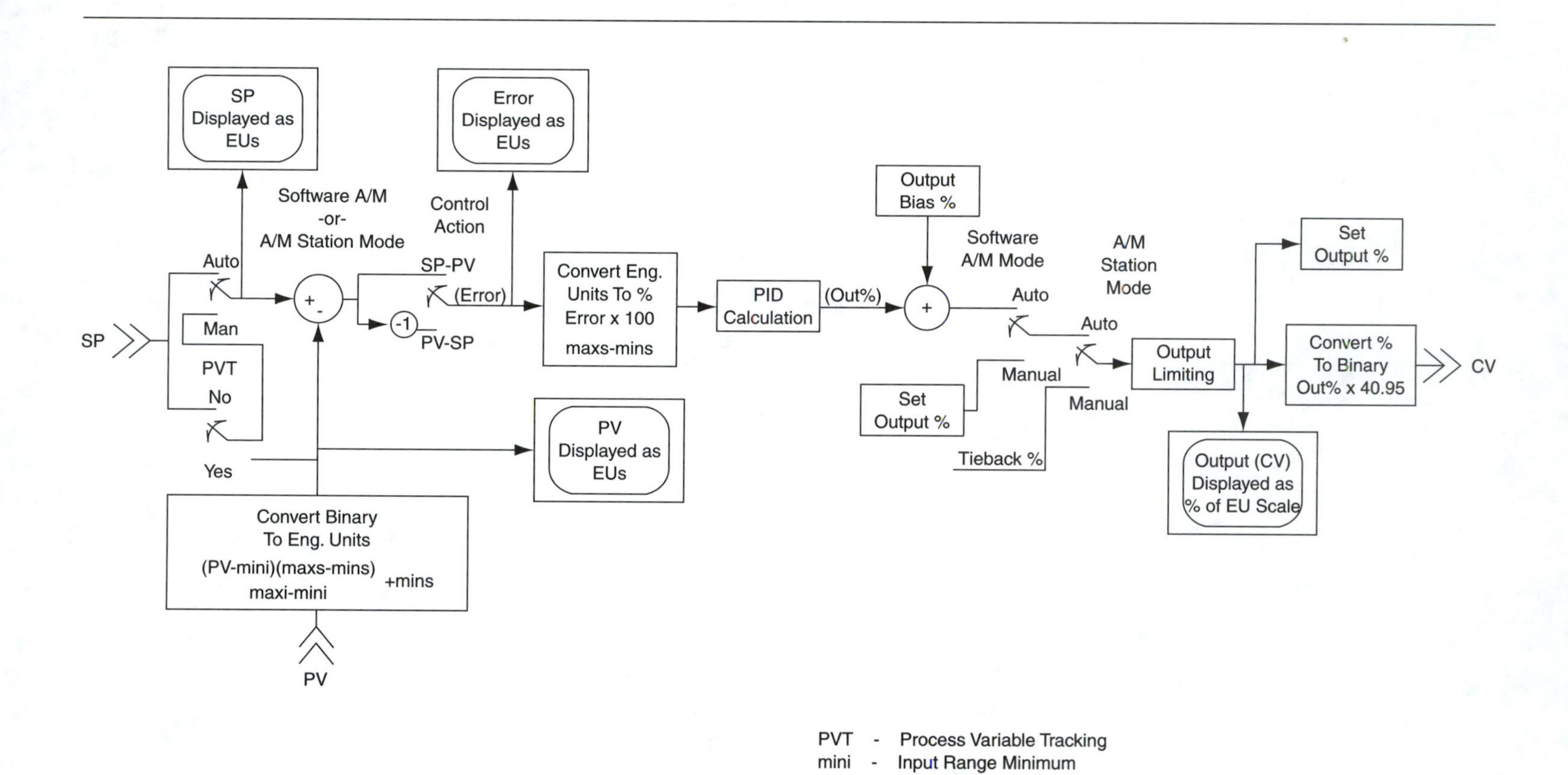

FIGURE 12.41
Operation of a PLC-5 PID instruction that is programmed to use a PID control block.

differences. PID control block mnemonics and integer control block offsets are included in the descriptions of data that a user-program might want to monitor or change. If the mnemonic or offset address is not included, that data should only be changed while the PLC is in program mode (tables here and in the Appendix give all addresses and mnemonics). *Italic type indicates that PID instructions in an SLC 500 PLC work differently from the PLC-5 descriptions here.*

The control blocks contain control bits, including:

1. *An* EQUATION *selection bit. Choose an independent gains equation or a dependent gains equation. The equation will include an output bias value which we didn't see when we examined the two equation types earlier.*

Independent gains equation:

$$\text{CV} = K_p(E) + K_i \int (E)\, dt + K_d \frac{d(E)}{dt} + \text{bias}$$

Dependent gains equation:

$$\text{CV} = K_c\left[E + \frac{1}{T_i} \int (E)\, dt + T_d \frac{d(E)}{dt}\right] + \text{bias}$$

2. A CONTROL ACTION selection bit, for selecting how to calculate error.
   - **(a) SP-PV (direct).** This is the traditional way to calculate error. If the setpoint (SP) is greater than the measured process variable (PV), the error will be positive, so the resulting control variable will be positive and will try to move the process variable forward to catch the setpoint.
   - **(b) PV-SP (reverse).** This method inverts the error sign, so that the PID calculation will result in an output that drives opposite to the direct equation's result. If, for example, you are using a PID instruction to control a fluid's temperature by controlling how much cold fluid is allowed into the mixture, and the current temperature (PV) is too low, we would want to decrease, not increase, the amount of cold fluid allowed in.
3. A DERIVATIVE INPUT VALUE selection bit.
   - **(a) Error.** This chapter has only shown a derivative component that was calculated as the rate of change of error and which was added to the proportional and integral components. Set the derivative action bit to ER (= 1) for this type of derivative action.
   - **(b) PV.** If this derivative action bit is cleared, the PID instruction will calculate the rate of change of the process variable instead of the rate of change of error. If the setpoint is rarely changed, calculating the derivative based on the process variable might result in better control of a process that is affected by external disturbances, because the PID calculation will act to cancel environmentally caused changes in the process variable. If this type of derivative action is chosen, and if SP-PV control action is selected, the derivative term will be subtracted, as follows:

Independent gains equation:

$$\text{CV} = K_p(E) + K_i \int |\, (E)\, dt \underset{\uparrow}{-} K_d \times \frac{d(\text{PV})}{dt \uparrow} + \text{bias}$$

Dependent gains equation:

$$CV = K_c\left[E + \frac{1}{T_i} * \int (E)\,dt - T_d\,\frac{d(PV)}{dt}\right] + \text{bias}$$

It is necessary to subtract a value proportional to the rate of change of a process variable in a formula with direct (SP-PV) control to avoid overshoot, because a rapidly changing process variable is likely to shoot right past the setpoint.

4. A **MODE** or **A/M SELECTION MODE** selection bit. (PD mnemonic = .MO, integer file offset = 0/1.) A user-program rung can monitor an operator-controlled mode switch and change this bit accordingly:

(a) In automatic mode (= 0), control variable values are calculated using the PID equation.

(b) *In manual mode (= 1), the control variable values come from the tieback address (an address or value must have been entered as part of the PID instruction). The user-program might read an operator-controlled dial attached to an analog input module to get a value for the tieback address. The tieback value is also copied to the integral term's storage location while in manual mode, so that when automatic mode restarts, the PID equation's output will begin at a value close to the manual mode's final value and there won't suddenly be a large control value change.*

5. *A* ***SET OUTPUT MODE*** *or* *SOFTWARE A/M MODE* *selection bit (PD mnemonic = .SWM, integer offset = 0/4). This bit may be controlled by the user-program. It performs a function very similar to the (automatic/manual) mode selection bit explained above.*

(a) *In software automatic mode (= 0) the PID instruction's control value is calculated normally.*

(b) *In software manual mode (= 1) the control variable value comes from the set output % address (explained later), which can contain data entered by an operator at a data monitor or supplied by another part of the PLC program. If the set output mode and the A/M selection bits are both set, the control variable value is controlled by the tieback value, not by the set output % value.*

The following options must be selected if an integer control block is used. These features are turned on automatically if a PID control block is used.

1. An **OUTPUT LIMITING** selection bit (offset = 0/3).

(a) When no (= 0), the output control variable is limited only by the 12-bit unsigned range of values, 0 to 4095.

(b) When yes (= 1), the PID instruction's maximum and minimum values are limited to the values entered for the upper CV limit % and lower CV limit % parameters (explained below).

Whether output limiting is selected or not, the PID instruction sets status bits in the control block when the calculated control value is above its high limit (PD mnemonic = .OLH, offset = 0/9) or below its low limit (.OLL, 0/10). The PID instruction also stops increasing the integral term while the calculated control variable is outside this range, to prevent "reset windup," which means that the integral term has been allowed to grow too big.

2. *A* SETPOINT DESCALING *selection bit.*
   (a) *If off (= 1), then descaling is inhibited. Setpoint values are simply truncated to 12 bits and treated as unsigned binary numbers. They are not adjusted other than by having their high 4 bits removed.*
   (b) *If on (= 0, default), values for setpoint (and the deadband if an integer control block is being used) can be supplied to the PID instruction using "engineering units" (a range of values other than the standard 12-bit value range of 0 to 4095). The upper and lower values for the engineering unit range of values must be entered into the control block's maximum and minimum scaled input or engineering unit parameters (described below) so that the PID instruction can convert engineering unit setpoint values into the data format used by the PID instruction.*
3. *A* RESUME LAST STATE *selection bit (offset = 0/7)*
   (a) *When off (= 0), the PID calculation's integrated value is reset to 0 after the PLC has been out of run mode.*
   (b) *When on (= 1), the control block's integrated value is not reset to zero when the PLC goes into run mode. If your analog output module is configured to hold its last value when the PLC goes into program mode or faults, you will probably want the calculation result to resume its last integrated component state when it resumes running.*[10]

The integer control block also contains:

4. *A* PID ENABLED *status bit (offset = 0/0) and a* PID DONE *status bit (offset = 0/1), which are useful since a* PID *instruction with an integer control block executes only once each time its control logic goes from false to true. The program in Figure 12.38 showed how to use the PID enabled bit in an STI program.*

The following options are available only if using a PD control block:

1. A **PROCESS VARIABLE TRACKING** selection bit (mnemonic = .PVT). If selected (= 1), the PID instruction copies the process variable to the setpoint value while the process is in manual mode. The PID calculation continues to execute while in manual mode even though its result isn't used, so this prevents the integral term from growing (if process variable = setpoint, then error = 0). Switching the process back into automatic can be done with minimized sudden change in control variable output.
2. A **CASCADED LOOP** selection bit and a **CASCADE TYPE** selection bit. In cascaded control loops, the control variable output of a master control loop becomes the setpoint for a slave control loop. If this PID loop is a slave, the master PID file number and the master PID element number must also be entered (mnemonic = .ADDR). (Cascade control is not discussed in this book.)

Data words that must be entered into a PID instruction's control block before using the PID instruction, and which can sometimes be changed while a control program executes, include:

---

[10] A PID instruction that is executing for the first time may be assigned a previously used memory for the integrated value, and the memory location may contain a nonzero value. Execute the PID instruction at least once before the RESUME LAST STATE selection bit is turned on, to clear the integrated value; otherwise, unpredictable, perhaps even dangerous, output values may result.

1. A **setpoint** value (PD mnemonic = .SP, integer offset word 2). This value is usually written into the control block by another rung in your PLC program, as shown in the first rung of the programs in Figures 12.38 and 12.39. Setpoint values must be between the maximum and minimum of the scaled engineering unit range. If scaling of setpoints is turned off during configuration of an integer control block, setpoint values must be between *0 and 4095.* If a setpoint value outside the allowable range is entered, the PID instruction sets a status bit (PD mnemonic = .SPOR, integer offset = 0/11).

2. Values for the **proportional, integral, and derivative gains** (K_p, K_i, and K_d); or values for **controller gain, reset time, and derivative time** (K_c, T_i, and T_d) (PD mnemonics = .KP, .KI, and .KD; integer offset words = 3, 4, and 5, respectively). Only positive values should be entered. Negative values will be assumed to be zero in the PID calculation. See Figures 12.34 and 12.37 for ranges and required number format. *Note that integer control block entries are assumed to be in units of hundredths or thousandths of the multiplier value entered. A method for selecting these values is discussed later.*

3. A **loop update time** or **update time,** in seconds (hundredths of a second in an integer control block, floating-point number in a PID control block). A nonzero loop update time value *must* be entered or the PLC will fault because the loop update time is used in calculating the integral and the derivative terms. The loop update time is used as the *dt* value in calculating the integral and the derivative terms[11]:

$$\text{new integral term} = (\text{old integral term}) + K_i(e_e\,dt)$$

$$\text{derivative term} = \frac{\text{change in } e_e \text{ since last calculation}}{dt}$$

Select an update time that is fast enough to control the process adequately but not so frequent that it disturbs the PLC's scanning. It should be no longer than one-fifth the period of the process's observed natural frequency (which is discussed later). It should be at least as long as five average scan cycles.

The programmer must ensure that the PID instruction is executed at a frequency matching the loop update time entered here. (Perhaps when you read this, PLCs will use their internal clocks to determine how much time has passed between PID instruction executions, but they can't now.) Entering an update time other than the actual calculation interval time will result in unintentional scaling of the derivative term and the integral term. The larger the value entered, the faster integral terms will build up and the less effect the derivative term will have.

4. A **deadband** value (PD mnemonic = .DB, integer offset = 9) entered in the same units as used in entering the setpoint. If the calculated error has gone through zero since the last time it was greater than the deadband value, but the error is less than the deadband value, the PID instruction will calculate its new control variable output as if the error were zero. The control variable will therefore be equal to the previously calculated integral component, but the integral term won't build up. The control variable value will probably be large enough to allow an actuator system (such as a lifting mechanism) to oppose forces that might cause error (gravity, for example), but won't change to correct the small outstanding error. If a system has physical

---

[11] The formulas shown here assume that the independent gains formula is used and that the derivative term is based on calculated error. They are slightly different otherwise. The formula shown for the derivative term is simplified; the Allen-Bradley PID calculation actually smooths the derivative term over 16 time intervals.

deadband characteristics (e.g., static friction), a programmed deadband should be used to prevent the integrator from causing sudden moves when it finally builds up enough force to overcome the physical deadband. A status bit (PD mnemonic = .EWD, integer offset 0/8) is set if error has gone through zero but is still within the deadband range.

5. Upper and lower **CV limit %** values, also called high and low **output limit %** values (PD mnemonics = .MAXO and .MINO, integer word offsets 11 and 12, respectively). Enter as a percentage between 0 and 100 (use floating point in a PD data file). Indicates the upper and lower limits of the 0 to 4096 range of control variable output that the PID instruction is allowed to generate. If the PID instruction generates a result outside this range, and if output limiting is selected, as described earlier, the PLC-5 will reduce the output value to the limiting value. A tieback value or a software-supplied set output value will also be limited. After being limited, the control variable is stored in the control block (PD mnemonic = .OUT, integer offset word = 16). This % value is then converted to a 12-bit number between 0 and 4095 before being written out as a control variable.

6. A **feedforward** or **output bias** value (PD mnemonic = .BIAS, integer word = 6) that is added to the PID calculation's output. Values between 0 and 4095 can be entered in an integer control block. *Floating-point values between −100% and +100% can be entered in a PD control block and correspond to values between the engineering unit range's minimum (−100%) and maximum (+100%). Scaling is described below.*

7. *A **set output** value (PD mnemonic = .SO, integer word offset = 10). Enter as a percentage between 0 and 100 (floating point in a PD data file). Indicates the control variable value that will be outputted as a percentage of the full scale of possible outputs before output limiting if the set output mode or software A/M mode selection bit is on (described above).*

8. *Values required to **scale setpoint, process variable, deadband,** and **error** values. A PID instruction using a PID control block requires different scaling values than a PID instruction using an integer control block, because floating-point math is used with a PID control block, and unsigned integer math is used with an integer control block.*

(a) *If the PID instruction has an integer control block, scaling parameters must be entered for:*

- ***Maximum** and **minimum scaled input** values. The values in the setpoint data word and deadband data word can then be entered in engineering units in the scaled input range instead of in the standard 0 to 4095 range (the setpoint descaling selection bit must be left at 0 to enable descaling). The PID calculation will adjust the setpoint and deadband engineering values to numbers between 0 and 4095 before using them in the PID equation. If, for example, a setpoint is intended to control a motor's speed within −200 to +350 rpm, "−200" and "350" would be entered as the minimum and maximum scaling values, and the PID instruction would adjust this 550 value span to fill the 0 to 4095 standard input value span. A new setpoint value of 75 would be recognized as halfway between −200 and +350, so it would be adjusted to 2047, which is halfway between 0 and 4095. The PID instruction will use the same scaling factors to convert the process variable and the error value from 12-bit numbers into engineering units for display purposes only. Offset word 14 contains the scaled process variable value, and offset word 15 contains the scaled error value. Other program rungs can copy these scaled values to an operator display screen.*

(b) *If the PID instruction has a PID control block, two sets of scaling parameters must be entered to allow entering of setpoint and deadband values in engineering units. The PID instruction will use the scaling factor to adjust the process variable into engineering units before comparing it with the setpoint to calculate error, and then to convert the error value from engineering units to percentage units for the PID calculation. The scaling values that must be entered include:*

- ***Maximum*** *and* ***minimum input range*** *values (mnemonics = .MAXI and .MINI). These floating-point values must reflect the maximum possible range of process variable values that can be supplied to the PID instruction. This range is typically defined by the analog input module, so is usually within a 12-bit unsigned integer range (0 to 4095). If the programmer wants to restrict the process control range to less than the full process variable measurement system's range, this is where reduced values should be entered.*
- ***Maximum*** *and* ***minimum engineering unit*** *scaling values (mnemonics = .MAXS and .MINS) can be entered. Enter upper and lower engineering unit values (in real numbers) that correspond to the maximum and minimum input range values entered. If .MAXS and .MINS are left zero, then setpoint and deadband values will be assumed to be in the range 0 to 4095. If a setpoint value outside this engineering unit range is later supplied to the PID instruction, the PID instruction sets a* SETPOINT OUT OF RANGE *status bit (mnemonic = .SPOR).*

  Assume, for example, that .MINS and .MAXS are entered as −300 and +300, so that rpm (revolutions per minute) speed setpoints in this range can be supplied to the PID instruction to control a motor's speed. Assume also that the process variable measurement system (sensor, amplifier, and analog input module) could measure speeds of between −600 and +600 rpm if the analog input module's entire 12-bit range (0 to 4095) was used. 1024 should be entered for .MINI, since it is the process variable value that represents the minimum speed of −300 rpm (.MINS), and 3092 should be entered as .MAXI because 3092 represents the +300 rpm maximum speed (the .MAXS value). The PID instruction will calculate the process variable value (mnemonic = .PV) and the error value (.ERR) in engineering units. A ladder logic program can copy these values to a display device.

The following configuration words can only be entered into a PID control block structure:

1. *A* ***process variable high alarm*** *value and* ***process variable low alarm*** *value (mnemonics = .PVH and .PVL). If the process variable is found to be outside these (engineering unit) limits, the PID instruction sets a* PV ALARM HIGH *bit* or a *PV* ALARM LOW *bit (mnemonics = .PVLA and .PVHA). To prevent the alarm bits from going on and off rapidly while the process variable oscillates across the .PVL or .PVH value, a* ***process variable alarm deadband*** *value can also be entered (in engineering units). This deadband effectively introduces hysteresis into the alarm. The alarm status bit turns on as soon as the process variable exceeds a .PVL or .PVH limit, but doesn't go back off until the process variable value returns to inside the .PVL to .PVH range by more than the PV alarm deadband specification.*

2. *An* ***error high alarm*** *value and an* ***error low alarm*** *value, called the (+)* ***deviation***

alarm *and the (−)* ***deviation alarm*** *values on the configuration screen (mnemonics = .DVP and .DVN, respectively). If the error value (in engineering units) exceeds the high or low limit, an error is high alarm or an error is low alarm bit is set (mnemonics = .DVPA and .DVNA, respectively) while those alarm limits are exceeded. An* ***error alarm deadband*** *value (called deviation alarm deadband in some places) can also be entered. Once an alarm bit is set, the bit will not be reset until the error returns from beyond the alarm limit and moves inside that alarm limit by an amount equal to the deadband.*

The PID control block structure also contains a ***tieback*** *% data value (mnemonic = .TIE), which is the 0 to 4095 tieback value converted to % output units. This value is available for display purposes.*

SLC 500 ALLEN-BRADLEY SLC 500 PID INSTRUCTIONS

The SLC 500 PID instruction, shown in Figure 12.42, calculates its result using 16-bit signed integer math. The PID instruction does not execute every scan cycle while its control logic is true, but instead, executes at repeatable timed intervals in either of two modes, as we will see later. The SLC 500 PID instruction always uses the dependent gains formula, as follows:

$$\text{CV} = K_c\left[E + \frac{1}{T_i}\int (E)\,dt + T_d * \frac{d(E)}{dt}\right] + \text{bias}$$

When entering an Allen-Bradley SLC 500 PID instruction, the user must enter:

1. An address for the process variable. Each time the PID instruction executes, it will read the feedback value from this address, and will calculate error, sum of error, and rate of change of error based on this value. *The PID instruction requires process variable values to be between 0 and 16,383, so the user-program may have to scale process variable values into this range. During configuration of the PID instruction (explained later), the programmer can set the PID instruction up so that it scales process variable values so that they can be displayed in engineering units, but the scaled values are not used in the PID calculation.*

2. An address for a control variable. After the PID instruction finishes executing it will place the result, *a value between 0 and 16,383,* into this address. The PID instruction can be

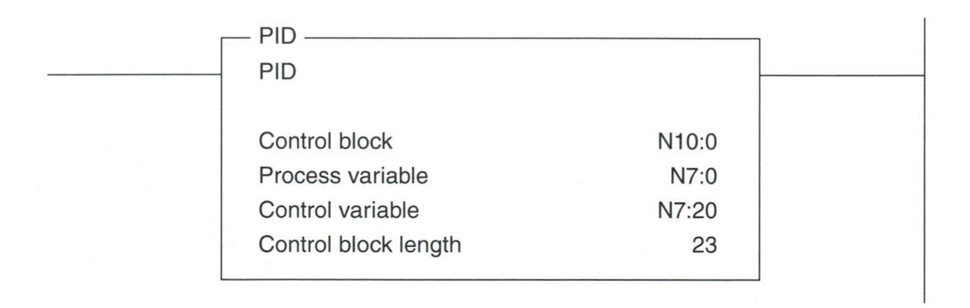

FIGURE 12.42
Allen-Bradley SLC 500 PID instruction.

```
        auto/manual:  MANUAL*                       time mode Bit:  1 TM
               mode:  TIMED *                     auto/manual bit:  1 AM
            control:  E=SP-PV *                  control mode bit:  0 CM
      setpoint (SP):        0          Ooutput limiting enabled bit:  0 OL
       process (PV):        0 *              reset and gain range:  0 RG
       scaled error:        0 *               scale setpoint flag:  0 SC
           deadband:        0             loop update time too fast:  0 TF
        output (CV):        0 %*            derivitive (rate) action:  0 DA
                                          DB, set when error is in DB:  0 DB
        loop update:        0 [.01 secs]   output alarm, upper limit:  0 UL
               gain:        0 [/10]        output alarm, lower limit:  0 LL
              reset:        0 [/10 m/r]        setpoint out of range:  0 SP
               rate:        0 [/100 min]    process var out of range:  0 PV
         min scaled:        0                               PID done:  0 DN
         max scaled:        0
  output (CV) limit:        NO *                         PID enabled:  0 EN
    output (CV) min:        0 %
    output (CV) max:        0 %
```

FIGURE 12.43
Data screen for an SLC 500 PID instruction.

configured to limit the output so that it doesn't use this full range, as we will see. If the output module requires numbers in another range, the user-program must perform a conversion.

3. The address of a control block. The control block is a set of 23 data words that contain values that the PID calculation needs to save from one execution to the next. This includes such values as:

- The setpoint value
- The multiplier constants (K_c, T_i, etc.)
- Interim calculated values the PID equation needs to store (e.g., the sum of error)
- Configuration bits for selection of the controller mode (manual or automatic) and selection of other optional features to be used in this PID calculation
- PID calculation result and status bits

The control block should be in integer memory. The fourth line in the instruction can't be changed; it reminds the programmer that a total of 23 consecutive data words will be used by the PID instruction. The programmer can enter values into the control block through the data screen or through instructions in the user-program.[12] The data screen looks like *the screen in Figure 12.43* and displays many (but not all) of the values a programmer may wish to monitor and change. Some entries include mnemonic names in parentheses following the entry's purpose [e.g., setpoint (SP)] or following the display of a bit's value (e.g., time mode bit: 1 TM). *The user-program can only address these values numerically, not by their mnemonic names. If, for example, the program needed to change the setpoint for the process, the programmer would have to know that the setpoint was the third data word in the control block (N10:2 if the control block starts at N10:0 as in the example above) and would have to program a move instruction as shown in Figure 12.44, to place a new setpoint value into N10:2. The structure of an SLC 500's PID instruction control block is shown in Figure 12.45.*

---

[12] Some data entries shouldn't be changed in run mode. Doing so could make the equipment that is being controlled perform in an unpredictable and even dangerous manner. See an Allen-Bradley instruction set manual.

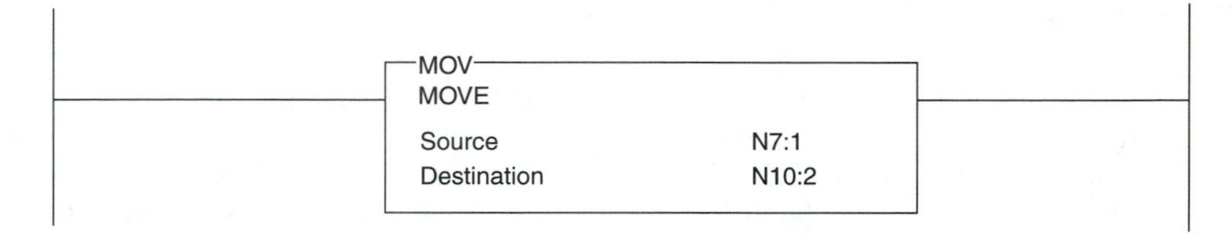

FIGURE 12.44
Using a move instruction to write the setpoint into an SLC 500 integer control block.

	②		②				

15	14	13	12	11	10	09	08	07 ②	06	05	04 ②	03	02	01	00	Word
EN		DN	PV	SP	LL	UL	DB	DA	TF	SC	RG	OL	CM	AM	TM	0

Contents	Word
* PID Sub Error Code (MSbyte)	1
* Setpoint SP	2
* Gain K_C	3
* Reset T_i	4
* Rate T_d	5
* Feed Forward Bias	6
* Setpoint Max (Smax)	7
* Setpoint Min (Smin)	8
* Deadband	9
INTERNAL USE DO NOT CHANGE	10
* Output Max	11
* Output Min	12
* Loop Update	13
Scaled Process Variable	14
Scaled Error SE	15
Output CV% (0-100%)	16
MSW Integral Sum 5/03 MSW Integral Sum	17
LSW Integral Sum 5/03 LSW Integral Sum	18
INTERNAL USE DO NOT CHANGE	19
	20
	21
	22

OL, CM, AM, TM ① (word 0, bits 03–00)

① You may alter the state of these values with your ladder program.
② Applies to the SLC 5/03 and SLC 5/04 processors.

FIGURE 12.45
Structure of an integer control block for an SLC 500 PID instruction.

Figure 12.46 shows an SLC 500 program with a PID instruction. *Notice that the PID instruction is programmed unconditionally. It will actually execute at timed intervals selected by the programmer during configuration of the PID instruction.* Immediate input and output instructions are shown, as if the analog input and analog output modules used values scaled to between 0 and 16,383, the same as the PID instruction. In fact, process variables and control variables often require scaling. The SLC 500 offers two instructions, **SCL** and **SCP,** either of which can be used to scale a number's magnitude and offset.

SLC 500 PID Control Block Data

In this section we explain the control block data entries. The descriptions include the offset for each data value that a user-program might want to monitor or change. If the offset is not in-

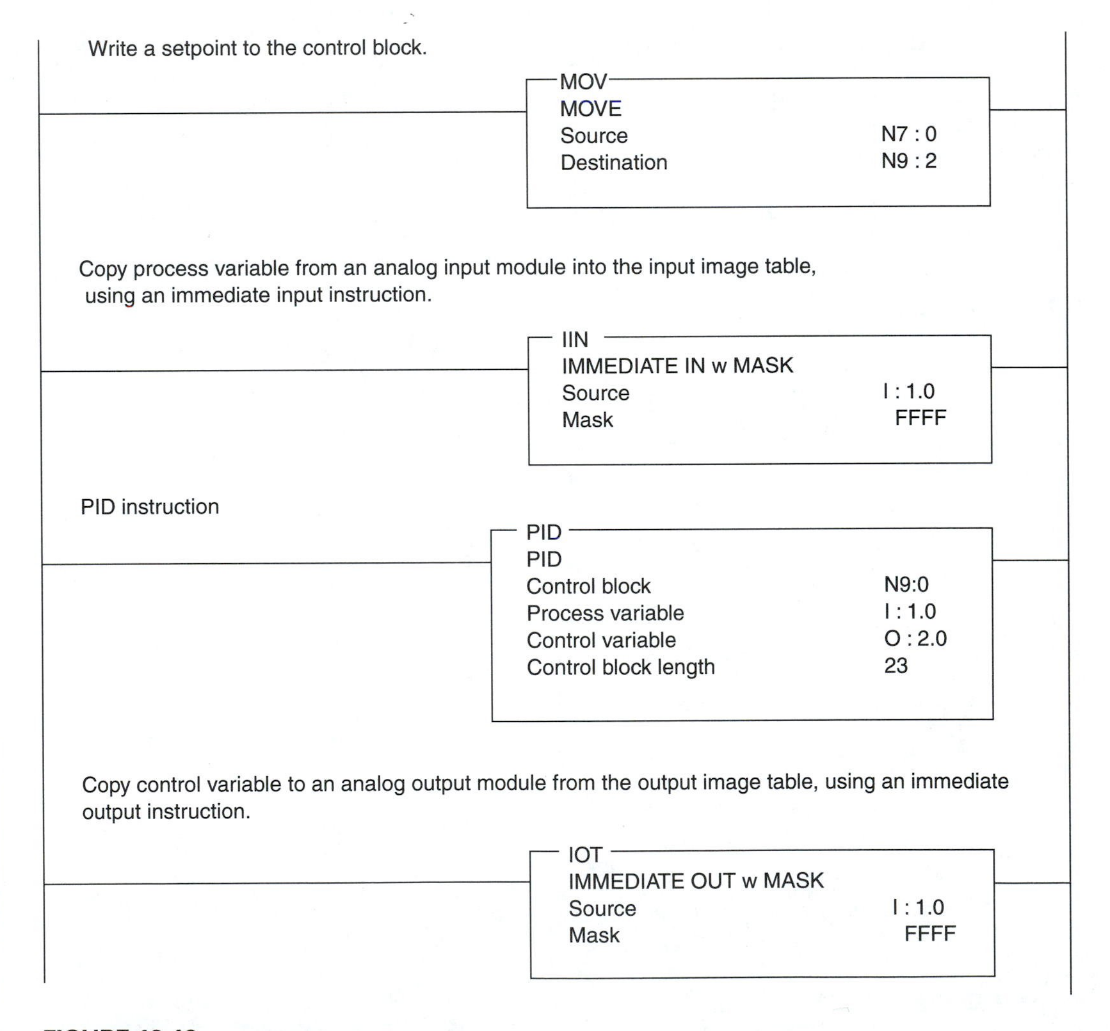

FIGURE 12.46
SLC 500 program with a PID instruction.

cluded, that data should only be changed while the PLC is in program mode, although the reader can use Figure 12.45 to find the other addresses if they are required. *Italic type will continue to mean that PID instructions in a PLC-5 work differently from the description here.*

The control bits include:

1. *A* **TIME MODE** *selection bit, for selecting how the PID instruction will be forced to execute at repeatable interval times.*

- **(a)** *STI mode. If = 0, the PID instruction will execute every time it is encountered in an STI (selectable timed interrupt) program. The programmer must ensure that the STI that contains the PID instruction executes at a rate equal to the loop update time entered in the PID instruction's control block, or the integral and derivative terms calculated in the result will be wrong. (Loop update time is described later.)*
- **(b)** *TIMED mode. If = 1, the PID instruction must be executed as part of the main scan (usually with unconditional logic as in Figure 12.46), but it will use an internal timer so that it only calculates a new process variable at the intervals defined in the control block's loop update time (described later).*

2. An **AUTO/MANUAL** selection bit (offset in control block = 0/1), for selecting the source of the control variable. A user-program rung can control this bit.

- **(a)** In automatic mode (= 0), control variable values are calculated using the PID equation.
- **(b)** *In manual mode (= 1), new control variables are not calculated by the PID instruction. The programmer must write another rung in the PLC program, which should execute only when the PID instruction is in manual mode, to write an alternative control variable to the control variable address.*

> *The value in the control variable address value is also used in Allen-Bradley's bumpless transfer feature. When the controlled process is in manual mode, the value for the PID instruction's integral term is copied from the control variable address each time the PID instruction executes, so that when the control system is switched back into automatic mode, the output due to the integrator starts at the value that was being used in manual mode, and there won't suddenly be a large change in the control variable value.*

3. A **CONTROL MODE** selection bit, for selecting how to calculate **error.**

- **(a)** E = SP − PV (= 0). This is the traditional way to calculate error. If the setpoint (SP) is greater than the measured process variable (PV), the error will be positive, so the resultant control variable will be positive and will try to move the process variable forward to catch the setpoint.
- **(b)** E = PV − SP (= 1). This method inverts the error sign, so that the PID calculation results in an output that drives opposite to the "direct" equation's result. If, for example, you use a PID instruction to control a fluid's temperature by controlling how much cold fluid is allowed into the mixture, and if the temperature (PV) is too low, we would want to decrease, not increase, the amount of cold fluid allowed in.

4. An **OUTPUT LIMITING ENABLED** selection bit (offset = 0/3).

- **(a)** When = 0, the output control variable is limited only by the standard range of values, 0 to 16,383.

(b) When = 1, the PID instruction's maximum and minimum values are limited to the values entered for the output CV min % and output CV max % parameters (explained below).

Whether or not output limiting is selected, the PID instruction sets status bits in the control block while the calculated control value is higher than its high limit (status bit at offset = 0/9) or is lower than its low limit (bit at offset = 0/10). The PID instruction also stops increasing the integral term while the calculated control variable is outside the min/max range, to prevent *reset windup,* which means that the integral term has been allowed to grow too big.

5. *A* **RESET AND GAIN RANGE** *selection bit (not available in the SLC 5/02).*

(a) *When = 0, the numbers entered as the gain multiplier and as the reset multiplier are divided by 10 before being used in the PID calculation. (Gain and reset multiplier parameters are explained below.)*

(b) *When = 1, the numbers entered as the gain multiplier and as the reset multiplier are divided by 100 before being used in the PID calculation. This enhances control precision. Since larger gain and reset values can be entered in SLC 500 models above the SLC 5/02, the programmer doesn't have to sacrifice multiplier size to get precision.*

6. A **DERIVATIVE (RATE) ACTION** selection bit. Use this bit to force the PID instruction to calculate the rate of change of:

(a) Error. If = 1, use a derivative term calculated as the rate of change of error and add the derivative component to the proportional and integral terms.

(b) Process variable. If = 0, the PID instruction will calculate the rate of change of the process variable instead of the rate of change of error. If the setpoint rarely changes, calculating the derivative based on the process variable might result in better control of a process affected by external disturbances, because the PID calculation will act to cancel environmentally caused changes in the process variable. If this method is chosen, and if SP-PV control action is selected, the derivative term will be subtracted, as follows:

Dependent gains equation:

$$CV = K_c\left[E + \frac{1}{T_i}\int (E)\,dt \underset{\uparrow}{-} T_d\frac{d\,(PV)}{dt\uparrow}\right] + \text{bias}$$

It is necessary to subtract a value proportional to rate of change of a process variable in a formula with direct (SP-PV) control to avoid overshoot, because a rapidly changing process variable is likely to shoot right past the setpoint.

The control block also contains several status bits, which a program can monitor. Most status bits are affected by other parameters that a programmer can enter, so they are explained with the parameters that affect them. Status bits that are not affected by other parameter entries include:

1. A **PROCESS VARIABLE OUT-OF-RANGE** bit (offset 0/12): set whenever the process variable value used most recently was outside the allowable range 0 to 16,383.

2. A **PID ENABLED** bit (offset 0/15): reflects the status of the control logic on the rung controlling this PID instruction.

3. A **PID DONE** bit (offset 0/13): turns on after each time the PID calculation is executed and turns off on the next scan.

Data words that must be entered into a PID instruction's control block before using the PID instruction, which can sometimes be adjusted while the control program executes, include:

1. A **setpoint** value (offset 2). This value is usually written into the control block by another rung in a user-program, as shown in the first rung of the programs in Figure 12.46. Setpoint values must be between the min scaled and the max scaled values (explained below). If scaling of setpoints is not set up during configuration of a control block, setpoint values must be between *0 and 16,383.* If a setpoint value outside the allowable range is entered, the PID instruction sets a SETPOINT OUT OF RANGE status bit (offset = 0/11).

2. Values for the **gain, reset, and rate** multipliers (K_c, T_i, and T_d) (offsets = 3, 4, and 5, respectively). *Values must be between 0 and 255 if programming an SLC 5/02, or between 0 and 32,767 when programming an SLC 5/03 or 5/04. Gain and reset values will be divided by 10 before being used in the PID calculation, and the rate value will be divided by 100 unless the reset and gain range enhancement bit has been set, in which case all three values are divided by 100 before being used.* A method for selecting these values is discussed later.

3. A **feedforward** or **bias** value (offset = 6). Although this parameter doesn't appear on the PID instruction data screen, it can be changed using other data screens or by an instruction in the program. The value from this address is added to the PID calculation's output. *Values between −16,383 and +16,383 can be entered.*

4. A **deadband** value (offset = 9). Entered in the same units as used in entering the setpoint, but only positive numbers are allowed. If the calculated error has gone through zero and is still less than the deadband value, the PID instruction will calculate its new control variable output as if error was zero. The control variable will therefore be equal to the previously calculated integral term, but the integral term won't build up. The control variable value will probably be large enough to allow the actuator system to oppose forces that might cause error (gravity, for example), but won't change to correct the small outstanding error. If the integral term was allowed to grow in systems that have physical deadband characteristics (e.g., static friction), the control variable value would eventually grow large enough to overcome the deadband and to cause sudden movements. A status bit, DB, SET WHEN ERROR IS IN DB (offset 0/8), is set if error has gone through zero but is still within the deadband range.

5. A **loop update** time. Enter times in hundredths of seconds. *Values between 1 and 255 (0.01 to 2.55 s) can be entered in a SLC 5/02, or between 1 and 1024 (0.01 to 10.24 s) in other SLC 500s. If too small a value is entered and your PLC is unable to execute the PID instruction at that rate, the LOOP UPDATE TIME TOO FAST status bit (offset 0/6) will be turned on. The loop update time is used in calculating the integral and the derivative terms, so the PID calculation result will be wrong if the TOO FAST status bit is on.* The loop update time is used as the *dt* value in calculating the integral and the derivative terms[13]:

---

[13] These formulas assume that the independent gains formula is used and that the derivative term is based on calculated error. They will be slightly different otherwise. The formula shown for the derivative term is simplified; Allen Bradley's PID calculation actually smooths the derivative term over 16 time intervals.

$$\text{new integral term} = (\text{old integral term}) + K_i\,(e_e\,dt)$$

$$\text{derivative term} = \frac{\text{change in } e_e \text{ since last calculation}}{dt}$$

Select an update time that is fast enough to control the process adequately, but not so frequent that it disturbs the PLC's scanning. It should be no longer than one-fifth the period of the process's observed natural frequency (which is discussed later). It should be at least as long as five average scan cycles.

6. *A **min scaled** value and a **max scaled** value. Used to allow setpoint values and deadband values to be entered in engineering units. The allowable range is −16,383 to +16,383 for an SLC 5/02, or −32,768 to +32,767 for an SLC 5/03 or 5/04. The PID instruction will use these values to scale the engineering unit values into the standard 0 to 16,383 value range it uses in the PID calculation. If scaling values are entered, the PID instruction also uses them to scale the process variable value and the calculated error value into engineering units. The scaled process variable and error values are not used in the PID calculation, but the PID instruction displays them on the PID data screen and stores them in the control block at offset word 14 (process variable) and offset word 15 (error). Other program rungs can copy the scaled values to display devices.*

If, for example, a setpoint is intended to control a motor's speed within −200 to +350 rpm, "−200" and "350" would be entered as the minimum and maximum scaling values, and the PID instruction would adjust this 550 value span to fill the 0 to 16,383 standard input value span. A new setpoint value of 75 would be recognized as halfway between −200 and +350, so it would be adjusted to 8191, which is halfway between 0 and 16,383. The PID instruction will use the same scaling factors to convert the process variable and the error value from the 0 to 16,383 number range into engineering units for display purposes only.

The scale setpoint flag status bit (offset = 0/5) will be set by the SLC 500 if scaling values have not been entered.

7. Output CV max. and **output CV min.** values (word offsets 11 and 12, respectively). Enter as a percentage between 0 and 100. Indicates the upper and lower limits of the 0 to +16,383 range of control variable output values the PID instruction is allowed to output. If the PID instruction generates a result outside this range, and if the OUTPUT LIMITING ENABLED bit has been set (as described earlier), the SLC 500 will reduce the output value to the limiting value. After limiting, the control variable percentage is stored in the control block at offset word = 16. This % value is converted to a number between 0 and 16,383 before being written out as a control variable.

Despite the state of the OUTPUT LIMITING ENABLED bit, if the calculated (or the operator-entered) control variable value is outside the output CV max. and output CV min. range, then either the OUTPUT ALARM, UPPER LIMIT status bit (offset = 0/9) or the OUTPUT ALARM, LOWER LIMIT status bit (offset = 0/10) will be set.

The second word in the control block is used as an error status word. If the SLC detects a major error as it tries to execute the PID instruction, the SLC 500 will:

1. Set the major fault bit, S:1/13. The PLC will halt unless a fault routine file exists and the fault routine includes instructions to clear this bit.

2. Write the error code H0036 into S:6 to indicate that the major fault was due to the PID instruction.

3. Write an 8-bit error code into the PID instruction's control block word at offset 1, indicating why the PID instruction failed to execute. Most error codes indicate that a control block parameter has been found to contain incorrect data, but one code indicates that the PID instruction has been interrupted too often to work properly. See your manual for a complete list of these error codes.

S5

SIEMENS STEP 5 PID INSTRUCTIONS

Siemens S5 PLCs don't have a PID instruction; they have a PID organization block: OB251. OB251 is preprogrammed to perform the PID calculation. To run the PID program in OB251, the user-program must call a data block containing the data that OB251 needs, then jump to OB251. The data block must contain setup parameters that are usually entered by the programmer during setup, a setpoint and a feedback value that are put into the data block by the user-program, and the PID calculation result, which is placed there by OB251. The user-program must read the result from the data block after OB 251 executes. An STL language program that does all of this would look like Figure 12.47.

In STEP 5 programs, OB251 can only be jumped to from a function block, so the section of code shown in Figure 12.47 would have to be in a function block that is called by the user-program. To ensure that OB251 executes at consistent intervals, it makes sense to call the function block from OB13, which is a block that STEP 5 reserves for timed interrupt programs (covered in Chapter 11). An S5 PLC can be configured to call OB13 at timed intervals specified by the value in reserved system word 97 (RS97). Writing to RS97 can only be done in a function block, which might be called by one of STEP 5's initialization organization blocks. A complete example program is shown in Figure 12.48.

Another STEP 5 feature that the programmer might wish to use is direct access addressing (covered in the immediate I/O section of Chapter 11). Direct access addressing causes large S5 PLCs to read and write data in the I/O modules and causes medium-sized S5 PLCs (e.g., S5-103U) to read/write the interrupt image table. Since medium-sized S5 PLCs update their interrupt image table only when they call an interrupt service routine such as OB13, direct access addressing can only be used in their interrupt service routine blocks such as OB13. A program using direct access addressing is shown in Figure 12.49. In this example, function block 2 has been modified to read values directly from an input module (or from the interrupt image table that contains data that was read from the input modules when OB13 was called) and

```
C DB4       ; Data block 4 contains data for the PID calculation
L IW72      ; Read a feedback value from an analog input module
T DW 22     ; and put it into a DW22 of DB004
JU OB251    ; Perform the PID calculation
L DW48      ; Read the result of the PID calculation from DW48
T QW112     ; and write it to an analog output module
```

FIGURE 12.47
Using the Siemens STEP 5 PID program.

OB21

Organizational block 21 is a STEP 5 initialization block. It runs automatically once each time the PLC is switched from ***stop*** to ***run*** mode. Here, it is programmed to call function block 1 to establish a timed interrupt interval.

```
JU FB1
     Init_Int
BE
```

FB1

This function block only sets up the timed interrupt to execute at intervals of 500 ms (50 standard 10 ms intervals)

```
NAME :   Init_Int
DECL  :

L KF + 50
T RS 97
BE
```

OB13

The timed interrupt organizational block, which will run at intervals defined by RS97 while the PLC is in ***run*** mode. It is programmed to call the function block that calls OB251.

```
JU FB2
     PID_exec
BE
```

FB2

This block updates the process variable (feedback) data word in a data block for the PID calculation, jumps to OB251 to do the PID calculation, then copies the control variable (the result) from the data block to an analog output.

```
NAME :   PID_exec
DECL  :

C   DB4
L   IW72
T   DW22
JU  OB251
L   DW48
T   QW112
BE
```

FIGURE 12.48
Using the Siemens STEP 5 PID instruction at timed intervals.

OB13

The timed interrupt organizational block, which runs at intervals defined by RS97. Unchanged from the previous example.

```
JU  FB2
        PID_exec
BE
```

FB2

The function block that calls OB251. Addresses IW72 and QW112 have been changed to PW72 and PW112.

```
NAME:  PID_exec
DECL:

C  DB4
L  PW72
T  DW22
JU  OB251
L  DW48
T  PW112
BE
```

FIGURE 12.49
Using STEP 5 direct access addressing.

has also been modified to write directly to an output module (or to the interrupt image table for transfer to the output module when OB13 ends).

STEP 5's PID calculation in OB251 can be used as a position algorithm or as a velocity algorithm. Its **position algorithm** form is a combination of the independent gains equation and the dependent gains equation:

$$\mathrm{CV} = K_c\left[K_p(E) + K_i \int (E)\,dt + K_d \frac{d(E)}{dt}\right] + \text{bias}$$

Siemens uses different notation for the calculations however, so the same equation looks different, as follows:

$$\mathrm{YA} = \mathrm{K}\left[\mathrm{R(XW)} + \mathrm{TI} \int \mathrm{(XW)} + \mathrm{TD} \frac{d\mathrm{(XW)}}{dt}\right] + \mathrm{Z}$$

where

YA = CV = command variable or controller output
XW = E = error
Z = bias or disturbance variable
K = K_c[14]

[14] Since K is entered in units of 1/1024 (or 1/1000), entering the value 1024 (or 1000) actually means 1 (1024/1024 = 1).

$$\mathrm{R} = K_p$$
$$\mathrm{TI} = K_i$$
$$\mathrm{TD} = K_d$$

Other variations of the independent gains PID formula can also be used:

1. The PID calculation can be configured to use the derivative of a random disturbance variable (XZ) instead of error. XZ could be the process variable or could come from anywhere else the programmer wants. This feature is described further later in this section.

2. TD can be entered as a negative value if the programmer wishes to cause the derivative value to be subtracted during the PID calculation.

3. If a negative value is entered for K, the PID calculation's direction will be reversed.

The **velocity algorithm** form of the PID equation generates a value that is proportional to the rate of change of the position algorithm's result, as in the following equation:

$$\mathrm{YA} = \frac{d(\mathrm{K}\{\mathrm{R(XW)} + \mathrm{TI}\int (\mathrm{XW})\,dt + \mathrm{TD}[d(\mathrm{XW})/dt]\} + \mathrm{Z})}{dt}$$

The velocity algorithm's output is zero while OB251's input values (setpoint, process variable, disturbance, and random variable) all remain constant. When one or more of those inputs changes, the velocity algorithm will output a number that is equal to how much the output of the position algorithm would have changed if the position algorithm were being used. The velocity algorithm's output might be used to maintain a constant output from a controlled system, by adjusting the signal that drives the controlled system whenever the system's measured characteristics are observed to change. An operator could, for example, manually set a conveyor belt to any desired speed, press a "set speed" switch, then let the velocity algorithm take over speed control. The velocity algorithm would generate correction factors whenever the actual speed, an external disturbance, or even the setpoint changed. Another program line can add the velocity algorithm's output value to the initial manually selected value to adjust it each time OB251 generates a result other than zero. This example program in shown in Figure 12.50.

While the conveyor speed remains constant, the result of the velocity algorithm in Figure 12.50 will be zero and the value at QW112 will not be changed. If, however, the setpoint increases due to an instruction somewhere else in the program, OB251 will generate a positive value that will be added to the value being outputted to the analog output module at address QW112. The higher analog output value will cause the motor to work harder until the conveyor accelerates. When speed increases, the change in the process variable value (read from IW72 and provided to OB251 via DW22) will result in the velocity algorithm generating a negative value. The negative value will be added to QW112, reducing the analog output back toward its original value. When the motor speed has increased by the same amount that the setpoint was increased, the setting at QW112 will have been returned to its initial value, which should still be adequate to maintain the new speed.

Configuring STEP 5's PID Calculation

The programmer selects the type of PID formula and provides multipliers for OB251 via entries in a data block. A programmer must enter configuration data into some of the 49 words of the

PB1

Program block 1 is called once each time the operator presses a "set speed" switch. It reads a manually selected setting and provides that value as an initial value for dc output to a motor.

```
L  IW72; Initial setting value for motor speed

T  QW112
BE
```

OB13

The timed interrupt block calling the function block that calls OB251.

```
JU  FB2
      PID_exec
BE
```

FB2

Updates the process variable data word for the PID calculation, jumps to OB251 to do the velocity algorithm PID calculation, then adjusts the analog output value by the result.

```
NAME  :   PID_exec
DECL  :

C   DB4         ;Includes configuring OB251 as velocity algorithm
L   I W72
T   DW22
JU  OB251
L   DW48
L   QW112       ;Reads the value currently outputted via the analog
+ F             ;  output module and adds the value from DW48 to it
T   QW112       ;  then writes the modified value back to the module
BE
```

FIGURE 12.50
Using STEP 5's velocity algorithm.

data block prior to running the program. Other data words contain settings that the user-program should change. For example, OB251 looks for the process variable value in data word 22, so the user-program should read the feedback sensor and write its value to DW22 before jumping to OB251. Figure 12.51 shows the data words that are important to the programmer and shows how OB251 uses those data words. The complete set of data words is explained below.

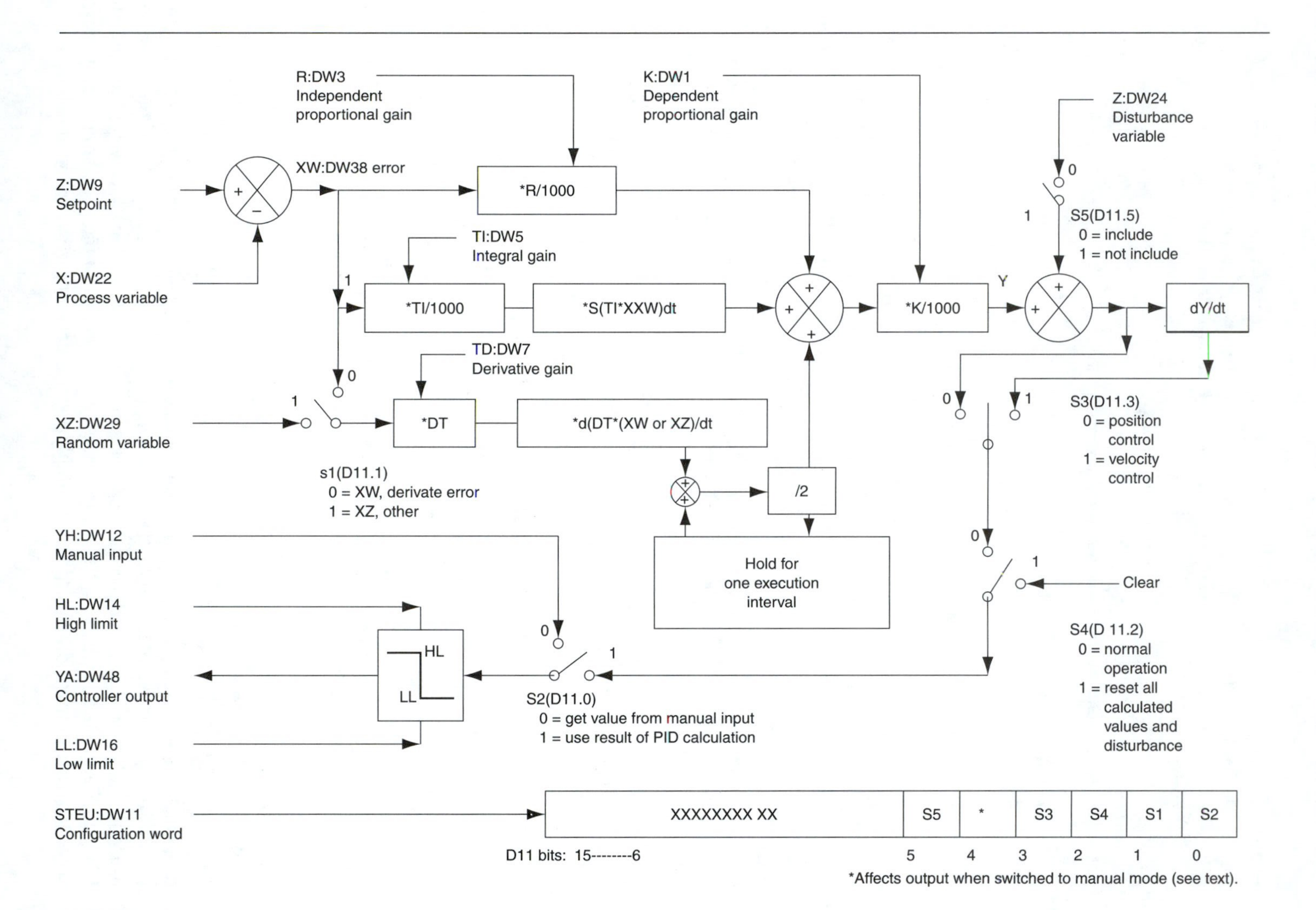

FIGURE 12.51
STEP 5's OB251, according to the manuals.

Figure 12.51 shows how OB251 operates according to the manual set that I received with my S5-103U PLC, but Figure 12.52 shows how OB251 operates according to my observations. There are a few significant differences, which means that there are probably at least two different versions of OB251 in the marketplace.[15] The differences are highlighted in Figure 12.52. I will cover both the "as documented" and the "as observed" versions here.

The values that must be entered into a data block for use by the STEP 5 PID calculation include configuration bits and data words. The configuration bits are in data word 11, which Siemens calls a **STEU** word. The low 6 bits of data word 11 include:

1. A MANUAL/AUTOMATIC mode selection bit (bit 0, called **S2** in the STEP 5 programming manual). In automatic mode (when S2 = 1), the value outputted via DW48 is the result of the PID calculation. In manual mode, OB251 copies the manual input value from DW12 to DW48 (subject to the setting of bits 3 and 4), reducing that value as required so that DW48 does not exceed the high or low limit values.

The user-program may monitor an operator-controlled input switch to control the S2 bit and can read an analog value from an operator-controlled dial to write to DW12. Use the STL set unconditional (SU) and reset unconditional (RU) instructions in a function block to control S2, because individual bits of data words cannot be changed by the standard Boolean logic instructions. SU and RU aren't available in ladder logic; you must use word logic instructions to turn S2 off in ladder logic, as shown in Figure 12.53a, or to turn it on as in Figure 12.53b.

2. A DERIVATIVE VALUE selection bit (bit 1, **S1**). When S1 = 0, the derivative term is calculated using the rate of change of error (XW), in the normal way. If S1 = 1, OB251 will calculate the rate of change of the random variable (XZ) in DW29, and use that result in the PID equation.

A programmer may want to use the derivative of the process variable instead of error, as in the formula

$$YA = K\left[R(XW) + TI\int (XW) \underset{\uparrow}{\mp} TD\frac{d(XZ)}{dt}\right] + Z$$

so the program must copy the process variable value from the input module to both the process variable data word (DW22) and to the random variable data word (DW29). To subtract the derivative term, a negative value must be entered for the derivative gain.

The random variable doesn't have to be the process variable value. Any value can be used here. The derivative of the setpoint is sometimes added to the output, so that the system will respond to rapid setpoint changes with more effort than it responds to slowly changing setpoints.

3. An ALGORITHM selection bit (bit 3, **S3**). If S3 = 0, the position algorithm is selected, or set S3 = 1 to select the velocity algorithm. The choice of position or velocity algorithm, examined earlier in this section, also affects what STEU bit 4 does.

4. A NO JERK selection bit (bit 4, the only configuration bit not identified as "S-something" in STEP 5 programming manuals). If OB251 is set up as a position algorithm and D11.4 = 0, the "no jerk" option is selected. Whenever the PID calculation is switched to manual,

---

[15] My PLC was actually a Westinghouse PC503, which is an S7 = 103U PLC manufactured by Siemens for Westinghouse to sell in North America.

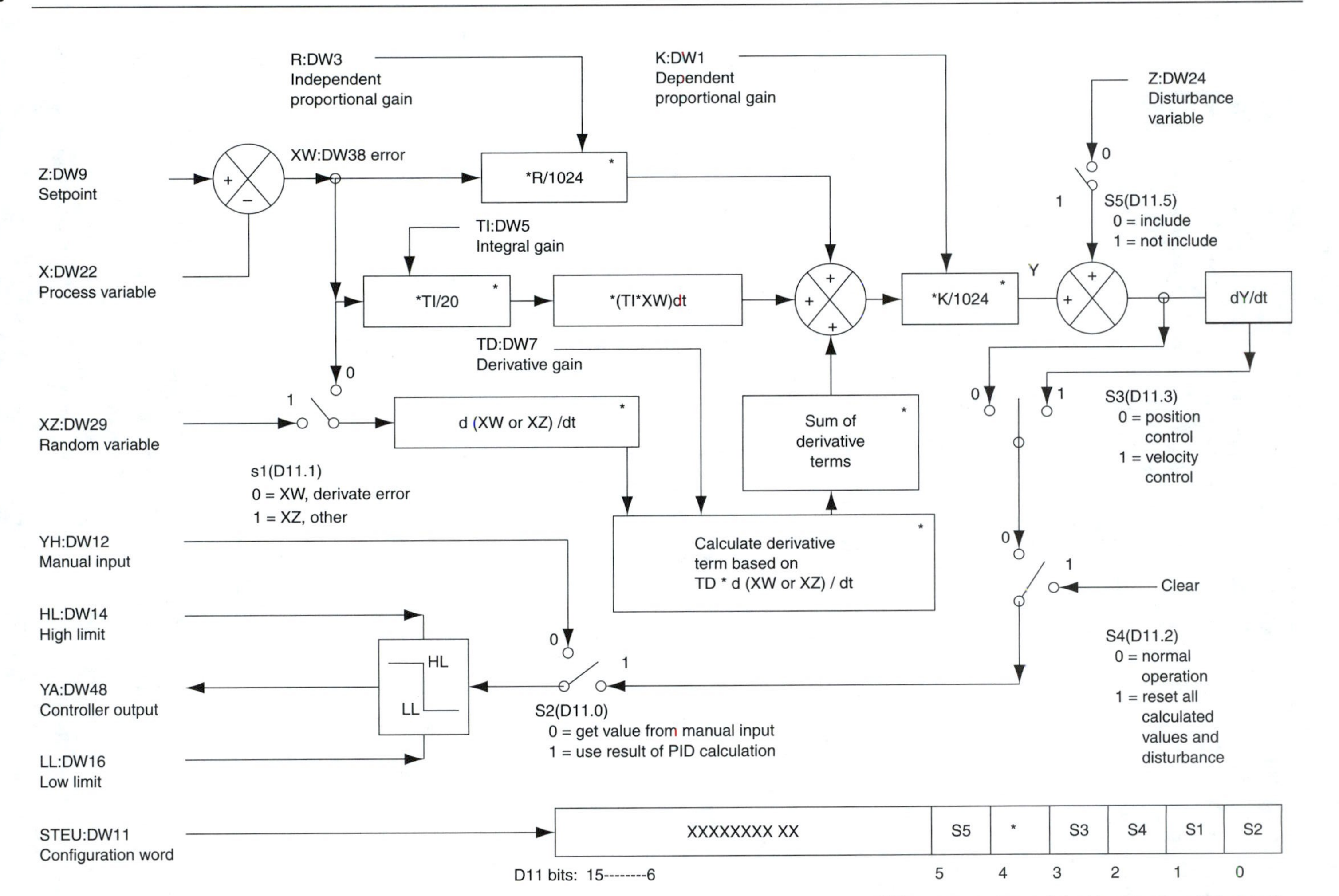

FIGURE 12.52
STEP 5's OB251, as observed.

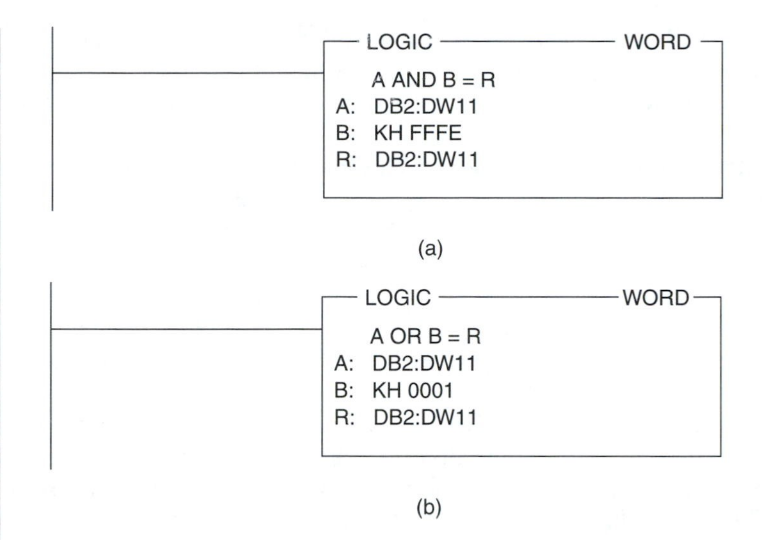

FIGURE 12.53
Masking to change a single data bit value in STEP 5: (a) using AND logic to turn bit D11.0 off; (b) using OR logic to turn bit D11.0 on.

OB251 will change its output to the value provided as a manual input (at DW12) in equal-sized changes over four calculation intervals. If bit 4 = 1, OB251 will hold its output at the last calculated value while in manual mode.

If OB251 is set up as a velocity algorithm and D11.4 = 0, OB251 will change its output value (while in manual mode) to the value provided at the manual input in the very first calculation interval. If bit 4 = 1, OB251's output (while in manual mode) will be zero.

5. A DISTURBANCE VARIABLE selection bit (bit 5, **S5**). The disturbance value in DW24 is included if S5 = 0.

6. A RESET bit (bit 2, **S4**). If the PLC program sets this bit to 1, then resets it to 0, all calculated values, including integrated error, derivative smoothing values, process variable value, and disturbance value, will be cleared. Setpoint, gain settings, output limits, and manual input values are not cleared.

Some data words must be entered into the data block before OB251 is called, to configure the PID calculation. Some of these values can be modified as the program executes, to adjust the way the calculation works. Other values (e.g., setpoint, process variable, control variable) must be read and written by the user-program. According to the manuals, the data words must have values between 2047 and −2047 (except as noted below). This range corresponds to 12-bit signed binary numbers. I have found that numbers from −32,768 to +32,767 work fine (the range for signed 16-bit numbers). There is increased danger of generating a result that is outside the 16-bit number range if very large numbers are used. OB251 does not perform scaling of input or output values, so the user-program may have to do scaling separately.

Of the 49 data words in OB251's data block, the user can manipulate the following:

1. W, the **setpoint** (in DW9). Usually placed into the data block by the user-program, as in earlier examples.

2. X, the **process variable** (DW22). Must be read from a feedback sensor and copied to the data block before each execution of OB251. The user-program may need to scale process variable values so that they have the same range as the setpoint values, before writing them to DW22.

3. K, the **proportional action coefficient** (DW1); sometimes called the **dependent proportional gain.** The manuals say that values between −32,768 and 32,767 can be entered, but they are divided by 1000 before being used, so a value of 2000 for K is actually a proportional gain of 2. (I have observed that the value is actually divided by 1024, so 2048 must be entered to get a gain of 2.[16]) Setting this value to zero makes the entire PID calculation's output zero, so OB 251's control variable output is equal to the disturbance value only. Setting DW1 to a negative value reverses the control action of OB251. If DW1 = 1000, the PID equation is an independent gains equation.

4. R, the **independent proportional gain** (DW3). As with the dependent proportional gain, values between −32,768 and 32,767 are allowed, but they are divided by 1000 (or by 1024, according to my observations) before being used. Setting DW3 = 1000 makes the PID equation a dependent gains equation.

5. TI, the **integral gain** (DW5). The manuals say that a value between 0 and 9999 can be entered, and it is divided by 1000 before being included in the PID calculation. (My observations show that the larger allowable input value range can be used, and values are divided by 20.) The integrated output value will not change while OB251's calculated output is greater than the high limit value or lower than the low limit value. (HL and LL are covered later.)

The PID calculation in OB251 assumes that the calculation executes every second, regardless of the actual interval time between executions of OB251. If, for example, OB251 executes every second for 5 s while the error remains constant at 30 and TI is 2000, then during those 5 s the output due to the integrator term will have increased by

$$
\begin{aligned}
\text{output}_{\text{integrator}} &= \frac{\text{TI}}{1000} \times \text{error} \times \text{iterations} \\
&= \frac{2000}{1000} \times 30 \times 5 \\
&= 300
\end{aligned}
$$

If, on the other hand, OB251 is actually executed every half-second, it will execute 10 times in 5 s and the increase in output due to the integral term will be

$$
\begin{aligned}
\text{output}_{\text{integrator}} &= \frac{\text{TI}}{1000} \times \text{error} \times \text{iterations} \\
&= \frac{2000}{1000} \times 30 \times 10 \\
&= 600
\end{aligned}
$$

---

[16] To check to see which version of OB251 you have, configure a position control algorithm that isn't connected to sensors or actuators, enter zeros for TI and TD, enter a setpoint and a process variable with a reasonably large difference, then try values of 1000 and 1024 to see which one causes the controlled variable (DW48) to be the same as the setpoint - PV difference.

so it is up to the programmer to ensure that OB251 executes exactly once per second, or to adjust the integrator term accordingly.

6. TD, the **derivative gain** (DW7). There is a big difference between how the manuals say that this term works and how I have observed it to work:

(a) The manuals say: Values between 0 and 999 can be entered. The derivative equation assumes that each interval between executions is 1 s. The programmer must ensure that OB251 executes once per second, or must adjust TD accordingly.

If, for example, OB251 executes every second while error is changing at a constant rate of 50 every second and TD is set to 2, the output of the derivative term would be

$$\begin{aligned} \text{output}_{\text{derivative}} &= \text{TD} \times \frac{\text{error change rate}}{\text{iteration time}} \\ &= 2 \times \frac{50\ \text{s}^{-1}}{1\ \text{s}} \\ &= 100 \end{aligned}$$

but if OB251 were executing at half-second intervals, only one-half as much error changes would be observed per second (25 per iteration), so TD would have to be doubled (to 2) to compensate:

$$\begin{aligned} \text{output}_{\text{derivative}} &= \text{TD} \times \frac{\text{error change rate}}{\text{iteration time}} \\ &= 4 \times \frac{25\ \text{s}^{-1}}{1\ \text{s}} \\ &= 100 \end{aligned}$$

The manual also says that the derivative term is smoothed by averaging each new derivative result with the value calculated previously, as shown in Figure 12.51. The result is used in the PID calculation and is also stored so it can be averaged with the next calculated derivative. The smoothing is also affected by the actual execution interval for OB251. If OB251 is executing every half-second, the effect of a sudden error change will decay twice as fast as if OB251 were executing every second.

(b) I have observed that numbers between −32,768 and 32,767 can be entered, and (as shown in Figure 12.52) the derivative term's output is smoothed in a manner that does not decay over time. Every time the error changes the output of the derivative term is changed by an increment that is related to the size of the recent error change (positive or negative) and to the size of the derivative gain value (TD). (The increment sizes never exceed the range represented by a signed 5-bit binary number.) While error continues to grow, the derivative term's output also continues to grow by accumulating more increments. If error is decreasing, negative increments are added to the derivative's accumulated output value, reducing it. If the rate of change of error is zero, smoothing increments are not added, so the derivative term does not become zero. Increment sizes generally increase with the increases in rate of change of error and/or with larger TD values, but the sizes of the increments are not perfectly linearly related to the rate of change of error or to

TD. The uneven effect usually disappears after a few executions of OB251, but programmers should be aware that the derivative term can accumulate a small but significant offset, so that it may not return to zero when it should. The program in Figure 12.54 is the same as previous examples, but it clears OB251's accumulated values if error remains small and hasn't changed since the last time OB251 exe-

```
FB211
This block executes OB251 as in earlier examples, but includes
a jump to a routine that corrects cumulative derivative error.

NAME: PID_exec
DECL :

C DB4
L IW72
T DW22
JU OB251
L DW48
T QW112
JU FB3
      Corrrection
BE

FB3
Corrects accumulated derivative error if error is small and stable.

NAME: Correction
DECL:

L DW38       ; Load OB251's most recent calculated error value
L KH0008     ;   and load a small positive value (8)
>F           ; Check if error is greater than 8
BEC          ;   and terminate execution of FB3 if it is

L DW38       ; Load the error value again
L KHFFF8     ;   and load a small negative value (–8)
<F           ; Check if error is less than –8
BEC          ;   and terminate FB3 if it is

L DW50       ; Load a word used by FB3 only (the previous error)
L DW38       ;   and the most recent error from OB251
T DW50       ;   (store the new error for next time FB3 runs)
><F          ; Check if new error is different from old error
BEC          ;   and terminate FB3 if it is

L KH000      ; Clear PID calculation result if error is small and
T DW44       ;   hasn't changed since last time FB3 executed
BE
```

FIGURE 12.54
Clearing accumulated derivative error in STEP 5.

cuted. (*Be careful:* This will also clear the integral term, which might create a problem.)

7. HL and **LL,** the **high limit** (DW14) and the **low limit** (DW16) of the output value. These are the maximum and minimum values OB251 will allow the output to go to. If the sum of the PID calculation plus the disturbance value (or the manual input value) is outside these limits, it is changed to the limiting value.

While output is at a limiting value, the integrator term is not allowed to grow. This feature prevents windup of the integrator term, which might otherwise lead to large overshoots. The programmer should consider setting HL and LL to reasonably low values, to make use of the antiwindup feature.

8. Z, the **disturbance variable** (DW24), sometimes called **bias** or **offset.** The disturbance value will be added to the result of the PID position algorithm calculation before calculating the velocity algorithm output if the STEU word configures OB251 to do so. OB251 actually calculates the change in the disturbance value from what it was when OB251 was switched into automatic mode, then adds that calculated value!

9. YH, the **manual input** (DW12). This value may be used as an output value instead of using the result of the PID calculation when OB251 is in manual mode (depending on how OB251 is configured).

10. XZ, the **random variable** (DW29). The value from DW29 is used in the derivative calculation instead of derivating the error value, if OB251 is configured to do so. The value that the user-program writes into this address may be a copy of the process variable if derivative of output is wanted, or can be a copy of the setpoint if derivative of setpoint is wanted.

11. YA, the **control variable** (DW48). This is where OB251 puts its final result. After each execution of OB251, the user-program should copy the value from this address and write it to an output module to control the process. It may be necessary to scale the output from DW48 before writing it to an analog output module.

Other data words in the 49-word data block are used by OB251 to store data between executions, and normally should not be changed by the programmer.

S7

SIEMENS STEP 7 PID INSTRUCTIONS

Only a few S7 PLCs come with a built-in PID control algorithm (the S7-215, S7-216, and the S7-314 IFM). In these PLCs, the PID algorithm is in a system function block, so it can be used to control multiple processes if each process has its own instance data block to store parameters and working values.

If you don't have a PLC with a built-in PID algorithm, you must purchase the optional function block or write it yourself. Since the built-in PID control algorithm in the S7 314 IFM is a good position control type of algorithm, which uses floating-point math for high accuracy, in this section we examine how to use it and how it works so that you can write a similar function block.

The built-in system function block in the S7-314 IFM for PID control is **SFB41 "CONT_C".** The process variable can be supplied to CONT_C as a real number (**PV_IN**) or as a signed 16-bit number (**PV_PER**) as the address of a peripheral input. CONT_C converts PV_PER to a real number, then scales it. CONT_C performs its calculation, then outputs the control variable as a real number (**LMV**) and also as a scaled 16-bit signed number (**LMV_PER**) suitable for sending to a peripheral output.

Figure 12.55 shows how CONT_C calculates a new control variable output each time it is executed with a new process variable. All input and output parameters are either floating-point (real) numbers or Boolean bits, except as noted. The following is a description of CONT_C from the STEP 7 manual set, supplemented with what the author knows about PID algorithms:

1. The process variable can be entered as an integer input parameter, **PV_PER.** PER_VAL is converted from a 16-bit signed binary number to a 32-bit floating-point (REAL) binary number, then scaled by being multiplied by a process variable scaling factor (**PV_FAC**). The result is adjusted by adding an offset value (**PV_OFF**) and then is used as the process variable (**PV**) if the **PVPER ON** bit is set; otherwise, the real number supplied as **PV_IN** is used as the process variable (**PV**).

2. The difference between the setpoint (**SP_INT**) and the process variable is calculated.

3. CONT_C examines the calculated error to see if its absolute value is greater than a deadband value (**DEADB_W**). If it isn't, CONT_C uses zero instead of the actual error in the following calculations, so that tiny steady-state errors can't cause integrator buildup.

4. The error value (now called **ER**) is then multiplied by a proportional **GAIN** value, and the result is included in the final PID result if the proportional output is enabled (if **P_SEL** is TRUE).

5. A copy of the newly calculated (GAIN * ER) value is divided by the integral time value (**TI,** entered in TIME format) and multiplied by the actual time since CONT_C last executed (**CYCLE,** entered in time format). If the integral term isn't frozen (**INT_HOLD** can be set to TRUE to freeze the integral term), this newly calculated value will be added to the integral (sum) of similar calculations from previous executions of CONT_C. The integrated result is included in the final PID calculation if the integral output is enabled (if **I_SEL** is TRUE).

6. The derivative (rate of change) of (GAIN * ER) is calculated by subtracting a copy of the (GAIN * ER) value from the (GAIN * ER) value from the previous execution of CONT_C, then dividing the difference by the actual time since CONT_C executed last (**CYCLE**). The result is multiplied by the derivative time value (**TD,** entered in time format), and then a time lag value (**TM_LAG,** entered in time format) is applied to smooth the effects of sudden changes. It isn't explained how CONT_C uses TM_LAG in smoothing, but if you are writing your own delay routine, you can use one of the following methods:

- **(a)** Store a reasonable number (10 or 20) of the most recently calculated derivative term values in an array, discarding the oldest value each time a new value is added. Calculate the average of the stored values as a sort of delayed output. The averaged result will immediately be affected by the latest derivative value, but the full effect would be spread over the 10 to 20 iterations.
- **(b)** Add the newly calculated derivative value to the saved result of the previous smoothed derivative calculation, and divide the result by 2. Output the result and save it for use in the next calculation. To allow adjustable smoothing, the old and new values can be weighted before being added. If larger weighting is used for the old value, the result is greater smoothing.

After the smoothing calculation is performed, the derivative result is included in the final PID calculation if the derivative output is enabled (if **D_SEL** is true).

7. The proportional, integral, and derivative results are added (**LMN_P** + **LMN_I** + **LMN_D**).

8. A disturbance variable value (**DISV**) is added to the PID result.

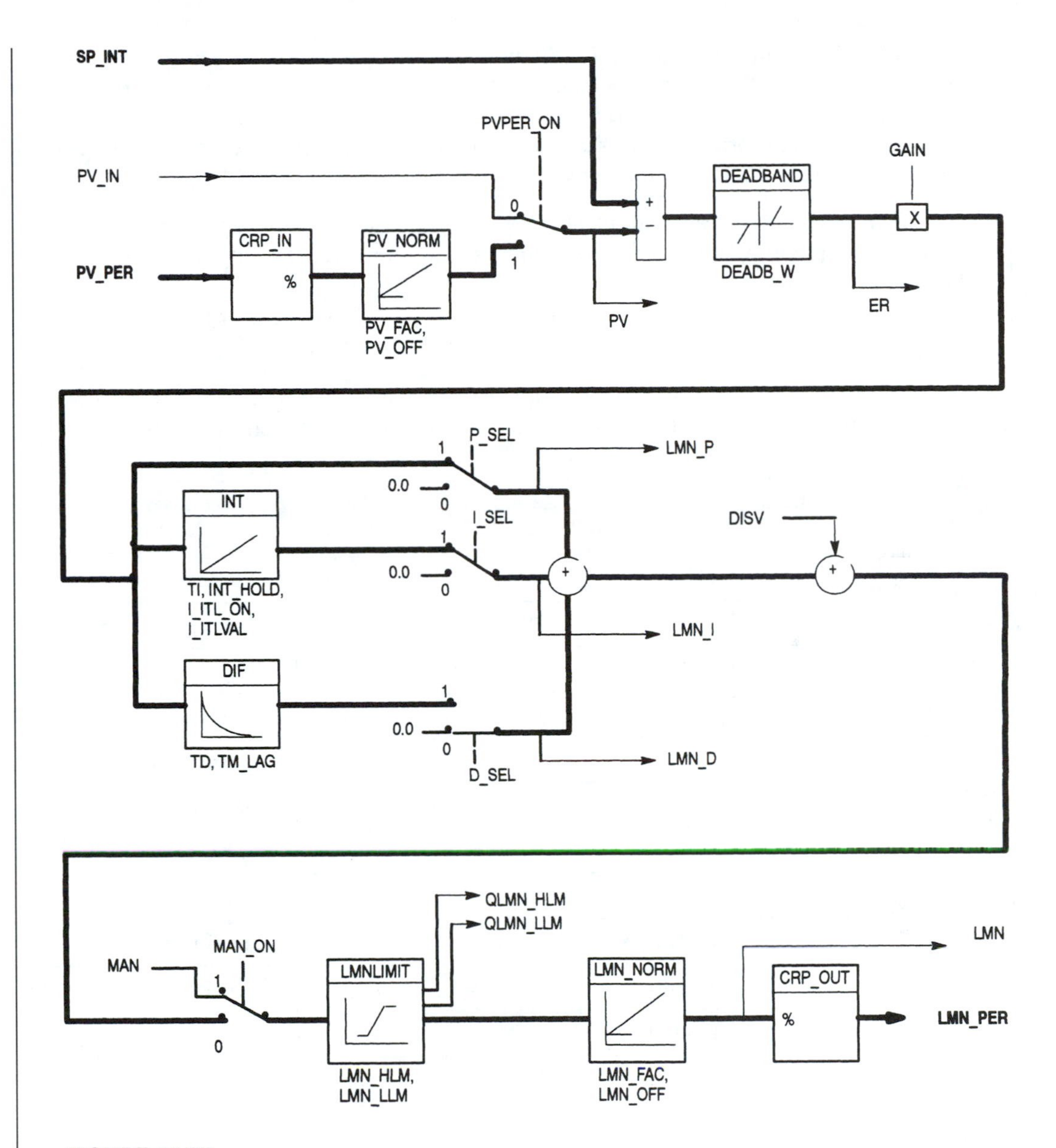

FIGURE 12.55
Block diagram of STEP 7's CONT_C.

9. If **MAN_ON** is true, the calculated result is ignored and a manual input value (**MAN**) is used in its place.

10. The result so far is compared against a high limit value (**LMN_HLM**) and a low limit value (**LMN_LLM**). If the result is outside these limits, it is discarded and the limit value is used in its place. If CONT_C makes this change, it also turns on one of two status bits (**QLMN_HLM** or **QLMN_LLM**).

11. The result is scaled by multiplying it with a scaling factor (**LMN_FAC**), then an offset value (**LMN_OFF**) is added to the result. The result of this calculation is the control variable in REAL number format and is available as CONT_C's **LMN** output parameter.

12. The real format result is converted to an integer format number and is available as CONT C's **LMN_PER** output parameter.

The sequence of calculations above is the equivalent of the dependent gains PID algorithm (if we ignore the number conversions, scaling, integral freezing, limiting, and time-delaying options):

$$\text{LMN} = \text{GAIN}\left[(\text{ER}) + \frac{\text{CYCLE}}{\text{TI}}\int(\text{ER})\ dt + \frac{\text{TD}}{\text{CYCLE}}\frac{d(\text{ER})}{dt}\right] + \text{DISV}$$

A function block to execute the same PID calculation that CONT_C executes would look like the program in Figure 12.56. The example program does not check status word bits to en-

Address	Declaration type	Name	Data type	Initial value	Comment
0.0	in	M_SP_IN	REAL	0.0	Setpoint
4.0	in	M_PV_IN	REAL	0.0	Process variable
8.0	in	M_GAIN	REAL	1.0	K_p
12.0	in	M_TI	REAL	0.0	In seconds
16.0	in	M_TD	REAL	0.0	In seconds
20.0	in	M_DISV	REAL	0.0	Disturbance
24.0	in	M_CYCLE	REAL	1.0	In seconds
28.0	out	M_LMN	REAL	0.0	Control variable
32.0	stat	M_OLD_P	REAL	0.0	Saved error
36.0	stat	M_LMN_I	REAL	0.0	Integral
0.0	temp	M_LMN_P	REAL	0.0	Proportional term
4.0	temp	M_LMN_D	REAL	0.0	Derivative term

Network 1

Calculate the error and generate the proportional term.

```
L #M_SP_IN      // Setpoint
L #M_PV_IN      // Minus the process variable
–R              // Equals error
L #GAIN         // Times dependent proportional gain
*R              //    = proportional term
T #M_LMN_P      // Saved for other networks to use
```

FIGURE 12.56
STEP 7 function block to perform PID control.

Network 2

Generate and save the integral term.

```
L #M_LMN_P          // Error, with dependent proportional gain
L #CYCLE            // times the interval since the last
*R                  //    calculation
L #M_PI             // divided by the integral time
/R                  // equals value to integrate
L #M_LMN_I          // plus previously integrated values
+R                  //    = integral term
T #M_LMN_I          // saved for network 4
```

Network 3

Generate the derivative term, then save the current error for future use.

```
L #M_LMN_P          // Error, with dependent proportional gain
L #M_OLD_P          // minus the same value from the previous
–R                  //    calculation
L #M_CYCLE          // divided by actual time since last
/R                  //    calculation
L #M_PD             // times the derivative time constant
*R                  //    = derivative term
T #M_LMN_D          // stored for network 4


L #M_LMN_P          // Currrent error, with dependent gain
T #M_OLD_P          // Save for next calculation interval
```

Network 4

Sum the P, I, and D terms.

```
L #M_LMN_P          // Proportional term
L #M_LMN_I          // plus integral term
+R
L #M_LMN_D          // plus derivative term
+R                  //    = PID calculation result
T #M_LMN            // as an output parameter
BE
```

FIGURE 12.56
(continued)

sure error-free operation, and leaves out the number conversion, scaling, integral freezing, limiting, and time-delaying features that CONT_C contains. (I have used the same parameter names that CONT_C uses, but prefixed each with an M_ to remind the reader that this is not CONT_C but just a similar program).

The S7-314 IFM also offers **SFB 42 "CONT_S",** which calculates a value proportional to the error between a setpoint and a process variable, and adds an integral term that is proportional to the difference between the accumulated past error and the accumulated past output of CONT_S. The result is a type of PI control in which the integral term is proportional to the integrated uncorrected error. CONT_S also converts the calculated result from a real number into two Boolean outputs that can be used to switch a positive and a negative power supply, for three-level (positive, zero, negative) pulse width modulation to control a dc actuator.

CONT_S automatically calls another system function block, **SFB 43 "PULSE_GEN",** which converts the calculated PI result into the 2-bit PWM output. A user-program can call PULSE_GEN to convert any real number (the output from CONT_C, for example) into a three-step PWM signal. When called by a user-program, PULSE_GEN can be set up to generate:

1. Symmetric three-step output, in which the entire range of real numbers is scaled linearly to the range of positive and negative PWM output.

2. Asymmetric three-step output, in which the positive PWM output has a different gain than the negative PWM output.

3. Bipolar two-step output, in which the entire range of input real numbers is scaled to match just the positive PWM output range.

4. Unipolar two-step output, in which positive real numbers are scaled to positive PWM output, and negative real numbers cause zero output.

5. Any of the above, but with a minimum pulse-duration time and no-pulse-duration time (to yield a deadband for real numbers near zero), and with an output boost that causes maximum PWM output if real numbers are near their maximum.

CQMI

OMRON CQMI PID INSTRUCTION

OMRON's PID instruction is available only in the CQM1 CPU4x PLCs, and it performs a PID calculation based on a slightly older version of the dependent gains equation than the one covered so far in this book. It uses a **proportional band** value instead of the unitless proportional gain multiplier for the proportional gain, and it doesn't add a bias offset value. It also differs from the equations we have seen so far in that it self-adjusts according to its execution interval. OMRON's PID formula looks like this:

$$\text{CV} = \frac{\mathbf{PB}}{\mathbf{100}}\left[E + \frac{1}{R_i\boldsymbol{T_n}}\int (E)\,dt + R_d\boldsymbol{T_n}\frac{d(E)}{dt}\right]$$

where

$$\text{PB} = \text{proportional band, in percentage (\%)}$$

$$= \frac{\text{change in proportional output}}{\text{change in error value}} \times 100$$

$$= K_c \times 100$$

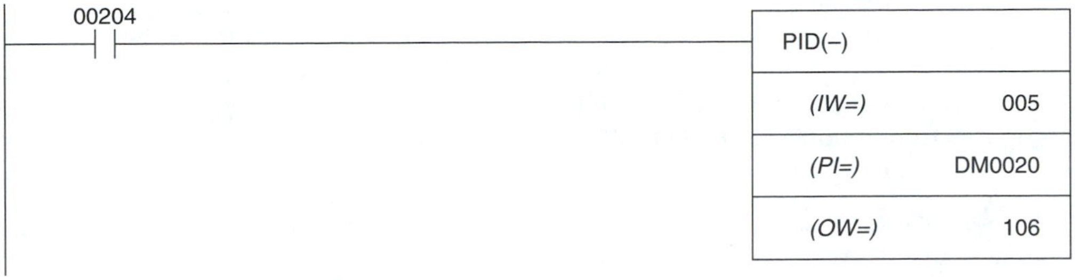

00204	PID(–)
(IW=)	005
(PI=)	DM0020
(OW=)	106

FIGURE 12.57
CQMI PID instruction.

Enter a value that is 100 times the K_c value that you would enter in a standard independent gains equation.

T_n is the sampling period (execution interval) of the PID calculation. Enter ratio values for R_i and R_d to indicate how many execution interval times to use as the integral time (T_i) and derivative time (T_d).

$$R_i = \frac{T_i}{T_n}$$

$$R_d = \frac{T_d}{T_n}$$

Increasing or decreasing T_n automatically changes the actual integral time and derivative time used in the calculation.

The PID ladder logic element, **PID(−),** is shown as it might be programmed in Figure 12.57. The PID(−) instruction does not have a differentiated version. If its control logic is true (e.g., if IR 00204 is on in Figure 12.47), the PID instruction executes. It will read a process variable value from the address entered in the IW parameter (IR 005 in the example program), perform the PID calculation as configured by the data in the 33 data words starting at the address specified as the P1 parameter (DM0020 to DM0052 in the example), and will write the result to the address specified in the OW parameter (IR 106 in the example). The configuration parameters will have been entered by the programmer before the PLC was put into run mode, and perhaps adjusted as the program ran.[17] One configuration parameter makes the PID(−) instruction recalculate its output at a fixed rate [controlled by a timer built into the PID(−) instruction] while the instruction's logic remains true. The execution interval will be slightly inaccurate, because the PID(−) instruction must execute to check the timer's accumulated value. If the accumulated value is greater than the preset value, the PID calculation is executed, and the timer is reset. The timer's accumulated value is not reset to zero but to the difference between its preset value and the (slightly later) time when the PID calculation actually executed. This causes the execution interval to remain approximately correct.

[17] Changes to the PID(−) instruction's configuration parameters are not recognized until the PID(−) instruction's logic goes from false to true. If you want to adjust the algorithm gains as the system runs, you must program control logic to allow acceptance of new parameters.

Because the PID(−) instruction must execute frequently to update to check its timer, PID(−) instructions should execute every scan cycle. They do not work well if they are in program segments that can be jumped past or inhibited by interlocks, if they are programmed in conditionally executed subroutines, or if they are in interrupt routines. *Note that OMRON's PID(−) instruction is not intended to be programmed into a timed interrupt.*

If the PID(−) instruction's control logic is false, the PID(−) instruction does not write values to the OW address, so the PLC program can include other instructions to write to this address, effectively operating the controlled system in manual mode. When the PID(−) instruction's control logic goes from false to true, the PID calculation begins to execute using the latest configuration parameters, but the value in the output address (specified by the OW parameter) is ramped to the PID calculation's result by the PID instruction. This prevents sudden jumps in the controlled system when it first starts to be controlled by the PID instruction.

The parameters that configure the PID instruction must be in the 33 consecutive addresses that start at the address specified in the P1 parameter. Most should not be changed by the programmer or the user-program, but the first seven contain values that can be changed. [Changes to these configuration parameters affect the PID calculation only after the PID(−) instruction's control logic goes from false to true.] The parameter values in the first seven data words include:

1. The **setpoint** value (SV), in the first data word in the P1 range. Must be entered as a value within the range of allowable engineering unit values as specified as the input range of values (covered below). Remember to program a false-to-true transition of the PID(−) instruction's control logic to force acceptance of a new setpoint.

2. The **proportional band** setting (PB), in the second data word. BCD numbers between 0001 and 9999 can be entered, and are interpreted as proportional band values between 0.1 and 999.9%.

3. The **integral time** ratio value (R_i), in the third data word. Entered as a BCD value between 0000 and 8191, this value is multiplied by the sampling period (T_n) to determine the integral time (T_i). If you enter 9999, the integral term is disabled.

4. The **derivative time** ratio value (R_d), in the fourth data word. Entered as a BCD value between 0000 and 8191, this value is multiplied by the sampling period (T_n) to determine the derivative time (T_d). If you enter 0000, the derivative term is disabled.

5. The **sampling period** (T_n), in the fifth data word. Entered as BCD values between 0001 and 1023 and interpreted as times from 0.1 to 102.3 s. This sets the preset value for the timer that is built into the PID(−) instruction, which controls execution of the PID calculation while the PID(−) instruction's control logic is true.

6. The **operation specifier,** in the low nibble (4 bits) of the sixth data word.

- **(a)** A value of 0 in this 4-bit number allows what OMRON calls *reverse action.* When reverse action is selected, error is calculated in what in this book we call the normal way by subtracting the process variable (from the IW address) from the setpoint (from the first data word). Reverse operation is used in situations such as when water temperature is being controlled by heating the water. If the temperature is too low, you would want to increase the heater output.
- **(b)** A value of 1 in this 4-bit number forces what OMRON calls normal action. When

normal action is selected, error is calculated in a way that most users would call reverse: the setpoint is subtracted from the process variable. Normal operation is used in situations such as when water temperature is being controlled by allowing cold water in. If the temperature is too low, you would want to reduce the amount of cold water allowed in.

7. The **input filter coefficient** in the high 3 nibbles (12 bits) of data word 6. Enter BCD values of 000 to 999, corresponding to time constants of 0.0 to 0.999 s. High filter coefficients help to reduce undesirable effects resulting from rapid changes (including electrical noise and spiking) in the process variable value. The PID calculation uses a calculated process control value, which is the actual process variable value (from the IW address) adjusted so that only 65% of any change is recognized in any time constant period. Entering 000 disables input filtering.

If you have electrical theory background, you will recognize "time constant" from when you observed the charge level at a capacitor in series with a resistor. When you changed the supply voltage, the observed charge at the capacitor changed by 65% of the difference every time constant.

8. The **output range,** in the low byte (8 bits) of word 7. BCD values of 00 to 08 cause the control variable output of the PID calculation to be scaled to between 8- and 16-bit values (engineering unit values) before being written to the address specified as the OW address.

9. The **input range,** in the high byte (8 bits) of word 7. BCD values of 00 to 08 cause the process variable from the address specified as the IW address to be scaled from an 8- to a 16-bit value (engineering unit values) before being used in the PID calculation. Setpoint values, from data word 1, are scaled in the same way as process control values.

Note that the PID(−) instruction's scaling capabilities are very limited and don't even allow for an offset. Other rungs in your PLC program may have to scale setpoint and process variable before they are used by the PID(−) instruction, and/or control variable values after the PID(−) instruction executes.

If an invalid parameter is discovered as the PID(−) instruction executes, it will set the **ER bit (SR 25603)** and the PLC will go into fault mode. If the PID instruction causes the PID calculation to execute, the **CY bit (SR 25604)** will be turned on. The PID(−) instruction turns this bit off if it doesn't execute the PID calculation.

Tuning PID Control Algorithms

It can be difficult to tune PID control algorithms, especially those implemented in digital controllers such as a PLC, because of the large number of factors that can affect a controlled system's behavior. The following procedure will help in making a first approximation, but the setup person will probably still have to make adjustments later.

1. Start with a proportional-only control algorithm that executes at consistent intervals (use an execution interval of three to five times the normal scan time of your PLC). Do not use output limiting yet, except as required for safety purposes.

2. Determine the controlled system's deadtime: In manual mode, change the control variable value by a value large enough to cause an obvious observable response. Time and record the delay between when the control variable change is made and when the observable response starts.

3. Determine the controlled system's natural cycle time (at its natural or resonant frequency): In automatic mode, with proportional gain low, enter a changed setpoint and observe the controlled system. The system should be driven toward the new setpoint, but weakly. Repeat with larger proportional gains until you observe that the response is strong enough to include significant overshoot. Continue repeating with larger proportional gains until changing the setpoint causes the system to oscillate without stopping. If the oscillations are so large that the limits of the output are reached, try with smaller setpoint changes. Observe and record the time of one full cycle (oscillation) of system response. Record this natural cycle time and the proportional gain setting that caused the system to resonate at its natural frequency.

4. If the natural cycle time is less than four times the length of the deadtime, you will have difficulty in tuning. Modify the system being controlled, if possible, to reduce the deadtime (preferably) or to increase the natural cycle time (if deadtime can't be reduced). Repeat steps 2 and 3.

5. Change the PID calculation time interval so that it will execute at least five times during each natural cycle, but not so frequently that other PLC control functions are affected. In general, the PID instruction shouldn't execute more frequently than every fifth scan cycle, or more than 10 times during the controlled system's natural cycle time. Remember, you may have to change a PID parameter containing the loop update time when you change the frequency with which the PLC program calls the PID instruction.

6. Calculate and enter the proportional, integral, and derivative multipliers as follows. Remember that some PID instructions divide the values you enter by a factor to calculate a final multiplier actually used in the calculation. Don't mix minutes and seconds!

(a) Set the proportional gain (K_c or K_p) to one-half the value it was at when the system resonated at its natural frequency (in step 3).

(b) Enter the integral multiplier in the form your PID equation requires:
- For integral time (T_i), enter the observed natural cycle time.
- Calculate the dependent gains equation's integral gain as $1/T_i$.
- Calculate the independent gains equation's integral gain (K_i) by dividing K_p by T_i.

(c) Enter the derivative multiplier in the form that your PID equation requires:
- For derivative time (T_d), use one-eighth of the natural cycle time (or one-eighth of the value actually used as the integral time in part b).
- Calculate the independent gains equation's derivative gain (K_d) by multiplying T_d by K_p.

Fuzzy Logic Fuzzy logic control algorithms don't generally require as much tuning as PID control algorithms do, yet they have been demonstrated to be good at controlling complex systems. Fuzzy logic as a control method has been slow to be accepted in North America, and as a result it wasn't commonly used in PLC control at the time that this book was being written. Some PLC suppliers do offer intelligent I/O modules that contain fuzzy logic routines, and some offer add-in software routines for fuzzy logic control, but none of the PLCs discussed in this book now offer preprogrammed fuzzy logic instructions. By the time you read this, some probably will, so in this section we discuss fuzzy logic concepts.

Fuzzy logic calculates the optimum control variable value in three steps: (1) fuzzification, (2) approximate reasoning, and (3) defuzzification.

Fuzzification The process variable is compared against the setpoint value and error is calculated. The amount of change in error since the previous execution is also calculated. After these calculations, the **error (E)** and **change in error (CE)** values are in what fuzzy logic users call **crisp** form (they are numerically equal to the actual error or change in error). They must be converted to fuzzy form before they can be used in the approximate reasoning step.

During fuzzification, crisp E (error) and CE (change in error) values are converted into **membership categories** and **degree of membership** fuzzy value form using membership function charts. Read the y-axis values for the crisp values along the x-axis of a membership function chart such as that shown in Figure 12.58. Crisp values for E can have values between 0 and 4095 in the example shown in Figure 12.58. A crisp E value (e.g., 976) can often be fuzzified into more than one fuzzy value (e.g., 976 has a 0.85 degree of membership in the MN membership category and a 0.23 degree of membership in the LN membership category).

Creation of the two membership function charts (one for E and one for CE) is one of the programmer's tasks. The programming software should display membership function charts graphically as in Figure 12.58.

1. There are usually between three and seven membership categories (five are shown in Figure 12.58), of which the central category (for small error values) is divided into two halves (called SN for *small negative error* and SP for *small positive error*).

2. Each category is typically assigned an easy-to-remember category name consisting (for example) of the letters L, M, and S for *large, medium,* and *small,* and the letters N and P for *negative* and *positive.*

3. Membership category shapes are typically triangular, but they can be trapezoidal (triangular with flat tops), as shown in the example. The programmer enters crisp values as endpoints for the triangles and for the flat tops during creation of the membership function charts.

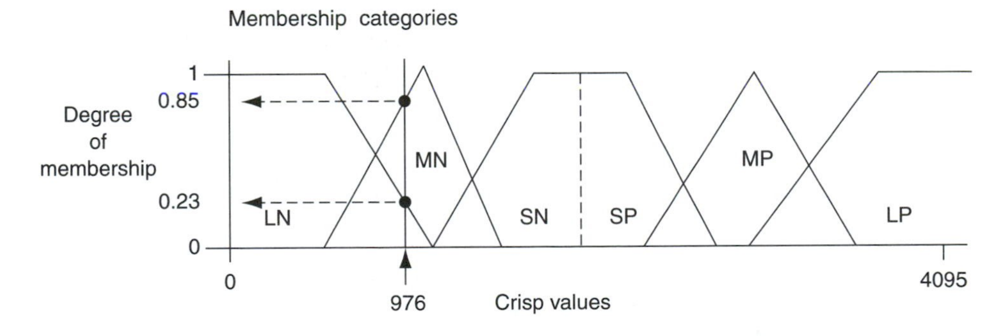

FIGURE 12.58
Membership function chart for error (E).

4. The categories usually overlap. If they don't overlap, each crisp value will only be fuzzified into a single fuzzy value, but if they do overlap (as in Figure 12.58), a crisp value may have two or more resultant fuzzy values. A fuzzy value such as "0.85 membership in MN category" can actually describe two crisp values (at the side of the MN category close to the LN category and at the side close to the SN category), so having other fuzzified results helps define the true input value.

Approximate Reasoning The programmer must enter a set of rules. There must be a rule for every possible combination of fuzzified E and CE values. The programming software will typically generate a chart containing all the combinations of E and CE membership categories. The finished chart might look like the one in Figure 12.59. For each row (one rule) the programmer has entered a fuzzy category name for the strength of the output (O) required. Some rules will be simple. Rule 1 says: If error is large negative (LN) but is decreasing rapidly (CE = LN), a medium positive (MP) output is required (to maintain the reduction in error). Other rules will require a little more care. Rule 5 says: If there is a medium negative error but the error is reducing rapidly, stop driving the output (and let inertia or built-up energy carry it to the setpoint).

During approximate reasoning, the controller will bond the rules that apply and then determine a degree of membership for each output (O) membership category. A common way of assigning an output degree of membership is to set it equal to the smaller of the degrees of membership in the E (error) or CE (change in error) categories. If, for example, rule 1 applied because E was LN with a 0.95 degree of membership, and CE was LN with a 0.45 degree of membership, the output would be MP with a 0.45 degree of membership. Averaging E and CE degrees of membership is another of the many methods used to calculate an O degree of membership.

Since membership function charts for E and/or for CE contain overlapping categories, ap-

FIGURE 12.59
Rule table for approximate reasoning.

Rule	E	CE	O
1	LN	LN	MP
2	LN	SN	LP
3	LN	SP	LP
4	LN	LP	LP
5	MN	LN	Z (Z means zero)
6	MN	SN	MP
7	MN	SP	LP
8	MN	LP	LP
9	SN	LN	LN
10	SN	SN	Z
11	SN	SP	SN
etc.			

proximate reasoning will find more than one rule that applies. Having more than one fuzzy output result helps in the next step: defuzzification.

Defuzzification The fuzzy output categories and degree of membership values must be converted to a single crisp (numerical) value so it can be sent to an analog output module. A programmer-created output (O) membership function chart like the one in Figure 12.60 is used. Note that in this example, a central membership category (Z) has been created with a minimum and a maximum crisp value of 2048. It is a triangular shape but looks like a vertical line. Membership function categories cannot have flat tops, because there should only be one crisp output value possible for each fuzzy output value. Overlapping of the categories is essential to prevent ambiguity. In Figure 12.60, approximate reasoning has yielded two fuzzy output values, 0.56 in the SP output category and 0.49 in the SM output categories. Both are plotted onto Figure 12.60. Note that there are two possible crisp values represented by each of the fuzzy logic values. On the SP category, there are two crisp values corresponding to a 0.56 degree of membership: one at the side close to Z and another at the side close to MP. Knowing that another rule generated a fuzzy logic result in the SN category tells the controller which side of the SP category to select. In some defuzzification routines, the fuzzy logic result with the highest degree of membership is used to determine the crisp value (2155). In other defuzzification routines, a formula is used to calculate the weighted average of the fuzzy logic results.

After determination of the crisp output value, defuzzification may include a calculation to multiply the current crisp output value by a proportional gain value. Sometimes the formula also sums (integrates) all the historical crisp output values, multiplies the sum by an integral gain, then adds the two terms, as follows:

$$\text{output} = G_pO + G_i \int (O)\, dt$$

PID control is actually used in fuzzy logic! Proportional gains show up as the degree of membership in E categories and in the final calculation of output. An inte-

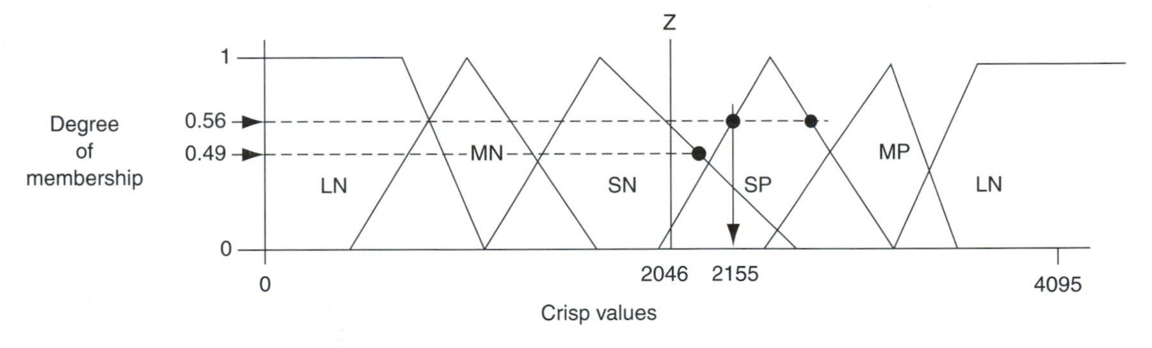

FIGURE 12.60
Membership function chart for output (O).

gral control component is present in the summing of the crisp output values. A derivative control component is there, too, as the change in error (CE) values.

Taking Manual Control of the System

Even the best-programmed control system must sometimes be taken over by a human operator. It may be necessary, for example, to manually control heating of a vat of liquid to its operating temperature before relinquishing control to an integral-only control program. An operator may be forced to take control if a component in the controlled system fails. In these circumstances, it is necessary for the operator to have access to a manual/automatic selector switch, and a dial or computer terminal to use to supply values for the control variable.

Preprogrammed PID instructions usually have optional manual configurations. The PLC program just has to change the PID instruction's configuration when the operator places the manual/automatic switch in the manual position. If the PLC programmer has written his or her own control program, it is easy to program the PLC to get another value to use for the control variable instead of using the PID calculation's result. The PLC program can be written to examine the manual/automatic switch and select a subroutine depending on the switch condition. One subroutine will contain the process control calculation and will execute if the manual/automatic switch is in the manual (off) position. The other subroutine executes if the manual/automatic switch is in the manual (on) position, and it reads an operator-controlled analog input value to get the value for the output.

For safety purposes, a PLC can be set up and programmed so that it will automatically release control of the process to a manual controller whenever it is in stop mode. Using an unconditional output instruction in the PLC program and a PLC-controlled relay switch will accomplish this. Figure 12.61 shows the extra program line that would be added to a user-program, and shows a relay that connects the PLC's analog output module to the system when the PLC is running and connects the operator-controlled dial when the PLC stops.

Sometimes a manual/automatic selector switch has to be wired to bypass the PLC entirely when the switch is set to manual. This is a book about PLC programming, so we will not discuss bypassing the PLC.

Forcing Different Methods of Calculating the Control Output Depending on Sensed Conditions

Control systems can be programmed to select their own control system programs or to modify their own control parameters. Such programs may be called **constraint control systems.** Constraint control is desirable, for example, to select between a control program that is tuned to control the temperature in a vat under normal conditions, and another control program tuned to bring the vat up to operating temperature at the start of the working day.

One way to change the way the control system operates is to change the gain parameters while the control program is working. A program to hold temperature constant could include instructions to monitor the error between the setpoint and the process variable value. When the error is large, the program would use a large proportional gain, but no integrator gain. The control system will work hard to correct error but won't build up a large integrated term while doing so. When the error becomes small, the program would change the proportional gain to zero

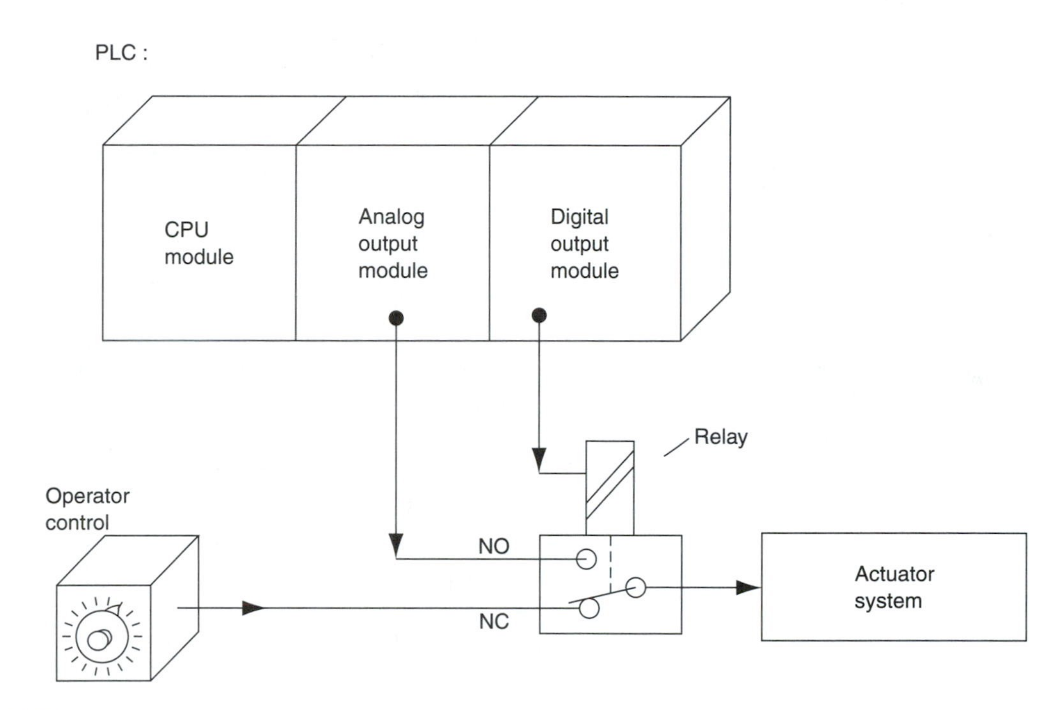

FIGURE 12.61
To release control of the process to an operator when a PLC stops.

and increase the integral gain to a large value. The control system will then work hard to hold the vat temperature despite gradual changes in the surrounding temperature.

Sometimes it is desirable to change the entire set of parameters that a PID instruction is using. The PID routines store their configurations and working data in sets of data words, sometimes in control blocks or data blocks. When you enter a call to the PID routine, you include a specification describing where the data is. The programmer can write a program that executes a different PID call (with a call to a different data set) to control the same process under a different set of conditions. A user-program controlling vat temperature, for example, might call the configuration data to bring the vat up to temperature, or the configuration data to maintain the operating temperature. In PLCs with indirect addressing, the program might only contain one PID routine call, but the PID routine can include indirect addressing to the data set it uses so that the indirect address can be changed under different circumstances. Remember that PID data sets are used to save integrated values, old values so that derivative terms can be updated, and derivative smoothing data. If your program suddenly starts reusing a PID data set after it hasn't been used for a while, these old values can lead to very unpredictable results. If your pro-

Save previous value of N7:0 to allow detecting when it changes.

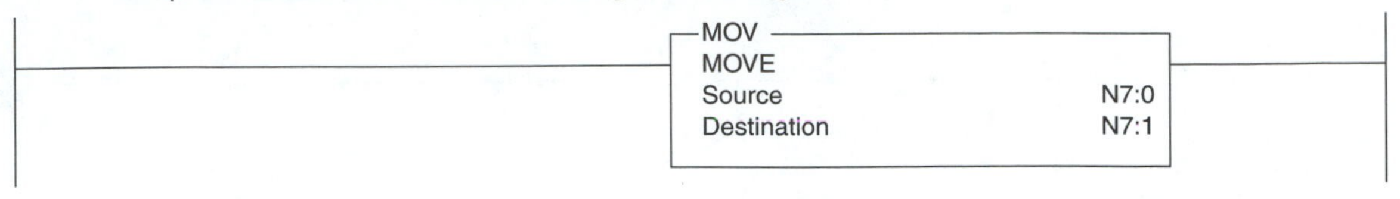

Determine what value to use for N7:0 this scan.

B3
0
MOV
MOVE
Source **10**
Destination **N7:0**

B3
0
MOV
MOVE
Source **11**
Destination **N7:0**

If N7:0 has changed, initialize data in integer control block N10:0 or N11:0 for the PID instruction.

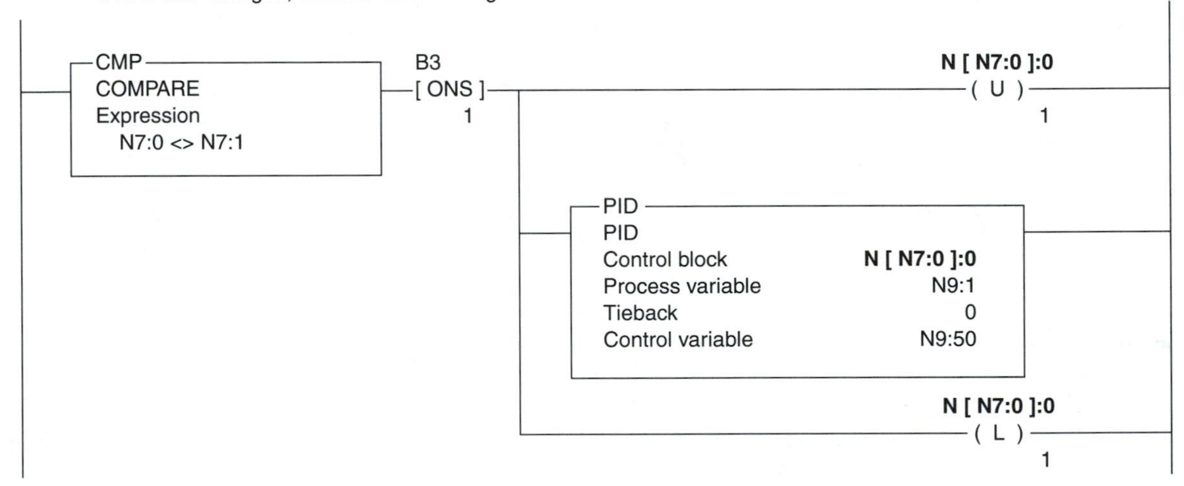

Normal execution of the PID instruction, using integer control block N10:0 or N11:0.

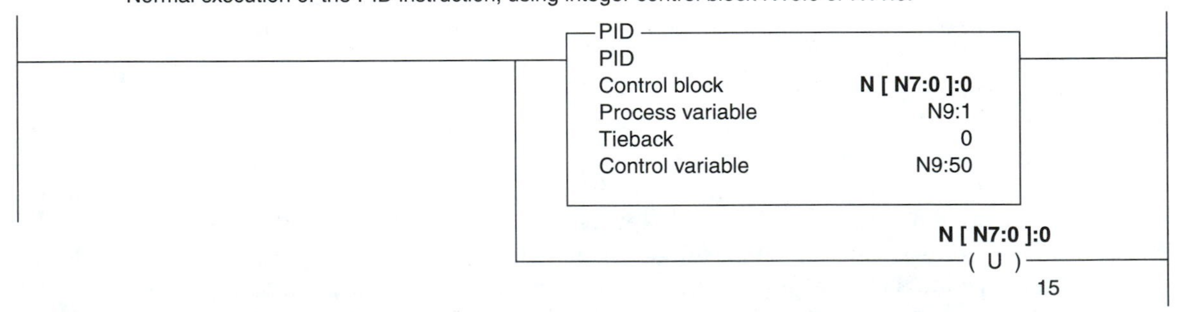

FIGURE 12.62
Using PLC-5 indirect addressing to select a control block for a PID instruction.

```
        C    DB1           ; Parameters best for normal operation,
        L    FW2           ; Load OB251's recently calculated error value
        L    KF   +100     ;   and load a small positive value (100)
        >=F                ; Check to see if system is in normal
                           ;   operation range (small positive error)
        JU   =M1           ;   and jump to where the PID routine is called
        C    DB2           ; Otherwise, change to startup parameters
M1      JU   OB251         ; Then jump to the PID routine,
        L    DW38          ;   and save the error calculation from this
        T    FW2           ;   execution to use selecting the next DB
```

FIGURE 12.63
Automatic selection of operation characteristics in an STL language program.

gram is written so that PID data sets are switched, you will probably want to include instructions to reset these old values each time you restart using a data set that wasn't used the last time the PID routine executed.

Figure 12.62 shows an Allen-Bradley PLC-5 program that uses N7:0 as a pointer for indirect addressing of integer files[18] N10 and N11. Bit (B3/2) changes the contents of N7:0, effectively selecting a control block. Figure 12.62 includes a one-shot that detects a changed value in N7:0 and executes the PID instruction once with the A/M selection bit (bit 1 in the first word of the control block) cleared to select manual mode, then immediately resets the bit. The next rung executes the PID instruction every scan cycle. Switching to manual, then back to automatic, clears old integrator and derivative data.

Figure 12.63 shows part of an example Siemens STEP 5 STL program that automatically selects whether to use data block 1, containing parameters optimized for maintaining a vat at a constant temperature (it has a high integrator gain) or data block 2, containing parameters better suited for bringing a vat up to temperature (it has a high proportional gain). The program compares the setpoint to the process variable to determine whether there is a large difference or a small difference, and selects the data block with the most appropriate parameters. The example program does not include the instructions that should be present to reset OB251's data block each time it is used again after the other data block has been in use for a while.

TROUBLESHOOTING

Process control calculations need careful tuning, and it is easy to detune them. *Always check the controlled system first if the system starts acting wrong.* Look for damaged or worn components. Use the measurement tools applicable to the system; vibration-analysis tools can help pinpoint malfunctioning components in a system with moving parts. Use electrical signal monitoring tools (voltmeters, ammeters, oscilloscopes, etc.) to verify that the electrical system outside the PLC is performing correctly.

Even if you are confident that the problem is in the PLC program, record the tuning parameters in use before you change them. Then look for the following problems.

[18] Indirect addressing cannot be used to address an Allen-Bradley PID control file.

1. If the system being controlled drives consistently in one direction instead of changing direction as it should:
   (a) Check to ensure that the setpoint value is correct as used in the process control calculation, and can be changed.
   (b) Check to see if the feedback value is changing and that it is being used in the process control calculation.
   (c) Ensure that the control direction is correct. Consider the effect of calculating error in the normal way (setpoint - PV) versus the reverse way (PV - setpoint).
   (d) Check that number ranges are scaled correctly. You may need an offset (or perhaps there is an offset that shouldn't be there).
   (e) Check that the process control calculation's result is being used, or is the control variable value coming from somewhere else (perhaps from a manual-mode input)?
   (f) Are output limits correct? Can the system drive the other way?
   (g) Have calculation errors accumulated during tuning? If using a preprogrammed PID instruction, switch it to manual (for at least one execution interval), then back to automatic to clear accumulated errors.
2. If the system is trying to correct error, but often overshoots the setpoint:
   (a) Ensure that the process control calculation is being performed at regular intervals, and frequently enough to control the process. Monitor the variables being used by the calculation. Do they say that an error continues to exist after an actual error has been eliminated? See the section on tuning PID control. To ensure that the process control calculation is being called at consistent intervals, program a timer to run during each interval, and store the timer's accumulated value once each time the process control calculation executes. If the process is reasonably slow, you should be able to monitor the time saved to see if it is reasonably consistent. If it is too fast to see, program a counter to count the number of times that the time value goes outside what you believe are acceptable limits.

   Remember: If you make a change to the execution interval of a PID instruction, you often have to change a loop update parameter in the data that the instruction uses, so that the instruction can calculate the integral and derivative terms correctly. If the instruction doesn't have a loop update parameter, you may have to recalculate integral and derivative multipliers based on the new interval. If using timed interrupts, the loop update time and timed interrupt intervals should match.

   (b) Is the process variable being updated immediately prior to each process control calculation, or does its age at the time of the calculation vary? Read the section on reducing scan-time delays.
   (c) Is the control variable written to the output module immediately after being calculated? Read about reducing scan-time delay.
   (d) Does the controlled process need different tuning for driving in one direction versus driving in the other? (A hydraulic cylinder, for example, uses more fluid to extend 1 in. than it does to retract 1 in., because the oil chamber in front of the piston is partly filled with a piston rod.)

(e) Is integral error growing too large during response to a setpoint change, causing overshoot? Consider increasing the proportional gain, perhaps reducing the output limits. Most preprogrammed PID instructions have an antiwindup feature that prevents integral buildup while the output is at its maximum limit. Some PID instructions have a deadband feature, which can be used when the process variable is very close to the setpoint value to prevent driving the controlled process away from the setpoint.

(f) Is the process control calculation calculating the derivative term based on the value you think it is using? Some formulas calculate the rate of change of error, while others use the rate of change of the process variable. Some equations allow you to select one of these or any other value to be derivated.

(g) Is derivative action too small to prevent overshoot? Is it so large that it stops (or even reverses) the output before the setpoint is reached? Read about tuning PID control.

(h) Is the deadtime (time between when an error first exists to the time of the first observable actuator action to correct it) long? Consider modifications to the controlled system to reduce delays. Consider methods of predicting the need for corrections earlier and using the predictions as a feedforward signal in the process control calculation.

(i) Consider increasing the energy absorbtion (e.g., friction) to get rid of the energy that causes overshoot.

3. If the system is trying to correct error, but stops short of the setpoint:

(a) Increase the gains, especially proportional and integral gain. See the discussion on the tuning of PID calculations.

(b) Reduce the energy absorption (e.g., friction) in the system being controlled.

4. If the system doesn't respond to setpoint changes or to process variable changes:

(a) Monitor the setpoint and process variable values that are being used in the PID equation. If they are changing, verify that they are being read before (not after) execution of the calculation.

(b) Check that the control variable value from the process control calculation is changing and that it is being written to the output module after (not before) the calculation.

(c) Check to see if the output address is also being written to by another rung in your program.

5. If the system suddenly stops working correctly after the program changes a calculation parameter or after switching which control calculation is used.

(a) Does your PLC require you to reset the PID instruction (e.g., by toggling it to manual and back) before it will recognize the newly changed parameters? Does changing tuning parameters as the process is being controlled in automatic mode lead to a buildup of error? These types of potential problems are often not documented, or the technical writer who wrote the manual may not have fully understood the process control calculation. Try switching the instruction out of automatic mode, then back into automatic. If this works, change your program to include the mode toggling each time it changes parameters.

QUESTIONS/PROBLEMS

1. Define *setpoint* (command variable), *feedback* (process variable), and *control variable.* Use an automobile speed-control (cruise-control) system for an example.
2. What are engineering units? Use an automobile's cruise-control system for an example.
3. Explain why and how output limiting might be required in an automotive cruise-control system.
4. Early cruise-control systems allowed a vehicle to move noticeably faster downhill than when driving uphill. Why? A sensor that detected the forward or backward inclination of the car might have been used to provide an offset (bias, disturbance, or feedforward variable) to correct. Explain.
5. The early cruise-control system exhibited more steady-state error when climbing hills. Why? What term in a PID control system would correct the steady-state error in this example? How?
6. When cresting a hill, the early cruise-control system might exhibit overshoot error (especially if PID control had corrected for steady-state error). Why? What term in a PID control system would correct the overshoot in this example? How?
7. Explain the difference between a dependent gains PID equation and an independent gains PID equation (i.e., what is dependent or independent?).
8. What is deadband, and why might it be desirable in a control system such as a cruise-control system?
9. Why might an antiwindup feature be useful in PID control?
10. Why might an antijerk feature be useful in PID control?
11. If you doubled the length of the interval times at which a PID equation executed, and the PID equation couldn't compensate automatically (most can't), what would the effect be on the size of the integral term? On the derivative term?
12. Create a fuzzy logic's approximate reasoning table for an automotive cruise-control system. Assume a very simple system that has only:
    - *Three membership regions for speed error:* ZE (zero error), TF (too fast, membership degree increases with error), and TS (too slow). Draw a membership function chart.
    - *Three membership functions for acceleration:* ZA (zero acceleration), P (positive = accelerating), and N (negative = deceleration).
13. If an automotive cruise-control system had a constraint-control selector switch that could be set to either *sensible* (for grandparents) or *fun* (for teenagers), how would the PID control algorithm used for "sensible" differ from those for "fun"? Assume the system is controlled by the PLC that you are learning to use.

SUGGESTED PLC LABORATORY EXERCISES

For a system with:

- Four control panel switches: inputs 0 to 3
- One position control actuator (e.g., a motor with speed control that accepts an analog signal, or a servovalve-controlled fluidic actuator)
- One position sensor to detect position of a load moved by the actuator (e.g., a potentiometer, or an encoder with pulse counter module)

On–Off Control of Analog Processes Position control using an actuator that moves at a speed relative to the analog signal it receives, and a linear position sensor or encoder position sensor:

Step 1 Program the PLC to read and write analog values (or read the encoder output). Write an MCP program to read and write data to each of the analog I/O modules continuously.

Step 2 Write a program to move the actuator through a repeating series of moves after the operator presses a start button, until a stop button is pressed.

(a) Determine the position setpoint values you need for position 1 and position 2. You may be able, for example, to move a linear sensor by hand while observing feedback.

(b) Write a PLC program that will:
- Initialize the actuator speed to 0 when the PLC goes into run mode.
- Restart the move sequence from the be-

ginning each time a momentary-contact switch (the start switch) is pressed, and stop the move sequence when a momentary-contact stop switch is pressed.

- Copy speed command setpoints from a data file (so that you can change them later without editing the program) to the analog output module so that the following sequence is executed:

  (1) The actuator retracts toward position 1.

  (2) After passing position 1, the actuator pauses for 2 s.

  (3) Moves toward position 2. Stops for another 2 s after passing position 2.

  (4) Repeat from step (1).

(c) With at least three different sets of speed setpoint values, observe and record the speed values and the position errors when the cylinder stops. Repeat observations for at least five cycles for each set of speed settings. (Note how much error there is. What effect do speed and direction have? Is error consistent?)

Step 3 Use timed interrupts to make the controlled process more deterministic but with minimized control overhead requirements.

(a) Copy the instructions that read sensor values and write actuator values to a new program. Change your processor configuration so that this new program file executes at repeatable timed intervals.

(b) Experiment with at least five different interval times, repeating the error observations made earlier. (Is error size or consistency affected by varying interval times?)

Proportional Control of Analog Processes Using a PID Instruction

Step 4 Modify your earlier program to use the PID instruction. (The following procedure order may have to be changed, depending on the PLC that you are using.)

(a) Enter a PID instruction in your timed interrupt program. It should get its process variable (sensed position signal) from the analog input module, and it should write its control variable (speed signal) to the analog output module.

(b) Enter configuration data into the data file or variables used by the PID instruction. Use a data screen if possible. Notice that the process variable, control variable, and some other PID values are displayed on the data screen while the PLC is in run mode.

(c) Change your MCP program so that it provides position setpoint values to the appropriate variable or word in the PID instruction's data block structure instead of writing directly to the analog output module. I suggest that your program should copy these setpoint values from data memory so they can be changed easily.

(d) Change your main program so that it recognizes that a move is finished when the position sensor's signal is acceptably close to the setpoint value (then waits during the 2-s delay before the next move). *Hint:* Your program should read an acceptably close value from data memory so that you can adjust it more easily.

(e) At each of two timed interrupt intervals (you may have to adjust a loop update time), experiment with at least two proportional gain values. For each, determine and record the smallest acceptably close values that won't cause the move sequence to get stuck. Record and observe the actual error in the stopping position for at least five cycles at each setting. Is the system more or less accurate than an on–off controller? More or less consistent?

Step 5 Introduce engineering units and deadband.

(a) Measure the actuator's movement range in the engineering units of your choice.

(b) Modify the PID instruction's scaling parameters so that setpoint values can be entered in engineering units.

(c) Add a deadband range that is at least the same size as the acceptable tolerance range your program used in the previous steps, but this time it is in engineering units.

(d) Change your program so that it provides setpoint values in engineering units and runs the 2-s delay timer between moves after the error enters the deadband.

(e) Determine and record the deadband range value that is the smallest value that will allow the cycle to stay running. You may also have to adjust gain values and limit values to minimize the deadband value.

Step 6 Use proportional–integral (PI) control to reduce steady-state error.

(a) Change your program so that each time a new setpoint is written, a one-shot reinitializes the PID instruction's accumulated integral value. (Pressing the start button sends the first setpoint.)
(b) Reduce the proportional gain slightly as you enter increasingly large integral gain (or reset time) values. Experiment to see how low you can get the deadband.
(c) Determine how small you can make the deadband value now.

Step 7 Use proportional–derivative (PD) control to accomplish better acceleration and deceleration.

(a) Set the integrator gain to zero and increase the deadband as necessary. Select a reasonable proportional gain and record how long it takes for one complete move cycle.
(b) Gradually increase the derivative gain or derivative time (depending on your chosen equation), reducing the proportional gain as you do so. Record how long a move cycle takes after you have finished tuning the system.

Step 8 Use proportional–integral–derivative (PID) control. Tune the gains and limits so that your move cycle uses a small deadband setting and still executes rapidly.

13

COMMUNICATIONS

OBJECTIVES

In this chapter you are introduced to:

- The communication requirements of a modern PLC.
- Network access methods, for shared communication channels.
- The communications capabilities of Allen-Bradley, Siemens, and OMRON PLCs.
- Configuration and instructions to use a PLC to communicate via serial ports and local area networks.
- Communication control from a user-program.

PLC COMMUNICATION CAPABILITIES

For the last 15 years (at least), experts have been predicting that programmable controllers were going to be replaced by personal computers. One of the most commonly quoted reasons was that PLCs had insufficient communications capacity. In this chapter we discuss the sophisticated level of PLC communication capabilities. (In Chapter 16 we examine the new list of reasons why PLCs are "dead.")

There are several ways in which PLCs use communications interfaces and communications software:

1. PLCs are programmed and configured through serial communication links to a **programming unit.** A PLC responds to programming software commands by accepting programs and data from the programming unit and by sending status information to the programming unit.

2. Some PLCs require that large blocks of data be moved between the CPU's data memory and **intelligent I/O modules.** The data must be moved via a parallel data bus that connects the CPU to I/O modules in its local rack.

3. Modern PLCs allow multiple remote I/O racks to be connected directly to the CPU in the host rack via a serial interface port. The user-program treats remote I/O modules in the same way that it treats I/O modules in its local rack: as input image table data and as output image table data. The PLC handles exchanging the data with the remote I/O racks. Each **remote rack** has a communication module instead of a CPU. The communication module reads data from input modules in its rack to send to the host CPU and receives output data from the host CPU to write to output modules in its rack.

4. Modern PLCs can be connected into **local area networks,** so the PLC can exchange data with other controllers (perhaps other PLCs) via a serial connection. The PLC must copy outgoing data from the PLC's memory onto the LAN link, and copy data from the LAN link into the PLC's memory.

5. As PLC control of manufacturing processes becomes more complex and the processes being controlled get farther away from the PLC's host rack, serial links are being used increasingly to move data over longer distances. **Sensor-actuator nets** allow PLCs to exchange data with individual sensors and actuators via a shared serial link. The PLC exchanges input and output image table data with the sensors or actuators, each of which has its own communication adaptor. The even more powerful **fieldbus** standard allows any of the multiple PLCs or other controllers connected to the serial link to read and write data anywhere in the fieldbus system. Each CPU must execute data exchange instructions in the user-program and respond to data exchange instructions initiated by other controllers on the fieldbus.

Modern PLC CPU modules now come with multiple microprocessors. The **main microprocessor** executes the standard program PLC scan cycle. The other microprocessors handle the serial exchange of data between the CPU module's RAM memory and other controllers. Some of the other controllers may be external to this PLC (e.g., personal computers, other PLCs, etc.), while some may be intelligent I/O modules under the PLC's control (e.g., a high-speed counter module, motion-control modules, etc.) Each **communication-handling microprocessor** comes preprogrammed specifically for communications and has some access to the main microprocessor's memory so that the user-program can read the communication handler's status and assign it new jobs.

A communication handler that is designed to exchange data with other controllers via a single set of conductors must share the conductors. PLCs use the same network access-control schemes that are used in other computer networks:

1. In **polling,** or **master–slave, systems,** a single computer is designated to be the master computer (usually node 0). It is programmed to send enquiry messages to the slave computers on the network, one slave at a time. A slave computer (e.g., node 12) can reply by requesting that the master pass a data packet on to another node (e.g., to node 5) or can ask the master to get data from node 5 and pass it back to node 12. All the PLCs covered in this book have master–slave communication capabilities.

OMRON's **host link** communications protocol is based on the master–slave scheme and is the master–slave system described in most detail in this book. Allen-Bradley PLCs have what they call **DF1** protocol. Siemens PLCs' **L1** networks and **Profibus-DP** networks include the master–slave scheme. (Profibus industrial communication networks are installed worldwide and are not limited to Siemens. Other versions of Profibus exist.)

2. In **token-passing systems,** as in a polling system, each computer on a token-passing network has a turn when it can send data or requests for data, but instead of being slaves, each computer is an equal, or a *peer*. In this peer-to-peer system, nodes can send data packets directly to other nodes, and each node has a scheduled turn to send data. When a computer has sent all the data it needs to send (or when its allotted time is over), it sends one last message. The last message is the *token,* and the computer that receives the token now has its turn to send data, data requests, or data in response to requests received previously. Some token-passing networks do have a master computer, but it just detects the loss of a token and starts a new token rotation. Token passing is faster than polling and preserves the dependability inherent when each node has a turn in a preassigned rotation.

In the 1970s, General Motors adopted the ISO's Open Systems Interconnect (OSI) model for computer communications. General Motor's completed version of the OSI standard, called **Manufacturing Automation Protocol (MAP),** became a model for following computer networks, although very few people every bought MAP products. MAP used a token-passing scheme for network access.

Allen-Bradley **Data Highway** networks, along with GE Fanuc's **Genius** networks and the popular **Modbus Plus** network, are de facto standards in North American plants. All use token passing.

3. The **carrier sensing multiple access** with **collision detection** network access-control method, better known as CSMA/CD or **Ethernet,** is the most popular network access-control method for office computers. In CSMA/CD, any computer can send on the network if the network isn't already in use. Collisions happen if two computers try to send simultaneously, but both computers sense the collision (the data on the network will be different from the data it is sending) and stop transmitting until after random delay times.

The Fieldbus Foundation, a group representing most industrial controller manufacturers, has recently adopted **Fast Ethernet** as the common network-access method for open industrial controllers of the future. All the PLC manufacturers covered in this book offer Ethernet communication modules now, and the next generation of PLCs will probably all have Ethernet ports in their CPU modules.

4. Several networks are actually **combinations of the foregoing methods.** In industrial control it is important for a supervisory controller (e.g., a PLC) to read remote sensors and write to remote actuators. The remote I/O only need to be slaves to the master controller and are sometimes called **passive** nodes on the network, while the controller is called an **active** node. To be more useful in a manufacturing plant, the same network must also allow multiple active nodes to be connected and to exchange data in a **peer-to-peer** way, using token-passing or CSMA/CD schemes.

OMRON's **host-link** master–slave communication scheme includes a pause between the polling of each slave during which the master will accept urgent service requests sent by slaves. **Profibus** is a master–slave system with peer-to-peer capabilities. Allen-Bradley offers an optional **control net** network with both master–slave and peer-to-peer communications.

PLC manufacturers and third-party suppliers offer several interface devices, complete with software, to translate from one network-access scheme to another. If network-access translation is their only function, the interfaces are properly called **bridges.** If the interface also adjusts data formats or performs data transmission control, the interface is properly called a **gateway.**

An **open network** is one that is not owned by a commercial interest. Anybody can freely manufacture and sell open network–compatible components, so users aren't forced to buy all their control components from a single supplier, nor are they limited by the options available from that supplier. Recently, several open standards have been developed and adopted by PLC manufacturers, and are actually in competition with each other. (Even when they agree to work together, industry can't stop competing.) Chapter 16 covers some of the more successful open standards.

PLC-5

SLC 500

COMMUNICATION AND THE ALLEN-BRADLEY PLCS

Allen-Bradley PLCs are easily programmed to communicate once they have been configured. Configuring an Allen-Bradley PLC to communicate is also fairly easy, but there are lots of options to select from. It is useful, therefore, to understand what happens when an Allen-Bradley PLC communicates. (The PLC-5 and SLC 500 are described individually in following sections.)

When a user-program executes an instruction requesting a data transfer via a communication port, the request is usually just added to the end of a queue containing other communication requests for the communication channel. During the I/O scan part of the scan cycle, the PLC's main processor services communication channels by checking to see if a communication handler has completed its previous task and is ready for a new task. If it is, the CPU copies newly arrived data from the communication handler's buffer memory into main memory, and/or copies new outgoing data from memory into the communication buffer. Communication tasks are performed one at a time (per channel), usually in the order in which the requests are received by the communication handler. Communication tasks are performed asynchronously with the scan cycle, which means that it may take several scan cycles between when a user-program executes an instruction requesting a data transfer and when the data transfer is complete. The user should write programs that don't try to evaluate incoming data in the main memory, or change outgoing data in the main memory, between when a communication-request instruction executes and when a communication status bit says that the data transfer has been executed. Some instructions can cause the PLC to interrupt the scan cycle to service the communication handler without waiting for the next I/O scan step, and there are some situations in which a new data transfer request will take precedence over other requests in the queue.

Allen-Bradley PLC-5s and SLC-500s have several communication channels through which they can respond to communications needs. Some communication activity requires little or no configuration or programming and is automatic enough that the programmer is often unaware of its use, although its performance can affect the PLC's performance. Other communication channels must be configured and are used only in response to user-program instructions. Allen-Bradley PLC communication channels are used for:

1. A **programming unit** interface. The PLC's response to programming unit commands cannot be modified, except that access to the PLC's memory can be limited by configuring privileges as described in Chapter 10. A PLC-5 can be connected to the RS-232C serial port of a personal computer for programming, but if a manufacturing plant has several PLCs, it makes more sense to use an Allen-Bradley interface card in the personal computer. The card connects to a plug in the faceplate of a PLC-5 CPU, and the programming unit actually connects into the data highway plus (DH+) local area network (as shown in Figure 13.1). Messages between the programming unit and the PLC are carried with other traffic on the DH+. If other nodes are using the DH+, program monitoring speed can be reduced, and if the programming unit requires

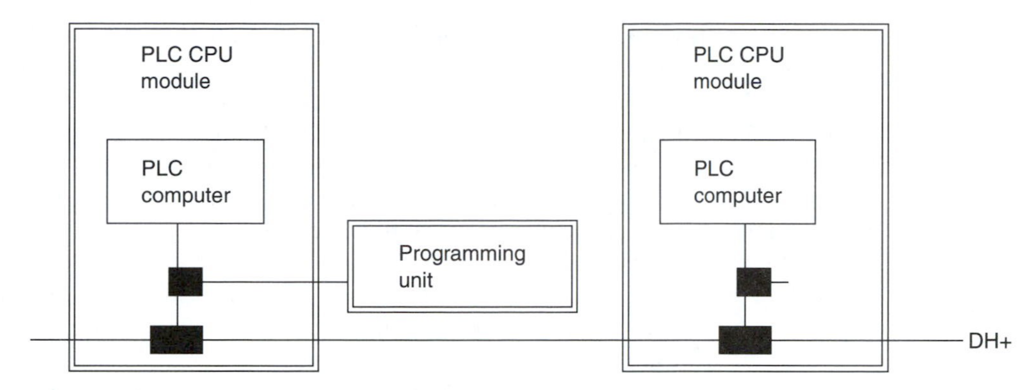

FIGURE 13.1
Connection of a programming unit to an Allen-Bradley PLC in a DH+.

large amounts of data, DH+ data exchanges will be slower. In Chapter 10 we described how to configure programming software so that it can connect to a PLC via the DH+. A programming unit connected to a DH+ network can connect to any of the PLCs on the DH+ network to monitor or change their memory contents.

2. **Local I/O** scanning, part of the main PLC scan cycle, reads and writes digital I/O modules and the status of other modules in the PLC's rack.

3. If an I/O interrupt is configured, the local rack communication handler has to read one input module periodically while the user-program is being executed. I/O interrupts are called **process input interrupts (PIIs)** in the PLC-5 and **discrete input interrupts** and **I/O interrupts** in the SLC 500.

4. **IMMEDIATE I/O** instructions **(IIN** and **IOT)** make use of the local rack communication handler to send or receive individual data words on the idle bus while the user-program runs.

5. Data is exchanged with other PLCs via the **data highway** channels (DH+ or the SLC500's DH485) by executing the **MESSAGE (MSG)** instructions in a user-program.

6. Other serial communications. An **integral serial port** can be configured to operate in:

(a) **System mode,** as a **DF1 communication channel,** to communicate with other devices via links such as modems. User-programs can communicate via a DF1 channel by executing a MESSAGE (MSG) instruction, or an ASCII WRITE (AWT) or an ASCII SET HANDSHAKE (AHL) instruction. A PLC-5 can be configured to use a DF1 port in:

- Point-to-point mode, linked to one other device (the default mode).
- DF1 slave mode, as one of up to 254 nodes connected to a local area network via interface modems or line drivers. Messages can be sent when a DF1 master allows the slave to send.
- DF1 master mode, polling each of the slaves on a DF1 network, and passing messages for them when requested to do so.

DF1 allows the connection of more nodes than DH+ and allows the user more options in selecting communication protocol but isn't as easy to configure as a DH+. Since DH+ networks can be connected together to get additional nodes and are easy to use and dependable without custom configuration, DH+ is used instead of DF1 in most places. DF1 will not be covered in detail here. Read about OMRON's host–link communications in this chapter for a description of how DF1 typically works, and see your Allen-Bradley manual set for details.

(b) **User mode,** as an **ASCII channel,** to exchange ASCII strings with peripherals such as bar code readers or serial printers. The user-program can use the ASCII instructions to send or receive data via a buffer in the CPU module's user mode channel. ASCII instructions to use a user mode channel are not covered in this book. Your manual set provides details on how to use ASCII instructions such as:

- ASCII WRITE (AWT) or ASCII WRITE APPEND (AWA) to write an ASCII string to the buffer for the user mode channel to transmit
- ASCII READ (ARD) or ASCII READ LINE (ARL) to read an ASCII string that the user mode channel has received into its buffer
- ASCII CHARACTERS IN BUFFER (ACB) or ASCII TEST FOR LINE (ABL) to determine how many characters are in the buffer or in the first message in the buffer
- ASCII HANDSHAKE LINE (AHL) to control the RTS and DTS handshake lines
- Other instructions for manipulating ASCII strings

ALLEN-BRADLEY PLC-5 COMMUNICATIONS

The PLC-5 uses some communication channels as described above for Allen-Bradley PLCs in general:

1. A **programming unit** interface. An interface card in the personal computer is generally used to connect to a DH+ via the PLC-5's programming port.

2. Local I/O scanning. I/O is read or written one group at a time via the bus conductors in the local chassis. PLC/5s with an "L" in their name (e.g., PLC-5/40L) have a parallel connector in channel 2, through which they also scan extended local racks. See your manuals for more information.

3. Process input interrupts (PIIs), if configured, cause the local chassis communication handler to read a specific input module periodically while the user-program is being executed.

4. Immediate I/O instructions **(IIN** and **IOT)** that address digital I/O modules in the local chassis cause the local chassis communication handler to send or receive individual data words on the chassis bus while the user-program runs.

5. A **data highway plus (DH+)** channel. Channel 1A is the default DH+ channel, although the defaults can be changed.

6. The **integral serial port.** Channel 0 is the default serial port channel. The serial port can be configured to use **RS-232C,** or **RS-423,** or **RS-422A.**

A PLC-5 has additional communication channel applications, including block transfer instructions, remote I/O, and Ethernet communications.

7. Block Transfer Instructions Block transfer instructions for BT modules in the local chassis are communication requests to the local chassis communication handler. The BLOCK

TRANSFER READ (BTR) and BLOCK TRANSFER WRITE (BTW) instructions request the communication handler to exchange blocks of data with I/O modules during the periods when the local I/O bus is idle. (See Chapter 7 for BTR and BTW.) Each block transfer request causes a small amount of block transfer identifier data to be placed at the end of the communication handler's wait queue. The identifier data includes the data file addresses for the source or destination of the data in the CPU. Data transfers are carried out one at a time, in the order the requests are received into the queue (except in PII or STI programs; see below), so it may take several scan cycles before the block transfer is completed. Each block transfer instruction reserves a block transfer control structure in memory, and the communication handler writes to status bits in this area of memory as it works. A user-program can examine the status bits to determine if a requested block transfer of data is still waiting its turn in the wait queue (.EW status bit), has been started (.ST), is done (.DN), or is in a less-desirable state (.ER is the error bit).

As it executes a block transfer, the communication handler copies data between the CPU's data files and the BT module one data word at a time. Outgoing data shouldn't be changed before the entire block of data has been sent, or partially changed data will be sent. Incoming data shouldn't be evaluated before the entire block has been received, or the program may be evaluating data that is only partially updated! Programs should be written to monitor block transfer control status bits (e.g., the .EW, .ST, and .DN bits) to protect data sets from changing or being used while block transfers are pending.

As noted in Chapter 11, if a BTR or a BTW is executed in an interrupt program (either a PII or an STI), the block transfer requested takes precedence over other block transfer requests in the queue and is executed immediately. The next instruction in the interrupt program is not executed until the block transfer has been completed, so there is no danger of an interrupt program using partially transferred blocks of data.

Block transfers can conflict with process input interrupts (PIIs). While a block transfer is being performed in the local chassis, the communication handler cannot read the input module, so signal changes that should cause PII interrupts may be missed. Allen-Bradley recommends against having BT modules in the local chassis if PII interrupts are used, but the two can work together if the user-program disables the PII interrupt [using the USER INTERRUPT DISABLE (UID) instruction] before issuing a BTR or a BTW, then reenables the interrupt [using the USER INTERRUPT ENABLE (UIE) instruction] after the block transfer is complete.

Block transfer can conflict with immediate I/O instructions. Block transfers that are executing take precedence over immediate I/O. If the program includes both block transfers and immediate I/O instructions for the local rack, the immediate I/O instruction should be conditional upon there being no active block transfers, which can be checked by examining the state of the .ST (start) bit for all block transfer instructions accessing BT modules in the local rack.

8. Remote I/O Remote I/O (RIO) racks are read and written to by a communication handler called the scanner in the PLC-5. Each RIO device is assigned input image and/or output image addresses by setting DIP switches at the RIO controller, and a user-program treats digital RIO devices the same way it treats digital I/O in its local rack. In Chapter 10 we described how to configure a communication channel to be used as a scanner channel. Channel 1B is the default scanner channel.

The PLC-5 CPU module can't scan remote I/O modules during the I/O part of its scan cycle. Instead, the PLC-5 reads and writes an area of its own remote I/O (RIO) buffer memory every scan cycle, as shown in Figure 13.2. The scanner communication handler executes its own scan cycle, during which it copies data between the RIO buffer memory and remote I/O (RIO) controllers via the serial port reserved for RIO scanning. The scanner cycle typically takes longer than the PLC's main scan cycle. At the RIO rack or other device, a third scan cycle is executing, during which the RIO controller writes output image data from the scanner to output modules, and reads input module data to pass on to the scanner the next time the scanner reads from this module. Although the scan cycle in the RIO controller is typically faster than the scanner cycle, it does add additional delay.

The accumulated delay means that it may take several PLC-5 scan cycles before a value in output image memory is actually written to a remote output module. Reading of remote inputs can similarly be delayed. In fact, if the PLC-5 turns an output image bit on, then off too quickly, the remote output contact may not be turned on at all, and if a remote input contact changes states too quickly, the PLC-5 may never see the change.

Block transfers to/from BT modules in remote I/O racks cause the PLC-5 to interrupt RIO scanning to execute the block transfer, so remote digital I/O status updating becomes slower. A single block transfer can easily double the RIO scan time. The PLC-5 limits RIO scanning delays by only executing one block transfer per remote rack per RIO scanner cycle.

Block transfers requested in interrupt programs (STIs and PIIs) for BT modules in remote I/O systems take precedence over all other block transfer requests in the channel's wait queue, as you might expect. In addition, all block transfer requests issued in an STI or a PII program are executed despite how many occur during a single RIO scanner cycle. In enhanced PLC-5 models, the MCP that was interrupted to execute the STI or the PII resumes operating while the interrupt program waits for a block transfer to be completed.

9. Ethernet Communications Ethernet communication is possible via channel 2, in CPU models with the letter "E" in their name (e.g., PLC-5/40E) or via channel 3 if an Ethernet control coprocessor is plugged into the CPU module at the side-mounted channel 3 port. (Configuration of a control coprocessor for Ethernet communication wasn't mentioned in Chapter 10 because no configuration is necessary.)

Configuring the PLC-5 for Communications

There is one communication setting that can be made as part of the CPU configuration, the communication time slice.

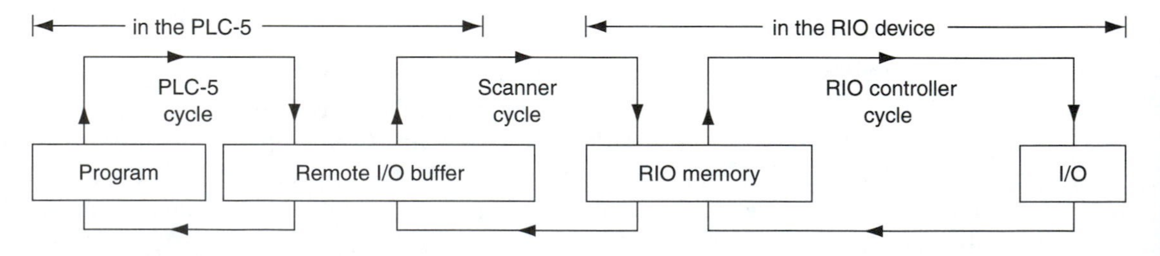

FIGURE 13.2
Three scan cycles involved in PLC-5 remote I/O.

The PLC-5 system performs serial communication servicing functions during a housekeeping step in the I/O part of the scan cycle. The amount of housekeeping time needed per scan cycle may vary, so scan time can vary if the amount of communication-channel servicing time is not controlled.

If a time of "0" is entered for the communication time slice, the PLC-5 will perform all the communication tasks required per scan, so scan time will be variable. If a time of other than 0 is entered, the PLC-5 will use that number of milliseconds, plus the basic 3.1 ms always required for housekeeping, for communications servicing every scan cycle, despite how many (or how few) communications tasks are actually required.

Several configuration changes can be made to communication channels, using the channel overview screen shown in Figure 13.3. This screen displays each channel's currently configured communication task and allows the programmer to select a channel to change the configuration or to monitor its communication status. Channels can be configured for data highway plus (DH+), remote I/O scanning, remote I/O adapter mode, or with other serial port protocol. More detailed configuration information is included in Chapter 10.

PLC-5 Message Instruction

The MESSAGE **(MSG)** instruction can be used to send data to another PLC or to read data from another PLC via a CPU port configured as a DH+ or as a DF1 channel. Messages can even be sent to a PLC on a remote network if a communication link exists to the remote network. DH+

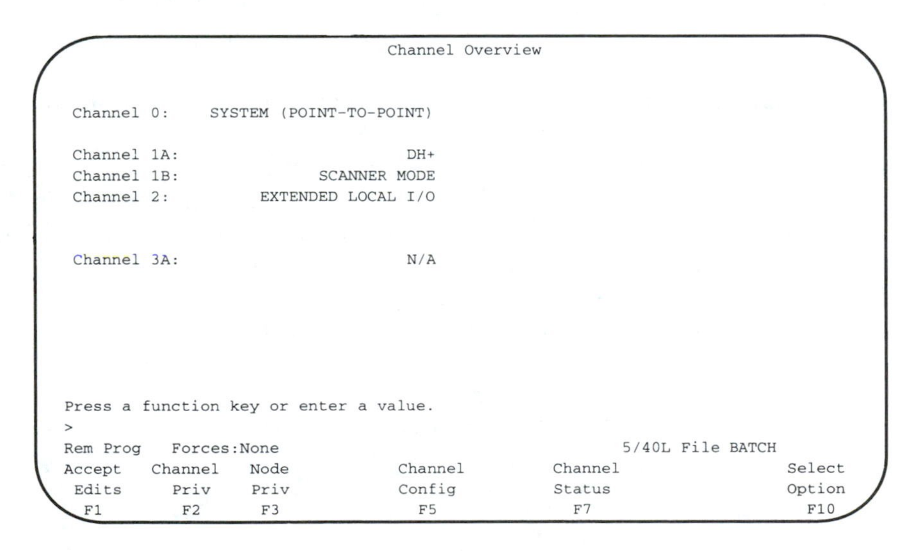

FIGURE 13.3
PLC-5 communication channel overview screen.

networks can be connected together via PLC-5s. Other networks can be connected via SLC 5/04s or other communication processors.

The MESSAGE (MSG) instruction for the PLC-5 can be programmed on a rung that controls its execution, and looks deceptively simple in a ladder logic program (see Figure 13.4). The MSG instruction's ladder logic symbol makes it appear as if only one piece of information is needed when entering a MSG instruction: the address for a control block. The programmer of an enhanced PLC-5 should enter a unique message control element address (e.g., MG10:1), but an integer address must be entered for classic PLC-5 processors and is allowed for some enhanced PLC-5 processors. A message control element contains 56 data words (integer control block size varies). When the programmer enters the control block address, an MSG data entry screen appears. The programmer must enter additional MSG parameters on this screen before the MSG instruction can be entered into the user-program. Yet another screen of status information can be accessed by invoking the data table while your programming unit's cursor is on the MSG instruction, and further MSG parameters can be entered using that screen. The data entry screen and data table monitor screens for a message control element are shown in Figure 13.5. (Screens for integer control blocks are only slightly different.)

Parameters that must be entered or selected through the data entry screen include:

1. **Communication command.** Select one of:
   (a) PLC-5 TYPED READ or PLC-5 TYPED WRITE to read data from another PLC-5 or write data to another PLC-5. Entire data structures will be copied as data elements if the data to be copied is structured data (e.g., floating-point numbers, counter elements, strings, etc.).
   (b) PLC-5 TYPED READ FROM SLC or PLC-5 TYPED WRITE TO SLC to copy whole data structures between this PLC-5 and an SLC-500 PLC.
   (c) SLC TYPED LOGICAL READ or SLC TYPED LOGICAL WRITE to copy data as individual 16-bit data words between this PLC-5's memory and an SLC 500. (Structures can be copied, but a floating-point structure, for example, would count as two 16-bit data words.)
   (d) PLC-3 WORD RANGE READ or PLC-3 WORD RANGE WRITE to copy 16-bit data words between this PLC-5 and the memory of a PLC-3.
   (e) PLC-2 UNPROTECTED READ or PLC-2 UNPROTECTED WRITE to copy 16-bit data words between a PLC-2 compatibility file in this PLC-5 and the memory of a PLC-2, or a controller that acts like a PLC-2. SLC 500s below the SLC 5/03 with OS 301 are designed to act like PLC-2 devices.

2. **PLC-5 data table address.** Enter the starting address, in this PLC-5, of the data to be written to the other PLC, or where data from the other PLC is to be copied to.

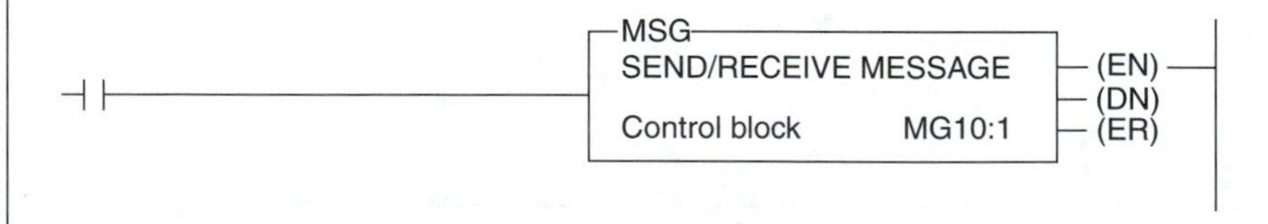

FIGURE 13.4
A PLC-5 message instruction in a Ladder Logic rung.

```
MESSAGE INSTRUCTION DATA ENTRY FOR CONTROL BLOCK MG10:10

          Communication Command               PLC-5 Typed Write
          PLC-5 Data Table Address:           N10:30
          Size in Elements:                   1
          Local/Remote:                       LOCAL
             Remote Station:                  N/A
             Link ID:                         N/A
             Remote Link Type:                N/A
          Local Node Address:                 00
          Destination Data Table Address:     N20:50
          Port Number                          1A

BLOCK SIZE = 56 WORDS
```

(a)

```
MESSAGE INSTRUCTION DATA MONITOR FOR CONTROL BLOCK MG10:10

Communication Command             PLC-5 Typed Write
PLC-5 Data Table Address:         N7:0              ignore if timed-out:  0 TO
Size in Elements:                 1                      to be retried:   0 NR
Local/Remote:                     LOCAL              awaiting execution:  0 EW
   Remote Station:                N/A                        continuous:  0 CO
   Link ID:                       N/A                             error:  0 ER
   Remote Link Type:              N/A                      message done:  0 DN
Local Node Address:               00               message transmitting:  0 ST
Destination Data Table Address:   N7:10                 message enabled:  0 EN
Port Number                       1A

ERROR CODE: 0

Press a function key to change a value.
MG10:10.TO=
Program     Forces:None      Data:Decimal     Addr:Decimal PLC-5/40 File DRILL1
               Toggle          Specify          Next    Prev   Next      Prev
                 Bit           Address          File    File   Element   Element
                 F3              F5              F7      F8      F9        F10
```

(b)

FIGURE 13.5
Data entry and data monitor screens for a PLC-5 message control element.

3. Size in elements. Enter a number indicating how many elements to copy. The PLC-5 can handle messages up to 1000 words, but an SLC 500 can only handle messages up to 120 words in length. If the communication command allows structures, remember that structures are made up of multiple words, so fewer elements can be transferred (except for ASCII elements, which need only one-half word each).

4. Local/remote

- **(a)** If LOCAL is selected, messages can only be sent to another PLC on the same (DH+) network, and for **local node address:** Enter the DH+ address of the other PLC.
- **(b)** If REMOTE is selected, up to four additional entries must be made, effectively describing how to route data to the other PLC on a remote network. See your manuals (and get a good technical advisor) for more information.

5. Destination table address. Enter the starting address in the other PLC where data is to be written to or copied from. For some combinations of command-type selections and types of other PLCs, you may have to include quotation marks or other formats. (See your manual for more details.) If the other PLC acts like a PLC-2, enter an offset number (10 or higher) indicating where the data will be written into the part of that other PLC's memory that acts as if it were in a PLC-2. In an SLC 500, file 9 is always used to act as a PLC-2-compatible memory space, called a 485CIF file.

6. Port number. Toggle to select the DH+ port. Unless you have modified the default configuration, port 1A is the DH+ port. If you select an Ethernet port, you will be asked to enter an **IP address** for the other PLC instead of entering LOCAL or REMOTE address specifications.

Most of the information on the MSG instruction data monitor screen is status information, but two message control element bits can be used for configuration:

1. The **CONTINUOUS OPERATION** bit **(.CO),** when set, causes the MSG instruction to repeat executing each time it finishes the previous execution, while the MSG instruction's control logic remains true.

2. The **TIMEOUT** bit **(.TO),** when set, causes the PLC-5 to abort attempting a message after between 30 and 60 s if the attempt isn't successful. The ERROR bit **(.ER)** will then be turned on and an error code will be put into the MESSAGE CONTROL ELEMENT **(.ERR)** and displayed on the data monitor screen.

Status bits can be examined in a user-program using the control element address and the bit mnemonic (e.g., MG10:1.ER), and some bits can be written to by a user-program (e.g., MG10:1.TO). The data screen shown in Figure 13.5 identifies the function of each bit. The reader will recognize the functions of most bits, knowing that messages are sent via serial ports and that communications handlers do serial port data exchanges asynchronously with the main PLC scan cycle.

SLC 500 ALLEN-BRADLEY SLC 500 COMMUNICATIONS

Some SLC 500s' communication channels can be used for the purposes listed previously for Allen-Bradley PLCs in general:

1. A port for the programming unit. RS-232C communication between a programming unit's serial port and a single SLC 500 is common, since SLC 500s are often used for

small jobs. You can also use an interface card in your personal computer programming unit to connect to a DH+ via the programming port of an SLC 5/04, or to a DH-485 network via other SLC 500 CPU programming ports.

2. I/O scanning (of course).
3. Discrete input interrupts (DIIs) and I/O interrupts can be configured in SLC 5/03 and above, so that the local rack's communication handler will periodically read an input module while the user-program is being executed.
4. IMMEDIATE I/O instructions (IIM and IOM) also make use of the local rack's communication handler to send or receive individual data words on the bus while the user-program runs.
5. A data highway plus (DH+) channel is available in SLC 5/04 and above. Channel 1 is the default DH+ channel.
6. An integral serial port. Only channel 0 can be used as the serial channel. DF1 or ASCII communication options are as described earlier.

In addition to the uses listed above, the SLC 500 PLC can also use:

7. A data highway 485 (DH-485) channel. Channel 0 can be configured for DH-485 communications instead of for serial communications. To connect channel 0 to the shared conductors of a DH-485 network, you need an interface adapter (a 1747-AIC).

Configuring the SLC 500 CPU for Communications

Communication configuration can be changed by making changes to the CPU's status file, either through the processor status screen or (sometimes) through the channel configuration screen. Both screens may access the same configuration memory.

The status file configuration options available through the processor status screen are described in detail in Chapter 10, where we discuss configuration of a DH-485 node address and/or a DH+ node address, the default baud rate, limiting the communication servicing time, and selection of whether to maintain a record of which stations are active on the DH+. Configuration also selects whether to use global status words. If global status words are enabled, this CPU will pass the contents of S:99 with the communications token when it passes the token to the next PLC on a DH+ network and will accept global status words from the other CPUs on the network and store them in S:100 to S:163 (for nodes 0 to 63, respectively).

Some communication configuration must be done using the channel configure screens. Figure 13.6 shows the channel configure screen for Allen-Bradley APS programming software. It shows the two channels, channel 0 and channel 1, which must be configured before they can be used. Function keys call up additional screens for selection of configuration options that are not shown.

Depending on the SLC 500 model you have, it may be possible to configure a channel as a DH+ system channel, a DH-485 system channel, an RS 232 DF1 system channel, or an ASCII user channel. See Chapter 10 for more detail.

The status file contains some bits that only contain status information (instead of configuration parameters), updated by the SLC 500 CPU as it runs. The user-program can examine these status bits to detect the presence of active nodes on:

- *Channel 0* (S:33/4 set if channel 0 is present). SLC 5/03s and 5/04s also maintain a bit map showing which DH-485 nodes are active (S:67 and S:68).

```
CHANNEL 0 CONFIGURATION
              Current Communication Mode:  SYSTEM
                      System Mode Driver:  DF1 FULL-DUPLEX
                        User Mode Driver:  GENERIC ASCII
                           Write Protect:  DISABLED
                            Mode Changes:  DISABLED
                Mode Attention Character:  /1b
                   System Mode Character:  S
                     User Mode Character:  U
        Edit Resource/File Owner Timeout:  60 (seconds)
                        Passthru Link ID:  1  (decimal)

CHANNEL 1 CONFIGURATION
                      System Mode Driver:  DH-485 MASTER
                           Write Protect:  DISABLED
        Edit Resource/File Owner Timeout:  60 (seconds)
                        Passthru Link ID:  2  (decimal)
```

FIGURE 13.6
SLC 500 channel configuration screen in the APS programming software.

- *Channel 1* (S:1/7 set if channel 1 is present). The SLC 5/03 also maintains a bit map showing which DH-485 nodes are active (S:9 and S:10). The SLC 5/04 maintains a bit map showing which DH+ nodes are active (S:83 to S:86).

SLC 500's DH+, DH-485, and the READ/WRITE MESSAGE (MSG) Instruction

There is a lot of variability among SLC 500 models and how they can use the **READ/WRITE MESSAGE (MSG)** command to exchange data:

1. The MSG command cannot be included in the program of a **fixed SLC 500** or an **SLC 5/01.** These PLCs can be connected to a DH-485 network, however, and MSG commands from other PLCs can be used to read or write the data files of these PLCs.

2. The MSG command can only be used to exchange data via a DH-485 in an **SLC 5/02** program.

3. The **SLC 5/03 with OS 300** can execute a MSG to communicate with a PLC-5 on a DH+ if the DH+ is connected to the SLC 5/03's DH-485 (via a 1785 KA5 module or via an SLC 5/04 configured as a pass-through link). The MSG command must select **485CIF** as the target device, and the PLC-5 must have a common interface file (CIF) to accept data from the SLC 5/03 or to hold data to be read by the SLC 5/03.

4. SLC 5/03 (with OS 301 and higher) has a **PLC-5** target device option in the MSG instruction, so it can communicate with a PLC-5 on a DH+ that is connected to the DH-485 (via a 1785 KA5 module or a pass-through SLC 5/04).

5. SLC 5/04 and higher can connect directly to a DH+ and have the PLC-5 target device option. These PLCs are described by Allen-Bradley as being Internet compatible. Since an SLC 5/04 "speaks" both DH+ and DH-485 protocols, it can be used as a pass-through link, translating messages from a PLC on one network so that they can be understood by a PLC on the other. When any SLC 500 receives a data packet destined for itself, it issues an ACK (acknowledgment) signal back to the PLC that provided the packet, indicating that the packet was received. When an SLC 5/04 receives a data packet to be passed through to another PLC on another network, the pass-through PLC does not issue an ACK but just forwards the packet on the other network. Eventually, the packet will reach its destination where an ACK will be issued, and the pass-through PLC must pass the ACK back to the originating PLC.

The **READ/WRITE MESSAGE** instruction **(MSG)** initiates a data exchange with another PLC via the DH+, the DH-485, or a DF1 channel. The other PLC does not need to be programmed to participate in the data exchange. The SLC 5/03 and 5/04 have a newer version of the MSG instruction that is slightly different from the earlier SLC 5/02 version. The newer version is discussed in this book and is shown in Figure 13.7. After entering the parameters shown in the figure, another screen is displayed on the programming unit, asking the programmer to enter additional information. The parameters that must be entered include:

1. Selection of WRITE, if you want to copy a block of data from this SLC 500 to the other PLC, or READ if you want to copy data from the other PLC.

2. Selection of a type of target device. If the other PLC is an SLC 500, select 500CPU. If it is a PLC-5, select PLC-5. (I/O data files can't be exchanged between SLC 500 and PLC-5 PLCs.) If you select 485CIF, the data will be transmitted as individual words or bytes (select using bit S:2/8) between this PLC and a common interface file in the other computer. Allen-Bradley sometimes calls this method a PLC-2 read/write. Another SLC 500 can be treated as a 485CIF device, in which case data will always be written to, or read from, its data file 9.

3. The integer file address for a control block. This SLC 500 will always maintain 14 data words, starting at the address entered, to hold the parameters and working data that it needs as it executes the message transfer, including status bits that the user-program can examine to determine how a message transfer is proceeding. Figure 13.8 shows the structure of the control block if 500 CPU or PLC-5 has been selected as the target device. The structure will be different if 485CIF is chosen and is only seven words long for an SLC 5/02's MSG instruction.

4. Selection of **LOCAL** if the other PLC is connected to the same network as this PLC, or **REMOTE** if it is on another network that is linked to this one.

(a) If LOCAL is selected, the following parameters must be entered:

- The channel (0 or 1) of this PLC that is connected to the network that contains the target PLC,
- Target node is the node number of the other PLC, as assigned when it was configured for communications.
- A destination or source file address (in this PLC), a corresponding source or des-

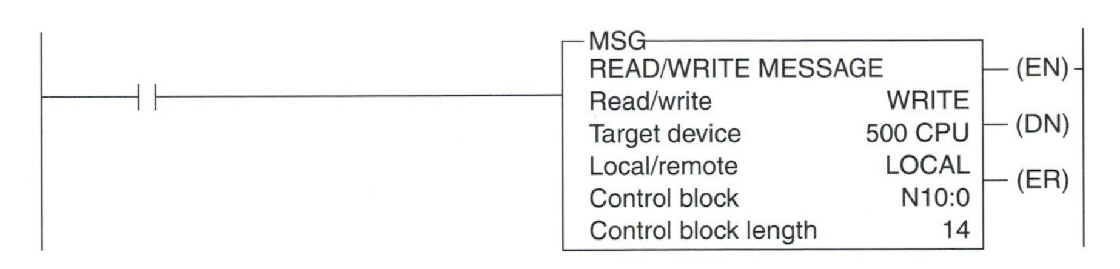

FIGURE 13.7
SLC 5/03 or SLC 5/04 MSG instruction.

15 14 13 12 11 10 09 08 07 06 05 04 03 02 01 00	Word
EN ST DN ER CO EW NR TO | Error Code	0
Node Number	1
Reserved for length in words	2
File Number	3
File Type (O, I, S, B, T, C, R, N, F, St, A)	4
Element Number	5
Subelement Number	6
Reserved (Internal Messaging Bits) | WQ	7
Message Timer Preset	8
Message Timer Scaled Zero	9
Message Timer Accumulator	10
Reserved (Internal use only)	11
Reserved (Internal use only)	12
Reserved (Internal use only)	13

FIGURE 13.8
Integer file control block layout for an SLC 5/03 or 5/04 MSG instruction if the target device is 500CPU or PLC-5.

tination target address (in the other PLC), and a message length in elements to describe the data to be copied between PLCs. If 485CIF is the target type, a target offset instead of a target address is specified to indicate where the data starts in the target computer's CIF file.

- A message timeout time in seconds (1 to 255) can be entered. If 0 is entered, the message won't ever "timeout."

(b) If REMOTE is selected, the PLC has to build slightly different data packets for transmission. These packets are called *gateway packets* in Allen-Bradley literature, and they require additional parameters to be entered for the MSG instruction. There are two ways to select the parameters, depending on how the DH+ and/or DH-485 networks are connected to each other:

- Networks can be connected via an SLC 5/04 if it is connected to both networks and configured as a pass-through link (sometimes called a *bridge* or a *gateway*[1]).

---

[1] Allen Bradley can be forgiven for its confusing mixture of network interconnection terminology, but only because most other manufacturers are just as confused and imprecise. In fact, according to the ISO, a bridge is only supposed to connect networks with different methods of control and perhaps different signaling and addressing methods but not with different data packet content requirements, whereas a gateway can connect totally incompatible networks by translating even the data in the packets if required. The ISO does nor define the term *link,* so it is a safe word to use. Translation between DH-485 and DH+ protocol appears to require just slightly more than a bridge, although one of the benefits of using a proprietary network such as those offered by Allen-Bradley is that users (like the author) don't need to know the technical details.

The pass-through SLC 5/04 must have its two channels configured so that each has a different link ID number (0 to 255) identifying the DH+ or DH-485 that each is connected to. The SLC 5/04 acts like a 1785-KA5 module. (1785-KA5 modules plug into the rack of a PLC-5 to connect data highways.) If the target PLC is on the other network, the following additional information must be included when programming the message instruction:

- The **local bridge node address** is the address of the SLC acting as the pass-through link, on the same DH+ or DH-485 as the PLC with the MSG instruction. (The pass-through link PLC has a separate node address for each of the two networks it is connected to.)
- The **remote bridge link ID** is the link ID number of the network that the target PLC is attached to.
- The **remote bridge node address** is a confusing name, especially in this two-network communication situation:

If the target PLC is an SLC 5/03, an SLC 5/04, or a PLC-5, the remote bridge node address is 0. Allen-Bradley calls the PLC-5, SLC 5/03, and SLC 5/04 Internet-capable devices.

If the target PLC is not Internet capable, its remote bridge node address is the same as the local bridge node address. This alerts the bridge that it may have to translate the message so that it can be interpreted by the lower-level processor.

- If the target PLC is on a more remote network, requiring the use of more than a single pass-through link and/or 1785-KA5 module, choice of REMOTE parameter values depends on the types of networks in the system. See your manual for specific addressing requirements.

How an SLC 500 Responds to a MSG Instruction

If the conditions for a MSG instruction are evaluated as true during program execution, a message request is placed into a wait queue for the communication handler for the appropriate communication channel. Newer SLCs immediately try to place the data to be sent (or the data request message) into a message packet in one of the message buffers. The PLC follows up by servicing the communication handler every time it executes the I/O scan, putting outgoing data packets into buffers as buffers become available, and copying incoming data packets from other buffers. Outstanding requests in the wait queue are executed in the order that they are put into the wait queue.[2] If even the wait queue is filled when a MSG instruction tries to write a request to it, the processor will try again the next time the MSG instruction is executed or the next time that communication servicing is performed.

Each PLC has a turn (called a *token*) to make use of the network by initiating data transfers. It can send data or requests for data to other PLCs. Other PLCs send data packets in response to these requests when they get the token. As communication handlers initiate data trans-

---

[2] An SLC 5/02 has only one message buffer, some 5/03s have four, while other 5/03s and all 5/04s have four per channel. The 5/04 and some 5/03s can hold 10 message requests in each channel's wait queue, other 5/03s can hold only 10 requests total in the queue, while the 5/02 is said to be able to hold "several" requests.

fers or respond to data transfer requests from an initiator, they write to status bits in the status file and in the MSG instruction's control block. These bits can be examined by a user-program. The status file bits include:

1. An OUTGOING MESSAGE COMMAND PENDING bit, which is set if there is a message request in the wait queue because a MSG instruction has been executed at this PLC but there is no data packet in the buffer to be sent (S:2/7 for channel 1, S:33/2 for channel 0). When the communication servicing occurs next, the oldest message request still in the wait queue will be examined, and a data packet will be assembled containing the data (or the request for data) that is to be sent and put into the buffer. If a packet can't be assembled as requested, the ERROR bit in the control block for the MSG instruction is set (see below), and the next message request in the wait queue is used.

2. A MESSAGE REPLY PENDING bit, which is set when another PLC has provided data asked for by a MSG instruction in this PLC (S:2/6 if data received on channel 1, S:33/1 if via channel 0). During the next communication servicing, the CPU will move the data from the incoming data buffer, put it into data memory, and will set the DONE bit in the control block for the MSG instruction that asked for the data from the other PLC.

3. An INCOMING COMMAND PENDING bit, which is set if another PLC on the network has sent data or a command requesting data due to a MSG instruction in its user-program (S:2/5 for incoming packets received on channel 1, S:33/0 for channel 0). During the next communication servicing, the newly received packet is examined:

> If it contains data the other PLC is writing to this PLC's memory, the data will be placed in memory, then a packet containing an "I have successfully performed your write request" will be put into the outgoing buffer or queue for outgoing messages. If the data cannot be written into this PLC's data memory, this PLC instead replies with an "I haven't performed your request, you are in error" message packet.
>
> If it contains a request from the other PLC to read data from this PLC's memory, the data requested is copied from data memory to an outgoing message buffer, complete with a "here is the data you requested" message (or the request is queued for eventual moving of data into a buffer). If the outgoing packet can't be assembled with the requested data, an "I haven't performed your request, you are in error" message packet is returned instead.

Each MSG instruction specifies a control block starting address, entered as an integer file word address, where the PLC will store 14 message control and status words used by the communication handler. Figure 13.8 shows the structure of a message control block. The control block status bits that the communication handler can change as it performs tasks requested by MSG instructions include the following, all of which are in the first word of the control block unless otherwise specified:

1. The ENABLED bit, **EN.** If EN (bit 15) is set, the message instruction logic is true.

2. The WAITING FOR QUEUE bit, **WQ.** If WQ (bit 0 of word 7) is set, the message request hasn't been added to the queue because the queue is already full. WQ remains set until a subsequent scan is successful in putting the request on the queue. The SLC 5/02 does not use this bit.

3. The **ENABLED AND WAITING** bit, **EW.** If EW (bit 10) is set, the data is in a buffer, waiting to be sent (and can't be modified).

4. The **START** bit, **ST.** If ST (bit 14) is set, the destination PLC has sent the ACK signal, indicating reception of the packet.

5. The **DONE** bit, **DN.** If DN (bit 13) is set, the destination PLC has sent a reply packet indicating successful reception of data from this PLC, or has sent the data requested by this PLC.

6. The **NO RESPONSE** bit, **NR.** If NR (bit 9) is set, the other PLC hasn't yet responded with an ACK signal to the data packet.

7. The **ERROR** bit, **ER.** If ER (bit 12) is set, this PLC cannot assemble the packet to be sent, or the NR bit has been set for a full scan duration, or the other PLC has reported that it cannot comply with a message request, or the message has timed out (see the TIME OUT bit below). The error is *not* set if the other PLC can't accept a packet because it is busy; instead, the message is kept in the queue for another attempt later.

Some bits in the control block can be written to by a programmer or by a program, to control the communication as follows:

8. the **CONTINUOUS OPERATION** bit, **CO.** Set CO (bit 11) to cause the message to be sent repeatedly, even if an error occurs.

9. the **TIME OUT** bit, **TO.** Set TO (bit 8) to cause the message to be cleared from the buffer and from the wait queue after the MSG instruction has been enabled without completing within the specified time. Enter a time (from 1 to 256 s) in word 8 of the control block.

If you do not set the TIME OUT bit, an outgoing packet that can't be sent will permanently occupy a buffer position. To remove the message from the buffer, you must set the time to 0 s, then set the TIME OUT bit and enable the instruction for one scan. The MSG instruction won't work again until you clear the TIME OUT bit and reenable the MSG instruction.

Other SLC 500 Instructions for Communication

SLC 500 models starting with the 5/02 offer instructions to cause the PLC to interrupt the user-program. The **SERVICE COMMUNICATIONS (SVC)** instruction causes servicing of the serial communications channels without waiting until the I/O scan. The **I/O REFRESH (REF)** instruction (discussed in Chapter 11) causes a complete I/O scan, including servicing of serial communication channels. SLC 500s have only limited communication buffer space (the SLC 5/02 has only enough buffer for a single data packet), so a user-program can monitor the **INCOMING COMMAND PENDING, MESSAGE REPLY PENDING,** and **OUTGOING MESSAGE COMMAND PENDING** status file bits (described earlier) and execute the SVC (or REF) commands as required to improve communication performance.

The 5/03 and higher allow the programmer to specify which of the two serial communications channels to service in response to the SVC or REF instruction. SVC and REF instructions cannot be programmed into interrupt service routines responding to I/O interrupts, DIIs, or STIs.

Several ASCII instructions are available to send and receive data via a channel configured as an ASCII user channel. The instructions are listed in the PLC-5 communications section of this chapter but are not covered in this book.

S5 S7

SIEMENS COMMUNICATIONS, WITH PROFIBUS

Siemens does not own Profibus, although a Siemens customer could be forgiven for thinking so. Siemens has played a big part in promoting Profibus networks.

There are four types of Profibus (process field bus) networks still in use, so connecting to a Profibus network requires the correct interface.

1. The oldest Profibus version you are likely to encounter is often just called Profibus, but sometimes the letters **FDL** (*fieldbus device level*) or the descriptor **part 1** are used to differentiate it from the others. This was primarily a master–slave network. The Siemens **L1** network was based on this early standard.

2. Profibus-DP (*distributed peripherals* or *decentralized periphery*) is the most common today and is used to interconnect active nodes so that they can send messages to each other as peers, and passive nodes which act as slaves to active nodes. Siemens had an earlier version of Profibus-DP, which they called an **L2** network. Siemens now uses Profibus-DP, which is compliant with the European norm standard, EN 50 170.

3. Profibus-PA (*process automation*) is the same as Profibus-DP except that it uses lower voltage and current levels. It is intended for intrinsically safe applications where computerized equipment can't generate sparks or electrical noise.

4. Profibus-FMS (*fieldbus Message service*) is the fastest and highest-level Profibus network, designed to meet EN 50 254, which German manufacturers hoped would be the worldwide standard for the open communication network approved by the Fieldbus Foundation. Now that another system (fast Ethernet) has been chosen, Profibus-FMS is scheduled to be dropped as a Fieldbus Foundation acceptable standard in about 2004. (Sometimes plans don't work out the way they were intended.)

S5

COMMUNICATIONS USING THE SIEMENS S5

S5 PLCs can communicate via the serial ports in their CPU modules, or via interface modules (IMs) or communication processor (CP) modules. The main serial port in an S5 CPU module is used to connect a programming unit, so it is usually called the programming port. Some S5-115U CPU modules have a second serial port, so that they can be connected to communicate using both channels. To use a serial port for communication with another PLC (instead of just for connecting to a programming unit) requires the CPU module to be configured to use parts of its memory for coordinating communication and to hold data packets. User-programs must write outgoing data to a SEND MAILBOX, and then initiate the transmission using the SEND COORDINATION byte. The user-program must monitor the RECEIVE COORDINATION byte to detect the reception of data, and must then copy the newly received data from the RECEIVE MAILBOX. An S5 CPU can use its CPU ports to communicate in three different methods:

1. In point-to-point communications, only two S5 PLCs are connected to exchange data via their mailboxes. Only a simple communication cable is required.

2. In L1 network communications, up to 30 S5 CPUs can be interconnected to act as slaves to an L1 master. The L1 master can be an L1 communication processor module in an S5-115U rack, or can be a personal computer with an L1 interface card and communication software, but CPU modules cannot be masters. The L1 master usually polls each CPU slave in turn, reading its SEND COORDINATION bytes to see if it wants to send data and then performing the requested data transfers. The L1 master can also be programmed to interrupt the handling of normal-priority messages to transmit high-priority messages in response to interrupt requests from selected CPU modules. Each CPU module needs an L1 interface module to translate current-loop protocol at the programming port to RS-485 protocol on the shared communication bus. Communication processor modules are also available so that S5 PLCs can communicate on the L1 network via I/O modules instead of via their programming port.

3. Some S5-115U CPU modules have ASCII drivers built into their CPU module. These CPU modules can use ASCII serial communications to exchange data packets with another computer, or with a serial device such as a bar-code reader or a serial printer. Besides the coordination bytes and mailboxes, memory must also be assigned (during CPU configuration) for an ASCII parameter set, which must contain parameters describing how the S5 CPU will communicate with the other computer or peripheral. Interface hardware may be required to exchange the current-loop signals at the CPU to the signal type required by the other computer/peripheral.

The S5-115U CPU can be connected into remote I/O systems using the appropriate interface (IM) modules in the central rack and in the expansion unit racks. S5 CPUs can be connected into Profibus networks, into MODBUS networks, or to RS-232C devices using point-to-point protocol. These communication processes require the appropriate communication processor (CP) modules. AS-I CP modules are available to connect actuator/sensor-interface networks so that they can be controlled by an S5 PLC.

S5 L1 and Point-to-Point Serial Communication

An S5 CPU must be configured to act as an L1 slave or to communicate in point-to-point mode. (L1 masters are covered later.) To configure an S5 PLC as an L1 slave or for point-to-point communications, you must enter network addresses and pointers into the CPU's system data area of memory, as described in Chapter 10. Configuration data that must be entered includes a unique L1 slave node number for the CPU and pointers to the memory areas that will be reserved for coordination bytes and mailboxes for data packets.

Coordination bytes contain control bits and status bits that a user-program can read and write, including:

Bit 7. Set this bit in a SEND COORDINATION BYTE (CBS) to cause data in the SEND MAILBOX to be sent, or in a RECEIVE COORDINATION BYTE (CBR) to allow data to be received into the RECEIVE MAILBOX. Bit 7 of the CBS and the CBR is automatically set by the operating system as the PLC goes into run mode.

While CBS bit 7 is set, the CPU is prevented from changing the contents of the SEND MAILBOX. While CBR bit 7 is set, the CPU is prevented from reading the contents of the RECEIVE MAILBOX.

The STEP 5 operating system resets bit 7 after a data transmission (in the CBS after sending data from the SEND MAILBOX, or in the CBR after receiving data into the RECEIVE MAILBOX). Your program should read this bit to know when the transmission has happened.

Bit 0. Set by the STEP 5 operating system if an error was detected during the most recently requested data transmission.

The following status bits in coordination bytes are only used in L1 networks:

Bit 4 can be set in the send coordination byte (CBS) to indicate that the data in the SEND MAILBOX should be treated as higher priority than other data transmission requests by other PLCs. The STEP 5 operating system sets bit 4 in the receive coordination byte (CBR) of the destination controller as it delivers the high-priority data.

Bit 1 of the CBR is set by the STEP 5 operating system if it detects that a node is turned off.

Bit 2 of the CBR is set by the STEP 5 operating system while the L1 network is in run mode.

Bit 3 of the CBS is set by the STEP 5 operating system to inhibit other data transmissions while a programming unit is using the L1 network.

A SEND MAILBOX can contain a data packet of up to 66 bytes. A RECEIVE MAILBOX should be large enough to receive the longest data packet that any other node can send. A data packet must contain:

In **byte 1,** a number indicating the length of the entire data packet (2 to 66 bytes). This number is transmitted with the rest of the packet.

In **byte 2,** a number indicating to which node(s) to transmit this packet. Node 0[3] is the L1 master, nodes 1 to 30 are other slave nodes, and number 31 means to broadcast this packet to all the nodes on the network. As the packet is written into a receiving node's memory, the STEP 5 operating system changes byte 2 to indicate the address of the sending node.

Bytes 3 to 66 contain the data that the sending node wishes to send. *Note:* Byte 1 indicated how many total bytes (including bytes 1 and 2) will actually be sent.

S5 ASCII Serial Communication

Some S5-115U CPU modules have ASCII drivers built into their CPU serial ports. To use ASCII communications, configuration (see Chapter 10) must include configuration to select the ASCII driver for the serial port and pointers to coordination bytes and mailboxes and to an ASCII parameter set.

Bit 7 of the **SEND COORDINATION BYTE (CBS)** initiates data transmissions, and bit 7 of the **RECEIVE COORDINATION BYTE (CBR)** allows data receptions when set by a user-program. (The STEP 5 operating system clears these bits after performing a data transmission.) **Bit 0** is set in both the CBS and the CBR if an error is detected during the data transmission.

The **SEND MAILBOX** and the **RECEIVE MAILBOX** can be up to 1 kilobyte in length and can contain only data. The ASCII parameter set indicates whether the data packet size is fixed, or indicates a specific control character to be interpreted as an end of data signal in the data set.

S5 Communication via Interface Modules and Communication Processor Modules

Interface (IM) modules perform signal conversion and some minor communication control functions to connect an S5 PLC to networks not compatible with their serial port signal types.

---

[3] Node 32 can also mean the L1 network master if the slave is connected to the L1 network via a communication processor module instead of its programming port.

Communication processor (CP) modules can be used to connect a PLC to a significantly different type of network, such as a Profibus network.

Some CP modules exchange information with S5-115U CPU modules using interface memory in the CPU module. (Interface memory was not discussed in Chapter 5 because it is used only for communication with CP modules.) Each 1 kilobyte section of interface memory in the CPU module is reserved for data that will be shared with one CP module.[4] A communication-handling processor in the CPU module automatically handles the transfer of data back and forth between CP modules and the interface memory. Each CP module has a unique interface number, assigned during hardware setting using DIP switches on the CP module, or held in EEPROM memory in that CP module. Interface numbers can sometimes be changed by writing a new number to the CP module in data sent to the module from the CPU. A user-program must execute data handling function blocks to read/write the interface memory in the CPU (thereby reading/writing CP module data). The S5-115U CPU module comes with preprogrammed data handling blocks (see Chapters 7 and 10).

Some CP modules use interprocessor communication flags to control communications between the CP module and the CPU module. In Chapter 10 we described how to configure your CPU to use interprocessor communication flags.

The Siemens L1 Network Master

An L1 network must have a master node, which reads each slave's SEND COORDINATION BYTE (CBS) in turn for data transfer requests, and performs the data transfers requested. As the description of CBS bits implied, the master will also respond to interrupt requests from slaves to perform high-priority data exchange operations. The L1 master can be a CP 530 communication processor module in an S5-115U PLC's rack, or may be a personal computer with an interface card.

Configuration of a CP 530 module was described in Chapter 10. Configuration consists of calling the data handling function blocks from a user-program to read and write data in an interface memory area of the CPU memory. The S5's communication handlers automatically copy interface memory data contents between the CPU and intelligent I/O modules, including communication processor modules, to exchange data with the CPU. After configuration (and perhaps after switching the L1 network on) a CP 530 module can operate as the L1 network master without any further configuration or control. Slaves connected to the L1 network can exchange data.

The same data handling function blocks that are used to configure a CP module can also be used in the L1 master's user-program to send and receive data on the L1 network. Although data handling function blocks were described in Chapters 7 and 10, we will review their use for communications here:

FB 246, "SYNCHRON", must be called once every time the PLC restarts, to synchronize communications between the CPU module and the CP module (will have been called during configuration).

FB 244, "SEND", used to write configuration data to the CP 530 module, but after config-

---

[4] It is also possible to use page framing instead of this linear method of assigning memory for each CP module. In page framing, the same 1 kilobyte of CPU memory is used with all CP modules. See your manual set.

uration SEND is also used to send control data and message data to a CP 530 module. Parameters that must be included with a call to send data on the L1 network include:

- **A-NR:** the **job number.** Enter as KY 0,y. where "y" is a job number indicating the purpose of this data transmission to the CP module. SEND must be executed with the following job numbers to send data via the L1 network after it is configured:
  - 1 to 31 — The data for the interface memory contains a SEND MAILBOX data set to be sent from this node (the master node) to node number 1 to 30, or to broadcast to all nodes (job 31). The SEND MAILBOX data set must consist of:
    - A SEND COORDINATION BYTE (CBS), then
    - An unused byte, then
    - Up to 64 bytes of data.

    The SEND COORDINATION BYTE can be used to switch the destination CPU into stop mode (if bit 7 is set) or into run mode (if bit 6 is set).
  - 51 to 81 — The data contains a SEND MAILBOX data set, for high-priority transmission of the slaves 1 to 30 (jobs 51 to 80) or for broadcasting (job 81).
- **ANZW:** the address for two words of condition code data. Enter a flag word address (e.g., FW6) or a data word address if a data block is open. The first of these two words will contain flag bits containing the status of the requested transmission of data to the CP module, and to another CPU via the L1 network. The second word will contain a number indicating how many data words have been exchanged between the CPU and the CP module for this request. (See Chapter 7 and your manuals.)
- **QLAE:** how many words or bytes of data to copy to the interface area of memory. Enter as KF x, where "x" is the number of data words or the number of flag bytes.

Remember that a **SEND MAILBOX** data set includes a two-byte header followed by the data.

FB 247, "CONTROL", to read the status of the CP 530 module. Parameters that must be entered include:

- **ANZW:** the address where two words of **condition code** data will be stored. The CONTROL function block uses these words to describe the status of communication with a specific slave if that slave is singled out by using the appropriate job number. A count of the words transferred is put into the high word. The low word includes these important status bits:
  - 0 Set when a CP module has received a data block from another PLC. Reset when a RECEIVE instruction finishes reading the data from the CP module.
  - 1 Set if a **SEND** function has been executed, supplying data for the CP module to forward to another PLC, or if a **FETCH** function has requested the CP module to get data from another PLC. Reset when the SEND data or the FETCH request has been sent to the other PLC.
  - 2 Set when a requested data SEND or RECEIVE has been completed without an error. Reset when a new SEND or RECEIVE function block is executed.

3 Set if an error was detected during a SEND or RECEIVE. An error code will also be placed in bits 8 to 11. Reset by a new SEND or RECEIVE.

4 Set while the data transfer between the CPU module and the CP module network is being performed.

5 Set by a SEND function block. Reset after data has been transferred to the CP module from the CPU.

6 Set by a RECEIVE function block. Reset after data has been transferred from the CP module to the CPU.

7 Can be set by a user-program to disable future SEND or RECEIVE function blocks from accessing the CP's interface memory, or reset to reenable them.

- **A-NR:** the job number for which condition code status data is requested. Any job number can be entered, but the use of job numbers for receiving data from slaves is perhaps the most useful application, to determine whether the CP 530 module contains a message sent from a slave to this master. Use job numbers:

  101–130 Job 101 reads the status of messages received from slave node 1, 102 from slave 2, etc.

FB 245, "RECEIVE", to read CP 530 module status and messages received by the CP 530 into CPU memory. Parameters that must be entered with RECEIVE include:

- **ANZW:** returns the same condition code data that CONTROL receives.
- **A-NR:** the **job number,** as described for SEND, but a different set of job numbers is used:

  101–130 to read RECEIVE MAILBOX data packets from the CP 530 module into CPU memory. Job 101 reads a packet from slave node 1, 102 from slave 2, etc. Each RECEIVE MAILBOX data set will consist of a 4-byte header and data:
  - The RECEIVE COORDINATION BYTE (CBR) is the first byte, followed by:
  - The number of data bytes in this message, then
  - The number of the node that sent this data, then
  - An unused byte, followed by
  - Up to 64 bytes of data.

  221 to read the full SYSTEM STATUS (SYSTAT) data set, which includes an error number and an error code for recent communication problems. (See your manual for error numbers and codes.)

  220 to read only 10 bytes of SYSTAT data, including a maximum of three error numbers and codes.

  201 to read a 4-byte-long listing of the node numbers of the slaves that did not respond during the last full cycle of polling.

  223 to read the current SYSID configuration data (data written using job 222).

  143 to read the current polling list configuration.

  144 to read the current interrupt list configuration.

  142 to read the current control byte.

FB 248, "RESET". Cancels any outstanding job and resets status for that job. Parameters for RESET are SSNR, PAFE, and A-NR, all described above.

S5 PLCs and the Profibus Local Area Network

An S5-115U CPU module can be connected into an RS-485 Profibus-DP or Profibus-FDL network using an interface module (IM 308-C). The S5-115U will be an active node on the Profibus-DP network, which means that it can contain a program that commands data exchanges. An S5-115U can be the Profibus master, to coordinate all Profibus-DP traffic, or a CP 5431 module can be used to act as a master on either a Profibus-DP or a Profibus-FMS, doing the controlling functions under the control of an S5-115U CPU.

An S5-100U or an S5-115U CPU module can be connected into a Profibus-DP or Profibus-FDL network via a communication processor (CP 541) module. The CP 541 connects to the S5 CPU via the programming port, not via the PLC's bus system. S5-100U CPUs can only be passive components in a Profibus-DP network, which means they can only send/receive data in response to commands from active nodes on the Profibus-DP.

A rack of S5-100U I/O modules can be controlled by an active controller via a Profibus-DP network. Instead of paying for an S5-100U CPU and a CP 541, install the I/O modules into a distributed I/O module (e.g., an ET200U with a CP 318 Profibus-DP interface module). Distributed I/O modules are always passive nodes.

Other ET200 distributed I/O modules, with CP modules, can also be connected to a Profibus network. Interfaces are available to connect actuator-sensor interface (AS-I) sensor nets as Profibus-DP slave nodes. Personal computers with Siemens-supplied interface cards can also be connected and can be active components.

S7

COMMUNICATIONS USING THE SIEMENS S7

Siemens S7 PLCs are designed to communicate via what Siemens calls *subnets.* A subnet is any point-to-point connection or network, and subnets can be parts of larger networks. The most common types of networks used as subnets for S7 PLCs include multiprocessor interface (MPI), Profibus (especially Profibus-DP), and Industrial Ethernet. In Chapter 10 we described how to set up a networked PLC system, and described in general how to use subnets in a project. In this chapter we look at network configuration in more detail and discuss how to use the various types of communication subnets.

Configuring S7 Communications

Configuring an S7 PLC for communications depends on how you want it to communicate with other nodes and via what types of network it is going to communicate:

1. You can configure global data circles within an MPI network so that selected data will be exchanged automatically between small numbers of S7 nodes.

2. User-programs can contain standard functions (SFCs) to exchange data with other nodes on an MPI network, without having to preconfigure connections between the nodes (sometimes called *communicating via nonconfigured connections*).

3. Standard function blocks (SFBs) are available for S7-400 CPUs to exchange data between nodes on any network type if the nodes have been configured with connections to each other (*communication via configured connections*).

4. When you configure a Profibus-DP network, you assign address space in the DP master's memory for each DP slave. The DP master can read or write up to 4 input bytes and 4 output bytes of the slave's data using peripheral I/O addressing, just as if the slave were in a slot in the DP master's rack system. The Profibus-DP system handles copying of data between the slave's memory and the master's memory.

5. A Profibus-DP master can use SFCs in user-programs to read or write data sets larger than the 4 input and 4 output bytes at DP slaves.

If an S7 CPU module detects a communication error, it will interrupt any program with a priority class of less than 35 and will execute OB 87, the communication error organization block. If OB 87 does not exist, the PLC will go into fault mode. When OB 87 is called, the STEP 7 operating system provides it with a 1-byte error code (symbolic name: OB87_FLT_ID_) and as many as two additional codes identifying the problem. See Chapter 11 for more information on fault interrupts, and your manuals for error codes.

S7 Global Data Circles in an MPI Network

To configure an S7 CPU to participate in global data circles in an MPI network:

1. You must have created and configured an MPI subnet in the project, and MPI configuration must have been entered for at least two modules for the subnet.

2. Select the MPI subnet, then select OPTIONS, then DEFINE GLOBAL DATA. A configuration table will open for configuration of global data for an MPI subnet. Figure 13.9 shows how that table might appear after you have entered some global data parameters, as in the following steps.

3. Double-click on a blank column header; a menu of the stations that have been configured as being connected to this subnet will appear. Select one of those stations. Repeat for the other stations that will participate in the global circle. Any one station can send data to as many as four other stations in a global data circle.

4. Each row of the table must describe where each PLC will store the global data packet that will be sent between participants in the global data circle. Select a cell in the first row, press F2, then enter the address(es) at that station that you want to dedicate to global data circle communication. Enter an address followed by a colon, then the number of bytes (maximum 22) that you want to be exchanged in a packet in this global data circle. Addresses can be any memory area except local (L) or peripheral I/O (PI and PQ). Any number and size of data unit can be specified, up to a total of 22 bytes, as long as the sender and receiver(s) are configured for the same amount of data.

If more than two columns contain addresses (as in row 1 of Figure 10.21), one of the stations must be identified as the sender by selecting SELECT AS SENDER when the cursor is in the appropriate cell. Row 1 of Figure 13.9 configures Stn 2 to send the contents of memory locations IB 4 to IB 19 (16 bytes) to Stn 1 and to Stn 3. At Stn 1 it will overwrite 16 bytes of bit memory starting at MB 100. At Stn 3, it will overwrite data starting at MB 20.

If exactly two columns contain addresses (as in row 2), a sender does not have to be identified. Whenever either station writes to the memory it uses for this GD packet, the corresponding memory in the other station will be changed.

	GD identifier	Stn 1/CPU 1	Stn 2/CPU 2	Stn 3/CPU 3
	GD 1.1.1	MB100 : 16	>>IB4 : 16	MB10 : 16
	GD 1.2.1		MW40 : 4	MW60:4

FIGURE 13.9
STEP 7 global data table configuration screen.

5. Repeat data packet definitions for the other rows. There can be up to four packets exchanged per global data circle.

6. Create additional global data circles by selecting GD TABLE–OPEN–GLOBAL DATA FOR SUBNET, then repeating as in steps 2 through 5. Depending on the S7 CPU model, each station may be configured to be a member of as many as 16 different global data circles (only four circles for an S7-300).

7. Select GD TABLE COMPILE. The programming software will create system data blocks (SDBs) from your global data configuration if you have entered data correctly, and will report when it has completed this phase 1 compiling step successfully.

8. It's optional, but you may want to set a scan rate and/or create GD status words. These must be done after a phase 1 compiling is finished (i.e., now).

(a) Select VIEW–SCAN RATE, then enter a scan rate such that the scan rate you select multiplied by the PLC's scan cycle time is greater than 60 ms for an S7-300, or 10 ms for an S7-400, when sending data. For receiving data, (scan rate) × (cycle time) should be shorter than the sending data rate. If you enter 0 as the scan rate, global data packets will not be sent automatically but will only be exchanged in response to execution of the SFCs that command global data sending or receiving (SFC 60, GD_SEND or SCF 61, GD_RCV).

(b) Select VIEW–GD STATUS, then enter the address where a global data status (GDS) double word can be stored, for each CPU where you want global data exchange status to be available. Receivers of global data packets do *not* respond, so without setting up a GDS status word, there is no way of verifying that a transmission of global data was successful. All bits of the GDS double word will be "0" when data exchanges have been successful. When using a programming unit to monitor global data exchanges, the programming unit ORs all GDS double words and displays a group status word (GST).

(c) You must select GD TABLE COMPILE a second time after entering scan rate and/or GD status configuration.

9. Save and download these configuration parameters after you have finished entering them!

The user-programs can now write to the memories that are copied into global data circle packets, and that data will be copied automatically to the other PLCs in the global data circles. The other PLCs should be programmed to read the global data they receive before the sender sends new data. Two standard functions (SFC 60, GD_SEND and SFC 61, GD_RCV) can be called in a user-program to initiate an exchange of global data packets without waiting for the cyclic exchange.

S7 SFCs for Nonconfigured Connections

To use SFCs for nonconfigured connections, subnet(s) must be configured, and each node must be configured to know which subnet it is on, as covered in Chapter 10. Whenever you execute one of these SFCs, the S7 CPU automatically establishes a connection to the other computer.

The following standard functions (SFCs) are preprogrammed into S7 CPU modules and can be called from a user-program:

SFC 67, "X_GET". Copies one data element from another CPU on the same MPI network. The other CPU does not have to be programmed to respond. (Siemens describes communications that don't need to be programmed at both PLCs as *unilateral.*)

- Input parameters of X_GET include:

  REQ. If this bit is on, X_GET requests that the communication handler get the data element if the job isn't already in progress from a previous scan. Rerequests the data if REQ is still true after the data element has been received.

  DEST_ID. This is a word indicating the MPI node number of the other CPU.

  VAR_ADDR. This is an ANY pointer to a data unit containing the data element to read from the other CPU. The data element cannot be a STRUCT, or an ARRAY of BOOL. If it is important that the data unit shouldn't be broken up for transmission, it should be no larger than 8 bytes if either CPU is an S7-300, 16 bytes between smaller S7-400 CPUs, or 32 bytes for S7-414 CPUs and above.

  CONT. If this bit is on, maintain the connection after the data has been sent, to save time next time data is sent to the same destination. If off, break the connection after sending, so that it is available for other communications.

- X_GET has three *output parameters:*

  RD. This is an ANY pointer indicating the memory area where the incoming data should be placed in this CPU. Must be as large as the size of VAR_ADDR.

  BUSY. This bit is set when the communication buffer accepts the job, or if the job has already been accepted from an earlier scan. BUSY is reset if the job can't be accepted.

  RET_VAL. This would contain an integer code describing the job status (including connection difficulties) or the reason the X_GET function can't work.

SFC 68, "X_PUT". To write a single data unit to another CPU on the MPI network. Like X_GET, X_PUT is unilateral, so the other PLC does not need to be programmed to accept the data unit. X_PUT's parameters include:

REQ, CONT, BUSY, and **RET_VAL:** as described for X_GET.

DEST_ID and **VAR_ADDR:** as described for X_GET, except that they now identify the MPI node and memory address where the data from this CPU will be written.

SD: an ANY pointer to the memory area containing the single data element to be written to the other CPU. Cannot be a STRUCT data unit, or an ARRAY of BOOL. If it is important that the data unit shouldn't be broken up for transmission, it should be no larger than 8 bytes if either CPU is an S7-300, 16 bytes between smaller S7-400 CPUs, or 32 bytes for S7-414 CPUs and above.

SFC 65, "X_SEND". Writes a block of data to the CPU's communication buffer. A communication processor takes over to send the data to the destination address. The CPU that the data is being sent to needs to execute an X_RCV to accept the data from its communication buffer, so X-SEND and X-RCV are used for bilateral communications. Parameters of X_SEND include:

REQ, CONT, BUSY, RET_VAL, and **DEST_ID:** as described above.

REQ_ID: a double word assigned by the programmer to identify a unique X_SEND job. Sent to the receiving computer, where the other computer can use it to identify the data's source.

SD: the ANY pointer to the memory area containing the data to be sent to the destination CPU. Cannot be a STRUCT data unit or an ARRAY of BOOL.

SFC 66, "X_RCV". Reads data from the communication buffer after an X_SEND instruction has sent the data from another S7 CPU. The communication buffer can hold more than one incoming block of data.

- Only one input parameter is available to control X_RCV:
  EN_DT: If this bit is:
  - Off: X_RCV uses its Boolean NDA output parameter to indicate whether there are any blocks of data in the incoming communications buffer. (NMA = 1 if there is.)
  - On: X_RCV copies the oldest block of data from the communication buffer.
- Output parameters from X_RCV include:
  NDA: a bit that can have either of two functions, depending on the state of EN_DT when X_RCV was executed. If:
  - EN_DT = 0, NDA = 1 means that there is data in the buffer.
  - EN_DT = 1, NDA = 1 means that a block of data has been copied from the buffer to the CPU's memory.

  REQ_ID: the unique double word identifying number for the oldest block of data in the buffer, as assigned by the X_SEND instruction that sent it.

  RD: an ANY pointer to the memory area where the data from the buffer is to be placed in this CPU.

  RET_VAL: an integer, will contain an error code describing why X_RCV can't be accepted, or the size (in bytes) of the oldest block of data in the buffer (if EN_DT was 0) or of the block of data just transferred to the CPU (if EN_DT was 1), or the hexadecimal value 7000 if the buffer is empty.

SFC 69, X_ABORT. Terminates a connection established with X_GET, X_PUT, or X_SEND. **REQ, BUSY, RET_VAL,** and **DEST_ID** parameters are required.

SFC 72, I_GET; SFC 73, I_PUT; and **SFC 74, I_ABORT,** are like **SFC 67, X_GET; SFC 68, X_PUT;** and **SFC 69, X_ABORT,** except that they are used to read/write data elements or abort connections with communication partners in the same station as the CPU that initiates the communication. The parameters are the same, except that the DEST_ID parameter is replaced by:

IORD: a byte to indicate whether the communication partner is addressed as a peripheral input module (IORD = hex: 54) or as a peripheral output (hex: 55).
LADDR: a word indicating the module's peripheral I/O address number.

S7-400 SFBs for Configured Connections

The S7-400 series of CPUs has preprogrammed standard function blocks (SFBs) for communications using any type of subnet(s). These SFBs require that connections be configured. In Chapter 10 we described how to configure an S7-400 PLC for configured connections. During configuration, the programming software assigns local ID numbers and sometimes partner ID numbers. These ID numbers will be needed as parameters for the SFB calls for communication in the user-program.

Once connections have been configured, a user-program in an S7-400 CPU can call communications system function blocks (SFBs)[5] to communicate with connected communication partners. Data that is exchanged can be segmented into smaller packets. The parameters for communication SFBs must be in instance data blocks. (User-programs can read and write variables in instance data blocks.) The communication SFBs and their parameters include:

SFB 14, GET. Copies a set of data from another CPU unilaterally (without the other CPU having to be programmed to provide the data, so it can be an S7-300 CPU.) Parameters include:

REQ: an input bit that initiates the GET if the GET isn't already in process. Unlike the SFCs for nonconfigured communications, the communication SFBs are initiated when the REQ input parameter changes from false to true between executions of the SFB.
NDR: an output bit that goes on after data has been received in this CPU's memory in response to the GET.
ID: a word where the programmer enters the local ID number that was assigned automatically for the connection to the other CPU when the connection was configured.
ADDR_1, ADDR_2, ADDR_3, and **ADDR_4:** ANY pointers to the memory area(s) in the other CPU that contain the data to be copied.
RD_1, RD_2, RD_3, and **RD_4:** ANY pointers to the memory area(s) where the data is to be copied into this CPU.
ERROR: an output bit that GET turns on if a communication error is detected.
STATUS: an output word containing a code describing any communication errors.

SFB 15, "PUT". Copies a set of data to another CPU unilaterally (without the other CPU having to be programmed to accept the data, so it can be an S7-300 CPU). Parameters include:

REQ, ID, ERROR, and **STATUS:** as described for GET.

---

[5] The STEP 7 manual set sometimes refers to the communication system function blocks as CFBs.

SD_1, SD_2, SD_3, and **SD_4:** ANY pointers to this CPU's memory area(s) containing the data to be copied to the other CPU.

ADDR_1, ADDR_2, ADDR_3, and **ADDR_4:** ANY pointers to the memory area(s) in the other CPU to which the data will be copied.

DONE: an output bit that PUT turns on after the other CPU acknowledges receiving the data.

SFB 8, USEND. Copies data to a partner CPU, which must execute a URCV to accept the data. Since both CPUs must be programmed, communication is bilateral and both CPUs must have a configured connection to the other CPU. USEND's parameters include:

REQ, ID, DONE, ERROR, STATUS, and **SD_1** to **SD_4:** as described above.

R_ID: a double word containing a user-assigned number. The BRCV must contain the same R_ID number so that the communicating CPUs can match the USEND to its matching URCV. (S5 PLCs connected to the network via CP 441 modules use the R_ID number for other purposes.)

SFB 9, URCV. Accepts data sent using BSEND. BRCV parameters include:

ID, R_ID, NDR, ERROR, STATUS, R_ID, and **RD_1** to **RD_4:** as described above.

EN_R: an input bit that allows new data to be received when set. If USEND is executed again at the other CPU, new incoming data can overwrite the data received previously. To prevent this, turn EN_R off after NDR goes on (indicating a successful data reception) and hold EN_R off until ready to receive new data.

SFB 12, BSEND. Copies up to 64 kilobytes of data to a partner CPU, which must execute a BRCV to accept the data. Since both CPUs must be programmed, communication is bilateral and both CPUs must be configured with a connection to the other CPU. USEND's parameters include:

REQ, ID, DONE, ERROR, STATUS, and **R_ID:** as described above.

SD_1: an ANY pointer to the memory area in this CPU, containing data to be sent to the other CPU.

LEN: a word indicating how many data bytes to send. BSEND automatically divides the data into smaller packets and waits for the receiver to acknowledge each packet before sending the next.

R: an input bit, which can be turned on to cancel the BSEND.

SFB 13, BRCV. Accepts data sent using BSEND. BRCV parameters include:

EN_R, ID, R_ID, NDR, ERROR, STATUS, and **LEN:** as described above. EN_R and NDR should be used to prevent the overwriting of received data, as described for URCV.

RD_1: an ANY pointer indicating the memory area where incoming data is to be stored.

Three SFBs can be used to control a remote S7 PLC:

SFB 19, START. Put a PLC into run mode via a complete restart.
SFB 20, STOP. Put a PLC into stop mode.
SFB 21, RESUME. Put an S7-400 into run mode via a restart.

START, STOP, and RESUME are unilateral. They require the following parameters:

REQ, ID, DONE, ERROR, and **STATUS:** as described above.
PI_NAME: An ANY pointer to an ASCII string containing an IEC-approved device name. For an S7 PLC, the name must begin with P_PROGRAM.
IOSTATE and **ARG:** a byte and an ANY pointer, respectively, that should be left blank. Eventually, the IEC will require their use.

There are two SFBs and one SFC that can be used to:

- Request status information from another CPU (**SFB 22, "STATUS"**).
- Allow another CPU to report mode changes (**SFB 23, "USTATUS"**).
- Request the status of a connection to another CPU (standard function: **SFC 62, "CONTROL"**).

See your manuals for more details.

SFC 16, PRINT. Can be used to send ASCII characters and formatting codes to a printer via a point-to-point connection. (See your manuals.)

S7 Communications via Profibus-DP

You can configure an S7 project with a Profibus subnet, and you can select modules capable of Profibus communications when you configure station hardware, as described in Chapter 10. During configuration of a DP master, you will have assigned 4 bytes of input image and 4 bytes of output image address space for each DP slave. The master can read or write I/O addresses in each slave via these I/O image addresses, using Ladder Logic MOVE or STL LOAD and TRANSFER instructions.

During configuration, you may also have configured the DP master and some DP slaves to maintain shared interface data areas for use by the DP network. The Profibus system copies data automatically between the master's interface memory and the corresponding interface memory in slaves. The interface data areas can be used to exchange more than the 4 input bytes and 4 output bytes of data with the DP master. The user-program in the DP master can call standard functions (SFCs) to exchange as much as 122 bytes of data with each slave by reading or writing the interface area in the master's own memory. The SFCs include:

SFC 14, "DPRD_DAT": to read slave data from the interface memory.
SFC 15, "DPWR_DAT": to write slave data via interface memory.

Parameters used by DPRD_DAT and DPWR_DAT include:

LADDR: a word identifying the number for the interface memory to use (thereby selecting the slave).
RECORD: an ANY-type pointer to another area of the CPU's memory, for DPRD_DAT to copy data to or DPWR_DAT to copy data from.
RET_VAL: an integer containing an error code written by the SFC if the SFC fails.

DP slaves must be configured with a diagnostic area of memory. DP master can read the slave's diagnostic memory using **SFC 13, "DPNRM_DG".** Active DP slaves can execute their own user-programs, which can include calling **SFC 7, "DP_PRAL",** to cause the DP master to interrupt and run its hardware interrupt routine, OB40. (See your manuals and Chapter 11.)

CQMI

COMMUNICATIONS USING THE OMRON CQM1

Most CQM1s have two serial communications ports: a peripheral port, and an RS-232C port.[6] The programming unit is usually attached at the peripheral port, but either port can be configured to communicate in one of three modes:

1. In one-to-one mode, two CQM1 PLCs are configured to share a portion of their link register (LR) memory with each other. One CQM1 is configured as the master and can write to the lowest addresses of the shared memory. The other CQM1 is the slave and can write to the high addresses of the shared memory. Whatever one PLC writes to its portion of link memory is sent to the other PLC during the originator's I/O scan step and is accepted and written into the other's link memory during the receiving PLC's I/O scan. The data copying process is completely automatic and there are no commands or configuration options that can specifically speed up or slow down the serial data exchange. One-to-one communication can be configured only via the RS-232C port, not via the peripheral port.

2. By default, both ports are in host link mode. The programming unit software is written to communicate with a CQM1 using this mode, but it is useful for local area network communications as well. In host link mode, a host computer (not a CQM1) can be connected to as many as 32 slave CQM1s (each configured with a different node number) via a shared serial link. The host computer can issue *C-mode commands* to any of the CQM1 PLCs. There are C-mode commands to allow the host to read or write any part of a CQM1's data, program, error code, or operating status memory. A CQM1's response is completely automatic and cannot be controlled from the CQM1's program. Host link communications using C-mode commands allow networking of CQM1 PLC in either of two ways:

(a) A host computer can be programmed to poll the CQM1 slaves by reading a specific area of each CQM1's memory, one CQM1 after the other. Each CQM1 would have to be programmed to place communication request messages into that area of its memory for the host to read. The master would have to be programmed to interpret the communication request messages from CQM1s and to execute the C-code commands necessary to perform the functions requested (such as copying data from one CQM1 to another). The host would also have to notify each CQM1 after a requested communication service has been accomplished, by writing status information to a separate area of that CQM1's memory. For an example of how such a system would work, read about Siemens' L1 networking for S5 PLCs in this chapter.

(b) Any CQM1 can be programmed to transmit data packets to the host using a **TRANSMIT, TXD(48)**, instruction in its user-program. The host computer can be programmed to interpret data packets from CQM1s as communication requests, and to respond by executing the required set of C-code commands. The CQM1 can therefore be programmed to initiate communications (via the host computer) without waiting to be polled.

---

[6] The CPU11-E has only a peripheral port.

C-mode commands sent by a host computer to a CQM1 have a standard format. The CQM1 automatically assembles data sets into the standard frame format as it transmits. A C-mode frame contains:

- A start code (the @ character)
- A destination node address
- A coded command
- Data for the CQM1 (with some commands)
- An error-checking data byte, then
- A termination code (a * followed by a carriage return)

C-mode replies from the CQM1 are similarly structured frames.

Every character in the frame is always transmitted in ASCII code format, using RS-232C asynchronous protocol. The CQM1 automatically translates data from hexadecimal to ASCII as it is sent, and back into hexadecimal as it is received.

Programming of a host computer is outside the scope of this book, but OMRON says that host computer interfaces and drivers are available from several third-party sources. If you want to write your own host computer program, OMRON's manuals include a complete description of C-mode frame formats, including the set of command codes and the format of preprogrammed CQM1 replies. A programmer can easily devise a set of communication requests that a CQM1 could issue and program the host computer to respond to those requests by issuing the appropriate C-code frames.

3. In RS-232C mode, the user-program can execute a **TRANSMIT, TXD (48)**, instruction to cause data to be sent out via the serial port, or a **RECEIVE, RXD (47),** instruction to cause the PLC to read data received through the serial port. Messages can be transmitted to a peripheral (such as a printer or another CQM1) or received from a peripheral (such as a bar code reader or another CQM1).

The CQM1 maintains RS-232C communication status bits in its additional register (AR) memory area (AR 08 to 10). Some communication status bits indicate the completed transmission of a message (but not whether the message was received by the peripheral). Other communication status bits indicate that an RS-232C message has been received into the communication buffer. CQM1 user-programs should check these status bits before executing a TXD (48) or an RXD (47) instruction.

RS-232C messages can be fixed-length sets of characters, or they can be variable-length character sets. CQM1s must be configured to use specific byte values for message start and end codes in variable-length messages.

Larger OMRON PLCs can be interconnected into high-speed token ring networks via SYSMAC NET interface modules, into high-speed token bus networks via SYSMAC LINK interface modules, into peer-to-peer networks using link adapters, into Ethernet networks via Ethernet units, or into host link networks as described here. Link adapters are required to connect to a host link network that uses the RS-422 protocol instead of RS-232C. Larger OMRON PLCs can also have remote I/O racks, using a SYSMAC BUS remote master module in the main rack, and SYSMAC BUS remote slave modules in each remote rack. RS-485 protocol is used in the SYSMAC BUS remote I/O system.

Configuring CQM1 Communication Channels

Configuration of CQM1 communications is done by entering configuration data into the CPU module's RS-232C port and peripheral port settings area of data memory (DM 6645 to 6654). This configuration data can only be changed using a programming unit, but changes can be made even while the PLC is in run mode. Changes are effective as soon as the changes are made.

If the CPU DIP switch 5 is on, the CQM1's RS-232C port is hardware-configured to be in host link mode, to have node number 0, and to transmit and receive data using the default standard RS-232C communication protocol, explained below. You must turn switch 5 off or the settings in DM 6645 to DM 6649 will have no effect.

To configure a port for RS-232C communications with a peripheral:

- **DM 6645** (for RS-232C port),
  DM 6650 (for peripheral port) must contain:
  - 1000 — for a "standard" asynchronous communication character protocol: 7 data bits, an even parity bit, 1 start bit, and 2 stop bits, at 9600 baud.
  - **1001** — for a special protocol selected using DM 6646 or DM 6651.
- **DM 6646** (for RS-232C port),
  DM 6651 (for peripheral port) must contain the following if a special protocol is selected:
  - **NN**0**P** — where NN is a code selecting a combination of data, parity, start, and stop bits. P is a code selecting a baud rate. (See your manual set.)
- **DM 6647** (for RS-232C port),
  DM 6652 (for peripheral port) must contain:
  - 0000 — for no delay time between characters
  - **NNNN** — to force a delay between character transmissions, enter a number for NNNN (0000 to 9999) indicating how many 10-ms time units to delay.
- **DM 6648** (for RS-232C port),
  DM 6653 (for peripheral port) must contain:
  - **0**000 — for no start codes or end codes in the character stream.
  - **x2**00 — use a CR/LF character to indicate the end of a character stream (end code).
  - **x1**00 — another byte value (defined below) indicates the end of a character stream (end code).
  - **1x**00 — another byte value (defined below) indicates the start of a character stream (start code).
- **DM 6649** (for RS-232C port),
  DM 6654 (for peripheral port) must contain either:

(a) End and start codes if DM 6648 or DM 6653 contains configuration requiring the programmer to specify them:
  - **NNPP** — where NN is a hexadecimal value from 00 to FF for an end code, and PP is a hexadecimal value from 00 to FF for a start code.

(b) A size for fixed data packets without start or stop codes if DM 6648 or DM 6653 does not call for start or end codes:

NN00 where NN is a hex value from 00 to FF indicating the fixed size of each data packet in bytes (from 0 to 258 bytes).

To configure a port for one-to-one communications with one other computer:

- **DM 6645** (for RS-232C port),
  DM 6650 (for peripheral port) must contain:
  x**N**00 where N selects how much memory in this CQM1's link register to share with the other CQM1.
  - If N = 0, use LR 00 to 63.
  - If N = 1, use LR 00 to 31.
  - If N = 2, use LR 00 to 15.

  2x00 to be the slave in the communications pair. The slave can write only to the highest addresses of the shared memory (i.e., LR 32–63, 16–31, or 08–15).

  3x00 to be the master in the communications pair. The master can write only to the low addresses of the shared memory (i.e., LR 00–31, 00–15, or 00–07).

To configure a port for host link communications in a local area network:

- **DM 6645** (for RS-232C port),
  DM 6650 (for peripheral port) must contain:
  00NN where NN is a node number for this CQM1 in the host link local area network. Nodes can be numbered from 00 (the default) up to 31.
- **DM 6645 to 6647** (for RS-232C port),
  DM 6650 to 6652 (for peripheral port) must be configured as for RS-232C configuration (described above).

Programming the CQM1 to Communicate

To use one-to-one communications once configuration is complete, a CQM1 program only has to write data to the portion of link memory (LM) that it is allowed to write to, using the instructions that are usually used to write to memory: OUT instructions for bits, **MOV (21)** instructions for words, and so on. Standard instructions are used to read data from the part of link memory that the other PLC is allowed to write to: AND for bits, **CMP (20)** for words, and so on. Remember that data written into one PLC's link memory does not instantaneously show up in the other PLC's link memory; it is moved during the I/O scans of the two PLCs. There may be a delay of as long as two scan cycles (one PLC's scan cycle, plus the other PLC's scan cycle) before the data from one PLC can be used by the other PLC.

To use *host link communications* in systems where the host computer is programmed to *poll each CQM1* slave, each CQM1 must be programmed to write communication requests into its own memory, in addresses where the host computer can find them, and to check the communication status bits periodically to know when the host has performed a communication service. The specific request format and memory requirements depend on how the host is programmed.

To use *host link communications* in systems where the host computer is programmed to accept data packets sent by any CQM1 and interpret them as requests for communication services, each CQM1 must be programmed to execute TRANSMIT instructions, TXD(48), to send data packets containing requests to the host computer. Data packets cannot exceed 122 bytes of data. Figure 13.10 shows a CQM1 instruction to send a data packet to a host computer. If a port is configured for host link communications, the CQM1 will automatically translate the data packet into a C-mode frame format with an instruction code to indicate that it is an unsolicited message from a slave, and the host must be programmed to interpret the message data and to determine which slave sent it. Whenever a CQM1 executes a TXD(48) instruction to send data via the RS-232C port, it automatically turns the communication status bit at AR 0805 off and holds it off while the data to be transmitted is in the communication buffer waiting to be sent. AR 0813 is used the same way if data is transmitted via the peripheral port. TXD(48) instructions should be programmed to execute once each time there is a new data set to send to the host, and only to transmit if AR 0805 (or AR 0813) is ON.

In the program of Figure 13.10:

1. The differentiated form of TXD(48) is used so that the message will only be sent once if:

- **(a)** The work bit IR 01204 is on. In rungs that are not shown, this bit is set whenever a new data set has been prepared, and reset when the TXD(48) instruction starts, turning bit AR 0805 off.
- **(b)** AR 0805 is on, indicating that the peripheral port is not already busy.

2. TXD(48)'s first parameter, S, the first source word, indicates that the data to be sent is in data memory starting at DM 0020. The third parameter, N, the number of bytes parameter, indicates that 18 bytes of data (nine words from DM 0020 to 0028) are to be transmitted. Data that the host computer can interpret must be in these nine data memory locations before TXD (47) executes.

3. TXD(48)'s second parameter, C, the control word, contains 0000 to indicate to use the RS-232C port ("1000" here would mean to use the peripheral port).

To use RS-232C communications in a CQM1 configured for RS-232C mode, the user-program must execute TRANSMIT, TXD(48), and/or RECEIVE, RXD(47), instructions. The TXD(48) instruction causes the PLC to copy data into the transmit communications buffer, and

FIGURE 13.10
OMRON CQM1 instruction to transmit an 18-byte message to a host computer.

the RXD(47) command is used to copy data from the receive communications buffer. The actual sending and receiving of serial data are performed by a separate communications-handling processor in the CQM1's CPU module. The TXD(48) instruction does not cause the PLC to wait until data is actually sent, nor does an RXD(47) cause the PLC to wait for data to be received. The communications handler controls status bits indicating when data has been received into the communications buffer and indicating when a requested transmission is complete. SR 25209 can be set to reset the RS-232C port, and SR 25208 resets the peripheral port. SR 25209 and SR 25208 reset themselves after the port has reset.

RXD(47) uses the following additional register (AR) and status register (SR) bits:

1. AR 0806 is set when data has been received via the RS-232C port, and AR 09 indicates how many bytes are in the receive buffer. Bits 0 to 4 and/or bit 7 of AR 08 will be set if data errors are detected in incoming data.

2. AR 0814 is set when data has been received via the peripheral port, and AR 10 indicates how many bytes are in the receive buffer. Bits 8 to 12 and/or bit 15 of AR 08 will be set if data errors are detected in incoming data.

All these bits are reset when RXD(47) is executed, so look for error bits first.

TXD(48) uses the following additional register bits:

1. AR 0805 goes off when a TXD(48) instruction writes data to the transmit communication buffer for the RS-232C port and goes back on again after the data has been sent.

2. AR 0813 goes off when a TXD(48) instruction writes data to the transmit communication buffer for the peripheral port and goes back on again after the data has been sent.

TROUBLESHOOTING

If your PLC system is failing to pass data between components as it should, you may have a hardware problem, a firmware problem, or a software problem. If the system is failing to pass data as you think it should, you may not be using the communication feature correctly.

Hardware problems should be looked for first, especially if the system has been communicating successfully but has recently started to fail:

- Are all the required communication components turned on? This includes the controllers that are trying to exchange data, the network master if there is one (a master may have to be powered-up before the other components), as well as any interface devices along the communication path.
- Are the conductors connected to the controllers properly, and is the wiring bus terminated correctly?
- If a common communication standard is used for the physical layer (e.g., RS-232C specifies connectors and conductor uses), you can purchase an inexpensive protocol checker to insert between the PLC and the rest of the network, to see if communication line states are changing. Manufacturers of proprietary networks usually offer network analyzers, which report network activity with other statistics on performance. Programming software often includes a channel monitor screen that shows statistics, including attempts at communications.

Firmware is the software that is preprogrammed into the communication equipment when you buy it. If the devices that you purchase are described by the manufacturer as compatible, you shouldn't have a firmware problem. Check to ensure that the components are all compatible with the same version of the network, because standards have evolved over the years.

There may be rules on using the preconfigured port that can lead to problems when you ignore them:

- If a channel is used for multiple purposes, the purposes may conflict. Allen-Bradley, for example, recommends that you not execute block transfers with PLC-5 modules in a chassis that has PII interrupts enabled. If you do, you will miss interrupt signals if they occur while a block transfer is executing.
- Excessive traffic can make any shared communication channel slow. Is your program issuing jobs to communication handlers unnecessarily? Some PLC programming software packages come with channel diagnostic screens. Check to see if there are more data packets being sent than you intended, and then check to see which instructions are so guilty. Try using a counter to count the number of times that suspect instructions issue job requests. Remember, programming units often use the shared communication channel as they monitor PLCs on a network, so a programming unit may be issuing requests for large quantities of data.

Software problems include incorrect configuration or user-programming.

- Check that the addresses that you have assigned to each network node is unique and within the allowable range of addresses. Some network systems allow you to change the node address range, so each node must be configured to use the same range. Does your user-program use the correct node address for the controller with which it is trying to communicate? When a programming unit is used to change the node address of a PLC, the programming unit must then reconnect to the node at that new address.
- Are the communication ports at all network nodes configured to use the same communication protocol? If one is configured to communicate at a baud rate different from the others, for example, it won't be able to communicate. Master–slave networks such as Allen-Bradley's DF1 network or OMRON's host link network have a lot of flexibility in how they can be configured, which also means that they have the greatest potential for incompatible protocol. Sometimes it takes only one node with incorrectly configured protocol to cause confusion in all the other nodes' communication handlers. If you use a programming unit to change the protocol at the port that the programming unit uses, you will then have to change the programming unit's protocol to match the new protocol.
- All PLCs report communication errors and reasons for errors somewhere. Determine what is available and use it to help identify the problem.
  - Each Allen-Bradley communication channel has a diagnostic file (set up during configuration) which holds statistical data on the channel's performance. Allen-Bradley message instructions make use of status bits in a message control structure, and other status bits in the status file, to report the progress of attempts at message sending. Allen-Bradley PLCs can also send MSG instruction error codes to a programming unit if one is installed and is monitoring the message instruction status.

- Siemens PLCs maintain control words in S5 PLCs communicating via L1 networks. Communication processor modules use interface flags or interface memory to report their status. S7 PLCs log all errors into a diagnostic buffer (which is covered in more detail in Chapter 15). Profibus-DP slaves maintain diagnostic data that can be read by a Profibus master.
- OMRON CQM1 PLCs maintain communication status bits in the AR area of memory, as well as logging up to 10 error messages into a diagnostics buffer (configuration affects which 10 are stored). Chapter 15 covers the diagnostic buffer in more detail.

Incorrect usage of a communication capability can cause undesirable results. If data is being transmitted (verify this by monitoring the receiving PLC's memory) but your program seems to use the wrong data, your user-program's timing may be at fault. Remember that a communication handler can't finish its assigned jobs as quickly as the user-program can execute an instruction assigning jobs. Ensure that your program executes the following steps in this sequence:

1. Write the data to be sent to the memory location(s) where the communication handler will look for data to send, or copy recently received data from the memory area(s) where the communication handler will place new data (if you don't want to lose the old data). Do not read or write these areas again until after the last step in this sequence of steps!
2. Check the communication handler's status to ensure that the communication handler is ready for a new job.
3. Execute the instruction that issues the job request to the communication handler.
4. Check communication handler status to determine when the communication handler has completed the job. (You may have to check other progress-reporting status bits in some applications, to ensure that the communication handler accepts the job, for example.)
5. Repeat.

QUESTIONS/PROBLEMS

1. (a) Explain the difference between the types of network access systems typically used in PLC local area networks: polling (master–slave), token passing, and CSMA/CD (Ethernet).

(b) Early PLCs were good at industrial control, in part because input sampling intervals were constant. Which method of network access to the shared network does not allow the sampling interval to remain constant?

2. What is an open network protocol?

3. What is a communication handler as found in modern CPU modules? How does it coordinate its activity with that of the main MPU (which is executing the user-program)?

4. An Allen-Bradley MSG (message) instruction can send or receive data via an Allen-Bradley Data Highway+ (DH+) network. Each MSG instruction needs a __________ __________ data element to control the message transmission. This data element contains status bits such as the EW bit, which indicates that the message is __________; the DN bit, which is written to by the __________ when the message has been sent successfully; and the ER bit, which indicates that

____________________. The control element also contains data that the programmer enters while entering the MSG instruction, including, for example:

(a) __________________________________
(b) __________________________________
(c) __________________________________

5. A Siemens L1 network (an early form of Profibus) is a master–slave system in which each PLC must set aside an area of its memory for the L1 network master to read and write. Why do the network and your PLC program need this area of memory?

SUGGESTED LABORATORY EXERCISE

1. Program your PLC to send an ASCII message to another PLC whenever switch 0 is on. Depending on the PLC you are using, you may have to use a message instruction, a send instruction, or you may have to configure your PLC to exchange messages between mailboxes in each PLC. Use the programming unit's data screen to enter messages into the sending PLC's memory and to observe the memory at the receiving PLC.

14

ROBOTICS, AUTOMATION, AND PLCS

OBJECTIVES

In this chapter you are introduced to:

- Robot components.
- Operating system differences between PLCs and robotic controllers. The application-specific strengths and weaknesses of each are identified.
- An automated workcell, described to show how a robot and a PLC can complement each other in a manufacturing system.

ROBOTS AND PLCS IN WORKCELLS

PLCs are used in automated workcells. In fact, a PLC is usually the main controller in a workcell. Even in a workcell with a robot, the robot is usually a slave to the far cheaper PLC. Why not use the more expensive robot to control the PLC? (Robot controllers always have digital I/O capabilities, along with assorted analog I/O and communications abilities.) PLCs are optimized for supervisory control, whereas robot controllers are optimized to control a single actuator: the robotic arm.

When programming a PLC to work with other controllers in a workcell, it is important to understand the strengths and weaknesses of each type of controller. While the PLC will usually be the master, some control functions are best programmed into other types of controllers. Robot controllers are typical of non-PLC controllers, so in this chapter we examine how to program a PLC to work with a robot.

DIFFERENCES BETWEEN ROBOT CONTROLLERS AND PLCS

When you purchase a complete **servo robot** (as opposed to a pick-and-place robot), it usually arrives in three parts: a robotic actuator, sometimes called the *arm;* a controller; and a pendant.

1. The **arm** of a robot has a set of beams that can move relative to each other so that an

end-effector at the end of the last beam can be positioned to do work. A new robot doesn't usually include an end-effector, so the user has to buy or build one to do the work that the robot was purchased to do. Each beam of an electric robot is moved by an electric motor and contains a position sensor.

2. The **controller** contains interface circuitry to allow it to read the arm's position sensors and to drive the arm's motors. The controller also contains a power supply to provide power for the arm's motors, and a computer that stores and executes programs to control the robot.

3. The **pendant** connects to the controller and has controls that are used to command the arm to move manually. Some pendants or controllers have additional displays and operator input capabilities so that they can be used to enter robot programs into the controller's memory, but programming is often done through a personal computer running robot-programming software and software to communicate with the robot controller.

So far, the robot appears to differ from a PLC in two ways: It includes its own actuators, and it includes built-in power supplies and interface circuitry for those actuators. But there is another, more important difference: A robot can only perform a **sequential** series of actions. Actions that a robot can perform include:

- Moving the arm. A move requires driving all the motors simultaneously and monitoring all the position sensors.
- Controlling actuators other than the arm's motors.
- Exchanging data with other controllers in the workcell.
- Testing input signals from sensors other than the arm position sensors.
- Branching forward or backward to bypass or repeat program segments.

While the actions themselves aren't any different from those a PLC can perform, the robot controller is designed to perform the actions sequentially and cannot begin to perform a new action until the previous action has been completed.[1] A robot program might take from a few minutes to several hours to execute just once. A PLC, on the other hand, executes a full scan cycle every few milliseconds, executing all its control actions almost simultaneously. You can program a PLC, for example, to detect tall parts on a moving conveyor and to push them off the conveyor every time they are detected, no matter when they are detected (if the conveyor isn't moving too fast). For a robot to do the same work, the tall parts can arrive only when the robot is executing the right step in the robot's sequential program. If they arrive at any other time, the robot won't be able to push them off.

Okay, you may ask, why not control the robot arm's movements using the PLC? A PLC can control electric motors and monitor position sensors, and a PLC can monitor and control unscheduled workcell activities better than a robot controller can, so why use a robot controller? There are a couple of answers:

---

[1] Some robot controllers can deviate slightly from their sequential nature in a couple of ways. Once an arm-move command has been issued, some controllers can execute the next few sequentially programmed actions without waiting for the move to be completed. Some robots can be programmed with interrupt capabilities so that they will interrupt one sequential program to jump to a different sequential program under specific conditions.

1. A robotic arm must be programmed to execute moves sequentially and in synchronization with other activities in the workcell. To program a PLC to move an arm through a series of sequential moves, you need to latch bits to keep track of the sequential step that is supposed to be active, or use a sequencer type of instruction (if the PLC has one), or use a programming language designed for sequencing operations (such as IEC's sequential flowchart language). It is much easier to program sequential operations in the robotic programming language, which inherently finishes one instruction before executing the next.

2. A robot controller is optimized to control the robot arm with which it is supplied. The robot controller contains programs in its ROM memory that coordinate the movements of all the motors so that all the beam segments reach their destination simultaneously. Preprogrammed ROM routines in the robot controller may allow the programmer to specify a preprogrammed path that the end-effector should move through, or to specify that the end-effector should move through a path for which the controller must calculate interim points. The robot controller's movement control programs automatically contain servo control and position error control routines, which the programmer doesn't have to write. The motion control programs often allow the programmer to specify a changed tool **offset,** and the robot controller will automatically recalculate the movements required to move an end-effector's tool center point through a previously programmed path (e.g., if the robot puts a grinding tool down and picks up a paintbrush). The built-in routines can control the arm using its maximum controllable velocity and acceleration. None of the above is impossible to program into a PLC, but the PLC program will be more complex, less flexible, and the arm will move much more slowly. You can buy PLC I/O modules for motion control, to achieve higher speeds and flexibility, but you can't buy I/O modules that are preconfigured to optimize any single robot's performance.

SIMILARITIES BETWEEN ROBOT CONTROLLERS AND PLCS

Like PLCs, robot controllers are computers with digital I/O capabilities and communication ports and have commands in their programming languages to use the I/O points and communication channels. Some robot controllers even have additional analog I/O capabilities, like a PLC.

Most robot controllers can be programmed to interrupt their main program. They can monitor an I/O point and immediately execute an interrupt service routine if the input condition changes, just as a PLC can be configured with I/O interrupt capabilities.

Like a PLC, robot controllers have retentive memory, in which they store configuration parameters, programs, and variable values while they are not running a program and even when they are disconnected from power. (Unlike in a PLC, a robot does not necessarily clear any of this memory when it starts running a program. This can lead to some surprising results when you restart a robot that was turned off midway through program execution!)

For more similarities, we would have to start looking at characteristics that aren't of much interest to the user, such as the computer architecture or signal processing capabilities. In fact, there are not many functional similarities between PLCs and robot controllers! There aren't many functional similarities between PLCs and any other type of computers. The standard PLC operating system (based on the scan cycle) makes it different.

PROGRAMMING A ROBOT AND A PLC TO WORK TOGETHER

Because of their differences, robot controllers and PLCs make ideal control partners in automated workcells.[2] You should assign each controller to control the types of processes that it is best at controlling. The robot controller can be programmed to monitor the environmental conditions that affect the robot's sequential operations and to control the outputs that should be synchronized with the robot actions. The PLC, on the other hand, can be programmed to control processes that can't be synchronized with the robot's activities or that don't need to be synchronized. Signals can be exchanged between the PLC and the robot so that some PLC operations can be synchronized to the robot's requirements.

A sample workcell is shown in Figure 14.1. The workcell contains a PLC and a robot. The workcell is designed to complete the assembly of two parts we will call part A and part B. A description of the workcell's operation follows (the workcell is for learning purposes only, so please ignore its obvious deficiencies):

1. The A parts arrive at the left of the workcell, where a human operator places a part into a fixture in one of the identical shuttles taking it to the robot, then presses a button to signal that a shuttle is loaded. (There is no control over which shuttle the operator selects if both are available.)

2. After the operator has pressed the button, the shuttle beside the button takes the fixture, with the A part, to where the robot can reach it. It is possible for both shuttles to be actuated simultaneously.

3. When a shuttle arrives near the robot, a lock engages to hold the A part firmly into a precise position for the following assembly operation.

4. A vibratory feeder feeds B parts onto a gravity feeder whenever a sensor on the gravity feeder detects fewer than 10 B parts, and continues until a second sensor detects that there are more than 20 parts on the gravity feeder. The vibratory bowl is supplied by a hopper containing B parts, and a warning lamp at the hopper turns on if the hopper needs refilling.

5. B parts slide down the gravity feeder into a fixture at the bottom. A sensor detects a part in the fixture. The B part in the fixture is *singulated* when the fixture moves sideways, so that the robot can pick the B part up without bumping into the other parts on the gravity feeder. No other B parts can slide down the gravity feeder into the fixture until the robot takes the B part and the singulator retracts.

6. The robot uses vacuum pressure to pick the B part up from the singulator. A vacuum sensor in the end-effector detects a loss of vacuum if a part isn't picked up or if it is dropped during the following operations.

7. Since the B parts are symmetrical, the vibratory feeder can orient them in either of two directions. The robot must place the B part into an orientation fixture with an inductive proximity sensor to detect the metal plate built into one face of the B part. The orientation fixture rotates the part 180 degrees if it is in the wrong orientation; then the robot can pick the B part up again.

8. The robot then assembles the B part onto the A part using a complex continuous motion that no robot programmer would want to program twice. Since there are two assembly sta-

---

[2] So ideal that at least one recently announced new robot comes with a controller that includes a coprocessor that executes a scan cycle like a PLC.

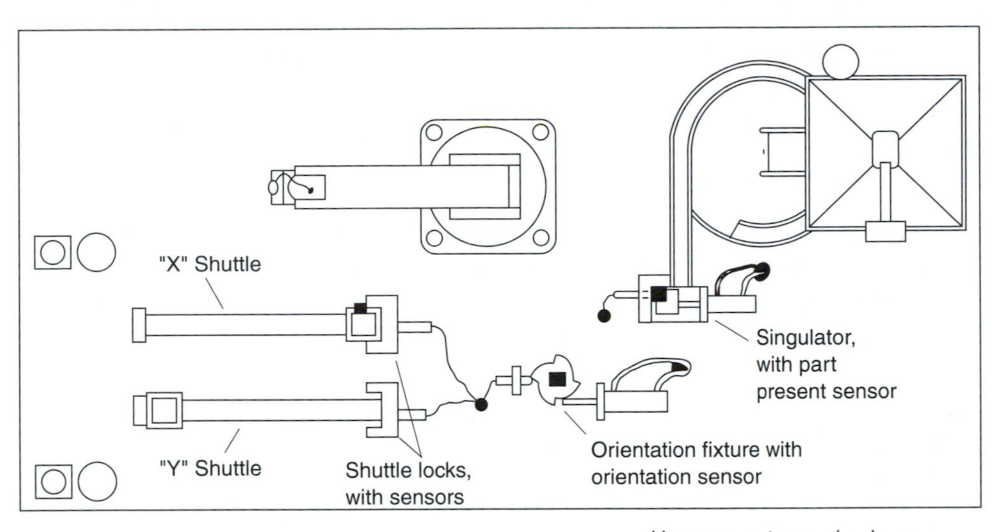

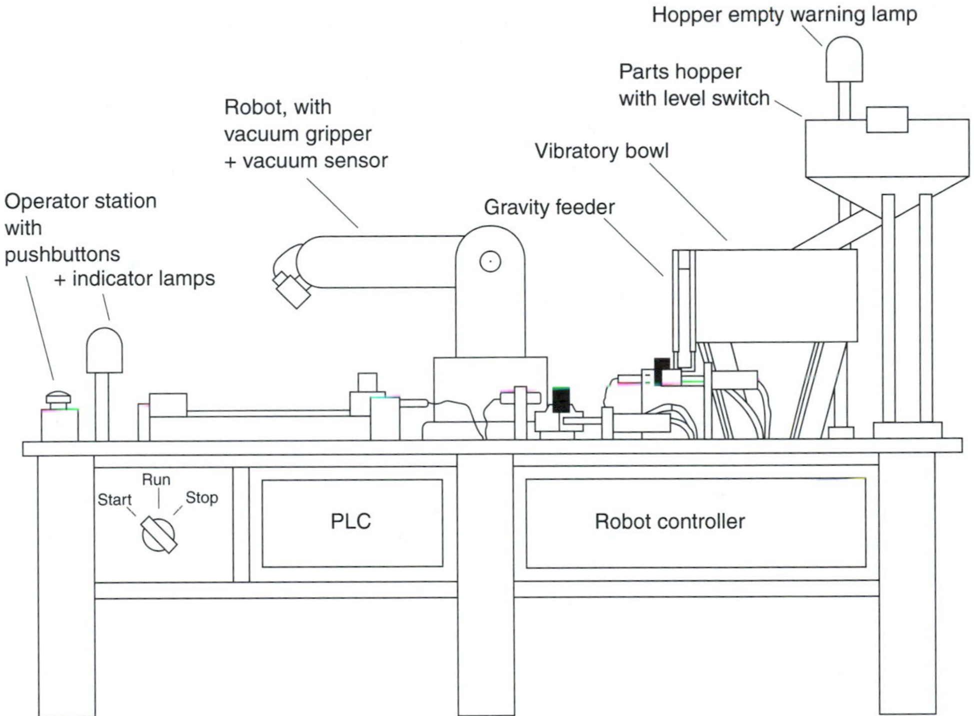

FIGURE 14.1
Workcell containing a PLC and a robot.

tions at the robot's end of the shuttles, an offset is used in the robot program to allow it to execute the same path motion at two locations.

9. After assembly is complete, the A part lock will release, the shuttle will retract, and a light will go on to notify the operator to remove the assembly amd to reload the shuttle with another A part to begin the cycle again.

At the start of the shift, there should be no A parts in the workcell, although there may be some completed assemblies still in the shuttle fixtures near the robot from the previous shift, and there should be plenty of B parts in the gravity feeder, bowl feeder, and hopper at the top right.

1. The operator must place a workcell control switch briefly in the START position, at which time shuttles with completed assemblies will return to the operator's station.

2. The operator then places the workcell control switch in RUN and leaves it there for the remainder of the shift.

3. At the end of the shift, the operator must place the workcell control switch in the STOP position. The shuttles will not move again until the next shift, but the robot must complete the assembly of any parts on shuttles still waiting near the robot.

There are some operations that *must* be synchronized with the robot, so they are best included in the robot program. These operations are sometimes called **internal operations,** to indicate that they *occupy time in the sequential production cycle.* Other operations should be performed as early as possible, to avoid having the robot wait for them to be completed. Operations that can be performed at the same time as operations in the sequential production cycle are sometimes called **external operations.** External operations should not add to the production cycle time. To increase the production rate, as many operations as possible should be made external.

The Robot Program

Let's look first at what has to be in the robot program. The robot must:

1. Initialize any variables, configuration parameters, and output signal states that need to be initialized at the start of a shift. Retentive memory may contain undesirable states from the last time the robot ran, especially if the work-cycle operation was terminated abnormally.

2. Move the arm to a position near the singulator and wait for a signal indicating that a B part is present. The vibratory bowl, gravity feeder, and singulator will be external operations controlled by the PLC, so the PLC must supply a B PART READY signal.

3. Perform the following sequential series of operations:

- **(a)** Turn on the vacuum generator.
- **(b)** Pick up the B part.
- **(c)** Briefly output a signal indicating that the B part has been taken from the singulator.
- **(d)** Actuate an interrupt condition that monitors the vacuum sensor to detect a dropped B part.
- **(e)** Move to the orientation fixture.
- **(f)** Deactivate the interrupt condition.

(g) Place the B part in the orientation fixture.

The vacuum generator and the vacuum sensor must be connected directly to digital I/O connections of the robot controller since they are sequentially controlled as part of the production cycle. One robot output is needed to send a B PART TAKEN signal to the PLC.

4. Should the orientation fixture be controlled by the PLC or the robot? Since the robot must wait while the orientation fixture works, it might as well be part of the robot's program. The inductive proximity sensor and orientation fixture rotator must be connected directly to the robot's digital I/O connections. The robot program will:

(a) Go to step 5 if the metal detector's signal is on, or

(b) Rotate the B part 180 degrees.

5. (a) Pick up the oriented B part from the orientation fixture.

(b) Reenable the interrupt condition that monitors the vacuum sensor.

(c) Move to a neutral position near the shuttles.

6. (a) Go to step 8 if the PLC is signaling that an A part is ready in shuttle X near the robot, or

(b) Go to step 7 if an A part is ready in shuttle Y near the robot, or

(c) Repeat step 6.

The PLC controls the shuttles, the locks, and the indicator lamps at the shuttles. Two signals from the PLC to the robot must indicate either X SHUTTLE READY or Y SHUTTLE READY.

7. (a) If the Y-position variable is SET, go to step 9; otherwise,

(b) Apply the offset needed to transform the path coordinates, so that assembly at the Y shuttle can be performed.

(c) SET the Y-position parameter.

(d) Go to step 9.

The path coordinates were taught at shuttle X, not at shuttle Y.

8. (a) If the Y-position variable is RESET, go to step 9; otherwise,

(b) Remove the offset that adjusted the path coordinates for assembly at the Y shuttle, so that assembly at the X shuttle can be performed.

(c) RESET the Y-position parameter.

9. (a) Perform the assembly path move.

(b) Disable the interrupt condition.

(c) Turn the vacuum off, releasing part B.

(d) If the Y-position parameter is SET, send the Y ASSEMBLY COMPLETE signal briefly; otherwise,

(e) Send the X ASSEMBLY COMPLETE signal briefly.

The X ASSEMBLY COMPLETE and Y ASSEMBLY COMPLETE signals go from the robot to the PLC.

10. (a) If the END OF SHIFT signal is on, *and* if *both* the X SHUTTLE READY and the Y SHUTTLE READY signals are RESET, go to step 11; otherwise,

(b) Go to step 2.

The PLC will monitor the workstation control switch and will tell the robot when the shift is over. (A better robot program would have checked the END OF SHIFT signal before starting at step 2.)

11. Stop.

The interrupt service routine that executes if a block is lost from the end-effector, or is not picked up properly, is not described here. The main program must manipulate variables for the interrupt service routines to examine to find where the robotic arm was in the sequenced process, so that the program can return and repeat only what needs to be repeated and can adjust any outputs that need to be changed.

The PLC Program

The PLC program's functions are obvious, since we wisely designed the internal operation program (the robot program) for the workcell first. Except for a few operations that are performed as sequencer operations, most of the PLC operations can be considered to be executing simultaneously. Only sequencer steps are numbered in the following description.

- If the workstation control switch is in the START position, unlatch both shuttle control outputs, so that the shuttles will retract to the operator position, and reset any bits used for the shuttle sequencer controls in the main part of this program.
- If the workstation control switch is in the STOP position, latch the END OF SHIFT signal to the robot on.
- If the workstation switch has just been put into the RUN position, use a one-shot to unlatch the END OF SHIFT signal.
- If the workstation control switch is *not* in the RUN position, do *not* execute the rest of this program.
- Execute the following as a sequencer program:
  1. When the operator presses the X shuttle pushbutton, latch the X shuttle output to move the X shuttle to the robot and unlatch the output that turns the X shuttle indicator lamp on.
  2. When the X shuttle is detected at the robot, latch the X shuttle part lock.
  3. After the lock closes, latch the X SHUTTLE READY signal to the robot.
  4. When the robot's X ASSEMBLY COMPLETE signal is received, unlatch the X shuttle lock.
  5. When the X shuttle lock has opened, unlatch the X shuttle control to return the shuttle to the operator.
  6. When the X shuttle has returned to the operator, latch the X shuttle indicator lamp on.
- Execute another sequencer program for shuttle Y, identical with the sequencer for shuttle X.
- While the sensor in the vibratory bowl detects low parts, turn the warning light on.
- If the (normally open) sensor on the gravity feeder that detects fewer than 10 parts is *off* for longer than 2 s, latch the vibratory feeder output to feed more B parts.
- If the (normally open) sensor that detects more than 20 parts at the gravity feeder is *on* for longer than 2 s, unlatch the vibratory feeder output.
- Execute the following as a sequencer program. Use retentive bits so that the sequencer position is retained even when the PLC is not running.
  1. When a B part arrives in the singulator fixture, latch the singulator output to singulate the part.

2. When the singulator is detected in the forward position, latch the B PART READY signal to the robot.
3. When the robot's B PART TAKEN signal is received, unlatch the B PART READY signal and unlatch the singulator output.

Figure 14.2 summarizes the inputs and outputs required in the two control programs, including the signals that are exchanged between the PLC and the robot controller.

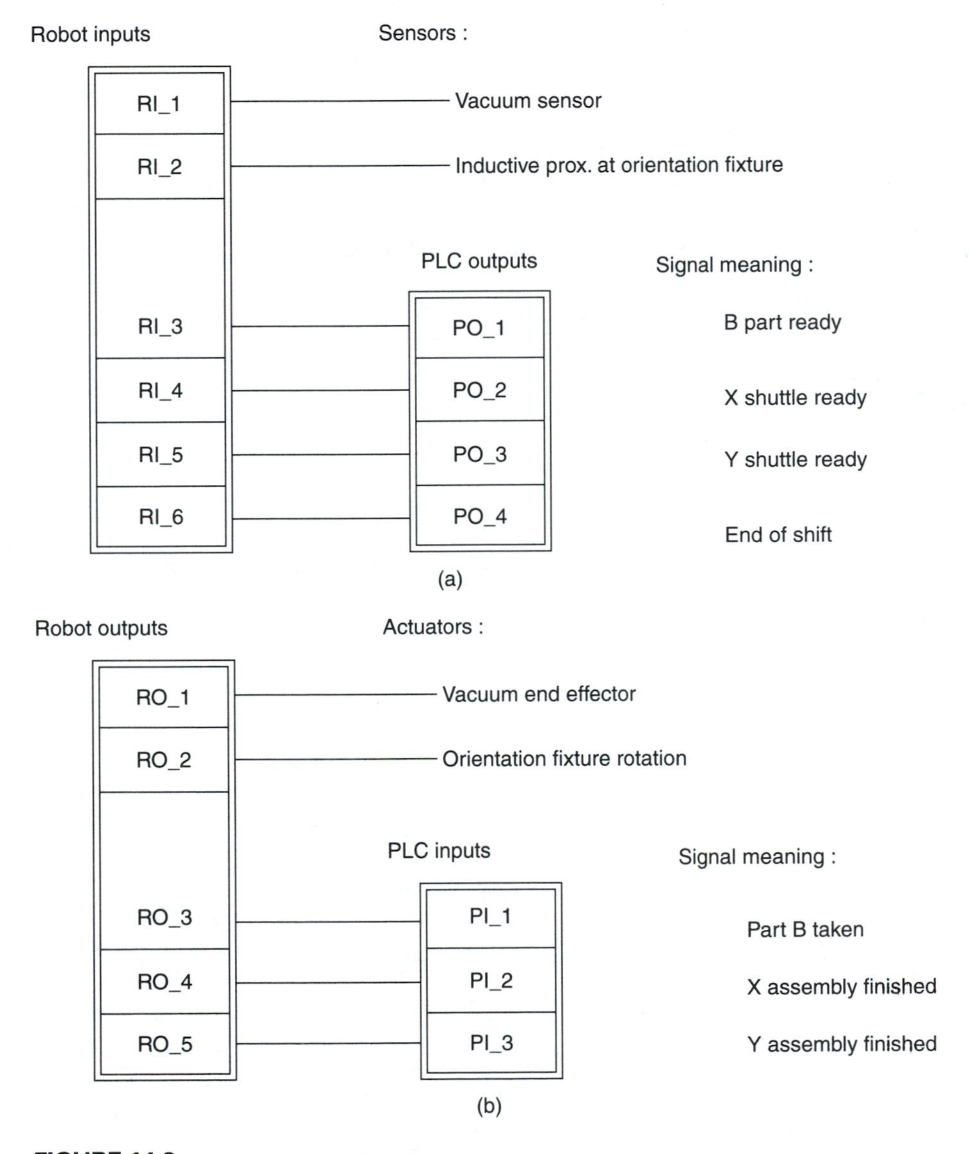

FIGURE 14.2
(a) Robot inputs; (b) robot outputs; *(continued)*

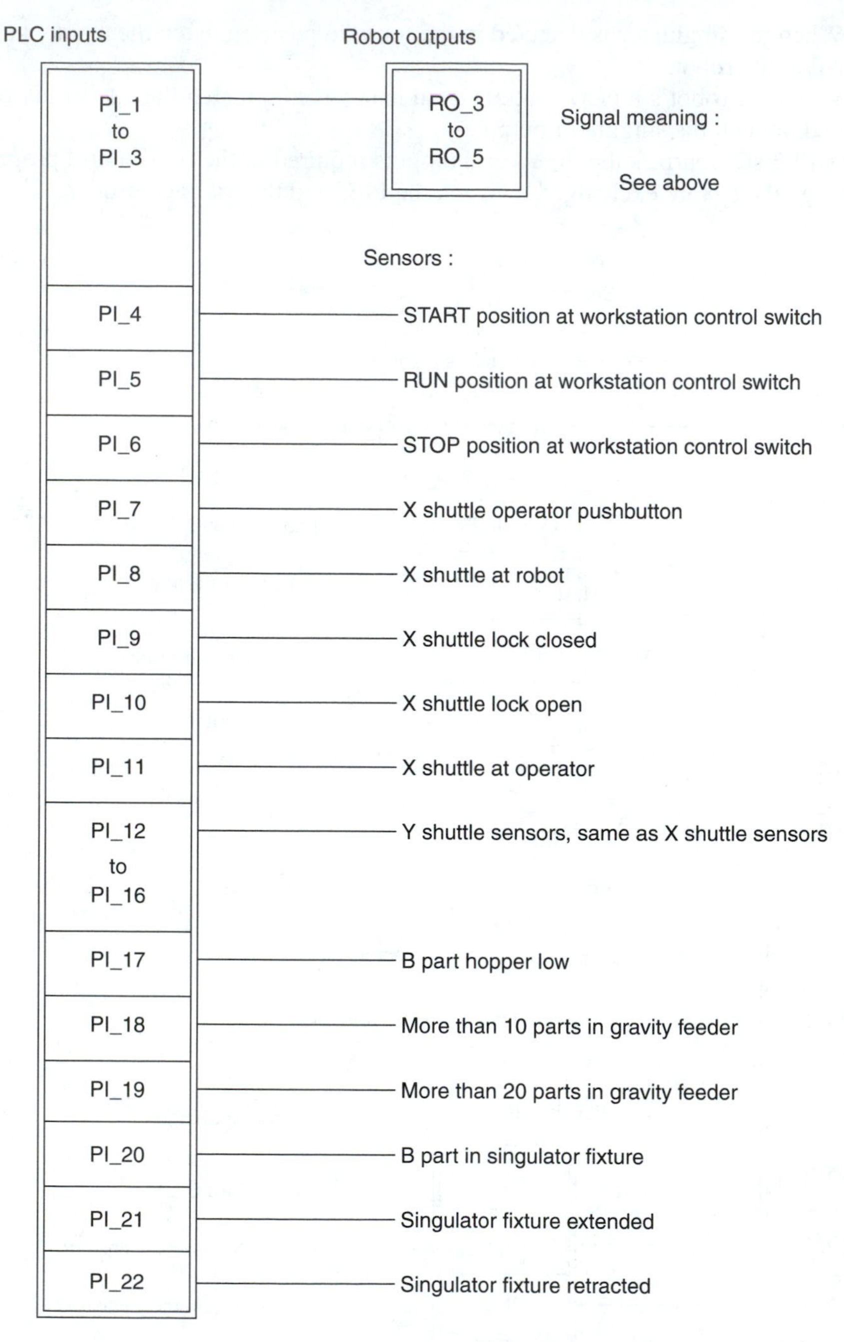

FIGURE 14.2
(continued) (c) PLC inputs.

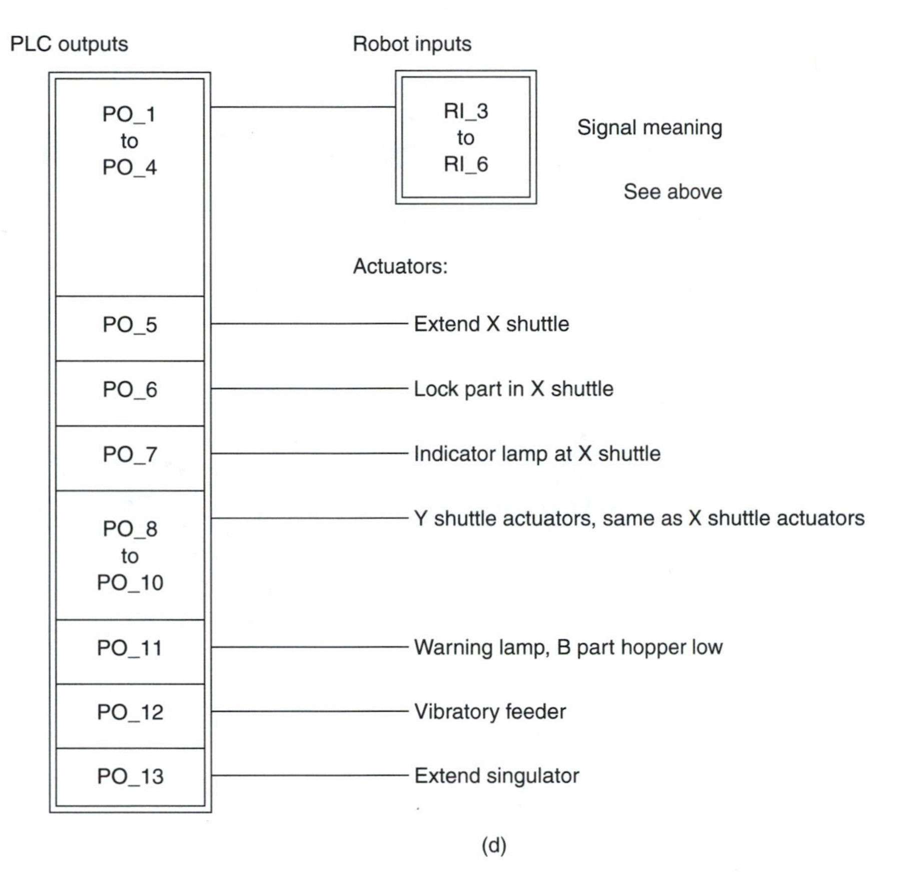

FIGURE 14.2
(continued) (d) PLC outputs.

15

TROUBLESHOOTING

OBJECTIVES

In this chapter you are introduced to:

- A systematic approach to troubleshooting.
- Troubleshooting hardware outside the PLC.
- Problems in the PLC that may be due to fatal errors; nonfatal errors, including math errors; or programming logic errors.
- Error codes, bits, status information, and other fault identification features of Allen-Bradley, Siemens, and OMRON PLCs.
- Some instructions and programming techniques that can help in detecting the causes of undesirable behavior.

SYSTEMATIC APPROACH

PLC-related problems should be solved using a systematic method:

1. Identify the problem.
2. Determine what the correct system behavior should be.
3. Consider the method(s) that can be used to achieve optimum system behavior.
4. Implement the best solution. (Return to earlier steps as required.)
5. Verify that the problem has been corrected and that the system behavior is correct.

The easiest solution isn't always the best solution, especially in today's complex manufacturing systems. A PLC can help you identify the problem, but be prepared to look outside the PLC for the solution. Managers in automated manufacturing facilities tell lots of stories about perfectly good PLC programs being changed to correct problems such as seized roller bearings or broken electrical signal wires!

Use the troubleshooting features built into your PLC to help identify the problem. Look for possible solutions by inspecting the hardware outside the PLC and the interfaces to your PLC. Next, check the hardware of the PLC itself, and check the PLC's configuration. Finally, look for improvements in the PLC's user-program. New user-programs often contain errors, but new control systems often contain faulty components or interconnections, too.

TROUBLESHOOTING HARDWARE OUTSIDE THE PLC

All PLCs have LEDs on their CPU modules, I/O modules, and sometimes on power supply modules. In general, a red LED means a problem and a green LED means OK. If a LED is flashing, it often means that a function is active or that the module is waiting for something. Correct interpretation of LED-indicated status can save you a lot of troubleshooting time. See the LED troubleshooting guide that all PLC manufacturers supply with their hardware.

1. If the PLC won't go into run mode, you can check in a couple of ways to see if the problem is hardware or software:

(a) Temporarily put an end-of-scan-cycle instruction as the first instruction in your user-program. If the PLC can be put into run mode now, the problem is probably in your user-program, not in the PLC's hardware. (Some PLCs require you to clear the error status before the PLC will go into run mode, even if the problem has been corrected.)

(b) Reset the PLC's memory (only if you have a backup copy of the PLC's memory contents or if you are willing to lose the program, data, and configuration). If the PLC can be put into run mode now, the problem was in the program or the configuration, or was due to corrupted memory.

2. If a PLC doesn't appear to be getting a signal from a sensor, as verified by monitoring the input image memory while you actuate the sensor, observe the LEDs on your PLC's input module to see if they change when the appropriate sensor is actuated:

(a) If the input LEDs do not change, use a multimeter to verify that the signal at the PLC's input module is changing state. If it isn't, disconnect the sensor from the PLC and test its action alone. Look for external power supply problems or dc polarity problems. Some PLC dc input modules are *current sinking,* so the sensor circuit should switch the positive terminal supply to the PLC's input contact. Other dc input modules are *current sourcing,* so a sensor circuit must switch the power supply's common terminal to the input contact.

If the sensor circuit checks out, replace the input module with a simulator module if one is available. If the PLC recognizes simulator switch changes, the input module was faulty (or you missed a sensor-circuit problem).

(b) If the LED does change, your program may be at fault. Insert an end-of-scan instruction as the first line in your program, then monitor the input image again as you run the program. If the bit changes with the LED now, the problem was in your program. Your program must be writing to the input image bit, changing its value.

If the input image bit still doesn't change with the LED status on the input module, the sensor circuit could be faulty. The circuit's current-carrying capacity may not be sufficient to change the input state, even if it can change the LED state. Check for an unusually small voltage change at the input contact.

3. If an actuator doesn't appear to be getting a signal that the PLC is trying to write to it, observe the LEDs on the output module to ensure that they change when the PLC changes the output state:

(a) If the LED is changing, use a multimeter to verify that the output module is providing signal changes adequate to drive the actuator, and at the right polarity. If so, disconnect the actuator from the PLC and try its action by itself. (Some output modules have fuses. Are they blown?)

(b) If the LED does not change, check the output circuit's power supply and its wiring to the output module. (Dc outputs can be sinking or sourcing.) If the wiring is correct, disconnect the actuator(s) from the output module to see if the LEDs will change when not driving the actuators.

Replace the output module with a simulator module if one is available. If the PLC can change simulated output states, the output module or wiring was faulty.

4. If the sensor or actuator circuit works, you can look for problems in your PLC or PLC program.

TROUBLESHOOTING PLC HARDWARE, CONFIGURATION, AND PROGRAMMING

Because PLCs are set up, configured, and programmed by human beings, ingenious errors might exist.[1] PLC manufacturers therefore provide a wide range of tools for finding those errors. Errors can be classified as:

1. Fatal errors, which cause the PLC to leave run mode and go into fault mode (similar to when a hardware fault is detected). Fatal errors can be caused by detection of a failed PLC component as the PLC performs a self-check when it is started, or during use of the component as a program executes. Some programming or configuration problems (e.g., watchdog timer times-out, an attempt to run a program file that doesn't exist) are also causes of fatal errors.

As the PLC goes into fault mode, it turns a fault LED on and turns all outputs off (or freezes some in their last states). The PLC also saves a fault code in memory. The programmer can read the fault code to determine the cause of the fault, solve the problem, and then clear the fault status and/or cycle the PLC's power off and on to put the PLC back into run mode.

Modern PLCs can often store detailed descriptions of several recent faults and may allow the programmer to write fault routines that execute in the event of a fatal error. Fault routines can examine the fault descriptions and respond appropriately. We described fault routines in Chapter 11. In this chapter we examine how to find and use fault codes.

2. Nonfatal errors, which the PLC can detect but which do not cause the PLC to leave run mode. Some detectable hardware problems, such as low memory backup battery power, only cause nonfatal errors. They can also be caused by configuration and program errors (e.g., a timed interrupt program's execution is delayed because a higher-priority program is running, or a math operation generates a result that is too large to be stored in the destination memory address).

---

[1] Make a system foolproof and the world will invent a better fool.

Nonfatal errors cause the PLC to set error status bits and/or to write error codes into memory. Data word manipulation instructions set math flags in memory when the resulting word is too big or small to be stored. The user-program must check those bits and/or codes and respond as necessary, because the PLC will continue executing the user-program as if no error were detected. Some nonfatal error status bits were discussed in earlier chapters. In this chapter we look at some others.

3. Programming or configuration logic errors, which the PLC cannot detect automatically but which can be detected using troubleshooting instructions in a program or program monitoring features in a programming unit.

Logic error examples include a user-program turning a bit off instead of on, two sections of a user-program competing for control of the same output, or a structured program bypassing a section of program that is needed.

Early PLCs were designed not to allow logical errors such as having two rungs that controlled the same output or having backward jumps in a program, but users demanded more flexibility, so even those safeguards were removed. Some programming units provide warning messages when a programmer writes a program with an obvious potential logical error, but the programmer can ignore the warning.

Programming languages include simple debugging tools: instructions to terminate a scan cycle early so that a program can be checked one section at a time; instructions to cause a fatal or a nonfatal error, to make the PLC stop immediately under specific conditions; and other instructions intended specifically for debugging a program. Standard instructions such as **counters** can be inserted into programs temporarily to record events that might otherwise be missed. Programming software also includes debugging tools: to monitor and to change data memory while a user-program executes; to force I/O image bits on or off to see how the program responds; to record changes at specific bits or words for short intervals, then display the changes in histogram form; to create cross-reference (X-ref) lists of all the places in a program where a specified bit or word is used; and a search tool, which will find each occurrence in a program of a specified address or instruction.

PLC-5

TROUBLESHOOTING THE ALLEN-BRADLEY PLC-5

PLC-5 Hardware Troubleshooting

Allen-Bradley manuals tell you how to interpret the status of the LEDs on the CPU modules and I/O modules. In general, green means OK, red means a problem, and a flashing LED means an incomplete operation. If, for example, an intelligent I/O module needs a block transfer of configuration or data before it can start working, it will flash a green LED. (This may indicate a programming problem.)

Some hardware problems that may be difficult to detect include:

1. DIP switch settings on analog output modules in remote I/O racks that select signal values when the rack faults (minimum, maximum, or middle of output range) can conflict with DIP switch settings in the chassis with some combinations of chassis and modules. Verify that the analog output signals do what you want them to, by disconnecting actuators from an analog output module, then causing a rack fault and measuring the analog outputs.

2. Analog inputs can be affected by signals in adjacent channels at the module. Look for voltage or current spikes in output signals. Reduce the probability of unused channels causing problems by connecting them to common.

3. Remember that analog input and output modules require occasional recalibration.

Configuring PLC-5 Startup Fault Detection

You can configure how a PLC-5 responds to some types of errors by configuring the PLC's startup characteristics. We discussed this in Chapter 10 and 11. We saw that:

1. Whenever a PLC-5 halts because of a major fault, the PLC cannot be put back into run mode until the major fault bit(s) in S:11 have been cleared by an operator using a programming unit or by a fault routine program.

2. If you set S:26/1, the PLC-5 will treat a power failure that occurs while it runs as a major fault, so it won't go directly into run mode when power is turned back on. Instead, it will look for and execute a fault routine, and will go into run mode only if the fault routine clears the major fault bit(s).

3. You can configure a PLC-5 always to restart its SFC program (if it has an SFC program) at the first step (the default) or at the step that was unfinished at the time that the PLC stopped running (if you set S:26/0).

PLC-5 Hardware Status

I/O Status Global status and rack control words in the status file indicate the status of I/O racks. Six words are needed for 24 racks. When you configure a PLC-5's remote I/O scanner channel, you specify an integer file (e.g., N9) to use as an I/O status file to store additional rack status (S:16 will contain the integer file number you enter). The I/O status file contains two status words for each I/O rack, so if there are 24 racks, it will contain 48 words. The global status, rack control, and I/O status words are used:

1. To indicate **rack or module faults.**

Global status words (S:7, S:32, and S:34) contain bits in their low bytes that are set if a rack fault is detected. Each bit represents one full rack, with the lowest addresses representing the lowest (octal) numbered racks (S:7/0 to S:7/7 for racks 00 to 07, S:32/0 for rack 10, etc.). If, for example, rack 03 faults, S:7 will contain the binary pattern:

S:7 ? ? ? ? ? ? ? ? **0 0 0 0 1 0 0 0**

I/O status file words contain 4 low bits in the first of the two status words, which are set if a fault occurs in one-fourth of a rack (bit 0 for groups 0 and 1, bit 1 for groups 2 and 3, etc.) If, for example, the I/O status file is N9, the I/O status file starts at word N9:0, and words N9:6 and N9:7 are status words for rack 03. If rack 03 faults because a module in its group 2 has faulted, N9:6 and N9:7 will contain the following pattern:

N9:6 ? ? ? ? ? ? ? ? **0 0 0 0 0 0 1 0**
N9:7 ? ? ? ? ? ? ? ? ? ? ? ? ? ? ? ?

The high bits of the low byte (N9:6/4 to N9:6/7 in this example) are always zero.

When a rack faults, it cannot be included in the I/O scan, and:

- The rack's outputs are placed in states as specified by setting the DIP switches at the chassis (reset or frozen), or overridden by DIP switch setting on the I/O module (e.g., max./min./mid analog range value).
- *Input images remain in their last state.* The user-program should write a set of default values to the input images for a rack that has faulted if their states can affect the controlled system dangerously.

2. To indicate that a rack's **block transfer request queue** is full.

   The **global status** words (S:7, S:32, and S:34) each contain 8 high bits that are set if there are enough block transfers requested for a rack to fill the block transfer request queue for that rack. Each bit represents one rack, with the lowest addresses representing the lowest rack numbers (S:7/8 to S:7/17 for racks 00 to 07, S:32/8 for rack 10, etc.). If, for example, there are too many block transfers requested for racks 03 and 04, S:7 will contain the binary pattern

   S:7 **0 0 0 1 1 0 0 0** ? ? ? ? ? ? ? ?

3. To indicate the **presence of I/O modules** in a rack.

   The **I/O status file** includes 4 low bits of the high byte in the first of the status words for a rack, which are set if that one-fourth of a rack contains I/O modules (bit 8 for groups 0 and 1, bit 9 for groups 2 and 3, etc.). If, for example, rack 03 does not have any I/O modules in groups 6 or 7, the words that describe rack 03's status will contain the following pattern:

   N9:6 **0 0 0 0 0 1 1 1** ? ? ? ? ? ? ? ?
   N9:7 ? ? ? ? ? ? ? ? ? ? ? ? ? ? ? ? ?

   The high bits of the high byte (N9:6/12 to N9:6/15 in this example) are always zero.

4. To **inhibit a rack or modules in a rack.**

   The **rack control** words (S:27, S:33, and S:35) each contain 8 low bits that a user-program (or an operator at a programming unit) can set to inhibit a rack, freezing its output states and the input images from that rack. Each bit represents one rack (S:27/0 to S:27/7 for racks 00 to 07, S:33/0 for rack 10, etc.). If, for example, you want to inhibit racks 02 and 03, S:27 must contain the binary pattern

   S:27 ? ? ? ? ? ? ? ? **0 0 0 0 1 1 0 0**

   The second **I/O status file** word contains 4 low bits that can each be set to inhibit two modules of a rack (bit 0 inhibits groups 0 and 1, bit 2 inhibits groups 2 and 3, etc.). If, for example, you only want to inhibit rack 03's groups 0 and 1, and groups 4 and 5, N9:6 and N9:7 must contain the following pattern:

   N9:6 ? ? ? ? ? ? ? ? ? ? ? ? ? ? ? ? ?
   N9:7 ? ? ? ? ? ? ? ? **0 0 0 0 0 1 0 1**

The high bits of the low byte (N9:7/4 to N9:7/7 in this example) are always zero.

5. To **reset a rack or modules in a rack.**

    The **rack control** words (S:27, S:33, and S:35) each contain 8 high bits that a user-program (or an operator at a programming unit) can set to reset a rack, turning its outputs off (or into their default states) until the scan cycle writes new values to them. Each bit represents one rack (S:27/7 to S:27/15 for racks 00 to 07, S:33/7 for rack 10, etc.). If, for example, you want to reset racks 00 and 03, S:27 must contain the binary pattern

    S:27 **00001001**????????

    The second **I/O status file** word's high byte contains 4 low bits that can each be set to reset two modules of a rack (bit 8 resets groups 0 and 1, bit 9 resets groups 2 and 3, etc.). If, for example, you want to reset rack 03's groups 0 and 1, and groups 4 and 5, N9:6 and N9:7 must contain the following pattern:

    N9:6 ????????????????
    N9:7 **00000101**????????

The high bits of the high byte (N9:7/12 to N9:7/15 in this example) are always zero.

PLC-5 Communication Channel Status When you configure a PLC-5's communication channel, you specify an integer file to be used as a diagnostic file that will contain status information that is relevant to the channel's type of communication. (Diagnostic file numbers are saved in the status file.) You won't need to interpret the diagnostic file's contents in its raw data form, and you should not write programs that change the file's contents. Use the programming unit's *channel status* display to display a channel's diagnostic file contents, complete with labels that describe what each value means. During configuration of a DH+ network channel, you also specify an integer file to be used as a global status file flag file. The DH+ WHO active screen displays global status file flag data to show you which DH+ nodes are active or to show you why they are inactive, and can also display the DH+ diagnostic file data when you select the DH+ WHO diagnostics screen. Ethernet WHO screens are also available for viewing Ethernet network status and diagnostics.

If you want your program to check whether a DH+ node is active, status words S:3 to S:6 contain individual bits that are on if the corresponding DH+ node is active. For example, S:3/4 is on if DH+ node 4 is available, and S:6/15 will be on if the highest node, node 63 (octal 77), is active.

PLC-5 CPU Status The PLC-5 status file contains various status information, including:

- Processor status flags (in S:1) to describe the current operating status of the CPU
- The positions of the DIP switches on the CPU (in S:2)
- The most recent scan time (S:8), the maximum scan time (S:9), and the scan times for individual MCPs, PII, and for extended local I/O scanning (if an extended local I/O chassis is connected via channel 2)

- A Checksum (in S:57) calculated for the current processor file, used to check for corrupted memory
- Fault information (discussed in the following section)
- Configuration and the status of interrupt programs

Most of the PLC status that is relevant to an operator for troubleshooting is accessible via the processor status screens, except for the more detailed channel status information.

PLC-5

SLC 500

Allen-Bradley Major Faults and Minor Faults

The Allen-Bradley PLC-5 and SLC 500 respond to major and minor faults in the same way, so in this section we discuss both types of PLC to avoid repetition. Status information is available in an SLC 500 and in a PLC-5 to help a programmer troubleshoot a PLC that is failing. Use of fault routines for error recovery is also discussed. A more detailed description of how an Allen-Bradley PLC responds to faults appears in the section on fault routine interrupts in Chapter 11.

Fault Bits and Codes A major fault can cause an Allen-Bradley PLC to exit run mode. Some major faults are recoverable, which means that when they are detected, the PLC will look for a fault routine file number in S:29 and execute the fault routine when the fault is detected. Fault routines are written by the user and can include instructions to clear the major fault bit(s). If a fault routine clears the major fault bits(s) before it ends, the PLC will return to run mode and will resume what it was doing before it detected the fault. Recoverable major faults are usually faults caused by the contents of a user-program, so Allen-Bradley literature often refers to them as recoverable major user faults.

Minor faults do not usually cause an Allen-Bradley PLC to exit run mode, although some SLC 500 minor fault bits (in the low 8 bits of **S:5**) trigger a major fault if they aren't cleared before the SLC 500 finishes executing its program scan (or when the user-program executes a REF instruction). Any user-program can include instructions to monitor minor fault bits and to correct minor errors, or can even include instructions to cause a major fault when a minor fault is detected. Arithmetic flags detect minor math errors that might lead to control problems. Arithmetic flags are set if an instruction that manipulates a number detects a carry or an overflow, and/or to indicate that the result is zero or is negative. Minor fault bits and arithmetic flag bits clear themselves.

You can find fault bits, codes, and arithmetic flags:

1. In a PLC-5:
   - S:23 contains the major fault bits (see Appendix H).
   - S:12 contains a major fault code (see Appendix H).
   - S:13 contains the number of the program file that was executing at the time of a major fault.
   - S:14 contains the number of the rung that was executing at the time of a major fault.
   - S:10 and S:17 contain the minor fault bits. (See Appendix H.)
   - S:0 contains the arithmetic flag bits. (See Chapter 6.)
2. In an SLC 500:
   - S:1/13 contains the only major fault bit.
   - S:6 contains the major fault code. (See Appendix I.)

- S:20 contains the number of the program file that was executing at the time of a major fault.
- S:21 contains the number of the rung that was executing at the time of a major fault.
- S:0 contains the arithmetic flag bits. (See Chapter 6.)
- S:5 contains the minor fault bits:
  (a) The low 8 bits of S:5 latch on under specific conditions and become major faults if not cleared by a program instruction before the user-program part of the scan cycle ends:
  - S:5/0 is on if a math overflow was detected (if S:0/1 was on).
  - S:5/2 is on if the error bit of the control element for a LIFO, a FILO, a bit shift, or a sequencer instruction was on.
  - S:5/3 is on if a new major fault occurred as the fault routine was executing. The fault code in S:6 describes this latest major fault.
  - S:5/4 is on if a program instruction tried to use data in the module memory (M0/M1 address space) of a module that was disabled.

  (Other low bits of S:5 have no function at this time.)

  (b) The other minor fault bits of S:5, which do not latch on (unless noted below) and do not become major faults:
  - S:5/8 is latched on if a memory module's contents were transferred into RAM at power-up (as required by the configuration).
  - S:5/9 or S:5/13 is on if a memory module's contents were *not* transferred into RAM at power-up (if configuration required it), for assorted reasons. (See your manual.)
  - S:5/10 or S:5/12 is on if an STI or a DII interrupt program was missed. (See Chapter 11.)
  - S:5/11 is on if the battery is low.
  - S:5/14 is on if a problem with a modem attached at channel 0 is detected. (See your manual.)
  - S:5/15 is on if an ASCII instruction has tried to manipulate a string over the allowable 82 characters in length.

After a PLC has faulted, the major fault bits and fault codes should be examined to determine the reason for the fault. You can use your programming unit to display the contents of the status file words that contain fault bits. The programming unit will also display a text message describing why each fault bit is set. Appendices H and I contain a list of the major fault bits, minor fault bits, and fault codes for faults that an SLC 500 and a PLC-5 can detect, and include the status file addresses where the user or the fault routine program can find them. After correcting the source of a major fault, the operator can put an SLC back into run mode by cycling the mode control switch to program mode, then to run mode, if the PLC has been configured to clear its own major fault bit. The operator must use the programming terminal to clear major faults in a PLC-5, in an SLC 500 that has its keyswitch at REM run, or in an SLC 500 not configured to clear its own major fault bit. After clearing the major fault bit(s), the PLC can be put back into run mode.

Watchdog Timer A program will fault if the watchdog timer has timing out. The current setting of the watchdog timer can be displayed and compared against the maximum, average, and most recent scan times. The current time can be displayed on the programming unit or can be read and changed in a user-program.

1. In a PLC-5:
   - S:28 contains the watchdog timer setting.
   - S:8 and S:9 contain the most recent and maximum scan times, respectively.
2. In an SLC 500:
   - S:3H (high byte of S:3) contains the watchdog timer setting.
   - S:22, S:23, and S:35 contain the maximum, average, and recent times, respectively.

I/O Status Maintained in I/O Modules I/O modules may contain status bits indicating fault conditions.

1. In a PLC-5, intelligent (BT) I/O module status is included in the data that can be read using a BLOCK TRANSFER READ (BTR) instruction. If the user-program has executed a BTR to read an I/O module's status, the I/O overview menu in a programming unit can be used to view the status information complete with descriptive labels. See the BT module manuals for BT file addresses for status information if you want to write a user-program to monitor I/O module status.

2. In an SLC 500, status bits can be read from the module's M0 and/or M1 memory. See the manuals for your specialty I/O module for status information availability and addresses.

We now return to PLC-5 specific troubleshooting.

PLC-5

PLC-5 Program Troubleshooting

Several PLC-5 instructions use control elements. A control element contains status bits and status words that are invaluable during programming debugging. Some but not all control element status bits and/or words are displayed (e.g., EN, DN, ER, LEN, etc.) on the Ladder Logic program display while you monitor the program. The other status information can be monitored and can sometimes be changed using the data screen or instructions in the user-program.

Besides the element called a *control element* (e.g., R6:0), several instructions contain status bits and words built into other data structures they use for control. These include timer and counter instructions (e.g., T4:0, C5:0), block transfer instructions (e.g., BT9:0), message instructions (e.g., MG10:0), the PID instruction (e.g., PD11:0), and structures for SFC program control (e.g., SC:12:0).

Some instructions are intended primarily for program debugging purposes. They include the instructions discussed below.

ALWAYS FALSE (AFI) If you insert an ALWAYS FALSE (AFI) instruction as the first element in a ladder logic rung, it holds the rung's Boolean result false, disabling or turning the rung's output off.

TEMPORARY END (TND) A TEMPORARY END (TND) is an output element. It can be programmed conditionally or unconditionally. When it executes, it immediately terminates the current scan cycle, even if it is programmed in a subroutine or in an interrupt. The I/O scan executes and then the next scan cycle starts.

DATA TRANSITION (DTR) The DATA TRANSITION (DTR) instruction is used to compare a single source data word (or selected bits of the source word) against a reference word containing expected bit values. It can be used as a Boolean logic element, in the way that a comparison instruction is used, except that:

- DTR only becomes true if the two words (or part-words) are *not* equal to each other.
- When the two values are found to be unequal, DTR copies the value from the source word into the reference word.
- A mask must be entered (as a hexadecimal value or as an address). If the mask contains 0 bit values, those bits of the source and reference words are not compared.

The DTR instruction can be used as a type of one-shot that goes true for one scan each time a changed source value is detected.

FILE BIT COMPARE (FBC) The FILE BIT COMPARE (FBC) instruction is shown in Figure 15.1. Use it in a program to compare a set of data words against a set of expected data word values, and to record the location of each bit that does not match, each time the DTR instruction's control logic goes from false to true. For learning purposes I have included dashed lines, which you won't see anywhere else, to separate the instruction parameters into three sets.

In the first set of FBC parameters, you enter three addresses, each preceded by the # symbol to indicate that these addresses are the starting addresses for three files. (Remember, a # causes the PLC-5 to use indexed addressing.) The three addresses indicate the starting word addresses of:

1. A *source file,* containing the data values you want to be compared against expected values, starting with the lowest bit of the first word. For diagnostic purposes, this might be a set of input or output images. (In the example, start at bit I:024/0.)

2. A *reference file,* containing the expected data values. This might be a pattern the programmer has entered into bit memory prior to executing the program. (In the example, start with bit B3:0.0.)

3. A *result file,* where the FBC instruction will write the *octal* addresses of the bits that don't match. If, for example, bits 1, 8, 9, and 17 do not match, then the octal numbers 0, 7, 10, and 20 will be stored into the first four words of the result file (in the example, into a result file that starts at N9:0).

The second set of compare parameters indicate how many bits to compare and supplies a compare control element to be used to keep track of where the PLC-5 is in the sequence of comparisons. The usual control element addresses can be read, written to, and monitored by the program or programmer, including the .EN (enabled) and .DN (done) status bits and the .LEN (length) value. Some comparison control element functions have special purposes in the FBC instruction:

1. The **.IN (inhibit)** bit can be set (by the programmer or by a program instruction) to cause the FBC instruction to only look for *one mismatch* each time the FBC's control logic goes

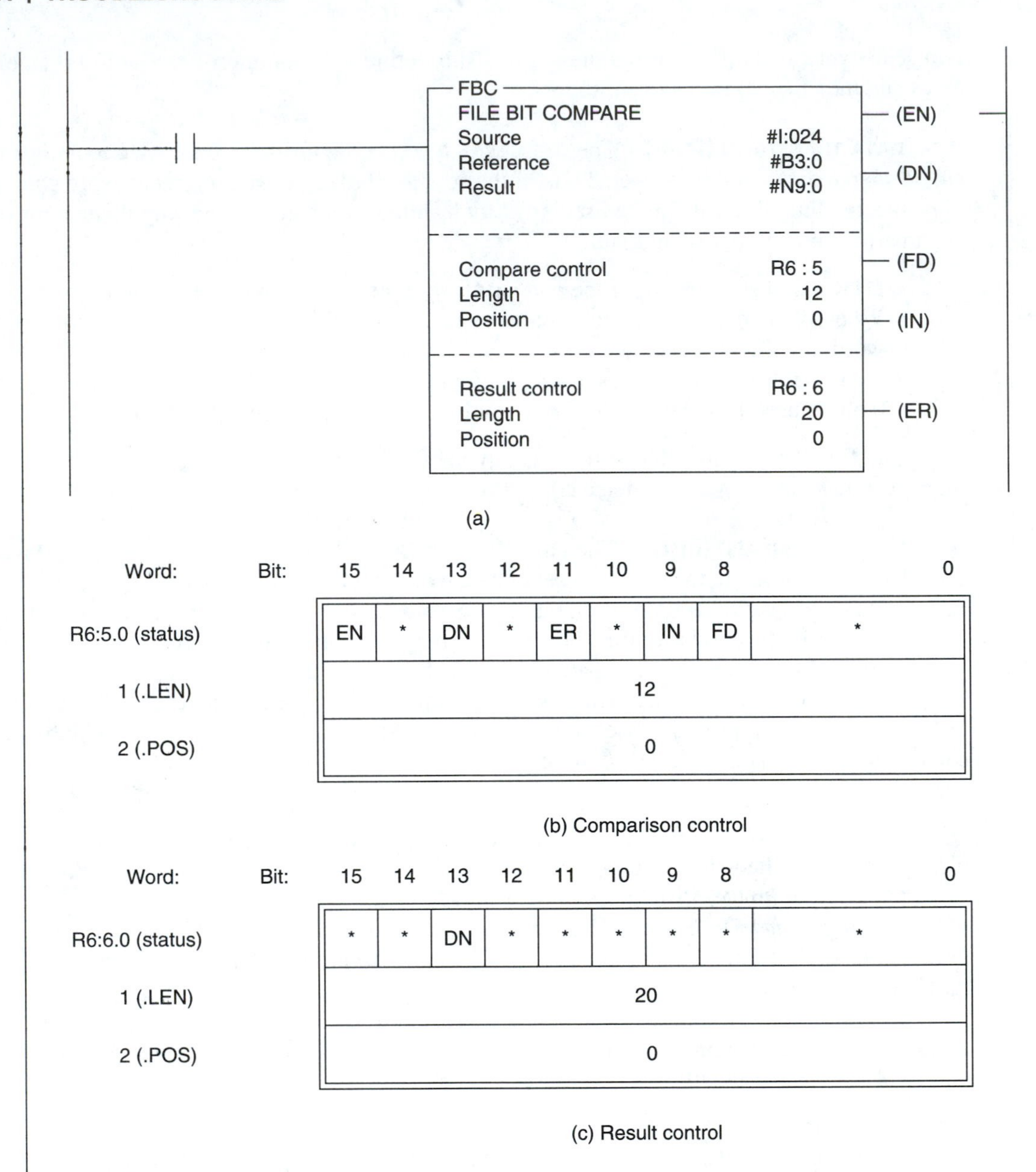

FIGURE 15.1
(a) PLC-5 FILE BIT COMPARE (FBC) instruction and the two control elements it uses: (b) comparison control element; (c) result control element.

from false to true. When the mismatch is found, the compare control element's **.FD (found)** bit will be set, and the bit position will be put in the **.POS (position)** word of the control element and displayed in the FBC instruction box. It will be stored in decimal, not in octal! When the FBC instruction's control logic goes from false to true again, the FBC instruction continues comparing from where it stopped, until the next mismatch is found. When all the comparisons have been performed, the FBC instruction does not restart another comparison set until another false-to-true logic transition occurs, even if no mismatches have been found.

2. If the .IN bit has not been set, the FBC instruction will compare all the source file bits with the reference file, recording all the mismatches into the result file, and will turn the .FD bit of the compare control element on only if *at least one mismatch* was found.

The third set of result parameters are used to specify a length for the result file and a result control element that the PLC-5 can use to keep track of how many mismatches have been entered into the result file. If more mismatches are found than there is room for, the .DN bit of the result control element will be set, the latest mismatch bit address will be stored in the first result file word, and the comparison will continue. When the comparison control element's .DN bit is on and the instruction's control logic goes from false to true, the result control element is reset so that it always restarts by placing new mismatch addresses into the first result file word.

DIAGNOSTIC DETECT (DDT) The DIAGNOSTIC DETECT (DDT) instruction looks like a FILE BIT COMPARE (FBC) instruction. It works the same as an FBC, too, except that every time the DDT instruction finds a mismatch between the source file bits and the reference file bits, the DDT instruction changes the reference bit value as it records the mismatch. DDT is therefore more appropriate for detecting changes in bit values.

User-Generated Major Faults Using JUMP TO SUBROUTINE (JSR) In Chapter 11 we describe how to use a JUMP TO SUBROUTINE (JSR) instruction to generate a user-generated major fault. Briefly, if you specify the fault routine file number as the subroutine program file number in a JSR instruction and enter a number as a single input parameter, then instead of just jumping to the fault routine as a subroutine, the fault routine is executed as if in response to a major fault (and it must clear the S:11/7 major fault bit that is set or the PLC will halt). The number entered as an input parameter is put into the major fault code word (S:12).

PLC-5 Programming Unit Features for Debugging

Allen-Bradley programming software supplies the usual program and data monitoring features, data modification features and I/O forcing features, and the search and cross-reference features. You can also use a PLC-5's programming unit to generate a histogram showing data value changes over time: Specify any bit or data value (not floating-point values at the time of this writing) and the programming unit will sample those PLC-5 memories and will display the value's transition times and new values as well as the total monitoring time. You can even specify a mask, so that only some bits of a data value are monitored and the others are ignored.

SLC 500

TROUBLESHOOTING THE ALLEN-BRADLEY SLC 500

SLC 500 Hardware Troubleshooting

See the section on PLC-5 hardware troubleshooting.

Configuring SLC 500 Startup Fault Detection

We discussed how to configure an SLC 500's startup characteristics in Chapter 10, and in Chapter 11 described how the SLC 500 responds to fault interrupts. We have seen that:

1. An SLC 500 clears its major fault bit (S:1/13) and the minor fault bits (S:5/0 to S:5/7) whenever power is restored if the PLC was in REM run mode when power failed, and if you have set S:1/8; otherwise, the SLC 500 will go into fault mode.

2. The SLC 500 can restart in REM run mode after a power failure, even if S:1/8 hasn't been set, if S:1/9 is set instead. The PLC will first look for and execute a fault routine. If the fault routine clears the major fault bit, the PLC will then return to REM run mode.

SLC 500 Hardware Status

I/O Status In the SLC 500, an I/O module fault is a fatal error, so S:1/13 will be set and (unless a fault routine executes and clears S:1/13) the PLC will halt. An error code will be written into S:6. The high byte of the error code will contain the number of the module that failed, and the low byte will be a code identifying the type of fault. (See Appendix I for the codes.) I/O slots can be individually disabled (freezing their outputs and the input image table) by clearing bits in S:11 or S:12. These bits are automatically set when the PLC starts, enabling all I/O modules.

Communication Channel Status The SLC 500 does not collect channel statistics into diagnostic files. Communication channel status can only be observed by examining the active node status words:

S:9 and S:10	for a DH 485 network connected at channel 1
S:67 and S:68	for a DH 485 network connected at channel 0
S:83 and S:86	for a DH+ network connected at channel 1

Each bit that is on indicates a node that has been participating in the passing of the tokens via the network, even if the node hasn't been passing data. Node 0 is represented by the lowest bit (S:9/0, S:67/0, or S:83/0) and the higher bits represent higher addressed nodes (up to DH-485 node 31 at S:10/15 or S:68/15, or DH+ node 63 at S:86/15). You can use a programming unit to display network and node activity by selecting WHO List or Station Diagnostics in DH-485 WHO Active or in DH+ WHO Active.

CPU Status If you can't see the SLC 500, but you can connect to it via a programming unit on a local area network, you can determine what type of SLC 500 you are working with. The SLC 500 holds information describing its model, memory, and a description of the user-program it contains, in status words S:58 to S:65.

Most of the PLC status that is relevant to an operator for troubleshooting is accessible via the processor status or the status file data screens, except for some detailed channel status information.

SLC 500 Major and Minor Faults

See the discussion in this chapter of Allen-Bradley major and minor faults in the section on troubleshooting the Allen-Bradley PLC-5.

SLC 500 Program Troubleshooting

Some instructions that are intended for use during debugging of a user-program include:

TEMPORARY END (TND). When the SLC 500 executes a TEMPORARY END (TAD) instruction (which can be programmed as an output element controlled by Boolean logic), it will terminate the current scan cycle even if it is encountered in a subroutine or in an interrupt. The PLC will then execute the I/O scan step and restart the user-program from its beginning.

SUSPEND (SUS). If, as shown in Figure 15.2, a SUSPEND (SUS) instruction is executed, the PLC immediately terminates scan cycling. The termination is not a major fault, although the processor status can be examined as it could be after a major fault. When you program the SUS instruction, you must enter a suspend ID number (number 91 is used in the example). The SUS instruction writes the suspend ID number to S:7 as it executes. A programmer can include several SUS instructions in a program, each with a different suspend ID, so that the programmer can read S:7 to determine which SUS instruction caused the PLC to halt. S:8 will contain the number of the program file that was executing when the SUS instruction executed.

SLC 500 User-Fault Routines To program an SLC 500 to fault under selected circumstances, write a program that sets the major fault bit, S:1/13.

SLC 500 Programming Unit Features for Troubleshooting Instructions for which you have entered control element addresses can be monitored as the PLC runs, through the data monitor screen at your programming unit. The control elements contain status bits that are written to by the instruction each time the instruction is executed in your program. Timer and counter elements also contain status bits. The PID instruction's control element also contains a 1-byte fault code. The status bits and fault codes for each instruction are covered in this book with each instruction.

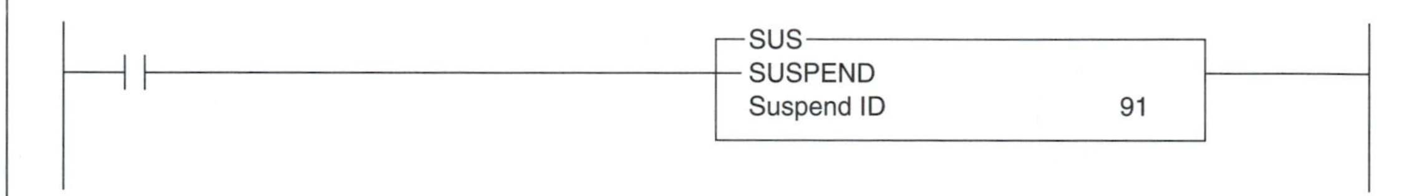

FIGURE 15.2
SLC 500 SUSPEND (SUS) instruction in a Ladder Logic rung.

Message instructions and some ASCII instructions display error codes in the instruction graphic as you monitor the program. These error codes cannot be monitored in any other way.

An SLC 500 program can be tested one rung at a time, in test single step mode. After selecting TEST MODE instead of run or program mode at the programming unit, select SINGLE STEP, then select EXECUTE STEP *once* to cause the PLC to execute one complete (initial) scan cycle, then select EXECUTE STEP as many more times as you choose, to execute one rung of your program each time, starting from the first rung. You can use the program monitoring and data monitoring features, including data value modification and I/O forcing, between single steps.

You can also test an SLC 500 program by causing it to run to a breakpoint in test single step mode. After selecting TEST MODE and SINGLE STEP, select SET END RUNG and enter a file number and rung number to which you want to test the program down. When you press ENTER, the PLC will execute one complete (initial) scan cycle, then a second scan down to the breakpoint. Each time you press ENTER again, the PLC will complete the unfinished scan cycle and execute the next scan down to the breakpoint. (You can change the breakpoint before pressing ENTER again.)

S5

TROUBLESHOOTING THE SIEMENS S5

The user cannot configure the S5's fault response (other than by writing error-handling OBs). After the reason for a fault has been eliminated, simply switch the S5 PLC into STOP, then back into RUN. There is no need to clear any error bits or codes.

S5 Fatal Errors and Nonfatal Errors

ISTACK The S5 CPU module maintains 32 bytes of status and fault information in two areas of system data memory. Siemens calls these 32 bytes the ISTACK, or interrupt stack memory area. Figure 15.3 shows the contents of these 32 bytes.[2] Figures 15.4 and 15.5 describe the purposes for each fault bit (which Siemens calls analysis bits) and the purpose of the other status information.

ISTACK information can be displayed on the programming unit while the PLC is in stop mode. The PLC will be in stop mode if it has encountered a fatal error that wasn't corrected by a user-written error-handling organization block. The bits that were set at the time of the fatal error will be indicated. You can cursor to each bit to see an explanation of the bit's meaning on the programming unit display. While the PLC is in run mode, the user can monitor some (the first 6 bytes) of ISTACK data.

BSTACK Besides the ISTACK, the S5 PLC also maintains status information in a BSTACK (block stack) area of memory. The programming unit can be used to display the contents of the BSTACK after a fatal error. The BSTACK contains a list of the logic blocks that are waiting to resume operation. The list will include the logic block(s) that contained jump in-

---

[2] One word address is not given. This table uses the SD prefix for system data instead of the RS prefix more commonly used. Some S5 CPUs don't maintain all of the bits shown, but newer S5 CPUs may have more than are shown.

Byte \ Bit	7	6	5	4	3	2	1	0
1			BST SCH	SCH TAE	ADR BAU			
2	CA-DA	CE-DA		REM AN				
3	STO ZUS	STO ANZ	NEU STA		BAT PUF		BARB	BARB END
4						AF		
5	ASP NEP	ASP NRA	KOPF NI		ASP NEEP			
6	KEIN AS	SYN FEH	NINEU					UR LAD
7	IRRELEVANT							
8	IRRELEVANT							
9	STOPS		SUF	TRAF	NNN	STS	STUEB	FEST
10	NAU	QVZ	KOLIF	ZYK	SYSFE	PEU	BAU	ASPFA
11								
12	ANZ1	ANZ0	OVFL		OR	STA TUS	VKE	ERAB
13	6th nesting level					OR	VKE	FKT
14	IRRELEVANT							
15	4th nesting level					OR	VKE	FKT
16	5th nesting level					OR	VKE	FKT

FIGURE 15.3
Siemens S5 ISTACK memory contents.

Byte \ Bit	7	6	5	4	3	2	1	0
17	2nd nesting level					OR	VKE	FKT
18	3rd nesting level					OR	VKE	FKT
19	Nesting depth (0 ... 6)							
20	1st nesting level					OR	VKE	FKT
21	Start address of the data block (high)							
22	Start address of the data block (low)							
23	Block stack pointer (high)							
24	Block stackpointer (low)							
25	STEP address counter (high)							
26	STEP address counter (low) [1]							
27	Statement register (high)							
28	Statement register (low)							
29	Accumulator 2 (high)							
30	Accumulator 2 (low)							
31	Accumulator 1 (high)							
32	Accumulator 1 (low)							

FIGURE 15.3
(continued)

Fault/Error	Fault/Error ID	Cause Defective program in CPU memory	Remedy
Cold Restart is not possible.	NINEU SYNFEH/ KOPFNI	Causes: - System start-up is faulty. - Compressing has been interrupted by a power failure. - Block transfer between programmer and PC was interrupted by a power failure. - Program error (TIR/TNB/DO FW).	Perform an Overall Reset. Reload the program.
	KOLIF	DB 1 is programmed incorrectly.	Rename DB 1.
	FEST	There is an error in the self-test routine of the CPU.	Replace the CPU.
Faulty submodule	ASPFA	The submodule ID is illegal	Plug in the correct submodule.
Battery failure	BAU	There is no battery or the battery is low and the retentive feature is required.	Replace the battery. Perform an Overall Reset. Reload the program.
I/Os not ready	PEU	The I/Os are not ready: - There has been a power failure in the expansion unit. - The connection to the expansion unit has been interrupted. - There is no terminator in the central controller.	- Check the power supply in the expansion unit. - Check the connection. - Install a terminator in the central controller.
Program scanning interrupted	STOPS	The mode selector is on STOP.	Put the mode selector on RUN.
	SUF	Substitution error: A function block was called with an incorrect actual parameter.	Correct the function block call.
	TRAF	Transfer error: - A data block statement has been programmed with data word number greater than the data block length. - A data block statement has been programmed without opening a DB first.	Correct the program error.
	STS	- Software stop by statement (STP) - STOP request from programmer - STOP request from L1 network master	
	NNN	- A statement cannot be decoded. - The nesting level has been exceeded. - A parameter has been exceeded.	Correct the program error.
	STUEB	Block stack overflow: - The maximum block call nesting depth of 16 has been exceeded.	Correct the program error.
	NAU	There has been a power failure.	
	QVZ	Time-out from I/Os: - A peripheral byte that was not addressed has been referenced in the program. - An I/O module does not acknowledge.	Correct the program error or replace the I/O module.
	ZYK	Scan time exceeded: The program scanning time is greater than the set monitoring time.	Check the program for continuous loops. If necessary, retrigger the scan time with OB 31 or change the monitoring time.

FIGURE 15.4
Siemens S5 ISTACK "analysis" (fault) bit meanings.

Control Bit Mnemonics	
ADRBAU	Construction of address lists
AF	Interrupt enable
ASPNEEP	Memory submodule is an EEPROM
ASPNEP	Memory submodule is an EPROM
ASPNRA	Memory submodule is a RAM
BARB	Program check
BARBEND	Request for end of program check
BATPUF	Battery backup okay
BSTSCH	Block shift requested
CA-DA	Interprocessor communication flag output address list available
CE-DE	Interprocessor communication flag input address list available
KEINAS	No memory submodule
KOPFNI	Block header cannot be interpreted
NEUSTA	PC in Cold Restart
NINEU	Cold Restart not possible
REMAN	0: all timers, counters, and flags are reset on Cold Restart 1: the second half of timers, counters, and flags are reset on Cold Restart
SCHTAE	Block shift active (function: COMP: PC)
SD	System data (from address $EA00_H$)
STOANZ	"STOP" display (internal request)
STOZUS	"STOP" state (external request)
SYNFEH	Synchronization error (blocks are incorrect)
URLAD	Bootstrapping required

Interrupt Display Mnemonics	
ANZ1/ANZ0	00: ACCUM 1 = 0 or 0 is shifted 01: ACCUM 1 > 0 or 1 is shifted 10: ACCUM 1 < 0
ASPFA	Illegal memory submodule
BAU	Battery failure
BEF-REG	Statement register
BST-STP	Block stack pointer
DB-ADR	Data block address
DB-NR	Data block number
ERAB	First scan
FEST	Error in the CPU self-test routine
FKT	0 : O(
KE1...KE6	Nesting stack entry 1 to 6 entered for A(and O(1 : A(
KOLIF	Interprocessor communication flag transfer list is incorrect.
NAU	Power failure
NNN	Statement cannot be interpreted in the PC-2000.
OB-NR	Organization block number
ODER	OR memory
OVF	Arithmetic overflow (+ or -) (set by "0" operation)
PEU	I/Os not ready: power failure in the I/O expansion unit; connection to the I/O expansion unit interrupted No terminator in the central controller
QVZ	Time-out from I/Os: A nonexistent module has been referenced.
REL-SAZ	Relative step address counter
SAZ	Step address counter
STATUS	STATUS of the operand of the last binary statement executed
STOPS	Mode selector on STOP
STS	Operation interrupted by a programmer STOP request or programmed STOP statements.
STUEB	Block stack overflow: The maximum block call nesting depth of 16 has been exceeded.
SUF	Substitution error
SYSFE	Error in the SYSID block
TRAF	Transfer error for data block statements: data word number > data block length.
UAW	Interrupt display word
VKE	Result of logic operation
ZYK	Scan time exceeded: The set maximum permissible program scan time of 0.5 sec. has been exceeded.

FIGURE 15.5
Siemens S5 ISTACK analysis bits and words.

struction(s) that called the logic block that was executing at the time of the fault. The list will also identify any logic block(s) that were waiting to resume operation after being interrupted. The BSTACK will also contain the address(es) of the data block(s) that were active in any suspended logic blocks when the logic blocks were suspended.

Error-Handling Organization Blocks Depending on the type of S5 CPU, the PLC's operating system may look for one of the following error-handling organization blocks and execute it instead of going into fault mode when certain types of otherwise fatal errors are encountered. After executing an error-handling OB, the S5 CPU will resume executing the program that was running at the time of the "fatal" error. The error-handling organization blocks include:

OB 23 executes if it takes longer than 160 microseconds to read or write an I/O module in response to an instruction using direct access addressing. The CPU writes the address of the I/O module into RS 103 before calling OB 23.

OB 24 executes if it takes longer than 160 microseconds to read or write an I/O module during the I/O scan or while updating interprocess communication flag memory. The CPU writes the address of the I/O module into RS 103 before calling OB 24.

OB 27 executes if the program calls a function block that has variable declarations that were changed after the call to the function block was programmed.

OB 32 executes if the program tries to use a data block that hasn't been called, or a data word that doesn't exist, or if the program tries to generate (create) a data block that is too big.

One organization block is called in response to a nonfatal error. The PLC will not fault if it isn't found, of course:

OB 34 executes if a battery-low condition is detected.

Watchdog Time Errors One final error-handling OB is used to avoid fatal errors, not to recover from them:

OB 31 can be called to reset the watchdog timer to avoid long scan times causing a fatal error.

Fatal errors due to exceeding the watchdog time setting can also be avoided by increasing the watchdog timer setting from its default 100 ms. The watchdog timer setpoint can be changed by writing a new value of between 1 and 255 to RS096. The values correspond to the number of 10-ms increments to be used as a setpoint, so the scan time can be set to values between 10 ms (too short for most programs) and 2.55 s. The following STL program segment shows how a function block can configure the PLC so that the watchdog timer won't cause a fault unless a program scan exceeds 300 ms. System data words, preceded with the letters RS, can be written to only by function block programs.

```
L KF30    ;  Load decimal value 30
T RS097   ;      and transfer it to RS097
BE        ;  . . that's all it takes!
```

Math Errors As in most other PLCs, the S5 can detect when it has performed a math operation that has generated a number that is too big a positive number or a negative number to be stored in memory. The PLC will set an overflow bit (OV), then continue to execute the user-program as if nothing were wrong (obviously, nonfatal). The programmer can use the (STL language only) instruction JUMP IF OVERFLOW (JO) to detect when this type of math error has occurred. The jump destination should include a routine to respond to the error. The other condition code bits can be used to detect whether the result was too large or too small. CC1 is set if the number was too large a positive number, and the JUMP IF POSITIVE (JP) instruction will cause a jump if CC1 is set. CC0 is set if the number was too large a negative number, and the JUMP IF MINUS (JM) instruction will cause a jump if CC0 is set. If CC1 and CC0 are both cleared when OV is set, the result of the math operation overflowed so far that the 16-bit result became zero!

STEP 5 Logical Error Debugging Tools

In keeping with Siemens' apparent belief that any programmer who can use the PLC's most powerful features can also program complex functions with simple tools, STEP 5 only offers simple STL language instructions for program debugging:

The **BLOCK END (BE)** instruction can only be placed at the very end of a logic block listing, so it isn't very useful for program debugging.

A **BLOCK END UNCONDITIONAL (BEU)** can be programmed anywhere in a logic block. If the PLC executes a BEU, the current logic block will end, and the logic block that is waiting to resume next (if any) will resume. If BEU is encountered in OB 001, the I/O scan step executes immediately, then OB 001 is restarted. You can program instructions to jump past a BEU instruction so that it won't execute.

The **BLOCK END CONDITIONAL (BEC)** can also be placed anywhere in a logic block. It works like the BEU instruction except that it can be controlled by a Boolean logic statement.

The **STS and STP** instructions can be used to force the PLC to terminate the scan cycle, at which time the programmer can examine the PLC's memory, CPU registers, ISTACK, and BSTACK to see their status at the time of the termination. Both instructions execute unconditionally, although a program can conditionally jump past them. The **STOP IMMEDIATELY (STS)** instruction effectively causes a fatal error so that the PLC will immediately leave run mode. The **STOP AT END OF PROGRAM SCANNING (STP)** instruction will allow the rest of the scan cycle to finish, including writing of output images to output modules, before it causes the CPU to leave run mode.

Common Programming Errors

There are some types of programming errors that are so common in STEP 5 that they deserve repeating here, although they have been covered elsewhere:

1. *Incorrect mixing of bytes and words.* When you load a byte from a memory address, the 8-bit value from that address is put into the low byte of the accumulator. When you load a word from a memory address, the 8-bit value from that address is put into the accumulator's high byte, and the 8-bit value from the next memory is put into the accumulator's low byte. A similar effect occurs in transferring data from accumulators to memory (see Chapter 5).

2. *Incorrect placement of timers and/or counters in structured programs.* Most timers reset themselves when they are executed with their control logic false. If the timer is in a logic block that is called only when that control logic is true, the timer can't reset itself and will run only once (the first time the logic block with the timer executes). Counters also need to be controlled by logic which must be *seen* to change (see Chapter 4).

3. *Incorrect use of analog values.* The digital-to-analog (DAC) converter in an analog output module ignores the low 4 bits of 16-bit data words and converts the high 12-bit value to analog. Some modules assume that the binary value is in signed binary format, so if the most significant bit is set, the analog output will be a negative value. The analog-to-digital (ADC) converter in an analog input module generates a 13-bit number (often in signed binary), which is stored into the high 13 bits of a 16-bit number. The low 3 bits are important! The lowest bit is set if the analog input value was outside the range of values that could be converted to binary. The next-lowest bit is set if the analog input value was outside the input module's nominal range but close enough that a correct digital value was generated. The third-lowest bit (in some analog input modules) is set if the input module is still in the process of converting analog values to binary, so the digital value isn't correct yet (see Chapter 2).

STEP 5 Programming Unit Features for Debugging

STEP 5 programming units can be used to create cross-reference lists showing where specific bits, words, or instructions are used; to view status variable screens showing values in the PLC's memory as the program executes (and allowing some values to be changed); and to force memory contents (including I/O images) to take on specific values while the program executes

STEP 5 also allows the user to force output module contacts on or off while the PLC is in stop mode. STEP 5 does not currently have a histogram feature (although Siemens is reportedly working on histograms for STEP 7).

S7

TROUBLESHOOTING THE SIEMENS S7

The S7 PLC has very sophisticated troubleshooting features, and it would take a whole book to cover them all. This section provides enough detail for the reader to be able to use most of the features and to know there are other features described in STEP 7 manuals.

When you connect to a PLC via a programming unit, graphics in the station hardware window display the operating status of each module. If a module isn't operating because of a configuration problem, there is a slash through the graphic for the module. An X indicates that the module has a fault, and modules with forced bits are shown with a red bar across their top. If you click on the graphic for a module in the station hardware window, a dialog box will let you view the diagnostic information stored in that module's memory.

The CPU maintains a list of *diagnostic events* (sometimes called *error events*) in its system status data. Diagnostic events are recorded as *diagnostic messages* in an area of memory called the *diagnostics buffer.* The user cannot erase entries in the diagnostic buffer, even by performing a complete memory reset. The diagnostic buffer and a related memory area used to store system status lists (SZL) are described in this section.

STEP 7 Programming Unit Trouble

If you can't get STEP 7 to start, Siemens' method for protecting STEP 7 from being illegally copied may be giving you trouble. An *authorization* must be present on the personal computer with STEP 7, where it can be damaged by such simple operations as allowing SCANDISK to fix files with lost clusters, or when a hard disk is compressed, or when a new operating system is loaded. Perhaps by the time you read this, Siemens will have realized that Microsoft's marketing goal is to get every user to buy a new operating system every three days and will have devised a better protection scheme.

S7 Hardware Troubleshooting

Each S7 module has a system fault (SF) LED, which is red if the module faults, and status LEDs, which are green if no circuit problem is detected. The CPU module has additional LEDs, each for a specific purpose (battery low, forces on, etc.). The S7 allows *hot-swapping* of modules and network connections (so you can exchange modules as the PLC runs), but be careful:

- You can't remove or insert a power supply, CPU, or interface module (IM) while the PLC is running.
- Changing the memory card in a CPU will cause the CPU to flash its stop LED, to tell you that a memory reset is required before the new card's contents will be loaded into the PLC's memory.
- It may take up to 2 s before the CPU recognizes a newly installed signal module requiring configuration data and writes configuration data to the module.
- You can't insert a module into a slot configured for a different type of module.
- The modules at both ends of a network segment must be powered (and must have their network terminator resistors switched on), or communication data may be lost.

S7 Configuration for Troubleshooting

There are only a few configuration changes that you can make to affect how an S7 PLC responds to faults or how it collects status information. Some configuration options useful in troubleshooting were discussed in Chapter 10, including those that cause the PLC to:

1. Check RAM memory and check the hardware configuration to verify that it is the same as the hardware that is actually connected, and fault if errors are found.
2. Limit the scan time, and fault if a scan cycle takes longer.
3. Send a message (the latest entry in the diagnostics buffer) via the MPI network each time the CPU goes into stop mode, to describe why the PLC stopped.
4. Write diagnostic data entries into the diagnostic buffer for an extended list of reasons in addition to the standard reasons.
5. Limit the time spent trying to write configuration parameters to modules, or waiting for replies, and fault after timing out.
6. Enable signal modules (SMs) with diagnostics capabilities to issue diagnostic interrupts to the CPU and to send extended diagnostics messages to the CPU when faults are detected.

7. Tell the PLC where to store DP-diagnostic information in a Profibus-DP master and in a Profibus-DP slave.

The S7 CPU can be configured to perform troubleshooting functions that were not described in Chapter 10:

1. Monitor the states of selected Boolean elements and send a message when they change state. Siemens calls this feature symbol-related (SCAN) messaging. You select a bit to monitor from those listed on the symbol table, then select STEP 7's EDIT–SPECIAL OBJECT PROPERTIES–MESSAGES to enter a message text, the time frame during which to monitor the bit, and other attributes. The message must then be compiled and downloaded to the CPU. After operating stations have been identified using INSERT–WINCC OBJECT–OPERATOR STATIONS, select the operating stations to receive the message(s) using OPTIONS–PLC-OS CONNECTION DATA–TRANSFER and the dialog screen that appears.

2. Message(s) to send as alarm messages. The block-related message texts and message destinations are also entered using EDIT–SPECIAL OBJECT PROPERTIES–MESSAGES, but the user-program must execute a message-sending system function (SFC 17 or SFC 18) or system function block (SFB 33 to SFB 37) to initiate sending the message.

3. Messages for user-defined diagnostic messages. The text for a user-defined diagnostic message is entered as above, but the program must execute SFC 52, WR_USMSG, to initiate sending it, and an input parameter for WR_USMSG must indicate where the message is to be sent. This type of message can only be sent to the diagnostic buffer and to the programming unit, not to operator stations.

S7 Status Information

S7 CPU modules and other modules with diagnostics capabilities maintain status information that can be read using a programming unit. Most of the same information is also readable using system functions in a user-program.

CPU Module Status In the CPU module, status information is stored in two places: in system status lists (SZL lists) and in a status word.

CPU Status Word The status word maintained in the S7 CPU module contains status bits that reflect the status of the CPU as it executes a user-program. Like other PLCs, the S7 won't fault just because your program has asked the PLC to perform a math operation that generates a number beyond the PLC's capability to handle. Like other PLCs, though, this nonfatal error is detected, and your program can detect it by checking the status bits that are affected. The status word contains 4 bits relevant to this purpose, 2 of which can be addressed by Boolean logic instructions:

OV (OVERFLOW) is set every time a result is generated that is outside the range of the numbering system in use. If this bit is set, the result is wrong! The next math operation will set or reset this bit again.

OS (OVERFLOW SET) is set like OV, but remains set until specifically reset, or until a specific instruction (JOS, CALL, or BE) executes. Because OS latches, it does not need to be checked after every math operation.

Two other status word bits, **CC1** and **CC0,** indicate the size of the result of an operation (or the result of a comparison). Although CC1 and CC0 cannot be addressed directly, they can be checked by Boolean logic instructions if one of the following codes is used instead of a bit address in a Boolean instruction:

==0	Set if result = zero (CC0 = CC1 = 0).
>0	Set if result was positive (CC1 = 1, CC0 = 0).
<0	Set if result was negative (CC1 = 0, CC0 = 1).
<>0	Set if result not zero (CC1 = 1 + CC0 = 0 or CC1 = 0 + CC0 = 1).
<=0	Set if result not negative (CC1 = 0 + CC0 = 0 or CC1 = 0 + CC0 = 1)
>=0	Set if result not positive (CC1 = 1 + CC0 = 0 or CC1 = 0 + CC0 = 0)

For example, the following STL program will add two numbers then will check the result. It will jump past the instruction that stores the result if the result is negative or is invalid because of an overflow.

```
      L      MW 10
      L      MW 14
      +I
      A <0
      O OV
      JC     Past
      T      MW 18
Past: BE
```

CPU System Status Lists (SZL) The system status list (SZL) area is divided into sublists of information. Each sublist has a hexadecimal SZL-ID number, and some sublists are further divided into sections with index numbers. When viewing SZL list information using a programming unit, you don't need to know the SZL-ID or index numbers, but they are required as input parameters when you use SFC 51, "RDSYSST", in a program to read SZL sublists. Figure 15.6 shows an overview of the SZL list information available in an S7-400 CPU and the hexadecimal (W#16#) SZL-ID numbers for each type of sublist. See your manual set (or use RDSYSST to output SZL-ID 0000) to see which system status records are available in your CPU. (A complete listing with descriptions would occupy over 60 pages!) As Figure 15.6 shows, SZL data can be classified as:

1. *System data,* which consists of descriptions of:
   - The SZL-ID sublists available in this CPU's system status list.
   - The CPU model and characteristics, including how much user and system memory it has.
   - Data describing the blocks that are programmed, their priority classes (configurable in some S7 models), and a list of the system data blocks (SDB2). (System data blocks contain the compiled configuration data you have downloaded to the CPU.)
   - The allowable I/O rack system size (S7-300 only).
   - The current LED status at the CPU and other modules.
   - The *start event* (also called *error event*) numbers assigned to cause organization blocks to interrupt lower-priority-class programs, and a list of the priority classes

	Partial List	SZL-ID
System Data	List of all the SZL-IDs of a module	W#16#xy00
	Module identification	W#16#xy11
	CPU characteristics	W#16#xy12
	User memory areas	W#16#xy13
	System areas	W#16#xy14
	Block types	W#16#xy15
	Priority classes	W#16#xy16
	List of the permitted SDBs with a number < 1000	W#16#xy17
	Maximum S7-300 I/O configuration	W#16#xy18
	Status of the module LEDs	W#16#xy19
	Interrupt / error assignment	W#16#xy21
	Interrupt status	W#16#xy22
	Priority classes	W#16#xy23
	Modes	W#16#xy24
Diagnostic Status Data	Communication capability parameters	W#16#xy31
	Communication status data	W#16#xy32
	Diagnostics: device logon list	W#16#xy33
	Start information list	W#16#xy81
	Start event list	W#16#xy82
	Module status information	W#16#xy91
	Rack / station status information	W#16#xy92
Diagnostic Buffer	Diagnostic buffer of the CPU	W#16#xyA0
Diagnostic Data from Modules and DP Slaves	Module diagnostic information (data record 0)	W#16#00B1
	Module diagnostic information (data record 1), geographical address	W#16#00B2
	Module diagnostic information (data record 1), logical address	W#16#00B3
	Diagnostic data of a DP slave	W#16#00B4

FIGURE 15.6
S7 system status list (SZL) topics.

that are disabled or masked because the user-program has executed an SFC to do so. (Read about SFC 36 to 42 in Chapter 11.)

- Status of interrupts and priority class, including the 20 bytes of parameter data passed to each organization block that is currently running or is interrupted, each OB's operating status, the available nesting level for interrupts, and the limits on start information and local stack data that OBs can use.
- The current operating mode, the most recent mode change, and status indicating how the PLC started or why it stopped.

2. *Diagnostic status data,* which consists of:
   - Extensive lists of communication configuration parameters and status, including baud rate settings, connections to communication partners, communication job activity, global data exchanges, and symbol-related (SCAN) and block-related message configuration and status (as described earlier in this section).
   - Lists of the start events that have initiated organization blocks, and the start information for each of those organization blocks. (This information is also stored in the diagnostics buffer.)
   - A list of the status of modules with diagnostics capabilities that are active in this PLC, including racks and including stations in DP networks connected to this PLC.

3. The *diagnostics buffer.* The S7 CPU writes an entry into the diagnostics buffer each time a diagnostic event occurs. Diagnostic events may be faults detected by the CPU or by modules that have their diagnostic interrupt capabilities enabled. Diagnostic events may simply be mode changes. These diagnostic events are treated as start events to cause organization blocks to execute or to cause the PLC to stop. User-defined diagnostic event messages can also be written to the diagnostic buffer by SFC 52, WR_USMSG. Each diagnostic event record includes:
   - The time and date of occurrence.
   - The start event identification number and the start information, which is passed to the organization block that this fault initiates. (Start event numbers are associated with text diagnostic messages and help information, all of which can be displayed on a programming unit.)

4. *Diagnostic data from modules or DP slaves.* This data is ordered in the system status list according to the location of the module or the diagnostic address of the DP slave. Diagnostic data in the SZL includes:
   - Record 0 diagnostics from a module with diagnostics capability. Record 0 is 4 bytes of fault bits in a format standardized for all modules. (See SZL-ID 00B1 in your manual set.)
   - Record 1 diagnostics from a module with diagnostics capability. Record 1 contains a data structure that is specific to the type of module. (See your manuals.)
   - The standard structured diagnostic information from a Profibus-DP slave, as specified in EN 50 170, consisting of 6 standard-format bytes and additional data depending on the DP slave type.

Module Status Signal modules (SMs), communication processor (CP) modules, interface modules (IMs), and DP slaves with diagnostic capabilities maintain their own diagnostic data information, which can be examined using a programming unit or can be read by calling SFC 59, "RD_REC", in a user-program. (RD_REC was covered in Chapter 7.) A module's diagnostic data is in the same format as it is when it is stored in the CPU's system status list (record 0 and record 1, or in standardized DP slave diagnostic structure). Diagnostic data in the module's memory may contain more information or more recent information than what is in the CPU's SZL list, especially if the module has not been configured to send all its diagnostic data to the CPU, or if there has been no reason for the module to transmit data recently.

Global Data Status Double Word When you configure your PLC, you can set up a global status double word to contain status bits describing the exchange of global data packets between PLCs. The status bits can be monitored through a programming unit, using VIEW–GD STATUS, or they can be read by instructions in the user-program. Global status bits that are set will remain set until specifically reset. One or more of the low 8 bits will be set if data has been lost or corrupted. Bit 3 of the second-lowest byte is set if the sender of global data has been restarted. The most significant bit is set when new data has been received. (None of the other bits currently have purposes.)

Faults Indicated by an SFC's RET_VAL All SFCs return a one-integer parameter with the symbolic name RET_VAL. RET_VAL contains status information describing how well the SFC worked. If the user-program includes an address for this output parameter, a programming unit can be used to display RET_VAL, or a user-program can evaluate it. If RET_VAL is equal to or greater than zero, the SFC worked. (Some SFCs use RET_VAL to return a positive result to the calling program if the SFC worked.) If RET_VAL's most significant digit is a 1, RET_VAL is negative, indicating that the SFC did not work. The low byte of RET_VAL will contain a code indicating the failure. If the rest of the bits of the high byte are all 0's, the error code is one that is specific to the SFC that generated it; otherwise, the code is a general error that could apply to any SFC. See your manual for RET_VAL error codes.

Fault-Response Organization Blocks

For some types of fatal errors, the S7 PLC looks for a fault-response organization block when the error is detected. If the fault-response OB isn't found, the PLC will go into stop mode after writing a diagnostic message to the diagnostic buffer to describe the fatal error.

Siemens S7 offers several organization blocks that are intended to be used as fault-response organization blocks. They include OB 80 to OB 87, which are called if an asynchronous error occurs, and OB 121 and 122, which are called in response to a synchronous error. Synchronous errors are usually caused by errors in a program (so the error occurs in synchronization with the program), whereas asynchronous errors are caused by events such as hardware failures and diagnostic interrupt signals from modules. In Chapter 11 we described the fault-response OBs in the "Fault Routine" section. We also described how to use SFC 36 to SFC 42 to disable, delay, or mask selected types of errors so that they can't cause interruptions of the program until reenabled or unmasked.

If a type of fatal error occurs that causes the S7 to look for a fault-response organization block, and the OB is found (because the programmer wrote a program for that block), the S7 will immediately interrupt the program currently executing, and will start the fault-response OB. The interrupted program will resume when the fault-response OB finishes. Fatal errors can thus be turned into nonfatal errors. A diagnostic message will still be entered into the diagnostic buffer, even if the PLC does not enter stop mode. SFC 46, "STP", can be called in a fault-response OB to prevent the interrupted program from resuming.

Even before a programmer creates a fault-response OB, each OB already has a 20-byte variable declaration table, complete with default symbolic names. The S7 operating system always passes data from the diagnostic buffer for the variables when it calls the fault-response OB. The programmer can add more (temporary) variable declarations and should write a program to examine the error-description parameters and to correct the error. If more status data is required by the OB, the OB can call SFC 51, "RDSYSST", to read any of the sublists in the system status list (SZL) in the CPU, or SFC 6, "RD_SINFO", to read status information describing how the PLC started last. SFC 59, "RD_REC" can be called to read data from an I/O module. SFC 13, "DPNRM_DG", can be used to read diagnostic data from a Profibus-DP slave. RDSYSST was covered in Chapter 11, and RD_REC was covered in Chapter 7. RD_SINFO and DPNRM_DG are not covered in this book.

Fault-response OBs that allow the interrupted program to resume (by not calling STP) often need to pass parameters back to the interrupted programs, indicating that they have taken some action, but organization blocks can only have temporary variables. There are several ways to write a fault-response OB to output data that other programs can read after the OB terminates: The OB can write to a globally defined variable declared in a symbol table or to a value to a data block, or SFC 58, "WR_REC", can be used to write data (not configuration data) to a module. Synchronous fault-response OBs share CPU registers with the programs that they interrupt, so any changes they make to the registers remain when the fault-response OB ends.

S7 Diagnostic Data As Seen Using a Programming Unit

After the PLC has stopped due to a fault, the programmer can use a programming unit to connect online to the faulted PLC to see diagnostic data (including the contents of the diagnostic buffer) describing the state of the PLC at the time of the fatal error. Select the faulted module (displayed with a diagnostics symbol) from the station hardware window, and a dialog box will appear offering the module's diagnostic data. If the faulted module is the CPU, the diagnostic data consists of data from the system status list (SZL).

Diagnostic data for modules that have not faulted can be displayed in the same way. This feature is not intended to allow you to observe intermittent events, because the screen does not update automatically as the PLC runs. An UPDATE button must be selected each time you want to capture a new set of data from the CPU to display.

Diagnostic data will not appear in the format(s) described above for SZL lists and module diagnostic data; instead, it will be arranged into the following dialog pages, depending on the type of module you have selected from the project window:

1. The **DIAGNOSTICS BUFFER** page shows a time, date, and text message for every entry in the diagnostics buffer. Cursor to a diagnostic event and (optionally) select HELP ON EVENT to see more detailed descriptions. Select OPEN BLOCK to jump immediately to a screen showing the logic block and instruction that was executing at the time of the diagnostic event.

2. **DIAGNOSTIC INTERRUPTS** shows a module's diagnostic data, on two pages, corresponding to record 0 and record 1.

3. The **STACKS** page can be displayed *only* if the PLC is in stop mode, and is useful in determining why a PLC suddenly went into stop mode. This screen shows the contents of:

(a) The **BSTACK,** which contains a list of the programs that are paused due to being interrupted (including the program that was interrupted by the stop) or are paused because they called another program. The BSTACK also indicates which part of each paused program is to be executed when it resumes.

(b) The **ISTACK,** which contains the information a PLC saves so that it can resume each program that the BSTACK says has been interrupted. The information includes the contents of the CPU registers at the time of the interrupt, and the priority class of the program that was interrupted. OPEN BLOCK can be selected to jump to a screen showing information on each interrupted program, including the program that was interrupted by the change to stop mode.

(c) The **LSTACK** shows the contents of the local stack (temporarily assigned) memory for each of the programs listed in the BSTACK. Cursor to a BSTACK entry and select LSTACK.

4. **DP SLAVE DIAGNOSTICS** shows a DP slave's diagnostic data, on subpages corresponding to EN 50 170's specifications for standard data and module-specific data. Diagnostic data for submodules of the DP slave (if any) is also available.

5. **SCAN CYCLE TIMES** shows the shortest, longest, and most recent scan cycle times and the time configured as the maximum scan time.

6. The **COMMUNICATIONS** page shows baud rates, the allowable number of communications connections this CPU can have, and the number of connections that are in use.

7. **GENERAL INFORMATION** describing the module type, characteristics, location, and its operating status.

8. **USER MEMORY UTILIZATION** shows how much of the CPU module's memory is used. (You can select COMPRESS to try to generate more free memory.)

9. **TIME INFORMATION** shows you the current time and date (the clock is set during configuration), and the elapsed operating hours for this CPU.

10. **PERFORMANCE DATA** shows the ranges of memory addresses available in each area of the CPU's memory. (Check this before trying to download a program that may require too much memory.) Select BLOCKS to see a list of the organization blocks this CPU uses, the allowable number and size for user-written program blocks, and a list of the SFC and SFBs available.

STEP 7 Logical Error Debugging Tools

Instructions and System Functions for Debugging Other than the block end instructions: BEC and BEU, Siemens does not offer any special instructions for debugging purposes. (BEC and BEU are described in the section describing STEP 5 instructions for program debugging.)

Several system functions (SFCs) perform operations that can be used for debugging programs. Some have already been discussed or mentioned in this or other chapters:

SFC 6, "RD_SINFO"	Read OB start information.
SFC 13, "DPNRM_DG"	Read diagnostic data of a DP slave.
SFC 17, "ALARM_SQ"	Generate acknowledgeable block-related message.
SFC 18, "ALARM_S"	Generate permanent acknowledgeable block-related message.

SFC 19, "ALARM_SC" Query acknowledgment of block-related message.
SFC 31, "QRY_TINT" Query time-of-day interrupt.
SFC 34, "QRY_DINT" Query time-delay interrupt.
SFC 46, "STP" Change the CPU to stop.
SFC 51, "RDSYSST" Read a system status list or partial list.
SFC 52, "WR_USMSG" Write user-defined message to diagnostic buffer.
SFC 59, "RD_REC" Read a data record.

Some additional SFCs, which are not discussed in this chapter, are useful in determining, correcting, or avoiding faults. These include:

SFC 38, "READ_ERR" Read error register.
SFC 43, "RE_TRIGR" Retrigger cycle time monitoring.
SFC 44, "REPL_VAL" Transfer substitute value to accumulator 1.
SFC 47, "WAIT" Delay execution of the user-program.

Programming Unit Features for Debugging CPU reference data can be generated by a programming unit to describe programs in a S7 CPU module. The reference data can include lists of the memories addressed in programs, the memories still available, and the addresses for which symbolic names don't exist. You can generate cross-reference lists describing where each memory is used and lists of the calls to other programs, and you can generate descriptions of program structure.

During program editing in STEP 7, the editor is usually in incremental mode by default, so it will highlight the detectable programming errors (such as the use of undefined symbols).

For monitoring of a program, the programmer can set up and save multiple variable tables, then retrieve the one needed to monitor a specific process. Each table can have any combination of variables, specified as absolute addresses or as symbols. The triggering of data capturing for a variable table can be set up to happen once only (the laboratory test environment), or once per scan (the process test environment), and the trigger point for capturing data can be either at the start of a scan cycle, at its end, or just when the PLC goes into stop mode. You can use the programming unit's debug menu to set breakpoints. The PLC will go into halt mode until you select CONTINUE, each time it reaches the breakpoint.

Variable values can be modified while the PLC is in run or stop mode, using a variable table. While the PLC is in stop mode VARIABLE–ENABLE PERIPHERAL OUTPUT (**PQ**) can be used to edit output values, then VARIABLE–ACTIVATE PERIPHERAL OUTPUT (**PQ**) can be selected to write the new output values.

CQMI

TROUBLESHOOTING THE OMRON CQM1

If a CQM1 faults, error codes must be cleared before the PLC will go into run mode again, even after the cause of the fault has been fixed.

CQM1 Configuration for Troubleshooting

You use DM 6655, as described in Chapter 10, to configure the CQM1 to store fatal and non-fatal error descriptions in an error log area of memory between DM 6569 and DM 6599. The

CQM1 can store descriptions of the most recent 10 errors (the default), or only the first 10, or not store any. DM 6655 can also be used to disable the CQM1 from setting SR 25309 each time the scan cycle takes longer than 100 ms[3] or to disable the CQM1 from setting SR 25308 when a low battery charge is detected.

CQM1 Fatal Errors and Nonfatal Errors

Reasons for fatal errors include things such as power failures, detection of faulty memory, errors in communicating with an I/O module, no END(01) instruction in a program, or scan time exceeding the cycle monitor time setting. Execution of the SEVERE FAILURE ALARM instruction: FALS(07) in a user-program will also trigger a fatal error. FALS(07) was covered in Chapter 11. In the event of a fatal error, the CQM1 immediately stops after storing the error information, turning all outputs off, and turns on an ERR/ALM indicator on the CPU module. The operator can examine the error information in memory, repair the problem and clear the codes,[4] then switch the PLC back into run mode.

Nonfatal errors include errors in transferring data between the CPU and a memory module, problems with the PLC setup data in the DM area of memory, cycle time exceeding 100 ms, a bad backup battery, or execution of a FAILURE ALARM instruction: FAL(06). FAL(06) was covered in Chapter 11. Problems encountered during use of a CPU port (RS-232C, peripheral, pulse, encoder) are also considered nonfatal errors. In the event of a nonfatal error, the CQM1 does not stop running after storing the error information. The ERR/ALM indicator on the CPU module starts flashing, but the scan cycle continues executing.

When a fatal or nonfatal error occurs, the CQM1 stores:

1. A two-digit BCD FAL error code or FALS error code[5] into the low byte of SR 253 (which OMRON calls the *FAL area of memory*). The CQM1 actually saves the three most recent FAL/FALS error codes, but only the most recent code is available in SR 253's low byte.

2. A description of the error in the error log (affected by configuration word DM 6655). The first memory in the error log (DM 6569) contains a pointer to where the next entry must be made into the log and is followed by a three-word record for each of up to 10 errors (in DM 6570 to DM 6599). Each record contains the FAL error code or the FALS error code and the time and day of occurrence.

Other status bits in the SR and AR areas of memory may also be set to further describe nonfatal errors, and the user-program should include instructions to detect and correct these problems. In addition, AR 26 contains the time of the longest scan cycle since the PLC was put into run mode, and AR 27 contains the time of execution of the current scan cycle so far.

The FAL and FALS error codes are described in Figures 15.7 and 15.8, and the other memory areas that may be affected are also described.

---

[3] SR 25309, the CYCLE TIME OVERRUN FLAG, is different from AR 2405, the LONG CYCLE TIME FLAG. AR 2405 is set if a scan cycle exceeds the minimum scan cycle setting in DM 6619.

[4] Executing FAL(06) with FAL number 00 deletes one of the three FAL error codes stored in the PLC's memory each time it is executed, and the next oldest is then moved into the low byte of SR 253. Turning bit SR 25214 on from a peripheral device clears the entire error log (bit SR 25214 then resets itself). Status bits in the SR and AR memory areas refresh their own state each scan cycle.

[5] FAL stands for FAILURE ALARM (a nonfatal error); FALS stands for FAILURE ALARM SEVERE (a fatal error).

Message	FALS No.	Meaning and appropriate response
Power interruption (no message)	None	Power has been interrupted for at least 10 ms. Check power supply voltage and power lines. Try to power-up again.
MEMORY ERR	F1	AR 1611 ON: A checksum error has occurred in the PLC Setup (DM 6600 to DM 6655). Initialize all of the PLC Setup and reinput.
		AR 1612 ON: A checksum error has occurred in the program, indicating an incorrect instruction. Check the program and correct any errors detected.
		AR 1613 ON: A checksum error has occurred in an expansion instruction's data. Initialize all of the expansion instruction settings and reinput.
		AR 1614 ON: Memory Cassette was installed or removed with the power on. Turn the power off, install the Memory Cassette, and turn the power on again.
		AR 1615 ON: The Memory Cassette contents could not be read at start-up. Check flags AR 1412 to AR 1415 to determine the problem, correct it, and turn on the power again.
NO END INST	F0	END(01) is not written anywhere in program. Write END(01) at the final address of the program.
I/O BUS ERR	C0	An error has occurred during data transfer between the CPU and an I/O Module. Determine the location of the problem using flags AR 2408 to AR 2415, turn the power off, check for loose I/O Modules or end covers, and turn on the power again.
I/O UNIT OVER	E1	The number of I/O words on the installed I/O Modules exceeds the maximum. Turn off the power, rearrange the system to reduce the number of I/O words, and turn on the power again.
SYS FAIL FALS** (see note)	01 to 99	An FALS(07) instruction has been executed in the program. Check the FALS number to determine the conditions that would cause execution, correct the cause, and clear the error.
	9F	The cycle time has exceeded the FALS 9F Cycle Time Monitoring Time (DM 6618). Check the cycle time and adjust the Cycle Time Monitoring Time if necessary.

FIGURE 15.7
CQM1 fatal error codes.

Notice that there are no FAL error codes for CPU port problems, but bits in AR 24 are set when communication problems occur at the RS-232C or peripheral port, and bits in AR 04 are set if pulse output ports or high-speed input counting ports exist and are not working (see your manual). If a CPU port problem occurs (always considered nonfatal), the indicator light at the port connector will stop flashing.

Data errors can be treated as nonfatal errors for which no error codes exist. Error bits are affected by instructions that manipulate data values. The user-program should monitor these arithmetic error bits following every data manipulation instruction that can result in an incorrect resultant value:

SR 25404 turns on if the result overflows the allowable positive number range for the binary number system in use, which means that the result that will be stored in the destination address is much smaller than the correct answer should be.

SR 25405 turns on if the result underflows the lowest allowable number for the range in the binary number system in use, which means that the result that will be stored in the destination address is much larger than the correct answer should be.

SR 25504 turns on if there was a carry bit generated (e.g., a 17-bit result is generated when using an instruction for manipulating 16-bit numbers), which means the result was too big for the destination address.

Message	FAL No.	Meaning and appropriate response
SYS FAIL FAL**	01 to 99	An FAL(06) instruction has been executed in the program. Check the FAL number to determine conditions that would cause execution, correct the cause, and clear the error.
	9D	An error has occurred during data transmission between the CPU and Memory Cassette. Check the status of flags AR 1412 to AR 1415, and correct as directed. AR 1412 ON: Switch to PROGRAM Mode, clear the error, and transfer again. AR 1413 ON: The transfer destination is write-protected. If the PLC is the destination, turn off the power to the PLC, be sure that pin 1 of the CPU's DIP switch is OFF, clear the error, and transfer again. If an EEPROM Memory Cassette is the destination, check whether the power supply is on, clear the error, and transfer again. If an EPROM Memory Cassette is the destination, change to a writeable Memory Cassette. AR 1414 ON: The destination has insufficient capacity. Check the source's program size in AR 15 and consider using a different CPU or Memory Cassette. AR 1415 ON: There is no program in the Memory Cassette or the program contains errors. Check the Memory Cassette.
	9C	An error has occurred in the pulse I/O function or in the absolute-type encoder interface function. Check the contents of AR 0408 to AR 0415 (two digits BCD), and correct as directed. (This error code applies only to CQM1-CPU43-E and CQM1-CPU44-E CPUs.) 01, 02: An error has occurred in the hardware. Turn the power off, and then power up again. If the error persists, replace the CPU. 03: The PLC Setup (DM 6611, DM 6612, DM 6643, DM 6644) settings are incorrect. Correct the settings. 04: CQM1 operation was interrupted during pulse output. Check to see whether the Module receiving the pulse output was affected.
SYS FAIL FAL**	9B	An error has been detected in the PLC Setup. Check flags AR 2400 to AR 2402, and correct as directed. AR 2400 ON: An incorrect setting was detected in the PLC Setup (DM 6600 to DM 6614) when power was turned on. Correct the settings in PROGRAM Mode and turn on the power again. AR 2401 ON: An incorrect setting was detected in the PLC Setup (DM 6615 to DM 6644) when switching to RUN Mode. Correct the settings in PROGRAM Mode and switch to RUN Mode again. AR 2402 ON: An incorrect setting was detected in the PLC Setup (DM 6645 to DM 6655) during operation. Correct the settings and clear the error.

FIGURE 15.8
CQM1 nonfatal error codes.

SR 25503 turns on if another error was detected. It often means that indirect addressing is incorrect or that a data value was not in the binary form a math instruction was intended to manipulate.

CQM1 Logical Error Debugging Tools

Several instructions exist for use during debugging of a user-program.

END(01) To terminate every scan cycle earlier than normal but to continue scanning, insert an END (01) instruction. This will allow you to check that the part of the program ahead of the END (01) works correctly.

FAILURE ALARM, FAL(06) and SEVERE FAILURE ALARM, FALS(07) To make the PLC stop its scan cycle at a specific point and enter fault mode, insert a **SEVERE FAILURE, FALS (07),** instruction. FALS(07) was described in the fault routine section of Chapter 11. A program can have several conditional FALS (07) instructions,[6] each with its own programmer-assigned FALS error code, so that the programmer can determine which FALS(07) instruction caused the program to stop, and when the stoppage occurred. The CPU state, including its memory contents, will be preserved in the state they were in when the FALS(07) conditions were evaluated as true.

One or more **FAILURE ALARM, FAL(06),** instructions can be used to store error messages with user-assigned FAL error codes and time of occurrence into the error log every time a FAL(06) instruction's execution condition is evaluated as true, without making the PLC stop executing its normal scan cycle. FAL(06) was described in Chapter 11.

MESSAGE DISPLAY, MSG(46) The MESSAGE DISPLAY, **MSG(46),** instruction can be used similarly to how the FAL(06) instruction is used. MSG(46) sends a message to the programming unit when the MSG(46) instruction's conditions are evaluated as true. The user can create any message of up to 16 extended-ASCII characters and store it in memory as ASCII codes, each entered as a two-digit BCD number, in eight consecutive memory locations. The first ASCII code goes into the high byte in the lowest memory address. Messages of fewer than 16 characters can be entered if the ASCII code 0D is entered following the last character. One parameter is included with the MSG(46) instruction: the **FM** (FIRST MESSAGE WORD) parameter, which is the address where the ASCII code sequence starts. At the time of this writing, programming units could only hold up to three messages, so OMROM has devised a message-priority system: Messages stored in LM memory are the highest priority and can bump lower-priority messages from the programming unit's memory. The rest of the priority sequence is: LM > IR > HR > AR > TC > DM. Within a single memory area, messages from lower-numbered addresses have higher priority than those from higher addresses.

FAILURE POINT DETECT, FPD(−) A FAILURE POINT DETECT, **FPD(−),** instruction can be used to determine which part of the Boolean logic statement in the following rung fails to become true within an acceptable time. The time is entered in tenths of seconds as the **MONITORING TIME (T)** parameter. After FPD(−) starts executing and the time expires without the logical statement on the following rung going true, the CQM1 will generate a nonfatal error and will examine each Boolean logic instruction on the following rung, in the same order that the CQM1 would normally evaluate the statement. The CQM1 will find the address that the first false statement examines (even if several are false).

FPD(−)'s timer doesn't have a DONE bit, but the CARRY flag (SR 25504) is set after the timer times out.

The CQM1 can automatically generate a time for T. See your manuals for details.

The **CONTROL (C)** parameter must be entered as a constant. The low byte of C must con-

[6] With only a few exceptions, OMRON output instructions *must* be programmed with conditions, although the ALLAYS ON bit, SR 25313, can be used to effectively make the output instruction execute unconditionally.

tain a unique FAL number that will be written to the error buffer and to the FAL area of memory. If the highest bit of the C parameter is a:

0 The CQM1 will write a 16-bit code into the first memory following the address entered as the **DESTINATION (D)** parameter. The code identifies the address examined by the false Boolean statement. The code identifies the bit's memory area and bit address (see your manual). Only the two high bits of the word *at* the D address are used: bit 15 is set if an address code has been written into the next word, and bit 14 is set if the false Boolean logic instruction is an EXAMINE OFF or NORMALLY CLOSED instruction.

1 The CQM1 will write the false Boolean statement's bit address in BCD ASCII codes to the three addresses following the memory address entered as the **DESTINATION (D)** parameter, and will write a code indicating normally open or normally closed to the next word. The CQM1 also writes to the high 2 bits of the word at the D address, as described above. FPD(−) then sends nine words, starting with the contents of the D address, via the peripheral port. The programmer (or the program) can enter data for a message in the last four of the nine addresses. The entire nine-address area is structured as follows:

- Addresses controlled by the FPD(−) instruction:

D	High two bits controlled by CQM1 as described above.
D+1	High byte: The ASCII code for a space ("20"). Low byte: MSD of the bit address.
D+2	Next two characters in the bit address.
D+3	The low characters in the bit address.
D+4:	High byte: The ASCII code for a "−" ("2D") Low byte: ▪ ASCII code for 0 ("30") if instruction was a normally open instruction. ▪ ASCII code for 1 ("31") if instruction was a normally closed instruction.

- Addresses controlled by programmer or program:

D+5 to D+8	can contain up to eight extended-ASCII codes, or fewer if the last ASCII code is 0D.

Figure 15.9 shows how an FPD(−) instruction could be used. IR 00204 triggers the start of the timer in the FPD(−) instruction. A constant of #0030 is entered as the T parameter, so if 3 s pass without the logical statement on the following rung going true, the CQM1 will generate a nonfatal error with the FAL number 55 (from the low byte of the constant #0055 entered as a control parameter). The CQM1 will then examine the Boolean logic instructions on the following rung and find the first false statement (even if more are false). If the first false instruction happens to be the EXAMINE OFF instruction with address 00214, and since the most significant bit of the C parameter (#0010) is a 0, the FPD(−) instruction will write the following two words to addresses DM 0000 and DM 0001:

▪ DM 0000:	1100 0000 0000 0000
bit 15	set because an address follows in DM 0001
bit 14	set because the instruction was an EXAMINE OFF, also called a NORMALLY CLOSED
other bits	not used

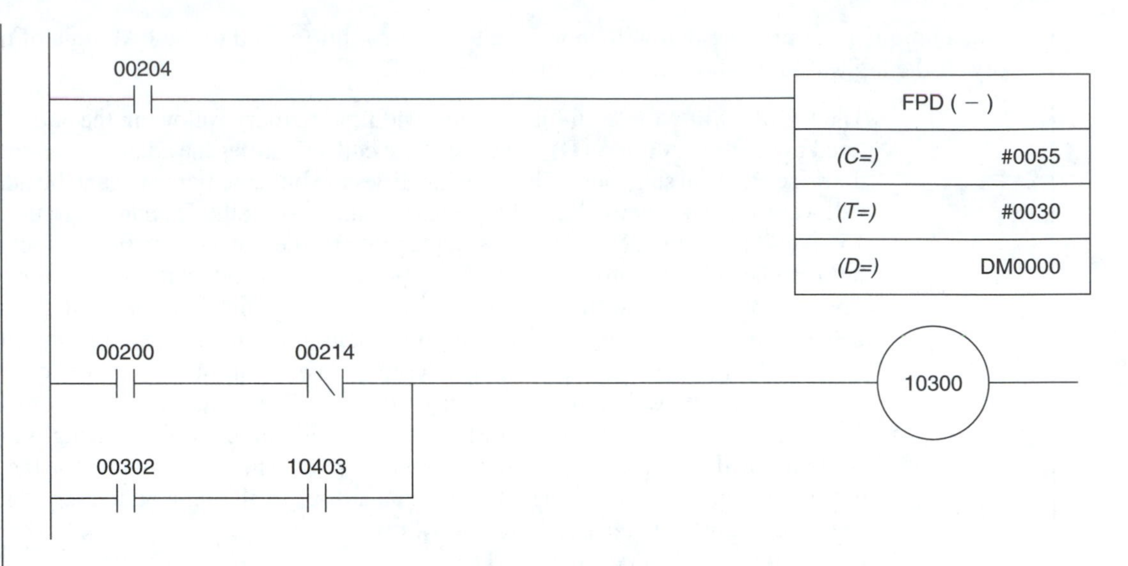

FIGURE 15.9
OMRON FAILURE POINT DETECT instruction in a program.

▪ DM 0001:	1000 0000 0010 1110
1000	code for IR address area (see manual)
0000 0010	binary for word address: 002
1110	binary for bit address: 14

FRAME CHECKSUM, FCS(−) The FRAME CHECKSUM, **FCS(−)**, instruction can be used to generate a checksum value from a series of data words.[7] It converts each nibble (4 bits) of the resulting checksum into a hexadecimal character, then converts each hex character into a 1-byte BCD ASCII code for that character. For example:

	1100	1111
Exclusive-ORed with:	1010	0101
binary result:	0110	1010
in hex:	6	A
ASCII coded:	36	41

The FCS(−) instruction can be executed again to generate a checksum number using the same data. If the resulting checksum is different, one or more of the values has changed. Checksums are typically used where data is most likely to be corrupted, such as when it is transmitted from one PLC to another. The checksum generated before transmission will be sent with the data, and the receiver should generate a new checksum to compare with the first.

[7] The checksum is generated by Exclusive ORing the first binary value with the second, then Exclusive ORing the result with the next value, until all values are included.

FCS(−)'s first parameter is a CONTROL (**C**) parameter. If bit 13 of C is set, the checksum is generated using bytes instead of words. If bit 12 is set, the least significant byte is used before the most significant byte in calculating the checksum. The BCD number in the low 3 bytes of the control parameter (000 to 999) indicates how many bytes or words to include in calculating the checksum. The second parameter is the FIRST WORD IN RANGE (**R_1**) address, which indicates the start of the data. The third parameter, DESTINATION ADDRESS (**D**), indicates where to write the two or four ASCII codes for the 8- or 16-bit result.

The CQM1 programming unit can generate a time chart to help in debugging a user-program. A CQM1 time chart histogram can simultaneously show the state of up to 16 bits and/or three data words over time. To generate a time chart histogram, use your programming unit to connect ONLINE to the CQM1, access the UTILITY menu and select TIME CHART, then EXECUTE. Enter a sampling interval (minimum is 0.3 s even if you enter 0.0) and select the bit(s) and word(s) to monitor. You must also specify a bit address and an edge type (rising or falling) on which to start the sampling, and you can enter a delay of up to 250 sampling intervals. You can enter a negative delay to record up to 249 samples before the trigger! When ready to start, select CONFIRM and answer Y.

Some OMRON PLCs are more powerful than the CQM1 and can be programmed to record bit and/or word values in their own trace memory at a higher frequency than is possible using a time chart.

SUMMARY

Modern PLCs store extensive status information, which can be accessed by a user to help in troubleshooting a PLC that is not behaving as it is expected to. Some PLCs can be programmed with fault-handling routines that execute automatically in the event of a major error before the PLC halts running. Nonfatal errors and math errors do not usually cause a PLC to fault, but user-programs can be written to detect these types of errors. Don't change a PLC's program or configuration to correct a problem until you are sure the problem can't be fixed by changing something outside the PLC!

QUESTIONS

1. What should you do before even considering changing a PLC's program or configuration if the system being controlled starts to behave poorly?
2. What fault-identification features does the PLC you are learning to use offer that will tell you why the PLC has faulted?
3. Does your PLC tell you what program was executing at the time that a fatal fault occurred? How do you access this data?
4. Does your PLC try automatically to execute a fault-recovery routine? Can you configure your PLC to do so?
5. What kinds of faults are recorded in your PLC but do not cause the PLC to stop?
6. What specific instructions are available to help you find logical errors in the PLC that you are learning to program?

16

THE FUTURE: WITHER THE PLC?

OBJECTIVES

In this chapter you are introduced to:

- Traits that have kept the PLC popular.
- Developments in open standards for integrating PLCs with other production equipment, from simple sensor-actuator nets to sophisticated fieldbus networks.
- Centralizing of control using supervisory control and data acquisition (SCADA) software.
- Soft logic: Will it replace the PLC or is it just another form of PLC?
- Simulation to test a PLC program without sensors or actuators.
- Reflective memory to allow fast exchange of data between controllers.
- Motion controllers and process controllers built into PLCs.

This chapter's title is a pun. "*Whither* the PLC?" would ask where the PLC is going, whereas "*Wither* the PLC?" asks if PLCs will shrink in importance. Predicting the future is risky. In 1984, I told a set of interviewers that I believed PLCs would disappear (wither away) and be replaced by personal computers. The interviewers must have felt I was a person of great foresight, because I got the job. Since then I have changed my mind, obviously, and none of the interviewers still work for the company that hired me.

It is still popular to predict that PLCs are going to be replaced by personal computers, but PLC sales continue to increase because PLC manufacturers continue to improve their product. The PLC of today is significantly different from the PLC of a decade ago, and the PLC of the next decade will be different from the one of today. Despite the changes, though, today's PLCs still have two important things in common with yesterday's PLCs: They still deliver *deterministic* control response, and they are still highly *dependable*. They are deterministic because their underlying scan cycle makes it relatively easy to determine how much delay there will be in responding to a sensed condition. They are dependable because they are designed to withstand unpleasant environmental conditions and to be connected to other pieces of equipment. (How deterministic and dependable are personal computers?)

TOMORROW'S PLC

What will the PLC of tomorrow look like? What will it be able to do that today's PLC doesn't do? One thing that you can be sure of is that some things about PLCs will not change radically: PLCs will still be *deterministic* and they will still be *dependable.* Another thing you can be sure of is that the changes will be those required by PLC users, not changes that PLC suppliers make just to sell new models. Industrial users are very cautious. If the PLC they are using is doing its job satisfactorily, they won't risk causing a production problem just for the sake of buying a new model with colorful advertising. If the industrial user becomes convinced that there is a better way of doing a job than the current way, however, they quickly seek out PLCs that can work in the new system. It is therefore a little easier to see into the future of PLCs than it would be for some other products. We can look at manufacturing trends without having to worry too much about fashion trends.

What trends do we see today in the world of production? Increased *flexibility,* greater *visibility* of the manufacturing process, better *control* from remote locations, and (of course) *reduced cost.* Reduced production cost is achieved through reduced inventory, higher productivity, better quality control, and better product design. All these trends require better and easier communication. From the point of view of a computer manufacturer (PLC manufacturers included), it means that computers must be able to exchange data, and they must have programs that can make use of data from other computers. Almost all the trends in this chapter are trends toward greater interconnectivity between PLCs and other computers, and many of the trends are related. PLC manufacturers are already making these improvements. Personal computer manufacturers are reacting more slowly, so it appears that the PLC is not in any immediate danger of "withering" away.

FIELDBUS AND SENSOR-ACTUATOR NETS

As controllers have become more powerful, they have become able to control production activity in increasingly large areas. This means that it is no longer practical to separately wire each sensor and actuator to the PLC controller. Remote I/O systems consisting of racks of I/O modules, with each rack controlled by a communication adapter that exchanges data with the single central PLC controller, is one solution, but the remote I/O solution still leaves us with racks wired to undesirably huge numbers of sensors and actuators.

Sensor/actuator nets, sometimes called just *sensor nets,* are becoming increasingly common. In a sensor/actuator net, small I/O modules are interconnected to each other and to a network control module via serial communication cable. One cable with as few as three conductors is all that you need between the widely separated I/O modules. Each I/O module can be connected to a handful of digital sensors and/or actuators, and each I/O module has its own node address. The network controller writes small packets of data to each I/O module to control its actuators, and reads small packets from I/O modules to read sensors. It is usually necessary to program individual I/O modules at setup (sometimes just by setting DIP switches) for such purposes as assigning each a unique node address. A sensor/actuator network controller can often be inserted directly into a PLC's central I/O rack, where it presents I/O data to the CPU as if it were a standard I/O module, so that the user-program doesn't have to be written differently to use remote sensor/actuator modules.

Sensor/actuator network interface circuitry is cheap enough that it is now economical to build interfaces into individual sensors or actuators. The problem is that a sensor or actuator with a built-in interface circuit for one type of actuator/sensor network can't be used in any other type of actuator/sensor network. Users have, therefore, been pushing for the adoption of an open sensor/actuator network standard. PLC manufacturers have been generally supportive, although proprietary sensor/actuator networks are still very popular. Two of the more successful open sensor/actuator network standards are DeviceNet, which Allen-Bradley designed based on the earlier open controller area network (CAN) standard and then "donated" to the general community, and As-i (actuator/sensor interface), which Siemens is associated with. Needless to say, even if there were only two types of sensor/actuator networks, that would still be one too many, and none of the manufacturers are anxious to support a standard that they think will give their competitor an edge.

Most sensor/actuator networks are only good at interfacing digital sensors and actuators to controllers, because the data packet sizes are generally limited to only a few bits per node. Even if an analog I/O value were digitized as a 10-bit number, the 10-bit number would be too big for most sensor/actuator nets to handle. Nonetheless, there are a few sensor/actuator networks that can handle larger data packets, or have variable data packet size, so they *can* be used to interface remote analog devices to a controller.

If the user is going to pay for an analog sensor or actuator with a built-in communications interface chip and a chip to convert between digital and analog, it makes economic sense to add a few more features on the chip, such as the ability to perform scaling and the ability to perform some self-testing. If you are going to connect the sensor or actuator into a network, why not use the network to write configuration data to the chip, and use the network to read its status? For that matter, why not build ROM into the chip to contain data describing the "intelligent" sensor or actuator device, so that the controller can read the ROM chip to see what type of sensor or actuator it is. If the controller can query the connected devices, the sensor/actuator network system can be designed to be self-configuring (plug-n-play), and you could even design the system to allow sensors and actuators to be hot-swapped (replaced without having to turn power off). Somewhere on the road from the earlier simple proprietary networks that could only read and write remote digital I/O to self-configuring networks with complex operating systems, the name *sensor/actuator network* has been replaced by the name *fieldbus.*

Several years ago, international standards organizations recognized that an open standard should be developed for transmission of digitized analog values for control purposes before too many proprietary systems were developed. After a good deal of infighting and revisions of the goal(s), a single group, comprised of industry representatives, has taken on responsibility to develop a single open fieldbus standard. This group is called the Fieldbus Foundation (FF), and the developing standard is a joint ISA/IEC standard identified as SP 50. SP 50 defines four layers of communication network requirements and defines what a PLC (in a fifth layer) has to do to use the fieldbus. The four-layer model contains three layers based on the ISO's OSI (open system's interconnect) seven-layer model for computer communications, plus a new user layer standard. SP 50 is actually a set of standards, so that a user can select SP 50–compliant equipment for the level of functionality required. Figure 16.1 shows the structure of a Fieldbus Foundation type of fieldbus.

Two computers are shown connected to the fieldbus in Figure 16.1. One of them is a PLC and the other is just a sensor! Sensors and actuators will have to contain embedded controller

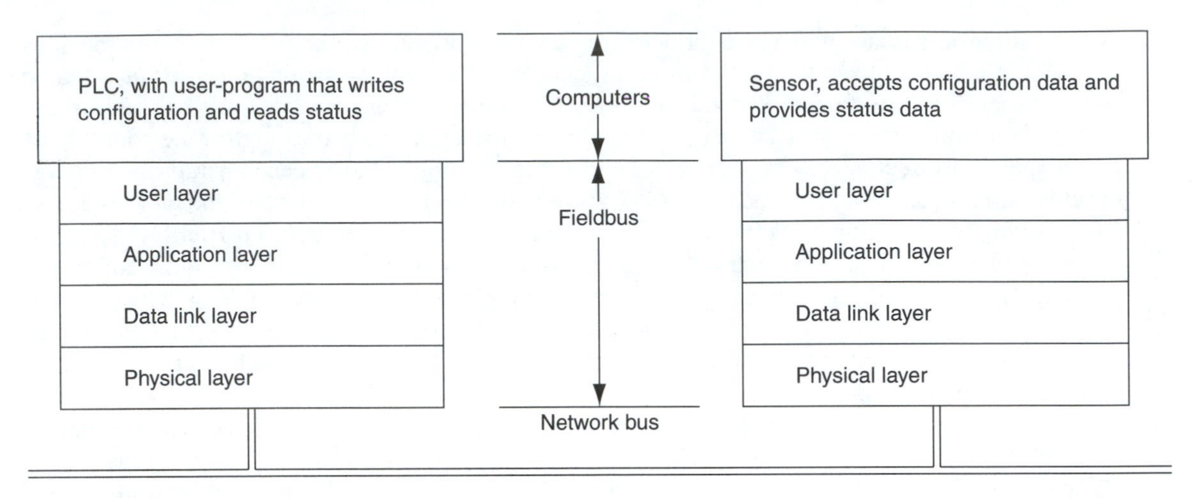

FIGURE 16.1
Two computerized nodes on an SP 50 fieldbus.

chips so that they will be smart enough to be able to communicate, use configuration data, report status, and more important, to execute control programs specific to the type of device. (SP 50 interface modules will also be available, so that dumb sensors can still be connected.) A low-level smart sensor might only have to accept scaling values from the PLC and calculate digitized output values for the PLC to read. A high-level smart actuator for motor control would have to accept a position setpoint from the PLC and might even contain its own servocontrol program (e.g., a PID function block). It would read process variable values from a separate sensor (e.g., a position sensor) on the fieldbus and would drive the motor to the required position using control parameters provided by the PLC or by another controller on the fieldbus.

You will be pleased to hear that the Fieldbus Foundation says that a user shouldn't have to know what the application and data link layers do. A little knowledge of the two versions (so far) of physical layers is required, however, because the user has to interconnect the fieldbus interfaces via network bus conductors. The H1 type of physical layer provides a level of activity that is adequate to replace most sensor/actuator nets, and the H2 type of physical layer provides speed and data-handling capabilities adequate for all the functionality described above for a complete fieldbus. It is the user layer that a PLC programmer must know the most about. The user layer will respond to user-program instructions by transmitting or requesting data via the fieldbus and will perform some fieldbus-required operations for the system automatically.

Some *user layer services* that a fieldbus system will perform (with the help of the application layer) include:

- Automatic detection and identification of devices on the fieldbus, at startup and as the system operates. Each device will contain a description of itself (installed by the manufacturer) in device description language (DDL) format. Fieldbus Foundation specifies a limited set of virtual device formats to limit the variability of device descriptions.
- Automatic address assignment and maintenance of an address directory, so that a user-program can be written using symbolic names.

- Cyclic reading and writing of inputs and outputs (as in a PLC) to update I/O images.
- Exchange of high-priority messages without waiting for cyclic refresh cycles. When a user-program (the application layer) executes an IEC 1131-5 communication function block, the fieldbus user layer will respond by performing the required data exchange. Some communications will require the fieldbus to interrupt another controller so that it can respond to an alarm message.
- Remote control, via execution of standard control function blocks in the user-program.
- Synchronization of all computer clocks at periodic intervals.

Will the Fieldbus Foundation's SP 50 standard be adopted universally? (Will you lose your job if you guess wrong?) SP 50 should do well, because it became available before all but a few proprietary standards and because industry desperately needs an open fieldbus standard. Some proprietary networks that could challenge SP 50 by acquiring a large installed base before SP 50 include Allen-Bradley's ControlNet, the Profibus networks supported by Siemens, and a few networks from companies not as large. Sensor/actuator nets already have a large-enough installed base that users may prefer simply to upgrade them as improvements are offered, instead of replacing them with an SP 50–compliant system.

SCADA SYSTEMS

Supervisory control and data acquisition (SCADA) systems allow the user to display data from controllers (such as PLCs) on a personal computer and to use the personal computer to change data values in the memory of the controller, as you might do using a programming unit. SCADA systems are more than just programming units, however, in a few ways:

- The SCADA software running on the personal computer can collect data from several controllers simultaneously.
- SCADA software can perform analysis of the acquired data.
- SCADA software offers graphics capabilities, so that the user can assemble operator display screens containing recognizable features (e.g., pump and tank graphics) and which make status interpretation easy (e.g., the tank can display its fill level and the pump can change color when it is on). If combined with a touch-sensitive screen and control graphics (e.g., a switch graphic), the operator can control the process without needing a keyboard or a mouse.

One data-analysis operation that SCADA software often performs using acquired data is to plot it against a time base. The graph, displayed on the operator screen, gives an operator a sense of the trend in a system's output. Further SCADA data manipulation sometimes includes SPC analysis of the graphed data, complete with triggering of operator alarms to warn of a process that is outside its control limits.

As networking of controllers expands, SCADA software will be required increasingly to allow supervisory personnel to monitor systems. Currently, the best SCADA packages are offered by third-party vendors (not the equipment supplier and not the equipment purchaser), but by the time you read this, PLC manufacturers will probably have begun a major effort to reclaim the SCADA market.

SOFT LOGIC

If a personal computer can contain software to simulate the scan cycle of a PLC, and if the computer can run a standard PLC program and perform I/O updates of remote I/O racks through an interface card, you don't need a PLC CPU module in your PLC system. You would have a *soft logic processor.* Soft logic software and interfaces for personal computers are the latest justification for the continued predictions that the PLC will die.

The growth trends in PLC technology make a strong case for the success of soft logic. Let's follow the growth cycle, starting with a typical small manufacturing operation:

- There are perhaps a dozen PLCs on the production floor. A programming unit consisting of a personal computer with PLC programming software and (at least) a serial interface cable is used. The personal computer isn't an industrially hardened PC, because it would have cost more, so the owners are worried about damaging the personal computer and look for a more dependable system.
- A slightly more sophisticated system can be assembled using the built-in local area network capabilities of the PLCs. An interface card can be added to the personal computer so that it can be connected into the PLC network. Now the personal computer can connect to any of the PLCs without leaving the engineering office.
- Since the personal computer is permanently connected to the PLC network now, it makes sense now to add some SCADA software so it can be used to monitor the entire controlled system. Some network analysis software could also be added.
- For improved integration of the entire system, the separate PLC controllers can be replaced with a single central controller rack and 11 remote I/O racks (or with remote I/O connected via an actuator/sensor net or a fieldbus). The wiring requirements are about the same, but you only have to write one program for the central PLC. You would probably need to buy a larger CPU for the central PLC to get the speed and memory that you need, but you gain the flexibility to modify the remote nodes more freely.
- You now realize that you actually have two central computers: the PLC's CPU module and the personal computer that is used to run the programming software, the SCADA software, and the network analysis software. The personal computer is probably faster and has more memory than the more expensive PLC CPU module. Why not run soft logic software in the personal computer instead of upgrading the CPU modules for the PLC system? It would allow the other software running in the personal computer to have faster and more complete access to the status of the control system. Even more important, it would mean that the PLC's memory, I/O capability, and communication channels capacity could be expanded easily by adding standard personal computer components.

Well, actually, there are some arguments for keeping a real PLC instead of using soft logic control. The arguments, predictably, have to do with dependability and control interval determinism. Suppliers of soft logic control software and interface hardware claim that these problems have been solved. Here are some arguments pro and con for soft logic. You decide for yourself:

Con: Personal computer motherboards and power supplies aren't good enough for the production environment. They fail too frequently.

Pro: Motherboards are no longer manufactured by small electronics suppliers. One supplier (Intel) has already captured more than half the worldwide market. Quality is better and more consistent. When was the last time you saw a failed motherboard (not counting connectors)? The next generation of motherboards are going to be sealed modules, which even PLC manufacturers will have to use! Power supplies are still a problem, but if you pay a little more for a better computer you will get a good one. You should have power-conditioning hardware in your plant anyway, so power problems such as spikes shouldn't get to a power supply, let alone through it.

Con: Personal computer operating systems aren't stable enough. They tend to hang, and hang all the software that is running at the same time, requiring the computer to be restarted.

Pro: Most reviewers claim that Windows operating systems from NT are very stable and can be trusted.

For those who refuse to trust an operating system, some soft logic systems run the PLC simulation software right on the interface card, so it will continue to operate even if the computer does hang, if the power supply isn't interrupted. In effect, you have a PLC CPU module, but it is in the personal computer's case, and other personal computer software has better access to it. If you need to turn the personal computer off to restart other computer programs, you still have a problem, but at least you can control when the control system stops and for how long.

Con: A personal computer's control interval isn't deterministic. A personal computer can really only execute one program at a time. If several "windows" have been opened and each is executing a program, the operating system has to switch back and forth between programs using methods that were not written to preserve deterministic behavior. There is no way of predicting the delay between when a situation needing response occurs, and when the soft logic controller can affect the required control outputs.

Pro: Modern operating systems usually contain features that can be called by well-written soft logic control programs to reduce or eliminate variability in control intervals. Even if these features aren't used, personal computers are becoming so fast that the variations in control intervals are becoming insignificant.

Some soft logic controller software is meant to be loaded before the operating system. The soft logic programs execute at repeatable intervals, and allow the operating system to use the computer only when the soft logic functions aren't needed. Variability depends only on the bias programs built into the computer (which are not necessarily optimized for the types of data exchange that a PLC requires).

If the soft logic controller is built into the interface card, its control interval is affected by the personal computer only if it needs to use a personal computer resource such as a communication channel in the personal computer or the personal computer's memory. If all these features are built into the interface card, the card becomes even more of a PLC-CPU-module-on-a-card.

If soft logic control does succeed in replacing the PLC, the soft logic PLC will still have to be programmed. Control languages will be dictated by the suppliers of the soft logic programs, and soft logic from the major PLC suppliers will probably be the most trusted. Rockwell Automation, for example, now offers an Allen-Bradley soft logic system called softLogix 5, which they describe as "a member of the PLC-5 family of processors." All the popular soft logic programs now being sold offer the IEC 1131-3 standard PLC programming languages, so programmers won't have to learn other computer programming languages. I/O modules will still be required, and the major PLC suppliers will still dominate that market.

In summary, even if soft logic does "kill" the PLC, most PLC users won't be able to tell the difference.

PROCESS SIMULATION

To help in the writing and debugging of PLC programs, several suppliers now offer software and interface cards for personal computers so that the personal computer can be programmed to simulate a system for the PLC to control. A PLC program can be tested by having the PLC program control the simulated system.

Siemens offers a process simulation software package called S7-PLCSIM. You load a user-program into S7-PLCSIM to test it, so you don't even need to have a PLC. Third-party vendors offer personal computer-based simulation systems that can be driven by a real PLC. SST (formerly, S-S Technologies), for example, offers a program called PICS and an interface card that allows a personal computer to be connected to a PLC-5 like a remote I/O module. The personal computer simulates the response of a system to the user-program running in the PLC-5.

REFLECTIVE MEMORY

As a complete replacement for a standard PLC, VMIC offers a soft logic program called IOWorks and sells remote I/O racks containing VMIC RTnet communication adapters and VMIC I/O modules. Data exchange between the central computer (running IOWorks) and the racks is via a reflective memory system. With reflective memory the network has direct memory access (DMA) to the memory of the RTnet adapters in the remote racks, so that I/O data is copied between the personal computer and the remote racks at rates of up to 29 Mbps.

OMAC MOTION AND PROCESS CONTROL

The name *OMAC (open modular architecture controller)* was proposed by the Big Three automakers in the United States, who are pushing for the definition of a standard for an open PLC controller. The OMAC controller definition will include a standard operating system (perhaps a modified version of Windows NT), a standard programming interface, standard programming languages (as defined by IEC 1131-3), and standard function blocks for motion control and other standard machine functions. Other controller users have joined the Big Three, including users of process control equipment. The process control industry representatives are pushing for the adoption of a continuous flowchart variation of the IEC function block diagram programming language, to incorporate standard process control operations.

PLC manufacturers are responding by offering new CPU modules for their PLCs, with motion control I/O capabilities and high-speed control algorithms built in. Siemens has a new S7-600-level CPU and Allen-Bradley has a new Logix 5 processor. Standards groups such as PLCOpen are working to develop additions to the IEC 1131 set of standards that will meet the requirements of continuous flowchart supporters without sacrificing the commonality that has already been agreed upon for the other IEC (6)1131-3 set of PLC programming languages.

No, PLCs aren't likely to wither away.

QUESTIONS/PROBLEMS

1. Instead of connecting sensors to a standard PLC I/O module, you could connect sensors to your PLC via a sensor bus such as DeviceNet (Allen-Bradley's favorite) or AS-i (Siemens' favorite). Your PLC would have to have a sensor bus interface module, of course, but the connections to the sensors would be different. Describe (or sketch a circuit showing) how you would connect eight sensors to the sensor bus interface module.

2. *Processes* in a fieldbus network synchronize their operations with each other with the help of a network controller. How does the network controller keep the system synchronized?

3. Processes in a fieldbus network communicate with each other as if they were separate computers. What is the difference between a process and a computer?

4. In addition to regularly scheduled data exchanges, the fieldbus also allows __________-type messages to be sent.

5. Every component connected to a fieldbus network must have a DDL (device description language) identifier. Why?

6. **(a)** Do you think that personal computers will replace PLCs on the plant floor in the foreseeable future?

(b) If personal computers do replace PLCs, do you think that PLC programmers will notice the difference? Explain.

Appendixes

ALLEN-BRADLEY PLC-5 STATUS FILE STRUCTURE

References to:	Include these Allen-Bradley Processors:
Classic PLC-5 processors	PLC-5/10, -5/12, -5/15, -5/25, and -5/VME processors.
Enhanced PLC-5 processors	PLC-5/11, -5/20, -5/30, -5/40, -5/40L, -5/60, -5/60L, and -5/80 processors. **Important:** Unless otherwise specified, Enhanced PLC-5 processors include Ethernet PLC-5 and VME PLC-5 processors.
Ethernet PLC-5 processors	PLC-5/20E, -5/40E, and -5/80E processors.
VME PLC-5 processors	PLC-5/V30, -5/V40, -5/V40L, and -5/V80 processors. See the PLC-5/VME VMEbus Programmable Controllers User Manual, publication 1785-6.5.9, for more information.

Table 12.A
Processor Status File Addresses

This Word of the Status File:		Stores:
S:0		Arithmetic flags • bit 0 = carry • bit 1 = overflow • bit 2 = zero • bit 3 = sign
S:1		Processor status and flags

This Word of the Status File:		Stores:
S:2		Switch setting information • bits 0 – 5: DH+ station number • bit 7: 1 = scanner; 0 = adapter • bit 11, 12: hardware addressing bit 12 bit 11 0 0 illegal 1 0 1/2-slot 0 1 1-slot 1 1 2-slot • bit 13: 1 = load from EEPROM • bit 14: 1 = RAM backup not configured • bit 15: 1 = memory unprotected
S:3 to S:6		Active Node table Word Bits DH+ Station # 3 0-15 00-17 4 0-15 20-37 5 0-15 40-57 6 0-15 60-77
S:7	(Classic PLC-5 and Enhanced PLC-5 processors)	Global status bits: • low 8 bits — rack fault bits for racks 0-7 • high 8 bits — rack queue-full bits for racks 0-7
S:32	(PLC-5/40, -5/40L, -5/40E, -5/V40, -5/V40B, -5/V40L,PLC-5/60, -5/60L, -5/80, -5/80E, PLC-5/V80)	• low 8 bits — rack fault bits for racks 10-17 (octal) • high 8 bits — rack queue-full bits for racks 10-17
S:34	(PLC-5/60, -5/60L, -5/80, -5/80E, -5/V80)	• low 8 bits — rack fault bits for racks 20-27 (octal) • high 8 bits — rack queue-full bits for racks 20-27
S:8		Last program scan (in ms)
S:9		Maximum program scan (in ms)
S:10		Minor fault (word 1)
S:11		Major fault
S:12		Fault codes
S:13		Program file where fault occurred
S:14		Rung number where fault occurred
S:15	PLC-5/VME processor	VME Status File
S:16		I/O Status File
S:17	Enhanced PLC-5 processors	Minor fault (word 2)
S:18		Processor clock year
S:19		Processor clock month

This Word of the Status File:		Stores:
S:20		Processor clock day
S:21		Processor clock hour
S:22		Processor clock minute
S:23		Processor clock second
S:24		Indexed addressing offset
S:25	(PLC-5/12, -5/15, -5/25)	I/O Adapter image file
S:26		User control bits • bits 0 and 1: startup procedure • bit 2: local rack address
S:27	(Classic PLC-5 and Enhanced PLC-5 processors)	Rack control bits: • low 8 bits — I/O rack inhibit bits for racks 0-7 • high 8 bits — I/O rack reset bits for racks 0-7
S:33	(PLC-5/40, -5/40L, -5/40E, -5/V40, -5/V40L, PLC-5/60, -5/60L, -5/80, -5/80E, -5/V80)	• low 8 bits — I/O rack inhibit bits for racks 10-17 • high 8 bits — I/O rack reset bits for racks 10-17
S:35	(PLC-5/60, -5/60L, -5/80, -5/80E, -5/V80)	• low 8 bits — I/O rack inhibit bits for racks 20-27 • high 8 bits — I/O rack reset bits for racks 20-27
S:28		Program watchdog setpoint
S:29		Fault routine file
S:30		STI setpoint
S:31		STI file number
S:46	(Enhanced PLC-5 processors)	PII program file number
S:47	(Enhanced PLC-5 processors)	PII module group
S:48	(Enhanced PLC-5 processors)	PII bit mask
S:49	(Enhanced PLC-5 processors)	PII compare value
S:50	(Enhanced PLC-5 processors)	PII down count
S:51	(Enhanced PLC-5 processors)	PII changed bits
S:52	(Enhanced PLC-5 processors)	PII events since last interrupt
S:53	(Enhanced PLC-5 processors)	STI scan time (in ms)
S:54	(Enhanced PLC-5 processors)	STI maximum scan time (in ms)
S:55	(Enhanced PLC-5 processors)	PII last scan time (in ms)
S:56	(Enhanced PLC-5 processors)	PII maximum scan time (in ms)
S:57	(Enhanced PLC-5 processors)	User program checksum
S:59	(PLC-5/40L, -5/60L)	Extended local I/O channel discrete transfer scan (in ms)
S:60	(PLC-5/40L, -5/60L)	Extended local I/O channel discrete maximum scan (in ms)
S:61	(PLC-5/40L, -5/60L)	Extended local I/O channel block-transfer scan (in ms)

This Word of the Status File:		Stores:
S:62	(PLC-5/40L, -5/60L)	Extended local I/O channel maximum block transfer scan (in ms)
S:77	(Enhanced PLC-5 processors)	Communication time slice for communication housekeeping functions (in ms)
S:78	(Enhanced PLC-5 processors)	I/O scan after MCP
S:79	(Enhanced PLC-5 processors)	MCP inhibit bits
S:80-S:127	(Enhanced PLC-5 processors)	MCP file number MCP scan time (in ms) MCP max scan time (in ms)

ALLEN-BRADLEY SLC 500 STATUS FILE STRUCTURE

The status file S: contains the following words:

Word	Function (applies to all processors)
S:0	Arithmetic Flags
S:1	Processor Mode Status/Control
S:2	STI Bits/DH485 Comms.
S:3L	Current/Last Scan Time
S:3H	Watchdog Scan Time
S:4	Free Running Clock
S:5	Minor Error Bits
S:6	Major Error Code
S:7, S:8	Suspend Code/Suspend File
S:9, S:10	Active Nodes (DH-485)
S:11, S:12	I/O Slot Enables
S:13, S:14	Math Register
S:15L	Node Address
S:15H	Baud Rate

Word	Function (applies to SLC 5/02, SLC 5/03, and SLC 5/04 processors)
S:16, S:17	Test Single Step – Start Step On – Rung/File
S:18, S:19	Test Single Step – Breakpoint – Rung/File
S:20, S:21	Test – Fault/Powerdown – Rung/File
S:22	Maximum Observed Scan Time
S:23	Average Scan Time
S:24	Index Register
S:25, S:26	I/O Interrupt Pending
S:27, S28	I/O Interrupt Enabled
S:29	User Fault Routine File Number
S:30	Selectable Timed Interrupt Setpoint
S:31	Selectable Timed Interrupt File Number
S:32	I/O Interrupt Executing

Word	Function (applies to SLC 5/03, and SLC 5/04 processors)
S:33	Extended Processor Status and Control Word
S:34	Passthru Disabled (SLC 5/04 only)
S:35	Last 1 ms Scan Time
S:36	Extended Minor Error Bits
S:37	Clock/Calendar Year
S:38	Clock/Calendar Month
S:39	Clock/Calendar Day
S:40	Clock/Calendar Hours
S:41	Clock/Calendar Minutes
S:42	Clock/Calendar Seconds
S:43	STI Interrupt Time (SLC 5/03 and SLC 5/04)
S:44	I/O Event Interrupt Time (SLC 5/03 and SLC 5/04)
S:45	DII Interrupt Time (SLC 5/03 and SLC 5/04)
S:46	Discrete Input Interrupt – FIle Number
S:47	Discrete Input Interrupt – Slot Number
S:48	Discrete Input Interrupt – Bit Mask

Word	Function (applies to SLC 5/03, and SLC 5/04 processors)
S:49	Discrete Input Interrupt – Compare Value
S:50	Discrete Input Interrupt – Preset
S:51	Discrete Input Interrupt – Return Mask
S:52	Discrete Input Interrupt – Accumulator
S:53 and S:54	Reserved
S:55	Last DII Scan Time
S:56	Maximum Observed DII Scan Time
S:57	Operating System Catalog Number
S:58	Operating System Series
S:59	Operating System FRN
S:60	Processor Catalog Number
S:61	Processor Series
S:62	Processor Revision
S:63	User Program Type
S:64	User Program Functionality Index
S:65	User RAM Size
S:66	Flash EEPROM Size
S:67 and S:68	Channel 0 Active Nodes

Word	Function (applies to SLC 5/04 processors)
S:69 to S:82	Reserved
S:83 to S:86	DH+ Active Nodes (Channel 1 SLC 5/04 only)
S:87 to S:98	Reserved
S:99	Global Status Word
S:100 to S:163	Global Status File

C

OMRON CQM1 SR AND AR MEMORY AREAS

SR Area

Word	Bit(s)	Function
SR 244	00 to 15	**Input Interrupt 0 Counter Mode SV** SV when input interrupt 0 is used in counter mode (4 digits hexadecimal, 0000 to FFFF). (Can be used as work bits when input interrupt 0 is not used in counter mode.)
SR 245	00 to 15	**Input Interrupt 1 Counter Mode SV** SV when input interrupt 1 is used in counter mode (4 digits hexadecimal, 0000 to FFFF). (Can be used as work bits when input interrupt 1 is not used in counter mode.)
SR 246	00 to 15	**Input Interrupt 2 Counter Mode SV** SV when input interrupt 2 is used in counter mode (4 digits hexadecimal, 0000 to FFFF). (Can be used as work bits when input interrupt 2 is not used in counter mode.)
SR 247	00 to 15	**Input Interrupt 3 Counter Mode SV** SV when input interrupt 3 is used in counter mode (4 digits hexadecimal, 0000 to FFFF). (Can be used as work bits when input interrupt 3 is not used in counter mode.)
SR 248	00 to 15	**Input Interrupt 0 Counter Mode PV Minus One** Counter PV-1 when input interrupt 0 is used in counter mode (4 digits hexadecimal).
SR 249	00 to 15	**Input Interrupt 1 Counter Mode PV Minus One** Counter PV-1 when input interrupt 1 is used in counter mode (4 digits hexadecimal).
SR 250	00 to 15	**Input Interrupt 2 Counter Mode PV Minus One** Counter PV-1 when input interrupt 2 is used in counter mode (4 digits hexadecimal).
SR 251	00 to 15	**Input Interrupt 3 Counter Mode PV Minus One** Counter PV-1 when input interrupt 3 is used in counter mode (4 digits hexadecimal).

Word	Bit(s)	Function
SR 252	00	**High-speed Counter 0 Reset Bit**
	01	**CQM1-CPU43-E: High-speed Counter 1 Reset Bit** Turn ON to reset PV of high-speed counter 1 (port 1). **CQM1-CPU44-E: Absolute High-speed Counter 1 Origin Compensation Bit** Turn ON to set origin compensation for absolute high-speed counter 1 (port 1). Automatically turns OFF when compensation value is set in DM 6611.
	01	**CQM1-CPU43-E: High-speed Counter 2 Reset Bit** Turn ON to reset PV of high-speed counter 2 (port 2). **CQM1-CPU44-E: Absolute High-speed Counter 2 Origin Compensation Bit** Turn ON to set origin compensation for absolute high-speed counter 2 (port 2). Automatically turns OFF when compensation value is set in DM 6612.
	03 to 07	Not used.
	08	**Peripheral Port Reset Bit** Turn ON to reset peripheral port. (Not valid when peripheral device is connected.) Automatically turns OFF when reset is complete.
	09	**RS-232C Port Reset Bit** Turn ON to reset RS-232C port. Automatically turns OFF when reset is complete.
	10	**PLC Setup Reset Bit** Turn ON to initialize PLC Setup (DM 6600 through DM 6655). Automatically turns OFF again when reset is complete. Only effective if the PLC is in PROGRAM mode.
	11	**Forced Status Hold Bit** OFF: Bits that are forced set/reset are cleared when switching from PROGRAM mode to MONITOR mode. ON: The status of bits that are forced set/reset are maintained when switching from PROGRAM mode to MONITOR mode.
	12	**I/O Hold Bit** OFF: IR and LR bits are reset when starting or stopping operation. ON: IR and LR bit status is maintained when starting or stopping operation.
	13	Not used.
	14	**Error Log Reset Bit** Turn ON to clear error log. Automatically turns OFF again when operation is complete.
	15	**Output OFF Bit** OFF: Normal output status. ON: All outputs turned OFF.
SR 253	00 to 07	**FAL Error Code** The error code (a 2-digit number) is stored here when an error occurs. The FAL number is stored here when FAL(06) or FALS(07) is executed. This word is reset (to 00) by executing a FAL 00 instruction or by clearing the error from a Peripheral Device.
	08	**Low Battery Flag** Turns ON when a CPU battery voltage drops.
	09	**Cycle Time Overrun Flag** Turns ON when a cycle time overrun occurs (i.e., when the cycle time exceeds 100 ms).
	10 to 12	Not used.
	13	**Always ON Flag**
	14	**Always OFF Flag**
	15	**First Cycle Flag** Turns ON for 1 cycle at the start of operation.

Word	Bit(s)	Function
SR 254	00	1-minute clock pulse (30 seconds ON; 30 seconds OFF)
	01	0.02-second clock pulse (0.01 second ON; 0.01 second OFF)
	02 to 03	Not used.
	04	**CQM1-CPU4□-E: Overflow (OF) Flag** Turns ON when the result of a calculation is above the upper limit of signed binary data.
	05	**CQM1-CPU4□-E: Underflow (UF) Flag** Turns ON when the result of a calculation is below the lower limit of signed binary data.
	06	**Differential Monitor Complete Flag** Turns ON when differential monitoring is complete.
	07	**STEP(08) Execution Flag** Turns ON for 1 cycle only at the start of process based on STEP(08).
	08	**HKY(—) Execution Flag** Turns ON during execution of HKY(—).
	09	**7SEG(88) Execution Flag** Turns ON during execution of 7SEG(88).
	10	**DSW(87) Execution Flag** Turns ON during execution of DSW(87).
	11 to 14	Not used.
	15	**CQM1-CPU43-E: Pulse I/O Error Flag (FALS: 9C)** Turns ON when there is an error in a pulse I/O function using port 1 or 2. **CQM1-CPU44-E: Absolute High-speed Counter Error Flag (FALS: 9C)** Turns ON when there is an error in an absolute high-speed counter using port 1 or 2.
SR 255	00	0.1-second clock pulse (0.05 second ON; 0.05 second OFF)
	01	0.2-second clock pulse (0.1 second ON; 0.1 second OFF)
	02	1.0-second clock pulse (0.5 second ON; 0.5 second OFF)
	03	**Instruction Execution Error (ER) Flag** Turns ON when an error occurs during execution of an instruction.
	04	**Carry (CY) Flag** Turns ON when there is a carry in the results of an instruction execution.
	05	**Greater Than (GR) Flag** Turns ON when the result of a comparison operation is "greater."
	06	**Equals (EQ) Flag** Turns ON when the result of a comparison operation is "equal," or when the result of an instruction execution is 0.
	07	**Less Than (LE) Flag** Turns ON when the result of a comparison operation is "less."
	08 to 15	Not used.

Note Writing is not possible for the following words: SR 248 through SR 251, and SR 253 through SR255.

AR Area

Word	Bit(s)	Function
AR 00 to AR 03	---	Not used.
AR 04	08 to 15	**CQM1-CPU43/44-E: Pulse I/O or Absolute High-speed Counter Status Code:** 00: Normal 01, 02: Hardware error 03: PLC Setup error 04: PLC stopped during pulse output
AR 05	00 to 07	**CQM1-CPU43/44-E: High-speed Counter 1 Range Comparison Flags** 00 ON: Counter PV is within comparison range 1 01 ON: Counter PV is within comparison range 2 02 ON: Counter PV is within comparison range 3 03 ON: Counter PV is within comparison range 4 04 ON: Counter PV is within comparison range 5 05 ON: Counter PV is within comparison range 6 06 ON: Counter PV is within comparison range 7 07 ON: Counter PV is within comparison range 8
	08	**CQM1-CPU43/44-E: High-speed Counter 1 Comparison Flag** OFF: Stopped ON: Comparing
	09	**CQM1-CPU43/44-E: High-speed Counter 1 Overflow/Underflow Flag** OFF: Normal ON: Overflow or underflow occurred.
	10 to 11	Not used.
	12 to 15	**CQM1-CPU43-E: Port 1 Pulse Output Flags** 12 ON: Deceleration specified. (OFF: Not specified.) 13 ON: Number of pulses specified. (OFF: Not specified.) 14 ON: Pulse output completed. (OFF: Not completed.) 15 ON: Pulse output in progress. (OFF: No pulse output.)
AR 06	00 to 15	**CQM1-CPU43/44-E: High-speed Counter 2/Port 2 Pulse Output Flags** Identical to the High-speed Counter 1/Port 1 Pulse Output Flags in AR 05.
AR 07	00 to 11	Not used.
	12	**DIP Switch Pin 6 Flag** OFF: CPU's DIP switch pin no. 6 is OFF. ON: CPU's DIP switch pin no. 6 is ON.
	13 to 15	Not used.

Word	Bit(s)	Function
AR 08	00 to 03	**RS-232C Communications Error Code** (1-digit number) The code will be "F" when a computer running LSS/SSS is connected to the Peripheral Port.
	04	**RS-232C Error Flag** Turns ON when an RS-232C communications error occurs.
	05	**RS-232C Transmission Enabled Flag** Valid only when host link, RS-232C communications are used.
	06	**RS-232C Reception Completed Flag** Valid only when RS-232C communications are used.
	07	**RS-232C Reception Overflow Flag** Valid only when RS-232C communications are used.)
	08 to 11	**Peripheral Device Error Code** (1-digit number) The code will be "F" when a computer running LSS/SSS is connected to the Peripheral Port.
	12	**Peripheral Device Error Flag** Turns ON when a peripheral device communications error occurs.
	13	**Peripheral Device Transmission Enabled Flag** Valid only when host link, RS-232C communications are used.
	14	**Peripheral Device Reception Completed Flag** Valid only when RS-232C communications are used.
	15	**Peripheral Device Reception Overflow Flag** Valid only when RS-232C communications are used.
AR 09	00 to 15	**RS-232C Reception Counter** 4 digits BCD; valid only when RS-232C communications are used.
AR 10	00 to 15	**Peripheral Device Reception Counter** 4 digits BCD; valid only when RS-232C communications are used.
AR 11	00 to 07	**High-speed Counter 0 Range Comparison Flags** 00 ON: Counter PV is within comparison range 1 01 ON: Counter PV is within comparison range 2 02 ON: Counter PV is within comparison range 3 03 ON: Counter PV is within comparison range 4 04 ON: Counter PV is within comparison range 5 05 ON: Counter PV is within comparison range 6 06 ON: Counter PV is within comparison range 7 07 ON: Counter PV is within comparison range 8
	08 to 15	Not used.
AR 12	00 to 15	Not used.
AR 13	00	**Memory Cassette Installed Flag** Turns ON if the Memory Cassette is installed at the time of powering up.
	01	**Clock Available Flag** Turns ON if a Memory Cassette equipped with a clock is installed.
	02	**Memory Cassette Write-protected Flag** ON when an EEPROM Memory Cassette if mounted and write protected or when an EPROM Memory cassette is mounted.
	03	Not used.
	04 to 07	**Memory Cassette Code** (1-digit number) 0: No Memory Cassette installed. 1: EEPROM, 4K-word Memory Cassette installed. 2: EEPROM, 8K-word Memory Cassette installed. 4: EPROM-type Memory Cassette installed.
	08 to 15	Not used.

Word	Bit(s)	Function
AR 14	00	**CPU to Memory Cassette Transfer Bit** Turn ON for transfer from the CPU to the Memory Cassette. Automatically turns OFF again when operation is complete.
	01	**Memory Cassette to CPU Transfer Bit** Turn ON for transfer from the Memory Cassette to the CPU. Automatically turns OFF again when operation is complete.
	02	**Memory Cassette Compare Flag** ON when the contents of the PLC and the Memory Cassette are being compared. Turns OFF automatically when comparison has completed.
	03	**Memory Cassette Comparison Results Flag** ON: Difference found or comparison not possible OFF: Contents compared and found to be the same.
	04 to 11	Not used.
	12	**PROGRAM Mode Transfer Error Flag** Turns ON when transfer could not be executed due to being in PROGRAM mode.
	13	**Write-protect Error Flag** Turns ON when transfer could not be executed due to write-protection.
	14	**Insufficient Capacity Flag** Turns ON when transfer could not be executed due to insufficient capacity at the transfer destination.
	15	**No Program Flag** Turns ON when transfer could not be executed due to there being no program in the Memory Cassette.
AR 15	00 to 07	**Memory Cassette Program Code** Code (2-digit number) indicates the size of the program stored in the Memory Cassette. 00: There is no program, or no Memory Cassette is installed. 04: The program is less than 3.2K words long. 08: The program is less than 7.2K words long.
	08 to15	**CPU Program Code** Code (2-digit number) indicates the size of the program stored in the CPU. 04: The program is less than 3.2K words long. 08: The program is less than 7.2K words long.
AR 16	00 to 10	Not used.
	11	**PLC Setup Initialized Flag** Turns ON when a checksum error occurs in the PLC Setup area and all settings are initialized back to the default settings.
	12	**Program Invalid Flag** Turns ON when a checksum error occurs in the UM area, or when an improper instruction is executed.
	13	**Instructions Table Initialized Flag** Turns ON when a checksum error occurs in the instructions table and all settings are initialized back to the default settings.
	14	**Memory Cassette Added Flag** Turns ON if the Memory Cassette is installed while the power is on.
	15	**Memory Cassette Transfer Error Flag** Turns ON if a transfer cannot be successfully executed when DIP switch pin no. 2 is set to ON (i.e., set to automatically transfer the contents of the Memory Cassette at power-up.)

Word	Bit(s)	Function
AR 17	00 to 07	"Minutes" portion of the present time, in 2 digits BCD (Valid only when a Memory Cassette with the clock function is installed.)
	08 to 15	"Hour" portion of the present time, in 2 digits BCD (Valid only when a Memory Cassette with the clock function is installed.)
AR 18	00 to 07	"Seconds" portion of the present time, in 2 digits BCD (Valid only when a Memory Cassette with the clock function is installed.)
	08 to 15	"Minutes" portion of the present time, in 2 digits BCD (Valid only when a Memory Cassette with the clock function is installed.)
AR 19	00 to 07	"Hour" portion of the present time, in 2 digits BCD (Valid only when a Memory Cassette with the clock function is installed.)
	08 to 15	"Date" portion of the present time, in 2 digits BCD (Valid only when a Memory Cassette with the clock function is installed.)
AR 20	00 to 07	"Month" portion of the present time, in 2 digits BCD (Valid only when a Memory Cassette with the clock function is installed.)
	08 to 15	"Year" portion of the present time, in 2 digits BCD (Valid only when a Memory Cassette with the clock function is installed.)
AR 21	00 to 07	"Day of week" portion of the present time, in 2 digits BCD [00: Sunday to 06: Saturday] (Valid only when a Memory Cassette with the clock function is installed.)
	08 to 12	Not used.
	13	**30-second Adjustment Bit** Valid only when a Memory Cassette with the clock function is installed.
	14	**Clock Stop Bit** Valid only when a Memory Cassette with the clock function is installed.
	15	**Clock Set Bit** Valid only when a Memory Cassette with the clock function is installed.
AR 22	00 to 07	**Input Words** Number of words for input bits (2 digits BCD)
	08 to 15	**Output Words** Number of words for output bits (2 digits BCD)
AR 23	00 to 15	**Power-off Counter** (4 digits BCD) This is the count of the number of times that the power has been turned off. To clear the count, write "0000" from a peripheral device.
AR 24	00	**Power-up PLC Setup Error Flag** Turns ON when there is an error in DM 6600 to DM 6614 (the part of the PLC Setup area that is read at power-up).
	01	**Start-up PLC Setup Error Flag** Turns ON when there is an error in DM 6615 to DM 6644 (the part of the PLC Setup area that is read at the beginning of operation).
	02	**RUN PLC Setup Error Flag** Turns ON when there is an error in DM 6645 to DM 6655 (the part of the PLC Setup area that is always read).
	03, 04	Not used.
	05	**Long Cycle Time Flag** Turns ON if the actual cycle time is longer than the cycle time set in DM 6619.
	06, 07	Not used.
	08 to 15	Code (2 digits hexadecimal) showing the word number of a detected I/O bus error 00 to 07: Correspond to input words 000 to 007. 80 to 87: Correspond to output words 100 to 107. FF: End cover cannot be confirmed.
AR 25	00 to 07	Not used.
	08	**FPD(—) Teaching Bit**
	09 to 15	Not used.

Word	Bit(s)	Function
AR 26	00 to 15	**Maximum Cycle Time** (4 digits BCD) The longest cycle time since the beginning of operation is stored. It is cleared at the beginning, and not at the end, of operation. The unit can be any of the following, depending on the setting of the 9F monitoring time (DM 6618). Default: 0.1 ms; "10 ms" setting: 0.1 ms; "100 ms" setting: 1 ms; "1 s" setting: 10 ms
AR 27	00 to 15	**Current Cycle Time** (4 digits BCD) The most recent cycle time during operation is stored. The Current Cycle Time is not cleared when operation stops. The unit can be any of the following, depending on the setting of the 9F monitoring time (DM 6618). Default: 0.1 ms; "10 ms" setting: 0.1 ms; "100 ms" setting: 1 ms; "1 s" setting: 10 ms

ALLEN-BRADLEY COMPARE INSTRUCTION OPERATORS

This Operation:	Using this Operator:
math binary	+, −, *, |
	OR, **
	AND, XOR
math unary	− (negate)
	LN
	FRD, TOD, DEG, RAD, SQR, NOT, LOG, SIN, COS, TAN, ASN, ACS, ATN
comparative	=, <, >
	<>, <=, >=

ALLEN-BRADLEY COMPUTE (CPT) INSTRUCTION OPERATORS AND PRECEDENCE

Valid Operations for Use in a CPT Expression

Type	Operator	Description	Example Operation
Copy	none	copy from A to B	enter source address in the expression enter destination address in destination
Clear	none	set a value to zero	0 (enter 0 for the expression)

Type	Operator	Description	Example Operation
Arithmetic	+	add	2 + 3 2 + 3 + 7 (Enhanced PLC-5 processors)
	–	subtract	12 – 5 (12 – 5) – 7 (Enhanced PLC-5 processors)
	*	multiply	5 * 2 6 * (5 * 2) (Enhanced PLC-5 processors)
	| (vertical bar)	divide	24 | 6 (24 | 6) *2 (Enhanced PLC-5 processors)
	–	negate	– N7:0
	SQR	square root	SQR N7:0
	**	exponential * (x to the power of y)	10**3
	LN	natural log *	LN F8:20
	LOG	log to the base 10 *	LOG F8:3
Trigonometric	ACS	arc cosine *	ACS F8:18
	ASN	arc sine *	ASN F8:20
	ATN	arc tangent *	ATN F8:22
	COS	cosine *	COS F8:14
	SIN	sine *	SIN F8:12
	TAN	tangent *	TAN F8:16
Bitwise	AND	bitwise AND	D9:3 AND D10:4
	OR	bitwise OR	D10:4 OR D10:5
	XOR	bitwise exclusive OR	D9:5 XOR D10:4
	NOT	bitwise complement	NOT D9:3
Conversion	FRD	convert from BCD to binary	FRD N7:0
	TOD	convert from binary to BCD	TOD N7:0
	DEG	convert radians to degrees *	DEG F8:8
	RAD	convert degrees to radians *	RAD F8:10

* Available in Enhanced PLC-5 processors only.

Order of Operation for CPT Expressions

<table>
<tr><th>Order</th><th>Operation</th><th>Description</th></tr>
<tr><td>1</td><td>**</td><td>exponential (X^Y)
Enhanced PLC-5 processors only</td></tr>
<tr><td rowspan="2">2</td><td>–</td><td>negate</td></tr>
<tr><td>NOT</td><td>bitwise complement</td></tr>
<tr><td rowspan="2">3</td><td>*</td><td>multiply</td></tr>
<tr><td>|</td><td>divide</td></tr>
<tr><td rowspan="2">4</td><td>+</td><td>add</td></tr>
<tr><td>–</td><td>subtract</td></tr>
<tr><td>5</td><td>AND</td><td>bitwise AND</td></tr>
<tr><td>6</td><td>XOR</td><td>bitwise exclusive OR</td></tr>
<tr><td>7</td><td>OR</td><td>bitwise OR</td></tr>
</table>

SIEMENS S7 STATUS BIT AFFECTED BY MATH AND LOGIC OPERATIONS

Evaluating the Bits in the Status Word

The math instructions affect the following bits of the status word:

- CC 1 and CC 0
- OV
- OS

Valid Result

A hyphen (–) entered in a bit column of the table means that the bit in question is not affected by the result of the integer math operation.

Signal State of Bits in the Status Word for Integer Math Result in Valid Range

Valid Range for an Integer (16 Bits) and Double Integer (32 Bits) Result	Bits of Status Word			
	CC 1	CC 0	OV	OS
0 (zero)	0	0	0	–
I: -32,768 ≤ Result < 0 (negative number) D: -2,147,483,648 ≤ Result < 0 (negative number)	0	1	0	–
I: 32,767 ≥ Result >0 (positive number) D 2,147,483,647 ≥ Result >0 (positive number)	1	0	0	–

Invalid Result

Signal State of Bits in the Status Word for Integer Math Result That Is Not in Valid Range

Range Not Valid for a Double Integer Result (32 Bits)	Bits of Status Word			
	CC 1	CC 0	OV	OS
I: Result > 32,767 (positive number) D: Result > 2,147,483,647 (positive number)	1	0	1	1
I: < –32,768 (negative number) D: Result < -2,147,483,648 (negative number)	0	1	1	1

Signal State of Bits on the Status Word for Double Integer Math Instructions + D, /D, and MOD92

Instruction	Bits of Status Word			
	CC 1	CC 0	OV	OS
+D: Result = -4,294,967,296	0	0	1	1
/D or MOD: Division by 0 has occurred.	1	1	1	1

Signal State of Bits in the Status Word for Floating-Point Math Result in Valid Range

Valid Range for a Floating-Point Result (32 Bits)	**Bits of Status Word**			
	CC 1	**CC 0**	**OV**	**OS**
+0, -0 (zero)	0	0	0	–
-3.402823E+38 < Result < -1.175494E–38 (negative number)	0	1	0	–
+1.175494E–38 < Result < 3.402823E+38 (positive number)	1	0	0	–

Signal State of Bits in the Status Word for Floating-Point Math Result That Is Not in Valid Range

Range Not Valid for a Floating-Point Result (32 Bits)	**Bits of Status Word**			
	CC 1	**CC 0**	**OV**	**OS**
-1.175494E–38 < Result < -1.401298E–45 (negative number) Underflow	0	0	1	1
+1.401298E–45 < Result < +1.175494E–38 (positive number) Underflow	0	0	1	1
Result < -3.402823E+38 (negative number) Overflow	0	1	1	1
Result > -3.402823E+38 (positive number) Overflow	1	0	1	1

Settings of Bits CC1 and CC0 after a Compare Instruction

Condition	Signal states of CC 1 and CC 0:		Possible check with the instructions
	CC 1	CC 0	A, O, X, AN, ON, XN
ACCU 2 > ACCU 1	1	0	>0
ACCU 2 < ACCU 1	0	1	<0
ACCU 2 = ACCU 1	0	0	==0
ACCU 2 <> ACCU 1	0 or 1	1 or 0	<>0
ACCU 2 >= ACCU 1	1 or 0	0 or 0	>=0
ACCU 2 <= ACCU 1	0 or 0	1 or 0	<=0

Settings of Bits of the Status Word after a Compare Real Instruction (IEEE) 32-bit Floating-Point Numbers

Condition	CC 1	CC 0	OV	OS
==	0	0	0	not applicable
<>	0 or 1	1 or 0	0	not applicable
>	1	0	0	not applicable
<	0	1	0	not applicable
>=	1 or 0	0 or 0	0	not applicable
<=	0 or 0	1 or 0	0	not applicable
UO	1	1	1	1

G

SIEMENS S7 SYSTEM FUNCTIONS (FC), SYSTEM FUNCTION BLOCKS (SFB), AND IEC FUNCTIONS (FC)

System Functions, Sorted Numerically

No.	Short Name	Function
SFC0	SET_CLK	Set System Clock
SFC1	READ_CLK	Read System Clock
SFC2	SET_RTM	Set Run-Time Meter
SFC3	CTRL_RTM	Start/Stop Run-Time Meter
SFC4	READ_RTM	Read Run-Time Meter
SFC5	GADR_LGC	Query Logical Address of a Channel
SFC6	RD_SINFO	Read OB Start Information
SFC9	EN_MSG	Enable Block-Related, Symbol-Related and Group Status Messages
SFC10	DIS_MSG	Disable Block-Related, Symbol-Related and Group Status Messages
SFC13	DPNRM_DG	Read Diagnostic Data of a DP Slave (Slave Diagnostics)
SFC14	DPRD_DAT	Read Consistent Data of a Standard DP Slave
SFC15	DPWR_DAT	Write Consistent Data to a DP Standard Slave
SFC17	ALARM_SQ	Generate Acknowledgable Block-Related Messages
SFC18	ALARM_S	Generate Permanently Acknowledged Block-Related Messages
SFC19	ALARM_SC	Query the Acknowledgment Status of the last ALARM_SQ Entering State Message
SFC20	BLKMOV	Copy Variables
SFC21	FILL	Initialize a Memory Area
SFC22	CREAT_DB	Create Data Block
SFC23	DEL_DB	Delete Data Block
SFC24	TEST_DB	Test Data Block
SFC25	COMPRESS	Compress the User Memory
SFC26	UPDAT_PI	Update the Process Image Update Table
SFC27	UPDAT_PO	Update the Process Image Output Table
SFC28	SET_TINT	Set Time-of-Day Interrupt
SFC29	CAN_TINT	Cancel Time-of-Day Interrupt
SFC30	ACT_TINT	Activate Time-of-Day Interrupt
SFC31	QRY_TINT	Query Time-of-Day Interrupt
SFC32	SRT_DINT	Start Time-Delay Interrupt
SFC33	CAN_DINT	Cancel Time-Delay Interrupt
SFC34	QRY_DINT	Query Time-Delay Interrupt
SFC35	MP_ALM	Trigger Multicomputing Interrupt
SFC36	MSK_FLT	Mask Synchronous Errors
SFC37	DMSK_FLT	Unmask Synchronous Errors
SFC38	READ_ERR	Read Error Register
SFC39	DIS_IRT	Disable New Interrupts and Asynchronous Errors
SFC40	EN_IRT	Enable New Interrupts and Asynchronous Errors

System Functions *(continued)*

No.	Short Name	Function
SFC41	DIS_AIRT	Delay Higher Priority Interrupts and Asynchronous Errors
SFC42	EN_AIRT	Enable Higher Priority Interrupts and Asynchronous Errors
SFC43	RE_TRIGR	Retrigger Cycle Time Monitoring
SFC44	REPL_VAL	Transfer Substitute Value to Accumulator 1
SFC46	STP	Change the CPU to STOP
SFC47	WAIT	Delay Execution of the User Program
SFC48	SNC_RTCB	Synchronize Slave Clocks
SFC49	LGC_GADR	Query the Module Slot Belonging to a Logical Address
SFC50	RD_LGADR	Query all Logical Addresses of a Module
SFC51	RDSYSST	Read a System Status List or Partial List
SFC52	WR_USMSG	Write a User-Defined Diagnostic Event to the Diagnostic Buffer
SFC55	WR_PARM	Write Dynamic Parameters
SFC56	WR_DPARM	Write Default Parameters
SFC57	PARM_MOD	Assign Parameters to a Module
SFC58	WR_REC	Write a Data Record
SFC59	RD_REC	Read a Data Record
SFC60	GD_SND	Send a GD Packet
SFC61	GD_RCV	Fetch a Received GD Packet
SFC62	CONTROL	Query the Status of a Connection Belonging to a Communication SFB Instance
SFC63	AB_CALL	Assembly Code Block
SFC64	TIME_TCK	Read the System Time
SFC65	X_SEND	Send Data to a Communication Partner outside the Local S7 Station
SFC66	X_RCV	Receive Data from a Communication Partner outside the Local S7 Station
SFC67	X_GET	Read Data from a Communication Partner outside the Local S7 Station
SFC68	X_PUT	Write Data to a Communication Partner outside the Local S7 Station
SFC69	X_ABORT	Abort an Existing Connection to a Communication Partner outside the Local S7 Station
SFC72	I_GET	Read Data from a Communication Partner within the Local S7 Station
SFC73	I_PUT	Write Data to a Communication Partner within the Local S7 Station
SFC74	I_ABORT	Abort an Existing Connection to a Communication Partner within the Local S7 Station
SFC79	SET	Set a Range of Outputs
SFC80	RSET	Reset a Range of Outputs

* SFC63 “AB_CALL” only exists for CPU 614. For a detailed description, refer to the corresponding manual.

System Function Blocks, Sorted Numerically

No.	Short Name	Function
SFB0	CTU	Count Up
SFB1	CTD	Count Down
SFB2	CTUD	Count Up/Down
SFB3	TP	Generate a Pulse
SFB4	TON	Generate an On Delay
SFB5	TOF	Generate an Off Delay
SFB8	USEND	Uncoordinated Sending of Data
SFB9	URCV	Uncoordinated Receiving of Data
SFB12	BSEND	Sending Segmented Data
SFB13	BRCV	Receiving Segmented Data
SFB14	GET	Read Data from a Remote CPU
SFB15	PUT	Write Data to a Remote CPU
SFB16	PRINT	Send Data to Printer
SFB19	START	Initiate a Complete Restart on a Remote Device
SFB20	STOP	Changing a Remote Device to the STOP State
SFB21	RESUME	Initiate a Restart on a Remote Device
SFB22	STATUS	Query the Status of a Remote Partner
SFB23	USTATUS	Receive the Status of a Remote Device
SFB29	HS_COUNT	Counter (high-speed counter, integrated function)
SFB30	FREQ_MES	Frequency Meter (frequency meter, integrated function
SFB32	DRUM	Implement a Sequencer
SFB33	ALARM	Generate Block-Related Messages with Acknowledgment Display
SFB34	ALARM_8	Generate Block-Related Messages without Values for 8 Signals
SFB35	ALARM_8P	Generate Block-Related Messages with Values for 8 Signals
SFB36	NOTIFY	Generate Block-Related Messages without Acknowledgment Display
SFB37	AR_SEND	Send Archive Data
SFB38	HSC_A_B	Counter A/B (integrated function)
SFB39	POS	Position (integrated function)
SFB41[1)]	CONT_C	Continuous Control
SFB42[1)]	CONT_S	Step Control
SFB43[1)]	PULSEGEN	Pulse Generation

[1]SFBs 41 "CONT_C", 42 "CONT_S", and 43 "PULSEGEN" only exist on the CPU 314 IFM.

IEC Functions, Sorted Numerically

No.	Short Name	Function
FC1	AD_DT_TM	Add duration to a time
FC2	CONCAT	Combine two STRING variables
FC3	D_TOD_DT	Combine DATE and TIME_OF_DAY to DT
FC4	DELETE	Delete in a STRING variable
FC5	DI_STRNG	Data type conversion DINT to STRING
FC6	DT_DATE	Extract the DATE from DT
FC7	DT_DAY	Extract the day of the week from DT
FC8	DT_TOD	Extract the TIME_OF_DAY from DT
FC9	EQ_DT	Compare DT for equal
FC10	EQ_STRNG	Compare STRING for equal
FC11	FIND	Find in a STRING variable
FC12	GE_DT	Compare DT for greater than or equal
FC13	GE_STRNG	Compare STRING for greater than or equal
FC14	GT_DT	Compare DT for greater than
FC15	GT_STRNG	Compare STRING for greater than
FC16	I_STRNG	Data type conversion INT to STRING
FC17	INSERT	Insert in a STRING variable
FC18	LE_DT	Compare DT for less than or equal
FC19	LE_STRNG	Compare STRING for less than or equal
FC20	LEFT	Left part of a STRING variable
FC21	LEN	Length of a STRING variable
FC22	LIMIT	Limit
FC23	LT_DT	Compare DT for less than
FC24	LT_STRNG	Compare STRING for less than
FC25	MAX	Select maximum
FC26	MID	Middle part of a STRING variable
FC27	MIN	Select minimum
FC28	NE_DT	Compare DT for unequal
FC29	NE_STRNG	Compare STRING for unequal
FC30	R_STRNG	Data type conversion REAL to STRING
FC31	REPLACE	Replace in a STRING variable
FC32	RIGHT	Right part of a STRING variable
FC33	S5TI_TIM	Data type conversion S5TIME to TIME
FC34	SB_DT_DT	Subtract two time values
FC35	SB_DT_TM	Subtract duration from a time
FC36	SEL	Binary selection
FC37	STRNG_DI	Data type conversion STRING to DINT
FC38	STRNG_I	Data type conversion STRING to INT
FC39	STRNG_R	Data type conversion STRING to REAL
FC40	TIM_S5TI	Data type conversion TIME to S5TIME

ALLEN-BRADLEY PLC-5 MAJOR AND MINOR FAULT BITS AND CODES

Status Word II
Response to Major Faults

This bit:	Indicates this fault:	And the fault is:
00	Corrupted program file	Recoverable—the fault routine can instruct the processor to clear the fault and then resume scanning the program.
01	Corrupted address in ladder program (see fault codes 10-19)	
02	Programming error (see fault codes 20-29)	
05	Start-up protection fault (see word 26, bit 1) Processor sets bit 5; if your fault routine does not reset this bit, the processor inhibits startup	
06 [1]	Peripheral device fault	
07	User-generated fault; processor jumped to fault routine (see fault codes 0-9)	
08	Watchdog faulted	
11 [1]	MCP does not exist or is not a ladder file	
12 [1]	PII file does not exist or is not a ladder file	
13	STI file does not contain ladder logic or does not exist	
03	Processor detected an SFC fault (see fault codes 74-79)	Not recoverable—the processor enters fault mode without scanning the fault routine.
04	Processor detected an error when assembling a ladder program file (see fault code 70)	
09	System is configured wrong; you installed a RAM cartridge but configured the system for an EEPROM or you violated 32-point I/O module placement rules for 1-slot addressing	
10	Non-recoverable hardware error	
14	Fault routine does not contain ladder logic or does not exist	
15	Fault routine program file does not contain ladder logic	

[1] This fault applies to PLC-5/11, -5/20, -5/30, -5/40, -5/40L, -5/60, -5/60L or -5/80 processors only.

Possible Major Faults

Status Word	If the Status Bit Values Are: 15.....8 7......0	Bit Number:	Then the Fault is:
S23	xxxxxxxx xxxxxxx1	0	Bad user program file (see fault codes 10-19)
S23	xxxxxxxx xxxxxx10	1	Illegal operand address (see fault codes 20-29)
S23	xxxxxxxx xxxxx100	2	Programming error (see fault codes 30-49)
S23	xxxxxxxx xxxx1000	3	SFC fault (see fault codes 71-79)
S23	xxxxxxxx xxx10000	4	Program assembly error (see fault code 70)
S23	xxxxxxxx xx100000	5	Powerup protection fault
S23	xxxxxxxx x1000000	6	Channel 3 device fault [1]
S23	xxxxxxxx 10000000	7	User generated fault (see fault codes 0-9)
S22	xxxxxxx1 00000000	8	Watchdog timer fault
S22	xxxxxx10 00000000	9	Bad system configuration (see fault codes 80-89)
S22	xxxxx100 00000000	10	Hardware error
S22	xxxx1000 00000000	11	MCP does not exist or is not ladder [1]
S22	xxx10000 00000000	12	PII file does not exist or is not ladder [1]
S22	xx100000 00000000	13	STI file does not exist or is not ladder [1]
S22	x1000000 00000000	14	Bad fault program
S22	10000000 00000000	15	Non ladder file

Each x indicates a bit that can be 0 or 1 for the status value described.

[1] For Enhanced PLC-5 processors only.

Possible Minor Faults in Word 1 (stored in S:10)

If the Status Bit Values Are: 15.....8 7......0	**Bit Number:**	**Then the Fault Is:**
xxxxxxxx xxxxxxx1	0	Battery is bad or missing
xxxxxxxx xxxxxx10	1	DH+ table changed
xxxxxxxx xxxxx100	2	STI overlap
xxxxxxxx xxxx1000	3	EEPROM transferred
xxxxxxxx xxx10000	4	Edits prevent SFC continuing
xxxxxxxx xx100000	5	Invalid I/O status file
xxxxxxxx x1000000	6	Memory cartridge battery low [1]
xxxxxxxx 10000000	7	No more command blocks exist [1]
xxxxxxx1 00000000	8	EEPROM too small, burn failed [1]
xxxxxx10 00000000	9	No MCP configured to run [1]
xxxxx100 00000000	10	MCP not allowed [1]
xxxx1000 00000000	11	PII word number isn't in local rack [1]
xxx10000 00000000	12	User PII routine overlap [1]
xx100000 00000000	13	No command blocks exist to get PII [1]
x1000000 00000000	14	Arithmetic overflow occurred [1].
10000000 00000000	15	SFC 'lingering' action overlap [1]

Each x indicates a bit that can be 0 or 1 for the status value described.

[1] For Enhanced PLC-5 processors only.

Possible Minor Faults—Enhanced PLC-5 only (Stored in S:17)

If the Status Bit Values in Word 2 Are: 15.....8 7......0	Bit Number:	Then the Fault Is:
xxxxxxxx xxxxxxx1	0	Queue full between local and remote I/O
xxxxxxxx xxxxxx10	1	Queue full servicing channel 1A
xxxxxxxx xxxxx100	2	Queue full servicing channel 1B
xxxxxxxx xxxx1000	3	Queue full servicing channel 2A
xxxxxxxx xxx10000	4	Queue full servicing channel 2B
xxxxxxxx xx100000	5	No modem on serial port
xxxxxxxx x1000000	6	• Remote I/O rack in local rack table or • Remote I/O rack is greater than the image size. This fault can also be caused by the local rack if the local rack is set for octal density scan and the I/O image tables are smaller than 64 words (8 racks) each.
xxxxxxxx 10000000	7	Error not defined
xxxxxxx1 00000000	8	ASCII instruction error
xxxxxx10 00000000	9	Duplicate node address
xxxxx100 00000000	10	DF1 Master poll list error
xxxx1000 00000000	11	Error not defined
xxx10000 00000000	12	Error not defined
xx100000 00000000	13	Error not defined
x1000000 00000000	14	Error not defined
10000000 00000000	15	Error not defined

Each x indicates a bit that can be 0 or 1 for the status value described.

Major Fault Codes

Code	Fault
00-09	Reserved for user-defined fault codes
10 [1]	Run-time data table check failed
11 [1]	Bad user program checksum
12	Bad integer operand type, restore new processor memory file
13	Bad mixed mode operation type, restore new processor memory file
14	Not enough operands for instruction, restore new processor memory file
15	Too many operands for instructions, restore new processor memory file
16	Corrupted instruction, probably due to restoring an incompatible processor memory file
17	Can't find expression end; restore new processor memory file
18	Missing end of edit zone; restore new processor memory file
19 [1]	Download aborted
20	You entered too large an element number in an indirect address
21	You entered a negative element number in an indirect address
22	You tried to access an undefined program file
23	You used a negative file number, you used a file number greater than the number of existing files, or you tried to indirectly address files 0, 1, or 2
24	You tried to indirectly address a file of the wrong type
30	You tried to jump to one too many nested subroutine files
31	You did not enter enough subroutine parameters
32	You jumped to an invalid (non-ladder) file
33 [1]	You entered a CAR routine file that is not 68000 code
34	You entered a negative preset or accumulated value in a timer instruction
35	You entered a negative time variable in a PID instruction
36	You entered an out-of-range setpoint in a PID instruction
37	You addressed an invalid module in a block transfer, immediate input, or immediate output instruction
38	You entered a return instruction from a non-subroutine file
39	FOR instruction with missing NXT
40	The control file is too small for the PID, BTR, BTW, or MSG instruction
41	NXT instruction with missing FOR
42	You tried to jump to a deleted label
43 [1]	File is not an SFC
44-69	Reserved

[1] Fault applies only to PLC-5/11, -5/20, -5/30, -5/40, -5/40L, -5/60, -5/60L, and -5/80 processors.

Major Fault Codes *(continued)*

Code	Fault
70	The processor detected duplicate labels
71 [1]	The processor tried to start an SFC subchart that is already running
72 [1]	The processor tried to stop an SFC subchart that is not running
73 [1]	The processor tried to start more than the allowed subcharts
74	SFC file error detected
75	The SFC has too many active functions
77	SFC missing file or of wrong type for step, action, transition; or Subchart is created but empty; or SC or timer file specified in SFC empty or too small
78	The processor cannot continue to run the SFC after power loss
79	You tried to download an SFC to a processor that cannot run SFCs; or This specific PLC does not support this enhanced SFC
80	You incorrectly installed a 32-point I/O module in a 1-slot configuration (PLC-5/15, -5/25) You have an I/O configuration error (PLC-5/11, -5/20, 5/30, -5/40, -5/40L, -5/60, -5/60L, -5/80)
81	You illegally set an I/O chassis backplane switch; either switch 4 or 5 must be off
82 [1]	Illegal EEPROM cartridge type for selected operation
83 [1]	User watchdog fault
84 [1]	Error in user-configured adapter mode block transfer
85 [1]	Bad memory cartridge
86 [1]	Memory cartridge is incompatible with host
87 [1]	Scanner rack list overlap
88 [1, 2]	Insufficient time to process remote I/O
90 [1]	Coprocessor extensive memory test failed
91 [1]	Coprocessor undefined message type
92 [1]	Coprocessor illegal pool index
93 [1]	Coprocessor illegal maximum pool size
94 [1]	Coprocessor illegal ASCII message
95 [1]	Coprocessor reported fault
96 [1]	Coprocessor present signal lost
97 [1]	Coprocessor illegal minimum pool size
98 [1]	Coprocessor first/last 16 bytes RAM test failed
99 [1]	Coprocessor data transfer faulted
100 [1]	Processor to coprocessor transfer failed
101 [1]	Coprocessor end of scan transfer failed
102 [1]	File number specified for raw data transfer through the coprocessor is an illegal value
103 [1]	The element number specified for raw data transfer through the coprocessor is an illegal value
104 [1]	The size of the transfer requested through the coprocessor is an illegal size
105 [1]	The offset into the raw transfer segment of the coprocessor is an illegal value

[1] Fault applies only to PLC-5/11, -5/20, -5/30, -5/40, -5/40L, -5/60, -5/60L, and -5/80 processors.
[2] Occurs when scanner channels overload the remote I/O buffer and the processor cannot process all of the data.

I

ALLEN-BRADLEY SLC 500 MAJOR FAULT CODES

Address	Error Code (Hex)	Powerup Errors	Fault Classification			Processor			
			Non-User	User					
				Non-Recov	Recov	Fixed 5/01	5/02	5/03	5/04
S:6	0001	NVRAM error.	X			•	•	•	•
	0002	Unexpected hardware watchdog timeout.	X			•	•	•	•
	0003	Memory module memory error. This error can also occur while going into the REM Run mode.	X				•	•	•
	0005	Reserved			X			•	•
	0006	Reserved			X			•	•
	0007	Failure during memory module transfer.	X					•	•
	0008	Internal software error.	X					•	•
	0009	Internal hardware error.	X					•	•

Address	Error Code (Hex)	Going–to–Run Errors	Fault Classification			Processor			
			Non-User	User					
				Non-Recov	Recov	Fixed 5/01	5/02	5/03	5/04
S:6	0010	The Processor does not meet the required revision level.	X			•	•	•	•
	0011	The executable program file number 2 is absent.	X			•	•	•	•
	0012	The ladder program has a memory error.	X			•	•	•	•
	0013	• The required memory module is absent or • S:1/10 or S:1/11 is not set as required by the program.			X	•	•	•	•
	0014	Internal file error.	X			•	•	•	•
	0015	Configuration file error.	X			•	•	•	•
	0016	Startup protection after power loss. Error condition exists at powerup when bit S:1/9 is set and powerdown occurred while running.			X		•	•	•
	0017	NVRAM/memory module user program mismatch.		X				•	•
	0018	Incompatible user program – Operating system type mismatch. This error can also occur during powerup.	X					•	•
	0019	Missing or duplicate label was detected.		X				•	•
	001F	A program integrity problem occurred during an online editing session.	X					•	•
	0004	Memory error occurred while in the Run mode.	X				•	•	•
	0020	A minor error bit is set at the end of the scan. Refer to S:5 minor error bits.			X	•	•	•	•

Address	Error Code (Hex)	Runtime Errors	Fault Classification			Processor			
			Non-User	User					
				Non-Recov	Recov	Fixed, 5/01	5/02	5/03	5/04
S:6	0021	Remote power failure of an expansion I/O chassis occurred. **Note**: *A modular system that encounters an over–voltage or over–current condition in any of its power supplies can produce any of the I/O error codes listed on pages B–42 and B–43 (instead of code 0021). The over–voltage or over–current condition is indicated by the power supply LED being off.* ⚠ **Fixed and FRN 1 to 4 SLC 5/01 processors – If the remote power failure occurred while the processor was in the REM Run mode, error 0021 will cause the major error halted bit (S:1/13) to be cleared at the next powerup of the local chassis.** **SLC 5/02 processors and FRN 5 SLC 5/01 processors – Power to the local chassis does not need to be cycled to resume the REM Run mode. Once the remote chassis is re–powered, the CPU will restart the system.**	X			•	•	•	•
	0022	The user watchdog scan time has been exceeded.		X		•	•	•	•
	0023	Invalid or non-existent STI interrupt file.		X			•	•	•

Address	Error Code (Hex)	Runtime Errors	Fault Classification			Processor			
			Non-User	User		Fixed, 5/01	5/02	5/03	5/04
				Non-Recov	Recov				
S:6	0024	Invalid STI interrupt interval (greater than 2550 ms or negative).		X			•	•	•
	0025	Excessive stack depth/JSR calls for STI routine.		X			•	•	•
	0026	Excessive stack depth/JSR calls for I/O interrupt routine.		X			•	•	•
	0027	Excessive stack depth/JSR calls for user fault routine.		X			•	•	•
	0028	Invalid or non-existent "startup protection" fault routine file value.	X				•	•	•
	0029	Indexed address reference outside of entire data file space (range of B3:0 through the last file). ⚠ **The SLC 5/02 processor uses an index value of zero for the faulted instruction following error recovery.**			X		•		
				X				•	•
	002A	Indexed address reference is beyond specific referenced data file.		X			•	•	•
	002B	The file number exists, but it is not the correct file type or the file number does not exist.			X			•	•

Address	Error Code (Hex)	Runtime Errors	Fault Classification			Processor			
			Non-User	User		Fixed, 5/01	5/02	5/03	5/04
				Non-Recov	Recov				
S:6	002C	The indirectly referenced element does not exist, but the file type is correct and it exists. For example, T4:[N7:0] N7:0=10, but T4 only goes to T4:9.			X			•	•
	002D	Either a subelement is referenced incorrectly or an indirect reference has been made to an M-file.			X			•	•
	002E	Invalid DII Input slot.			X			•	•
	002F	Invalid or non-existent DII interrupt file.		X				•	•

I/O Errors

ERROR CODES: The characters xx in the following codes represent the slot number, in hex. If the exact slot cannot be determined, the characters xx become 1F.

RECOVERABLE I/O FAULTS (SLC 5/02, SLC 5/03, and SLC 5/04 processors only): Many I/O faults are recoverable. To recover, you must disable the specified slot, xx, in the user fault routine. If you do not disable slot xx, the processor will fault at the end of the scan.

Note: *An I/O card that is severely damaged may cause the processor to indicate that an error exists in slot 1 even though the damaged card is installed in a slot other than 1.*

SLOT NUMBERS (xx) IN HEXADECIMAL

Slot	xx	Slot	xx	Slot	xx	Slot	xx
0	00	8	08	16	10	24	18
1	01	9	09	17	11	25	19
2	02	10	0A	18	12	26	1A
** 3	03	11	0B	19	13	27	1B
4	04	12	0C	20	14	28	1C
5	05	13	0D	21	15	29	1D
6	06	14	0E	22	16	30	1E
7	07	15	0F	23	17	*	1F

* This value indicates that the slot was not found (SLC 5/01, SLC 5/02, SLC 5/03, and SLC 5/04 processors).

** This value indicates that the slot was not found (500 fixed controller).

Address	Error Code (Hex)	User Program Instruction Errors	Fault Classification			Processor			
			Non-User	User					
				Non-Recov	Recov	Fixed, 5/01	5/02	5/03	5/04
S:6	0030	Attempt was made to jump to one too many nested subroutine files. This code can also mean that a program has potentially recursive routines.		X		•	•	•	•
	0031	An unsupported instruction reference was detected.		X		•	•	•	•
	0032	A sequencer length/position parameter points past the end of a data file.			X	•	•	•	•
	0033	The length of LFU, LFL, FFU, FFL, BSL, or BSR instruction points past the end of a data file.			X	•	•	•	•

Address	Error Code (Hex)	User Program Instruction Errors	Fault Classification			Processor			
			Non-User	User		Fixed, 5/01	5/02	5/03	5/04
				Non-Recov	Recov				
S:6	0034	A negative value for a timer accumulator or preset value was detected.			X	•	•	•	•
		Fixed processors with 24 VDC inputs only: A negative or zero HSC preset was detected in a HSC instruction.			X	•			
	0035	TND, SVC, or REF instruction is called within an interrupting or user fault routine.		X			•	•	•
	0036	An invalid value is being used for a PID instruction parameter.			X		•	•	•
	0038	A RET instruction was detected in a non-subroutine file.	X			•	•	•	•
	xx3A	An attempt was made to write to an indirect address located in a file that has constant data file protection.			X			•	•
	1f39	Invalid string length was detected in a string file.			X			•	•
	xx50	A chassis data error is detected.			X	•	•	•	•
	xx51	A “stuck” runtime error is detected on an I/O module.		X		•	•	•	•
	xx52	A module required for the user program is detected as missing or removed.			X	•	•	•	•

Address	Error Code (Hex)	I/O Errors	Fault Classification			Processor			
			Non-User	User					
				Non-Recov	Recov	Fixed, 5/01	5/02	5/03	5/04
S:6	xx53	When going-to-run, a user program declares a slot as unused, and that slot is detected as having an I/O module inserted. This can also mean that an I/O module has reset itself.			X	•	•	•	•
		An attempt to enter the run or test mode was made with an empty chassis.			X			•	•
	xx54	A module required for the user program is detected as being the wrong type.			X	•	•	•	•
	xx55	A discrete I/O module required for the user program is detected as having the wrong I/O count. This code can also mean that a specialty card driver is incorrect.			X	•	•	•	•
	xx56	The chassis configuration specified in the user program is detected as being incorrect.	X			•	•	•	•
	xx57	A specialty I/O module has not responded to a lock shared memory command within the required time limit.			X	•	•	•	•
	xx58	A specialty I/O module has generated a generic fault. The card fault bit is set (1) in the module's status byte.		X		•	•	•	•
	xx59	A specialty I/O module has not responded to a command as being completed within the required time limit.			X	•	•	•	•
	xx5A	Hardware interrupt problem.			X		•	•	•

Address	Error Code (Hex)	I/O Errors	Fault Classification			Processor			
			Non-User	User					
				Non-Recov	Recov	Fixed, 5/01	5/02	5/03	5/04
S:6	xx5B	G file configuration error – user program G file size exceeds capacity of the module.			X		•	•	•
	xx5C	M0–M1 file configuration error – user program M0–M1 file size exceeds capacity of the module.			X		•	•	•
	xx5D	Interrupt service requested is not supported by the processor.			X		•	•	•
	xx5E	Processor I/O driver (software) error.			X		•	•	•
	xx60 to xx6F	Identifies an I/O module specific recoverable major error. Refer to the user manual supplied with the specialty module.			X		•	•	•
	xx70 to xx7F	Identifies an I/O module specific non-recoverable major error. Refer to the user manual supplied with the specialty module.		X			•	•	•
	xx90	Interrupt problem on disabled slot.		X			•	•	•
	xx91	A disabled slot has faulted.		X			•	•	•
	xx92	An invalid or non-existent module interrupt subroutine (ISR) file.		X			•	•	•
	xx93	Unsupported I/O module specific major error.		X			•	•	•
	xx94	In the REM Run or REM Test mode, a module has been detected as being inserted under power. This can also mean that an I/O module has reset itself.		X			•	•	•

J

ALLEN-BRADLEY PLC-5 PID CONTROL BLOCK

Parameter	Address Mnemonic:	Description:
Setpoint	.SP	Enter a floating-point number in the same engineering units that are on the PID Configuration screen. Valid range is $-3.4\ E^{+38}$ to $+3.4\ E^{+38}$.
Process Variable	.PV	Displays data from the analog input module that the instruction scales to the same engineering units that you selected for the setpoint.
Error	.ERR	Displays one of the following error values: Reverse acting: Error = SP-PV Direct acting: Error = PV-SP
Output %	.OUT	Displays the PID algorithm control output value (0-100%).
Mode	.MO .MO=0 .MO=1 .SWM=1	Displays operating mode: AUTO - automatic PID control MANUAL - control from a manual control station SW MANUAL - simulated manual control from the data monitor or ladder program
PV Alarm	 .PVHA=1 .PVLA=1	Displays whether the PV is within or exceeds the high or low alarm limits you selected on the PID Configuration screen. Displays one of the following: NONE - PV within alarm limits HIGH - PV exceeds high alarm limit (used with deadband) LOW - PV exceeds low alarm limit (used with deadband)
Deviation Alarm	 .DVPA=1 .DVNA=1	Displays whether the error is within or exceeds the high or low deviation alarms you selected on the PID Configuration screen. Displays one of the following: NONE - error within deviation alarm limits POSITIVE - error exceeds high alarm (used with deadband) NEGATIVE - error exceeds low alarm (used with deadband)
Output Limiting	 .OLH=1 .OLL=1	Displays whether or not the instruction clamps the output at the high and low limiting values (.MAXO and .MINO) you selected on the PID Configuration screen. Displays one of the following: NONE - output not clamped HIGH - output clamped at the high end (.MAXO) LOW - output clamped at the low end (.MINO) The PID algorithm has a anti-reset-windup feature that prevents the integral term from becoming too large when the output reaches the high or low alarm limits. If the limits are reached, the algorithm stops calculating the integral term until the output comes back into range.
SP Out of Range	.SPOR=0 .SPOR=1	Displays whether or not the setpoint is out of the range of engineering units you selected on the PID Configuration screen. Displays one of the following: NO - SP within range YES - SP out of range **Important:** A major processor fault occurs if the SP is out of range when the instruction is first enabled.

Parameter	Address Mnemonic:	Description:
Error Within DB	 .EWD=0 .EWD=1	Displays whether the error is within or exceeds the deadband value you enter on this screen. The deadband is zero crossing. Displays one of the following: RESET - Error exits the deadband zone SET - Error crosses the deadband centerline
PID Initialized	 .INI=0 .INI=1	Each time you change a value in the control block, the PID instruction takes over twice as long to execute (until initialized) on the first scan. Displays one of the following: NO - PID instruction not initialized after you changed control block values YES - PID instruction remains initialized because you did not change any control block values **ATTENTION:** Do not change the range of input or engineering units when running. If you must do this, then you must reset this bit to re-initialize. Otherwise, the instruction will malfunction with possible damage to equipment and injury to personnel.
A/M Station Mode	 .MO=0 .MO=1	Enter whether you want automatic (0) or manual (1) PID control. Displays one of the following: AUTO (0) - automatic PID control MANUAL (1) - manual PID control Manual control specified that an output from a manual control station overrides the calculated output of the PID algorithm. **Important:** Manual overrides Set Output Mode.
Software A/M Mode	 .SWM=0 .SWM=1	Enter whether you want automatic PID (0) control or Set Output Mode (1), for software-simulated control. Displays one of the following: AUTO (0) - automatic PID control SW MANUAL (1) - software-simulated PID control You can simulate a manual control station with the data monitor when you program a single loop. To do this, set .SWM to SW MANUAL and enter a Set Output Percent value. You can simulate a manual control station with ladder logic, pushwheels, and pushbutton switches when you program several loops. To do this, set .SWM to SW MANUAL and move a value into the set output element .SO.
Status Enable	 .EN=0 .EN=1	Enter whether to use (1) or inhibit (0) this bit which displays the rung condition so you can see whether the PID instruction is operating. Displays one of the following: 0 - instruction not executing 1 - instruction executing
Proportional Gain	.KP	Enter a floating-point value. Valid range for independent or standard gains is 0 to 3.4 E^{+38} (unitless).
Integral Gain	.KI	Enter a floating-point value. Valid range for independent gains is 0 to 3.4 E^{+38} inverse seconds; valid range for standard gains is 0 to 3.4 E^{+38} minutes per repeat.

Parameter	Address Mnemonic:	Description:
Derivative Gain	.KD	Enter a floating-point value. Valid range for independent gains is 0 to 3.4 E^{+38} seconds; valid range for standard gains is 0 to 3.4E^{+38} minutes.
Deadband	.DB	Enter a floating-point value in the engineering units you selected on the PID Configuration screen. Valid range is 0 to 3.4 E^{+38}. See Error Within DB value (.EWD).
Output Bias %	.BIAS	Enter a value (-100 to +100) to represent the percentage of output you want to feed forward or use as a bias to the output. The bias value can compensate for steady-loss of energy from the system. The ladder program can enter a feedforward value to bump the output in anticipation of a disturbance. This value is often used to control a process with transportation lag.
Tieback %	.TIE	Displays a number (0 to 100) representing the percent of raw tieback (0 - 4095) from the manual control station. The PID algorithm uses this number to achieve bumpless transfer when switching from manual to auto mode.
Set Output %	.SO	Enter a percent (0 to 100), from this screen or a ladder program, to represent the software-manually controlled output. When you select software-simulated control (.SWM=1), the PID instruction overrides the algorithm with the set output value (0 - 4095) for transfer to the output module, and copies it to .OUT for display as a percent. The transfer to software-simulated control is bumpless because .SO (under your control) starts with the last automatic algorithm output. Do not vary .SO until after the transfer. To achieve bumpless transfer when changing from software-simulated control to automatic control, the PID algorithm changes the integral term so that the output is equal to the set output value.

Parameter:	Address Mnemonic:	Description:
PID Equation	 .PE=0 .PE=1	Enter whether you want to use independent (0) or dependent (1) gains. Displays one of the following: INDEPENDENT (0) - for independent gains DEPENDENT (1) - for dependent gains Use dependent gains when you want to use standard loop tuning methods. Use independent gains when you want the three gain constants (P, I, and D) to operate independently.
Derivative of	 .DO=0 .DO=1	Enter whether you want the derivative of the PV (0) or the error (1). Displays one of the following: PV (0) - for PV derivative ERROR (1) - for error derivative Select the PV derivative for more stable control when you do not change the setpoint often. Select the error derivative for fast responses to setpoint changes when the algorithm can tolerate overshoots.
Control Action	 .CA=0 .CA=1	Enter whether you want reverse (0) or direct acting (1). Displays one of the following: REVERSE (0) - for reverse acting (E = SP–PV) DIRECT (1) - for direct acting (E = PV–SP)
PV Tracking	 .PVT=0 .PVT=1	Enter whether you do not (0) or do (1) want PV tracking. Displays one of the following: NO (0) - for no tracking YES (1) - for PV tracking Select no tracking if the algorithm can tolerate a bump when switching from manual to automatic control. Select PV tracking if you want the setpoint to track the PV in manual control for bumpless transfer to automatic control.
Update Time	.UPD	Enter an update time (greater than or equal to .01 seconds) at 1/5 to 1/10 the natural period of the load (load time constant). The load time constant should be greater than: 3ms(algorithm) + block transfer time (ms) Periodically enable the PID instruction at a constant interval equal to the update time. When the program scan time is close to the required update time, use an STI to ensure a constant update interval. When the program scan is several times faster than the required update time, use a timer. **ATTENTION:** If you omit an update time or enter a negative update time, a major fault occurs the first time the processor runs the PID instruction.
Cascade Loop	 .CL=0 .CL=1	Enter whether this loop is not (0) or is (1) used in a cascade of loops. Displays one of the following: NO (0) - not used in a cascade YES (1) - used in a cascade
Cascade Type	 .CT=0 .CT=1	If this loop is part of a cascade of loops, enter whether this loop is the master (1) or a slave (0). Displays one of the following: SLAVE (0) - for a slave loop MASTER (1) - for a master loop

Parameter:	Address Mnemonic:	Description:
Master to this Slave	.ADDR	If this loop is a slave loop in a cascade, enter the control block address of the master. Tieback is ignored in the master loop of a cascade. When you change cascaded loops to manual control, the slave forces the master into manual control. When PV tracking is enabled, the order of events is: Slave.SP > Master.TIE > Master.OUT > Slave.SP When you return to automatic control, change the slave first, then the master.
Engineering Unit Max	.MAXS	Enter the floating-point value in engineering units that corresponds to the analog module's full scale output. Valid range is $-3.4\ E^{+38}$ to $+3.4\ E^{+38}$. **ATTENTION:** Do not change this value during operation because a processor fault might occur.
Engineering Unit Min	.MINS	Enter the floating-point value in engineering units that corresponds to the analog module's zero output. Valid range is $-3.4\ E^{+38}$ to $+3.4\ E^{+38}$ (post-scaled number). **ATTENTION:** Do not change the maximum scaled value during operation because a processor fault might occur.
Input Range Max	.MAXI	Enter the floating-point number ($-3.4\ E^{+38}$ to $+3.4\ E^{+38}$) that is the unscaled maximum value available from the analog module. For example, use 4095 for a module whose range is 0-4095.
Input Range Min	.MINI	Enter the floating-point number ($-3.4\ E^{+38}$ to $+3.4\ E^{+38}$) that is the minimum unscaled value available from the analog module. For example, use 0 for a module whose range is 0-4095.
Output Limit High %	.MAXO	Enter a percent (0-100) above which the algorithm clamps the output.
Output Limit Low %	.MINO	Enter a percent (0-100) below which the algorithm clamps the output.
PV Alarm High	.PVH	Enter a floating-point number ($-3.4\ E^{+38}$ to $+3.4\ E^{+38}$) that represents the highest PV value that the system can tolerate.
PV Alarm Low	.PVL	Enter a floating-point number ($-3.4\ E^{+38}$ to $+3.4\ E^{+38}$) that represents the lowest PV value that the system can tolerate.
PV Alarm Deadband	.PVDB	Enter a floating-point number ($0\text{-}3.4\ E^{+38}$) that is sufficient to minimize nuisance alarms. This is a one-sided deadband. The alarm bit (.PVH or .PVL) is not set until the PV crosses the deadband and reaches the alarm limit (DB zero point). The alarm bit remains set until the PV passes back through and exits from the deadband.
Deviation Alarm (+)	.DVP	Enter a floating-point number ($0\text{-}3.4\ E^{+38}$) that specifies the greatest error deviation above the setpoint that the system can tolerate.
Deviation Alarm (–)	.DVN	Enter a floating-point number ($-3.4\ E^{+38}\text{-}0$) that specifies the greatest error deviation below the setpoint that the system can tolerate.
Deviation Alarm Deadband	.DVDB	Enter a floating-point number ($0\text{-}3.4\ E^{+38}$) that is sufficient to minimize nuisance alarms. This is a one-sided deadband. The alarm bit (.DVP or .DVN) is not set until the error crosses the deadband and reaches the alarm limit (DB zero point). The alarm bit remains set until the error passes back through and exits from the deadband.

INDEX

Non-proprietary topics

IEC (6)1131 topics

Allen-Bradley PLC-5 topics

Allen-Bradley SLC 500 topics

Allen-Bradley MICROLOGIX 1000 topics

Siemens S5 topics

Siemens S7 topics

OMRON topics